Plant-derived Natural Products

Anne E. Osbourn • Virginia Lanzotti

Plant-derived Natural Products

Synthesis, Function, and Application

Springer

Editors
Anne E. Osbourn
Head, Department of Metabolic Biology
John Innes Centre
Colney Lane
Norwich NR4 7UH, UK
anne.osbourn@bbsrc.ac.uk

Virginia Lanzotti
University of Molise
DISTAAM
Via F. De Sanctis
86100 Campobasso
Italy
lanzotti@unimol.it
lanzotti@unina.it

ISBN 978-0-387-85497-7 e-ISBN 978-0-387-85498-4
DOI 10.1007/978-0-387-85498-4
Springer Dordrecht Heidelberg London New York

Library of Congress Control Number: 2009926301

Printed on acid-free paper

Springer is part of Springer Science+Business Media (www.springer.com)

Preface

Plants produce a huge array of natural products (secondary metabolites). These compounds have important ecological functions, providing protection against attack by herbivores and microbes and serving as attractants for pollinators and seed-dispersing agents. They may also contribute to competition and invasiveness by suppressing the growth of neighbouring plant species (a phenomenon known as allelopathy). Humans exploit natural products as sources of drugs, flavouring agents, fragrances and for a wide range of other applications. Rapid progress has been made in recent years in understanding natural product synthesis, regulation and function and the evolution of metabolic diversity. It is timely to bring this information together with contemporary advances in chemistry, plant biology, ecology, agronomy and human health to provide a comprehensive guide to plant-derived natural products.

Plant-derived natural products: synthesis, function and application provides an informative and accessible overview of the different facets of the field, ranging from an introduction to the different classes of natural products through developments in natural product chemistry and biology to ecological interactions and the significance of plant-derived natural products for humans. In the final section of the book a series of chapters on new trends covers metabolic engineering, genome-wide approaches, the metabolic consequences of genetic modification, developments in traditional medicines and nutraceuticals, natural products as leads for drug discovery and novel non-food crops.

Anne E. Osbourn
Virginia Lanzotti

Contents

Contributors

Asaph Aharoni
Department of Plant Sciences, Weizmann Institute of Science, Israel
asaph.aharoni@weizmann.ac.il

Stephen Barnes
Department of Pharmacology and Toxicology, Room 452 McCallum Research Building, University of Alabama at Birmingham, 1918 University Boulevard,
Birmingham, AL 35294
Center for Nutrient-Gene Interaction, Birmingham, AL 35294
University of Alabama at Birmingham, and Purdue University-University of Alabama Botanicals Center for Age-Related Disease, Birmingham, AL 35294
sbarnes@uab.edu

Giuliano Bonanomi
Dipartimento di Arboricoltura, Botanica e Patologia vegetale,
University of Naples Federico II, Via Università 100, 80055 Portici, Napoli, Italy
giuliano.bonanomi@unina.it

Dianna J. Bowles
Centre for Novel Agricultural Products, Department of Biology,
University of York,
York YO10 5DD, UK
djb32@york.ac.uk

Esperanza J. Carcache de Blanco
Division of Medicinal Chemistry and Pharmacognosy, College of Pharmacy, The Ohio State University, Columbus, OH 43210, USA
carcache-de-blan.1@osu.edu

Ian J. Colquhoun
Institute of Food Research, Norwich Research Park, Norwich NR4 7UA, UK
ian.colquhoun@bbsrc.ac.uk

Franck E. Dayan
Natural Products Utilization Research Unit, Agricultural Research Service, United States Department of Agriculture, P.O. Box 8048, University, MS 38677, USA
fdayan@olemiss.edu

Richard A. Dixon
Plant Biology Division, Samuel Roberts Noble Foundation, Ardmore, OK 73401, USA
radixon@noble.org

Natalia Dudareva
Department of Horticulture and Land scape Architecture, Purdue University, West Lafayette, IN 47907, USA
dudareva@purdue.edu

Stephen O. Duke
Natural Products Utilization Research Unit, Agricultural Research Service, United States Department of Agriculture, P.O. Box 8048, University, MS 38677, USA
sduke@olemiss.edu

Robert Edwards
School of Biological & Biomedical Sciences, Durham University,
South Road, Durham, DH1 3LE, UK
robert.edwards@durham.ac.uk

Jürgen Ehlting
Department of Plant Metabolic Responses, Institute for Plant Molecular Biology (IBMP), C.N.R.S., 28 rue Goethe, 67000 Strasbourg, France
Current address: Centre for Forest Biology, University of Victoria, P.O. Box 3020 STN CSC, Victoria BC V8W 3N5, Canada
je@uvic.ca

Robert A. Field
Department of Biological Chemistry, John Innes Centre, Colney Lane, Norwich NR4 7UH, UK
rob.field@bbsrc.ac.uk

Jean-François Ginglinger
Department of Plant Metabolic Responses, Institute for Plant Molecular Biology (IBMP), C.N.R.S., 28 rue Goethe, 67000 Strasbourg, France
jean-francois.ginglinger@ibmp-ulp.u-strasbg.fr

Ruth Gordon-Weeks
Department of Biological Chemistry, Centre for Pest and Disease Management, Rothamsted Research, West Common, Harpenden, Hertfordshire, AL5 2JQ, UK
ruth.gordon-weeks@bbsrc.ac.uk

Erich Grotewold
Department of Plant Cellular and Molecular Biology, The Ohio State University, Columbus, Ohio, USA
grotewold.1@osu.edu

Björn Hamberger
Michael Smith Laboratories, University of British Columbia, 2185 East Mall, Vancouver BC V6T 1Z4, Canada
bjoernh@interchange.ubc.ca

Alison Harner
The National Non-Food Crops Centre, Biocentre, York Science Park, Innovation Way, Heslington, York, YO10 5DG, UK
a.harner@nnfcc.co.uk

Adrian P. Higson
The National Non-Food Crops Centre, Biocentre, York Science Park, Innovation Way, Heslington, York, YO10 5DG, UK
a.higson@nnfcc.co.uk

Lionel Hill
Department of Metabolic Biology, John Innes Centre, Norwich Research Park, Colney, Norwich NR4 7UH, UK
lionel.hill@bbsrc.ac.uk

Efstathia Ioannou
Department of Pharmacognosy and Chemistry of Natural Products, School of Pharmacy, University of Athens, Panepistimiopolis Zografou, Athens 15771, Greece
eioannou@pharm.uoa.gr

Maxim Itkin
Department of Plant Sciences, Weizmann Institute of Science, Israel
maxitkin@yahoo.com

A. Douglas Kinghorn
Division of Medicinal Chemistry and Pharmacognosy, College of Pharmacy, The Ohio State University, Columbus, OH 43210, USA
kinghorn.4@osu.edu

Markus Klein
Zurich Basel Plant Science Center, University of Zurich, Plant Biology, Zollikerstrasse 107, CH-8008 Zurich, Switzerland
Martin-Luther University Halle-Wittenberg, Pharmaceutical Biology, Molecular Cell Biology, Kurt-Mothes-Str. 3, D-06120 Halle, Germany
markus.klein@botinst.uzh.ch

Daniel J. Kliebenstein
Department of Plant Sciences, University of California, Davis, One Shields Avenue, Davis, CA 95616, USA
kliebenstein@ucdavis.edu

Toni M. Kutchan
Donald Danforth Plant Science Center,
975 North Warson Road,
St. Louis, MO 63132,
USA
tmkutchan@danforthcenter.org

Gwénaëlle Le Gall
Institute of Food Research,
Norwich Research Park,
Norwich NR4 7UA,
UK
gwenaelle.legall@bbsrc.ac.uk

Eng-Kiat Lim
Centre for Novel Agricultural Products,
Department of Biology, University of York,
York YO10 5DD, UK
ekl100@york.ac.uk

Michael C. Long
Department of Horticulture and Landscape
Architecture, Purdue University, West Lafayette,
IN 47907, USA
longmc@purdue.edu

Johan Memelink
Institute of Biology, Leiden University, Clusius
Laboratory, Wassenaarseweg 64, 2333 AL Leiden,
The Netherlands
j.memelink@biology.leidenuniv.nl

Richard F. Mithen
Institute of Food Research, Colney Lane,
Norwich, NR4 7UA, UK
richard.mithen@bbsrc.ac.uk

Luzia V. Modolo
Plant Biology Division, Samuel Roberts Noble
Foundation, Ardmore, OK 73401, USA
lamodolo@noble.org

John P. Morrissey
Microbiology Department, University College
Cork, Ireland
j.morrissey@ucc.ie

Florence Negre-Zakharov
Department of Plant Sciences,
University of California, Davis,
CA 95616, USA
fnegre@plantsciences.ucdavis.edu

Sarah O'Connor
Massachusetts Institute of Technology,
Department of Chemistry, 18-592, 77
Massachusetts Ave, Cambridge,
MA 02139, USA
soc@mit.edu

Li Pan
Division of Medicinal Chemistry and
Pharmacognosy, College of Pharmacy, The Ohio
State University, Columbus, OH 43210, USA
pan.106@osu.edu

John A. Pickett
Department of Biological Chemistry,
Centre for Pest and Disease Management,
Rothamsted Research, West Common, Harpenden,
Hertfordshire, AL5 2JQ, UK
john.pickett@bbsrc.ac.uk

Lucille Pourcel
Department of Plant Cellular and Molecular
Biology, The Ohio State University,
Columbus, Ohio, USA
pourcel.1@osu.edu

Jeevan Prasain
Department of Pharmacology and Toxicology,
Birmingham,
AL 35294 University of Alabama at Birmingham,
and Purdue University-University of Alabama
Botanicals Center for Age-Related Disease,
Birmingham, AL 35294
jeevan.prasain@ccc.uab.edu

Angelika I. Reichert
Plant Biology Division, Samuel Roberts Noble
Foundation, Ardmore, OK 73401, USA
aireichert@noble.org

Werner Roos
Martin-Luther University Halle-Wittenberg, Pharmaceutical Biology, Molecular Cell Biology, Kurt-Mothes-Str. 3, D-06120 Halle, Germany
werner.roos@pharmazie.uni-halle.de

Vassilios Roussis
Department of Pharmacognosy and Chemistry of Natural Products, School of Pharmacy, University of Athens, Panepistimiopolis Zografou, Athens 15771, Greece
roussis@pharm.uoa.gr

Felice Scala
Dipartimento di Arboricoltura, Botanica e Patologia vegetale, University of Naples Federico II, Via Università 100, 80055 Portici, Napoli, Italy
felice.scala@unina.it

Laetitia Shintu
Institute of Food Research, Norwich Research Park, Norwich NR4 7UA, UK
laetitia.shintu@bbsrc.ac.uk

Monique S.J. Simmonds
Royal Botanic Gardens, Kew, Richmond Surrey UK, TW9 3AB
m.simmonds@kew.org

Karin Springob
Donald Danforth Plant Science Center, 975 North Warson Road, St. Louis, MO 63132, USA
karin.springob@gmx.de

Maria Traka
Institute of Food Research, Colney Lane, Norwich, NR4 7UA, UK
maria.traka@bbsrc.ac.uk

Fabián E. Vaistij
Centre for Novel Agricultural Products, Department of Biology, University of York, York YO10 5DD, UK
fv4@york.ac.uk

Francesco Vinale
Dipartimento di Arboricoltura, Botanica e Patologia vegetale, University of Naples Federico II, Via Università 100, 80055 Portici, Napoli, Italy
frvinale@unina.it

Trevor L. Wang
Department of Metabolic Biology, John Innes Centre, Norwich Research Park, Colney, Norwich NR4 7UH, UK
trevor.wang@bbsrc.ac.uk

Danièle Werck-Reichhart
Department of Plant Metabolic Responses, Institute for Plant Molecular Biology (IBMP), C.N.R.S., 28 rue Goethe, 67000 Strasbourg, France
daniele.werck@ibmp-ulp.u-strasbg.fr

Brenda S.J. Winkel
Department of Biological Sciences, Virginia Tech, Blacksburg, VA 24061, USA
winkel@vt.edu

Hongtao Zhang
Institute of Biology, Leiden University, Clusius Laboratory, Wassenaarseweg 64, 2333 AL Leiden, The Netherlands
h.zhang@biology.leidenuniv.nl

Part I

The Chemical Composition of Plants

Introduction to the Different Classes of Natural Products 1

Karin Springob and Toni M. Kutchan

Abstract Plants produce an enormous variety of natural products with highly diverse structures. These products are commonly termed "secondary metabolites" in contrast to the "primary metabolites" which are essential for plant growth and development. Secondary metabolites were formerly regarded as "waste products" without physiological function for the plant. With the emergence of the field of chemical ecology about 30 years ago, it became evident, however, that these natural products fulfill important functions in the interaction between plants and their biotic and abiotic environment. They can serve, for example, as defense compounds against herbivores and pathogens, as flower pigments that attract pollinators, or as hormones or signal molecules. In addition to their physiological function in plants, natural products also have a strong impact on human culture and have been used throughout human history as condiments, pigments, and pharmaceuticals.

This chapter provides an overview about the diversity of secondary metabolites in plants, their multiple biological functions and multifaceted cultural history. The compounds are classified into four different groups according to their biosynthetic origin: alkaloids, phenylpropanoids, polyketides, and terpenoids. Since more than 200,000 structures of natural products from plants are known, only selected groups and compounds are presented.

K. Springob and T.M. Kutchan (✉)
Donald Danforth Plant Science Center,
975 North Warson Road, St. Louis, MO 63132, USA
e-mail: karin.springob@gmx.de;
tmkutchan@danforthcenter.org

Nitrogen-Containing Natural Products

The term alkaloid is derived from the Arabic word "al-qali" that refers to potassium carbonate-containing ashes from plant material. Traditionally, alkaloids are defined as heterocyclic nitrogen compounds biosynthesized from amino acids. Many other substances, however, that do not exactly match this rule are classified as alkaloids, either for historical reasons or due to their bioactivities. With currently more than 12,000 known structures, alkaloids represent one of the biggest groups of natural products. Due to this large number and the high structural diversity, it is impossible to give a comprehensive summary of all different types of alkaloids, and only some classes will be introduced. In addition to alkaloids, benzoxazinoids, glucosinolates, and cyanogenic glucosides will be presented. Like the alkaloids, these metabolites contain nitrogen and are derived from amino acids.

A.E. Osbourn and V. Lanzotti (eds.), *Plant-derived Natural Products*,
DOI 10.1007/978-0-387-85498-4_1, © Springer Science+Business Media, LLC 2009

Purine Alkaloids

Purine alkaloids are nitrogen containing compounds derived from nucleoside metabolism (Ashihara and Crozier, 2001). The purine backbone is synthesized from several small molecules of primary metabolism that include L-aspartic acid, L-glutamine, L-glycine, and formate. Cytokinins, plant hormones that control, e.g., stem growth and differentiation, apical dominance, and senescence, are derived from the same pathway. Purine alkaloids are produced in a variety of taxonomically unrelated plant species, e.g., coffee (*Coffea arabica* and other *Coffea* species, Rubiaceae), tea (*Camellia sinensis*, Theaceae), cacao (*Theobroma cacao*, Sterculiaceae), maté (*Ilex paraguariensis*, Aquifoliceae), guaraná (*Paullinia cupana*, Sapindaceae), and cola (*Cola nitida*, Sterculiaceae). The most abundant purine alkaloid is caffeine, followed by theobromine and some minor purines, e.g., theophylline and paraxanthine (Fig. 1). Coffee seeds ("beans") contain ca. 1% caffeine, young tea leaves 2-3% (Ashihara and Suzuki, 2004).

Since caffeine is accumulated in higher amounts than the other purines, its function in plants has been investigated. It may serve as defense against herbivores (Hollingsworth et al., 2002) and as autotoxin, because it inhibits the germination of coffee seedlings (Friedmann and Waller, 1985).

Caffeine is a central stimulant and widely consumed in beverages like coffee, tea, and sodas, but also in cold medicine and analgesics. The average daily caffeine consumption of adults is 280 mg; one cup of filter coffee contains ca. 140 mg caffeine, one cup of black tea ca. 80 mg (Lovett, 2005). Besides caffeine, theophylline is of interest, since it has found application in the therapy of asthma due to its bronchodilatory effect.

The predominant mode of action of caffeine and other purine alkaloids is the blockade of adenosine receptors resulting in the release of neurotransmitters (Benowitz, 1990). In higher concentrations, phosphodiesterase, the enzyme that hydrolyzes the second messenger cAMP, is inhibited. However, these blood concentrations are normally not reached by consumption of caffeine-containing beverages. More recently, the attention towards caffeine increased because coffee drinkers show a reduced risk for Parkinson's disease (Ascherio et al., 2004).

C. arabica originated from Ethiopia, where the fruits were first used as food by nomads. Roasted coffee seeds ("beans") were brewed in Arabia around AD 1000 to prepare a drink called "qahwah", and it was introduced into Europe as "kahveh" after AD 1600 coffee and coffee houses became soon popular in Europe. Johann Sebastian Bach's "Coffee Cantata" (BMV 211), which he composed for a text written by Picander in the beginning of the eighteenth century, reflects this growing popularity as well as the controversy on the assumed dangerous health effects of coffee at that time.

Tea is prepared from fermented (black tea) or unfermented (green tea), dried leaves of *C. sinensis*. The earliest records on tea drinking come from China in the first millennium BC. From there, it was introduced into Japan in the

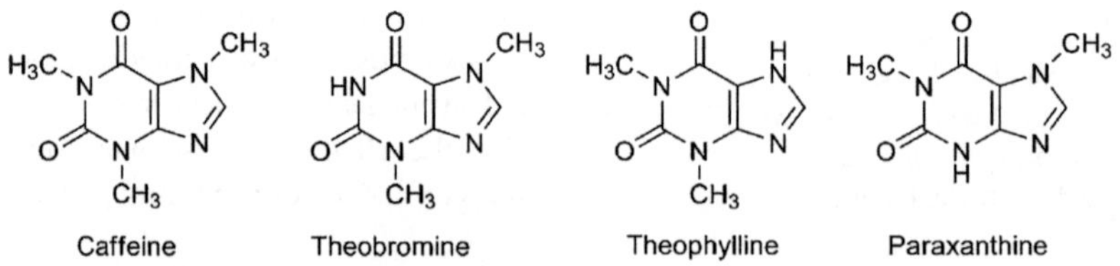

Fig. 1 Structures of the purine alkaloids caffeine, theobromine, theophylline, and paraxanthine

eighth century AD by Buddhist monks. Tea was first shipped to Europe by the Dutch East India Company in 1606.

The cacao tree (*T. cacao*, Sterculiaceae) originates from the Amazon Basin, but it was cultivated by the Mayas in Mesoamerica. Its seeds ("beans") contain theobromine and caffeine. Mayas and Azteks used roasted cacao seeds together with chili peppers and other spices to prepare a drink, which the Aztecs called "xocoatl". According to Aztec belief, cacao was given to humanity by the god Quetzalcoatl. The Swedish botanist Carl Linnaeus named the cacao tree after the Aztec tradition; *Theobroma* means "food of the gods" in Greek. The first cacao beans were brought to Europe by the Spanish Conquistador Hernán Cortés.

Tropane Alkaloids

Tropane alkaloids originate from the amino acids ornithine and/or arginine. They all have in common the bicyclic tropane skeleton that consists of a seven-membered ring with an *N*-bridge between C-1 and C-5, the nitrogen being methylated. Nortropanes lacking the *N*-methylation and *seco*-tropanes with a dissected *N*-bridge have been described, too (Griffin and Lin, 2000). Many tropane alkaloids are esters of the alcohols tropine (tropane-3α-ol) or pseudotropine (tropane-3β-ol) (Fig. 2) with aliphatic or aromatic acids. Tropane alkaloids were isolated first from the nightshade family (Solanaceae). Many structurally diverse tropanes, however, have been discovered in the related family Convolvulaceae, like the Solanaceae a member of the order Solanales, and in some species from the unrelated plant families Brassicaceae, Euphorbiaceae, Erythroxylaceae, Proteaeceae, and Rhizophoraceae (Griffin and Lin, 2000).

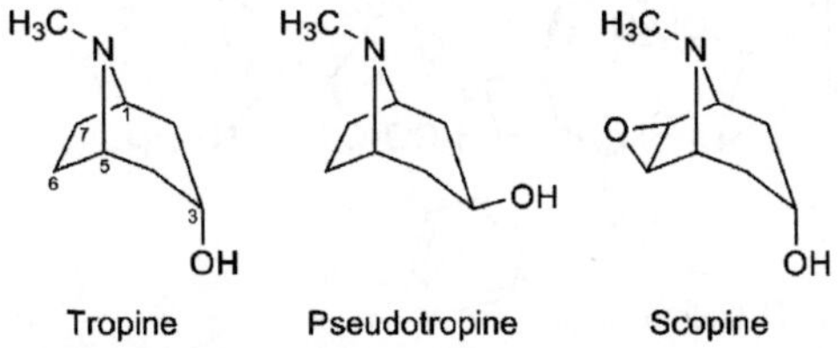

Fig. 2 Tropane amino alcohols

Hyoscyamine and Scopolamine

(*S*)-Hyoscyamine and (*S*)-scopolamine are esters of the amino alcohols tropine and scopine with (*S*)-tropic acid, which is derived from phenylalanine (Fig. 3). The two alkaloids occur exclusively in the Solanaceae family. They act as antagonists of muscarinic acetylcholine receptors (parasympatholytics) and lead to an increase in pulse rate, relaxation of smooth muscles, e.g., in the gastrointestine and the bronchial tract, reduction of salivary, bronchial, gastric, and sweat gland secretion. While hyoscyamine is a central stimulant, scopolamine depresses the central nervous system.

Atropine, the racemate of (*S*)- and (*R*)-hyoscyamine, is formed during the extraction of plant material. Although (*S*)-hyoscyamine is more effective than the (*R*)-enantiomer, atropine is more widely used for traditional reasons. In medicine, atropine is used against spasms during a biliary colic, as antidote against intoxication with organophosphorous insecticides, and as pre-medication before surgery to decrease salivation and respiratory secretion. (*S*)-Scopolamine

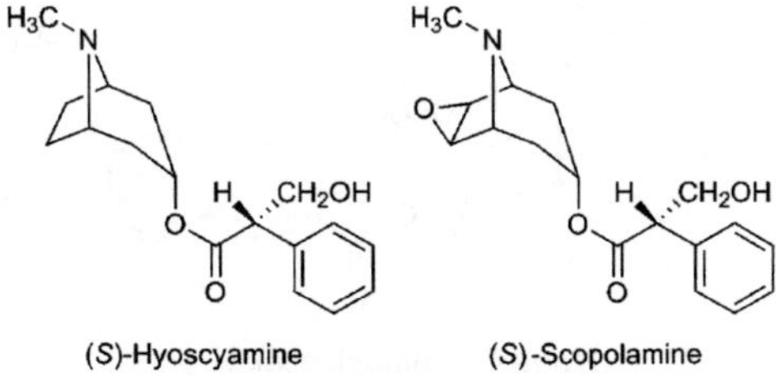

Fig. 3 The tropane alkaloids (*S*)-hyoscyamine and (*S*)-scopolamine occur only in the Solanaceae family

1

is applied as treatment of motion sickness. Derivatives of hyoscyamine or scopolamine are used as mydriatics for eye examinations, as treatment for asthma and chronic obstructive bronchitis, and against gastrointestinal spasms.

Plants containing hyoscyamine and scopolamine have been used throughout history as psychoactive drugs, poisons, aphrodisiacs, and for the preparation of analgesic and sleeping potions. Famous examples include the deadly nightshade (*Atropa belladonna*), thorn-apple (*Datura* species), henbane (*Hyoscyamus niger*), and mandrake (*Mandragora officinarum*). All plants are toxic, and, for example, five to ten fruits of *A. belladonna* are lethal in an adult. This is reflected in the name "Atropa", the Greek goddess of destiny who cuts the thread of life. "Bella donna" (Italian for "beautiful woman") refers to the custom of Renaissance ladies who dilated their pupils with extracts of the deadly nightshade. Solanaceous plants with tropanes are often mentioned in literature, for example, in Homer's Odyssey and several pieces by William Shakespeare.

Atropine was first isolated from *A. belladonna* (Mein, 1833), and (*S*)-hyoscyamine was extracted from *H. niger* (Geiger and Hesse, 1833). In the late nineteenth century, (*S*)-scopolamine was detected by Ladenburg (1881) and Schmidt (1892). Today, (*S*)-hyoscyamine and (*S*)-scopolamine are obtained from *Duboisia leichhardtii* and *Duboisia myoporoides*, trees native to Australia, and hybrids of the two species.

Cocaine

Cocaine is the benzoic acid ester of the tropane base methylecgonine. Only *Erythroxylum coca* and *Erythroxylum novogranatense*, shrubs or small trees native to the Andes, contain substantial amounts of cocaine in their leaves, i.e., up to 1% of their dry mass (Plowman and Rivier, 1983). If the coca leaves are dried or stored improperly, the cocaine content decreases rapidly. The two *Erythroxylum* species contain also other ecgonine derivatives, e.g., *cis*- and *trans*-cinnamoylcocaine and the truxillins, esters of methylecgonine with dimeric cinnamic acid (Fig. 4) (Griffin and Lin, 2000).

Cocaine is a highly addictive central stimulant that inhibits the re-uptake of the neurotransmitters dopamine and norepinephrine at synapses,

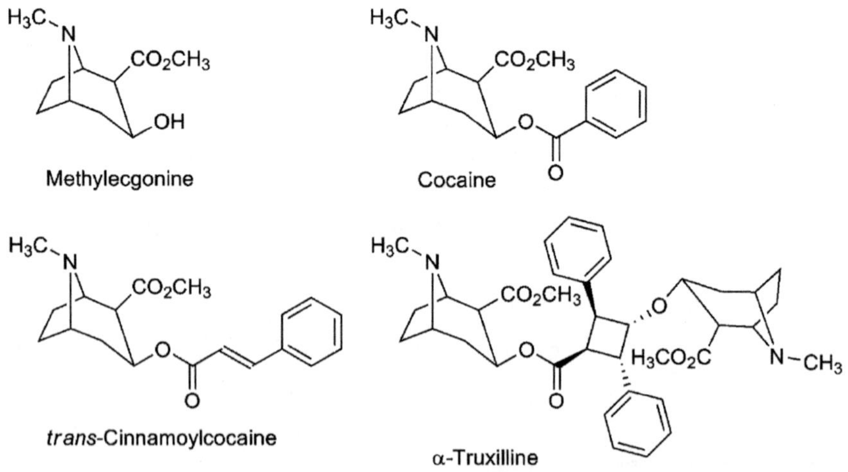

Fig. 4 *Erythroxylum* alkaloids

and inhibits monoamine oxidase, the enzyme that degrades dopamine, epinephrine, and norepinephrine. This leads to euphoria, hyperactivity, suppression of hunger and fatigue. Peripheral effects include increased heart rate and blood pressure, dilation of pupils, hyperglycaemia, and hyperthermia (White and Lambe, 2003). If cocaine is applied on mucous membranes, it blocks Na^+ channels leading to local anaesthesia. Therefore, cocaine is used as local anaesthetic in surgeries of the eye, ear, nose, and throat.

As an illegal drug, cocaine occurs in different forms. The hydrochloride is soluble in water and can be injected, sniffed, or chewed. "Freebase" is cocaine base, which is gained by extraction of cocaine from alkaline solutions with ether. It evaporates at high temperatures and can be inhaled by smoking. Another smokable form of the cocaine base is "crack", which is obtained by precipitating cocaine hydrochloride from solution with baking soda.

In South America, the chewing of coca leaves has a long tradition and dates back to 3000 B.C. It is used to overcome exhaustion, hunger, and thirst, and presumably does not have the addictive potential of cocaine. The leaves are chewed together with an alkaline agent like plant ash or sodium bicarbonate, which converts the alkaloids to their free bases. Only a small portion of the cocaine is hydrolyzed to methylecgonine (Rivier, 1981).

Coca leaves were brought to Europe by the Spanish conquistadores, and cocaine was isolated from the leaves in the 1860s. In 1863, the French chemist Angelo Mariani created the tonic "Vin Mariani", an extract of coca in Bordeaux wine. The non-alcoholic version "Coca-Cola" was invented in 1886 by the American Pharmacist John Pemberton, who mixed extracts of coca leaves and caffeine-containing cola nuts with soda. With the introduction of the first anti-drug laws in the USA in 1906, however, only decocainized leaves were used for the production of Coca-Cola.

Calystegines

Calystegines contain the nortropane skeleton with three to five hydroxyl groups. In contrast to most other tropane alkaloids, the hydroxyl groups are not esterified, but they can be glycosylated. The bridgehead C-1 of calystegines is hydroxylated, only in calystegine N_1 it is linked to an amino group instead (Dräger, 2004). The structures of the three most widespread calystegines are shown in Fig. 5. Calystegines were discovered in roots of *Calystegia sepium* (Tepfer et al., 1988), and the first structures were determined in 1990 (Goldmann et al., 1990). Since then, they have been isolated from numerous members of the Brassicaceae, Convolvulaceae, Erythroxylaceae, Solanaceae, and from two species of the Moraceae (Biastoff and Dräger, 2007). One important reason for their late discovery is their high hydrophily due to which they cannot be extracted with organic solvents like other alkaloids.

Calystegines are sugar-mimicking glycosidase inhibitors. Due to their structural similarity to sugars, they compete with polysaccharides for binding at the active site of the glycosidase. Therefore, one possible medicinal application for this group of metabolites is the prevention of post-prandial glucose peaks in patients with type II diabetes. In addition, calystegines might

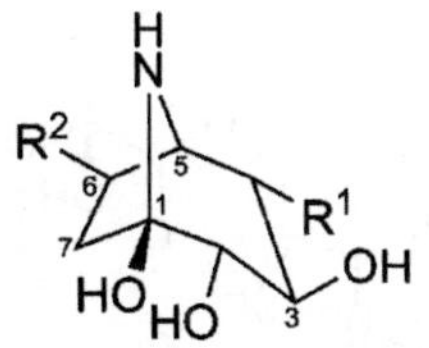

	R^1	R^2
Calystegine A_3	H	H
Calystegine B_1	H	OH
Calystegine B_2	OH	H

Fig. 5 Structures of calystegines A_3, B_1, and B_2

1

become useful for the therapy of Morbus Gaucher, a lysosomal storage disease that is caused by a mutation in the gene encoding glucocerebrosidase. Calystegines were reported to act as chaperones on the mutated enzyme thus preventing its misfolding and degradation. Nevertheless, it is still unclear whether these hydrophilic polyhydroxylated alkaloids can be absorbed in the intestines and transported to the lysosomes of target cells (reviewed by Biastoff and Dräger, 2007).

Pyrrolizidine Alkaloids

The backbone of pyrrolizidine alkaloids is composed of a hydroxymethylpyrrolizidine (necine base) that is mostly esterified with branched aliphatic mono- or dicarboxylic acids (necic acids). The necine base is biosynthesized from spermidine and putrescine, which in turn originates from arginine (Hartmann et al., 1988). The origin of the necic acids has been investigated only for pyrrolizidine alkaloids of the senecionine and lycopsamine type; they are derived from amino acid metabolism (Stirling et al., 1997; Weber et al., 1998). The major structural types of pyrrolizidine alkaloids are depicted in Fig. 6. In plants, these alkaloids are usually stored and transported as polar *N*-oxides. Pyrrolizidines occur mainly in the plant families Asteraceae, Boraginaceae, Fabaceae, and Orchidaceae, although scattered occurrences in other plant families have also been described (Hartmann and Ober, 2000).

Many pyrrolizidine alkaloids are hepatotoxic, mutagenic, and carcinogenic. They can cause veno-occlusive disease of the liver that may lead to cirrhosis and eventually liver failure. The main reasons for intoxications with pyrrolizidines are contamination of cereals with

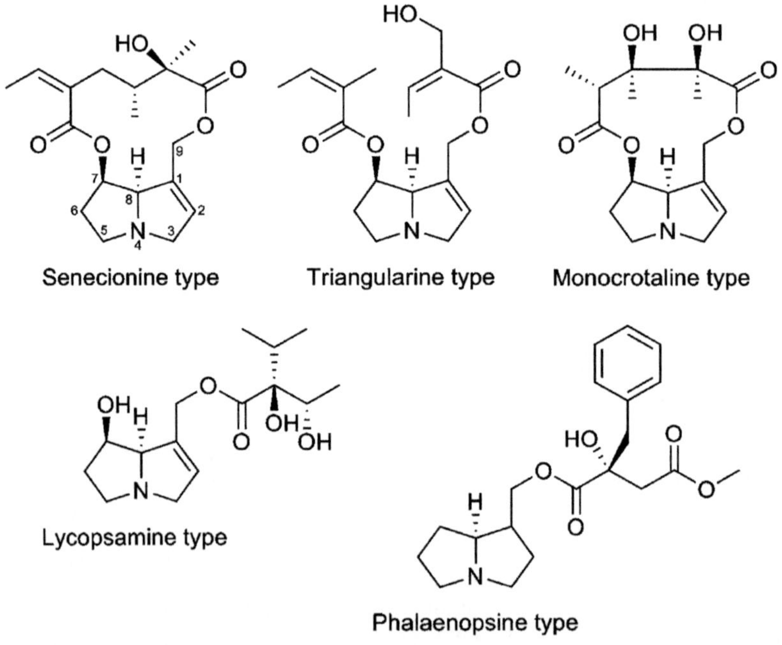

Fig. 6 The five major structural types of pyrrolizidine alkaloids. In plants, these alkaloids occur mostly in form of their *N*-oxides

pyrrolizidine-containing plants and the ingestion of herbal medicines containing these alkaloids.

The structural features responsible for the genotoxicity are a double bond in the necine base between C-1 and C-2, presence of hydroxy groups at C-7 and C-9, and esterification of at least one of these hydroxy groups with a branched carbon chain (Frei et al., 1992). In most vertebrates and insect herbivores, the alkaloid *N*-oxides are reduced in the gut to their free bases. The reduced alkaloids are then taken up and bioactivated by cytochrome P450-dependent monooxygenases of the liver to highly reactive dehydropyrrolizidine alkaloids that react with nucleophilic groups of proteins and DNA (Röder, 1995).

Although pyrrolizidines are toxic, many insects feed on pyrrolizidine-containing plants. Several butterflies and moths (Lepidoptera) and some Chrysomelid leaf beetles (Coleoptera) are even able to sequester pyrrolizidine alkaloids as defense compounds against predators. Adult members of the Lepidoptera selectively ingest plants with pyrrolizidines, a behaviour called pharmacophagy. In the gut of adapted Lepidoptera, the *N*-oxides are reduced and taken up as free bases. In the hemolymph, however, they are detoxified by oxidation to the water-soluble *N*-oxides, which do not serve as substrates for bioactivating cytochrome P450 enzymes. In addition to their function in chemical defense in adapted butterflies, pyrrolizidine alkaloids also play an important role in the mating process. Male moths utilize pyrrolizidines to synthesize the pheromone hydroxydanaidal in order to signal their alkaloid load to the females. During courtship, male moths of the species *Utetheisa ornatrix* transfer sequestered pyrrolizidine alkaloids as a nuptial gift to the female. The female moth transfers her own pyrrolizidines and the alkaloids acquired during mating to the egg mass to protect the offspring (Eisner and Meinwald, 1995).

Quinolizidine Alkaloids

Quinolizidine alkaloids are biosynthesized from lysine *via* cadaverine. Apart from the bicyclic lupinine, most other compounds of this group are tri- or tetracyclic. Some representative structures are shown in Fig. 7. Quinolizidine alkaloids occur abundantly in the Fabaceae, but also in several unrelated taxa, e.g., Berberidaceae, Chenopodiacae, Ranunculaceae, Rubiaceae, and Solanaceae (Wink, 2002). Traces of quinolizidines were found in elicited cell cultures of species that normally do not produce these metabolites (Wink and Witte, 1983). This finding, together with the occurrence of quinolizidines alkaloids in taxonomically unrelated species, has lead to the hypothesis that the genes for the biosynthesis of quinolizidines are widely distributed in the plant kingdom, but are actively transcribed only in a few species that use them as feeding deterrents against herbivores.

The function of quinolizidines as defense compounds can be observed in the example of sweet lupins, an alkaloid-free breeding form. In contrast to the alkaloid-containing wild form,

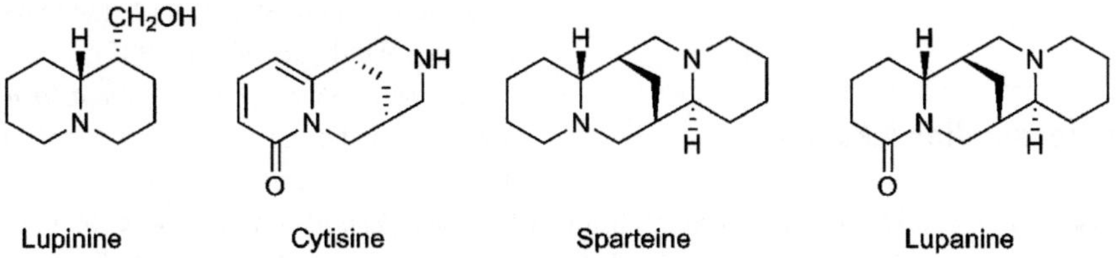

Fig. 7 Four representative structures of quinolizidine alkaloids

the bitter lupin, sweet lupins are more susceptible to herbivores (Wink, 2003). In addition to their function as defense compounds, a minor function of quinolizidines is nitrogen transport in the phloem and probably storage of nitrogen in seeds (Wink and Witte, 1984, 1985).

Quinolizidines have antiarrhythmic, CNS-depressant, hypotensive, and hypoglycemic effects. Their toxicity and some of their pharmacological properties can be explained through inhibition of Na^+ and K^+ channels and interaction with nicotinic and muscarinic acetylcholine receptors. Sparteine from broom (*Cytisus scoparius*) is used as antiarrhythmic. However, its medicinal use is declining and restricted because about 10% of all patients are unable to metabolize this alkaloid and suffer from intoxication (Wink, 2003).

Only few insects have adapted to quinolizidine alkaloids and sequester them as defense compounds, e.g., some aphids and larvae of the pyralid moth *Uresiphita reversalis* (Wink and Witte, 1991; Montllor et al., 1990). This is in contrast to pyrrolizidines, which are utilized by a large number of butterflies and beetles.

Alkaloids Derived from Tyrosine

The amino acid tyrosine is a precursor of numerous alkaloids. The largest group is formed by the benzylisoquinoline alkaloids. In addition, several other alkaloid classes originate from tyrosine, for example, the Ipecac alkaloids and the Amaryllidaceae alkaloids. The benzylisoquinolines, Ipecac- and Amaryllidaceae alkaloids will be reviewed.

Benzylisoquinoline Alkaloids

Benzylisoquinoline alkaloids are derived from two molecules of tyrosine. The central intermediate in their biosynthesis, (*S*)-reticuline, can undergo various rearrangements and modifications to yield the different structural classes of benzylisoquinolines (Fig. 8). At present, this diverse group of alkaloids comprises about 2,500 known structures. Benzylisoquinolines occur mainly in basal angiosperms, e.g., in members of the Berberidaceae, Fumariaceae, Papaveraceae, Menispermaceae, and Ranunculaceae, but also in other taxa.

Some benzylisoquinoline alkaloids have powerful pharmacological activities and have therefore found application in medicine, e.g., the analgesic morphine, the antitussive and analgesic codeine, the muscle relaxant tubocurarine, and the antimicrobial and anti-inflammatory sanguinarine. As can be expected from their highly diverse structures, these compounds have different mechanisms of action. Morphine and codeine are agonists at the μ-, δ-, and κ- opioid receptors, which are normally targeted by endorphines, enkephalines and dynorphines as endogenous ligands (Schiff, 2002). The bis-benzylisoquinoline alkaloid tubocurarine is the active principle of arrow poison from the liana *Chondrodendron tomentosum*, which is native to the Amazon Basin. Tubocurarine is an antagonist at nicotinic acetylcholine receptors on the neuromuscular end plate of skeletal muscle and is used to induce complete muscle relaxation before surgeries (Howland et al., 2005). Due to its quaternary nitrogen it is absorbed poorly and has to be injected intravenously.

Like tubocurarine, the quarternary benzophenanthridine sanguinarine is absorbed badly. It reacts with negatively charged and nucleophilic groups of proteins and inhibits several enzymes, e.g., Na^+/K^+-ATPase (Straub and Carver, 1975). In addition, it intercalates DNA due to its planar structure (Nandi and Maiti, 1985).

The opium poppy (*Papaver somniferum*) is an important medicinal plant with a colorful history. Opium, the dried latex of unripe capsules of *P. somniferum*, contains more than 80 isoquinoline alkaloids. The main alkaloids in opium are morphine (4-21%), followed by codeine,

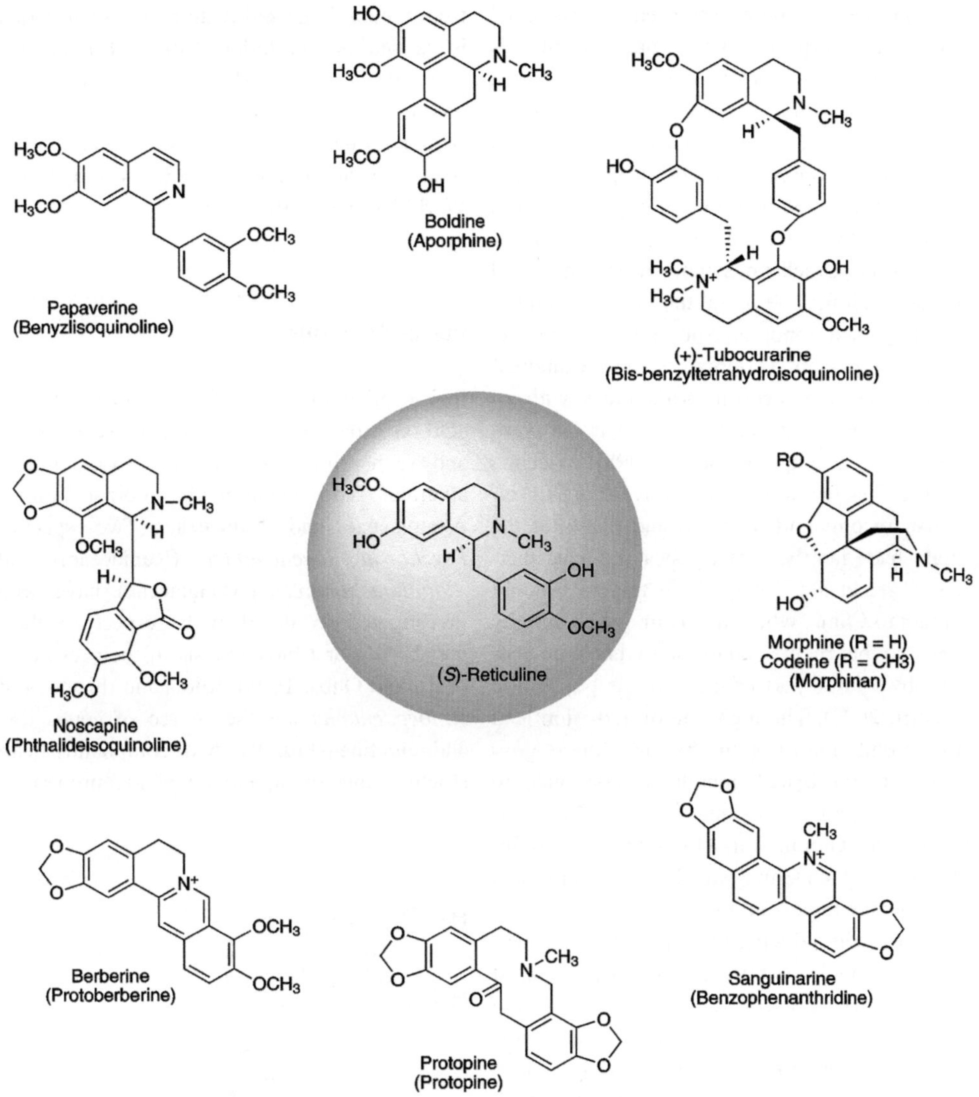

Fig. 8 (*S*)-Reticuline is the precursor for the various classes of benzylisoquinoline alkaloids

thebaine, papaverine, noscapine, and narceine (Dewick, 2002). The alkaloid concentrations vary strongly, and depending on the *P. somniferum* cultivar, also other alkaloids can occur in substantial amounts, e.g., oripavine in poppy from Tasmania (Frick et al., 2005). The only other plant species that accumulates the morphinans morphine and codeine is *Papaver setigerum*.

The opium poppy originated from the Mediterranean area. Earliest mention of the opium poppy, its cultivation, and the harvest of poppy latex are found on Sumerian clay tablets dating back to 3000 BC (Schiff, 2002). In ancient Greece, opium was used for medicinal and ritual purposes. The word "opium" is derived from the ancient Greek "opos" which means "milky juice of plants" (Askitopoulou et al., 2000). The twin brothers Hypnos and Thanatos, the Greek gods of sleep and death, are often depicted with opium poppies. Opium was mentioned by Homer in his "Iliad", and by the famous Greek physicians Hippocrates and Galen (Schiff, 2002). In the Roman empire, opium gained importance not only as medicine, but also as poison. Agrippina, emperor Claudius' wife, killed her stepson Brittannicus with an overdose of opium, so that her own son Nero could become emperor (Booth, 1998). Avicenna (980-1037), the famous Arab physician and scientist, recommended opium and plants of the nightshade family as analgesics and anaesthetics (Aziz et al., 2000). Arab traders brought opium to China, where it was first used only by the elite, but by the end of the seventeenth century by a large part of the Chinese population (Schiff, 2002). The high rate of addiction lead to a prohibition of opium by the Chinese government, but British merchants continued to smuggle opium into China. The conflict escalated in the Opium Wars (1839-1842 and 1856-1880), in which China was defeated and forced to allow the import of opium.

The sleep-inducing principle of opium was identified in 1806 by the German pharmacist Friedrich Sertürner. He succeeded in isolating crystalline morphine, which he named "morphium" after the Roman god of sleep "Morpheus" (Zenk and Jünger, 2007). It took more than 100 years, until the chemical structure of morphine was elucidated (1924-1925) by Gulland and Robinson. Total synthesis of morphine turned out to be extremely difficult due to its five stereo centers, and it was achieved by Gates and Tschudi (1952).

Since no cost-efficient synthesis of morphine and codeine has been developed, these alkaloids are isolated from "poppy straw", which consists of the entire plant tops (Dewick, 2002), and from opium. Legal cultivation of *P. somniferum* is carried out in India, Turkey, Russia, and Australia. Today, nearly all illegally produced opium (93%) originates from Afghanistan; smaller amounts are produced in South East Asia and South America (Sanderson, 2007; World Drug Report, 2007).

Ipecac Alkaloids

Ipecac alkaloids are derived from the amino acid tyrosine and the monoterpene secologanin and are therefore termed terpenoid-isoquinoline alkaloids. They occur in the eudicot families Alangiaceae and Rubiaceae. Two species, *Psychotria ipecacuanha* (Rubiaceae) and *Alangium lamarckii* (Alangiaceae), have been investigated in detail with respect to their metabolites and biosynthesis of their alkaloids (Fujii and Ohba, 1998). Roots and rhizomes of *P. ipecacuanha* are the source of cephaeline and emetine (Fig. 9), two compounds with emetic, expectorant, and amebicidal properties.

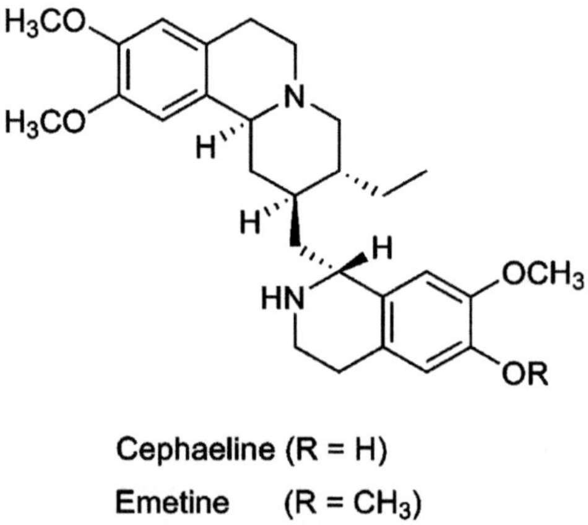

Fig. 9 Cephaeline and emetine, two alkaloids with emetic properties from roots of *Psychotria ipecacuanha*

The emetic effects are presumably mediated by 5-hydroxytryptamine 3 ($5HT_3$) receptors (Hasegawa et al., 2002). Ipecacuanha syrup is used to induce emesis after accidental ingestion of poisons. In lower doses, the extract of *P. ipecacuanha* roots is used as expectorant.

P. ipecacuanha occurs in the rainforests of Meso and South America. It was traditionally used in the Brazilian folk medicine. In the seventeenth century, the plant was brought by traders to France, and soon it found application in Europe as treatment against dysentery. The British physician Thomas Dover invented a special preparation *P. ipecacuanha* that was named Dover's powder after him. It consisted of Ipecacuanha root, opium, and potassium sulphate and was used as diaphoretic and medicine against cold and fever.

Amaryllidaceae Alkaloids

The Amaryllidaceae alkaloids are restricted to the monocot family that coined their name. They are derived from one molecule of tyrosine and protocatechuic aldehyde, which originates from phenylalanine. The central intermediate of their biosynthetic pathway is norbelladine. Nearly 500 structures of Amaryllidaceae alkaloids are known, and some of them possess significant pharmacological activities (Jin, 2007) (Fig. 10). For example, the isocarbostyrils pancratistatin from the spider lily (*Hymenocallis littoralis*) and narciclasine from *Narcissus* species show promising antineoplastic properties (Dumont et al., 2007; McLachlan et al., 2005). Lycorine that occurs, e.g., in *Clivia*, *Crinum* and *Galanthus*

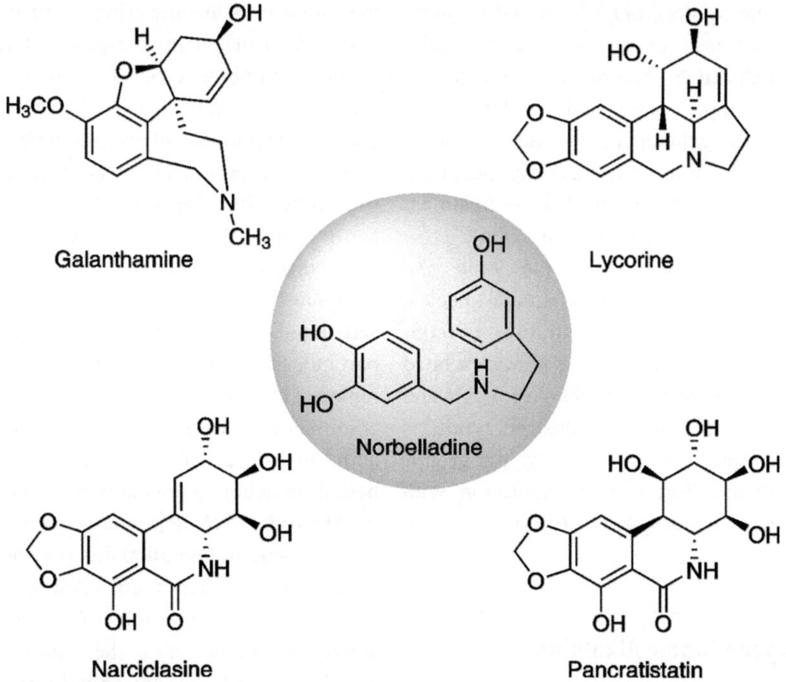

Fig. 10 Exemplary structures of Amaryllidaceae alkaloids

species exhibits antiviral activity (Ieven et al., 1983; Szlávik et al., 2004).

Galanthamine is the only alkaloid of this class that has already found application in medicine. It is approved for the symptomatic treatment of Alzheimer's disease in Europe and the United States. Its mode of action consists in a competitive and reversible inhibition of acetylcholinesterase, which leads to an increased concentration of acetylcholine at neuronal synapses. In addition, galanthamine acts as an allosteric modulator on nicotinic acetylcholine receptors. Since a characteristic feature of Alzheimer's disease is the loss of acetylcholinergic neurons concomitant with decreased levels of acetylcholine, galanthamine can, at least partially, compensate the damage and thus enhance cognitive functions in Alzheimer's patients.

Galanthamine was isolated first from the Caucasian snowdrop (*Galanthus woronowii*) in the early 1950s. Most of the early research on galanthamine was carried out in Bulgaria and the USSR during the Cold War. Initially, galanthamine was used to reverse neuromuscular blockade induced by muscle relaxants and for the treatment of post-polio paralysis. After it was discovered that galanthamine passes the blood brain barrier, the interest in this drug increased, and it was eventually established as treatment for Alzheimer's disease (Heinrich, 2004).

Galanthamine occurs in the bulbs of *Galanthus*, *Narcissus*, and *Leucojum* species, where it accumulates in concentrations of 0.05-0.2% (Dewick, 2002). Initially, it was isolated from these plant species. In 1999, a feasible and economic protocol for the industrial synthesis of galanthamine was developed by the groups of Fröhlich and Jordis in collaboration with Sanochemia (Küenburg et al., 1999).

Monoterpene Indole Alkaloids

This class of alkaloids is biosynthesized from tryptophan and secologanin via the central intermediate $3\alpha(S)$-strictosidine. Over 2,000 structurally diverse monoterpene indole alkaloids are known, and among them are several pharmacologically valuable compounds (O'Connor and Maresh, 2006). Some representative structures of the major classes of monoterpene indole alkaloids are depicted in Fig. 11. These alkaloids are mainly found in the plant families Apocynaceae, Loganiaceae, Nyssaceae, and Rubiaceae. The following section will focus on some representative alkaloids with significant pharmacological activities.

Rauwolfia Alkaloids

The Indian snakeroot, *Rauwolfia serpentina* (Apocynaceae), is a shrub that grows in southern and southeast Asia. The root of this plant has been used traditionally in Ayurvedic medicine (Sanskrit name: "Sarpagandha") to treat hypertension and mental disorders. In the 1950s, the blood pressure lowering agent was identified as reserpine, a monoterpene indole alkaloid of the yohimbine class. Reserpine inhibits a proton pump that is responsible for maintaining a high concentration of protons in neuronal vesicles. If the proton pump is inhibited, neurotransmitters like dopamine and norepinephrine can no longer be stored in the vesicles. Thus, the neurons are depleted of their transmitters, which leads to a decrease in blood pressure and sedation. Due to serious negative side effects, in particular depression, reserpine has mainly been replaced by other drugs, but is still used in some combinations with other antihypertonics. In addition, extracts of *Rauwolfia* roots are used in herbal remedies against hypertension.

The roots of *R. serpentina* contain 0.7–2.4% monoterpene indole alkaloids (Dewick, 2002). In addition to their main alkaloid reserpine, they also produce other pharmacologically active alkaloids, e.g., the antihypertensive ajmalicine and the antiarrhythmic ajmaline. Ajmaline blocks Na^+ channels in the heart and thus prolongs intraventricular conduction times.

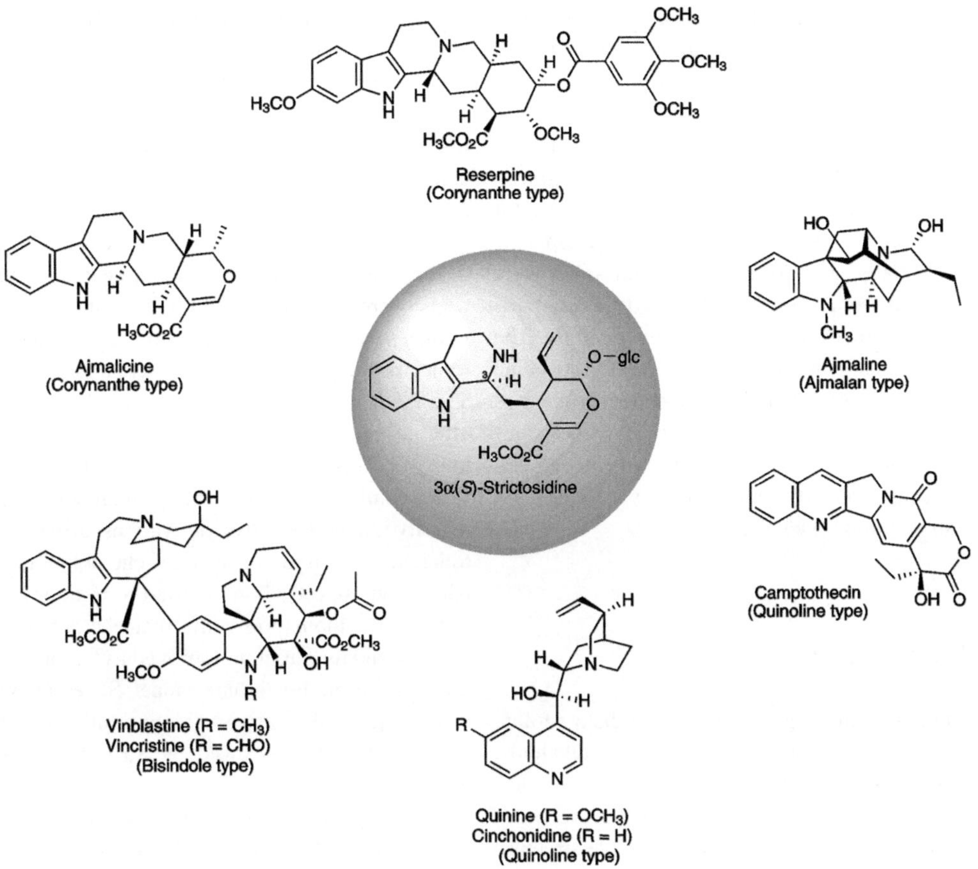

Fig. 11 Different classes of monoterpene indole alkaloids

Catharanthus Alkaloids

The Madagascar periwinkle (*Catharanthus roseus*, Apocynaceae, formerly known as *Vinca rosea*) is a small subshrub or herbaceous plant native to Madagascar. It contains about 130 monoterpene indole alkaloids of different subclasses (van der Heijden et al., 2004). Nowadays, *C. roseus* occurs worldwide in subtropical and tropical regions. It is cultivated as an ornamental plant, but it has also found application in the folk medicine of various countries. Because the plant was used as an antidiabetic in Jamaica, it was screened for hypoglycaemic activity by Eli-Lilly, USA, and the Cancer Research Center, Canada, in the late 1950s. Although plant extracts proved to be ineffective against diabetes, scientists of both institutes independently discovered the anticancer activity of several bisindole alkaloids, in particular, vinblastine and vincristine (reviewed by Noble, 1990). These compounds bind tubulin and inhibit its polymerization, thus hindering the formation of the mitotic spindle and causing cell cycle arrest at metaphase in dividing cells (Jordan et al., 1991). Vinblastine is used as therapy against Hodgkin's disease and non-Hodgkin's lymphomas. Its major side effect is a suppression of the bone marrow. Vincristine is more powerful but also more neurotoxic than vinblastine. It is used to treat acute lymphatic leukaemia, non-Hodgkin's lymphomas, rhabdomyosarcoma, and Wilms tumor. Two semisynthetic analogs with less side effects than the two Vinca alkaloids, vinorelbine and vindesine, are used as to treat non-small cell lung cancer and breast cancer. Vindesine is also used as chemotherapeutic for lymphoma, acute leukaemia, and melanoma.

Initially, vincristine and vinblastine were isolated from leaves of the Madagaskar periwinkle, however, the yield was very low. The plant tissue contains only 0.0002% vinblastine (Noble, 1990), and the vincristine content is even lower. Therefore, a partial synthesis for the dimeric indole alkaloids was developed starting from the monomers vindoline and catharanthine (Dewick, 2002).

Camptothecin

Camptothecin (Fig. 11) belongs to the quinoline class of the monoterpene indole alkaloids. Although it lacks the indole ring, feeding studies proved that it originates from tryptamine and a monoterpene precursor, and the indole structure undergoes rearrangements to a quinoline heterocycle (Hutchinson et al., 1974; Sheriha and Rapoport, 1976). The alkaloid occurs in several unrelated eudicot species, e.g., *Camptotheca acuminata* (Nyssaceae), *Ophiorrhiza pumila* (Rubiaceae), *Ervatamia heyneana* (Apocynaceae), and *Nothapodytes foetida* (Icacinaceae).

Camptothecin is unique in its mechanism of action. It binds to the cleavable complex of topoisomerase I and covalently attached DNA and stabilizes it (Hsiang et al., 1985). This non-degradable DNA/topoisomerase I complex arrests the replication fork and thus kills cells by inhibition of DNA synthesis (Hsiang et al., 1989). Camptothecin and its derivatives are therefore also termed topoisomerase "poisons".

Camptothecin was isolated first in 1966 (Wall et al., 1966) from *Camptotheca acuminata*, a tree native to China and Tibet, also known as "Happy Tree" (chinese "xi shu"). Despite the promising anticancer activities, the poor solubility of the alkaloid presented a major obstacle to clinical application. Water-soluble derivatives were prepared by opening the lactone ring. During clinical trials, however, it became apparent that the anticancer activity of these analogs was greatly decreased, and the trials were abandoned. Only later it became known that the anticancer activity of camptothecin is dependent on the intact lactone ring. The interest in camptothecin returned in 1985 after its unique mechanism of action became known. This encouraged the synthesis of water-soluble analogs that retained the activity. At present, two derivatives of camptothecin are used in cancer chemotherapy. Irinotecan (syn. CPT-11) is used to treat colon cancer in combination with other chemotherapeutics; and topotecan is approved as therapy of ovarian and small-cell lung cancer. Several new camptothecin derivatives are currently tested in clinical trials (reviewed by Sirikantaramas et al., 2007).

Camptothecin derivatives are produced semisynthetically using the alkaloids extracted from intact plants of *C. acuminata* or *N. foetida*. Alternatives to this limited resource have been suggested. For example, young leaves of *C. acuminata* accumulate high levels of alkaloids (4-5 mg/g dry weight; López-Meyer et al., 1994) and can be harvested repeatedly without killing the trees. In addition, the clonal propagation of elite cultivars by shoot and bud culture of *C. acuminata* (Vincent et al., 1997) or hairy roots of *Ophiorrhiza pumila* (Sudo et al., 2002) might present a suitable solution to overcome the shortage in plant material. Recently, camptothecin production was detected in the endophytic fungi *Entrophospora infrequens* of *N. foetida* (Puri et al., 2005; Amna et al., 2006), and this may open a new source for the commercial production of the antineoplastic alkaloid.

Cinchona Alkaloids

The genus *Cinchona* (Rubiaceae) comprises about 25 species of tall, evergreen trees that grow in South America. The bark of these trees accumulates quinoline alkaloids that are, like camptothecin, derived from tryptophan and secologanin. Cinchona alkaloids are also found in the genus *Remijia* of the Rubiaceae family.

The *Cinchona* bark was called "Quina-Quina", which means "bark of barks", in the native Indian language Quechua. It was discovered by Spanish monks in Peru around 1630. They either learned to use the bark against fevers from the indigenous Indian population or discovered its application by themselves (Bruce-Chwatt, 1988). *Cinchona* bark was introduced to Europe by Jesuits, where it became known as "Jesuits' powder" and was used to cure malaria, which was then widespread in Europe. However, initially the use of *Cinchona* powder was controversial. Due to its approval by the Vatican, Protestants refused to take it. In addition, healing of malaria with a hot, bitter drink form a powdered bark contradicted the humoral theory of the antique Greek philosophers, which was at that time still the base for medicinal treatment. Moreover, often bark of bad quality or adulterated bark was sold that proved to be inefficient. *Cinchona* bark became only widely accepted after the English apothecary apprentice Robert Talbor applied it successfully as a secret formula to cure many members of European royalty from malaria, among them the English King Charles II and the son of the French King Louis XIV. After Robert Talbor's death it was disclosed that his secret remedy was based on *Cinchona bark* (Bruce-Chwatt, 1988; Kaufman and Rúveda, 2005). Quinine (Fig. 11) was isolated from the bark of *Cinchona* trees in 1820 by the French pharmacists Pelletier and Caventou, and its molecular formula was established in 1854 by Adolf Strecker.

For two centuries, *Cinchona* bark was obtained solely from South America. It was particularly valuable for the colonial powers because malaria was frequent in Asia and Africa. Due to the high demand for the drug and the dwindling natural resources, efforts were taken to cultivate the trees outside South America. In the middle of the eighteenth century, the Dutch and English succeeded in growing *Cinchona* trees in Java and India. Shortage of quinine in World Wars I and II due to trade embargos encouraged the development of synthetic analogs, e.g., the 4-aminoquinolines chloroquine and mefloquine, and the 8-aminoquinoline primaquine. A formal synthesis of quinine was achieved by Woodward and Doering (1944), however, due to the four stereocenters in the quinine molecule it is a very complex synthesis and commercially not feasible.

Today, three *Cinchona* species are cultivated for the production of quinine. *C. succirubra* yields the "red bark", *C. legderiana* the "brown bark", and *C. calisaya* the "yellow bark" (Dewick, 2002). In addition to quinine, the barks contain significant amounts of three other quinoline alkaloids: quinidine, the diastereomer of quinine, which is used as an antiarrhythmic, and 6-demethoxy analogs of the two alkaloids, cinchonine and cinchonidine.

Quinine and its analogs act on erythrocytic stages of *Plasmodium falciparum*, the causative agent of malaria. It is assumed that these antimalarial agents inhibit the polymerization of haematine, which is released upon the degradation of haemoglobin and is toxic for the parasite, into non-toxic hemozoin (Chou et al., 1980; Sullivan et al., 1996).

Since quinine is extremely bitter, gin was added to make it easier to drink, giving rise to the cocktail "gin and tonic" that nowadays contains only minute amounts of the alkaloid. In addition to tonic water, quinine is also an ingredient of other beverages, e.g., bitter lemon or vermouth.

1

Benzoxazinones

Benzoxazinones are derived from indole-3-glycerol phosphate, a molecule that is also the direct precursor of tryptophan. In the literature, acronyms derived from the substitution pattern of the benzoxazinone ring are often used to distinguish individual members of this class. For example, DIMBOA is 2,4-dihydroxy-7-methoxy-1,4-benzoxazin-3-one. Benzoxazinones occur mainly in the monocot Poaceae family, but also in some families of eudicot plants, e.g., Ranunculaceae, Acanthaceae, and Plantaginaceae. At present, it is still being investigated whether the pathway developed only once or several times independently after the divergence of moncots and dicots (Sicker et al., 2000).

Two characteristic structural features are found in all benzoxazinones: a cyclic hemiacetal in combination with a cyclic hydroxamic acid or a cyclic lactam (Fig. 12); the hydroxy function of the hemiacetal is usually glucosylated. In the intact plant tissue, these benzoxazinone glucosides are stored in the vacuole, whereas a specific glucosidase is located in plastids. Only after injury of the plant tissue, glucosides and glucosidases are released, and free aglucones are formed. Those benzoxazinone aglucones that contain a hydroxamic acid function are chemically instable, and their tautomeric open-ring form can be converted to benzoxazolinones under loss of one carbon as formic acid (Fig. 12).

Benzoxazinones act as pre-formed defense and possess antibacterial, antifungal, and antialgal properties (Bravo and Lazo, 1993, 1996). In addition, they serve as feeding deterrents and reduce the vitality of pests. In particular, these metabolites confer resistance to one of the major corn pests, the European corn borer (*Ostrinia nubialis*) (Grombacher et al., 1989). Only benzoxazinones with hydroxamic acid function show these bioactivities. An electron-donating hydroxy or methoxy group at C-7 increases the reactivity. The mode of action of benzoxazinones can be explained by modification of amino and thiol groups of biomolecules. The aldehyde function of the tautomeric open-ring form can react as an electrophile with NH_2 groups and form Schiff bases (Pérez and Niemeyer, 1989). Thiol groups can be oxidized by the cyclic hydroxamic acid form, which is in turn reduced to a lactam. Structural prerequisite for this oxidation is an electron-donating substitution at C-7 of the benzoxazinone skeleton (Atkinson et al, 1991). Benzoxazinoids that have been bio-activated by *N*-acetylation may act as alkylating agents towards nucleic acids and proteins (Hashimoto and Shudo, 1996).

Due to their toxicity, benzoxazinones can also function as allelochemicals (Sicker et al., 2000) and are therefore discussed as natural herbicides.

Cyanogenic Glycosides

Cyanogenic glucosides are β-glucosides of α-hydroxynitriles (syn. cyanohydrins), which

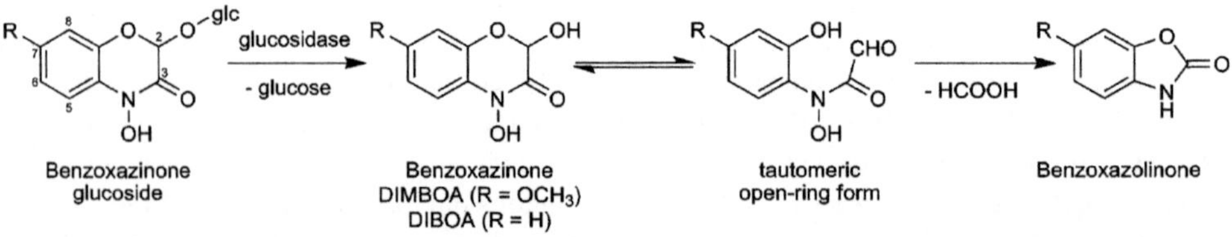

Fig. 12 Enzymatic and chemical degradation of benzoxazines with hydroxamic acid function (Sicker et al., 2000)

are derived from the five proteinogenic amino acids phenylalanine, tyrosine, valine, isoleucine, leucine and the non-proteinogenic amino acid cyclopentenyl-glycine. Because of the stereo-center in the α-hydroxynitrile function, (*R*)- and (*S*)-forms of several cyanogenic glucosides exist. About 2,500 different plant species including ferns, gymnosperms, and angiosperms produce cyanogenic glucosides (Hegnauer, 1986; Seigler, 1991). Despite their widespread occurrence, these natural products are found predominantly in the families Araceae, Asteraceae, Euphorbiaceae, Fabaceae, Passifloraceae, Poaceae, and Rosaceae (Dewick, 2002). Some of the most abundant molecules are amygdalin (Rosaceae), linamarin and lotaustralin (Fabaceae), and the epimers dhurrin and taxiphyllin in the genus *Sorghum* (Seigler, 1991).

Like the benzoxazinones, cyanogenic glucosides belong to the preformed defense of the plant and are stored in the vacuole. Upon disruption of the plant tissue, they are degraded by β-glucosidases to the corresponding α-hydroxynitriles, which are hydrolyzed by α-hydroxynitrile lyases to aldehydes or ketones and toxic hydrogen cyanide (HCN) (Fig. 13). Since the α-hydroxynitriles are unstable, they can also decompose spontaneously, but the enzyme catalyzed reaction proceeds up to 20 times faster (Selmar, 1999). In the gut of herbivores, the β-glucosidic bond can also be hydrolyzed by intestinal bacteria. The toxicity of hydrogen cyanide can be explained by its affinity to metal ions. Cyanide ions complex iron (III) in the active site of cytochrome oxidase and thus inhibit the respiratory chain. Mammals can consume small amounts of cyanogenic glucosides and detoxify them, mainly *via* the liver enzyme rhodanese that converts cyanide to thiocyanate. Chronic intake of non-lethal amounts, however, can result in paralysis of legs (Konzo) or neurological disorders due to cyanide intoxication or iodine deficiency caused by accumulation of the iodine antagonist thiocyanate (Selmar, 1999). Intoxications by cyanogenic glucosides are often observed in populations that live on a diet poor in protein with insufficient supply of sulfur-containing amino acids, which are required for the detoxification of cyanide.

Cyanogenic glucosides act as feeding deterrents. Herbivores are probably rejected by the keto or aldehyde compound that arises after cleavage rather than by the cyanide (Jones, 1988). By transferring all genes required for the

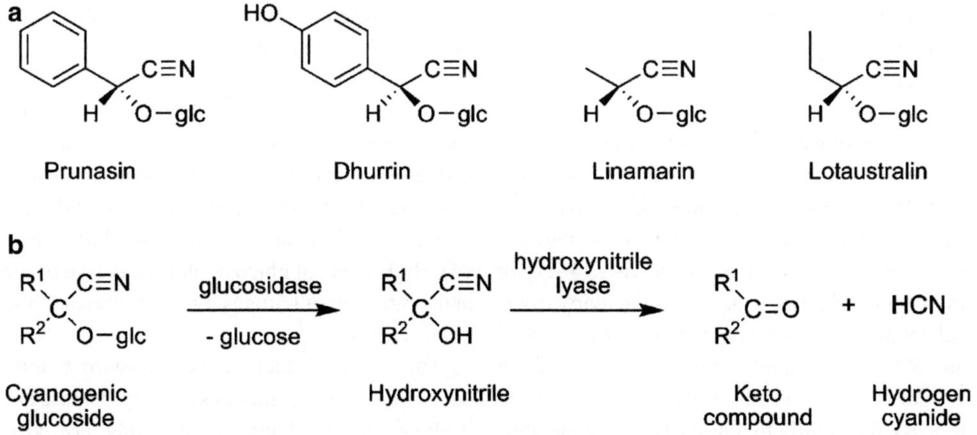

Fig. 13 Representative structures of cyanogenic glucosides (**a**) and degradation of cyanogenic glucosides with concomitant release of toxic hydrogen cyanide (**b**)

formation of the cyanogenic glucoside dhurrin from *Sorghum bicolor* into *Arabidopsis*, Tattersall et al. (2001) proved that cyanogenic glucosides play a role in plant defense. In comparison with wild-type leaves, leaves of transgenic *Arabidopsis* plants producing dhurrin were hardly consumed by the yellow-striped flea beetle (*Phyllotreta nemorum*). Despite the toxicity of the cyanogenic glucosides, several herbivores, especially insects, are able to feed on plants containing these natural products, and in this case the toxic compounds may act as phagostimulants. Some species of beetles, centipedes, and millipedes, but particularly many moths and butterflies sequester cyanogenic glucosides as defense compounds. The compounds are either taken up by feeding on cyanogenic plants or synthesized *de novo* by endogenous enzymes. In contrast to vertebrates, these arthropod species detoxify cyanide not by rhodanese, but mostly *via* β-cyanoalanine synthesis, a mechanism that is also used by plants (reviewed by Zagrobelny et al., 2008).

It has been postulated that cyanogenic glucosides also serve as storage compounds for reduced nitrogen and sugar (Selmar et al., 1988; Sánchez-Pérez et al., 2008). This has been deduced from the observation that cyanogenic glucosides are degraded during seed development or germination.

Interestingly, many important food crops accumulate cyanogenic glucosides (Jones, 1998), but usually not in the portion of the plant that is consumed. Some plants, however, contain high levels of these toxic constituents in the parts that are eaten, e.g., bamboo, cassava, lima beans, and sorghum. This problem is particularly serious in the case of cassava (*Manihot esculenta*), which is a major crop in many tropical countries. Cassava roots contain between 10 and 500 mg of cyanide equivalents per kg fresh weight (O'Brien et al., 1991) and have to be processed carefully to remove the toxic metabolites. Unfortunately, this treatment usually results in loss of protein, minerals, and vitamins. Various approaches to produce transgenic cassava with reduced content of cyanogenic glucosides in roots are currently underway (Jørgensen et al., 2005; Siritunga and Sayre, 2007).

Glucosinolates

Glucosinlates are β-thioglucosides of (Z)-*N*-hydroximinosulfate esters (Fig. 14). They are derived from the aliphatic amino acids alanine, isoleucine, leucine, methionine, and valine or from the aromatic amino acids phenylalanine, tryptophan, and tyrosine and share the first steps of cyanogenic glucoside biosynthesis. About 120 different structures of glucosinolates are known (Fahey et al., 2001). They occur exclusively in sixteen eudicot families, the majority belonging to the order Brassicales. Many of these plants are cultivated and consumed as vegetables or spices, e.g., cabbage (*Brassica oleraceae*, Brassicaeae), capers (*Capparis spinosa*, Capparidaeae), mustard (*Sinapis alba*, Brassicaeae), and wasabi (*Wasabia japonica*, Brassicaeae). The strong or pungent flavor of these plants can be explained by the presence of glucosinolates. If the plant tissue is damaged, the glucosinolates are hydrolyzed by myrosinase, a thioglucosidase that is spatially separated in the undamaged tissue. The resulting unstable thiohydroximate-*O*-sulfate intermediates undergo non-enzymatic loss of sulfate and spontaneous rearrangement to various bioactive products like isothiocyanates, nitriles, oxazolidine-2-thiones, epithionitriles, and thiocyanates (Halkier and Gershenzon, 2006) (Fig. 14). Hydrolysis of glucosinolates can also occur in the intestine of humans by myrosinases of the gut microflora.

The main product of the "mustard bomb" consisting of glucosinolates and myrosinase are isothiocyanates. These compounds are also responsible for many of the biological effects of glucosinolates, e.g., antibacterial, antifungal,

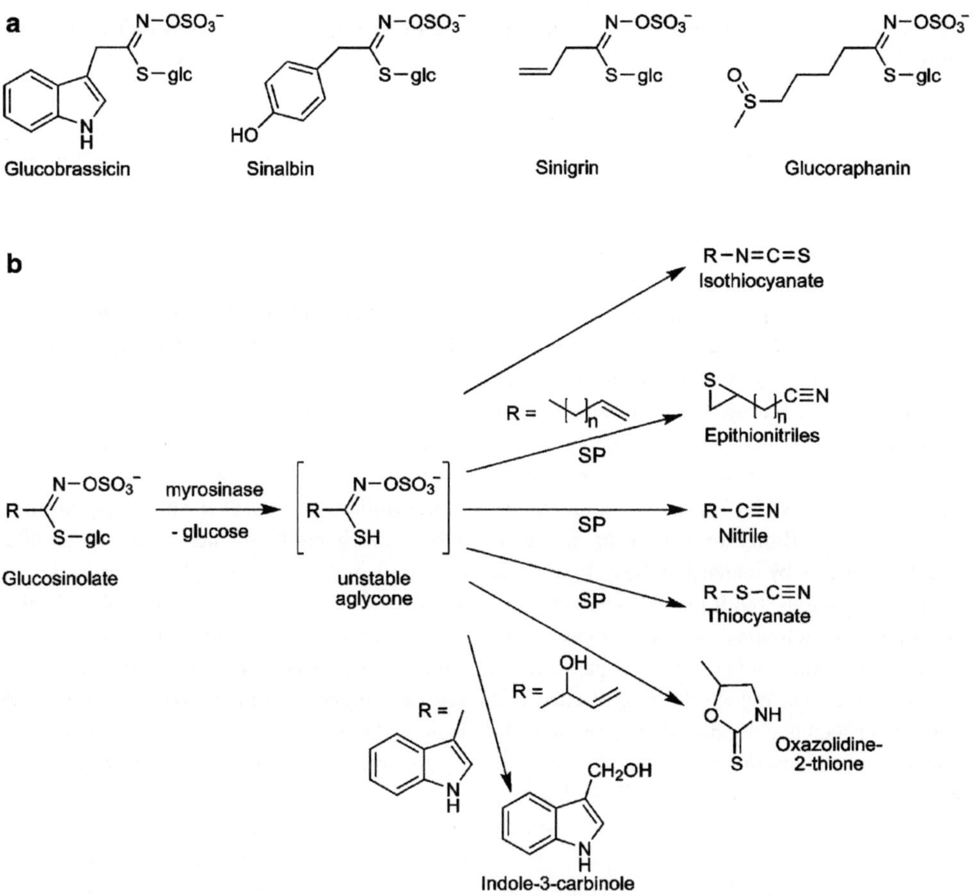

Fig. 14 Exemplary structures of glucosinolates (**a**) and hydrolysis of glucosinolates by myrosinase and rearrangement to various products (**b**) Isothiocyanates are the predominant degradation products. For the formation of epithionitriles, nitriles, and thiocyanates, specifier proteins (SP) are required

nematicidal, and feeding deterrent activities (Fahey et al., 2001). The formation of hydrolysis products distinct from thiocyanates depends on the structure of the glucosinolates, pH, and the presence or absence of Fe^{2+} ions or specifier proteins. Specifier proteins do probably not possess catalytic activity, but may modulate myrosinase activity allosterically to yield nitriles, epithionitriles, or thiocyanates as degradation products instead of thiocyanates (Wittstock and Burow, 2007). Epithionitriles are produced only from glucosinolates with a terminal double bond. Indolyl glucosinolates are hydrolyzed to unstable indole isothiocyanates that give rise to the alcohol indole-3-carbinol and a variety of other products. Hydrolysis of β-hydroxyalkenyl glucosinolates yields oxazolidine-2-thiones that can cause goitre by inhibiting the incorporation of iodine into thyroid hormones. A crop particularly rich in goitrogenic glucosinolates is rape (*Brassica napus*), an important source of vegetable oil. To make the protein rich seed cake that remains after extraction of the oil suitable as animal foodstuff, rape plants with low levels of glucosinolates have been developed by breeding efforts (Fahey et al, 2001).

Glucosinolates from Brassicaceae vegetables (e.g., broccoli, cauliflower, Chinese cabbage, kale, kohlrabi, mustard) are discussed as cancer preventive agents. In particular, sulforaphane and indole-3-carbinol, the degradation products of glucoraphanin and glucobrassicin, respectively, show promising activities, e.g., stimulation of apoptosis. Sulforaphane enhances the excretion of cancerogenous compounds by inducing phase II detoxification enzymes like glutathione-*S*-transferase, UDP-glucuronosyl transferase, and NADPH quinone oxidoreductase. Indole-3-carbinol may prevent estrogen-sensitive cancers by increasing the ratio of weak to strong estrogens, but it may also possess cancer-promoting activities. Although some epidemiological studies suggest that a diet rich in glucosinolates can reduce the risk of cancer, it has not yet been unambiguously proven. In addition, the bioavailability of glucosinolates may be influenced by genetic polymorphisms that lead to a slower excretion of these compounds (Higdon et al., 2007).

Although glucosinolates act as feeding deterrents, many insect herbivores feed on plants containing these natural products. Two very different mechanisms for the detoxification of glucosinolates are known from two insect species. The diamond-black moth (*Plutella xylostella*) produces a sulfatase that cleaves the sulfate from glucosinolates and converts them to compounds that cannot be degraded by myrosinases (Ratzka et al., 2002). The cabbage white butterfly (*Pieris rapae*) contains a specifier protein that transforms glucosinolates in the presence of myrosinase to nontoxic nitriles that are excreted with the feces (Wittstock et al., 2004). Both detoxifying enzymes are expressed in the gut of the insects, the organ in which the glucosinolates would normally be degraded to isothiocyanates. Other insect herbivores sequester glucosinolates and use them for their own defense. This requires either an endogenous myrosinase that is spatially separated from the glucosinolates in the insects or myrosinases from the gut microflora of their enemies (Halkier and Gershenzon, 2006).

Natural Products Derived from the Shikimate Pathway and Phenylpropanoids

The shikimate pathway provides the precursors for benzoic acid derivatives and phenylpropanoid compounds in plants (Fig. 15). Shikimate is biosynthesized from D-erythrose-4-phosphate and phosphoenolpyruvate, two metabolites derived from the pentose phosphate cycle and glycolysis, respectively. Shikimate is further converted to chorismate by addition of a C_3 unit from phosphoenolpyruvate; and chorismate serves as precursor of the aromatic amino acids L-phenylalanine, L-tyrosine, and L-tryptophan.

An intermediate of the shikimate pathway, most likely 5-dehydroshikimate, is the precursor of gallic acid and the gallotannins (Werner et al., 1997), which are esters of glucose with several molecules of gallic acid. Gallotannins have been used for centuries for the tanning of hides and for the preparation of ink from ferrous sulfate and oak gall extract.

Since the shikimate pathway occurs only in plants and microorganisms, L-phenylalanine, L-tyrosine, and L-tryptophan are essential for animals and have to be taken up with food. L-Phenylalanine, and in monocots also L-tyrosine, are the precursors of the phenylpropanoids. This class comprises cinnamic acid derivatives, lignin, lignans, phenylpropenes, and coumarins, which all share the basic C_6-C_3 skeleton. Phenylpropanoids are aromatic compounds, often with a hydroxy group in the *para* position. If more than one hydroxy group is present at the aromatic ring, the new hydroxy function is usually positioned next to the first hydroxy group (*ortho* position).

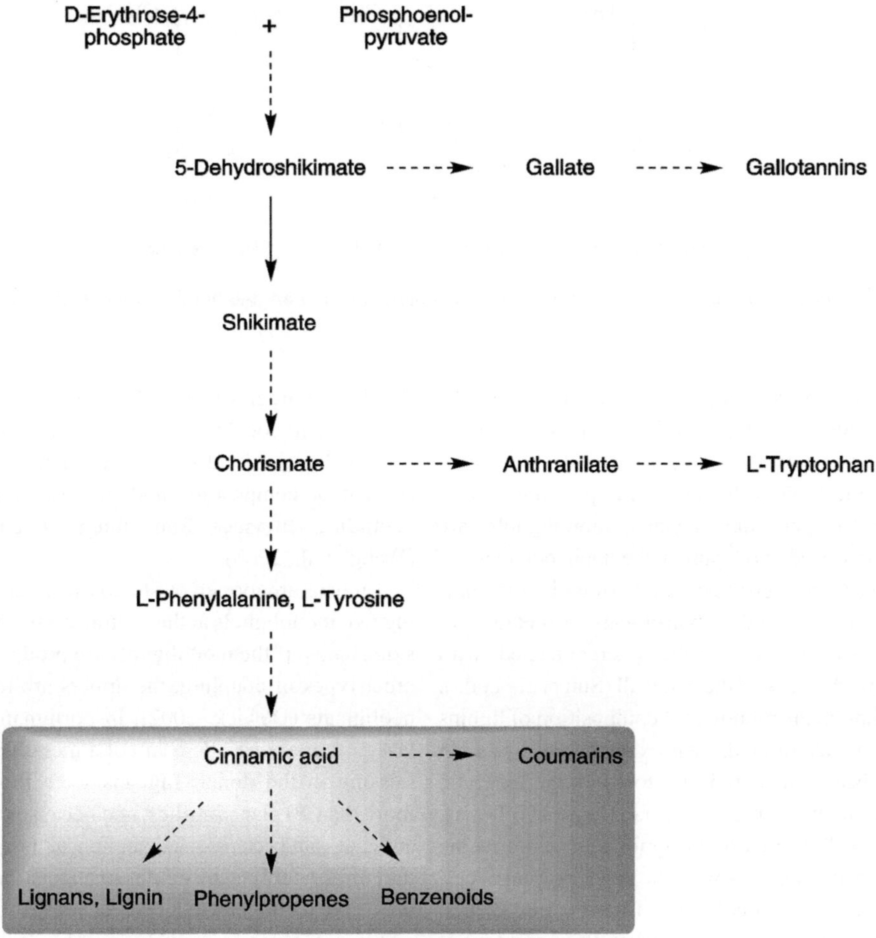

Fig. 15 Schematic overview of shikimate and phenylpropanoid biosynthesis. *Arrows* with dashed lines indicate multiple biosynthetic reactions. Boxed compounds are phenylpropanoids

Phenylpropanoids with additional carbons derived from acetate units, e.g., the flavonoids, will be discussed together with the polyketides.

Lignans and Lignins

Lignans and lignins are both composed of the hydroxy cinnamic alcohols (monolignols) *p*-coumaryl alcohol, coniferyl alcohol, and sinapyl alcohol (Fig. 16). Lignans are formed by stereoselective coupling of two hydroxy cinnamic alcohols units and lignins are polymers of monolignols.

After incorporation into the polymer lignin, the monolignols *p*-coumaryl alcohol, coniferyl alcohol, and sinapyl alcohol are also referred to as H (*p*-hydroxyphenyl), G (guaiacyl), and S (syringyl) units, respectively.

Lignin from gymnosperms consists mainly of G units and low levels of H units. Eudicots

1

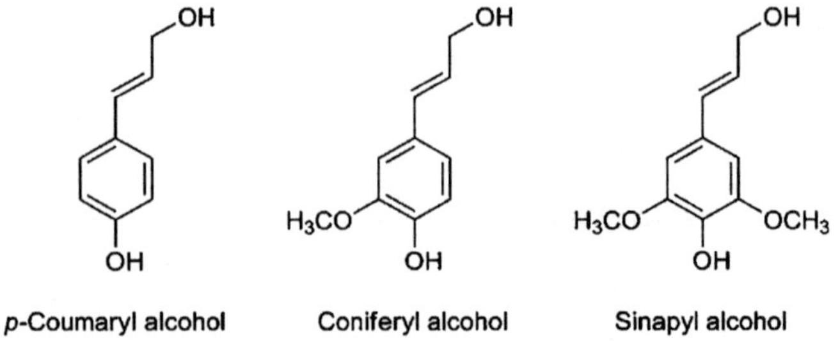

Fig. 16 *p*-Coumaryl alchol, coniferyl alcohol, and sinapyl alcohol are the building blocks of lignins and lignans

and monocots utilize all three monolignols, although lignin from eudicots consist mainly of G and S units (Boerjan et al., 2003). Recently, it became evident that also other phenolic monomers, in particular acylated monolignols, are incorporated into lignin. In the lignin polymer, the alcohols are connected by various bonds comprising ether and carbon-to-carbon linkages. In addition, lignin can be interconnected with hemicelluloses of the cell wall (Sun et al., 2005). Although the monolignol composition of lignins can be determined, their exact structure has not yet been elucidated due to the large size and complexity of the polymers (Davin and Lewis, 2005). The function of lignin is to reinforce the cell walls together with the sugar polymers cellulose and hemicellulose. Lignification of cell walls is required to strengthen the vascular tissue and evolved ca. 400 million years ago in the Silurian with the emergence of the first vascular plants. After cellulose, lignin is the second most abundant biopolymer on earth. From an economic point of view, lignin is important for the quality of wood, but it is an undesirable component for the paper industry, because its oxidation leads to yellowing of paper. Lignin cannot be digested by ruminants and therefore decreases the digestibility of forage and the absorption of nutritients (Boerjan et al., 2003; Rouhi et al., 2000). In addition, lignin has to be removed from lignocellulose-containing plant material prior to the production of biofuels, because it hinders the degradation and extraction of cellulose. Possible strategies to improve biofuel production from lignocellulose are the generation of genetically modified crops with altered lignin content or composition and the use of lignin degrading enzymes from fungi or bacteria (Weng et al., 2008).

Lignans are formed by stereoselectively linking two monolignols at the central atoms of their side chains. If the monolignols are produced by other types of coupling, the dimers are termed neolignans (Dewick, 2002). In norlignans, the last carbon of one side chain of a monolignol is missing in the dimer. Lignans were found in more than 70 plant families, and because of their antiviral, antibacterial, and antifungal properties they presumably act as defense against herbivores and pathogens (Saleem et al., 2005). In addition, they occur in many plant foods like oil seeds, whole cereals, fruits and vegetables. A particular rich source of lignans with more than 0.3 g/100 g are flaxseed and sesame seed with secoisolariciresinol (Fig. 17) and sesamine as main constituents, respectively (Adlercreutz, 2007; Milder et al., 2005). These latter and several other lignans can be converted by the intestinal microflora to the mammalian lignans enterodiol and enterolactone (Fig. 17). The two enterolignans are weak phytoestrogens; they increase the concentrations of sex hormone binding globulin in the plasma and modulate steroid hormone concentrations by competing for their metabolizing enzymes (Adlercreutz, 2007). It is assumed that consumption of a diet

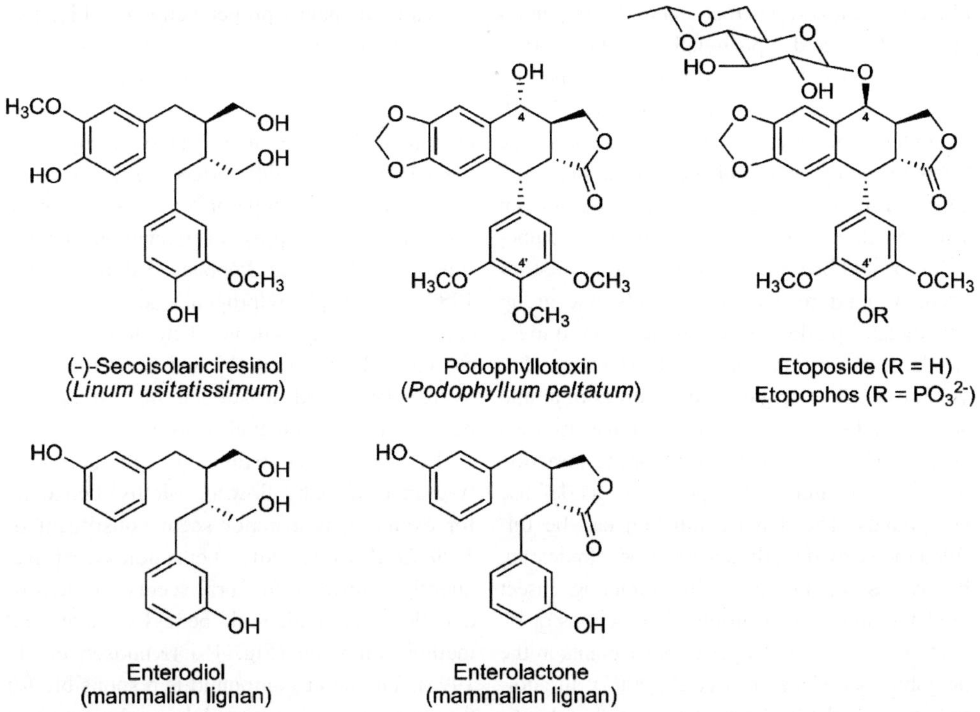

Fig. 17 Lignans of various origin

rich in lignans benefits health and reduces the risk for colon and breast cancer. Since lignan-rich food usually contains other health-promoting ingredients like fibers or other polyphenols, it is difficult, however, to attribute the beneficial effects solely to lignans.

A strong cytotoxic lignan is podophyllotoxin (Fig. 17) from *Podophyllum peltatum* or *P. hexandrum* (Berberidaceae). Podophyllotoxin has the same mechanism of action as the terpene indole alkaloids vinblastine and vincristine; it inhibits cell division by binding to tubulin and preventing its polymerization. This lignan is used as treatment for warts, but is too toxic for systemic application. Derivatives with reduced toxicity are etoposide, etopophos, and teniposide. They are prepared semi-synthetically from 4'-demethylpodophyllotoxin and have an inverted stereochemistry at C-4. These changes lead to a new anticancer mechanism, the stabilization of topoisomerase-DNA complexes, which is similar to that of camptothecin. Unlike camptothecin, however, the podophyllotoxin analogues attack topoisomerase II, not topoisomerase I. Etoposide and its prodrug teniposide are used in combination with other drugs to treat small cell lung cancer, testicular cancer, and certain lymphomas. Teniposide is used as therapy for childhood acute lymphocytic leukemia.

Phenylpropenes and Benzenoids

Phenylpropenes are derived from cinnamic acid and share the first steps of lignin/lignan biosynthesis. Shortening of the cinnamic acid side chain by two carbons leads to compounds with a C_6-C_1 skeleton called benzenoids. Other volatile

1

phenylpropanoid-related compounds are phenylacetaldehyde and 2-phenylethanol. They originate from phenylalanine which is shortened by one carbon (Pichersky and Dudareva, 2007). Phenylpropene and benzenoid volatiles are lipophilic compounds with a characteristic scent. They constitute the second largest group of plant volatiles after terpenoids. In many plants, phenylpropanoid/benzenoid and terpenoid volatiles occur as mixture, although usually one group accumulates predominantly. In addition to these two large groups of natural products, many other volatiles of plant origin are derivatives of amino acids and fatty acids. Each plant has its own cocktail of volatiles that is used for the attraction of pollinators and seed dispersers or as defense compounds. The defense function can be fulfilled either by directly deterring or intoxicating herbivores, or indirectly by attracting insect predators in case of tritrophic interactions or by communicating the danger to other plants in the neighborhood (Dudareva et al., 2006). In addition, many volatiles have antibacterial and antifungal properties (Kalemba and Kunicka, 2003).

Essential oils with phenylpropenes are found, e.g., in the Apiaceae, Lauraceae and Myrtaceae families. Many of these phenylpropene-containing plants have been employed by humans since antiquity as condiments and herbal remedies. Cloves, the unopened flower buds of the evergreen clove tree (*Syzygium aromaticum*, Myrtaceae) native to the Maluku islands, are used as spice, but also as anaesthetic and antiseptic in dentistry. The active ingredient and major component of essential oil from cloves is the phenylpropene eugenol (Fig. 18). Another evergreen tree of the tropics, *Cinnamomum ceylanicum* (Lauraceae) from Sri Lanka, is the source of cinnamon bark with *trans*-cinnamaldehyde as main flavor component. Cinnamon was highly priced in the antique world. The roman emperor Nero is said to have spent a year's supply of cinnamon for the funeral of his wife Poppaea Sabina (Klein, 1987). Some phenylpropenes are potentially carcinogenic, e.g., safrole, methyleugenol, and estragole. They require bioactivation including hydroxylation and sulfation at the side chain to become toxic (Zhou et al., 2007).

Benzenoids contribute to characteristic fragrance of many flowers. Methyl benzoate, for example, is a major scent constituent of *Petunia* flowers. Other benzenoids that frequently contribute to floral scents are benzaldehyde, benzyl alcohol, benzyl acetate, and methyl salicylate (Fig. 18) (Knudsen et al., 1993). The latter compound is responsible for the characteristic smell and the analgesic effect of wintergreen (*Gaultheria procumbens*, Ericaceae) (Dewick, 2002).

Polyketides

Polyketides are synthesized from two-carbon units derived from activated acetate in the form of acetyl-CoA and malonyl-CoA. Unlike fatty acids, which also originate from these precursors, polyketides retain all or most of their oxygen

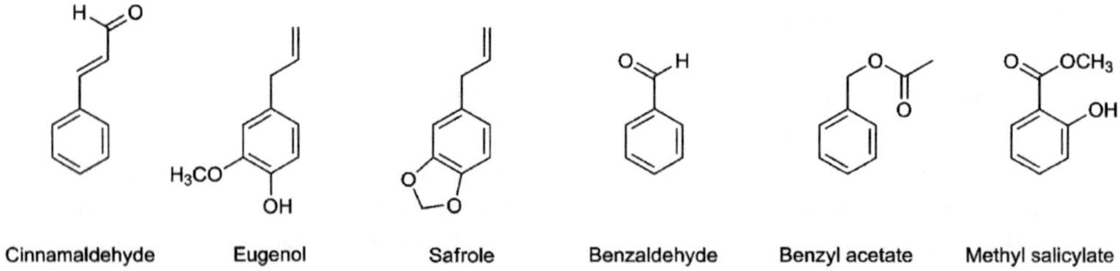

Fig. 18 Phenylpropanoid and benzenoid volatiles

functions. In the course of polyketide biosynthesis, highly reactive poly-β-keto intermediates are formed, which often undergo cyclization to six-membered aromatic or 2-pyrone rings. Many polyketides are phenolics like the phenylpropanoids, but the two groups can be distinguished by the substitution pattern of the aromatic ring. While polyketides usually contain oxygen functions on alternate carbons (*meta* position), phenolics derived from the phenylpropanoid pathway show an *ortho* oxygenation pattern. Many polyketides are glycosylated and may carry acyl substitutents on the sugar unit.

Polyketides occuring in plants are not always exclusively synthesized from acetate units, but often are of mixed biosynthetic origin. Phenylpropanoid or terpenoid building blocks or sometimes both can be connected with the acetate-derived backbone. In addition, parts of the carbon skeleton can be derived from fatty acids or amino acids. This mixed assembly principle results in a plethora of structurally diverse compounds. Polyketide alkaloids obtained when nitrogen or nitrogen-containing precursors are incorporated into the polyketide backbone will also be discussed.

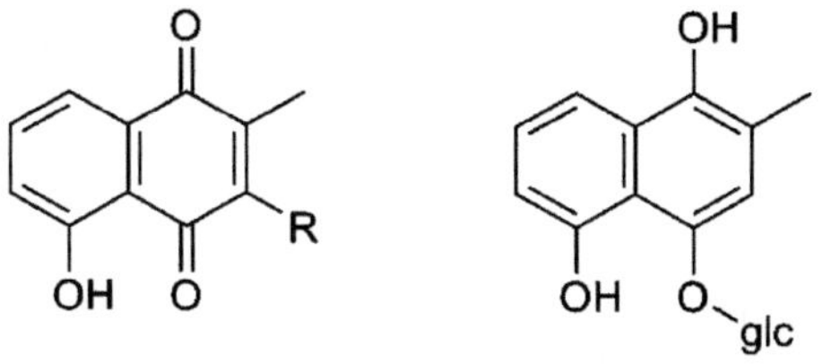

Plumbagin (R = H)
Droserone (R = OH)

Hydroplumbagin glucoside

Fig. 19 Naphthoquinones and a naphthohydroquinone glucoside

Polyketides Derived Exclusively from Acetate Units

Naphthoquinones of polyketide origin occur in a few taxonomically related families of the order Caryophyllales (Heubl et al., 2006), e.g., the Droseraceae, Nepenthaceae, Plumbaginaceae, and Polygonaceae, but also in the unrelated genus *Diospyros* (Ebenaceae) (Thomson, 1987). These compounds are built from six acetate units, and one carbon is lost in the course ot their biosynthesis, thus yielding an eleven carbon skeleton with two six-membered rings (Fig. 19). Plants that accumulate naphthoquinones often also contain naphthohydroquinone glycosides and dimeric naphthoquinones. Naphthoquinones, e.g., plumbagin and droserone, are strongly colored and lipophilic substances. Plumbagin is accumulated predominantly in roots and acts as an allelopathic compound. It has also antifeedant effects on insects (Tokunaga et al., 2004) and antimicrobial properties (Didry et al., 1994). Due to their structural similarity to ubiquinone, naphthoquinones may interfere with mitochondrial electron transport.

Structurally related to the naphthoquinones are the anthraquinones, which originate from eight acetate units and are composed of three six-membered rings. As in the case of the naphthoquinones, one of the sixteen carbons is lost by decarboxylation. Anthrones are reduced anthraquinones and their biosynthetic precursors. Dimeric anthrones are called dianthrones. All three groups of acetate-derived anthranoids in plants have two hydroxy groups at C-1 and C-8, a third hydroxy goup can occur at C-3. They are conjugated with sugar, mostly glucose, as *O*- and *C*-glycosides (Fig. 20). The presence of a carbonyl group in conjugation with an aryl is a strong chromophore, and consequently, anthranoids show yellow, orange or red pigmentation. In contrast to naphthoquinones of polyketide origin, anthranoids occur in several unrelated plant families, e.g., the eudicot Fabaceae (*Cassia*), Rhamnaceae (*Rhamnus*), Polygonaceae (*Rheum*), and the monocot Asphodelaceae (*Aloe*). Anthranoids have antibacterial and antifungal properties (Srinivas et al., 2007) and are therefore probably used as defense compounds.

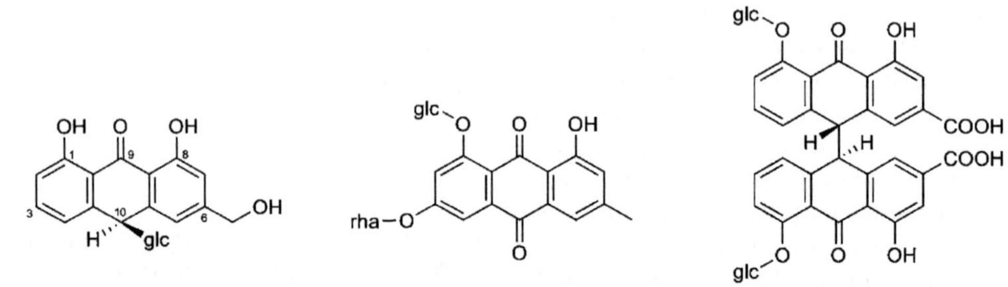

Aloin A
Anthrone *C*-glycoside
(*Aloe barbadensis, A. ferox*)

Glucofranguline A
Anthraquinone *O*-glycoside
(*Rhamnus frangula*)

Sennoside A
Dianthrone *O*-glycoside
(*Cassia senna, C. angustifolia*)

Fig. 20 Anthranoid glycosides from various plant species. Abbreviations for sugars: glc, glucose; rha; rhamnose

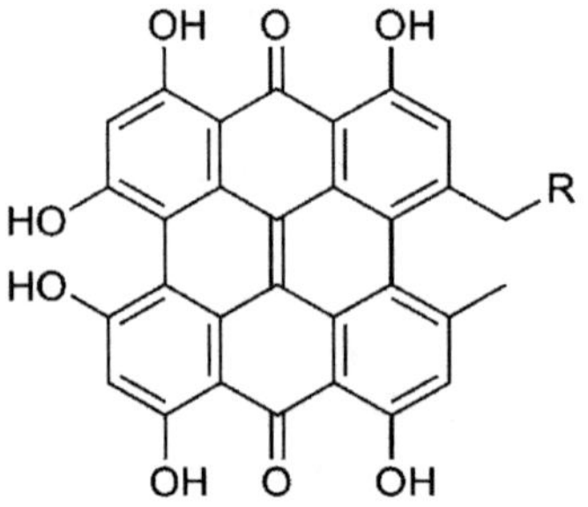

Hypericin (R = H)
Pseudohypericin (R = OH)

Fig. 21 Naphthodianthrones of *Hypericum perforatum*

Plants containing anthranoids are used as laxatives in case of habitual constipation. After ingestion, anthraquinones and dianthrones are converted to anthrones, the active metabolites, by bacteria of the large intestine. Anthrones stimulate the movement of the large intestine and increase secretion of water and electrolytes into the intestinal lumen. Frequent use of athranoid laxatives can lead to dehydration, depletion of minerals, and reversible pigmentation of the intestinal mucosa (*pseudomelanosis coli*). Emodin, a 1,3,8-trihydroxy-6-methylanthraquinone, proved to be mutagenic in *in vitro* assays. However, more recent *in vitro* studies suggest that it has also anti-tumor activities, e.g., by induction of apoptosis, inhibition of cell cycle and angiogenesis (Srinivas et al., 2007).

Hypericin and pseudohypericin (Fig. 21) are naphthodianthrones and occur in St. John's wort (*Hypericum perforatum*, Clusiaceae). These dark-red compounds are accumulated in glands on the margin of *H. perforatum* leaves and petals. It was assumed for a long time that they represent the antidepressant principle in extracts of St. John's wort. Recently, however, it became evident that the prenylated acylphloroglucinols hyperforin and adhyperforin are responsible for most of the pharmacological effects.

Anthraquinones and naphthoquinones are not exclusively formed via the polyketide biosynthetic route. They can also be derived from the shikimate pathway, e.g., the allelopathic naphthoquinone juglone from the walnut tree (*Juglans regia*) and the anthraquinone alizarin from madder root (*Rubia tinctorium*), which was used as purple dye.

Polyketides with Phenylpropanoid Moieties

In plants, phenylpropanoid and shikimate derived compounds can be combined with one to three C_2 units derived from malonyl-CoA to yield polyketides of various structures. The largest group of these compounds are the flavonoids with more than 6,000 known compounds (Harborne and Baxter, 1999). Other polyketides of phenylpropanoid origin are less widely distributed, e.g., stilbenes, styrylpyrones, and curcuminoids.

In the structure of flavonoids, the phenylpropanoid C_6-C_3 backbone is extended with three C_2 units that form a second aromatic ring. This basic C_6-C_3-C_6 skeleton of flavonoids can be modified by hydroxylations, methylations, prenylation and, in the case of isoflavonoids, by aryl migration. Flavonoids occur mostly in glycosylated form and are often accumulated in the vacuole. The central intermediates in flavonoid biosynthesis are chalcones, which serve as precursors of all other subgroups (Fig. 22). Flavones, flavonols, flavan-3,4-diols (leucoanthocyanidins), anthocyanins, and proanthocyanidins (polymerized flavan-3-ols, condensed tannins) occur nearly ubiquitously in higher plants. Other subgroups of flavonoids are restricted to certain taxa. Isoflavonoids are produced mainly in leguminous plants (Fabaceae). Aurones are yellow pigments with widespread occurrence, e.g., in snapdragon (*Antirrhinum majus*), dahlia (*Dahlia variabilis*), and tickseed (*Coreopsis* species). Flavan-4-ols or 3-deoxyanthocyanidins are precursors of the phlobaphene polymers, which are red pigments and occur in

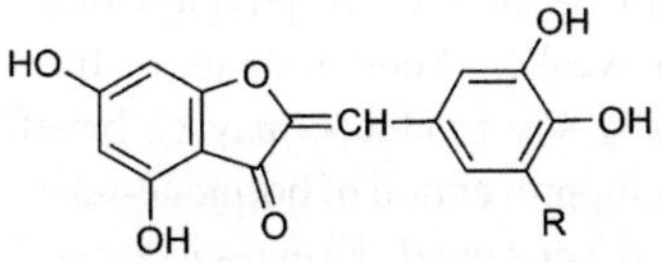

Areusidin (R = H)
Bracteatin (R = OH)
(Aurones)

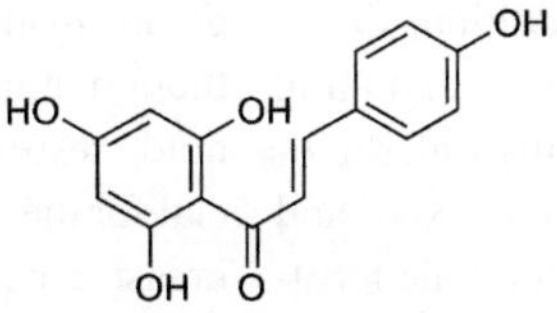

Naringenin chalcone
(Chalcone)

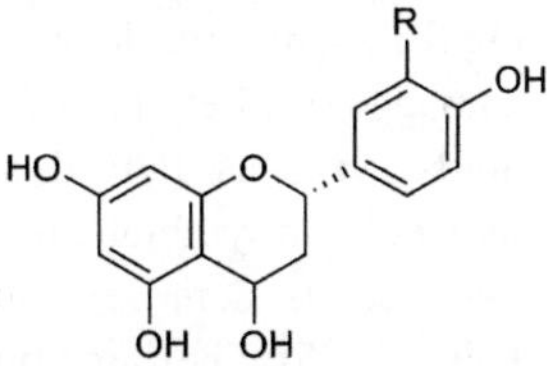

Apiferol (R = H)
Luteoforol (R = OH)
(Flavan-4-ols)

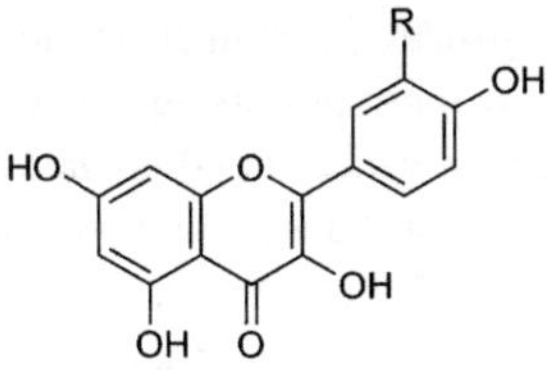

Kaempferol (R = H)
Quercetin (R = OH)
(Flavonols)

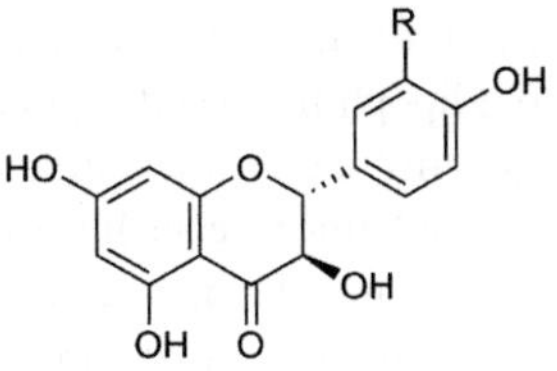

Dihydrokaempferol (R = H)
Dihydroquercetin (R = OH)
(Dihydroflavonols)

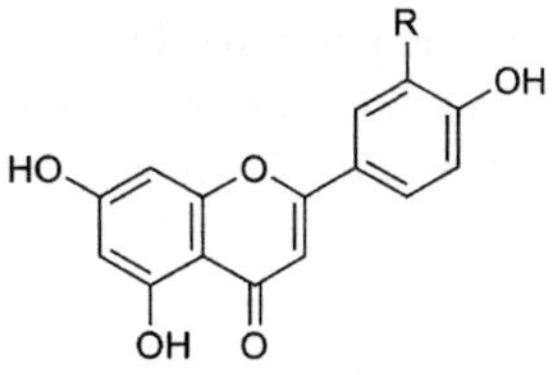

Apigenin (R = H)
Luteolin (R = OH)
(Flavones)

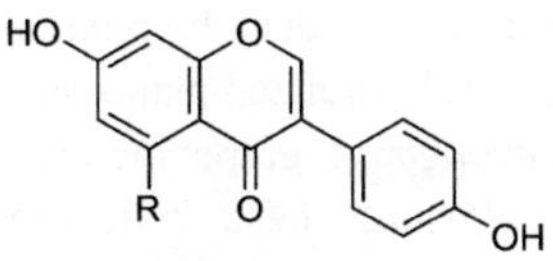

Daidzein (R = H)
Genistein (R = OH)
(Isoflavones)

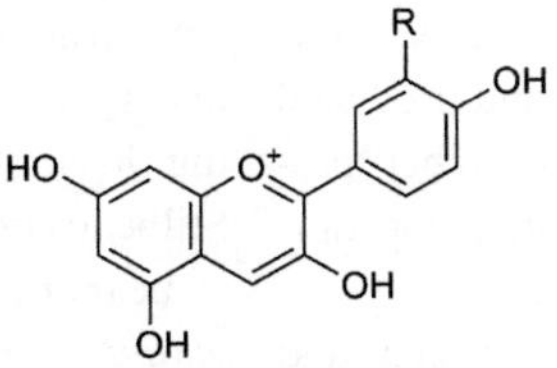

Pelargonidin (R = H)
Cyanidin (R = OH)
(Anthocyanidins)

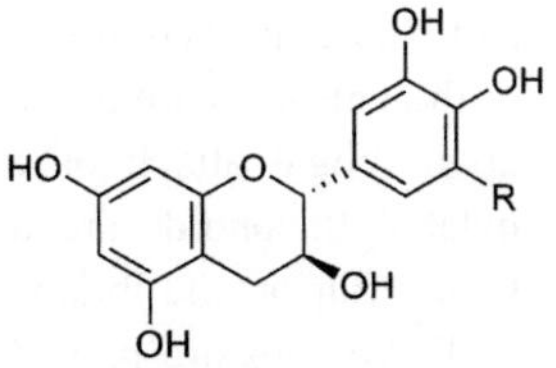

(+)-Catechin (R = H)
Gallocatechin (R = OH)
(Flavan-3-ols)

Fig. 22 Diverse classes of flavonoids

some Poaceae (*Sorghum bicolor*, *Zea mays*) and gloxinia (*Sinningia cardinalis*) (Winkel-Shirley, 2001).

Flavonoids fulfill several physiological functions in plants. Anthocyanins serve as pigments to attract pollinators and seed dispersers. They are responsible for the red, pink, purple, and blue coloration of many flowers, fruits, and leaves. Different hues and shades are achieved by glycosylation and acylation as well as by complexation with flavones or metal ions and by variation of vacuolar pH. Flavonoids have two absorption maxima in the ultraviolet (UV) range and therefore protect plant tissue from damage by UV radiation (Harborne and Williams, 2000). Anthocyanins absorb also light in the visible range and may provide protection for chlorophyll in senescing leaves from photooxidative damage (Feild et al., 2001). In addition, flavonoids serve as radical scavengers, signaling molecules in symbiotic relationships with rhizobia, and as defense compounds (Dixon and Paiva, 1995). They are important for male fertility (Mo et al., 1992) and modulate the transport of the phytohormone indole-3-acetic acid (Peer and Murphy, 2007). Many naturally occuring flavonoids carry one or two prenyl moieties. This modification renders them more lipophilic than other flavonoids, and consequently, they show higher affinities to biological membranes and often possess antibacterial and antifungal properties (Botta et al., 2005). These phytochemicals are of limited distribution in the plant kingdom and occur in several unrelated families, e.g., in the Fabaceae, Moraceae, Asteraceae, and Rutaceae. While prenylated isoflavonoids of the Fabaceae are phytoalexins and produced upon fungal attack or elicitation, other prenylated flavonoids are accumulated constitutively (Barron and Ibrahim, 1996).

Flavonoids are part of the human diet, and high levels occur in fruits, vegetables, wine, tea, and cocoa. It is suggested that dietary intake of flavonoids has beneficial health effects due to their antioxidant and radical-scavenging properties (Heim et al., 2002). By inhibiting the oxidation of low density lipoprotein (LDL), flavonoids may prevent coronary heart disease. This mechanism has been postulated for the cardio-protective effect of red wine polyphenolics including resveratrol. Two prenylflavonoids from hops (*Humulus lupulus*) with interesting bioactivities are 8-prenylnaringenin, the strongest phytoestrogen known so far, and the promising cancer chemopreventive agent xanthohumol (Stevens and Page, 2004). Xanthohumol *in vitro* inhibits phase I enzymes that activate procarcinogens and induces detoxifying phase II enzymes. Moreover, it has radical scavenging activities and inhibits cyclooxygenases 1 and 2, which participate in the biosynthesis of prostaglandins, inflammation mediators that promote carcinogenesis (Gerhäuser et al., 2002). Like 8-prenylnaringenin, many isoflavonoids are phytoestrogens, though much weaker. Food rich in isoflavonoids, especially soy products, may be beneficial for the chemoprevention of hormone-related breast cancer if consumed during adolescence. However, intake of purified genistein, the major isoflavonoid in soy together with daidzein, cannot be regarded as safe, because it increases breast cancer cell growth in animal models (Duffy et al., 2007). A more promising chemopreventive agent against cancer is (-)-epigallocatechin 3-gallate, the most abundant polyphenol in green tea (Nagle et al., 2006). Various mechanisms are discussed, e.g., induction of apoptosis, cell cycle arrest, inhibition of angiogenesis.

Stilbenes originate from the same biosynthetic precursors as flavonoids, but have a different structure, since the polyketide portion undergoes a different type of cyclization including loss of one carbon by decarboxylation. Stilbenes occur in several unrelated plants such as peanut (*Arachis hypogaea*), grapevine (*Vitis vinifera*), rhubarb (*Rheum*), false hellebores (*Veratrum*) and pine (*Pinus*) species. These compounds have antifungal properties. They are induced upon stress, injury, and fungal infection and can therefore be classified as phytoalexins.

Stilbene production can be conferred to a plant by transformation with a single gene, stilbene synthase. Several transgenic plants producing stilbenes or stilbene glucosides were described, and some showed enhanced resistance against fungal pathogens (Hain et al., 1993; Thomzik et al., 1997; Zhu et al., 2004).

Resveratrol (Fig. 23), a stilbene found in many food sources, e.g., peanuts and red wine, is assumed to have multiple benefits on human health. Most attention has been received by the so-called "French paradox", the low occurrence of cardiovacular disease in populations living on a diet high in saturated fats, but consuming red wine. The protective effect of red wine is attributed to its proanthocyanidin and resveratrol contents. Possible mechanisms disscussed are the inhibition of oxidation of LDL cholesterol and platelet aggregation. Resveratrol may also increase longevity by activation of sirtuins, NAD^+-dependent protein deacetylases involved in aging, which respond to oxidative stress and are induced by a low-calory diet. Resveratrol mimics the effects of a low-calory diet and extends the lifespans of baker's yeast (*Saccharomyces cerevisiae*), fruit flies (*Drosophila melanogaster*), and roundworms (*Caenorhabditis elegans*) (Howitz et al., 2003; Wood et al., 2004). In addition, resveratrol increases the survival of mice fed on a high-calory diet (Baur et al., 2006). Whether similar beneficial effects can be reproduced in humans will depend on the pharmacokinetics and long-term toxicity of resveratrol in humans.

In styrylpyrones, the phenylpropanoid is extended by two C_2 units derived from malonyl-CoA, and a lactone heterocycle is formed that contains four carbons of polyketide origin and one carbon of the phenylpropanoid side chain. These natural products occur, e.g., in horsetail (*Equisetum*) and kava (*Piper methysticum*). Kava grows on the Pacific islands of Melanesia, Micronesia, and Polynesia, and its roots and rhizomes were used to prepare an intoxicating drink named "Kava-Kava" (Briskin, 2000). An extract from the rhizomes of kava is used as medication against anxiety and tension, the styrylpyrones, also called kavapyrones, being the active ingredients (Fig. 23). Kavapyrones interact

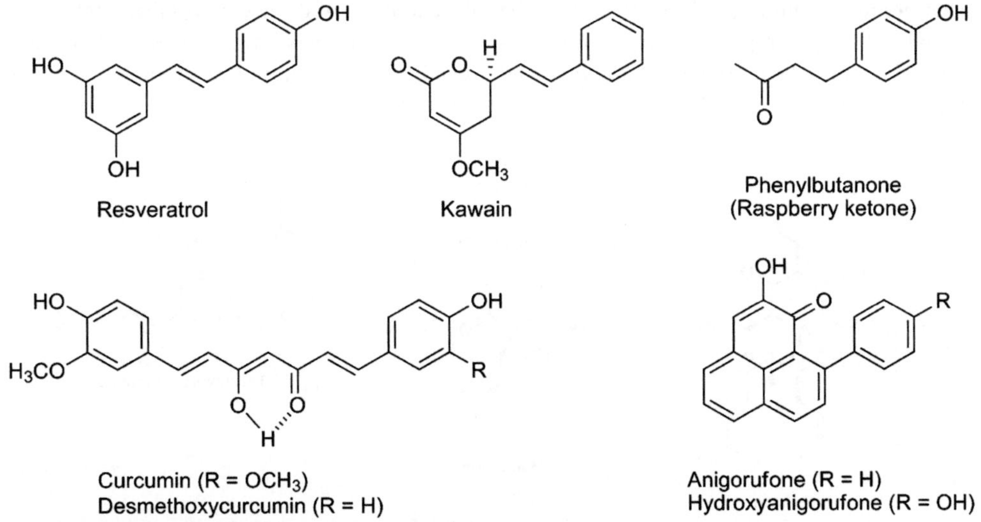

Fig. 23 Various polyketides containing one or two phenylpropanoid units

1

with the gamma-aminobutyric acid (GABA) receptor (Boonen and Häberlein, 1998), which mediates mainly inhibitory effects in vertebrates. Due to the liver toxicity of kava preparations, this phytomedicine was banned from the European market.

Polyketides with one or two phenylpropanoid moieties and only one carbon derived from malonyl-CoA are the raspberry aroma *p*-hydroxyphenylbutan-2-one, the curcuminoids of turmeric (*Curcuma*), and phenylphenalenones from *Musa* and *Anigozanthos* species (Fig. 23). Curcuminoids and phenylphenalenones contain two phenylpropanoid moieties on both sides of the acetate-derived carbon and are therefore classified as diarylheptanoids. These two groups of natural products are pigments due to their conjugated system of π-electrons. Curcuminoids are widely used as spices and also have antiinflammatory, antioxidative, and anti-cancer properties (Joe et al., 2004).

Polyketides with Terpenoid Building Blocks

This group of natural products comprises the prenylated acylphloroglucinols, e.g., bitter acids from the hop plant (*Humulus lupulus*) and hyperforin from St. John's wort (*Hypericum perforatum*), as well as cannabinoids from Indian hemp (*Cannabis sativa*) (Fig. 24). These compounds are lipophilic due to their terpenoid moiety and are often produced or stored in special glands or glandular trichomes.

Hop bitter acids and hyperforin are both derived from three different building blocks: a branched short-chain CoA ester derived from amino acid metabolism and three C_2 units derived from malonyl-CoA constitute the acyl phloroglucinol core, which carries two to several isopentenyl side chains that originate from terpenoid metabolism (Adam et al., 2002; Drawert and Beier, 1976; Goese et al., 1999; Karppinen et al., 2007).

Hyperforin accumulates in translucent glands of leaves of St. John's wort (Soelberg et al., 2007) and represents the major antidepressive principle of this plant. It inhibits the re-uptake of the neurotransmitters serotonin, noradrenaline, dopamine, and GABA, thus increasing their concentrations at the synapses of the brain. This is achieved by an unprecedented mechanism. By elevating the intracellular sodium concentration, hyperforin inhibits the sodium-driven proton gradient required for the transport of neurotransmitters from the synaptic gap into the axoplasm of the neuron (Müller, 2003).

Female inflorescences (cones) of hop carry glandular trichomes that contain volatiles derived from terpenes and fatty acids and a

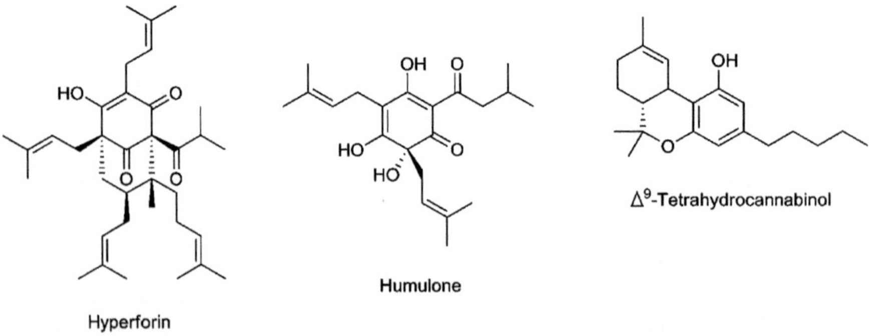

Fig. 24 Polyketides with terpenoid components

resin consisting of the hop bitter acids with humulone (α-resin) and lupulone (β-resin) as lead compounds. Hop cones are an important ingredient for beer production and contribute to its flavor, in particular its bitter taste. In addition, the hop bitter acids act as foam stabilizers and prevent the growth of bacteria due to their antimicrobial properties. Extracts of hop cones are also used as a mild sedative and sleep inducer.

Cannabinoids are derived from hexanoyl-CoA, three molecules of malonyl-CoA and the C_{10} terpenoid geranyldiphosphate. They occur only in Indian hemp, which belongs to the Cannabaceae family like hop. *C. sativa* plants accumulate a resin in glandular trichomes that are more abundant on female inflorescences than on male. The glandular resin contains more than sixty cannabinoids, e.g., δ^9-tetrahydrocannabinol (THC) and cannabidiol (CBD) (Turner et al., 1980). Cannabis has been used as psychoactive drug since prehistoric times. Its use is nowadays prohibited in most parts of the world, but cultivation of *C. sativa* plants with low THC contents is permitted for production of fibers and oil-rich seeds in serveral countries, though not in the USA. The dried flowering parts of female hemp plants are termed marihuana, while hashish (Arabic for "grass") is their concentrated resin. These cannabis preparations are usually smoked, but they can also be ingested orally, e.g., as cakes or cookies. THC, the major psychoactive compound in cannabis, binds to cannabinoid receptors that occur in brain, spinal cord and immune cells. The consumption of cannabis products leads to a feeling of euphoria and relaxation, but repeated use may cause addiction (Dewick, 2002). Nevertheless, THC has also useful medicinal properties, e.g., antiemetic, analgesic, and appetite-inducing. It is used for the prevention of nausea during radiotherapy and chemotherapy and as treatment of the wasting syndrome in AIDS patients.

Polyketide Alkaloids

The most famous compounds of this group are probably the piperidine alkaloids of poison hemlock (*Conium maculatum*), which were used to execute the Greek philosopher Socrates. While piperidine alkaloids are usually synthesized from the amino acid L-lysine, the carbon skeleton of the piperidine alkaloids in *C. maculatum* originates from four acetate units (Leete, 1963, 1964). Only the nitrogen is derived from L-alanine by transamination (Roberts, 1971). Hemlock alkaloids are accumulated in all plant parts, however, highest levels are found in unripe fruits (1.6%) (Dewick, 2002). The two major hemlock alkaloids are γ-coniceine and coniine (Fig. 25). Piperidine alkaloids like coniine occur

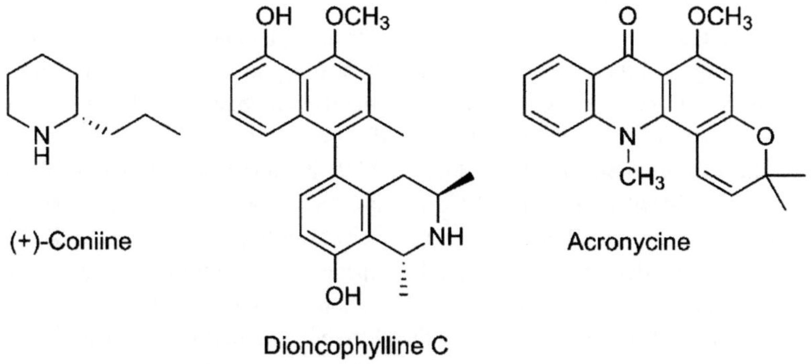

Fig. 25 Different types of polyketide alkaloids

not only in *C. maculatum*, but also in several *Aloe* species (Reynolds, 2005). These compounds are neurotoxic and cause paralysis, muscular tremor and death by respiratory paralysis.

Although the carbon skeleton of naphthylisoquinoline alkaloids is much more complex than that of the coniine alkaloids, it is composed of the same building blocks, C_2 units derived from acetyl-CoA and malonyl-CoA. Each part the the naphthylisoquinoline skeleton, naphthalene and isoquinoline moiety, are derived from six acetate units. Nitrogen is incorporated only in the isoquinoline part and most likely derived from amino acid metabolism (Bringmann and Feineis, 2001). Naphthylisoquinoline alkaloids occur only in the plant families Ancistrocladaceae and Dioncophyllaceae that comprise lianas from southeast Asia and Africa. They were shown to possess fungicidal activities and antifeedant properties towards insects, which might relate to their physiological function. Michellamine B, a dimeric naphthylisoquinoline from the liana *Ancistrocladus korupensis*, was discovered during a screening by the U.S. National Cancer Institute. It showed promising anti-HIV acitivity by inhibiting the viral reverse transcriptase and by blocking the fusion of virus particles with the human cell membrane (McMahon et al., 1995). Other alkaloids of this group, e.g., dioncophylline C (Fig. 25) and ancistrocladinium A and B, have antiplasmodial and antileishmanial properties, respectively (François et al.,1997; Ponte-Sucre et al., 2007).

In contrast to the previous two groups of alkaloids, the nitrogen in the backbone of acridone alkaloids is not acquired by transamination from an amino acid. Instead, the nitrogen and part of the carbon skeleton of acridones originate from *N*-methylanthraniloyl-CoA, which is derived from the shikimate pathway. In addition, three C_2 units derived from malonyl-CoA are incorporated. The basic acridone skeleton can be modified by prenylation with dimethylallyldiphosphate (DMAPP), which can be followed by the formation of an additional heterocyclic five- or six-membered ring. In plants, acridone alkaloids occur abundantly in the Rutaceae family. Due to their planar aromatic structure, acridones can intercalate DNA. Acronycine (Fig. 25) from *Acronychia baueri* showed promising anticancer activity, but did not yield convincing results in clinical studies. Its new derivative S23609-2, a benzoacronycine, is a potent DNA-alkylating agent and currently undergoes phase I clinical trials (Léonce et al., 2006).

Terpenoids

Terpenoids, also named isoprenoids, are the largest class of natural products in plants and comprise more than 40,000 different structures. They are derived from five-carbon isoprene units, and according to the number of isoprene molecules incorporated, they can be classified into hemiterpenes (C_5), monoterpenes (C_{10}), sesquiterpenes (C_{15}), diterpenes (C_{20}), triterpenes (C_{30}), tetraterpenes (C_{40}), and polyterpenes such as rubber (Dewick, 2002). In plants, terpenoids originate from two different biosynthetic routes: the cytosolic mevalonic acid (MVA) pathway and the plastid-located desoxyxylulose phosphate (DXP) pathway (also called methylerythritol phosphate or MEP pathway). Both biosynthetic routes yield the activated isoprene units dimethylallyl diphosphate (DMAPP) and isopentenyl diphosphate (IPP), which are joined by head-to-tail or tail-to-tail linkage and subsequently can undergo cyclization and other modifications, e.g., oxidations or rearrangements. While hemiterpenes, monoterpenes, diterpenes, and tetraterpenes are derived from the DXP pathway, triterpenes, steroids, and certain sesquiterpenes originate from mevalonic acid (Fig. 26). Although MVA and DXP pathway are located in different compartments, there is an exchange between the two biosynthetic routes, especially from the plastidial to the cytosolic pathway

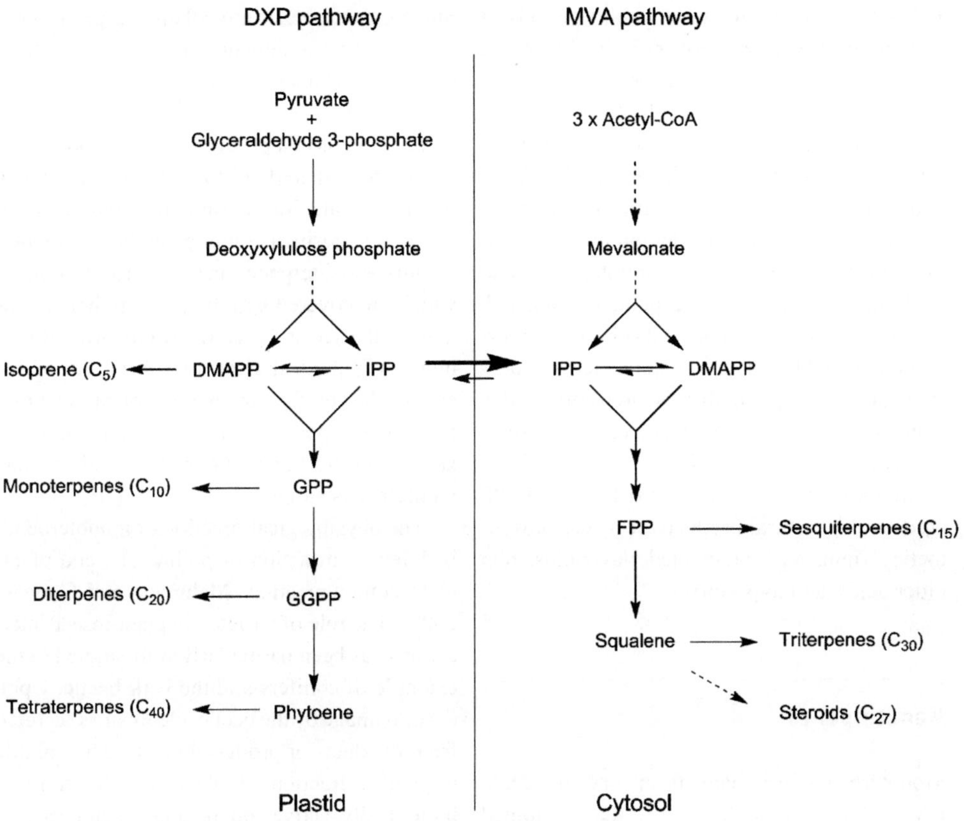

Fig. 26 Schematic overview of terpene biosynthesis in plants. DMAPP, dimethylallyl diphosphate; DXP, desoxyxylulose phosphate; FPP, farnesyl diphosphate; GGPP, geranylgeranyl diphosphate; GPP, geranyl diphosphate; IPP, isopentenyl diphosphate; MVA, mevalonate

(Laule et al., 2003; Schuhr et al., 2003). This has become particularly evident in the case of several sesquiterpenes, which are synthesized from DMAPP and IPP units provided by the DXP pathway, but not from MVA (Dudareva et al., 2005; Piel et al., 1998).

Hemiterpenes

The most abundant true hemiterpene from plants is isoprene (Fig. 27), a volatile compound synthesized from DMAPP. Production and emission of isoprene is distributed very widely in the plant kingdom, and species that synthesize this compound are found among mosses, ferns, gymnosperms, and angiosperms. Many isoprene-emitting species are trees, particularly poplar and aspen, and plants from the humid tropics. Isoprene is emitted into the atmosphere

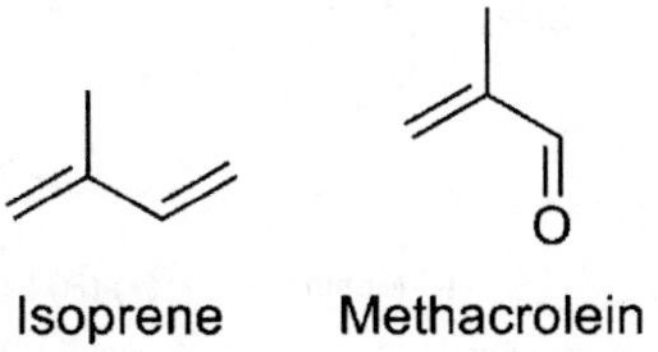

Fig. 27 Hemiterpenes

1

and protects leaves to survive short periods of high temperature. Moreover, it increases the plant's tolerance towards ozone and reactive oxygen species. (Sharkey et al., 2008).

Hemiterpenes may also act as signaling molecules. Leaves of sagebrush (*Artemisia tridentata*) emit the highly volatile hemiterpene methacrolein (Fig. 27) in addition to other volatile compounds like hexenal, monoterpenes, and methyljasmonate when the plant is damaged. This is perceived by plants in its close neighborhood and enables them to react faster to a possible attack. A plant that is prepared in this manner, is less likely to be damaged by herbivores (Baldwin et al., 2006).

In addition, C_5 units derived from DMAPP are found in natural products of mixed biosynthetic origin, e.g., prenylated flavonoids, hop bitter acids, and hyperforins.

Monoterpenes

Monoterpenes originate from one molecule DMAPP and one molecule IPP that are joined in most cases head-to-tail, yielding all-*trans* geranyldiphosphate (GPP) (Fig. 28). GPP can be folded into mono-, bi- and tricyclic structures and may undergo other modifications to yield more than 1,000 different monoterpenes. Monoterpenes are lipophilic volatile compounds that occur in defensive resins of conifers, essential oils, and floral scents and contribute to the characteristic flavor or aroma of many plants. Since monoterpenes are volatile, large amounts can only be accumulated in specialized structures. Several plant families, e.g., the Lamiaceae and Asterace, have glandular trichomes with secretory cells that produce terpenes and secrete them into a shared subcuticular storage cavity (Croteau et al., 2005). Similarly, conifers accumulate oleoresin, a complex mixture of mono-, sesqui-, and diterpenes, in resin blisters or ducts, which are covered by a layer of epithelial cells that synthesize and secrete the terpenes into the lumen (Trapp and Croteau, 2001). As in the case of the conifers, many other plants accumulate monoterpenes in mixtures containing the larger sesqui- and diterpenes, rather than monterpenes alone.

The physiological function of monoterpenes is defense, attraction of pollinators, and plant-plant communication (Mahmoud and Croteau, 2002). The role of terpenes in plant-insect-interactions has been particularly well-studied in the example of conifers and the bark beetle. Upon tissue damage by the beetle, oleoresin is secreted from the ducts or produced newly. The volatile turpentine fraction of oleoresin consisting of biologically active mono- and sesquiterpenes, e.g., limonene and pinene, kills the beetles and associated pathogenic fungi. After evaporation of turpentine, the remaining non-volatile rosin fraction consisting of diterpene resin acids solidifies, thus trapping the predators and sealing the wound (Philipps and Croteau, 1999). Despite their toxicity, monoterpenes in oleoresin serve as olfactory signals that help the bark beetles to find their host. Ingested monoterpenes

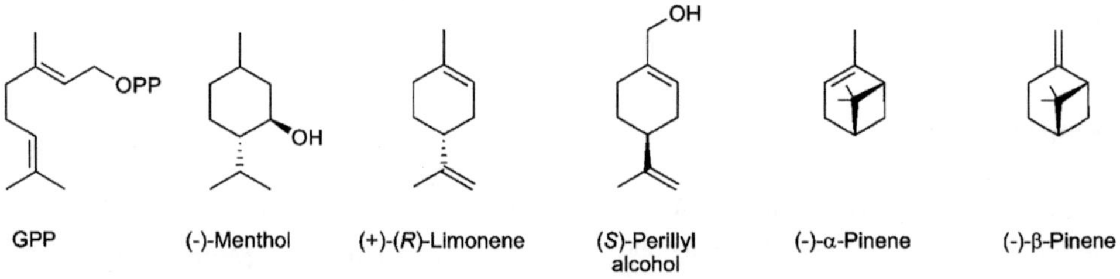

Fig. 28 Mono- and bicyclic monoterpenes derived from geranyldiphosphate (GPP)

are converted by the beetles to pheromones that either attract more beetles or serve as anti-aggregation signals. In addition, conifer monoterpenes take part in tritrophic interactions and attract insect predators that feed on bark beetles (Trapp and Croteau, 2001).

Many monoterpenes have found application in perfumery, aromatherapy, as cosmetics and insecticides. Menthol, a constituent of essential oils from *Mentha* species, is the most widely used monoterpene. It is contained in pharmaceuticals, oral health care products, chewing gums, and tobacco products (Croteau et al., 2005). More than 7,000 t of menthol are produced every year either by total synthesis or from the steam-distilled essential oil of corn-mint (*Mentha arvensis* var. *piperascens*). The cooling sensation stimulated by menthol is caused by excitation of cation channels that serve as thermal receptors (Jordt et al, 2003).

Two monoterpenes with promising anticancer effects are perrillyl alcohol and (+)-(*R*)-limonene (Mo and Elson, 2004). The two compounds induce apoptosis and suppress translation of 3-hydroxy-3-methylglutaryl-CoA (HMG-CoA) reductase, an enzyme of the MVA pathway. This enzyme is a promising target for anti-tumor compounds, because many proteins involved in cell growth are prenylated, and tumor cells have elevated HMG-CoA reductase levels. Suppression of HMG-CoA reductase is sufficient to decrease terpene biosynthesis in humans, since animals lack the alternative DXP pathway.

Iridoids are monoterpenes with a six-membered oxygen heterocycle annealed to a cycplopentane ring. The hydroxy group of the oxygen containing heterocycle (dihydropyrane) is glucosylated, thus converting the enol-hemiacetal into an acetal (Fig. 29). Cleavage of the cyclopentane ring of the iridoid skeleton yields the secoiridoids, which are biosynthetic building units of the Ipecac alkaloids and the monoterpene indole alkaloids. Iridoids are named after ants of the genus *Iridomyrmex* that produce these metabolites as defense compounds. In plants, iridoids are chemotaxonomic markers of the genera *Plantago* (Plantaginaceae), *Galium*

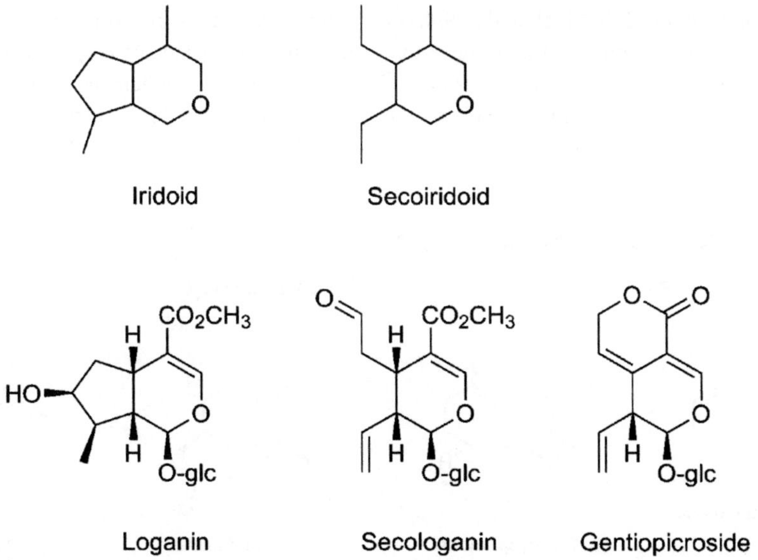

Fig. 29 Iridoids

1

(Rubiaceae), and *Scrophularia* (Scrophulariaceae) and also occur frequently in the Gentianaceae, Oleaceae, and Verbenaceae (Dinda et al., 2007a, b). Many iridoids have an intense bitter taste and therefore act as feeding deterrents (Seigler, 1998). On the other hand, plants with bitter tasting iridoids, e.g., gentian (*Gentiana lutea*) with its bitter principle gentiopicroside and amarogentine, are used for the preparation of tonics against anorexia and dyspepsia.

Sesquiterpenes

Sesquiterpenes contain three isoprene units and are formed by condensation of DMAPP with two molecules IPP. The central C_{15} intermediate farnesyldiphosphate (FPP) can be folded into mono-, bi- or tricyclic systems. In general, sesquiterpenes are less volatile than monoterpenes (Dewick, 2002). Initially, it was assumed that all sesquiterpenes are produced via cytosolic MVA pathway. Recent studies, however, revealed that certain sesquiterpenes originate from isoprene units provided by the DXP pathway (Dudareva et al., 2005; Piel et al., 1998) or by both biosynthetic routes (Adam and Zapp, 1998). This can be explained by transport of isoprenoid precursors from the plastids to the cytosol (Bick and Lange, 2003).

Abscisic acid is a sequiterpene phytohormone that is induced by drought and promotes stomatal closure and seed dormancy. Other sesquiterpenes take part in tritrophic plant-herbivore-parasite interactions (reviewed by Dudareva et al., 2006). In maize infested with lepidopteran larvae, the sesquiterpenes (*E*)-β-farnesene and the (*E*)-α-bergamotene (Fig. 30) attract the parasitic wasp *Cotesia marginiventris* (Schnee et al., 2006). Maize roots release (*E*)-β-caryophyllene (Fig. 30) upon attack of larvae of the beetle *Diabrotica virgifera* to attract the parasitic nematode *Heterorhabditis megidis* (Rasmann et al., 2005).

Many sesquiterpenes contain a pentacyclic lactone group and are therefore referred to as sesquiterpene lactones. These compounds occur abundantly in the family Asteraceae. Because of their bitter taste sesquiterpene lactones presumably serve as feeding deterrents of herbivores (Heinrich et al., 1998). Pharmacologically active sesquiterpene lactones often show anti-inflammatory effects due to inhibition of the transcription factor NF-κB that mediates immunological responses and inflammation (Lyß et al., 1998). Sesquiterpenes with such activities occur, for example, in chamomile (*Matricaria recutita*), one of the most popular medicinal plants. Antimigraine action of some sesquiterpene lactones, e.g., parthenolide from feverfew (*Tanacetum parthenium*), is mediated by inhibition of platelet aggregation

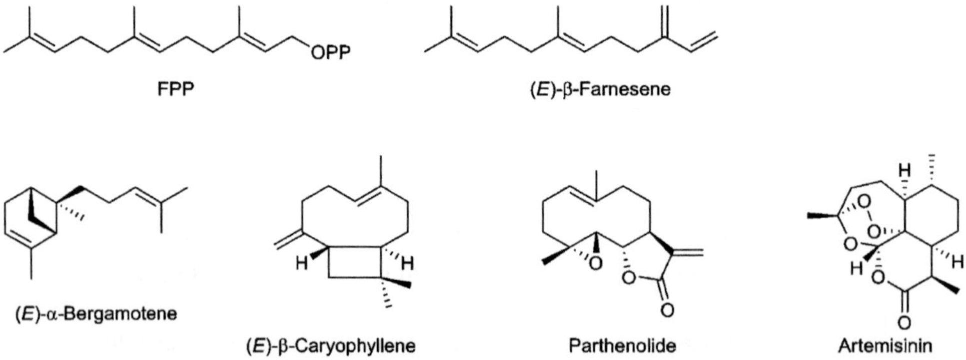

Fig. 30 Linear and cyclic sesquiterpenes

and serotonin secretion (Dewick, 2002). Structural prerequisite for the biological activities of sesquiterpene lactones is an α,β-unsaturated lactone that acts as nucleophile and alkylates proteins, particularly at their thiol groups. On the other hand, alkylation of proteins is the reason for the allergenicity and cytotoxicity of sesquiterpene lactones with an α,β-unsaturated lactone.

Artemisinin is a novel promising agent against malaria. Structurally, it is a tetracyclic sesquiterpene with a six-membered lactone ring and an unusual 1,2,4-trioxane ring (Fig. 30). It occurs in *Artemisia annua* (*qinghao*), which has been used for centuries in Traditional Chinese Medicine to treat fevers including malaria. The starting point of the discovery of artemisinin was a request of Ho Chi Minh, the president of North Vietnam, to the Chinese government for a cure against malaria to support his troops in the malaria-infested jungles during the American/Vietnamese war (Hsu, 2006). A screening of plants used in Traditional Chinese medicine revealed the antimalarial activity of an ether extract from *A. annua* in 1971, and artemisinin (*qinghaosu*) was isolated as the active principle in the late 1970s. The mode of action of artemisinin is still being investigated. Most likely, it interferes with a sarco-endoplasmic reticulum calcium ATPase (SERCA) of *Plasmodium falciparum*, but other mechanisms, e.g., alkylation of biological macromolecules or the production of reactive oxygen species are discussed as well (White, 2008). The peroxide bridge is a necessary structural feature required for antimalarial activity. In contrast to quinine, artemisinin kills already young erythrocytic forms of the parasite *Plasmodium*, thus curing malaria at an early stage. Two semisynthetic analogs, artemether and artesunate, with superior efficiency in comparison to artemisinin were developed and are now used as first-line therapy against malaria in combination with other antimalarial drugs like the quinine analogs mefloquine and lumefantrine. This combination of two drugs tends to prevent resistances of *Plasmodium*. The success of artemisinin and its analogs has triggered a huge demand that cannot be covered at low cost by extraction of the sesquiterpene from the plant, because *A. annua* contains only 0.01-1.5% of artemisinin (Covello et al., 2007). Approaches to provide the powerful drug at affordable price for the people in malaria-endemic areas are either breeding of *A. annua* plants with elevated artemisinin levels or biotechnological production of the artemisinin precursor artemisic acid by cloning the biosynthetic genes from *A. annua* (Covello et al., 2007) and engineering the pathway into the bacterium *Escherichia coli* or yeast (Chang et al., 2007; Ro et al., 2006).

Diterpenes

Diterpenes originate from the plastdic DXP pathway and are synthesized from DMAPP and three molecules IPP yielding the C_{20} metabolite geranylgeranyl diphosphate (GGPP). GGPP is the precursor of the lipophilic phytyl side chain of chlorophyll and plastoquinone. Like the smaller terpenes, GGPP can also undergo cyclization and rearrangements to many different structures.

Gibberellins are tetracyclic diterpenes that act as phytohormones and promote shoot elongation, flowering and seed germination (Bishopp et al., 2006). Diterpenes like abietic- and levopimaric acid (Fig. 31) are constitutents of conifer oleoresin and function as defense against herbivores and pathogens. After removal of mono- and sesquiterpenes (turpentine) from oleoresin by distillation, the solid diterpene fraction (rosin) is called colophonium and used on the bows of string instruments. The mono- and sesquiterpene containing distillate is used as oil of turpentine for the thinning of paints and varnishes.

The powerful cytostatic compound paclitaxel (Taxol®) is a diterpene with an *N*-benzoyl

1

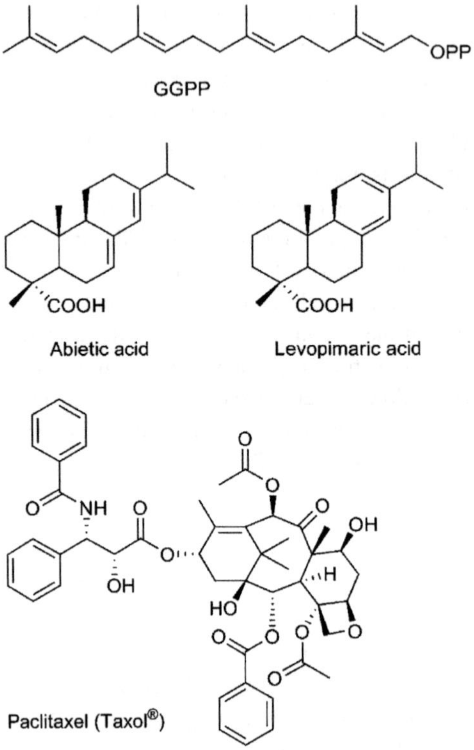

Fig. 31 Diterpenes

phenylisoserinoyl side chain derived from two molecules phenylalanine (Fig. 31). Paclitaxel was first isolated from the bark of the Pacific yew (*Taxus brevifolia*) by bioactivity-directed fractionation in 1966, and its structure was elucidated 5 years later (Wani et al., 1971). Its anti-cancer activity is based on a unique mechanism. Paclitaxel binds to microtubules, stabilizes them against depolymerisation and thus blocks cell proliferation (Schiff and Horwitz, 1980). Paclitaxel is used in the therapy of breast, ovarian and lung cancers, cancers of head and neck and Kaposi's sarcoma. Since paclitaxel occurs only in relatively low amounts in the bark of *T. brevifolia* (0.01-0.02%) and the trees grow slowly, other sources had to be found to supply enough of the diterpene for industrial production. Today, paclitaxel is obtained either from tissue cultures of various *Taxus* species or by semisynthesis from baccatin III and 10-deacetylbaccatin III, which can be extracted in sufficient amounts from leaves and twigs of the common yew (*T. baccata*), a tree that grows much faster than *T. brevifolia*.

Triterpenes and Steroids

Triterpenes are synthesized via the MVA pathway from two molecules of FPP that are joined by tail-to-tail condensation to squalene. Cyclization of its metabolite 2,3-oxidosqualene followed by rearrangements and methyl shifts yields various structures, mostly tetra- or pentacyclic. 2,3-Oxidosqualene is also the precursor of plant steroids. In this case, it is cyclized to the triterpene cycloartenol, which is then converted to the C_{27} compound cholesterol with the loss of three methyl groups. The oxygen of 2,3-oxidosqualene is usually retained as hydroxy group at C-3 in both triterpenes and steroids.

In contrast to animals, where cholesterol is the major sterol, many plant sterols are methylated or ethylated at C-24 of the side chain, e.g., campesterol and stigmasterol (Fig. 32). These phytosterols are constituents of biomembranes in plants and influence their permeability. Phytosterols inhibit the absorption of cholesterol in animals. Since they are more lipophilic than cholesterol, they are more readily incorporated into the micelles involved in fat digestion. Esters of phytosterols are therefore used as cholesterol-lowering food additives (Dewick, 2002). Brassinosteroids are a group of plant hormones that derive from campesterol. They regulate various biological processes, e.g., stem elongation, leaf expansion, seed germination, and xylem differentiation (Bishopp et al., 2006).

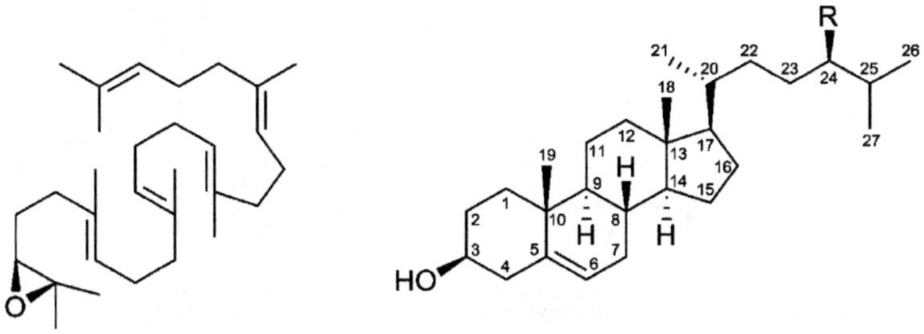

2,3-Oxidosqualene

Cholesterol (R = H)
Campesterol (R = CH_3)
Sitosterol (R = C_2H_5)

Fig. 32 Sterols derived from 2,3-oxidosqualene

Saponins

Triterpene saponins are widely distributed among eudicot plants, for example, in the Araliaceae, Caryophyllaceae, Fabaceae, and Primulaceae families. Monocots, instead, preferably accumulate steroidal saponins, which are abundant in the Agavaceae, Dioscoraceae, and Yuccaceae. Triterpenoid saponins often contain the pentacyclic α-amyrin (ursane), β-amyrin (oleanane) or lupane skeleton or the tetracyclic dammarane backbone as aglycone. This aglycone is linked with one to three carbohydrate chains containing up to six sugar molecules or uronic acids (Dewick, 2002; Hostettmann and Marston, 1995). The first sugar chain is attached to the hydroxy group at C-3 of the triterpene backbone. If two or more carbohydrate chains are present, they are usually connected with hydroxy or carboxy groups at C-28 or C-30.

Steroid saponins can be classified into two groups, spirostanols and furostanols. In furostanols, the side chain of cholesterol is used to form a tetrahydrofuran ring, and the hydroxy group at C-26 is glycosylated. Upon cleavage of this sugar moiety, a second oxygen-containing heterocycle is formed, thus yielding a spirostanol (Fig. 33). Steroidal glycoalkaloids have the same structure like spirostanol saponins, except that the oxygen in the six-membered heterocycle of the spiro function is replaced by nitrogen. As in the case of the triterpene saponins, steroidal saponins carry a sugar chain at the C-3 hydroxy group.

The name saponin is derived from the Latin word “sapo”, soap. This refers to the properties of saponins, which consist like soaps of a lipophilic moiety (triterpenoid or steroid aglycone, also called sapogenin) and a hydrophilic moiety (sugars) and produce foam when shaken in aqueous solution. Plants like soapwort (*Saponaria officinalis*) and soapbark tree (*Quillaia saponaria*) were therefore used as detergents. Today, extracts of saponin-containing plants or isolated saponins are used in cosmetics, as detergents and as foaming agents in soft drinks (Güçlü-Üstündag and Mazza, 2007). Saponins lyse red blood cells, a process called hemolysis, because they complex sterols of the plasma membrane and thus increase membrane permeability. This membrane-permeabilizing effect is also responsible for the antimicrobial and antifungal activities of saponins and their function as defence compounds in plants. In general, saponins with only one sugar chain (monodesmosides) show stronger hemolytic and antifungal effects than saponins with two oligosaccharide chains (bisdesmosides). Like cyanogenic glucosides or glucosinolates, bisdesmosidic saponins

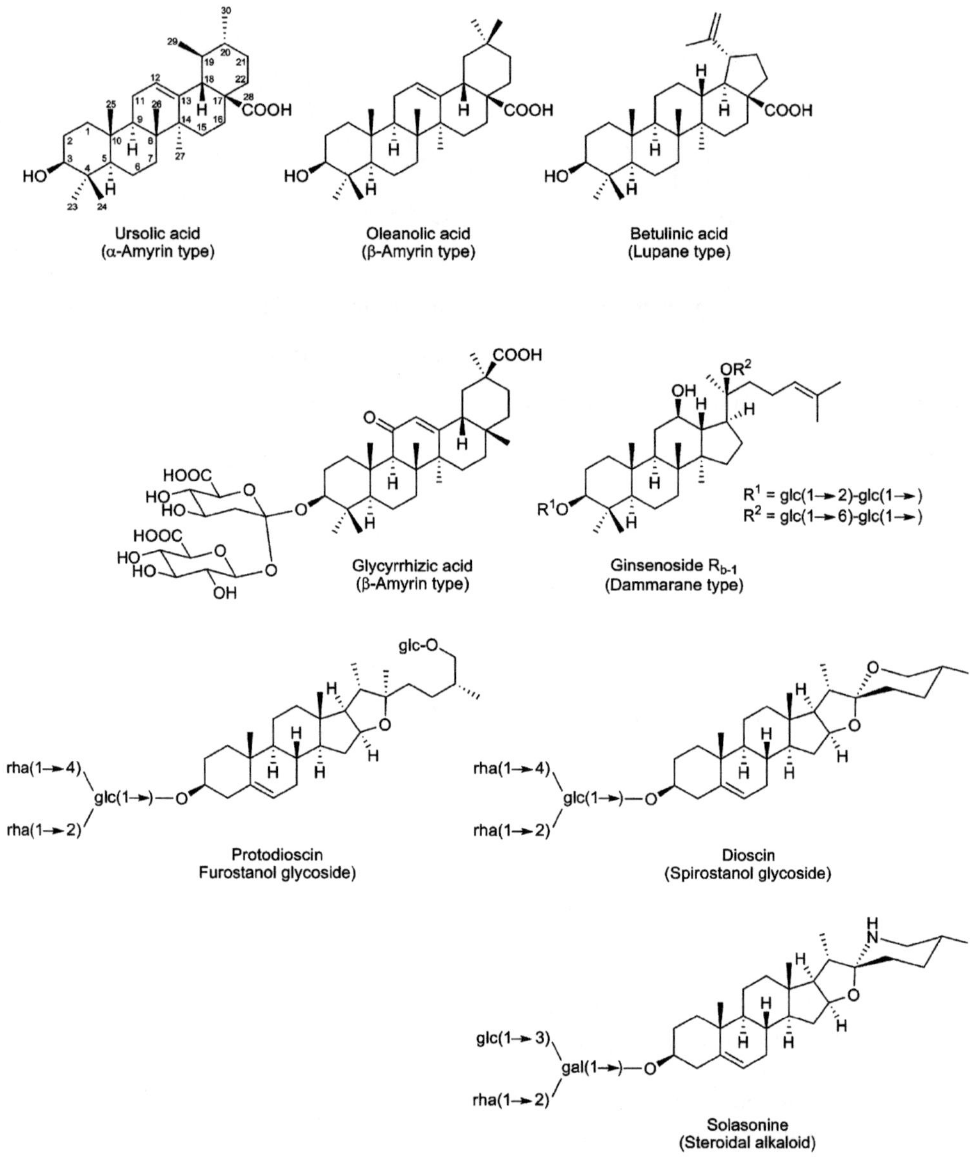

Fig. 33 Triterpene and steroid saponins

can be thought of as prodrugs that are cleaved in case of wounding by a specific hydrolase normally located in a different compartment and converted into active defence compounds (Osbourn, 1996). However, one sugar chain is required for the biological activity of saponins, whereas other natural products usually lose their activity when glycosylated.

Taken orally, saponins are not toxic because they are poorly absorbed and the sugar chain important for their hemolytic properties is hydrolyzed. In contrast, saponins are toxic to fish, since they damage the membranes of the gills (Hostettmann and Marston, 1995). Plant material rich in saponins has therefore been used to poison and stupefy fish. The fish can then be caught without difficulty and are not toxic to humans.

Most saponins have a bitter taste, but some sweet saponins are known as well. Glycyrrhizic acid from licorice root (*Glycyrrhiza glabra*), a β-amyrine type triterpene linked to two molecules of glucuronic acid (Fig. 33), is 50 times sweeter than sucrose (table sugar). Licorice extracts are used to prepare candies (licorice) and as a sweetener. They are also used as mild expectorant and as anti-inflammatory agent. The anti-inflammatory effect is caused by inhibition of an enzyme that inactivates cortisol. On the other hand, this may also lead to side effects such as sodium retention, excretion of potassium, water retention and increased blood pressure.

The roots of ginseng (*Panax ginseng*) have been used in the traditional medicine of Korea, China and Japan for several thousand years. It is used as adaptogen to help the body to cope with stress, to improve performance and during convalescence (Radad et al., 2006). Active ingredients are saponins, mainly of the dammarane type, containing two or three sugar side chains (Fig. 33) (Dewick, 2002). The ginseng aglycones protopanaxadiol and protopanaxatriol show promising anticancer activities (Güçlü-Üstündag and Mazza, 2007), and a preparation containing ginseng aglycones has been given conditional approval in China for the therapy of various tumors as single agent or in combination with paclitaxel.

The triterpene sapogenins betulinic acid, oleanolic acid and ursolic acid show cytotoxic and anti-inflammatory effects, and based on their structures novel chemopreventive and anticancer agents are being developed (Liby et al., 2007). A derivative of betulinic acid, bevirimat, is the first member of a new class of anti HIV therapeutics, maturase inhibitors. These compounds inhibit the processing of the HIV Gag protein, the precursor of the capsid, and lead to defective and non-infectious virus particles (Li et al., 2003).

Tetraterpenes

Tetraterpenes comprise only one group of compounds, the carotenoids. They are synthesized from two molecules GGPP by tail-to-tail addition. Double bonds are inserted to yield an extended conjugated system with all-*trans* configuration that is responsible for the yellow, orange and red color of the carotenoids. Either one or both ends of the tetraterpene chain are cyclized to a six-membered ring. Carotenoids with hydroxy or epoxy functions are classified as xanthophylls (Dewick, 2002).

Carotenoids fulfill important physiological functions in plants, since they are part of the light harvesting complex and act as accessory pigments of chlorophyll. In addition, they quench triplet chlorophyll and singlet oxygen in case of excess light energy and thus protect the plant from photo-oxidative damage. As pigments of flowers and fruits, carotenoids attract pollinators and seed dispersers (Howitt and Pogson, 2006).

Carotenoids are essential for human health. α-carotene, β-carotene (Fig. 34), and β-cryptoxanthine are precursors of vitamin A. They are taken up with food, cleaved in the intestinal mucosa and converted in the liver to vitamin A, which serves as pigment of the light receptors of human eyes. To overcome vitamin A deficiency

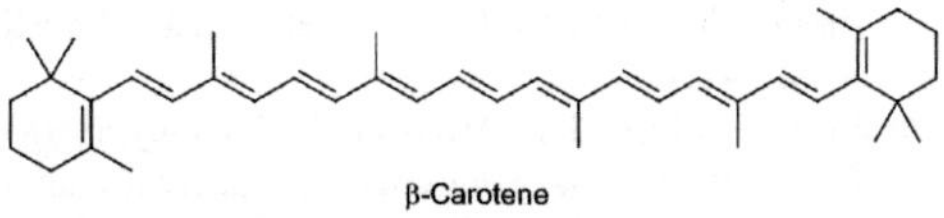

Fig. 34 β-Carotene

1

in areas with malnutrition, a transgenic rice termed, golden rice "was developed that expresses high levels of carotenoid biosynthetic enzymes in the endosperm and accumulates elevated levels of carotenoids (Ye et al., 2000). Due to their anti-oxidant and radical scavenging properties, a diet with fruits and vegetables rich in carotenoids is assumed to decrease the risk of cardiovascular disease or cancer. However, the intake of carotenoids as supplement probably has no health promoting effects (Riccioni et al., 2007; U.S. Preventive Service Task Force, 2003).

References

Adam P, Arigoni D, Bacher A, Eisenreich W (2002) Biosynthesis of hyperforin in Hypericum perforatum. J Med Chem 45:4786–4793

Adam KP, Zapp J (1998) Biosynthesis of the isoprene units of chamomile sesquiterpenes. Phytochemistry 48:953–959

Adlercreutz H (2007) Lignans and human health. Crit Rev Clin Lab Sci 44:483–525

Amna T, Puri SC, Verma V, Sharma JP, Khajuria RK, Musarrat J, Spiteller M, Qazi GN (2006) Bioreactor studies on the endophytic fungus Entrophospora infrequens for the production of an anticancer alkaloid camptothecin. Can J Microbiol 52:189-196

Ascherio A, Weisskopf MG, O'Reilly EJ, McCullough ML, Calle EE, Rodriguez C, Thun MJ (2004) Coffee consumption, gender, and Parkinson's disease mortality in the cancer prevention study II cohort: the modifying effects of estrogen. Am J Epidemiol 160:977-984

Ashihara H, Suzuki T (2004) Distribution and biosynthesis of caffeine in plants. Front Biosci 9:1864–1876

Ashihara H, Crozier A (2001) Caffeine: a well known but little mentioned compound in plant science. Trends Plant Sci 6:407–413

Askitopoulou H, Ramoutsaki IA, Konsolaki E (2000) Analgesia and anesthesia: etymology and literary history of related Greek words. Anesth Analg 91:486–91

Atkinson J, Morand P, Arnason JT, Niemeyer HM, Bravo HR (1991) Analogs of the cyclic hydroxamic acid 2,4-dihydroxy-7-methoxy-2H-1,4-benzoxazin-3-one (DIMBOA): decomposition to benzoxazolinones and reaction with beta-mercaptoethanol. J Org Chem 56:1788–1800

Aziz E, Nathan B, McKeever J (2000) Anesthetic and analgesic practices in Avicenna's Canon of Medicine. Am J Chin Med 28:147–151

Baldwin IT, Halitschke R, Paschold A, von Dahl CC, Preston CA (2006) Volatile signaling in plant-plant interactions: "talking trees" in the genomics era. Science 311:812–815

Barron D, Ibrahim RK (1996) Isoprenylated flavonoids - a survey. Phytochemistry 43:921–982

Baur JA, Pearson KJ, Price NL, Jamieson HA, Lerin C, Kalra A, Prabhu VV, Allard JS, Lopez-Lluch G, Lewis K, Pistell PJ, Poosala S, Becker KG, Boss O, Gwinn D, Wang M, Ramaswamy S, Fishbein KW, Spencer RG, Lakatta EG, Le Couteur D, Shaw RJ, Navas P, Puigserver P, Ingram DK, de Cabo R, Sinclair DA (2006) Resveratrol improves health and survival of mice on a high-calorie diet. Nature 444:337–342

Benowitz NL (1990) Clinical pharmacology of caffeine. Annu Rev Med 41:277–288

Biastoff S, Dräger B (2007) Calystegines. In: Cordell GA (ed) The Alkaloids, vol 64. Academic, New York

Bick JA, Lange BM (2003) Metabolic cross talk between cytosolic and plastidial pathways of isoprenoid biosynthesis: Unidirectional transport of intermediates across the chloroplast envelope membrane. Arch Biochem Biophys 415:146–154

Bishopp A, Mähönen AP, Helariutta Y (2006) Signs of change: hormone receptors that regulate plant development. Development 133:1857–1869

Boerjan W, Ralph J, Baucher M (2003) Lignin biosynthesis. Annu Rev Plant Biol 54:519–546

Boonen G, Häberlein H (1998) Influence of genuine kavapyrone enantiomers on the GABA-A binding site. Planta Med 64:504–506

Booth M (1998) Opium: a history. MacMillan, New York

Botta B, Vitali A, Menendez P, Misiti D, Delle Monache G (2005) Prenylated flavonoids: pharmacology and biotechnology. Curr Med Chem 12:717–739

Bravo HR, Lazo W (1993) Antimicrobial activity of cereal hydroxamic acids and related compounds. Phytochemistry 33: 569–591

Bravo HR, Lazo W (1996) Antialgal and antifungal activity of natural hydroxamic acids and related compounds. J Agric Food Chem 44:1569–1571

Bringmann G, Feineis D (2001) Stress-related polyketide metabolism of Dioncophyllaceae and Ancistrocladaceae. J Exp Bot 52:2015–2022

Briskin DP (2000) Medicinal Plants and phytomedicines. Linking plant biochemistry and

physiology to human health. Plant Phys 124:507–514
Bruce-Chwatt LJ (1988) Cinchona and its alkaloids: 350 years. N Y State J Med 88:318–322
Chang MCY, Eachus RA, Trieu W, Ro DK, Keasling JD (2007) Engineering Escherichia coli for production of functionalized terpenoids using plant P450s. Nat Chem Biol 3:274–277
Chou AC, Chevli R, Fitch CD (1980) Ferriprotoporphyrin IX fulfils the criteria for identification as the chloroquine receptor of malaria parasites. Biochemistry 19:1543-1549
Covello PS, Teoh KH, Polichuk DR, Reed DW, Nowak G (2007) Functional genomics and the biosynthesis of artemisinin. Phytochemistry 68:1864-1871
Croteau RB, Davis EM, Ringer KL, Wildung MR (2005) (-)-Menthol biosynthesis and molecular genetics. Naturwissenschaften 92:562–577
Davin LB, Lewis NG (2005) Lignin primary structures and dirigent sites. Curr Opin Biotechnol 16:407–415
Dewick PM (2002) Medicinal natural products: a biosynthetic approach, 2nd edn. Wiley, Chichester
Didry N, Dubreuil L, Pinkas M (1994) Activity of anthraquinonic and naphthoquinonic compounds on oral bacteria. Pharmazie 49:681–683
Dinda B, Debnath S, Harigaya Y (2007a) Naturally occurring iridoids. A review, Part 1. Chem Pharm Bull 55:159–222
Dinda B, Debnath S, Harigaya Y (2007b) Naturally occurring iridoids. A review, Part 2. Chem Pharm Bull 55:689–728
Dixon RA, Paiva NL (1995) Stress-induced phenylpropanoid metabolism. Plant Cell 7:1085–1097
Dräger B (2004) Chemistry and biology of calystegines. Nat Prod Rep 21:211–223
Drawert F, Beier J (1976) Aminosäuren als Vorstufe der Acylseitenkette der Hopfenbitterstoffe. Phytochemistry 15:1693–1694
Dudareva N, Andersson S, Orlova I, Gatto N, Reichelt M, Rhodes D, Boland W, Gershenzon J (2005) The nonmevalonate pathway supports both monoterpene and sesquiterpene formation in snapdragon flowers. Proc Natl Acad Sci USA 102:933–938
Dudareva N, Negre F, Nagegowda DA, Orlova I (2006) Plant volatiles: recent advances and future perspectives. CRC Crit Rev Plant Sci 25:417–440
Duffy C, Perez K, Partridge A (2007) Implications of phytoestrogen intake for breast cancer. CA Cancer J Clin 57:260–277
Dumont P, Ingrassia L, Rouzeau S, Ribaucour F, Thomas S, Roland I, Darro F, Lefranc F, Kiss R (2007) The Amaryllidaceae isocarbostyril narciclasine induces apoptosis by activation of the death receptor and/or mitochondrial pathways in cancer cells but not in normal fibroblasts. Neoplasia 9:766–76
Eisner T, Meinwald J (1995) The chemistry of sexual selection. Proc Natl Acad Sci USA 92:50–55
Fahey JW, Zalcmann AT, Talalay P (2001) The chemical diversity and distribution of glucosinolates and isothiocyanates among plants. Phytochemistry 56:5–51
Feild TS, Lee DW, Holbrook NM (2001) Why leaves turn red in autumn. The role of anthocyanins in senescing leaves of red-osier dogwood. Plant Physiol 127:566–574
François G, Timperman G, Eling W, Assi LA, Holenz J, Bringmann G (1997) Naphthylisoquinoline alkaloids against malaria: evaluation of the curative potentials of dioncophylline C and dioncopeltine A against Plasmodium berghei in vivo. Antimicrob Agents Chemother 41:2533–2539
Frei H, Lüthy J, Brauchli J, Zweifel U, Würgler FE, Schlatter C (1992) Structure/activity relationships of the genotoxic potencies of sixteen pyrrolizidine alkaloids assayed for the induction of somatic mutation and recombination in wing cells of Drosophila melanogaster. Chem Biol Interact 83:1-22
Frick S, Kramell R, Schmidt J, Fist AJ, Kutchan TM (2005) Comparative qualitative and quantitative determination of alkaloids in narcotic and condiment Papaver somniferum cultivars. J Nat Prod 68:666–673
Friedmann J, Waller GR (1985) Caffeine hazards and their prevention in germinating seeds of coffee (Coffea arabica L.). J Chem Ecol 9:1099–1106
Fujii T, Ohba M (1998) The ipecac alkaloids and related bases. In: Cordell GA (ed) The alkaloids, vol 51. Academic, New York
Gates M, Tschudi G (1952) The synthesis of morphine. J Am Chem Soc 74:1109-1110
Geiger PL, Hesse K (1833) Darstellung des Atropins. Ann Pharm 5:43–81
Gerhäuser C, Alt A, Heiss E, Gamal-Eldeen A, Klimo K, Knauft J, Neumann I, Scherf HR, Frank N, Bartsch H, Becker H (2002) Cancer chemopreventive activity of xanthohumol, a natural product derived from hop. Mol Cancer Ther 1:959–969
Goese M, Kammhuber K, Bacher A, Zenk MH, Eisenreich W (1999) Biosynthesis of bitter acids in hops. A ^{13}C-NMR and ^{2}H-NMR study in the building blocks of humulone. Eur J Biochem 263:447-454
Goldmann A, Milat ML, Ducrot PH, Lallemand JY, Maille M, Lepingle A, Charpin I, Tepfer D (1990)

1

Tropane derivatives from Calystegia sepium. Phytochemistry 29:2125-2128
Griffin WJ, Lin GD (2000) Chemotaxonomy and geographical distribution of tropane alkaloids. Phytochemistry 53:623–637
Grombacher AW, Russell WA, Guthrie WD (1989) Resistance to first-generation European corn borer (Lepidoptera: Pyralidae) and DIMBOA concentration in midwhorl leaves of the BS9 maize synthetic. J Kans Entomol Soc 62:103–107
Güçlü-Üstündag Ö, Mazza G (2007) Saponins: properties, applications and processing. Crit Rev Food Sci Nutr 47:231–258
Hain R, Reif HJ, Krause E, Langebartels R, Kindl H, Vornam B, Wiese W, Schmelzer E, Schreier PH, Stöcker RH, Stenzel K (1993) Disease resistance results from foreign phytoalexin expression in a novel plant. Nature 361:153–156
Halkier BA, Gershenzon J (2006) Biology and biochemistry of glucosinolates. Ann Rev Plant Biol 57:303–333
Harborne JB, Baxter H (1999) The handbook of natural flavonoids, vols 1 and 2. Wiley, Chichester
Harborne JB, Williams CA (2000) Advances in flavonoid research since 1992. Phytochemistry 55:481–504
Hartmann T, Ober D (2000) Biosynthesis and metabolism of pyrrolizidine alkaloids in plants and specialized insect herbivores. Top Curr Chem 209:207–243
Hartmann T, Sander H, Adolph R, Toppel, G (1988) Metabolic links between the biosynthesis of pyrrolizidine alkaloids and polyamines in root cultures of Senecio vulgaris. Planta 175:82–90
Hasegawa M, Sasaki T, Sadakane K, Tabuchi M, Takeda Y, Kimura M, Fujii Y (2002) Studies for the emetic mechanisms of ipecac syrup (TJN-119) and its active components in ferrets: involvement of 5-hydroxytryptamine receptors. Jpn J Pharmacol 89:113-119
Hashimoto Y, Shudo K (1996) Chemistry of biologically active benzoxazinoids. Phytochemistry 43:551–559
Hegnauer R (1986) Chemotaxonomie der Pflanzen. Vol 7. Birkhäuser Verlag, Basel
Heim KE, Tagliaferro AR, Bobilya DJ (2002) Flavonoids antioxidants: chemistry, metabolism and structure-activity relationships. J Nutr Biochem 13:572–584
Heinrich M (2004) Snowdrops: the heralds of spring and a modern drug for Alzheimer's disease. Pharmaceutical J 273:905–906
Heinrich M, Robles M, West JE, Ortiz de Montellano BR, Rodriguez E (1998) Ethnopharmacology of Mexican Asteraceae (Compositae). Annu Rev Pharmacol Toxicol 38:539–565
Heubl G, Bringmann G, Meimberg H (2006) Molecular phylogeny and character evolution of carnivorous plant families in Caryophyllales - revisited. Plant Biol 8:821–830
Higdon JV, Delage B, Williams DE, Dashwood RH (2007) Cruciferous vegetables and human cancer risk: epidemiologic evidence and mechanistic basis. Pharmacol Res 55:224–236
Hollingsworth RG, Armstrong JW, Campbell E (2002) Caffeine as a repellent for slugs and snails. Nature 417:915–916
Hostettmann KA, Marston A (1995) Saponins. Cambridge University Press, Cambridge
Howitt CA, Pogson BJ (2006) Carotenoid accumulation and function in seeds and non-green tissues. Plant Cell Environ 29:435–445
Howitz KT, Bitterman KJ, Cohen HY, Lamming DW, Lavu S, Wood JG, Zipkin RE, Chung P, Kisielewski A, Zhang LL, Scherer B, Sinclair DA (2003) Small molecule activators of sirtuins extend Saccharomyces cerevisiae lifespan. Nature 425:191–196
Howland R, Mycek MJ, Harvey RA, Champe PC (2005) Lippincott's Illustrated Reviews: Pharmacology, 3rd edn. Lippincott William & Wilkins, Philadelphia, PA
Hsiang YH, Hertzberg R, Hecht S, Liu LF (1985) Camptothecin induces protein-linked DNA breaks via mammalian DNA topoisomerase I. J Biol Chem 260:14873–14878
Hsiang YH, Lihou MG, Liu LF (1989) Arrest of replication forks by drug-stabilized topoisomerase I-DNA cleavable complexes as a mechanism of cell killing by camptothecin. Cancer Res 49:5077–5082
Hsu E (2006) Reflections on the 'discovery' of the antimalarial qinghao. Br J Clin Pharmacol 61:666–670
Hutchinson CR, Heckendorf AH, Daddona PE, Hagaman E, Wenkert E (1974) Biosynthesis of camptothecin. I. Definition of the overall pathway assisted by carbon-13 nuclear magnetic resonance analysis. J Am Chem Soc 96:5609-5611
Ieven M, van den Berghe DA, Vlietinck AJ (1983) Plant antiviral agents. IV. Influence of lycorine on growth pattern of three animal viruses. Planta Med 49:109–114
Jin Z (2007) Amaryllidaceae and Sceletium alkaloids. Nat Prod Rep 24:886–905
Joe B, Vijaykumar M, Lokesh BR (2004) Biological properties of curcumin - cellular and molecular

mechanisms of action. Crit Rev Food Sci Nutri 44:97–111
Jones DA (1988) Cyanogenesis in animal-plant interactions. Ciba Found Symp 140:151–170
Jones DA (1998) Why are so many food plants cyanogenic? Phytochemistry 47:155–162
Jordan MA, Thrower D, Wilson L (1991) Mechanism of inhibition of cell proliferation by Vinca alkaloids. Cancer Res 51:2212–2222
Jordt SE, McKemy DD, Julius D (2003) Lessons from peppers and peppermint: the molecular logic of thermosensation. Curr Opin Neurobiol 13:487–492
Jørgensen K, Bak S, Busk PK, Sørensen C, Olsen CE, Puonti-Kaerlas J, Møller BL (2005) Cassava plants with a depleted cyanogenic glucoside content in leaves and tubers. Distribution of cyanogenic glucosides, their site of synthesis and transport, and blockage of the biosynthesis by RNA interference technology. Plant Phys 139:363–374
Kalemba D, Kunicka A (2003) Antibacterial and antifungal properties of essential oils. Curr Med Chem 10:813–829
Karppinen K, Hokkanen J, Tolonen A, Mattila S, Hohtola A (2007) Biosynthesis of hyperforin and adhyperforin from amino acid precursors in shoot cultures of Hypericum perforatum. Phytochemistry 68:1038–1045
Kaufman TS, Rúveda EA (2005) The quest for quinine: those who won the battles and those who won the war. Angew Chem Int Ed 44:854–885
Klein RM (1987) The green world: an introduction to plants and people. HarperCollins, New York
Knudsen JT, Tollsten L, Bergström G (1993) Floral scents: a checklist of volatile compounds isolated by head-space techniques. Phytochemistry 33:253–28
Küenburg B, Czollner L, Fröhlich J, Jordis U (1999) Development of a pilot scale process for the anti-Alzheimer drug (-)-galanthamine using large-scale phenolic oxidative coupling and crystallisation-induced chiral conversion. Org Process Res Dev 3:425–431
Ladenburg A (1881) Die natürlich vorkommenden mydriatisch wirkenden Alkaloide. Justus Liebigs Ann Chem 206:274–307
Laule O, Fürholz A, Chang HS, Zhu T, Wang X, Heifetz PB, Gruissem W, Lange M (2003) Crosstalk between cytosolic and plastidial pathways of isoprenoid biosynthesis in Arabidopsis thaliana. Proc Natl Acad Sci USA 100:6866–6871
Leete E (1963) The biosynthesis of coniine from four acetate units. J Am Chem Soc 85: 3523–3524
Leete E (1964) Biosynthesis of the hemlock alkaloids. The incorporation of acetate-1-C^{14} into coniine and conhydrine. J Am Chem Soc 86:2509–2513
Léonce S, Kraus-Berthier L, Golsteyn RM, David-Cordonnier MH, Tardy C, Lansiaux A, Poindessous V, Larsen AK, Pierré A (2006) Generation of replication-dependent double-strand breaks by the novel N2-G-alkylator S23906-1. Cancer Res 66:7203–7210
Li F, Goila-Gaur R, Salzwedel K, Kilgore NR, Reddick M, Matallana C, Castillo A, Zoumplis D, Martin DE, Orenstein JM, Allaway GP, Freed EO, Wild CT (2003) PA-457: a potent HIV inhibitor that disrupts core condensation by targeting a late step in Gag processing. Proc Natl Acad Sci USA 100:13555–13560
Liby KT, Yore MM, Sporn MB (2007) Triterpenoids and rexinoids as multifunctional agents for the prevention and treatment of cancer. Nat Rev Cancer 7:357–5369
López-Meyer M, Nessler CL, McKnight TD (1994) Sites of accumulation of the antitumor alkaloid camptothecin in Camptotheca acuminata. Planta Med 60:558–560
Lovett R (2005) Coffee: the demon drink? New Scientist, 24 September
Lyß G, Knorr A, Schmidt TJ, Pahl HL, Merfort I (1998) The anti-inflammatory sesquiterpene lactone helenalin inhibits the transcription factor NF-κB by directly targeting p65. J Biol Chem 273:33508–33516
Mahmoud SS, Croteau RB (2002) Strategies fro transgenic manipulation of monoterpene biosynthesis in plants. Trends Plant Sci 7:366–373
McLachlan A, Kekre N, McNulty J, Pandey S (2005) Pancratistatin: a natural anti-cancer compound that targets mitochondria specifically in cancer cells to induce apoptosis. Apoptosis 10:619-630
McMahon JB, Currens MJ, Gulakowski RJ, Buckheit RW Jr, Lackman-Smith C, Hallock YF, Boyd MR (1995) Michellamine B, a novel plant alkaloid, inhibits human immunodeficiency virus-induced cell killing by at least two distinct mechanisms. Antimicrob Agents Chemother 39:484–488
Mein K (1833) Ueber die Darstellung des Atropins in weißen Krystallen. Ann Pharm 6:67–72
Milder IE, Arts IC, van de Putte B, Venema DP, Hollman PC (2005) Lignan contents of Dutch plant foods: a database including lariciresinol, pinoresinol, secoisolariciresinol and matairesinol. Br J Nutr 93:393-402
Mo H, Elson CE (2004) Studies of the isoprenoid-mediated inhibition of mevalonate synthesis

1

applied to cancer chemotherapy and chemoprevention. Exp Biol Med 229:567–585
Mo Y, Nagel C, Taylor LP (1992) Biochemical complementation of chalcone synthase mutants defines a role for flavonols in functional pollen. Proc Natl Acad Sci USA 89:7213–7217
Montllor CB, Bernays EA, Barbehenn RV (1990). Importance of quinolizidine alkaloids in the relationship between larvae of Uresiphita reversalis (Lepidoptera: Pyralidae), and a host plant Cytisus monspessulanus. J Chem Ecol 16:1853–1865
Müller WE (2003) Current St. John's wort research from mode of action to clinical efficacy. Pharmacol Res 47:101–109
Nagle DG, Ferreira D, Zhou YD (2006) Epigallocatechin-3-gallate (EGCG): chemical and biomedical perspectives. Phytochemistry 67:1849–1855
Nandi R, Maiti M (1985) Binding of sanguinarine to deoxyribonucleic acids of differing base composition. Biochem Pharmacol 34:321–324
Noble RL (1990) The discovery of the Vinca alkaloids - chemotherapeutic agents against cancer. Biochem Cell Biol 68:1344–1351
O'Brien GM, Taylor AJ, Poulter NH (1991) Improved enzymatic assay for cyanogens in fresh and processed cassava. J Sci Food Agric 56:277–289
O'Connor SE, Maresh JJ (2006) Chemistry and biology of monoterpene indole alkaloid biosynthesis. Nat Prod Rep 23:532–547
Osbourn A (1996) Saponins and plant defence - a soap story. Trends Plant Sci 1:4–9
Peer WA, Murphy AS (2007) Flavonoids and auxin transport: modulators or regulators? Trends Plant Sci 12:556–563
Pérez FJ, Niemeyer HM (1989) Reaction of DIMBOA with amines. Phytochemistry 28:1831–1834
Phillips MACroteau, (1999) Resin-based defenses in conifers. Trends Plant Sci 4:184–190
Pichersky E, Dudareva N (2007) Scent engineering: toward the goal of controlling how flowers smell. Trends Biotechnol 25:105–110
Piel J, Donath J, Bandemer K, Boland W (1998) Mevalonate-independent biosynthesis of terpenoid volatiles in plants: induced and constitutive emission of volatiles. Angew Chem Int Ed 37: 2478–2481
Plowman T, Rivier L (1983) Cocaine and cinnamoylcocaine content of Erythroxylum species. Ann Bot 51:641–659
Ponte-Sucre A, Faber JH, Gulder T, Kajahn I, Pedersen SE, Schultheis M, Bringmann G, Moll H (2007) Activities of naphthylisoquinoline alkaloids and synthetic analogs against Leishmania major. Antimicrob Agents Chemother 51:188–194
Puri SC, Verma V, Amna T, Qazi GN, Spiteller M (2005) An endophytic fungus from Nothapodytes foetida that produces camptothecin. J Nat Prod 68:1717–1719
Radad K, Gille G, Liu L, Rausch WD (2006) Use of ginseng in medicine with emphasis on neurodegenerative disorders. J Pharmacol Sci 100:175–186
Rasmann S, Kollner TG, Degenhardt J, Hiltpold I, Toepfer S, Kuhlmann U. Gershenzon J, Turlings TCJ (2005) Recruitment of entomopathogenic nematodes by insect-damaged maize. Nature 434:732–737
Ratzka A, Vogel H, Kliebenstein DJ, Mitchell-Olds T, Kroymann J. 2002. Disarming the mustard oil bomb. Proc Natl Acad Sci USA 99: 11223–11228
Reynolds T (2005) Hemlock alkaloids from Socrates to poison aloes. Phytochemistry 66:1399–1406
Riccioni G, Bucciarelli T, Mancini B, Corradi F, Di Ilio C, Mettei PA, D'Orazio N (2007) Antioxidant vitamin supplementation in cardiovascular diseases. Ann Clin Lab Sci 37:89–95
Rivier L (1981) Analysis of alkaloids in leaves of cultivated Erthroxylum and characterization of alkaline substances used during coca chewing. J Ethnopharmacol 3:313–335
Ro DK, Paradise EM, Quellet M, Fisher KJ, Newman KL, Ndungu JM, Ho KA, Eachus RA, Ham TS, Kirby J, Chang MC, Withers ST, Shiba Y, Sarpong R Keasling JD (2006) Production of the antimalarial drug precursor artemisinic acid in engineered yeast. Nature 440:940-943
Roberts MF (1971) The formation of γ-coniceine from 5-ketooctanal by a transaminase of Conium maculatum. Phytochemistry 10:3057–3060
Röder E (1995) Medicinal plants in Europe containing pyrrolizidine alkaloids. Pharmazie 50:83–98
Rouhi AM, Washington C, Washington EN (2000) Lignin and lignan biosynthesis. Chem Eng News 78:29–32
Saleem M, Kim HJ, Ali MS, Lee YS (2005) An update on bioactive plant lignans. Nat Prod Rep 22:696-716
Sánchez-Pérez R, Jørgensen K, Olsen CE, Dicenta F, Møller BL (2008) Bitterness in almonds. Plant Physiol 146:1040–1052
Sanderson K (2007) Opiates for the masses. Nature 449:268–269
Schiff P (2002) Opium and its alkaloids. Am J Pharm Edu 66
Schiff PB, Horwitz SB (1980) Taxol stabilizes microtubules in mouse fibroblast cells. Proc Natl Acad Sci USA 77:1561–1565
Schmidt E (1892) Ueber Scopolamin. Arch Pharm 230:207–231

Schnee C, Kollner TG, Held M, Turlings TCJ, Gershenzon J, Degenhardt J (2006) The products of a single maize sesquiterpene synthase form a volatile defense signal that attracts natural enemies of maize herbivores. Proc Natl Acad Sci USA 103:1129–1134

Schuhr CA, Radykewicz T, Sagner S, Latzel C, Zenk MH, Arigoni D, Bacher A, Rohdich F, Eisenreich W (2003) Quantitative assessment of crosstalk between the two isoprenoid biosynthesis pathways in plants by NMR spectroscopy. Phytochemistry Rev 2:3–16

Seigler DS (1991) Cyanide and cyanogenic glucosides. In: Rosenthal GA, Berenbaum MR (eds) Herbivores: their interactions with secondary plant metabolites. Academic, San Diego, CA

Seigler DS (1998) Iridoid monoterpenes. In: Plant secondary metabolism. Kluwer, Dordrecht

Selmar D (1999) Biosynthesis of cyanogenic glycosides, glucosinolates and non-protein amino acids. In: Wink M (ed) Biochemistry of plant secondary metabolism. Sheffield Acad Press, England

Selmar D, Lieberei R, Biehl B (1988) Mobilization and utilization of cyanogenic glycosides: the linustatin pathway. Plant Physiol 86:711–716

Sharkey TD, Wiberley AE, Donohue AR (2008) Isoprene emission from plants: why and how. Ann Bot (Lond) 101:5–18

Sheriha GM, Rapoport H (1976) Biosynthesis of Camptotheca acuminata alkaloids. Phytochemistry 15:505–508

Sicker D, Frey M, Gierl A (2000) Role of natural benzoxazinones in the survival strategy of plants. Int Rev Cytol 198:319–346

Sirikantaramas S, Asano T, Sudo H, Yamazaki M, Saito K (2007) Camptothecin: therapeutic potential and biotechnology. Curr Pharm Biotechnol 8:196–202

Siritunga D, Sayre R (2007) Transgenic approaches for cyanogen reduction in cassava. AOAC Int. 90:1450–1455

Soelberg J, Jørgensen LB, Jäger AK (2007) Hyperforin accumulates in the translucent glands of Hypericum perforatum. Ann Bot 99:1097–1100

Srinivas G, Babykutty S, Sathiadevan PP, Srinivas P (2007) Molecular mechanism of emodin action: transition from laxative Ingredient to an antitumor agent. Med Res Rev 27:591–608

Stevens JF, Page JE (2004) Xanthohumol and related prenylflavonoids from hops and beer: to your good health! Phytochemistry 65:1317–1330

Stirling IR, Freer IKA, Robins DJ (1997) Pyrrolizidine alkaloid biosynthesis. Incorporation of 2-aminobutanoic acid labelled with ^{13}C or ^{2}H into the senecic acid portion of rosmarinine and senecionine. J Chem Perkin Trans I 5:677–680

Straub KD, Carver P (1975) Sanguinarine, inhibitor of Na-K dependent ATP'ase. Biochem Biophys Res Commun 62:913–922

Sudo H, Yamakawa T, Yamazaki M, Aimi N, Saito K (2002) Bioreactor production of camptothecin by hairy root cultures of Ophiorrhiza pumila. Biotechnol Lett 24:359–363

Sullivan DJ Jr, Gluzman IY, Russell D G, Goldberg DE (1996) On the molecular mechanism of chloroquine's antimalarial action. Proc Natl Acad Sci USA 93:11865–11870

Sun XF, Sun RC, Fowler P, Baird MS (2005) Extraction and characterization of original lignin and hemicelluloses from wheat straw. J Agric Food Chem 53:860-870

Szlávik L, Gyuris A, Minárovits J, Forgo P, Molnár J, Hohmann J (2004) Alkaloids from Leucojum vernum and antiretroviral activity of Amaryllidaceae alkaloids. Planta Med 70:871–873

Tattersall DB, Bak S, Jones PR, Olsen CE, Nielsen JK, Hansen ML, Høj PB, Møller BL (2001) Resistance to an herbivore through engineered cyanogenic glucoside synthesis. Science 293:1826–1828

Tepfer D, Goldmann A, Pamboukdjian N, Maille M, Lepingle A, Chevalier D, Dénarié J, Rosenberg C (1988) A Plasmid of Rhizobium meliloti 41 encodes catabolism of two compounds from root exudate of Calystegium sepium. J Bacteriol 170:1153–1161

Thomson RH (1987) Naturally occurring quinones, vol III, Chapman & Hall, London

Thomzik JE, Stenzel K, Stöcker R, Schreier PH, Hain R, Stahl DJ (1997) Synthesis of a grapevine phytoalexin in transgenic tomatoes (Lycopersicon esculentum Mill.) conditions resistance against Phytophthora infestans. Physiol Mol Plant Pathol 51:265–278

Tokunaga T, Takada N, Ueda M (2004) Mechanism of antifeedant activity of plumbagin, a compound concerning the chemical defense in carnivorous plant. Tetrahedron Lett 45:7115–7119

Trapp S, Croteau R (2001) Defensive resin biosynthesis in conifers. Annu Rev Plant Physiol Plant Mol Biol 52:689–724

Turner CE, Elsohly MA, Boeren EG (1980) Constituents of Cannabis sativa L. XVII. A review of the natural constituents. J Nat Prod 43:169–234

U.S. Preventive Services Task Force (2003) Routine vitamin supplementation to prevent cancer and cardiovascular disease: recommendations and rationale. Ann Intern Med 139:51-55

Van der Heijden R, Jacobs DI, Snoeijer W, Hallard D, Verpoorte R (2004) The Catharanthus alkaloids: pharmacognosy and biotechnology. Curr Med Chem 11:607–628

Vincent RM, López-Meyer M, McKnight TD, Nessler CL (1997) Sustained harvest of camptothecin from the leaves of Camptotheca acuminata. J Nat Prod 60:618–619

Wall ME, Wani MC, Cook CE, Palmer KH, McPhail AT, Sim GA (1966) Plant antitumor agents. I. The isolation and structure of camptothecin, a novel alkaloidal leukemia and tumor inhibitor from Camptotheca acuminata. J Am Chem Soc 88:3888–3890

Wani MC, Taylor HL, Wall ME, Coggin P, McPhail AT (1971) Plant antitumor agents. VI. Isolation and structure of taxol, a novel antileukemic and antitumor agent from Taxus brevifolia. J Am Chem Soc 93:2325–2327

Weber S, Eisenreich W, Bacher A, Hartmann T (1998) Pyrrolizidine alkaloids of the lycopsamine type: biosynthesis of trachelanthic acid. Phytochemistry 50:1005–1014

Weng XL, Bonawitz ND, Chapple C (2008) Emerging strategies of lignin engineering and degradation for cellulosic biofuel production. Curr Opin Biotechnol 19:166–172

Werner I, Bacher A, Eisenreich W (1997) Retrobiosynthetic NMR studies with ^{13}C-labeled glucose. Formation of gallic acid in plants and fungi. J Biol Chem 272:25474–25482

White NJ (2008) Qinghaosu (Artemisinin): the price of success. Science 320:330–334

White SM, Lambe CJT (2003) The pathophysiology of cocaine abuse. J Clin Forensic Med 10:27–39

Wink M (2002) Production of quinolizidine alkaloids in in vitro cultures of legumes. In: Nagata, Ebizuka Y (eds) Biotechnology in agriculture and forestry, vol 51: medicinal and aromatic plants XII. Springer, Heidelberg

Wink M (2003) Allelochemical properties of quinolizidine alkaloids. In: Macías F, Galindo JC, Molinillo JM, Cutler HG (eds) Allelopathy: chemistry and mode of action of allelochemicals. CRC Press, Boca Raton, FL

Wink M, Witte L (1983) Evidence for a wide-spread occurrence of the genes of quinolizidine alkaloid biosynthesis. FEBS Lett 159:196–200

Wink M, Witte L (1984) Turnover and transport of quinolizidine alkaloids: diurnal variation of lupanine in the phloem sap, leaves and fruits of Lupinus albus L. Planta 161:519–524

Wink M, Witte L (1985) Quinolizidine alkaloids as nitrogen source for lupin seedlings and cell suspension cultures. Z Naturforschung 40c:767–775

Wink M, Witte L (1991) Storage of quinolizidine alkaloids in Macrosiphum albifrons and Aphis genistae (Homoptera: Aphididae). Entomol Gen 15:237–254

Winkel-Shirley B (2001) Flavonoid biosynthesis. A colorful model for genetics, biochemistry, cell biology, and biotechnology. Plant Physiol 126:485–493

Wittstock U, Agerbirk N, Stauber EJ, Olsen CE, Hippler M, Mitchell-Olds T, Gershenzon J, Vogel H (2004) Successful herbivore attack due to metabolic diversion of a plant chemical defense. Proc Natl Acad Sci USA 101:4859–48564

Wittstock U, Burow M (2007) Tipping the scales - specifier proteins in glucosinolate hydrolysis. IUBMB Life 59:744–751

Wood JG, Rogina B, Lavu S, Howitz KT, Helfand SL, Tatar M, Sinclair DA (2004) Sirtuin activators mimic caloric restriction and delay ageing in metazoans. Nature 430:686–689

Woodward RB, Doering WE (1944) The total synthesis of quinine. J Am Chem Soc 66:849

World Drug Report of the United Nations (2007)

Ye X, Al-Babili S, Klöti A, Zhang J, Lucca P, Beyer P, Potrykus I (2000) Engineering the provitamin A (beta-carotene) biosynthetic pathway into (carotenoid-free) rice endosperm. Science 287:303–305

Zagrobelny M, Bak S, Møller BL (2008) Cyanogenesis in plants and arthropods. Phytochemistry 69:1457–1468

Zenk MH, Jünger M (2007) Evolution and current status of the phytochemistry of nitrogenous compounds. Phytochemistry 68:2757–2772

Zhou SF, Xue CC, Yu XQ, Wang G (2007) Metabolic activation of herbal and dietary constituents and its clinical and toxicological implications: an update. Curr Drug Metabol 8:526–553

Zhu YJ, Agbayani R, Jackson MC, Tang CS, Moore PH (2004) Expression of the grapevine stilbene synthase gene VST1 in papaya provides increased resistance against diseases caused by Phytophthora palmivora. Planta 220:241–

Natural Products from Seaweeds

2

Efstathia Ioannou and Vassilios Roussis

Abstract Marine biota, even though the oceans are covering two thirds of earth's surface, remains an unexplored source of new and exciting chemical structures. Systematic investigations on marine organisms started only forty years ago, but the results have already proven the impact of the significantly diverse conditions and the distinct evolution on their biosynthetic pathways that frequently yield complex molecules with no counterparts in the terrestrial environment. Seaweeds are among the first marine organisms chemically analyzed, with more than 3,600 articles published describing 3,300 secondary metabolites from marine plants and algae, and they still remain an almost endless source of new bioactive compounds. In this chapter, some of the major classes of seaweed metabolites which find applications in the industrial sector, such as carotenoids, phycocolloids, polyunsaturated fatty acids and sterols, isolated either from aquacultures or wild harvesting, are presented. The ecological roles of a number of metabolites, as well as their potential application on the prevention of biofouling are described. The bioactive metabolites that target the pharmaceutical market, along with the spectrum of biological activities, are classified according to the class of producing seaweeds. The current status and the potential of seaweed metabolites for industrial exploitation is briefly discussed.

Introduction

Nature has provided mankind with folk medicines for centuries and continues to be the richest source of bioactive chemicals for the development of modern drugs. Terrestrial plants in particular were used as the basis of sophisticated traditional pharmacopoeias from as early as 2600 BC, and some of the earliest documentations come from inscriptions from Mesopotamia. Records from the ancient Egyptians and Chinese show that plants were used for the preparation of hundreds of drugs, covering a most impressive array of health problems and diseases.

In the ancient Western world, the philosopher and naturalist Theophrastus (~300 BC) compiled in his nine books entitled *History of Plants* the botanical characteristics and medicinal properties of herbs. Later, in AD 100, the physician Dioscorides, following the Roman armies, developed and recorded a wealth of complex prescriptions and formulas using medicinal herbs.

E. Ioannou and V. Roussis (✉)
Department of Pharmacognosy and Chemistry of Natural Products, School of Pharmacy, University of Athens, Panepistimiopolis Zografou, Athens, 15771, Greece
e-mails: eioannou@pharm.uoa.gr; roussis@pharm.uoa.gr

A.E. Osbourn and V. Lanzotti (eds.), *Plant-derived Natural Products*,
DOI 10.1007/978-0-387-85498-4_2, © Springer Science+Business Media, LLC 2009

Even though medicinal and pharmaceutical sciences, through the development of technology, have created milestones, plant-based systems continue to play an essential role in the healthcare of many communities. It was estimated by the World Health Organization that approximately 80% of the world's inhabitants rely mainly on traditional medicines for their primary healthcare [1]. For the remaining 20% of the world's population, mainly residing in developed countries, nature is equally important since approximately 25% of the prescribed drugs contain extracts or plant metabolites and an additional significant percentage of the market drugs have been developed through studies employing natural products as the lead molecules [2].

In the ancient days, marine organisms were mainly known from the Phoenicians that employed the chemical secretions from marine mollusks to produce purple dyes for woollen cloth. Still, there was a number of traditional drugs that incorporated seaweeds mainly for respiratory problems. Even today, in Eastern Africa the use of the roots of the seagrass *Enhalus acoroides* is very popular among the fishermen as a remedy against stings of different kinds of rays, stone, lion, scorpion (Scorpaenidae) and rabbit fish (Siganidae). This seagrass is also used for muscle pains, wounds, stomach problems and in the form of "mafusho" against fever. "Mafusho" is the smoke produced from a mixture of plants and herbs when burned, vapourized with water or prepared as incense. *Thalassia hemprichii* is mainly used as "mafusho" against fever and malaria. Species of the seagrass *Halophila* have also been reported as potent medicine ("mafusho") against malaria. The patients inhale the vapours mainly in order to lower body temperature. *Thalassodendron ciliatum* is used as "mafusho" and as treatment to relieve smallpox. *Halophila* species are also used for skin diseases and are also reported to be very effective in early stages of leprosy [3].

Taking into consideration that up to now only 5–15% of the approximately 250,000 species of higher plants have systematically been studied, and that the biodiversity of the marine environment is barely untapped, it is clear that nature will remain for a long time an abundant source of novel bioactive compounds.

The world's oceans, covering more than 70% of the earth's surface, represent an enormous resource for the discovery of potential chemotherapeutic agents. Taking higher taxonomic levels as an estimate of biodiversity, more phyla are found in the oceans than on land. Of the 33 known phyla of extant animals, only one is exclusive of land, while as many as 21 phyla are exclusive of the sea [4]. Prior to the development of reliable scuba diving techniques, some 40 years ago, the collection of marine organisms was limited to those obtainable by skin diving. Subsequently, depths from approximately 3-35 m became routinely attainable, and the marine environment is being increasingly explored as a source of novel bioactive agents [5].

Since 1975, three areas of research in marine natural products have emerged: bioactive metabolites, biotoxins and chemical ecology. Till today more than 15,000 novel compounds have been isolated from marine organisms [6]. Due to the wide range of competitive environments they survive in, marine organisms have developed unique defense strategies and bioactive compounds that, in some cases, are unparalleled by their terrestrial counterparts [7, 8]. The exploration of this chemical diversity for pharmaceutical purposes has revealed important chemical prototypes for the discovery of new agents, stimulating the use of sophisticated spectroscopic methods and development of new synthetic methodology.

Focusing on bioproducts, recent trends in drug research from natural sources suggest that algae are a promising group to furnish novel bioactive substances. Moreover, algae have potential to provide not only novel biologically active substances, but also essential compounds for human nutrition [9-14]. Several seaweeds are used as soil fertilizers and there have been reports showing that they are particularly good for the growth of specific plants, such as coconut trees.

Macro- and Microalgae

Seaweeds are heterogeneous groups of attached (benthic) photosynthetic plants characterized more by the lack of structural similarities with higher green plants than by their commonality of character [15]. Marine higher plants, such as seagrasses and salt marsh grasses, are often included with seaweeds due to similarities in habitat and function in nearshore marine ecosystems.

Algae are divided in two general categories – macroalgae, such as red, green and brown algae, and microalgae, such as blue-green algae [16]. Although according to Chen and Jiang [17], algae number only about 50,000 species worldwide (approximately 10% of the plant kingdom) [18], they may well be the most diverse group of organisms alive today. Diversity exists not only in morphological and reproductive features (typical of higher green plants), but also in complex life history phenomena and in exotic physiological and biochemical properties. Thus, it is perhaps not surprising that in most classifications algae constitute one third of the divisions within the plant kingdom [19].

Microalgae are the most primitive and simply organized members of the plant kingdom, with the majority existing as small cells of about 3-20 mm, and a few species organized into simple colonies. This group of microorganisms is extremely diverse and it constitutes a rich source of bioactive ingredients, such as vitamins, pigments, fatty acids, sterols, and polysaccharides [20-23].

Marine microalgae compose the majority of living species found in the oceans. There is no definite estimate for the total number of the existing species. New species are being discovered constantly, and the number is ever increasing. Currently, more than 10,000 known species are divided into five major divisions of marine microalgae: Chlorophyta (green algae), Chrysophyta (golden-brown, yellow algae, and diatoms), Pyrrhophyta (dinoflagellates), Euglenophyta, and Cyanophyta (blue-green algae). The phylogenetic positions and physiological characteristics of the organisms are important to consider when studying their metabolism and biochemistry. However, the taxonomy and phylogenetic relationships of microalgae are subjects on which taxonomists have never agreed [24].

In many countries, the food industries utilise a wide range of algae, which are well known to have high contents of fibers, minerals, vitamins and different antioxidants. In the last few decades, emphasis has moved from wild harvesting to farming and controlled cultivation for the production of valuable new products on a large scale.

The aquaculture sector produces large amounts of seaweeds, such as *Laminaria*, *Porphyra*, and *Gracilaria*, and microalgae, including *Dunaliella* and *Spirulina*. There is great potential in the use of microalgae for production of food ingredients, as they are photoautrophic microorganisms that can grow on a very simple culture medium containing seawater, nitrate, phosphate, trace amounts of certain metals, and carbon dioxide [25].

The production of algae-derived colloids (phycocolloids) such as algin, agar, and carrageenan has been developed into an important industrial sector [26]. Global production of aquatic plants for 2002 was 11.6 million tons, generating US$6.2 billion, with the highest production coming from the Japanese kelp *Laminaria japonica* (4.7 million tons), followed by Nori (*Porphyra tenera*, 1.3 million tons) [27]. Excluding the bioactive secondary metabolites that target the pharmaceutical market, the most common compounds from algae having already commercial applications besides phycocolloids include polyunsaturated fatty acids, steroids, carotenoids and lectins.

Carotenoids

Carotenoids are natural pigments derived from five-carbon isoprene units which are polymerized enzymatically to form regular, highly conjugated 40-carbon structures [6]. Carotenoids, which are present in all plants and many photosynthetic bacteria, represent photosynthetic pigments in the red, orange, or yellow wavelengths [16]. Animals lack the ability to synthesize carotenoids endogenously and thus obtain these compounds through their diet. Nature's most widespread pigments, carotenoids are linear polyenes that function both as light energy harvesters and as antioxidants that inactivate reactive oxygen species formed by exposure to light and air [28]. Of the approximately 600 known carotenoids, about 50 have been shown to exhibit some provitamin A activity, which is their primary beneficial role in the diet of humans and animals [28, 29]. As potent antioxidants and vitamin A precursors, carotenoids have been suggested to have protective activity against cancer, aging, ulcers, heart attack, and coronary artery disease [17]. Carotenoids are commonly used in food products as food-coloring or nutraceutical agents and they can be either produced synthetically or isolated from natural sources. Microalgal production of carotenoids, such as β-carotene and astaxanthin, is an attractive area of research, as they are valuable bioactive ingredients and can be present at relatively high concentrations in some algal cells.

β-Carotene (**1**) is mainly found in plant tissues and exhibits the highest provitamin A activity [28, 29]. The concentration of β-carotene among marine plants and algae is species-specific and is affected by stress factors, such as high light intensity, limited nitrates, and high salt concentrations. One of the major marine producers of β-carotene is the halophilic microalga *Dunaliella salina*, which is the highest β-carotene-enriched eukaryotic organism known [30, 31]. β-Carotene derived from *Dunaliella* has been marketed in several forms, such as extract in edible oils (1.5–30% β-carotene) and dried *Dunaliella* powder in capsules or tablets (approximately 5% β-carotene).

Astaxanthin (**2**) is a red pigment common to several aquatic organisms including microalgae, seagrasses, shrimps, lobsters and fish, such as salmon and trout. Crustaceans are unable to synthesize carotenoids *de novo* and require astaxanthin or its precursors in their diet in order to acquire the desired color for the seafood market [32]. Astaxanthin, with an antioxidant activity 10 times stronger than other carotenoids (e.g. β-carotene, canthaxanthin and lutein) [33], provides protection against cancer, inflammation, and UV light [34]. The beneficial properties of astaxanthin, in combination with the strong coloring ability, make it an important ingredient in the nutraceutical, cosmetics, food and animal feed industries [33, 34].

Phycocolloids

Hydrocolloids are carbohydrates that when dissolved in water form viscous solutions. The phycocolloids are hydrocolloids extracted from algae and represent a growing industry, with more than 1 million tons of seaweeds extracted annually for hydrocolloid production [27]. Polysaccharides are polymers of simple sugars linked together by glycosidic bonds. It has been proposed that they are involved in recognition mechanisms between seaweeds and pathogens [35]. Although several polysaccharides have been described with antioxidant, antiviral, antitumor and anticoagulant activities [11, 12, 36–38], the polysaccharides most commonly produced from seaweeds are agar, carrageenan and alginate due to their extensive use in food and cosmetics industries. Agar (**3**) and carrageenan (**4**) are polysaccharides mainly extracted from Rhodophyceae, while alginate (**5**) is extracted from Phaeophyceae. The wide use of these compounds is based on their gelling, viscosifying and emulsifying properties, which have generated an increasing commercial and scientific interest.

Phycocolloids have numerous commercial applications and commercial products containing algal constituents among others include: (1) Paints with alginates from the cell walls of brown algae (to smooth out pouring, brushing, and drying); (2) infant formulas with carrageenans from red algae (to thicken ingredients); (3) antacids with alginates from brown algae (to neutralize stomach acids); (4) food products with β-carotene from green algae (to improve color); (5) bread with carrageenans from red algae (to enhance browning); (6) paper with coatings containing alginates (to improve ink acceptance); (7) toothpastes with carrageenans (to increase water absorbance); (8) dental adhesives with alginate; (9) food products (e.g. salad dressings, ice-cream, pancake syrup, fruit juice, peanut butter, tomato sauce) with carrageenans or alginates (to improve thickness); (10) pasta products with carrageenans (to improve uniformity of dough and extrusion); (11) textiles printed with dyes containing alginates keep fine lines (to thicken print paste); (12) canned meat with agar from the cell walls of red algae (to gelify watery solutions at low concentrations); (13) cosmetic products with alginates (to improve protective coating and retain moisture); (14) air freshener gels with carrageenans (to stabilize gels and release odors gradually); (15) de-icers with carrageenans (to bind and stabilize ingredients for melting) [26, 27].

Fatty Acids

Algae are believed to be the primary producers of ω-3 polyunsaturated fatty acids (PUFAs) in the marine food chain and the only plant source for eicosapentaenoic acid (EPA, **6**) and docosahexaenoic acid (DHA) [39, 40]. PUFAs are essential for normal cell function, and after elucidation of their biological role in certain clinical conditions, such cardiovascular diseases [41, 42], they entered the biomedical and nutraceutical markets. PUFAs are important in cellular and tissue metabolism, including the regulation of membrane fluidity, electron and oxygen transport, as well as thermal adaptation [43]. EPA and DHA, found at various levels in marine micro- and macroalgal species, have shown relatively high oxidative stability compared to fish oils [44, 45] (Fig. 1). Recently, public preference for healthy food has increased the market visibility of PUFAs and in particular the interest for the ω-3 family of EPA, DHA and α-linolenic acid (**7**).

Sterols

Sterols are among the most important chemical constituents of micro- and macroalgae and a major nutritional component in the diet of aquacultured organisms. Microalgae are important components in the diet of many hydrobionts, especially bivalves [46]. A mixture of several algal species typically improves larval development, presumably by avoiding deficiency resulting from a unialgal diet [47]. The qualitative and quantitative sterol composition of microalgae used in hatcheries has direct implications in the phytosterol and cholesterol (**8**) composition of bivalve larvae and can affect their growth performance.

Chemical Ecology

Studies in chemical ecology have shown that natural products drive complex ecological interactions at all stages of marine plant and animal life cycles [48] (Fig. 2). Research in marine chemical ecology continues to focus on predator-prey, competitive interactions, settlement cues, and potential defenses against infection by microorganisms. Several comprehensive reviews of natural products and chemical ecology of macroalgae have been published recently and the medicinal and pharmaceutical uses of seaweed natural products that show the broad range of bioactivities of macroalgal metabolites has been reviewed by Smit [38].

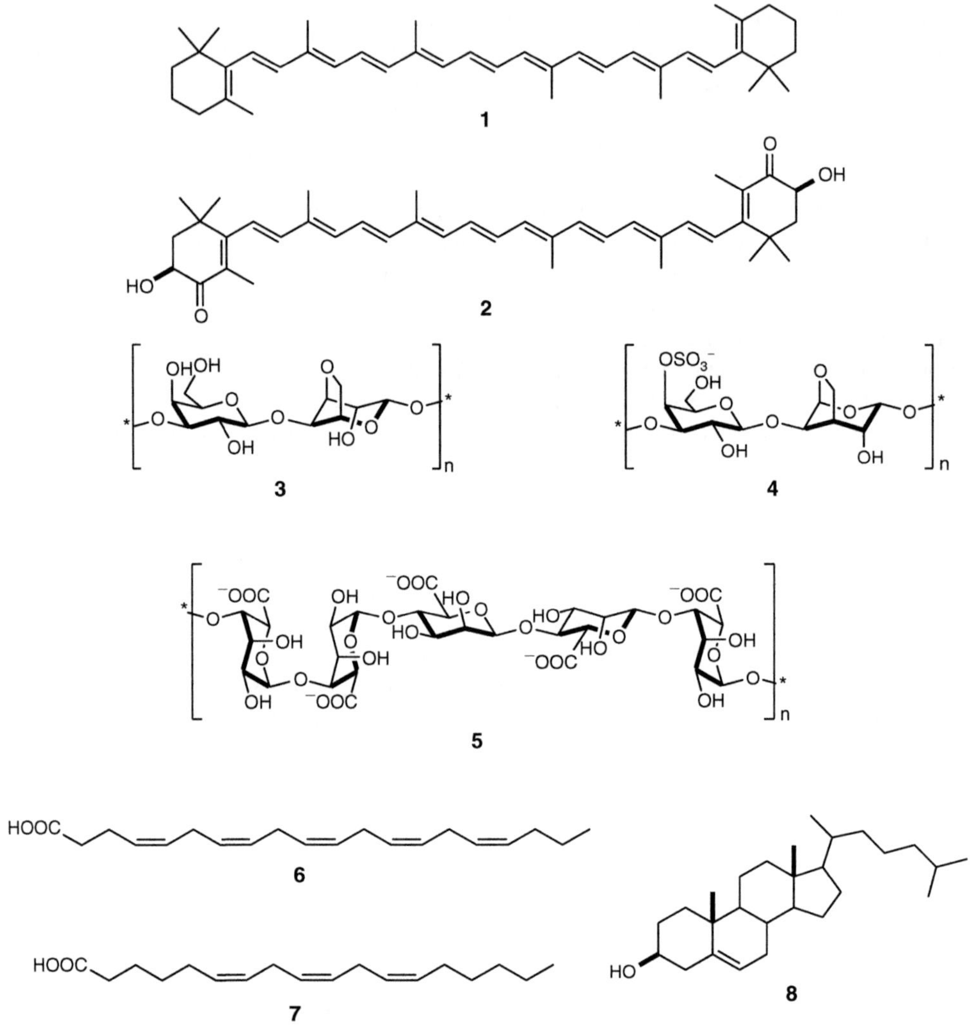

Fig. 1 Micro- and macroalgal metabolites with commercial use

Modern analytical techniques for the characterization and quantification of macroalgal chemical defenses were reported by La Barre and colleagues [49] and another review on marine metabolites implicated in antifouling, larval settlement and metamorphosis was published by Fusetani [50] (Fig. 3).

Several terpenoid compounds from brown algae have been proven to function as chemical defenses against herbivores [48]. The crude extract of *Dictyota pfaffii* and the major diterpene **9** deterred feeding of the sea urchin *Lytechinus variegatus* and generalist herbivore fishes in field assays, but did not influence the feeding of the crab *Pachygrapsus transversus* [51, 52].

The brown alga *Stypopodium zonale* produces diterpenes of mixed biogenesis that vary

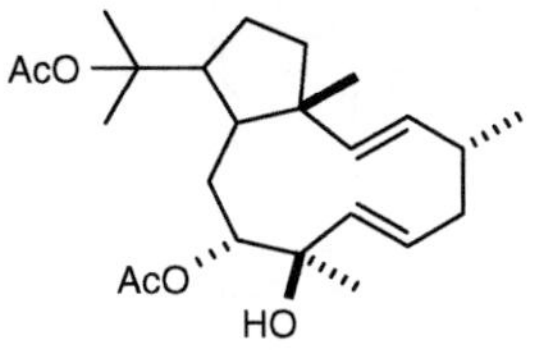

9

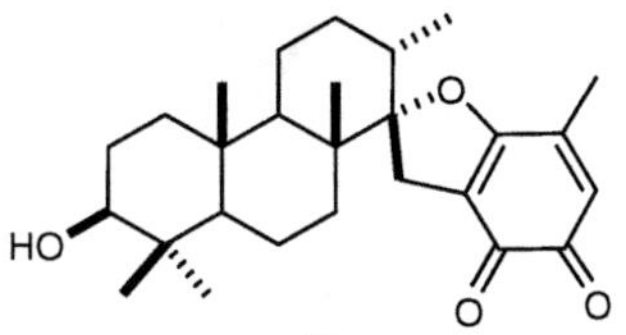

10

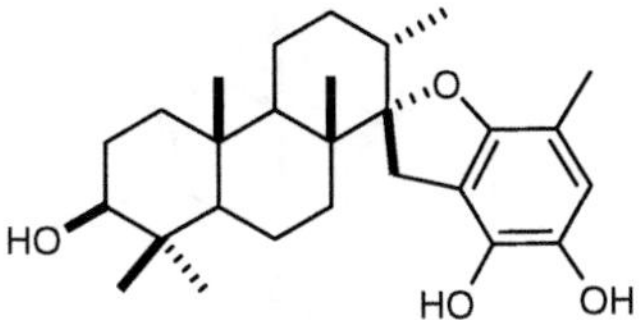

11

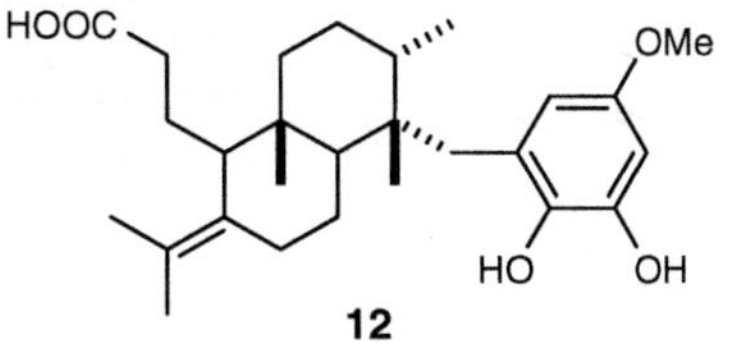

12

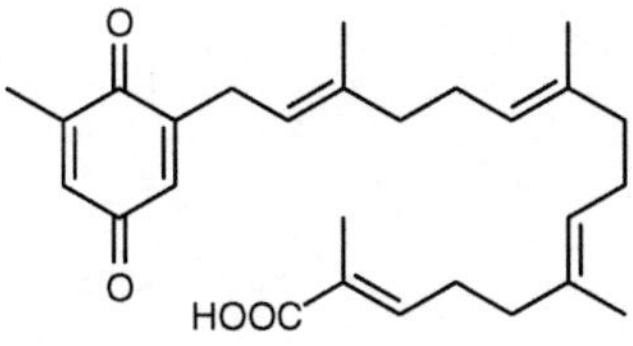

13

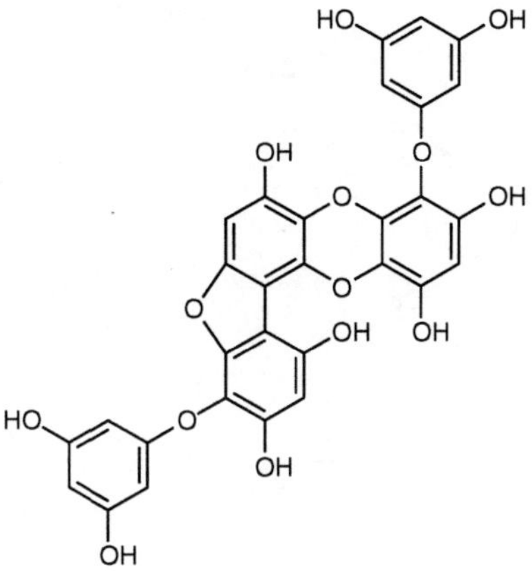

15

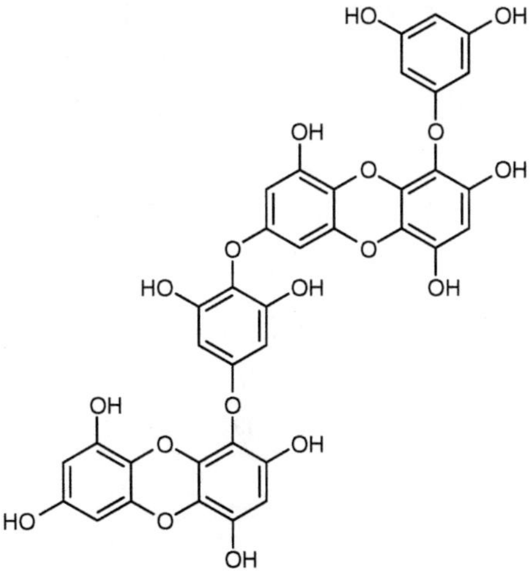

14

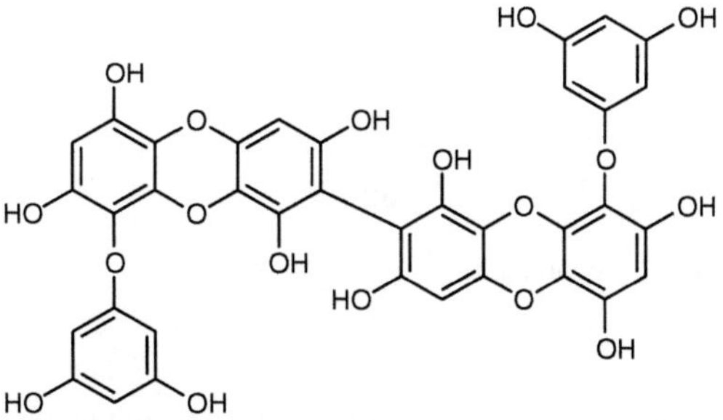

16

Fig. 2 Macroalgal metabolites with various ecological roles

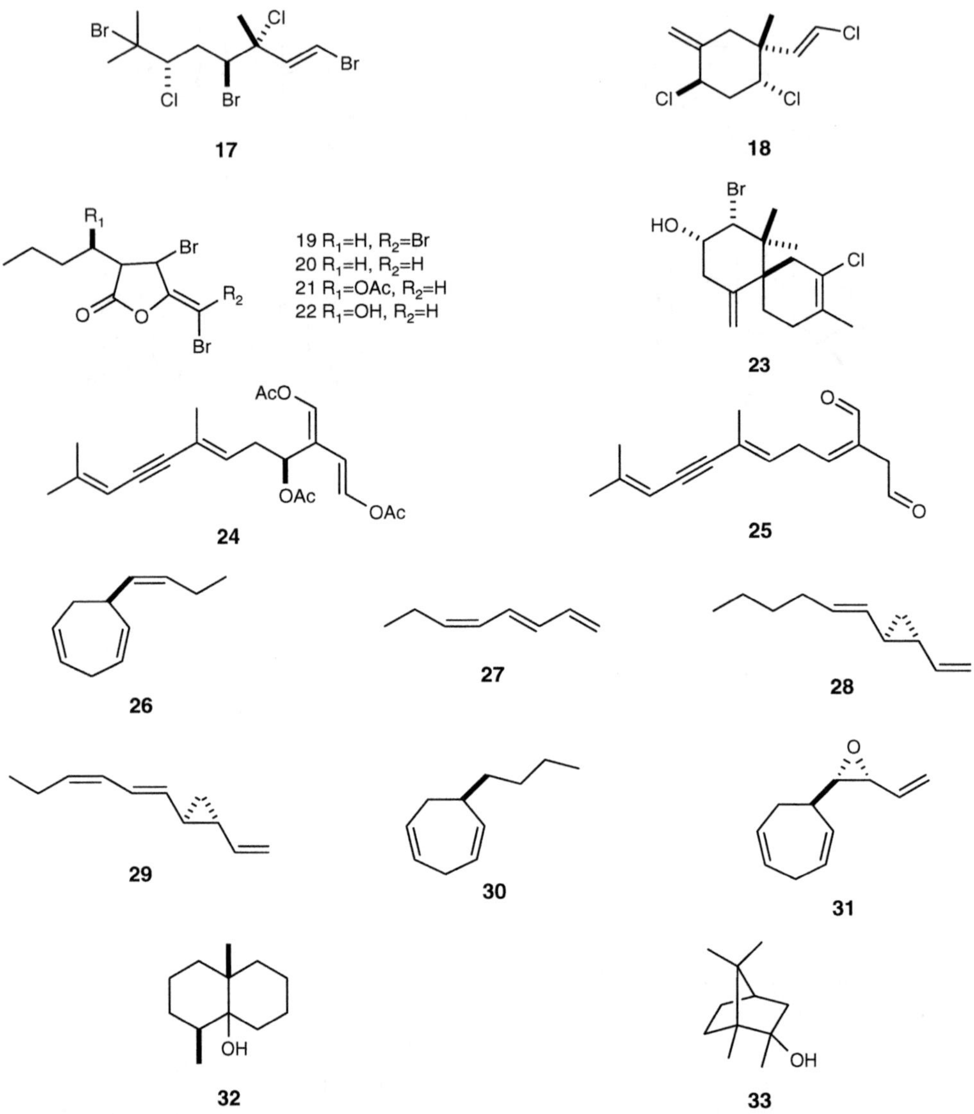

Fig. 2 (continued)

qualitatively and quantitatively depending on the collection site [53]. In a geographic variation study undertaken in Brazilian coasts Soares and coworkers [53] reported two distinct chemotypes of *S. zonale*. The specimens collected from the northern coasts contained stypoldione (**10**), which is the air-oxidation product of stypotriol (**11**), as the major metabolite, whereas collections from the southern coasts contained atomaric acid (**12**) as the major metabolite. Chemical defense studies of the two extracts showed that both chemotypes inhibited feeding

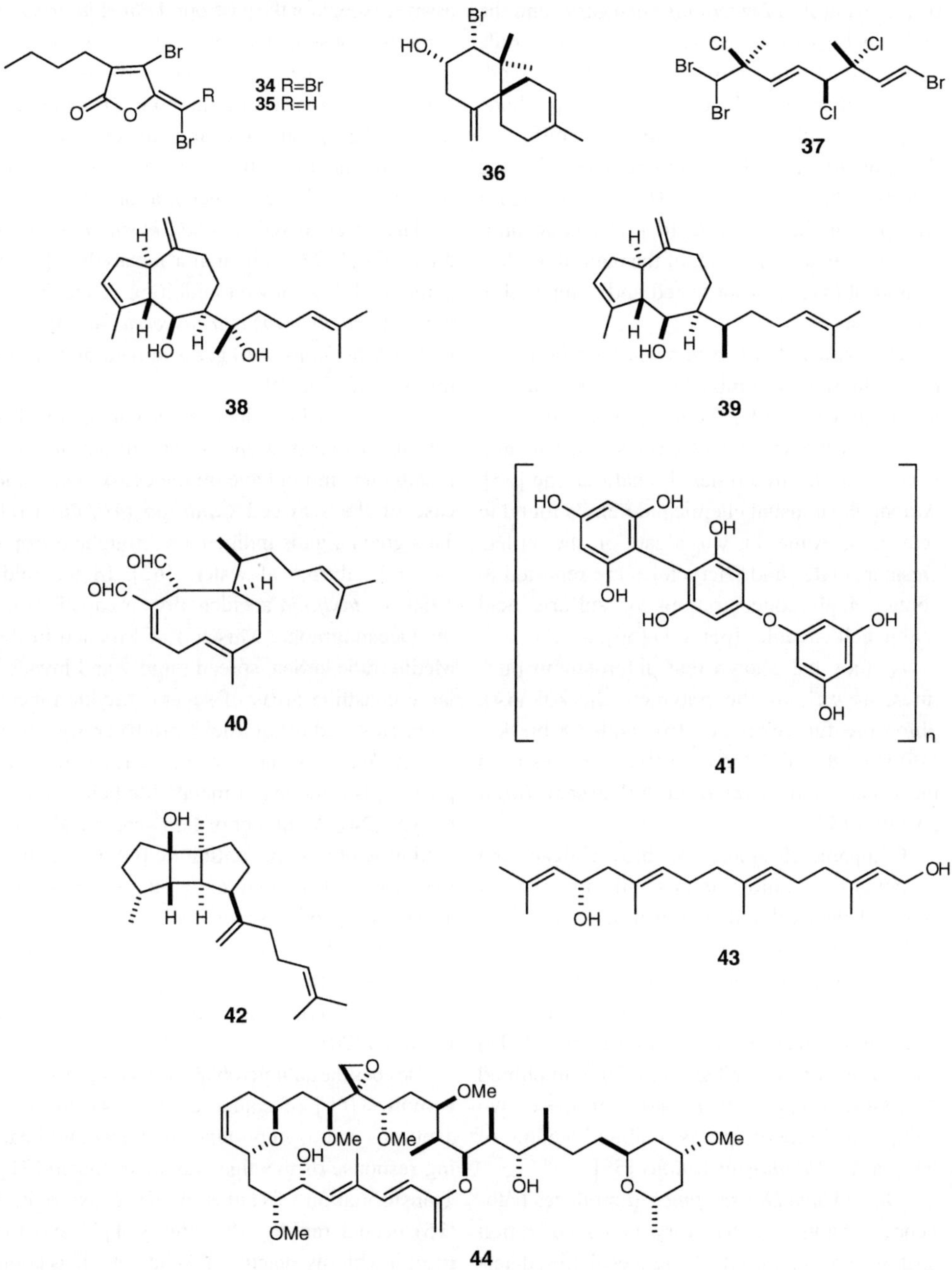

Fig. 3 Macroalgal metabolites with antifouling activity

of the sea urchin *Lytechinus variegatus* and the crab *Pachygrapsus transversus*, but the crude extract containing atomaric acid deterred feeding more than the extract containing stypoldione [54]. Tests with the pure compounds confirmed that atomaric acid (**12**) is a more potent feeding deterrent than stypoldione (**10**), even though it was present in *S. zonale* in lower concentrations. The deterrent effect of the natural product stypotriol (**11**) was not tested and compared in these assays.

The brown alga *Desmerestia menziesii* contained diterpenes of mixed biogenesis including menzoquinone (**13**). Menzoquinone deterred feeding of the sea star *Odontaster validus* at a concentration three times the natural one [55]. Among the unusual chemical defenses found in seaweeds, some brown algae of the orders Desmarestiales and Dictyotales are reported to contain high concentrations of sulfuric acid within cell vacuoles (pH < 1) [56].

A study has shown that phlorotannin mixtures, as well as the polymers dieckol (**14**), phlorofucofuroeckol A (**15**), and 8,8-bieckol (**16**) strongly inhibited digestive enzymes from the viscera of the herbivorous turban snail *Turbo cornutus* [57].

Compared to studies on brown algae, and particularly on phlorotannins, only a few recent studies have explored chemical defenses in red and green macroalgae.

Among the halogenated monoterpenes isolated from the red alga *Plocamium cartilagineum*, anverene (**17**) and *epi*-plocamene D (**18**) have proven to deter feeding of the amphipod *Gondogeneia antarctica*. None of these two halogenated monoterpenes inhibited feeding of the sea star *Odontaster validus* [55].

The red alga *Delisea pulchra* produces halogenated furanones that vary in concentration spatially and temporally. Wright et al. [58] determined the heritability of the four major furanones **19–22** and then determined the effects of different concentrations on feeding by herbivores. All six common herbivores used in the feeding assays, except for the gastropod *Phasianotrochus eximius*, consumed *D. pulchra* at lower rates than other macroalgae. The herbivores were deterred by extracts and furanone **21** at concentrations that spanned the range of concentrations found in the field, but were occasionally not deterred at the lowest concentrations [58].

The Brazilian red alga *Laurencia obtusa* produces elatol (**23**) as its major metabolite, [59] a compound that showed inhibition of feeding of the crab *Pachygrapsus transversus* and the sea urchin *Lytechinus variegatus* at natural concentrations [48, 59, 60].

An impressive, albeit environmentally disastrous demonstration of the influence algal metabolites might have on the ecosystem is the case of the seaweed *Caulerpa taxifolia* [61]. This green alga is indigenous throughout tropical and subtropical waters [62]. In the mid-1980s *C. taxifolia* accidentally "escaped" from the Oceanographic Museum of Monaco in the Mediterranean Sea, spread rapidly and invaded large coastline areas of several Mediterranean countries reducing their biodiversity [63]. *C. taxifolia* contains several terpenoid compounds [14], the major metabolite being caulerpenyne (**24**). A number of *Caulerpa* metabolites exhibit ichthyotoxic, antibiotic [64, 65], neurotoxic [66, 67] and cytotoxic [68, 69] properties. The presence of these metabolites offers protection to the alga and only a few autochthonous predators, such as the sea slugs *Oxynoe olivacea* and *Lobiger serradifalci*, can feed on *Caulerpa* [70].

Besides the antimicrobial and feeding-deterrent activities [60] of caulerpenyne (**24**), the biosynthesis and its importance in the wound-healing response of the alga was investigated [71]. Transformation of caulerpenyne to oxytoxin 2 (**25**) occurs rapidly after injury, [72] and the resulting highly reactive 1,4-dialdehyde is completely consumed in less than 4 min [71]. The decay kinetics of oxytoxin 2 (**25**) matched those of the formation of the external wound plug of *C. taxifolia* [71].

Volatile organic compounds (VOCs), in terrestrial ecosystems, are recognised as an important group of infochemicals, especially in the plant–herbivore interactions. Plants produce volatile compounds to attract pollinators and seed dispersers [73, 74]. Intraspecific interactions frequently are mediated by VOCs. In particular, many insects are able to recognise even single molecules of volatile pheromones released from the opposite sex for mate-finding [74, 75]. Additionaly, plants are able to signal an attack by herbivores to conspecifics, which then respond with upregulation of their defensive genes and increased defenses against the approaching predators [76]. The most sophisticated use of volatile semiochemicals is their role as guiding cues for predatory insects in so-called 'tritrophic interactions' [77, 78].

Volatile organic compounds from aquatic organisms have multiple functions – in intraspecific (pheromones) and interspecific (kairomones) communication, as well as in their defense system. An excellent review on the concepts and recent experimental studies on the ecological functions of such VOCs in aquatic ecosystems was recently reported by Fink [79].

Among the first ecological functions that were demonstrated for VOCs in aquatic ecosystems was that of sex pheromones in marine brown algae. Various genera of macroalgae, such as *Ectocarpus*, *Fucus*, *Dictyopteris* and *Laminaria*, produce volatile pheromones which are active at extremely low concentrations [80, 81]. From these species cyclic and acyclic compounds, e.g. ectocarpene (**26**), fucoserratene (**27**), dictyopterenes (**28–30**) and lamoxirene (**31**), have been identified in the excretions of female gametes that function as sperm attractants [81, 82]. The thermolabile cyclopropyl-pheromones **28** and **29** degrade rapidly via a spontaneous Cope-rearrangement to the inactive forms of **30** and **26**, controlled only by the environmental temperature [81]. Worth noting is that some of the chemicals that act as pheromones of brown algae occur also in freshwater diatoms [80, 83]. *Asterionella formosa* produces fucoserratene (the pheromone of *Fucus vesiculosus*) from EPA [84], while *Gomphonema parvulum* produces the brown algal pheromones hormosirene and two dictyopterenes from EPA and arachidonic acid (ARA) [81]. However, the functions of these volatile hydrocarbons in freshwater diatoms are unknown.

A recent example of volatile foraging kairomones are VOCs released by benthic green algae and diatoms [85, 86] that are utilised as food-finding cues by the freshwater gastropods *Radix ovata*. When the cells are damaged, the benthic green alga *Ulothrix fimbriata* releases a variety of volatile fatty acid and carotenoid degradation products. Several of these volatiles are attractive for *R. ovata* when present in blends.

Dimethyl sulfide (DMS) released from algae after cell damage might function as a volatile chemical cue in multitrophic interactions [87]. However, the details of such a multitrophic interaction are still unknown. While pheromones are actively excreted [82, 88], the liberation of VOCs from algae and cyanobacteria requires cell damage, e.g. from herbivory [89]. DMS release from the haptophyte *Emiliania huxleyi* was increased by dinoflagellate grazing on the alga. DMS produced by marine algae is released into the atmosphere and oxidised to sulphate. In higher atmospheric strata, the sulphates can then serve as cloud condensation nuclei. It is hypothesised that VOCs of marine microalgae may thereby play an important role in large-scale meteorological processes [90].

Frequently, VOCs produced by algae and cyanobacteria are a major problem in water processing, since aquatic primary producers are the reason for regularly encountered taste and odour problems in drinking water. The removal of the two most common cyanobacterial VOCs, geosmin (**32**) and 2-methylisoborneol (**33**) is difficult and therefore expensive, since these compounds can not be oxidized by chlorination [91].

Biofouling

Biofouling is one of the most serious problems that maritime domain currently faces and it has been estimated that the growth of marine fouling organisms costs the shipping and other marine industries over US$6.5 billion per year [92]. Biofouling is considered to have four distinct stages, the first one starting from the moment a man-made object is immersed in water. The surfaces of these objects quickly accumulate dissolved organic matter and molecules, such as polysaccharides and protein fragments. Gradually, bacteria and single-cell diatoms sense the surface and start settling on it, forming a microbial film [93]. Subsequently, the adhesive substances and rough irregular microbial colonies trap more particles and organisms. Spores of algae, e.g. species of *Enteromorpha intestinalis*, *Ulothrix zonata*, marine fungi and ciliate protozoa soon appear on the film [94]. In the final stage, other marine organisms, such as barnacles, tunicates, mussels, bryozoans, polychaetes and tubeworms, settle on the submerged surfaces [95].

Efficient antifouling paints are based on copper-containing compounds and booster biocides that when submerged, release toxic compounds causing adverse environmental effects [96–101]. New environmental regulations generate further restrictions on the use of biocides in industrial formulations [102] for the protection of rivers, lakes, estuaries, coastal waters and groundwater from further deterioration and for protection of biodiversity. There is clearly a need to develop new non-toxic or environmentally benign antifouling alternatives that would be efficient against the most severe fouling organisms, such as barnacles, blue mussels, bryozoans and algae [103]. On this direction there is an increasing interest for the antifouling potential exploration of natural compounds especially those derived from marine organisms, since through evolution, nature has already developed mechanisms for their gradual biodegradation.

Marine algae, as well as other benthic organisms, are frequently relatively free from settlement by fouling organisms [104–107] due to the production of biogenic compounds that possess antibacterial, antialgal, antifungal, antiprotozoan and antimacrofouling properties.

Many marine algae species have been evaluated for their antifouling properties and a significant number of their secondary metabolites have already been found active against a diverse spectrum of fouling organisms. Several excellent reviews on this topic have been published covering all aspects of fouling [50, 92].

From the Australian benthic marine macroalga *Delisea pulchra* (Class Rhodophyceae), which is free of fouling by macroorganisms, an unusual class of halogenated furanones (**34**, **35**) was isolated and found to act as a specific antagonist of the acylated homoserine lactone (AHL) regulatory system (quorum sensing) present in bacteria, thereby inhibiting bacterial colonization through a non-toxic and non-growth mechanism [108–113]. Furanones **34**, **35** inhibit AHL-regulated phenotypes in a wide range of gram-negative bacteria by shutting down the biofilm development process [109, 114, 115]. These halogenated furanones have been tested against fouling species, like the barnacle *Balanus amphitrite* [116], the macroalga *Ulva lactuca* [117] and marine bacteria [118]. The furanones were most potent in *B. amphitrite* cyprid assay (EC_{50} = 20 ng/mL), with activity 20 times higher than $CuSO_4$ [119]. These *D. pulchra* 2(5H)-furanones **34**, **35** being non-specific intercellular signal antagonists could be used for controlling bacterial biofilm formation without any toxicity [120, 121]. Additionaly, it has been shown that these furanones can inhibit the growth of *Bacillus subtilis* biofilm formation and its swarming motility in a concentration-dependent way [118].

Among the polar metabolites isolated from the extracts of the red algae *Polysiphonia morrowii*, *P. lanosa* and *Monostroma fuscum*, 5-bromo-3,4-dihydroxybenzaldehyde, 2,3-dibromo-

4,5-dihydroxyphenyl-ethylamine and 3,4-dihydroxyphenyl-ethylamine were found to be toxic to a variety of fouling unicellular algae [122].

The two well known antifeedant spiro-sesquiterpenes elatol (**23**) and deschlorelatol (**36**), isolated from the red alga *Laurencia rigida*, were found to exhibit strong settlement inhibition against the larvae of the bryozoan *Bugula neritina* and the barnacle *Balanus amphitrite* at 10 ng/cm^2 [123, 124]. The crude extract of *L. obtusa* significantly reduced fouling relative to controls after 5 weeks, when incorporated into phytagel TM and placed in the field [59].

A highly halogenated monoterpene (**37**) isolated from the Tasmanian red alga *Plocamium costatum* inhibited barnacle settlement at 1 μg/cm^2 [125, 126]. Similarly, dictyol E (**38**), pachydictyol A (**39**) and dictyodial (**40**), antifeedant diterpenes from the brown alga *Dictyota* spp. showed antifouling activity against the bryozoan *Bugula neritina* larvae. Dictyol E was lethal to larvae at 5 μg/mL, while pachydictyol A and dictyodial caused abnormal development and reduced rates of growth [128, 129]. Dictyodial also inhibited larval settlement and metamorphosis of the hydroid *Eudendrium carneum*.

Phlorotannins (**41**), polymers of phloroglucinol, isolated from the Australian brown algae *Ecklonia radiata* and *Sargassum vestitum*, have been shown to inhibit the settlement and growth of propagules of the fouling green alga *Ulva* sp. at concentrations of 100 mg/L [127]. The spatane diterpene **42** from the brown alga *Dilophus okamurai* inhibited metamorphosis of abalone larvae at 5–10 ppb [130].

The extracts of the brown alga *Bifurcaria bifurcata* that had shown seasonal variation were examined as settlement inhibitors of two marine bacteria and the barnacle *Balanus amphitrite* [131]. The increased activity during April–June corresponded to the seasonal maximum concentration of the major diterpene eleganediol **43** found in the crude extracts [132], but isolated compounds were not individually tested for their antifouling activity [131].

From the brown alga *Lobophora variegata* extract the new macrolide lobophorolide (**44**) has been identified and this 22-membered cyclic lactone showed promising antifungal activities against marine fungi implicated in biofouling [104].

Compounds belonging to the class of acetylene sesquiterpenoid lipid esters have been isolated along with caulerpynene (**24**) from the green alga *Caulerpa prolifera* [133]. The substantial inhibitory effects that the organic extract of the alga showed against a wide range of marine bacteriaand the marine microalga *Phaeodactylum tricornatum* implies a possible antifouling role for these metabolites in the alga.

Bioactive Metabolites from Macroalgae

Pharmaceutical compounds constitute one of the most important potential markets for algal products. Prior to 1950, the use of seaweed extracts and microalgae as drugs or drug sources was restricted to folk medicine. Still, up to date, there has been little commercial development of algal products as pharmaceutical agents. The vermifuge α-kainic acid from the red alga *Digenea simplex* was marketed in the past, but has now been withdrawn in western countries. However, there is a tremendous potential for the development of algae as sources of pharmaceutical compounds, since in the recent years researchers have described a wide range of biological activities for metabolites produced by algae.

Isolation of pharmacologically active compounds from marine algae has been the subject of many intensive investigations and comprehensive account of such studies in the field is given in numerous reviews [5, 9, 11, 12, 14, 36, 37, 134–137]. *In vitro* screens for the detection of a wide spectrum of biological activities for the initial evaluation of algal extracts and subsequent bioassay-guided fractionations for the isolation of the active compounds constitute one of the most common methods of investigation.

2

According to MARINLIT [138], 3,554 scientific articles have been published until 2007 describing the isolation of new natural products from algae and the biological activities of algal metabolites or extracts. More than 3,280 structures from algae have been reported in the literature, amounting to approximately 30% of the total number of natural products isolated from marine organisms. As it can be seen (Figs. 4 and 5) from the number of published articles and isolated metabolites, investigations on algae escalated between 1980 and 1989 and continue to attract the interest of scientists involved in natural products research. Among macroalgae, significantly more rich in secondary metabolites appear the brown and red algae, with the latter being the top producers of halogenated metabolites.

In Fig. 6, based on data from MARINLIT [138], it is clear that scientists are exploring, besides the easily accessible coastal areas, extreme and remote ecosystems in search of

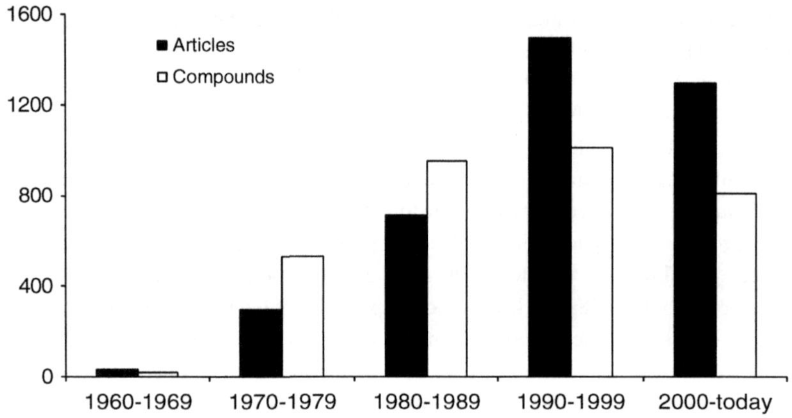

Fig. 4 Articles on marine algae published between 1960 and 2007

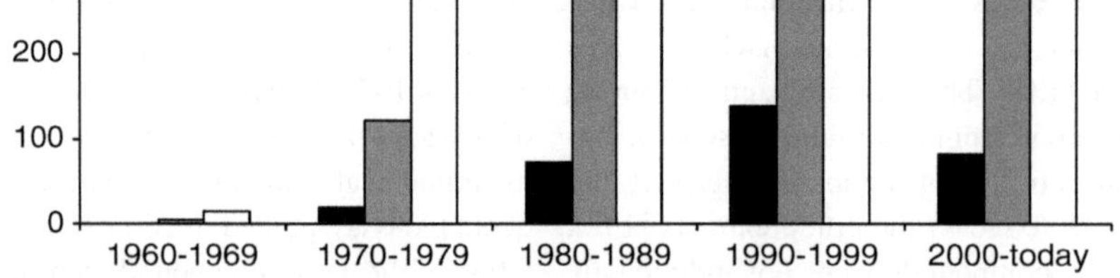

Fig. 5 Algal metabolites reported between 1960 and 2007

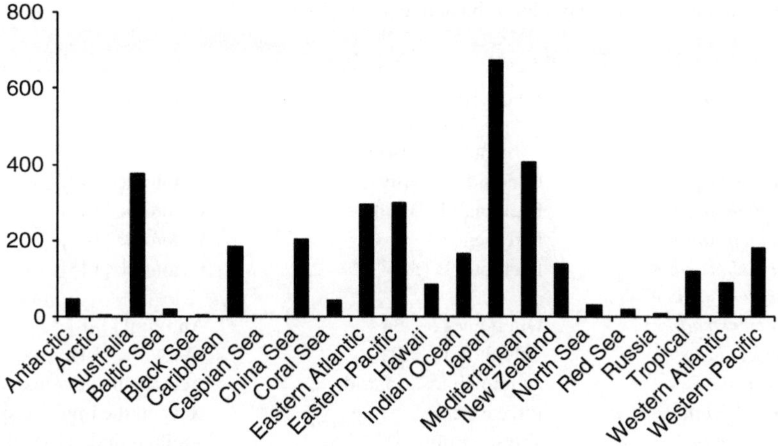

Fig. 6 Algal metabolites isolated from different geographic regions

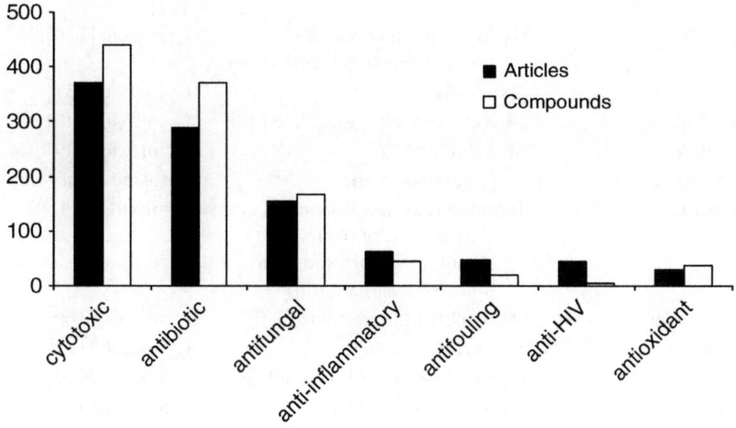

Fig. 7 Distribution of activity evaluation of algal metabolites

chemical diversity and unique metabolites. The number of metabolites reported from the mentioned geographic areas is directly connected to the biodiversity of marine life and the number of scientific groups involved in marine natural products research in these regions.

Algal metabolites have exhibited a wide spectrum of activity, with the majority of them having been evaluated for their cytotoxic and antibiotic properties (Fig. 7). The development of high-throughput screening systems and the enhanced sensitivity of pharmacological assays allow now extensive biological studies on compounds that are isolated in sub-milligram quantities.

Algal metabolites that have been evaluated for and have exhibited significant pharmacological activity are compiled in Tables 1–3 and Figs. 8–10 according to the taxonomic classification of the producing organisms.

2

Table 1 Brown algae that have produced bioactive metabolites

Producing organism	Metabolite	Activity
Cystoseira mediterranea	Mediterraneol A (**46**), mediterraneone (**47**), cystoseirol (**48**)	Cytotoxic [141]
Cystoseira usneoides	Usneoidone E (**49**)	Cytotoxic [142]
Dictyota dichotoma	Dictyotalide B (**50**)	Cytotoxic [143]
Dictyota dichotoma	Diterpene **51**	Cytotoxic [144]
Dictyota dichotoma & *Pachydictyon coriaceum*	Dictyotin B (**52**)	Cytotoxic [145]
Dictyota menstrualis	Diterpenes **53**, **54**	Antiviral [146]
Dilophus ligulatus	Dilopholide (**55**)	Cytotoxic [147, 148]
Ecklonia kurome	Acidic oligosaccharide	Acts on the nervous system [149]
Laminaria digitata	Phycarine	Acts on the immune system [150]
Notheia anomala	Tetrahydrofuran	Antihelmintic [151]
Padina pavonica	24-Hydroperoxy-24-vinyl-cholesterol (**56**)	Cytotoxic [152]
Sargassum horneri	Sulfated polysaccharide	Antiviral [153]
Sargassum macrocarpum	Sargaquinoic acid (**57**)	Acts on the nervous system [154, 155]
Sargassum tortile	Hydroxysargaquinone (**58**), sargasal-I (**59**), sargasal-II (**60**)	Cytotoxic [156]
Spatoglossum schmittii	Spatol (**61**)	Cytotoxic [157]
Stypopodium flabelliforme	14-Keto-stypodiol diacetate (**45**)	Cytotoxic [140]
Stypodium zonale	Stypoldione (**10**)	Cytotoxic [139]
Stypopodium zonale	Stypoquinonic acid	Tyrosine kinase inhibitor [158]
Taonia atomaria	Taondiol (**62**), isoepitaondiol (**63**), stypodiol (**64**), stypoldione (**10**), sargaquinone (**65**), sargaol (**66**)	Antioxidant [159]
Turbinaria conoides	Oxygenated fucosterol (**67**)	Cytotoxic [160]
Turbinaria ornata	Turbinaric acid (**68**)	Cytotoxic [161]
Turbinaria ornata	Hydroperoxy-sterols (**56**, **69**, **70**)	Cytotoxic [162]
Undaria pinnatifida	Hexadecatetraenoic acid, octadecatetraenoic acid	Anti-inflammatory [163]
Several species	Sulfated fucans	Anticoagulant [164]
Several species	Fucoidan	Anticoagulant [165]

Metabolites from brown algae have exhibited mainly cytotoxicity, but so far none of them has advanced in clinical or preclinical trials. One of the most potent cytotoxic compounds derived from brown algae is stypoldione (**10**), which was isolated from *Stypodium zonale*. Stypoldione inhibits microtubule polymerization and sperm motility, in contrast to the properties of other microtubule assembly inhibitors. This metabolite seems to prolong the survival time of mice injected with tumor cells, showing relatively little cytotoxicity itself. Actually, using tumor cells derived from P-388 lympholytic leukemia cells injected into BDF1 or CDF1 mice and drug treatment up to 30 days, a 42% increase in survival time in mice treated with stypoldione was observed [139].

Table 2 Red algae that have produced bioactive metabolites

Producing organism	Metabolite	Activity
Amphiroa zonata	Palmitic acid (**73**)	Cytotoxic [168]
Botryocladia occidentalis	Sulfated galactans (**74**, **75**)	Anticoagulant [169, 170]
Ceratodictyon spongiosum	Ceratospongamide	Anti-inflammatory [171]
Chondria atropurpurea	Chondriamide C	Antihelmintic [172]
Galaxaura marginata	Oxygenated desmosterols (**76**–**79**)	Cytotoxic [173]
Gigartina tenella	Sulfoquinovosyldiacylglycerol	Antiviral [174]
Jania rubens	16β-Hydroxy-5α-cholesta-3,6-dione (**80**)	Cytotoxic [175]
Laurencia calliclada	Callicladol (**81**)	Cytotoxic [176]
Laurencia cartilaginea	Chamigranes (**23**, **82**–**86**)	Cytotoxic [177]
Laurencia intricata	Laurenditerpenol (**87**)	Cytotoxic [178]
Laurencia obtusa	C-15 acetogenins (**88**, **89**)	Insecticidal [179]
Laurencia pannosa	Pannosanol (**90**), pannosane (**91**)	Antibacterial [180]
Laurencia viridis	Dehydrothyrsiferol (**71**)	Cytotoxic [166, 181]
Laurencia sp.	Ma'iliohydrin (**92**)	Cytotoxic [182]
Laurencia sp.	Lembyne-A (**93**)	Antibacterial [183]
Plocamium hamatum	Haloganted monoterpene (**94**)	Antituberculosis [125]
Portieria hornemanii	Halomon (**72**)	Cytotoxic [167]
Portieria hornemanii	Halomon- related monoterpenes (**72**, **95**–**97**)	Cytotoxic [184–186]
Rhodomela confervoides	Bromophenol **98**	Antibacterial [187]
Tricleocarpa fragilis	Sulfated triterpene **99**	Cytotoxic [188]

Table 3 Green algae that have produced bioactive metabolites

Producing organism	Metabolite	Activity
Bryopsis sp.	Kahalalide F (**100**)	Cytotoxic [189, 190]
Caulerpa taxifolia	Caulerpenyne (**24**)	Cytotoxic [69, 191, 192], pancreatic lipase inhibitor [193]
Codium arabieum	Clerosterol (**101**)	Cytotoxic [194]
Codium fragile	Sulfated polysaccharide	Acts on the immune system [195]
Codium iyengarii	Iyengaroside-A (**102**)	Antibacterial [196]
Codium pungniformis	Proteoglycan	Anticoagulant [197]
Cymopolia barbata	Cymopol, avrainvilleol	Antioxidant [198]
Tydemania expeditionis	Sulfated cycloartanol (**103**)	Cytotoxic [199]
Ulva lactuca	3-O-β-glucopyranosyl clerosterol	Anti-inflammatory [200]

14-Keto-stypodiol diacetate (SDA, **45**), isolated from *S. flabelliforme*, induces mitotic arrest of DU-145 human prostate tumor cells, an effect that could be associated to alterations in the normal microtubule assembly process. It also disrupts the normal organization of the microtubule cytoskeleton in the DU-145 cell line, as revealed by immunofluorescence studies, and affects protease secretion and the *in vitro* invasive capacity, two properties of cells from metastases. The microtubule assembly inhibition of SDA, together with its cellular effects in arresting mitosis and blocking protease secretion mechanisms and cell invasion

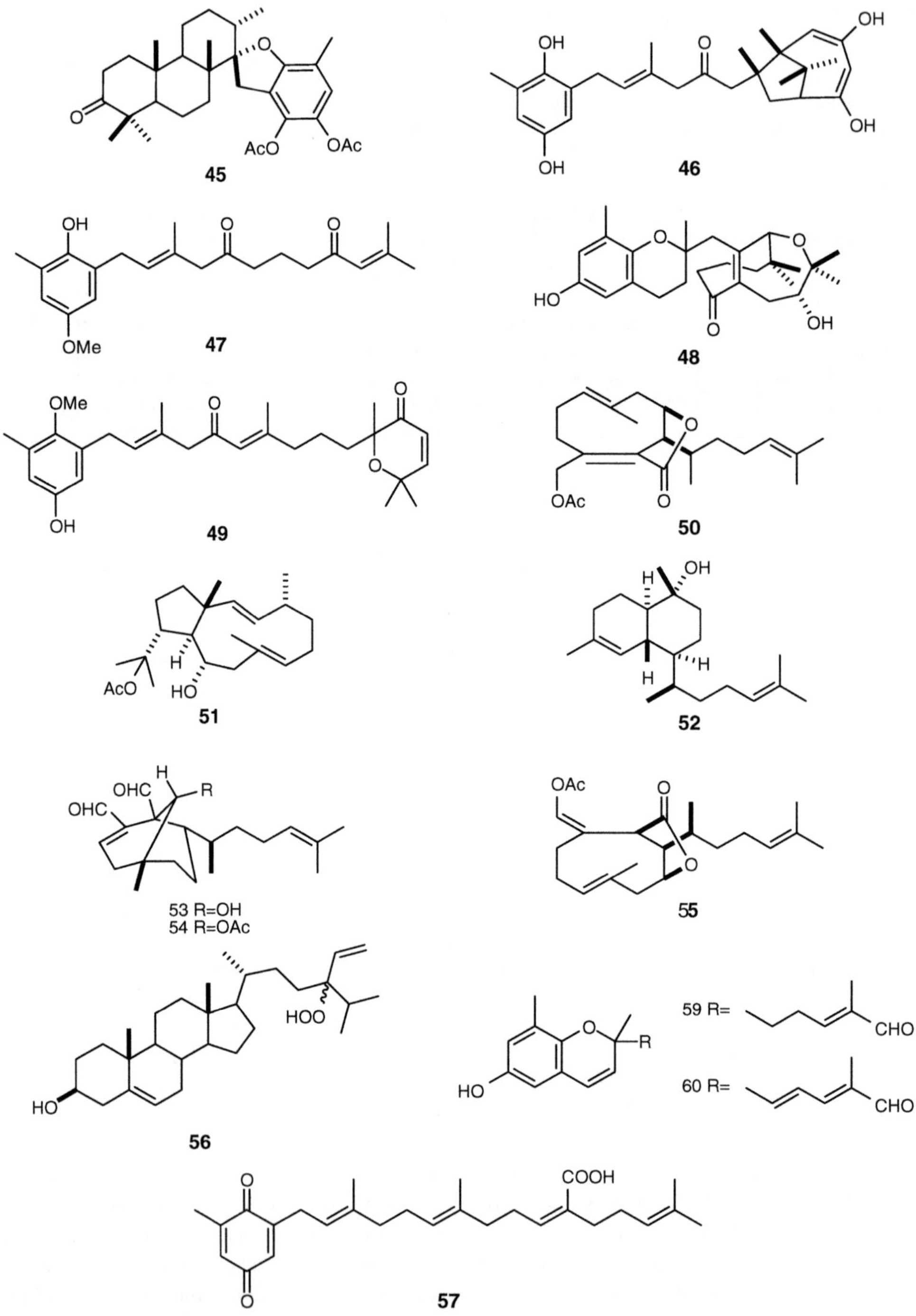

Fig. 8 Bioactive metabolites isolated from brown algae

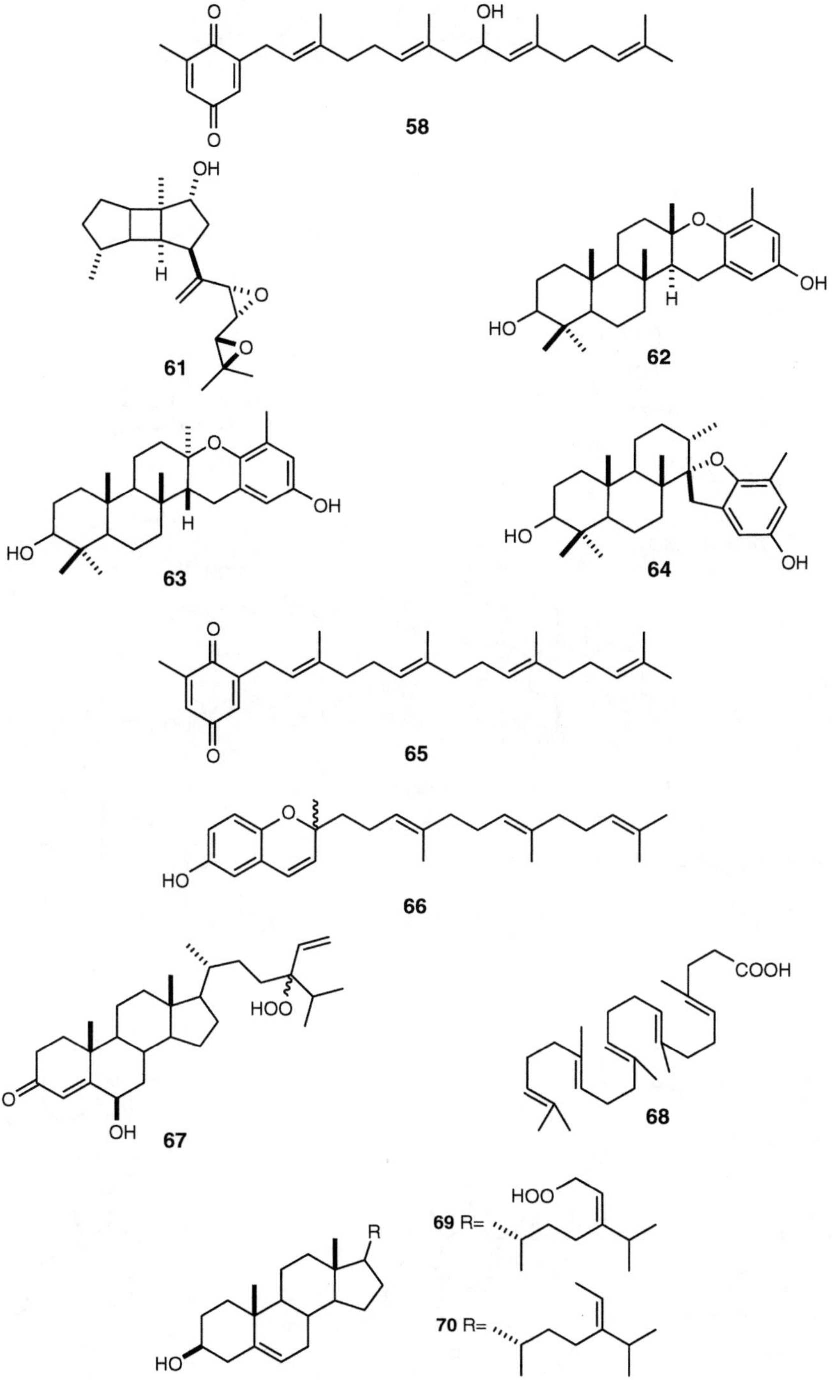

Fig. 8 (continued)

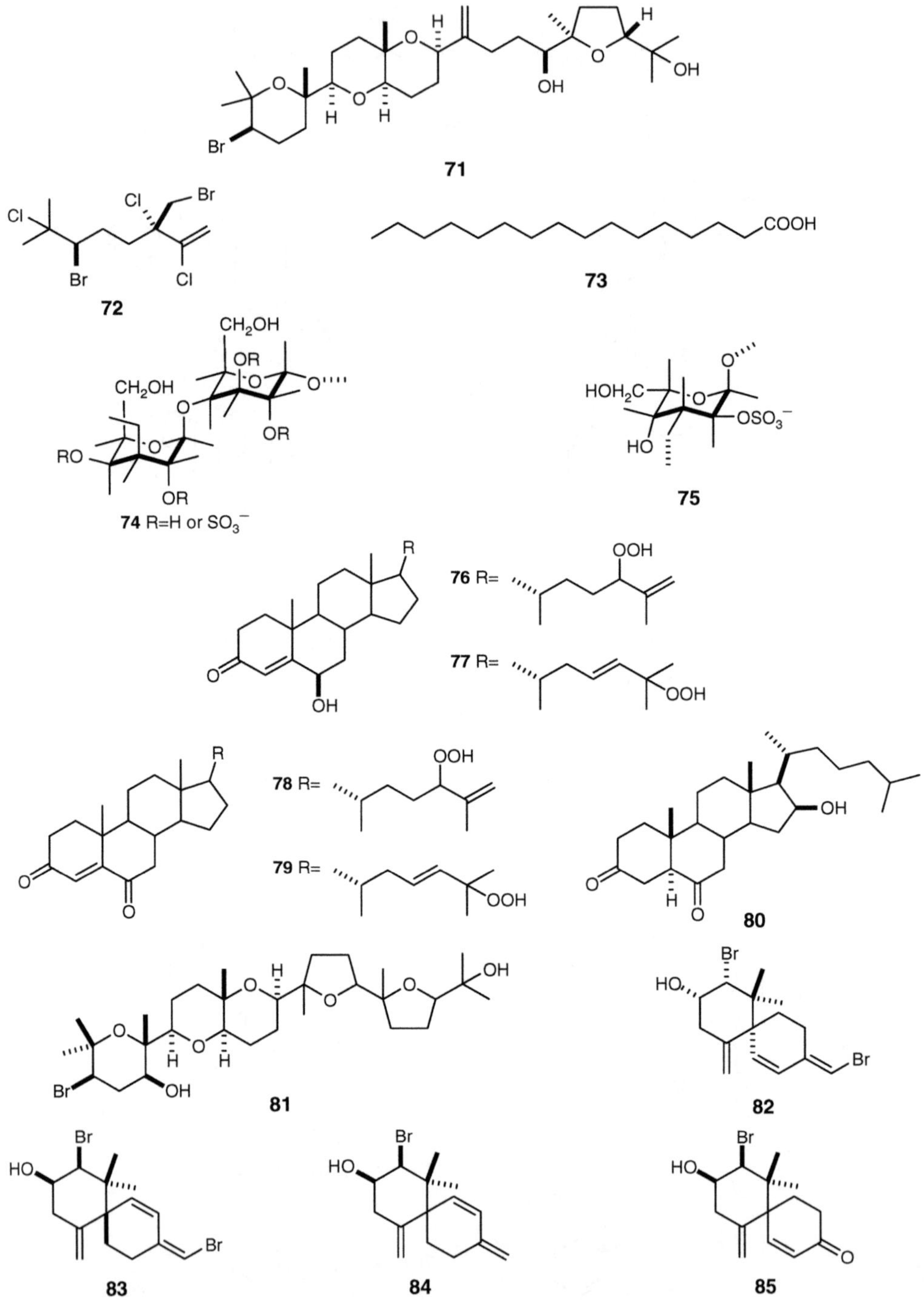

Fig. 9 Bioactive metabolites isolated from red algae

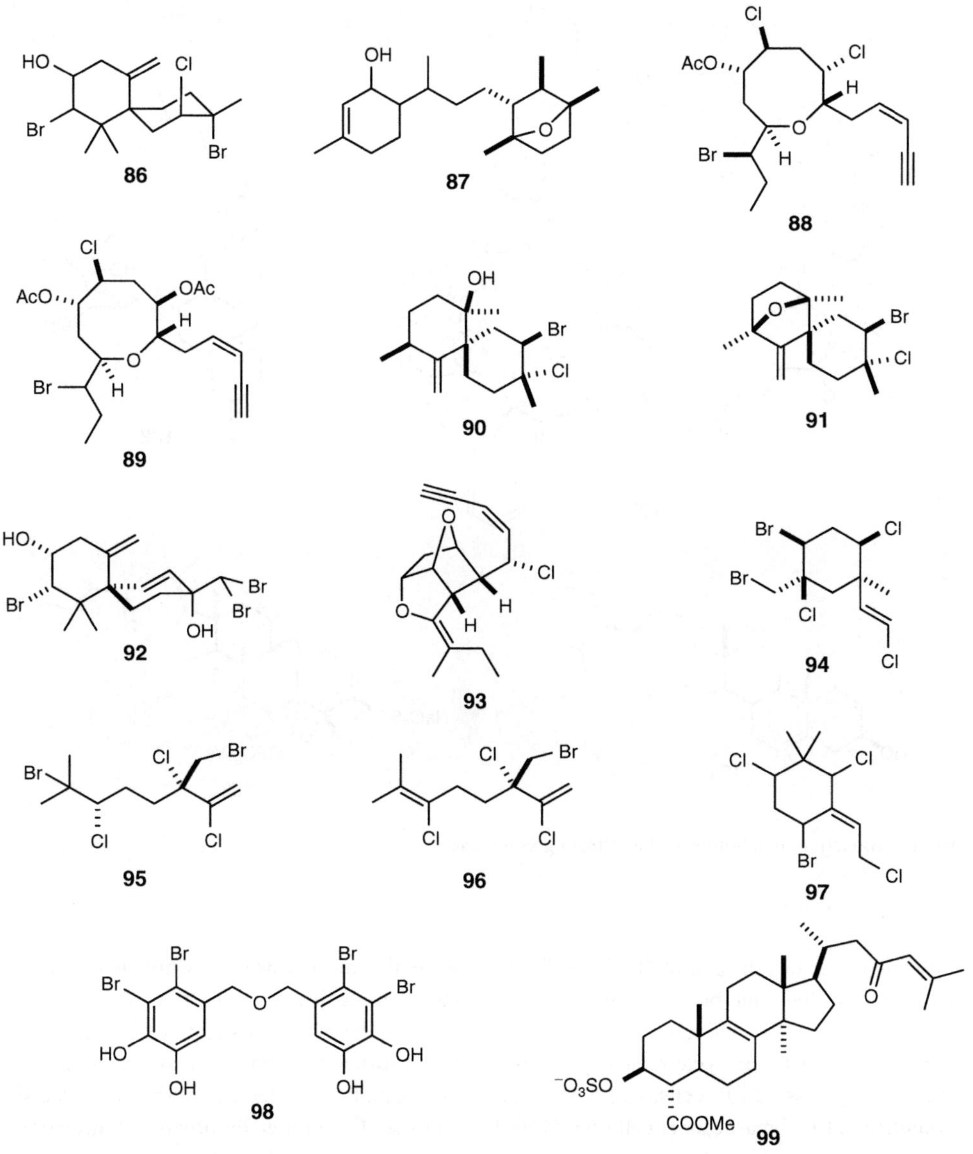

Fig. 9 (continued)

suggest that it interferes with the tumoral activity of these prostatic cancer cells [140].

Among the bioactive red algal metabolites, dehydrothyrsiferol (DT, **71**), a triterpenoid polyether isolated from a Canary island collection of *Laurencia viridis*, has exhibited promising cytotoxic activity. Therefore, the cytotoxic effect of DT on human estrogen receptor ER^+ and ER^- breast cancer cell lines was studied in a preclinical pharmacological evaluation. Although it was

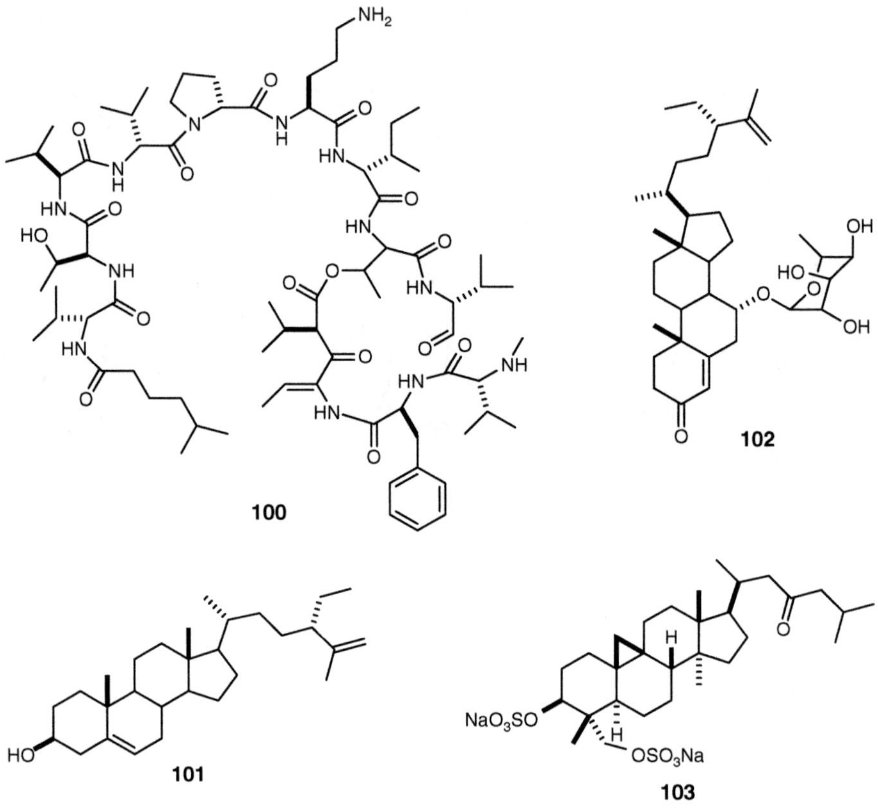

100 101 102 103

Fig. 10 Bioactive metabolites isolated from green algae

possible to exclude the possibility that DT functions as a mitosis inhibitor, researchers noted that apoptosis "was induced more efficiently and with distinct cell cycle-related patterns in the more aggressive ER- cells, while being less complete in ER+ breast cancer cell lines" [166].

Another natural product derived from red algae that entered preclinical evaluation was the polyhalogenated acyclic monoterpene halomon (**72**), obtained as the major constituent of *Portieria hornemannii*. Halomon exhibited highly differential cytotoxicity against the NCI's *in vitro* human tumor cell line screening panel. Brain tumor, renal, and colon tumor cell lines were the most sensitive, while leukemia and melanoma cell lines were relatively less sensitive. On the basis of its unprecedented cytotoxicity profile in the NCI primary screening, this compound was selected by the NCI Decision Network Committee for preclinical drug development. The preclinical work was hampered severely though, due to the inability to procure a collection of the organism which consistently produced halomon. Bioavailability studies using halomon showed that it was widely distributed to all tissues, but that it was concentrated and persistent in the fatty tissues. Another observation was that halomon persisted at the site of injection due to its low aqueous solubility. The problems of supply and bioavailability led to further clinical trials being delayed [167].

Currently, the most promising algal metabolite is the depsipeptide kahalalide F (KF, **100**), originally isolated from the sea slug *Elysia rufescens* and later found in the ethanolic extract of the green alga *Bryopsis* sp. which is the common diet of the mollusk [189]. KF showed a potent cytotoxic activity against a panel of human prostate and breast cancer cell lines, with an IC_{50} ranging from 0.07 to 0.28 μM. Importantly, non tumor human cells were 5–40 times less sensitive to the drug (IC_{50} = 1.6–3.1 μM). KF's cytotoxicity did not correlate with the expression level of the multidrug resistance MDR1 and of the tyrosine kinase HER2/NEU, and only slightly by the anti-apoptotic BCL-2 protein. KF's action was triggered rapidly by short pulse treatments. Neither a general caspase inhibitor (Z-VAD-fmk), nor transcription or translation inhibitors (actinomycin D, cycloheximide) blocked KF action. Flow cytometry analysis revealed that KF induced neither cell-cycle arrest, nor apoptotic hypodiploid peak. Using mitochondrial (JC-1)- and lysosomal (LysoTracker Green, Acridine Orange)-specific fluorophores, researchers detected loss of mitochondrial membrane potential and of lysosomal integrity following KF treatment. Confocal laser and electron microscopy revealed that KF-treated cells underwent a series of profound alterations including severe cytoplasmic swelling and vacuolization, dilation and vesiculation of the endoplasmic reticulum, mitochondrial damage, and plasma membrane rupture. In contrast, the cell nucleus showed irregular clumping of chromatin into small, condensed masses, while chromatin disappeared from other nuclear domains, but the nuclear envelope was preserved and no DNA degradation was detected. Together, these data indicate that KF induces cell death via oncosis preferentially in tumor cells. KF is currently in Phase II clinical trials in hepatocellular carcinoma, non-small cell lung cancer (NSCLC) and melanoma and is also under evaluation for the treatment of severe psoriasis [190].

Seaweeds have afforded to date the highest number of compounds within a single group of marine organisms. A high percentage of recent reports concerns bioactive metabolites with interesting biological properties. The reported in this review biological properties have focused on a diverse spectrum of activities. The effective discovery and development of novel drugs requires close international and multidisciplinary collaboration between scientific groups. This involves disciplines ranging from botany, marine biology and microbiology, through cell and molecular biology and synthetic chemistry, to pharmacology, toxicology and clinical trials. The continuing threat to biodiversity through the destruction of terrestrial and marine ecosystems lends urgency to the need to expand the collaborative exploration of these resources.

Even though algae were among of the first marine organisms that were investigated and proven to be rich sources of extraordinary chemical structures, still until today only a small percentage of algae has been studied and the fact that many species exhibit geographic variation in their chemical composition shows the huge potential algae hold as sources of interesting bioactive metabolites. Also, since some of the investigations on the algal chemistry preceeded the development of many of the current pharmacological bioassays, examination of these algal metabolites might prove very profitable.

As it can be seen on the basis of the reviewed literature, the potential of algae as sources of pharmaceutical molecules is far from been exhausted and further advancement of sophisticated spectroscopic techniques and development of new faster and more efficient pharmacological evaluation assays will stimulate, in the future, intense studies for the isolation of new bioactive algal metabolites.

2

References

1. Farnsworth, N. R. et al. (1985) Medicinal plants in therapy. *Bull. World Health Organ.* 63, 965–981
2. Cragg, G.M. and Newman, D.J. (2005) International collaboration in drug discovery and development from natural sources. *Pure Appl. Chem.* 77, 1923–1942
3. de la Torre-Castro, M. and Ronnback, P. (2004) Links between humans and seagrasses: an example from tropical East Africa. *Ocean Coast. Manage.* 47, 361–387
4. Pietra, F. (2002) Defining biodiversity. In *Biodiversity and natural product diversity.* (Pietra, F., eds.), p. 4, Pergamon, Amsterdam
5. Cragg, G.M. and Newman, D.J. (2005) Biodiversity: a continuing source of novel drug leads. *Pure Appl. Chem.* 77, 7–24
6. Cardozo, K.H.M. et al. (2007) Metabolites from algae with economical impact. *Comp. Biochem. Physiol. C.* 146, 60–78
7. Puglisi, M.P. et al. (2004) Capisterones A and B from the tropical green alga *Penicillus capitatus*: unexpected anti-fungal defenses targeting the marine pathogen *Lindra thallasiae. Tetrahedron.* 60, 7035–7039
8. Barros, M.P. et al. (2005) Rhythmicity and oxidative/nitrosative stress in algae. *Biol. Rhythm. Res.* 36, 67–82
9. Tringali, C. (1997) Bioactive metabolites from marine algae: recent results. *Curr. Org. Chem.* 1, 375–394
10. Burja, A.M. et al. (2001) Marine cyanobacteria-a profilic source of natural products. *Tetrahedron.* 57, 9347–9377
11. Mayer, A.M.S. and Hamann, M.T. (2005) Marine pharmacology in 2001-2002 marine compounds with anthelmintic, antibacterial, anticoagulant, antidiabetic, antifungal, anti-inflammatory, antimalarial, antiplatelet, anti-protozoal, antituberculosis, and antiviral activities; affecting the cardiovascular, immune and nervous systems and other miscellaneous mechanisms of action. *Comp. Biochem. Physiol C.* 140, 265–286
12. Mayer, A.M.S. et al. (2007) Marine pharmacology in 2003-4: marine compounds with anthelmintic, antibacterial, anticoagulant, antifungal, anti-inflammatory, antimalarial, antiplatelet, antiprotozoal, antituberculosis, and antiviral activities; affecting the cardiovascular, immune and nervous systems, and other miscellaneous mechanisms of action. *Comp. Biochem. Physiol. C.* 145, 553–581
13. Singh, S. et al. (2005) Bioactive compounds from cyanobacteria and microalgae: an overview. *Crit. Rev. Biotechnol.* 25, 73–95
14. Blunt, J.W. et al. (2007) Marine natural products. *Nat. Prod. Rep.* 24, 31–86 and earlier reviews in this series
15. Vadas, R.L. (1979) Seaweeds: an overview; ecological and economic importance. *Experientia.* 35, 429–433
16. Rasmussen, R.S. and Morrissey, M.T. (2007) Marine biotechnology for production of food ingredients. *Adv. Food Nutr. Res.* 52, 237–292
17. Li, H.B. and Chen, F. (2001) Preparative isolation and purification of astaxanthin from the green microalga *Chlorococcum* sp. by high-speed counter-current chromatography. In *Algae and Their Biotechnological Potential* (Chen, F. and Jiang, Y., eds.), pp. 127–134, Kluwer, Dordrecht/Boston, MA/London

18 Dawson, E.Y. (1966) *Marine Botany: An Introduction*, 371 p., Holt, Rinehart/Winston

19. Bold, H.C. (1967) *Morphology of Plants*, 2nd edn, 541 p., Harper & Row, London
20. Borowitzka, M.A. (1993) Products from microalgae. *INFOFISH Int.* 93, 21–26
21. Grobbelaar, J.U. (2004) Algal biotechnology: real opportunities for Africa. *S. Afr. J. Bot.* 70, 140–144
22. Kay, R.A. (1991) Microalgae as food and supplement. *Crit. Rev. Food Sci. Nutr.* 30, 555–573
23. Yap, C.Y. and Chen, F. (2001) Polyunsaturated fatty acids: Biological significance, biosynthesis, and production by microalgae and microalgae-like organisms. In *Algae and Their Biotechnological Potential* (Chen, F. and Jiang, Y., eds.), pp. 1–32, Kluwer, Dordrecht
24. Roussis, V. et al. (2004) Cytotoxic metabolites from marine algae. In *Plants That Fight Cancer* (Kintzios, S.E. and Barberaki, M.G., eds.), pp. 195–241, CRC Press, Boca Raton, FL
25. Luiten, E.E. et al. (2003) Realizing the promises of marine biotechnology. *Biomol. Eng.* 20, 429–439
26. Tseng, C.K. (2001) Algal biotechnology industries and research activities in China. *J. Appl. Phycol.* 13, 375–380
27. FAO (2004) *The State of the World Fisheries and Aquaculture 2004 (SOFIA)*, FAO, Rome, http://www.fao.org/sof/sofia/index_en.htm

28. von Elbe, J.H. and Schwartz, S.J. (1996) Colorants. In *Food Chemistry* (Fennema, O.R., ed.), pp. 651–722, Marcel Dekker, New York
29. Gregory, J.F., III (1996) Vitamins. In *Food Chemistry* (Fennema, O.R., ed.), pp. 531–616, Marcel Dekker, New York
30. Ben-Amotz, A. (1993) Production of beta-carotene and vitamins by the halotolerant alga *Dunaliella*. In *Marine Biotechnology, Volume 1: Pharmaceutical and Bioactive Natural Products* (Attaway, D.H. and Zaborsky, O.R., eds), pp. 411–417, Plenum, New York
31. El Baz, F.K. et al. (2002) Accumulation of antioxidant vitamins in *Dunaliella salina*. *J. Biol. Sci.* 2, 220–223
32. Meyers, S.P. and Latscha, T. (1997) Carotenoids. In *Crustacean Nutrition, Advances in World Aquaculture* (D'Abramo, L.R. et al., eds.), vol. 6, pp. 164–193, World Aquaculture Society, Baton Rouge, LA
33. Miki, W. (1991) Biological functions and activities of animal carotenoids. *Pure Appl. Chem.* 63, 141–146
34. Guerin, M. et al. (2003) *Haematococcus* astaxanthin: applications for human health and nutrition. *Trends Biotechnol.* 21, 210–216
35. Potin, P. et al. (1999) Oligosaccharide recognition signals and defence reactions inmarine plant-microbe interactions. *Curr. Opin. Microbiol.* 2, 276–283
36. Mayer, A.M.S. and Gustafson, K.R. (2004) Marine pharmacology in 2001–2: antitumour and cytotoxic compounds. *Eur. J. Cancer*. 40, 2676–2704
37. Mayer, A.M.S. and Gustafson, K.R. (2006) Marine pharmacology in 2003–2004: anti-tumour and cytotoxic compounds. *Eur. J. Cancer*. 42, 2241–2270
38. Smit, A.J. (2004) Medicinal and pharmaceutical uses of seaweed natural products: a review. *J. Appl. Phycol.* 16, 245–262
39. Ackman, R.G. et al. (1964) Origin of marine fatty acids. Analysis of the fatty acids produced by the diatom *Skeletonema costatum*. *J. Fish Res. Bd. Can.* 21, 747–756
40. Ohr, L.M. (2005) Riding the nutraceuticals wave. *Food Technol.* 59, 95–96
41. Gill, I. and Valivety, R. (1997) Polyunsaturated fatty acids: Part 1. Occurrence, biological activities and application. *Trends Biotechnol.* 15, 401–409
42. Sayanova, O.V. and Napier, J.A. (2004) Eicosapentaenoic acid: biosynthetic routs and the potential for synthesis in transgenic plants. *Phytochemistry*. 65, 147–158
43. Funk, C.D. (2001) Prostaglandins and leukotrienes: advances in eicosanoids biology. *Science*. 294, 1871–1875
44. Bajpai, P. and Bajpai, P.K. (1993) Eicosapentaenoic acid (EPA) production from microorganisms: a review. *J. Biotechnol.* 30, 161–183
45. Sijtsma, L. and de Swaaf, M.E. (2004) Biotechnological production and applications of the omega-3 polyunsaturated fatty acid docosahexaenoic acid. *Appl. Microbiol. Biotechnol.* 64, 146–153
46. Ponomarenko, L.P. et al. (2004) Sterols of marine microalgae *Pyramimonas* cf. cordata (prasinophyta), Atteya ussurensis sp. nov. (Bacollariophyta) and a spring diatom bloom form Lake Baikal. *Comp. Biochem. Physiol. B*. 138, 65–70
47. Delaunay, F. et al. (1993) The effect of monospecific algal diets on growth and fatty acid composition of *Pectenmaximus* (L.) larvae. *J. Exp. Mar. Biol. Ecol.* 173, 163–179
48. Paul, V.J. et al. (2006) Marine chemical ecology. *Nat. Prod. Rep.* 23, 153–180
49. La Barre, S.L. et al. (2004) Monitoring defensive responses in macroalgae-limitations and perspectives. *Phytochem. Rev.* 3, 371–379
50. Fusetani, N. (2004) Biofouling and antifouling. *Nat. Prod. Rep.* 21, 94–104
51. Barbosa, J.P. et al. (2003) A dolabellane diterpene from the Brazilian brown alga *Dictyota pfaffii*. *Biochem. Syst. Ecol.* 31, 1451–1453
52. Barbosa, J.P. et al. (2004) A dolabellane diteprene from the brown alga *Dictyota pfaffii* as chemical defense against herbivores. *Bot. Mar.* 47, 147–151
53. Soares, A.R. et al. (2003) Variation on diterpene production by the Brazilian alga *Stypopodium zonale* (Dictyotales, Phaeophyta). *Biochem. Syst. Ecol.* 31, 1347–1350
54. Pereira, R.C. et al. (2004) Variation in chemical defenses against herbivory in southwestern Atlantic *Stypopodium zonale* (Phaeophyta). *Bot. Mar*. 47, 202–208
55. Ankisetty, S. et al. (2004) Chemical investigation of predator-deterred macrolagae from the Antarctic Peninsula. *J. Nat. Prod.* 67, 1295–1302
56. Amsler, C.D. and Fairhead, V.A. (2005) Defensive and sensory chemical ecology of brown algae. *Adv. Bot. Res.* 43, 1–91
57. Shibata, T. et al. (2002) Inhibitory activity of brown algal phlorotannins against glycosidases from the viscera of the turban shell *Turbo cornotus*. *Eur. J. Phycol.* 37, 493–500

2

58. Wright, J.T. et al. (2004) Chemical defense in a marine alga: heritability and the potential for selection by herbivores. *Ecology*. 85, 2946–2959
59. Pereira, R.C. et al. (2003) Ecological roles of natural products of the Brazilian red seaweed *Laurencia obtusa*. *Braz. J. Biol.* 63, 665–672
60. Paul, V.J. et al. (2001) Chemical mediation of macroalgal-herbivore interactions: ecological and evolutionary perspectives. In *Marine Chemical Ecology* (McClintock, J.B. and Baker, B.J., eds.), pp. 227–265, CRC Press, Boca Raton, FL
61. Gross, H. and König, G.M. (2006) Terepenoids from marine organisms: unique structures and their pharmacological potential. *Phytochem. Rev*. 5, 115–141
62. Phillips, J.A. and Price, I.R. (2002) How different is Mediterranean *Caulerpa taxifolia* (Caulerpales: Chlorophyta) to other populations of the species. *Mar. Ecol. Prog. Ser*. 238, 61–71
63. Bellan-Santini, D. et al. (1996) The influence of the introduced tropical alga *Caulerpa taxifolia*, on the biodiversity of the Mediterranean marine biota. *J. Mar. Biol.* 76, 235–237
64. Paul, V.J. and Fenical, W. (1986) Chemical defense in tropical green algae, order Caulerpales. *Mar. Ecol. Prog. Ser*. 34, 157–169
65. Paul, V.J. and Fenical, W. (1987) Natural products chemistry and chemical defense in tropical marine algae of the phylum Chlorophyta. In *Bioorganic Marine Chemistry* (Scheuer, P.J., ed.), pp. 1–29, Springer, Berlin
66. Brunelli, M. et al. (2000) Neurotoxic effects of caulerpenyne. *Prog. Neuro-Psychopharmacol. Biol. Psychiat*. 24, 939–954
67. Mozzachiodi, R. et al. (2001) Caulerpenyne, a toxin from the seaweed *Caulerpa taxifolia*, depresses after polarization in invertebrate neurons. *Neuroscience*. 107, 519–526
68. Parent-Massin, D. et al. (1996) Evaluation of the toxicological risk to humans of caulerpenyne using human hematopoietic progenitors, melanocytes, and keratinocytes in culture. *J. Toxicol. Environ. Health*. 47, 47–59
69. Barbier, P. et al. (2001) Caulerpenyne from *Caulerpa taxifolia* has an antiproliferative activity on tumor cell line SK–N–SH and modifies the microtubule network. *Life Sci.* 70, 415–429
70. Thibaut, T. and Meinesz, A. (2000) Are the Mediterranean ascoglossan molluscs *Oxynoe olivacea* and *Lobiger serradifalci* suitable agents for a biological control against the invading alga *Caulerpa taxifolia*. *C. R. Acad. Sci. (Ser. 3)*. 323, 477–488
71. Adolph, S. et al. (2005) Wound closure in the invasive green alga *Caulerpa taxifolia* by enzymatic activation of a protein cross-linker. *Angew. Chem. Int. Ed*. 44, 2806–2808
72. Jung, V. and Pohnert, G. (2001) Rapid wound-activated transformation of the green algal defensive metabolite caulerpenyne. *Tetrahedron*. 57, 7169–7172
73. Metcalf, R.L. (1987) Plant volatiles as insect attractants. *CRC Crit. Rev. Plant Sci*. 5, 251–301
74. Harborne, J.B. (1994) *Introduction to Ecological Biochemistry*, 318 p., Academic, London
75. Kaissling, K.E. and Priesner, E. (1970) Die Riechschwelle des Seidenspinners. *Naturwissens-chaften*. 57, 23–28
76. Arimura, G. et al. (2000) Herbivory-induced volatiles elicit defence genes in lima bean leaves. *Nature*. 406, 512–515
77. Turlings, T.C.J. et al. (1990) Exploitation of herbivore-induced plant odors by host-seeking parasitic wasps. *Science*. 240, 1251–1253
78. Kessler, A. and Baldwin, I.T. (2001) Defensive function of herbivore-induced plant volatile emissions in nature. *Science*. 291, 2141–2144
79. Fink, P. (2007) Ecological functions of volatile organic compounds in aquatic systems. *Mar. Freshwater Behav. Physiol*. 40, 155–168
80. Boland, W. (1995) The chemistry of gamete attraction – chemical structures, biosynthesis, and (a)biotic degradation of algal pheromones. *Proc. Natl. Acad. Sci. USA*. 92, 37–43
81. Pohnert, G. and Boland, W. (2002) The oxylipin chemistry of attraction and defense in brown algae and diatoms. *Nat. Prod. Rep*. 19, 108–122
82. Müller, D.G. et al. (1971) Sex attractant in a brown alga - chemical structure. *Science*. 171, 815
83. Wendel, T. and Jüttner, F. (1996) Lipoxygenase-mediated formation of hydrocarbons and unsaturated aldehydes in freshwater diatoms. *Phytochemistry*. 41, 1445–1449
84. Hombeck, M. and Boland, W. (1998) Biosynthesis of the algal pheromone fucoserratene by the freshwater diatom *Asterionella formosa* (Bacillariophyceae). *Tetrahedron*. 54, 11033–11042
85. Fink, P. et al. (2006) Volatile foraging kairomones in the littoral zone: attraction of an herbivorous freshwater gastropod to algal odors. *J. Chem. Ecol*. 32, 1867–1881
86. Fink, P. et al. (2006) Oxylipins from freshwater diatoms act as attractants for a benthic herbivore. *Arch. Hydrobiol*. 167, 561–574

87. Steinke, M. et al. (2002) Trophic interactions in the sea: an ecological role for climate relevant volatiles? *J. Phycol.* 38, 630–638
88. Zeeck, E. et al. (1991) Sex-pheromones in a marine polychaete - biologically-active compounds from female *Platynereis dumerilii. J. Exp. Zool.* 260, 93–98
89. Wolfe, G.V. and Steinke, M. (1996) Grazing-activated production of dimethyl sulfide (DMS) by two clones of *Emiliania huxleyi. Limnol. Oceanogr.* 41, 1151–1160
90. Gabric, A. et al. (2001) Modeling the biogeochemical cycle of dimethylsulfide in the upper ocean: a review. *Chemosphere Glob. Change Sci.* 3, 377–392
91. Ellis, J. and Korth, W. (1993) Removal of geosmin and methylisoborneol from drinking water by adsorption on ultrastable Zeolite-Y. *Water Res.* 27, 535–539
92. Bhadury, P. and Wright, P.C. (2004) Exploitation of marine algae: biogenic compounds for potential antifouling applications. *Planta.* 219, 561–578
93. Costerton, J.W. et al. (1995) Microbial biofilms. *Ann. Rev. Microbiol.* 49, 711–745
94. Abarzua, S. and Jakubowski, S. (1995) Biotechnological investigation for the prevention of biofouling. I. Biological and biochemical principles for the prevention of biofouling. *Mar. Ecol. Prog. Ser. II.* 123, 301–312
95. Clare, A.S. (1996) Marine natural product antifoulants: status and potential. *Biofouling.* 9, 211–229
96. Bellas, J. (2006) Comparative toxicity of alternative antifouling biocides on embryos and larvae of marine invertebrates. *Sci. Total Environ.* 367, 573–585
97. Braithwaite, R.A. and Fletcher, R.L. (2005) The toxicity of Irgarol 1051 and Sea-Nine 211 to the non-target macroalga *Fucus serratus* with the aid of an image capture and analysis system. *J. Exp. Mar. Biol. Ecol.* 322, 111–121
98. Fernandez-Alba, A.R. et al. (2002) Toxicity evaluation of single and mixed antifouling biocides measured with acute toxicity bioassays. *Anal. Chim. Acta.* 456, 303–312
99. Kobayashi, N. and Okamura, H. (2002) Effects of new antifouling compounds on the development of sea urchin. *Mar. Pollut. Bull.* 44, 748–751
100. Kwok, K.W.H. and Leung, K.M.H. (2005) Toxicity of antifouling biocides to the intertidal harpacticoid copepod *Trigriopus japonicus* (Crustacea, Copepoda): effects of temperature and salinity. *Mar. Pollut. Bull.* 51, 830–837
101. Myers, J.H. et al. (2006) Effects of antifouling biocides to the germination and growth of the marine macroalga, *Hormosira banksii* (Turner) Desicaine. *Mar. Pollut. Bull.* 52, 1048–1055
102. Di Landa, G. et al. (2006) Occurrence of antifouling paint booster biocides in selected harbors and marinas inside the Gulf of Napoli: a preliminary survey. *Mar. Pollut. Bull.* 52, 1541–1546
103. Dahlström, M. et al. (2000) Surface active adrenoreceptor compounds prevent the settlement of cyprid larvae of *Balanus improvisus. Biofouling.* 16, 191–198
104. Kubanek, J. et al. (2003) Seaweed resistance to microbial attack: a targeted chemical defense against marine fungi. *Proc. Natl. Acad. Sci. USA.* 100, 6916–6921
105. Hellio, C. et al. (2002) Screening of marine algal extracts for anti-settlement activities against microalgae and maroalgae. *Biofouling.* 18, 205–215
106. Steinberg, P.D. and de Nys, R. (2002) Chemical mediation of colonization of seaweed surfaces. *J. Phycol.* 38, 621–629
107. Hay, M.E. (1992) The role of seaweed chemical defenses in the evolution of feeding specialization and in the mediation of complex interactions. In *Ecological Roles of Marine Natural Products*(Paul, V.J., ed.), pp. 93–118, Cornell University Press, Ithaca, NY/London
108. de Nys, R. and Steinberg, P.D. (2002) Linking marine biology and biotechnology. *Curr. Opin. Biotechnol.* 13, 244–248
109. Manefield, M. et al. (2002) Halogenated furanones inhibit quorum sensing through accelerated LuxR turnover. *Microbiology.* 148, 1119–1127
110. Hentzer, M. et al. (2002) Inhibition of quorum sensing in *Pseudomonas aeruginosa* biofilm bacteria by a halogenated furanone compound. *Microbiology.* 148, 87–102
111. Steinberg, P.D. et al. (2001) Chemical mediation of surface colonisation. In *Marine Chemical Ecology* (McClintock, J.B. and Baker, B.J., eds), pp. 355–387, CRC Press, Boca Raton, FL
112. Kjelleberg, S. et al. (1997) Do marine natural products interfere with prokaryotic AHL regulatory systems? *Aquat. Microb. Ecol.* 13, 85–93
113. Kjelleberg, S. and Steinberg, P.D. (2001) Surface warfare in the sea. *Microbiol. Today.* 28, 134–135
114. Manefield, M. et al. (1999) Inhibition of LuxR-based AHL regulation by halogenated

2

furanones from *Delisea pulchra*. *Microbiology*. 145, 283–291
115. Givskov, M. et al. (1996) Eukaryotic interference with homoserine lactone-mediated prokaryotic signalling. *J. Bacteriol*. 178, 6618–6622
116. Steinberg, P.D. et al. (1998) Chemical inhibition of epibiota by Australian seaweeds. *Biofouling*. 12, 227–244
117. Maximilien, R. et al. (1998) Chemical mediation of bacterial surface colonisation by secondary metabolites from red alga *Delisea pulchra*. *Aquat. Microb. Ecol*. 15, 233–246
118. Ren, D. et al. (2002) Inhibition of biofilm formation and swarming of *Bacillus subtilis* by (5Z)-4-bromo-5-(bromomethylene)-3-butyl-2(5H)-furanone. *Lett. Appl. Microbiol*. 34, 293–299
119. de Nys, R. et al. (1995) Broad spectrum effects of secondary metabolites from the red alga *Delisea pulchra* in antifouling assays. *Biofouling*. 8, 259–271
120. Rasmussen, T.B. et al. (2000) How *Delisea pulchra* furanones affect quorum sensing and swarming motility in *Serratia liquefaciens* MG1. *Microbiology*. 12, 3237–3244
121. Ren, D. et al. (2001) Inhibition of biofilm formation and swarming of *Escherichia coli* by (5Z)-4-bromo-5-(bromomethylene)-3-butyl-2-(5H)-furanone. *Environ. Microbiol*. 3, 731–736
122. McLachlan, J. and Craigie, J.S. (1966) Antialgal activity of some simple phenols. *J. Phycol*. 2, 133–135
123. König, G.M. and Wright, A.D. (1997) *Laurencia rigida*: chemical investigations of its antifouling dichloromethane extract. *J. Nat. Prod*. 60, 967–970
124. de Nys, R. et al. (1996) The need for standardised broad scale bioassay testing: a case study using the red algae *Laurencia rigida*. *Biofouling*. 10, 213–224
125. König, G.M. et al. (1999) *Plocamium hamatum* and its monoterpenes: chemical and biological investigations of the tropical marine red alga. *Phytochemistry*. 52, 1047–1053
126. König, G.M. et al. (1999) Halogenated monoterpenes from *Plocamium costatum* and their biological activity. *J. Nat. Prod*. 62, 383–385
127. Jennings, J.G. and Steinberg, P.D. (1997) Phlorotannins vs. others factors affecting epiphyte abundance on the kelp *Ecklonia radiata*. *Ocealogia*. 109, 461–473
128. Schmitt, T.M. et al. (1995) Constraints on chemically mediated coevolution: multiple functions for seaweed secondary metabolites. *Ecology*. 76, 107–123
129. Schmitt, T.M. et al. (1998) Seaweed secondary metabolites as antifoulants: effects of *Dictyota* spp. diterpenes on survivorship, settlement, and development of marine invertebrate larvae. *Chemoecology*. 8, 125–131
130. Taniguchi, K. et al. (1989) Inhibitory effects of the settlement and metamorphosis of the abalone *Haliotis discus hannai* veligers by the methanol extracts from the brown alga *Dilophus okamurai*. *Nippon Suisan Gakkaishi*. 55, 1133–1137
131. Maréchal, J. et al. (2004) Seasonal variation in antifouling activity of crude extracts of the brown alga *Bifurcaria bifurcata* (Cystoseiraceae) against cyprids of *Balanus amphitrite* and the marine bacteria *Cobetia marina* and *Pseudoalteromonas haloplanktis*. *J. Exp. Mar. Biol. Ecol*. 313, 47–62
132. Culioli, G. et al. (2002) Seasonal variations in the chemical composition of *Bifurcaria bifurcata* (Cystoseiraceae). *Biochem. Syst. Ecol*. 30, 61–64
133. Smyrniotopoulos, V. et al. (2003) Acetylene sesquiterpenoid esters from the green alga *Caulerpa prolifera*. *J. Nat. Prod*. 66, 21–24
134. Pietra, F. (1990) *A Secret World – Natural Products of Marine Life*, 288 p., Birkhauser Verlag AG, Basel
135. McConnell, O.J. et al. (1994) The discovery of natural products with therapeutic potential. *Biotechnology*. 26, 109–174
136. Riguera, R. (1997) Isolating bioactive compounds from marine organisms. *J. Mar. Biotechnol*. 5, 187–193
137. Kerr, R.G. and Kerr, S.S. (1999) Marine natural products as therapeutic agents. *Exp. Opin. Ther. Patents*. 9, 1207–1222
138. MarinLit database, Department of Chemistry, University of Canterbury, http://www.chem.canterbury.ac.nz/marinlit/marinlit.shtml
139. Mori, K. and Koga, Y. (1992) Synthesis and absolute configuration of (-)-stypoldione. *Bioorg. Med. Chem. Lett*. 2, 391–394
140. Depix, M.S. et al. (1998) The compound 14-keto-stypodiol diacetate from the algae *Stypopodium flabelliforme* inhibits microtubules and cell proliferation in DU-145 human prostatic cells. *Mol. Cell. Biochem*. 187, 191–199
141. Fadli, M. et al. (1991) Novel meroterpenoids from *Cystoseira mediterranea*: use of the Crown-Gall bioassay as a primary screen for lipophilic antineoplastic agents. *J. Nat. Prod*. 54, 261–264

142. Urones, J.G. et al. (1992) Meroterpenes from *Cystoceira usneoides*. *Phytochemistry*. 31, 179–182
143. Ishitsuka, M.O. et al. (1990) Antitumor xenicane and norxenicane lactones from the brown algae Dictyotaceae. *J. Org. Chem*. 53, 5010–5013
144. Durán, R. et al. (1997) New diterpenoids from the alga *Dictyota dichotoma*. *Tetrahedron*. 53, 8675–8688
145. Ishitsuka, M.O. et al. (1990) Bicyclic diterpenes from two species of brown algae of the Dictyotaceae. *Phytochemistry*. 29, 2605–2610
146. Pereira, H.S. et al. (2004) Antiviral activity of diterpenes isolated from the Brazilian marine alga *Dictyota menstrualis* against human immunodeficiency virus type 1 (HIV-1). *Antivir. Res*. 64, 69–76
147. Bouaicha, N. et al. (1993) Bioactive diterpenoids isolated from *Dilophus ligulatus*. *Planta Med*. 59, 256–258
148. Bouaicha, N. et al. (1993) Cytotoxic diterpenoids from the brown alga *Dilophus ligulatus*. *J. Nat. Prod*. 56, 1747–1752
149. Hu, J. et al. (2004) Acidic oligosaccharide sugar chain, a marine-derived acidic oligosaccharide, inhibits the cytotoxicity and aggregation of amyloid beta protein. *J. Pharm. Sci*. 95, 248–255
150. Vetvicka, V. and Yvin, J.C. (2004) Effects of marine beta-1,3 glucan on immune reactions. *Int. Immunopharmacol*. 4, 721–730
151. Capon, R.J. et al. (1998) Marine nematocides: tetrahydrofurans from a southern Australian brown algae, *Notheia anomala*. *Tetrahedron*. 54, 2227–2242
152. Ktari, L. and Guyot, M. (1999) A cytotoxic oxysterol from the marine alga *Padina pavonica* (L.) Thivy. *J. Appl. Phycol*. 11, 511–513
153. Hoshino, T. et al. (1998) An antivirally active sulfated polysaccharide from *Sargassum horneri* (Turner) C. Agardh. *Biol. Pharm. Bull*. 21, 730–734
154. Kamei, Y. and Tsang, C.K. (2003) Sargaquinoic acid promotes neurite outgrowth via protein kinase A and MAP kinases-mediated signaling pathways in PC12D cells. *Int. J. Dev. Neurosci*. 21, 255–262
155. Tsang, C.K. and Kamei, Y. (2004) Sargaquinoic acid supports the survival of neuronal PC12D cells in a nerve growth factor-independent manner. *Eur. J. Pharmacol*. 488, 11–18
156. Numata, A. et al. (1992) A cytotoxic principle of the brown alga *Sargassum tortile* and structures of chromenes. *Phytochemistry*. 31, 1209–1213
157. Gerwick, W.H. et al. (1980) Isolation and structure of spatol a potent inhibitor of cell replication from the brown seaweed *Spatoglossum schmittii*. *J. Am. Chem. Soc*. 102, 7991–7993
158. Wessels, M. et al. (1999) A new tyrosine kinase inhibitor from the marine brown alga *Stypopodium zonale*. *J. Nat. Prod*. 62, 927–930
159. Nahas, R. et al. (2007) Radical-scavenging activity of Aegean Sea marine algae. *Food Chem*. 102, 577–581
160. Sheu, J.H. et al. (1999) New cytotoxic oxygenated fucosterols from the brown alga *Turbinaria conoides*. *J. Nat. Prod*. 62, 224–227
161. Asari, F. et al. (1989) Turbinaric acid, a cytotoxic secosqualene carboxylic acid from the brown alga *Turbinaria ornata*. *J. Nat. Prod*. 52, 1167–1169
162. Sheu, J.H. et al. (1997) Cytotoxic sterols from the Formosan brown alga *Turbinaria ornata*. *Planta Med*. 63, 571–572
163. Ishihara, K. et al. (1998) Inhibition of icosanoid production in MC/9 mouse mast cells by n-3 polyunsaturated fatty acids isolated from edible marine algae. *Biosci. Biotechnol. Biochem*. 62, 1412–1415
164. Pereira, M.S. et al. (1999) Structure and anticoagulant activity of sulfated fucans. Comparison between the regular, repetitive, and linear fucans from echinoderms with the more heterogeneous and branched polymers from brown algae. *J. Biol. Chem*. 274, 7656–7667
165. Thorlacius, H. et al. (2000) The polysaccharide fucoidan inhibits microvascular thrombus formation independently from P- and L-selectin function *in vivo*. *Eur. J. Clin. Invest*. 30, 804–810
166. Pec, M.K. et al. (2003) Induction apoptosis in estrogen dependent and independent cancer cells by the marine terpenoid dehydrothyrsiferol. *Biochem. Pharmacol*. 65, 1451–1461
167. Jaspars, M. (1998) Pharmacy of the deep – marine organisms as a source of anticancer agents. In *Advances in Drug Discovery Techniques* (Harvey, A., ed.), p. 77, Wiley, Chichester
168. Harada, H. et al. (2002) Antitumor activity of palmitic acid found as a selective cytotoxic substance in a marine red alga. *Anticancer Res*. 22, 2587–2590
169. Farias, W.R.L. et al. (2000) Structure and anticoagulant activity of sulfated galactans – isolation of a unique sulfated galactan from the

red algae *Botryocladia occidentalis* and comparison of its anticoagulant action with that of sulfated galactans from invertebrates. *J. Biol. Chem.* 275, 29299–29307
170. Melo, F.R. et al. (2004) Antithrombin-mediated anticoagulant activity of sulfated polysaccharides: different mechanisms for heparin and sulfated galactans. *J. Biol. Chem.* 279, 20824–20835
171. Tan, L.T. et al. (2000) Cis,cis- and trans,trans-ceratospongamide, new bioactive cyclic heptapeptides from the Indonesian red alga *Ceratodictyon spongiosum* and symbiotic sponge *Sigmadocia symbiotica*. *J. Org. Chem.* 65, 419–425
172. Davyt, D. et al. (1998) A new indole derivative from the red alga *Chondria atropurpurea*. Isolation, structure determination, and anthelmintic activity. *J. Nat. Prod.* 61, 1560–1563
173. Sheu, J.H. et al. (1997) Study on cytotoxic oxygenated desmosterols isolated from the red alga *Galaxaura marginata*. *J. Nat. Prod.* 60, 900–903
174. Ohta, K. et al. (1998) Sulfoquinovosyldiacylglycerol, KM043, a new potent inhibitor of eukaryotic DNA polymerases and HIVreverse transcriptase type 1 from a marine red alga, *Gigartina tenella*. *Chem. Pharm. Bull.* 46, 684–686
175. Ktari, L. et al. (2000) 16β-Hydroxy-5α-cholestane-3,6-dione, a novel cytotoxic oxysterol from the red alga *Jania rubens*. *Bioorg. Med. Chem. Lett.* 10, 2563–2565
176. Suzuki, M. et al. (1995) Callicladol, a novel bromotriterpene polyether from a Vietnamese species of the red algal genus *Laurencia*. *Chem. Lett.* 1045-1046
177. Juagdan, E.D. et al. (1997) Two new chamigranes from an Hawaiian red alga, *Laurencia cartilaginea*. *Tetrahedron.* 53, 521–528
178. Mohammed, K.A. et al. (2004) Laurenditerpenol, a new diterpene from the tropical marine alga *Laurencia intricata* that potently inhibits HIF-1 mediated hypoxic signaling in breast tumor cells. *J. Nat. Prod.* 67, 2002–2007
179. Iliopoulou, D. et al. (2002) C15 acetogenins from the red alga *Laurencia obtusa*. *Phytochemistry.* 59, 111–116
180. Suzuki, M. et al. (2001) Novel halogenated metabolites from the Malaysian *Laurencia pannosa*. *J. Nat. Prod.* 64, 597–602
181. Pec, M.K. et al. (2002) Dehydrothyrsiferol does not modulate multidrug resistance-associated protein 1 resistance: a functional screening system for MRP1 substrates. *Int. J. Mol. Med.* 10, 605–608
182. Francisco, M.E.Y. and Erickson, K.L. (2001) Ma'iliohydrin, a cytotoxic chamigrene dibromohydrin from a Philippine *Laurencia* species. *J. Nat. Prod.* 64, 790–791
183. Vairappan, C.S. et al. (2001) Antibacterial halogenated metabolites from the Malaysian *Laurencia* species. *Phytochemistry.* 58, 291–297
184. Fuller, R.W. et al. (1994) Isolation and structure/activity features of halomon – related antitumor monoterpenes from the red algae *Portieria hornemanii*. *J. Med. Chem.* 37, 4407–4411
185. Egorin, M.J. et al. (1996) Plasma pharmacokinetics, bioavailability, and tissue distribution in CD_2F_1 mice of halomon, an antitumor halogenated monoterpene isolated from the red algae *Portieria hornemanii*. *Cancer Chemother. Pharmacol.* 39, 51–60
186. Egorin, M.J. et al. (1997) *In vitro* metabolism by mouse and human liver preparations of halomon, an antitumor halogenated monoterpene. *Cancer Chemother. Pharmacol.* 41, 9–14
187. Xu, N. et al. (2003) Antibacterial bromophenols from the marine red alga *Rhodomela confervoides*. *Phytochemistry.* 62, 1221–1224
188. Horgen, F.D. et al. (2000) New terpenoid sulfates from the red alga *Tricleocarpa fragilis*. *J. Nat. Prod.* 63, 210–216
189. Hamann, M.T. and Scheuer, P.J. (1993) Kahalalide F: a bioactive depsipeptide from the sacoglossum mollusk *Elysia rufescens* and the green alga *Bryopsis* sp. *J. Am. Chem. Soc.* 115, 5825–5826
190. Suárez, Y. et al. (2003) Kahalalide F, a new marine-derived compound, induces oncosis in human prostate and breast cancer cells. *Mol. Cancer Ther.* 2, 863–872
191. Fischel, J.L. et al. (1994) Mise en évidence d'effets antiprolifér ants de la caulerpényne (de *Caulerpa taxifolia*). Expérience sur cellules tumorales humaines en culture. *Bull. Cancer.* 81, 489
192. Fischel, J.L. et al. (1995) Cell growth inhibitory effects of caulerpenyne, a sesquiterpenoid from the marine algae *Caulerpa taxifolia*. *Anticancer Res.* 15, 2155–2160
193. Bitou, N. et al. (1999) Screening of lipase inhibitors from marine algae. *Lipids.* 34, 441–445
194. Sheu, J.H. et al. (1995) Oxygenated clerosterols isolated from the marine alga *Codium arabieum*. *J. Nat. Prod.* 58, 1521–1526

195. Nika, K. et al. (2003) Specific recognition of immune cytokines by sulphated polysaccharides from marine algae. *Eur. J. Phycol.* 38, 257–264
196. Ali, M.S. et al. (2002) Steroid and antibacterial steroidal glycosides from marine green alga *Codium iyengarii* Borgesen. *Nat. Prod. Lett.* 16, 407–413
197. Matsubara, K. et al. (2000) An anticoagulant proteoglycan from the marine green alga, *Codium pungniformis*. *J. Appl. Phycol.* 12, 9–14
198. Takamatsu, S. et al. (2003) Marine natural products as novel antioxidant prototypes. *J. Nat. Prod.* 66, 605–608
199. Govindan, M. et al. (1994) New cycloartanol sulfates from the alga *Tydemania expeditionis*: inhibitors of the protein tyrosine kinase pp60^{v-src}. *J. Nat. Prod.* 57, 74–78
200. Awad, N.E. (2000) Biologically active steroid from the green alga *Ulva lactuca*. *Phytother. Res.* 14, 641–643I

Use of Secondary Metabolite Variation in Crop Improvement

3

Daniel J. Kliebenstein*

Abstract Plants contain significant levels of natural genetic and phenotypic variation between individuals within a species for traits ranging from development to metabolism to pathogen resistance. This intra-specific variation is a foundation for research by evolutionary and ecological biologists interested in understanding plant fitness as well as by plant biologists focused on increasing the fitness or yield of agricultural plants. An important component of intra-specific variation for both research groups is the secondary metabolite complement present within a plant. Variation in these compounds controls important ecological and agronomic traits such as resistance to insect herbivores and benefit to human health. This intraspecific variation means that secondary metabolite diversity can be an important tool in crop improvement. This chapter will focus on sources of intra-specific variation in plant secondary metabolites and the potential use of this natural variation in crop improvement and potential pleiotropic consequences of this natural variation.

D.J. Kliebenstein (✉)
Department of Plant Sciences,
University of California, Davis,
One Shields Avenue, Davis, CA 95616, USA
e-mail: kliebenstein@ucdavis.edu

*The author has no conflict of interest to report with regards to this review.

Ecology Versus Agriculture

Secondary metabolites are believed to aid plant fitness via diverse biological mechanisms requiring secondary metabolites to have highly diverse biological activities. Generating these diverse biological activities has led plants to accumulate a vast catalogue of compounds of at least several hundred thousand secondary metabolites [1]. Plants have simplified the generation of diverse structures by relying on differential modification of common backbone structures. Often times networks of enzymes with broad substrate specificity can be shuffled to generate modular systems capable of producing metabolic diversity [2]. Differential modification of common backbone structures can alter the biological activity of numerous plant hormones and secondary metabolites including auxins, glucosinolates, gibberellins and phenylpropanoids [3–7]. The repeated modification of common backbone structures produces approximately 12,000 known alkaloid structures [5].

A plant research field that has extensively utilized the diversity and biology of secondary metabolites is crop breeding. This previously

A.E. Osbourn and V. Lanzotti (eds.), *Plant-derived Natural Products*,
DOI 10.1007/978-0-387-85498-4_3, © Springer Science+Business Media, LLC 2009

3

Table 1 Current or previous secondary metabolite breeding targets

Biology	Crop	Target	Goal	Potential Consequences
Flavour	Apple	Polyphenol	Decrease	Increased pest pressure
	Crucifers	Glucosinolate	Decrease	Increased pest pressure
	Carrot	Polyacetylenes	Decrease	Increased nematode pressure
Nutrition	Cassava	Cyanogenic glycosides	Decrease	Increased herbivore pressure
	Crucifers	Glucosinolate	Increase	Decreased flavour
	Potato	Glycoalkaloids	Decreased	Increased pest pressure
	Solanum	Phenolics	Increased	Decreased flavour
	Tomato	Lycopene	Increased	Altered phytohormone balance
Biotic stress	Crucifers	Glucosinolate	Increase	Decreased flavour
Resistance	Maize	Maysin	Increase	
	Lettuce	Sesquiterpenes	Increase	Decreased flavour
Abiotic stress				
Resistance				

and currently focuses on four areas of secondary metabolite biology (Table 1). In terms of crop plants, these improvements via secondary metabolites are focused on altering the flavour profile of the crops, altering its resistance to biotic and abiotic stresses and more recently directly targeting the nutritional quality of a crop. This chapter will focus on the biotic interactions of secondary metabolites in flavor, nutrition and biotic pest resistance as the specific manipulation of secondary metabolites for abiotic stress is less developed.

One indication of the importance of secondary metabolite changes in crop development comes about by a comparison of wild and domestic species. Many flavor related changes in secondary metabolites were involved in the early domestication of crops such as alterations in the polyphenolic content of Apples and Tomatoes, decreased glucosinolates in crucifers and lowered glycoalkaloid accumulation in most Solanum crops [8–12]. Thus, secondary metabolites have long been a focus of breeding. Interestingly, these changes geared towards improving human acceptance of the crop has typically decreased the content of secondary metabolites generating bitter flavors [10, 13, 14]. Unfortunately, these same bitter flavors are important biotic stress resistance mechanisms and the improved flavor has led to lowered resistance to numerous biotic stresses in comparison to the wild species. Thus, while secondary metabolites are important in crop breeding, changing their concentration or structure can have pleiotropic and detrimental consequences.

Another field focused on plant secondary metabolism is ecological and evolutionary studies where secondary metabolite diversity is believed to allow selection to quickly favor plants with new defenses in the presence of insects or other pests that have evolved the ability to overcome existing defenses [8]. New defensive compounds can be synthesized by structurally modifying a toxic compound to evade the pest's counter-defense while either maintaining or changing the compounds deterrent activity. As such, studying the sources of intraspecific variation in secondary metabolites and the biological consequence of this diversity is an important focus of ecological and evolutionary studies [9–17]. This has shown that the intraspecific variation is frequently under positive or balancing selection where a secondary metabolite has positive and negative consequences on fitness due to effects that range across multiple trophic levels [18–23]. For instance, genetic variation in glucosinolate metabolism alters neighboring plant species fitness and secondary parasitoid populations [24, 25].

Frequently the crop breeding and ecology/evolution fields are studying similar metabolites with comparative variation yet there exists minimal cross-communication. With future crop breeding studies increasingly focusing on the manipulation of distinct chemical classes with a specifically targeted plant phenotype, i.e. secondary metabolite mediated resistance to lepidopteran herbivory (Table 1), it seems imperative that the two fields begin to coalesce to a degree. This argument comes about from these targeted secondary metabolite changes frequently being associated with unintended consequences that could have been foretold by the incorporation of ecological studies as important guides to crop breeders. Frequently, ecological studies show that attempts to modify one trait will only succeed if the potential ramifications on all biotic levels that may influence crop productivity are also understood. As such, while this chapter will show that there is ample genetic variation in plant secondary metabolites to allow for specific crop modifications that will have unintended secondary consequences. However, these secondary consequences may have already been identified in ecological/evolutionary studies using related wild species. As such, before making any modification the two disparate literature resources should be thoroughly investigated. Throughout this chapter, I will attempt to utilize citations from both literature sources to illustrate this concept. To fully detail this in all crops and all secondary metabolite pathways would require at the very least a book, as such, I will focus on specific model pathways with a strong link to underlying mechanism.

Intraspecific Variation in Plant Secondary Metabolism

Plant breeding studies frequently investigate or use interspecific variation in plant secondary metabolites or transgenic manipulation of secondary metabolites to improve a crop. Given the diverse activities of secondary metabolites, this may be done to improve either agronomic or nutritional traits. Transgenic manipulation is typically done on the assumption that there is more variation available to the researcher when introducing germplasm from another species. However, secondary metabolites frequently have extensive intraspecific variation in both structure and content that is easily accessible. Using this intraspecific variation requires a deep analysis of all available germplasm within a given crop or species to identify the desired secondary metabolite traits. Almost every crop tested to date has significant variation in secondary metabolites ranging from carrot to potato to apple (e.g. [8, 9, 13]). Because an adequate description of the extensive work on secondary metabolism variation in crops would require an entire book, we will focus on two model pathways, glucosinolates and monoterpenes. This sections focus on these two pathways will allow a focus on how genomics information may be used to rapidly identify candidate genes for controlling secondary metabolite variation in a crop.

Monoterpenoid variation in a large number of species: A broad class of flavor related secondary metabolites that show great potential for interspecific variation and breeding applications are the monoterpenes. These are small isoprenoids that are frequently associated with changes in taste and smell of fruit and vegetable crops [15]. An excellent example of how interspecific variation in monoterpenes has been applied to breeding is the presence of numerous basil (*Basil basilicum*) varieties whereby the presence or absence of specific monoterpenes fundamentally changes the flavor of the basil variety[16–18]. Citrus is another prime example of how interspecific monoterpene variation can be modified to alter flavor quality in that the presence of limonene is a predominant component of lemon flavor [19, 20]. In addition, there are high levels of interspecific monoterpene diversity in crucifers and monocots mainly in non-consumed tissues [21, 22]. Finally, interspecific variation exists in tomato monoterpenes that are involved in determining tomato flavor, linalool and nerolidol [23-26]. However, the relationship between this monoterpene variation and differences in flavor qualities between domesticated tomatoes varieties remains to be tested. Given this, there is likely interspecific monoterpene variation in most fruit and vegetable crop and the specific use of this to breed for altered fruit and vegetable flavor quality may be a successful avenue for integrating secondary metabolites into breeding programs.

Glucosinolate structural variation in a large number of species: Glucosinolates are amino acid derived thioglucosides specific to the order Capparales including all cruciferous crops [27]. Crucifers utilize a three part biosynthetic pathway to produce glucosinolates from various amino acids [28, 29]. This involves carbon chain elongation of a protein amino acid, entrance of the amino acid into the core pathway and finally side chain modification [30–36]. The final glucosinolates form a bipartite defense system such that when the cell is disrupted, a myrosinase enzyme cleaves the sugar from the glucosinolate, and a series of toxic compounds are released that alter insect herbivory [37–41], plant-plant interactions [42], nematode survival [43], fungal resistance [44, 45], human cancer susceptibility [46] and human ulcer susceptibility [47]. As such, the glucosinolates are an important target for crucifer breeders due to their agronomic and nutritional value.

Glucosinolates in most cruciferous crops and model plants show high levels of interspecific variation in both structure and content due to numerous quantitative trait loci (QTLs) [2, 48–54]. In contrast to phenylpropanoids and isoprenoids that each have tens of thousands of structures, glucosinolate structural variation does not appear open ended. Instead, species within the order Capparales appear to resample the same set of ~150 glucosinolate structures suggesting that interspecific variation is more useful in crop improvement than intraspecific variation [55–57].

Within most crucifer crops four major loci, *GSL-AOP*, *-ELONG*, *-OH* and *-OX*, control the methionine derived glucosinolate structural variation and the nutritional and agronomic potential of a crop [50–52, 58, 59]. These four loci function epistatically with each other to generate a modular genetic system capable of producing 14 different structural profiles. One profile that produces progoitrin(2-hydroxy-but-3-enyl glucosinolate) is prone to causing goiter in humans where as a different profile producing 4-methylsulfinyl glucosinolate is beneficial in that it can increase anti-cancer defenses within humans [46]. A third profile will lend the crop a peppery flavor similar to arugula/ruccola and provide altered insect herbivory. Shifts between any of these profiles only require modification of the four structural loci allowing a breeder to quickly shift any crop between these profiles depending upon the desired final trait [2].

Interspecific variation and integrating genomics into breeding programs: Recent attempts to integrate genomics methodologies into breeding programs focus on increasing the speed and efficiency in generating new crop germplasm. Analysis of interspecific variation in the Arabidopsis glucosinolate system has suggested some conceptual approaches that may help the integration of genomics into breeding programs focused on secondary metabolites. These approaches facilitate the identification of candidate genes and linked markers controlling the secondary metabolite variation that will be available for breeding. Recent studies cloning Arabidopsis glucosinolate QTLs as well as investigating variation in the aliphatic glucosinolate metabolism have shown that there is a bias towards these secondary metabolite QTLs being controlled by differential gene expression [34, 35, 38, 40, 53, 60, 61]. Further, the variation in glucosinolate content is directly related to the variation in gene expression of the biosynthetic genes [53]. For these secondary metabolite QTLs, this differential gene expression is easily detectable in a comparison of the parents possibly due to secondary metabolite diversity being under different selective pressures in comparison to the rest of metabolism [62–65]. As such, it may be possible to use transcriptomics comparing two parental genotypes to rapidly identify candidate genes controlling secondary metabolite variation. This is supported by extensive variation in gene expression amongst terpene cyclases within *Arabidopsis thaliana* [64, 65].

The above approach is primarily focused on the analysis of individual defined mapping populations. Interest in using direct surveys of available germplasm to associate genetic polymorphisms with phenotypic variation using linkage disequilibria has been growing [66–69]. It may be possible to extend the above gene expression variation approach to a genomic survey of genomic diversity in a crop. An analysis of genomic diversity in gene expression within different Arabidopsis germplasm showed that the secondary metabolite pathways were under selection for increased genetic diversity [62]. As such, a genomic survey of gene expression diversity within a crop could quickly yield a list of candidate genes controlling secondary metabolite variation. Thus, a way to introduce genomics into any breeding project focused on secondary metabolite modification may be to conduct a simple transcriptomics survey of the available germplasm and compare this to the measured secondary metabolite diversity. This could quickly yield a list of genes and markers that could be used to improve the desired secondary metabolite related trait. This approach however would not identify genetic variation that is related to differential enzymatic activity or post-transcriptional regulation and the inclusion of these sources of variation would require other technologies such as proteomics which is currently being utilized to understand changes underlying flavor variation in Basil secondary metabolism [70].

Natural Variation/Breeding and Plant/Biotic Interactions

The identification of causal linkages between secondary metabolites and agronomically important traits such as plant resistance to insect or fungal pathogens is generating a desire to modify these for crop improvement. A hope is that these secondary metabolites traits may provide more stable resistance phenotypes or lead to decreased pesticide utilization. However, ecological and crop data shows that plant secondary metabolites mediate complex interactions with numerous biotic pests with differing fitness consequences. For instance in the manipulation of carrots for altered root knot nematode resistance has pleiotropic consequences upon flavor/bitterness [13, 71–73]. This negative correlation between flavor and biotic resistance is present in a wide range of crop species but is typically limited to a correlational relationship. To better delve into the mechanistic basis of the potential relationship between secondary metabolites and biotic resistance, this section will focus the glucosinolate model system which has extensive mechanistic information in the ecological literature on how genetic variation in secondary metabolites may help to guide breeding uses of plant secondary metabolites in crop improvement.

Analysis of ecological costs and benefits in glucosinolates: Glucosinolates are commonly described as providing resistance against insect herbivores [38, 74, 75]. However, this is typically limited to generalist insects that do not specialize upon glucosinolate containing plants [37, 38]. In contrast, there are groups of specialist insects that have evolved specific defenses that ameliorate or detoxify the glucosinolates [76, 77]. These specialist insects actually use the glucosinolates to identify the crucifers and this establishes a situation where generalists and specialist insects may generate opposing pressures upon glucosinolates [78–80]. In this case, the generalists should put a pressure favoring higher glucosinolate plants while the specialists would lead to lower glucosinolate plants being favored. Recent field work with *Brassica nigra* bred for different glucosinolate levels showed that this was in fact the case. In the presence of generalist herbivores the high glucosinolate lines were favored while in the presence of specialist herbivores, the moderately low glucosinolate lines were preferred [37]. Thus, any modification of glucosinolate levels either high or low may not alter total herbivory resistance but instead modify the insect herbivore populations affecting the crop itself.

The *Brassica nigra* lines bred for different glucosinolate levels were further examined for their ability to interact with other plants. This showed that the high glucosinolate lines led to other plant species having difficulties with germination and seedling establishment apparently from decreased mycorhizal success in the high glucosinolate plots [42]. In contrast, the low glucosinolate lines did not have the same effect. This ecological study suggests that any breeding program focused on modifying glucosinolate content in a crucifer needs to factor in potential crop rotations within which the crucifer may exist as breeding for increased glucosinolate content in the crucifer may impact the potential productivity of ensuing crop cycles. As such, the use of genetic variation in secondary metabolites by both breeding and ecological projects can cross-feed each field and improve the goals of breeding projects and how secondary metabolites may fit into these goals [81].

Analysis of costs and benefits of secondary metabolites: In addition to secondary metabolites having an ecological cost related to the presence of specialist herbivores and pathogens there is also the allocation cost related to the production of the secondary metabolite [82, 83]. For instance, glucosinolate levels are positively associated with fitness in the presence of generalist herbivory but show direct allocation costs, in the form of negative correlations with fitness in the absence of herbivory [84, 85]. In *Nicotiana attenuata*, the ability to induce secondary metabolites is beneficial in the presence of herbivory but the costs of this induction are shown by decreased fitness of plants with constitutively high metabolite levels when there is no herbivory [86, 87]. This cost/benefit relationship for secondary metabolites would suggest that breeding for altered secondary metabolite accumulation may lead to unexpected yield consequences.

One solution to this cost of resistance has likely been a pressure for the plants to make the compound inducible such that the plant only incurs the cost of producing a given metabolite when it is needed [82, 83, 88]. However, the capacity to induce a given secondary metabolite also shows interspecific genetic variation in numerous plant species [89–92]. Thus, when breeding for secondary metabolite concentrations, it will be critical to understand if the metabolite is inducible and if so what is the signal? This information will then have to be integrated into the final breeding design depending upon if high or low metabolite concentrations are required.

Presence of resistance variation in the pests: Some interest has been expressed in the horizontal transfer of novel secondary metabolites into crops in the hopes that these novel secondary metabolites may provide stable resistance against pests or pathogens. This horizontal transfer would typically involve transgenic or intraspecific breeding projects to move novel metabolic capacity into the desired crop. While this approach typically focuses on modulating resistance against specialist pathogens or pests that may have not encountered the novel secondary metabolite, it may run into difficulties when applied to generalist pests of pathogens. These generalists usually have very broad host-ranges that were believed facilitated by general defense mechanisms effective against numerous toxins. However, work on broad-host range pathogen resistance to specific toxins is showing that these generalist pathogens can contain specific resistance mechanisms targeted towards individual compounds [93–96]. For instance, the broad host range fungus *Botrytis cinerea* has enzymatic defenses against the tomato metabolite tomatine, the Arabidopsis metabolite camalexin and the grape metabolite resveratrol [93, 94, 97–99]. The broad host range pathogens counter any potential costs of having these specific resistance mechanisms by making these mechanisms genetically variable such that only a limited percent of the population has any given resistance [94]. This allows the species to have the resistance mechanism but not require every individual to maintain the resistance. This capacity of generalists to have specialist defenses against phylogenetically limited secondary metabolites is even extended to a generalist herbivore, locust, containing a specialist defense against the glucosinolate secondary metabolites [100]. Thus, while the introduction of novel metabolites into crops may provide resistance against specialist pathogens that have never encountered the metabolite, the novel metabolites are unlikely to provide stable resistance against generalist pests and pathogens.

Natural Variation/Breeding and Human Diet

Given the above potential difficulties inherent in using secondary metabolites to breed for altered agronomic traits the use of secondary metabolites for altering the nutritional quality of a crop may be more direct. Numerous secondary metabolites are already sold in a purified form using human health claims. In the health food section of any grocery store one can find purified secondary metabolites for sale such as isoflavones, flavonols, ginkolide, etc. As such, there may be a market for modulation of these same compounds within fruits and vegetables to allow the same compounds to be consumed as a food rather than a pill. This would especially be useful when a purported health beneficial metabolite is also a positive component of flavor. However knowing the identity of the beneficial metabolite is critically important in the proper design of any breeding project.

An example of how important it is to know the specific metabolite arises from the glucosinolate pathway. In glucosinolates, it is a specific isothiocyanate breakdown product of the 4-methylsulfinylbutyl glucosinolate that provides an anti-cancer and anti-ulcer benefit [46, 47] (Fig. 1). Other breakdown products of the 4-methylsulfinylbutyl glucosinolate do not provide this benefit (Fig. 1). Originally, it was thought that breeding for the 4-methylsulfinylbutyl glucosinolate would be sufficient to increase the anti-cancer activity in cruciferous vegetables such as broccoli. However, recent work identified genetic variation in crucifers that controls how the glucosinolate is broken down [38, 40, 101]. This was then expanded upon to show that it was more efficient to increase anti-cancer activity by breeding for the appropriate glucosinolate breakdown structure rather than simply increasing the level of the metabolite [102]. Thus, the efficiency of any secondary metabolite-breeding program meant to improve the nutritional quality of a crop requires knowing the important metabolite and all potential sources of genetic variation controlling the given metabolite. Finally, this glucosinolate example also displays the potential complexity of how any targeted breeding designed against a specific metabolite may ramify through multiple phenotypic levels (Fig. 1).

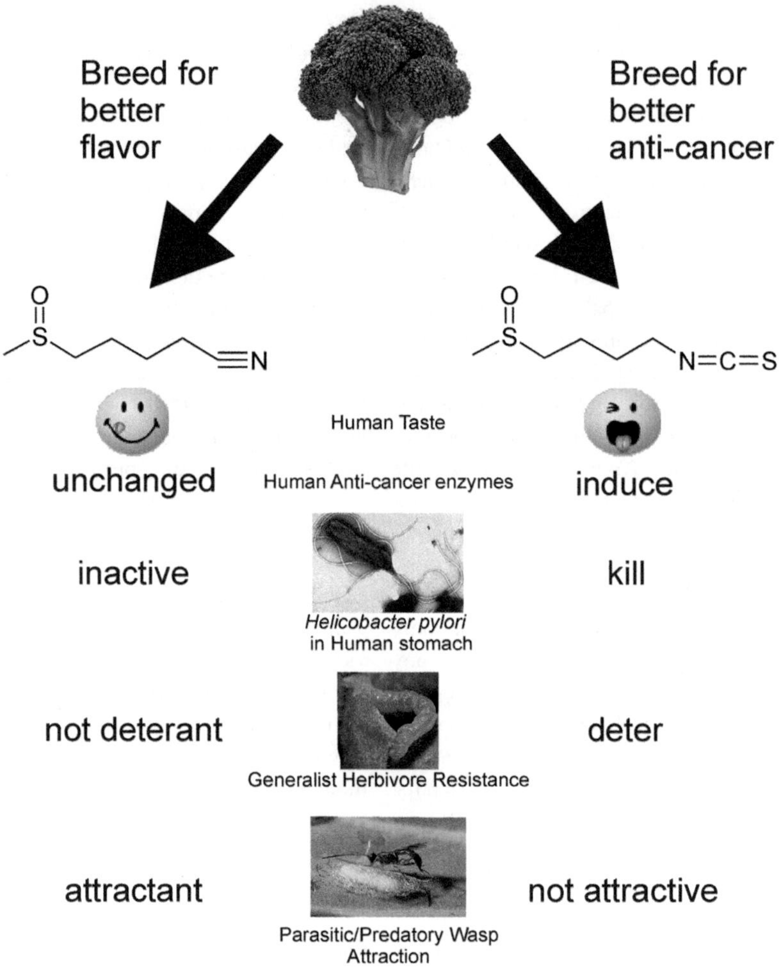

Fig. 1 Pleiotropic consequences inherent in Broccoli breeding. A theoretical shift of broccoli towards either the nitrile (*left*) or isothiocyanate (*right*) breakdown structure of 4-methylsulfinylbutyl glucosinolate is shown. The consequences of this breeding decision on human desirability of the crop (taste, cancer and *H. pylori*) and on the crops biotic resistance both direct (generalist herbivore) and indirect (wasp attraction) are shown

Conclusion

Natural genetic variation for secondary metabolites within a given species can be an important tool for the breeder looking to alter a plants metabolic signature. This can be targeted towards improving the agronomic or quality of a crop. Recent improvements in speeding chemical analysis of plants and genomic analysis of candidate genes can allow for rapid utilization of this germplasm for the targeted manipulation of secondary metabolites in crops. However, it is advised that any program utilize as much information from as many sources as possible to avoid any unexpected complications of these breeding programs upon the crop itself.

Glossary

Intraspecific variation – Genetic or phenotypic variation between members of a species

Interspecific variation – Genetic or phenotypic variation between species

References

1. Wink M (1988) Plant Breeding: Importance of plant secondary metabolites for protection against pathogens and herbivores. *Theor Appl Genet* 75: 225–233.
2. Kliebenstein DJ, Kroymann J, Brown P, Figuth A, Pedersen D, et al. (2001) Genetic control of natural variation in *Arabidopsis thaliana* glucosinolate accumulation. *Plant Physiol* 126: 811–825.
3. Du YJ, Vanloon JJA, Renwick JAA (1995) Contact chemoreception of oviposition-stimulating glucosinolates and an oviposition-deterrent cardenolide in two subspecies of *Pieris napi*. *Physiol Entomol* 20: 164–174.
4. Kobayashi M, Gasking P, Spray C, Suzuki Y, Phinney B, et al. (1993) Metabolism and biological activity of gibberellin A-4 in vegetative shoots of *Zea mays*, Oryza sativa, and *Arabidopsis thaliana*. *Plant Phys* 102: 379–386.
5. Facchini P (2001) Alkaloid biosynthesis in plants: Biochemistry, cell biology, molecular regulation and metabolic engineering applications. *Annu Rev Plant Physiol Plant Mol Biol.* 52: 29–66.
6. Morton LW, Abu-Amsha Caccetta R, Puddey IB, Croft KD (2000) Chemistry and biological effects of dietary phenolic compounds: Relevance to cardiovascular disease. *Clin Exp Pharmacol Physiol* 27: 152–159.
7. Salah N, Miller NJ, Paganga G, Tijburg L, Bolwell GP, et al. (1995) Polyphenolic flavanols as scavengers of aqueous phase radicals and as chain-breaking antioxidants. *Arch Biochem Biophys* 322: 339–346.
8. Friedman M (2006) Potato glycoalkaloids and metabolites: Roles in the plant and in the diet. *J Agric Food Chem* 54: 8655–8681.
9. Sanoner P, Guyot S, Marnet N, Molle D, Drilleau JF (1999) Polyphenol profiles of French cider apple varieties (Malus domestica sp.). *J Agric Food Chem* 47: 4847–4853.
10. Friedman M, McDonald GM (1997) Potato glycoalkaloids: Chemistry, analysis, safety, and plant physiology. *Crit Rev Plant Sci* 16: 55–132.
11. Flanders KL, Hawkes JG, Radcliffe EB, Lauer FI (1992) Insect resistance in potatoes – Sources, evolutionary relationships, morphological and chemical defenses, and ecogeographical associations. *Euphytica* 61: 83–111.
12. Cipollini ML, Levey DJ (1997) Why are some fruits toxic? Glycoalkaloids in solanum and fruit choice by vertebrates. *Ecology* 78: 782–798.
13. Kreutzmann S, Christensen LP, Edelenbos M (2008) Investigation of bitterness in carrots (Daucus carota L.) based on quantitative chemical and sensory analyses. *Lwt-Food Sci Technol* 41: 193–205.
14. Friedman M, Henika PR, Mackey BE (1996) Feeding of potato, tomato and eggplant alkaloids affects food consumption and body and liver weights in mice. *J Nutr* 126: 989–999.
15. Mahmoud SS, Croteau RB (2002) Strategies for transgenic manipulation of monoterpene biosynthesis in plants. *Trends Plant Sci* 7: 366–373.
16. Klimankova E, Holadova K, Hajslova J, Cajka T, Poustka J, et al. (2008) Aroma profiles of five basil (Ocimum basilicum L.) cultivars grown under conventional and organic conditions. *Food Chem* 107: 464–472.
17. Iijima Y, Davidovich-Rikanati R, Fridman E, Gang DR, Bar E, et al. (2004) The biochemical and molecular basis for the divergent patterns in the biosynthesis of terpenes and phenylpropenes in the peltate glands of three cultivars of basil. *Plant Physiol* 136: 3724–3736.
18. Iijima Y, Gang DR, Fridman E, Lewinsohn E, Pichersky E (2004) Characterization of geraniol synthase from the peltate glands of sweet basil. *Plant Physiol* 134: 370–379.
19. Allegrone G, Belliardo F, Cabella P (2006) Comparison of volatile concentrations in hand-squeezed juices of four different lemon varieties. *J Agric Food Chem* 54: 1844–1848.
20. Choi HS, Sawamura M (2001) Volatile flavor components of ripe and overripe Ki-mikans

3

(Citrus flaviculpus Hort. ex Tanaka) in comparison with Hyuganatsu (Citrus tamurana Hort. ex Tanaka). *Biosci Biotechnol Biochem* 65: 48–55.

21. Schnee C, Kollner TG, Held M, Turlings TCJ, Gershenzon J, et al. (2006) xThe products of a single maize sesquiterpene synthase form a volatile defense signal that attracts natural enemies of maize herbivores. *Proc Natl Acad Sci USA* 103: 1129–1134.
22. Kollner TG, Schnee C, Gershenzon J, Degenhardt J (2004) The variability of sesquiterpenes cultivars is controlled by allelic emitted from two Zea mays variation of two terpene synthase genes encoding stereoselective multiple product enzymes. *Plant Cell* 16: 1115–1131.
23. Baldwin EA, Goodner K, Plotto A, Pritchett K, Einstein M (2004) Effect of volatiles and their concentration on perception of tomato descriptors. *J Food Sci* 69: S310–S318.
24. Lewinsohn E, Schalechet F, Wilkinson J, Matsui K, Tadmor Y, et al. (2001) Enhanced levels of the aroma and flavor compound S-linalool by metabolic engineering of the terpenoid pathway in tomato fruits. *Plant Physiol* 127: 1256–1265.
25. Moco S, Capanoglu E, Tikunov Y, Bino RJ, Boyacioglu D, et al. (2007) Tissue specialization at the metabolite level is perceived during the development of tomato fruit. *J Exp Bot* 58: 4131–4146.
26. Tikunov Y, Lommen A, de Vos CHR, Verhoeven HA, Bino RJ, et al. (2005) A novel approach for nontargeted data analysis for metabolomics. Large-scale profiling of tomato fruit volatiles. *Plant Physiol* 139: 1125–1137.
27. Bones AM, Rossiter JT (1996) The myrosinase-glucosinolate system, its organisation and biochemistry. *Physiologia Plantarum* 97: 194–208.
28. Halkier BA, Du L (1997) The biosynthesis of glucosinolates. *Trends Plant Sci* 2: 425–431.
29. Hogge LR, Reed DW, Underhill EW, Haughn GW (1988) HPLC separation of glucosinolates from leaves and seeds of *Arabidopsis thaliana* and their identification using thermospary liquid chromatography-mass spectometry. *J Chrom Sci* 26: 551–556.
30. Graser G, Schneider B, Oldham NJ, Gershenzon J (2000) The methionine chain elongation pathway in the biosynthesis of glucosinolates in *Eruca sativa* (Brassicaceae). *Arch Biochem Biophys* 378: 411–419.
31. de Quiros HC, Magrath R, McCallum D, Kroymann J, Schnabelrauch D, et al. (2000) α-Keto acid elongation and glucosinolate biosynthesis in Arabidopsis thaliana. *Theor Appl Genet* 101: 429–437.
32. Wittstock U, Halkier BA (2000) Cytochrome P450 CYP79A2 from *Arabidopsis thaliana* L. catalyzes the conversion of L-phenylalanine to phenylacetaldoxime in the biosynthesis of the benzylglucosinolate. *J Biol Chem* 275: 14659–14666.
33. Kahn RA, Fahrendorf T, Halkier BA, Moller BL (1999) Substrate specificity of the cytochrome P450 enzymes CYP79A1 and CYP71E1 involved in the biosynthesis of the cyanogenic glucoside dhurrin in *Sorghum bicolor* (L.) Moench. *Arch Biochem Biophys* 363: 9–18.
34. Kliebenstein D, Lambrix V, Reichelt M, Gershenzon J, Mitchell-Olds T (2001) Gene duplication and the diversification of secondary metabolism: side chain modification of glucosinolates in *Arabidopsis thaliana*. *Plant Cell* 13: 681–693.
35. Hansen BG, Kliebenstein DJ, Halkier BA (2007) Identification of a flavin-monooxygenase as the S-oxygenating enzyme in aliphatic glucosinolate biosynthesis in Arabidopsis. *Plant J* 50: 902–910.
36. Textor S, de Kraker JW, Hause BG, Tokuhisa JG (2007) MAM3 catalyzes the formation of all aliphatic glucosinolate chain lengths in Arabidopsis. *Plant Physiol* 144: 60–71.
37. Lankau RA (2007) Specialist and generalist herbivores exert opposing selection on a chemical defense. *New Phytologist* 175: 176–184.
38. Zhang Z-Y, Ober JA, Kliebenstein DJ (2006) The gene controlling the quantitative trait locus *EPITHIOSPECIFIER MODIFIER1* alters glucosinolate hydrolysis and insect resistance in Arabidopsis. *Plant Cell* 18: 1524–1536.
39. Li Q, Eigenbrode SD, Stringham GR, Thiagarajah MR (2000) Feeding and growth of Plutella xylostella and *Spodoptera eridania* on *Brassica juncea* with varying glucosinolate concentrations and myrosinase activities. *J Chem Ecol* 26: 2401–2419.
40. Lambrix V, Reichelt M, Mitchell-Olds T, Kliebenstein D, Gershenzon J (2001) The Arabidopsis epithiospecifier protein promotes the hydrolysis of glucosinolates to nitriles and influences *Trichoplusia ni* herbivory. *Plant Cell* 13: 2793–2807.
41. Barth C, Jander G (2006) Arabidopsis myrosinases *TGG1* and *TGG2* have redundant function in glucosinolate breakdown and insect defense. *Plant J* 46: 549–562.

42. Lankau RA, Strauss SY (2007) Mutual feedbacks maintain both genetic and species diversity in a plant community. *Science* 317: 1561–1563.
43. Donkin SG, Eiteman MA, Williams PL (1995) Toxicity of glucosinolates and their enzymatic decomposition products to *Caenorhabditis elegans*. *J Nematol* 27: 258–262.
44. Mithen RF, Lewis BG, Fenwick GR (1986) In vitro activity of glucosinolates and their products against *Leptosphaeria maculans*. *Trans Br Mycol Soc* 87: 433–440.
45. Tierens K-J, Thomma B, Brower M, Schmidt J, Kistner K, et al. (2001) Study of the role of antimicrobial glucosinolate-derived isothiocyanates in resistance of Arabidopsis to microbial pathogens. *Plant Physiol* 125: 1688–1699.
46. Fahey JW, Zhang YS, Talalay P (1997) Broccoli sprouts: An exceptionally rich source of inducers of enzymes that protect against chemical carcinogens. *Proc Natl Acad Sci USA* 94: 10367–10372.
47. Fahey JW, Haristoy X, Dolan PM, Kensler TW, Scholtus I, et al. (2002) Sulforaphane inhibits extracellular, intracellular, and antibiotic-resistant strains of *Helicobacter pylori* and prevents benzo[a]pyrene-induced stomach tumors. *Proc Natl Acad Sci USA* 99: 7610–7615.
48. Kliebenstein DJ, Gershenzon J, Mitchell-Olds T (2001) Comparative quantitative trait loci mapping of aliphatic, indolic and benzylic glucosinolate production in *Arabidopsis thaliana* leaves and seeds. *Genetics* 159: 359–370.
49. Parkin I, Magrath R, Keith D, Sharpe A, Mithen R, et al. (1994) Genetis of aliphatic glucosinolates. II. Hydroxylation of alkenyl glucosinolates in Brassica napus. *Heredity* 72: 594–598.
50. Mithen R, Clarke J, Lister C, Dean C (1995) Genetics of aliphatic glucosinolates.III. Side-chain structure of aliphatic glucosinolates in *Arabidopsis thaliana*. *Heredity* 74: 210–215.
51. Magrath R, Bano F, Morgner M, Parkin I, Sharpe A, et al. (1994) Genetics of aliphatic glucosinolates. I. Side chain elongation in *Brassica napus and Arabidopsis thaliana*. *Heredity* 72: 290–299.
52. Giamoustaris A, Mithen R (1996) Genetics of aliphatic glucosinolates.IV. Side-chain modification in *Brassica oleracea*. *Theor Appl Genet* 93: 1006–1010.
53. Wentzell AM, Rowe HC, Hansen BG, Ticconi C, Halkier BA, et al. (2007) Linking metabolic QTL with network and *cis*-*e*QTL controlling biosynthetic pathways. *PLOS Genetics* 3: e162.
54. Toroser D, Thormann C, Osborn T, Mithen R (1995) RFLP mapping of quantitative trait loci controlling seed aliphatic glucosinolate content in oilseed rape (*Brassica napus* L.). *Theor Appl Genet* 91: 802–808.
55. Rodman J (1980) Population variation and hybridization in sea-rockets (*Cakile, Cruciferae*): Seed glucosinolate characters. *Am J Bot* 67: 1145–1159.
56. Rodman JE, Kruckeberg AR, Alshehbaz IA (1981) Chemotaxonomic diversity and complexity in seed glucosinolates of *Caulanthus* and *Streptanthus* (Cruciferae). *Syst Bot* 6: 197–222.
57. Daxenbichler ME, Spencer GF, Carlson DG, Rose GB, Brinkler AM, et al. (1991) Glucosinolate composition of seeds from 297 species of wild plants. *Phytochemistry* 30: 2623–2638.
58. Mithen R, Toroser D (1995) Biochemical genetics of aliphatic glucosinolates in Brassica and Arabidopsis. *Soc Exp Biol Semin Ser* 56: 261–275.
59. Parkin I, Magrath R, Keith D, Sharpe A, Mithen R, et al. (1994) Genetics of aliphatic glucosinolates. II. Hydroxylation of alkenyl glucosinolates in *Brassica napus*. *Heredity* 72: 594–598.
60. Hansen BG, Halkier BA, Kliebenstein DJ (2008) Identifying the molecular basis of QTLs: eQTLs add a new dimension. *Trends Plant Sci* 13: 72–77.
61. Sønderby IE, Hansen BG, Bjarnholt N, Ticconi C, Halkier BA, et al. (2007) A systems biology approach identifies a R2R3 MYB gene subfamily with distinct and overlapping functions in regulation of aliphatic glucosinolates. *PLos ONE* 2: e1322.
62. Kliebenstein DJ (2008) A role for gene duplication and natural variation of gene expression in the evolution of metabolism. *PLos ONE* 3: e1838.
63. West MAL, Kim K, Kliebenstein DJ, van Leeuwen H, Michelmore RW, et al. (2007) Global eQTL mapping reveals the complex genetic architecture of transcript level variation in Arabidopsis. *Genetics* 175: 1441–1450.
64. Van Leeuwen H, Kliebenstein DJ, West MAL, kim KD, van Poecke R, et al. (2007) Natural variation among *Arabidopsis thaliana* accessions for transcriptome response to exogenous salicylic acid. *Plant Cell* 19: 2099–2110.
65. Kliebenstein DJ, West MAL, Van Leeuwen H, Kyunga K, Doerge RW, et al. (2006) Genomic survey of gene expression diversity in *Arabidopsis thaliana*. *Genetics* 172: 1179–1189.

66. Zhao KY, Aranzana MJ, Kim S, Lister C, Shindo C, et al. (2007) An Arabidopsis example of association mapping in structured samples. *Plos Genetics* 3: e4.
67. Nordborg M, Borevitz JO, Bergelson J, Berry CC, Chory J, et al. (2002) The extent of linkage disequilibrium in Arabidopsis thaliana. *Nat Genet* 30: 190–193.
68. Belo A, Zheng PZ, Luck S, Shen B, Meyer DJ, et al. (2008) Whole genome scan detects an allelic variant of fad2 associated with increased oleic acid levels in maize. *Mol Genet Genomics* 279: 1–10.
69. Jung M, Ching A, Bhattramakki D, Dolan M, Tingey S, et al. (2004) Linkage disequilibrium and sequence diversity in a 500-kbp region around the adh1 locus in elite maize germplasm. *Theor Appl Genet* 109: 681–689.
70. Xie Z, Kapteyn J, Gang DR (2008) A systems biology investigation of the MEP/terpenoid and shikimate/phenylpropanoid pathways points to multiple levels of metabolic control in sweet basil glandular trichomes. *Plant J* 54: 349–361.
71. Vieira JV, Charchar JM, Aragao FAS, Boiteux LS (2003) Heritability and gain from selection for field resistance against multiple root-knot nematode species (Meloidogyne incognita race 1 and M-javanica) in carrot. *Euphytica* 130: 11–16.
72. Simon PW, Matthews WC, Roberts PA (2000) Evidence for simply inherited dominant resistance to Meloidogyne javanica in carrot. *Theor Appl Genet* 100: 735–742.
73. Wang M, Goldman IL (1996) Resistance to root knot nematode (*Meloidogyne hapla* Chitwood) in carrot is controlled by two recessive genes. *J Heredity* 87: 119–123.
74. Kliebenstein DJ, Pedersen D, Mitchell-Olds T (2002) Comparative analysis of insect resistance QTL and QTL controlling the myrosinase/glucosinolate system in *Arabidopsis thaliana*. *Genetics* 161: 325–332.
75. Halkier BA, Gershenzon J (2006) Biology and biochemistry of glucosinolates. *Annu Rev Plant Biol* 57: 303–333.
76. Wittstock U, Agerbirk N, Stauber EJ, Olsen CE, Hippler M, et al. (2004) Successful herbivore attack due to metabolic diversion of a plant chemical defense. *Proc Natl Acad Sci USA* 101: 4859–4864.
77. Ratzka A, Vogel H, Kliebenstein DJ, Mitchell-Olds T, Kroymann J (2002) Disarming the mustard oil bomb. *Proc Natl Acad Sci USA* 99: 11223–11228.
78. Renwick J, Chew F (1994) Oviposition behavior in Lepidoptera. *Annu Rev Entomol* 39: 377–400.
79. Chew FS, Renwick JAA (1994) Host plant choice in *Pieris* butterflies. In: Bell RTCaWJ, editor. Chemical Ecology of Insects II. New York: Chapman & Hall, pp. 214–238.
80. Renwick JAA, Haribal M, Gouinguene S, Stadler E (2006) Isothiocyanates stimulating oviposition by the diamondback moth, Plutella xylostella. *J Chem Ecol* 32: 755–766.
81. Lankau RA, Strauss SY (2008) Community complexity drives patterns of natural selection on a chemical defense of Brassica nigra. *Am Naturalist* 171: 150–161.
82. Karban R, Baldwin IT (1997) Induced Responses to Herbivory. Chicago, IL: University of Chicago Press.
83. Heil M, Baldwin IT (2002) Fitness costs of induced resistance: Emerging experimental support for a slippery concept. *Trends Plant Sci* 7: 61–67.
84. Mauricio R (1998) Costs of resistance to natural enemies in field populations of the annual plant Arabidopsis thaliana. *Am Naturalist* 151: 20–28.
85. Mauricio R, Rausher MD (1997) Experimental manipulation of putative selective agents provides evidence for the role of natural enemies in the evolution of plant defense. *Evolution* 51: 1435–1444.
86. Baldwin IT, Hamilton W (2000) Jasmonate-induced responses of Nicotiana sylvestris results in fitness costs due to impaired competitive ability for nitrogen. *J Chem Ecol* 26: 915–952.
87. Baldwin IT (1998) Jasmonate-induced responses are costly but benefit plants under attack in native populations. *Proc Natl Acad Sci USA* 95: 8113–8118.
88. Kant MR, Baldwin IT (2007) The ecogenetics and ecogenomics of plant-herbivore interactions: Rapid progress on a slippery road. *Curr Opin Genet Dev* 17: 519–524.
89. Glawe GA, Zavala JA, Kessler A, Van Dam NM, Baldwin IT (2003) Ecological costs and benefits correlated with trypsin protease inhibitor production in Nicotiana attenuata. *Ecology* 84: 79–90.
90. Kliebenstein DJ, Figuth A, Mitchell-Olds T (2002) Genetic architecture of plastic methyl jasmonate responses in *Arabidopsis thaliana*. *Genetics* 161: 1685–1696.
91. Strauss SY, Irwin RE, Lambrix VM (2004) Optimal defence theory and flower petal colour

predict variation in the secondary chemistry of wild radish. *J Ecol* 92: 132–141.
92. Byrne PF, McMullen MD, Wiseman BR, Snook ME, Musket TA, et al. (1998) Maize silk maysin concentration and corn earworm antibiosis: QTLs and genetic mechanisms. *Crop Sci* 38: 461–471.
93. Kliebenstein DJ, Rowe HC, Denby KJ (2005) Secondary metabolites influence Arabidopsis/Botrytis interactions: Variation in host production and pathogen sensitivity. *Plant J* 44: 25–36.
94. Quidde T, Osbourn AE, Tudzynski P (1998) Detoxification of alpha-tomatine by *Botrytis cinerea*. *Physiol Mol Plant Pathol* 52: 151–165.
95. Pedras MSC, Khan AQ (1997) Unprecedented detoxification of the phytoalexin camalexin by a root rot pathogen. *Bioorg Med Chem Lett* 7: 2255–2260.
96. Pedras MSC, Ahiahonu PWK (2002) Probing the phytopathogenic stem rot fungus with phytoalexins and analogs: unprecedented glucosylation of camalexin and 6-methoxycamalexin. *Bioorg Med Chem* 10: 3307–3312.
97. Adrian M, Rajaei H, Jeandet P, Veneau J, Bessis R (1998) Resveratrol oxidation in Botrytis cinerea conidia. *Phytopathology* 88: 472–476.
98. de Waard MA, Andrade AC, Hayashi K, Schoonbeek HJ, Stergiopoulos I, et al. (2006) Impact of fungal drug transporters on fungicide sensitivity, multidrug resistance and virulence. *Pest Manag Sci* 62: 195–207.
99. Schoonbeek H, Del Sorbo G, De Waard MA (2001) The ABC transporter BcatrB affects the sensitivity of Botrytis cinerea to the phytoalexin resveratrol and the fungicide fenpiclonil. *Mol Plant-Microbe Interact* 14: 562–571.
100. Falk KL, Gershenzon J (2007) The desert locust, Schistocerca gregaria, detoxifies the glucosinolates of Schouwia purpurea by desulfation. *J Chem Ecol* 33: 1542–1555.
101. Matusheski NV, Swarup R, Juvik JA, Mithen R, Bennett M, et al. (2006) Epithiospecifier protein from broccoli (*Brassica oleracea* L. *ssp italica*) inhibits formation of the anticancer agent sulforaphane. *J Agric Food Chem* 54: 2069–2076.
102. Mithen R, Faulkner K, Magrath R, Rose P, Williamson G, et al. (2003) Development of isothiocyanate-enriched broccoli, and its enhanced ability to induce phase 2 detoxification enzymes in mammalian cells. *Theor Appl Genet* 106: 727–734.I

Approaches to the Analysis of Plant-Derived Natural Products

4

Lionel Hill and Trevor L. Wang

Introduction

The term "plant-derived natural product" is extremely broad and the scope of this chapter is determined by the nature of lower-abundance secondary metabolites rather than storage proteins, starch, cell walls, and lipids. In some instances, however, similar techniques can used to measure both groups of compounds. The bioactivity of secondary metabolites underlines their importance in human nutrition, health, pharmacy and plant defence mechanisms, and is the basis for their commercial value. Consequently, these two features are the driving force behind the continuing development of techniques for their analysis. The most important recent advance has been the advent of metabolomics. The metabolome, by analogy to the genome, proteome and transcriptome, is the entire small-molecule complement of the plant. The study of the metabolome, metabolomics, cannot be achieved by any single method and is largely a consequence of recent improvements to technology permitting high throughput analyses and data-handling. Interest in understanding details of natural product biosynthetic pathways as well as a desire to measure the bioactive end-product means, however, that it is necessary to cover methods of sufficient sensitivity to detect low-abundance intermediates as well as those methods that can investigate metabolite structure. Hence this chapter represents an introduction and overview of the fundamentals that underlie the wide range of methods used in the quantification of plant-derived natural products and a brief introduction to metabolomics. It is hoped that a strong understanding of the fundamentals will allow the reader to judge from the plethora of manufacturers' brochures, primary literature, and on-line resources, which technologies and approaches best fit their situation. The process of analysis will be followed through its stages, from extraction of material to detection of analytes. Methods have been chosen not only with the chemistry of the analyte in mind, but also with a firm idea of the biological question that is to be answered. There will be an unashamed bias towards chromatography and mass spectroscopy since plants are complex systems containing many interesting chemicals, often in low amounts. Chromatography can simplify the process of addressing a complex mixture, and mass spectroscopy yields rich information from low-abundance analytes. For a brief list of methods for the main secondary metabolites, consult Table 1.

L. Hill and T.L. Wang (✉)
Department of Metabolic Biology,
John Innes Centre, Norwich Research Park,
Colney, Norwich NR4 7UH, UK
e-mails: trevor.wang@bbsrc.ac.uk;
lionel.hill@bbsrc.ac.uk

A.E. Osbourn and V. Lanzotti (eds.), *Plant-derived Natural Products*,
DOI 10.1007/978-0-387-85498-4_4, © Springer Science+Business Media, LLC 2009

Table 1 Recent and/or relevant methods in analysis of plant products. Articles referenced in this table have been chosen to illustrate the breadth of methodology currently in use, the breadth of analytes, and the reasons for analysis. Note that papers range from identifications of novel products to routine analyses of useful products. Many products are pharmaceutical (e.g. artemisinin), or of pharmaceutical relevance (atractyloside, a poisonous component of some traditional remedies). Others are pathway intermediates, hormones, or involved in defence against pathogens. Note also that many methods can be applied with success to the same analyte (see especially the phenolic compounds)

Analyte		Method	Reference
Phenolics	Isoflavone glycosides, identification of glycosylation	LC-fractionation followed by NMR	Shimoda et al. (2008)
	Acylated flavonol glycosides, identification of novel	LC-SPE-NMR (online)	Lee et al. (2008)
	Scopoletin and Scopolin	Micellar capillary electrophoresis	Wang et al. (2007)
		HPLC	Kai et al. (2008)
	Isoflavones and glycosides	Ion chromatography-electrochemical detection	Otieno and Shah (2007)
	Isoflavones and phenolic acids	UPLC-diode array detector (UV)	Klejdus et al. (2008)
	Flavonol profiling	HPLC-MS	Yonekura-Sakakibara et al. (2008)
	Flavanone enantiomers and glucoside diastereoisomers	CE and HPLC, both with diode array detector (UV)	Pan et al. (2008)
	Flavonoid pigments, anthocyanins anthocyanins	HPLC-UV	Edwards et al. (2008)
		HPLC-diode array detector (UV)	Lätti et al. (2008)
Polyprenylated xanthones		HPLC-ESI-Qtof-MS	Zhou et al. (2008)
Shikimate		Ion chromatography-MS/MS	Doehlemann et al. (2008)
Tocopherols		GC-MS	Kobayashi and DellaPenna (2008)
Glutathione and related compounds		LC-ESI-MS	Rellán-álvarez et al. (2006)
Diterpenes	Identification of novel	Normal phase LC-fractionation followed by NMR	Chien et al. (2008)
	Atractyloside	GC-MS	Laurens et al. (2001)

Triterpenes	Avenacins	LC-fluorescence detection	Papadopoulou et al. (1999)
Sesquiterpenes, monoterpenes		GC-MS	Davidovich-Rikanati et al. (2008)
	Monoterpenes (enantiomers separated)	Chiral GC-MS	Larkov et al. (2008)
Soya saponins		LC-evaporative light scattering	Rupasinghe et al. (2003)
Ginsenosides		Reverse phase HPLC-Pulsed amperometry	Kwon et al. (2008)
Alkaloids	Identification of novel	LC-fractionation followed by NMR	Alali et al. (2008)
Isoprenoids and precursors (dolichols)		NMR, LC-ESI-MS and ^{13}C labelling	Skorupinska-Tudek et al. (2008)
Polyunsaturated fatty acids		LC-APCI-MS	Řezanka et al. (2008)
Triacyl glycerols		2d-LC-UV, evaporative light scattering, APCI-MS	van der Klift et al. (2008)
		HPLC-charged aerosol detection (CAD)	Lísa et al. (2007)
		LC-APCI	Lísa and Holčapek (2008)
C6-aldehydes		Derivatisation-HPLC-UV	Kishimoto et al. (2008)
Oxylipins		GC-MS	Schulze et al. (2006)
Sugars and oligosaccharides		Ion-chromatography-electrochemical detection	Giannoccaro et al. (2008)
		Capillary electrophoresis	Linhardt (1994)
		Porous graphitic carbon LC-ESI-MS	Antonio et al. (2008)
	Also sugar phosphates	Porous graphitic carbon LC-ESI-MS	Antonio et al. (2007)
Glucosinolates		LC-APCI-MS	Reichelt et al. (2002)
Artemisinin		GC-MS, GC-FID	Ma et al. (2008)
		GC-electron capture detection (compared to FID)	Liu et al. (2008)
Hormones	Cytokinins	SPE-UPLC-ESI-triple quad MS/MS	Novák et al. (2008)
	ABA	Immunoassay	Zhang et al. (2008)
	IAA and other auxins	GC-MS	Barkawi et al. (2008)
	Jasmonates	UPLC-ESI-ToF	Glauser et al. (2008)

4

Metabolomics

Under the definition originally outlined in the pioneering work of Fiehn (2002), metabolomics is the quantitative and qualitative analysis of all metabolites in an organism. For plants, this has never been achieved or even closely approximated and is a situation that has been recognised as such (Allwood et al., 2008). In this respect, plant metabolomics may be regarded as an art rather than a science. It is impossible at present to identify all the metabolites since, by definition, metabolites are dynamic and there is no one method that will extract them all. The art is to generate large datasets and extract as much information from the data as is possible so that two or more samples can be distinguished and biological relevance for the distinction can be ascertained. Contrast this with genomics where there are a finite number of genes in a genome, all of which can be annotated even though many may have unknown functions. Metabolomics has now established itself firmly, however, as part of the - omics revolution (Guy et al., 2008) including its application to secondary metabolites (e.g. Goossens et al., 2003). In truth, most publications under the umbrella of metabolomics are concerned with metabolite profiling where a limited number of metabolites have been measured and this is exemplified in a recent journal volume dedicated to the subject (Guy et al., 2008). Analyses of secondary metabolites by virtue of the fact such analyses exclude primary metabolites and are frequently targeted, fall into the category of profiling and thus detailed consideration will not be presented in this section.

NMR has the greatest potential in metabolomics since it can 'see' all the metabolites in a single sample, but its sensitivity and resolution are poor compared to MS as already mentioned. Many metabolites in plants are well below current instrument detection limits (Allwood et al., 2008), and this is especially so for secondary metabolites. ^{1}HNMR has, however, been used extensively for natural products and was recently reviewed by Pauli et al. (2005). It has also been developed for 'fingerprinting' of metabolites in plants (Ward et al., 2003; Krishnan et al., 2005) and there are also some very good examples where it has been used for examining abundant secondary metabolites (see Pauli et al., 2005). Metabolite fingerprinting ignores the identification of individual peaks in the spectrum and uses multivariate statistics to determine whether two samples are similar or different (Krishnan et al., 2005; Ward et al., 2007). Fingerprinting (by NMR or FTIR), therefore, can be a useful prelude to targeted profiling by MS because there is no need to identify individual metabolites and samples can be classified rapidly without purification. NMR technology for natural products has been reviewed (e.g. Vogler and Setzer, 2006) and fingerprinting is dealt with in detail in the Chapter by Colquhoun and colleagues, including details of the use of NMR for metabolomics and especially for food metabolomics.

Direct injection (infusion) MS of unpurified samples has also been used as a metabolomics technique since it uses a single unpurified extract to examine all ions in a sample to produce a single spectrum. Although it has some drawbacks, such as ionisation suppression, it is considered useful for reducing data complexity especially when used with accurate mass, as long as the resolution of the instrument is sufficient to distinguish ions with the same nominal, but different monoisotopic mass (Dunn et al., 2005).

In any consideration of metabolomics, one must not forget that data handling is of paramount importance, both primary data and metadata. Collection of large datasets generated by the instrumentation required for metabolomics necessitates both comprehensive databases and sophisticated statistics to extract maximum biological relevance from the data. For metabolomics, unsupervised multivariate statistics (Fiehn, 2002; Ward et al., 2007; Weckwerth, 2008), most commonly principal component analysis, and supervised techniques such as genetic algorithms (Goodacre, 2005) have both been applied successfully. The major issue remaining is that experimenters and manufacturers collect human- and machine-dependent data that is often difficult to compare or collate. This has led to a framework for metabolomics experiments (Jenkins et al., 2004), to proposals for minimum reporting standards (Sumner et al., 2007) and to the launch of a reporting standards initiative (Fiehn et al., 2007).

The major drawback of any – omics approach to metabolites in plants, is our lack of knowledge of the vast number of compounds present (many of which are secondary metabolites) and an even greater lack of reference standards (Weckwerth, 2008). Technologies, such as accurate mass measurement and NMR sensitivity, will need to improve substantially in the future if we are to get close to identifying the total metabolite content of a plant and true metabolomics.

Extraction

Plants produce an enormous range of chemicals; the plant kingdom probably produces some hundreds of thousands of different chemicals (Hall et al., 2002) with very wide-ranging physical and chemical properties. It is obvious that a single solvent will not be suitable for both the very hydrophilic phosphates and sulphates, and the very hydrophobic membrane lipids and cuticle waxes. Instead it is necessary to choose a solvent system that will suit the intended purpose, and to know its limitations, both as an extractant and as a storage medium. In metabolomics there is great interest in the development of methods that will extract, reliably and reproducibly, as wide a range of low-molecular weight components as possible. Few plant-specific investigations have been carried out, but investigations exist for fish (Lin et al., 2007), erythrocytes (Sana et al., 2008), and for many examples of body fluids (Zhang et al., 2007). These studies are of relevance to plant work since the range of physicochemical properties of the analytes is similar. However, plant systems present additional problems because of the cell wall. Fortunately, the growth of metabolomics has created a demand for standardised methods for broad-scale extraction of plant secondary metabolites (Jones and Klinghorn, 2005; Weckwerth, 2007; Weckwerth et al., 2004; http://www.mpimp-golm.mpg.de/fiehn/forschung/blatt-protokoll-e.html).

In order to understand whether an extractant is appropriate for a particular analyte, it is necessary to test different methods using an internal standard or spike, where a small amount of a product is added at as early a stage as possible, to trace the recovery of the endogenous compound. This is particularly important in the analysis of intermediates in synthetic pathways, which may be much less abundant than the final product (ap Rees, 1980). There are two basic approaches. Either a non-native analogue of the product is chosen, with chemical and physical characteristics as close to those of the product as possible, or the product itself can be used. Isotopically labelled compounds are the most convenient to use, but are frequently unavailable. Using the analogue or isotopologue avoids the need to compensate for the product naturally present. If the natural product is used, it is necessary to carry out parallel extractions with and without the spike. Those without will reveal the amount of material naturally present, and the difference between these and the spiked samples will reveal the amount of spike recovered. If it is substantially less than the original addition, then the method is not suitable. These approaches are discussed more specifically in the context of high performance liquid chromatography (HPLC) and solid phase extraction in Snyder et al. (1997). There can be occasions where it is necessary to analyse a product that is not available as a standard. The best that can be hoped for is an indication that the analyte has increased or decreased in abundance in response to a particular experimental treatment. In order to exclude the possibility of analytical error, it is possible to use a variation on the internal standard spike method, known as a mixing experiment. In this, a high-abundance sample and a low-abundance sample are mixed, and the resulting value for the analyte is compared with the expected value from combination of the two individual measurements of the pure samples. If the low-abundance sample is low because it contains something that interferes with extraction or analysis of the target analyte, then it will also reduce the efficiency of analysis in the mixed sample, and the result will be unreasonably low.

Separation

Detection and measurement of natural products has always utilised chromatography in conjunction with selective detectors, liquid and gas chromatography employing different detection systems. Capillary electrophoresis can also be used as a method of separation with a variety of detectors, many similar to those used in liquid chromatography (Issaq, 1999). Separation is usually necessary because plant extracts are likely to contain a large number of analytes, and very few detectors have the specificity to analyse a complex mixture. Exceptions exist; it is not always necessary to use separation when carrying out analysis with high-resolution mass spectrometry or with NMR. Indeed, in high throughput metabolomic studies, where there is a need to collect data from as many samples as possible, the extra time required for separation can be unacceptable.

Chromatography is generally used to add specificity to a method. The classic methods of detection (UV, fluorescence, etc.) are not in themselves specific enough to quantify the individual analytes in a mixture. Supplementing the detector with chromatography, however, is only one way to increase specificity. The other has been to use enzyme coupled assays, in which the analyte is irreversibly converted to a product by a specific enzyme, the conversion being linked to a change in the amount of a cofactor that is easy to detect. These methods have their history in primary metabolism (e.g. Jellito et al., 1992) but have been less useful in secondary metabolism, where the enzymes can be hard to obtain. They can, however, be exquisitely sensitive and selective, especially when cycling assays are used (Gibon et al., 2002). Where enzymes of a pathway of secondary metabolism are being expressed heterologously and purified for other purposes, it is worth considering their use in measurement of the intermediates of the pathway.

In contrast to enzymatic methods, chromatography can be applied readily to any chemical that can be made to partition between two different phases, a stationary phase (the column) and a mobile phase (gas or liquid flowing through the column). There are numerous forms of binding depending on different properties of the analyte, and numerous types of column. In LC the most common columns are reverse phase columns. The term reverse phase means that the column is hydrophobic and the solvent hydrophilic (a historical "reversal", since the earliest columns were materials such as pure silica – hydrophilic – and were used with hydrophobic solvents). The analyte binds to a reverse phase column by hydrophobic interactions, and is eluted when the solvent (often methanol or acetonitrile) is sufficiently hydrophobic to attract at least some proportion of the analyte to detach itself from the stationary phase and be swept along by the solvent flow. Reverse phase methods have been applied to a huge range of biological analytes (see Chapters 4 and 9, Sadek, 2002); for reverse phase method development see Snyder et al. (1997).

Reverse phase column chemistry is a complex area that changes very rapidly. The catalogues and literature of the major manufacturers are the best source of up-to-date information, and trade journals such as LC-GC frequently provide updates on novel technologies and applications. Although columns differ, it is frequently quite possible to substitute a similar column when following a literature method with only slight changes to relative retention times. Conversely, it is worth understanding the characteristics of a column, since a change in column (or occasionally solvent) may allow two co-eluting analytes to be separated. For instance, the authors have found columns with phenyl groups a useful fall-back for phenolic compounds that coelute with an interfering metabolite in C18 columns. These columns still have a reverse-phase mechanism, but additionally phenolics can interact with the phenyl group on the column via π–π interactions, and are more strongly retained. This effect is solvent specific, being disrupted by solvents that themselves have lots of π-bonding (e.g. acetonitrile).

Reverse phase methods are appropriate for any analyte that has hydrophobic groups. Some analytes, such as glucosinolates, are too hydrophilic to interact well with reverse phase columns, and these provide a good example of alternative forms of chromatography. HILIC chromatography (Hydrophilic interaction chromatography; Wade et al., 2007) uses a column that is effectively immobilised water, into which the glucosinolates partition, and from which they are eluted by a gradient of *decreasing* organic solvent (contrast with reverse phase, where the gradient is of *increasing* organic). Ion pairing chromatography is a modification of reverse phase, and involves adding to the solvent a reagent that will bind to very polar parts of the analyte. The ion pair reagent might merely neutralise a charged group (e.g. glucosinolates; Mellon et al., 2002), or it might itself have a strongly non-polar feature that can interact with a reverse phase column (e.g. perfluorinated carboxylic acids as ion-pair reagent for amino acids; De Person et al., 2008).

In addition to chromatography, capillary zone electrophoresis (CE) can be used to separate ions, the most mobile ions emerging first. This is clearly applicable to anything that is charged in solution. With care, even analytes that at first appear unlikely candidates can be converted to ions by complexing or by using an extreme pH (e.g. oligosaccharides; Linhardt, 1994). CE has considerable advantages in resolution over HPLC. When liquids are driven down a tube by pressure, they undergo laminar flow, in which the flow nearest the walls of the tube is slower than the flow in the middle. This effect causes band-broadening in HPLC. Electroosmotically driven flow suffers much less from this effect, with the result that the analyte front as it moves down the tube is flat rather than domed. This allows greater theoretical plate count and narrower peaks than in HPLC, and excellent resolution. Capillary electrophoresis can even be applied to typical reverse phase chromatography situations by generating micelles. In micellar electrokinetic chromatography (Matsubara and Terabe, 1996), a detergent is added, which forms micelles. These are charged, and move in one direction, while the bulk electrophoretic flow tends to be in the other direction. An analyte can bind to the micelle if it is sufficiently hydrophobic, or remain in the aqueous phase, if it is not. Hence its rate of movement depends on its hydrophobic partition between two phases. This has been applied to plant phenols such as scopoletin, which are otherwise excellent candidates for reverse phase HPLC (Wang et al., 2007).

The essential feature of all methods is that the analyte we wish to measure must not come out at the same time (co-elute) as anything else that would also be detected and confused with the authentic analyte. Unfortunately co-elution is statistically hard to avoid. Many people know the paradox that if 23 or more people meet in a room, it is more likely than not that two will share a birthday. Translated into chromatography terms, if 23 chemicals are separated at random down a chromatogram that is 365 peak-widths long, it is likely that two will coelute. De Hoffmann and Stroobant (2002) point out that a mere 14 peaks in a chromatogram capable of holding 1,000 corresponds to a 10% chance of a peak containing two chemicals. The problem is exacerbated in that a typical HPLC run is unlikely to have anything like this resolution, and analytes will rarely be distributed evenly along the chromatogram. Simple HPLC should therefore be treated with great caution, *except* where the ingredients of the mixtures are already known, and the method has been validated by testing that there are no co-eluting analytes. Although in a totally unknown mixture (e.g. a new species of plant) it is impossible to exclude the possibility of co-elution, the chance can be much reduced by increasing resolution. GC often has better resolution than HPLC, and more recently, HPLC at high pressure using columns with small particle size (UPLC) has sparked considerable interest amongst manufacturers in the optimisation of the relationship between particle size and resolution.

Detection

The detector is determined by the chemistry of the analyte and the available technology; the same analytes are frequently measured by a wide range of different detectors (Lesney, 2004). This section, therefore, is restricted to a description of the basic principles behind detectors with an appropriate selection of examples of their use with natural products. Further information on specific detectors can be found in an online e-book (Scott).

HPLC systems are almost universally fitted with fixed wavelength or diode array light detectors (UV or DAD detectors) that measure the absorbance of the eluting analytes. These are most applicable to analytes with obvious chromophores, for instance, plant pigments such as anthocyanins (Lätti et al., 2008). UV detection can be applied to anything that absorbs at wavelengths from about 200 nm upwards, but the chemical groups that absorb in the far UV tend to be generic, and found in many analytes (for instance short unsaturated units, phenolic groups). At these wavelengths, UV offers little selectivity.

Fluorescence detectors are also common and offer better selectivity, since they ignore analytes that absorb at an appropriate wavelength, but fail to fluoresce, or emit light at the wrong wavelength. These detectors are also more sensitive as they measure a slight increase in light above zero, rather than a slight decrease of light. Although frequently used with fluorescent-tagged molecules, they can be used with naturally-fluorescing analytes (e.g. oat avenacins, Papadopoulou et al., 1999).

To deal with those classes of analyte that neither absorb at a reasonable wavelength, nor fluoresce, a wide range of other detectors has been developed. For instance, sugars can be detected in HPLC systems by measuring the refractive index of the sample as it elutes, or by electrochemistry (Giannoccaro et al., 2008). Refractive index suffers from lack of sensitivity, and from baseline drift in gradient methods, since the solvents themselves have a refractive index. Electrochemical detectors offer conductance and amperometry. Conductance measures whether ions are present in the eluant by applying an alternating voltage. Amperometry detects the analyte by its ability to become reduced or oxidised (i.e. a redox reaction) on application of a constant voltage (albeit often for a short time as part of a wave-form designed to preserve the electrode). Amperometry can be more specific because a voltage can be chosen at which the analyte undergoes a redox change, but other chemicals will not. Although classically favoured for sugars, amperometry has been used on other plant products (e.g. isoflavones and their glycosides; Otieno and Shah, 2007). While single-electrode detectors can only measure redox change at a single potential, electrochemical array detectors contain an array of electrodes at different potentials. Since the curve of extent of redox change versus potential is characteristic of the analyte, the array detector allows confirmation of the analyte by comparison of its redox properties with those of the standard. This is analogous to confirmation by spectral comparison in a UV diode array detector.

The development of liquid chromatography (LC) has been marked by a search for the universal, sensitive detector. The current claimant is probably evaporative light scattering. In its simplest form, the sample is merely evaporated to remove solvent, and allowed to enter a tube. Light is passed through the tube, and would normally not be deflected. If non-volatile materials were present, they will scatter the light, which can be detected. More complex forms of this detector can yield extra physical information about the sample. These detectors are compatible with nearly any volatile solvent, and universal in that they detect anything which is not volatile. Soya saponins are an example of a plant product investigated in this way (Rupasinghe et al., 2003).

Universal detection has not been an issue for GC, where the major limitation is the range of chemicals that can be rendered volatile and passed successfully through the column. The most common detector is the flame ionisation detector. The sample is ionised in a small flame of hydrogen, leading to an increase in conductivity. This is almost a universal detector since virtually everything can be ionised in this way.

The most powerful general detector for both LC and GC, however, is the selective mass detector, where we attach the chromatography system to a mass spectrometer. Mass spectroscopy (MS) systems require analytes to be ionised and the carrier medium to be removed. In this respect, LC-MS faces greater technical challenges than GC-MS since it is more difficult to remove the liquid carrier solvent; a chromatography flow of 1 mL min^{-1} aqueous solvent represents about 1.3 L min^{-1} of gas at room temperature and pressure, a substantial volume to remove. Methods of ionisation in both GC-MS and LC-MS are described in detail in standard textbooks (e.g. De Hoffmann and Stroobant, 2002), so will only be presented in summary here. The most commonly used in LC-MS are electrospray and atmospheric pressure chemical ionisation, whereas GC-MS uses electron ionisation.

Since many analytes naturally exist as ions in solution (carboxylic acids form R-COO$^-$ ions, amine bases form R-NH$_3^+$ ions), electrospray ionisation (ESI, Fig. 1a) is concerned mainly with removal of solvent from the flow of dissolved ions. The key feature of ESI is the application of a large potential difference between the spray needle and the surrounding instrument. In nano-spray sources, this potential gradient alone is enough to draw liquid out of the end of the needle, and to distort drops sufficiently for smaller drops to break off. The resulting spray of small droplets can dry rapidly to leave ions in the gas phase. Where ESI is interfaced to chromatography at larger flow rates (approximately 10 μL min^{-1} and upwards) it is usual to add gas flows to aid the formation of a spray (a nebulizer gas), and to improve drying. The exact design of spray chambers and associated ion desolvation systems (lenses, octopoles, etc.) varies greatly as manufacturers use voltage gradients and gas flows to maximise ion transmission and minimise contamination by neutral material. Unfortunately, it depends as much on the need to avoid patent infringement as on physics. In all cases, however, ions are removed from solvent and recovered as gas-phase ions in a drying gas (usually nitrogen), and subsequently the gas is pumped off.

In atmospheric pressure chemical ionisation (APCI; Fig. 1b), the solvent is first dried by a strong heater and drying gas, before entering a small spark where the analyte will encounter ionised solvent (and other ions and radicals) that can pass on a charge. ESI works for polar analytes,

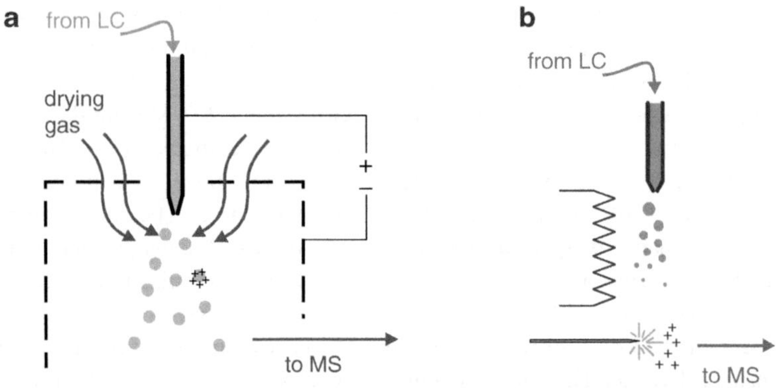

Fig. 1 Comparison of the "soft" ionisation methods of LC-MS. (**a**) Electrospray ionisation (ESI). The analyte already exists as a dissolved ion, and must be freed from solvent. The sample is sprayed with a nebulising gas, forming droplets, which are dried by a further stream of inert gas. The ions are driven to the outside of the droplet by an electric field, where their like charges repel. There are two theories about what happens next. Some ions may be expelled from the surface. Alternatively, the ions may remain, but as the droplets dry and shrink, the repulsive forces grow stronger until they overcome the surface tension of the droplet. The droplet breaks into smaller droplets, with a bigger surface area, which dry more quickly. The result in either case is a cloud of free ions floating in inert drying gas, which may be drawn into the mass spectrometer by a combination of electric fields and the vacuum in the instrument. (**b**) Atmospheric pressure chemical ionisation (APCI). Again the sample is nebulised, but the solvent is evaporated by a powerful heater before reaching a corona discharge. Here the solvent will be ionised, and can pass on its charge to analyte molecules

which include many alkaloids, phenolics, and, in fact, the majority of plant products. APCI is generally reckoned to outperform ESI on slightly less polar analytes, but neither method can handle completely non-polar alkanes. APCI gives very good sensitivity where it works, becoming, for example, the most widely-used choice for glucosinolates (Reichelt et al., 2002).

GC is usually interfaced with electron ionisation (EI), where electrons are removed from the analyte by an electron beam. This has the advantage of being highly reproducible, thus making possible libraries such as NIST (www.nist.gov). EI is less appropriate for analysis of large molecules, but these are in any case rarely suitable for GC. In addition to EI, GC-MS systems frequently offer chemical ionisation (CI), in which the electron beam is used to ionise a gas (e.g. methane or ammonia), which then ionises the analyte. This indirect method transfers less energy to the analyte, giving greater prominence to the molecular ion, and less fragmentation.

ESI and APCI in LC-MS are less universal than the electron ionisation of GC-MS, but they have the advantage that they work with the native analyte which does not need to be made gaseous. This avoids the difficulty of incomplete derivatisation, and renders them suitable for larger products. LC-MS ionisation methods are, however, less satisfactory for quantification, because the conditions in the ionisation chamber tend to be rather variable, and the ionisation process can be greatly affected by solvents and co-eluting analytes, so there is no guarantee that the efficiency of ionisation will be the same in different runs of different samples. Internal standards are therefore particularly useful, and external standard quantification, where the standard is run separately from the samples of interest (see later), must be treated with caution.

The ultimate universal detector is NMR, which is beyond the scope of this chapter. In the context of natural product identification, it has been reviewed recently (Vogler and Setzer, 2006) and elsewhere in this volume. While giving very valuable structural information (see below) the instruments are much more expensive to buy and operate than MS and are much less sensitive, requiring at least a thousand times more material for analysis.

Hyphenation

It is possible to improve resolution by separating chemicals in two "orthogonal" ways analogous to two-dimensional paper chromatography or two-dimensional gel electrophoresis. In this context orthogonal means that the second approach is independent of the first. Frequently the detector itself has a resolving power totally different to that of the column.

Attaching different instruments (chromatography systems, detectors, etc.) in series, to achieve orthogonal selectivity, is called hyphenation (hence LC-MS, GC-MS, LC-NMR, etc.). Hyphenation can also provide different types of data about the same peak (for instance, UV and MS spectra), allowing better proof that it is the correct analyte, and that it is pure. Some hyphenations are difficult and expensive. An NMR detector is a very valuable addition to an LC-MS system when unknown analytes must be identified in a complex mixture, but the addition is technically challenging. NMR requires a much more concentrated sample than the MS (ideally five orders of magnitude more material) and gives a better signal if the sample remains in the detector for many minutes, while an HPLC method is frequently optimised for chromatographic peaks only a few seconds wide. This can be solved in various ways including the accumulation and storage of peaks from LC on solid phase extraction media before their elution into NMR. The identification of a novel class of medicinally useful divanilloylquinic acids in a tropical root bark provides an example of the power of LC-MS-NMR in structural elucidation (Ouattara et al., 2004).

The principle benefit of hyphenation is the reduction in the chance of two different chemicals behaving identically, and thus being indistinguishable. A typical single-quadrupole MS can measure with 1 amu resolution over several thousand amu, offering, in theory, up to 2,000 data bins. Alone, it cannot resolve all products as plants contain more than 2,000 products, and many have the same mass (all isomers, but also many that are merely approximately isobaric but of different empirical formula). A GC- or LC-MS system might offer a chromatogram 100 peaks long, which is obviously inadequate, but coupled to a MS we obtain 200,000 data bins, enough that it is unlikely two reasonably dissimilar analytes will appear in the same bin, despite expected non-random distribution of peaks in both dimensions. Note, however, that the uneven distribution of chemicals in both dimensions strengthens the case for a hyphenated approach, and increases the desire for high resolution in both dimensions. Worse, certain analytes, particularly optical isomers, will remain particularly intransigent, because they are unresolvable in a large number of otherwise orthogonal methods of separation.

NMR and high resolution MS now offer such very high resolutions that some researchers have chosen to miss out the chromatography. This is the basis of direct injection MS for metabolomic screening (see above). While chromatography is time-consuming, the dangers of ignoring it are obvious. Currently the ultimate in MS is FT-MS where the masses of the ions are measured as the frequency of their movement in a magnetic field (and more recently, in Thermo's "Orbitrap", an electric field). FT-MS can measure to sub

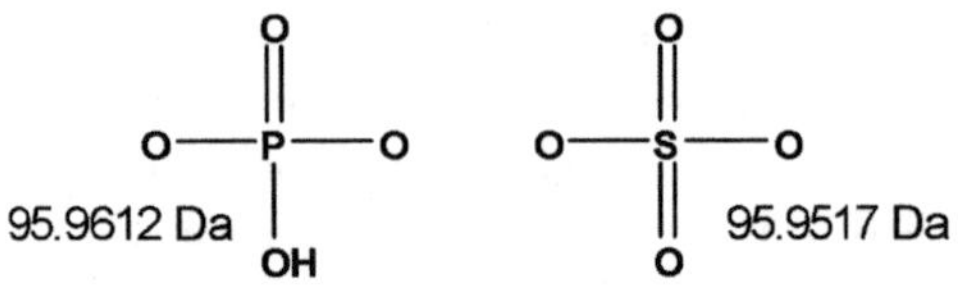

Fig. 2 Exact masses of a phosphate and a sulphate group as they appear as part of a molecule. Nominally (to the nearest integer) the two groups have the same mass because phosphorus weighs 1 Da less than sulphur, but has a valency one less than sulphur, and therefore needs an additional hydrogen. However, because the elements do not weigh exact integers, the exact masses of phosphate and sulphate groups are slightly different. This difference can be detected by a high-accuracy MS instrument

parts-per-million accuracy. This is relevant because while ^{12}C, by definition, weighs exactly 12 Da, no other isotope of any element weighs an exact integer. Thus although phosphate and sulphate have the same mass to 1 Da, they differ in the decimal places, and an "exact mass" MS can tell them apart (Fig. 2). This approach, however, cannot distinguish true isomers, because these have the same empirical formula and exactly the same mass. Like many tools, direct injection, high-resolution MS must be employed with understanding.

Detecting and Quantifying Single, Known Compounds

This and following sections examine how the design of analysis is determined by the biological question. Typical examples are presented from LC-MS.

Detection and quantification of a single analyte is the simplest biological requirement. Issues of selectivity have been discussed in the previous section. Simple HPLC or GC may be enough provided there is no risk of co-elution. Thus, LC-MS offers the opportunity to supplement chromatography by use of selected ion monitoring (SIM), where the MS is set up to detect only the mass of interest. In effect, resolution is increased by using a second, orthogonal dimension, except that data are collected only for points in the second dimension that are of interest. In many instruments (for instance, all quadrupoles), SIM increases sensitivity since the entire cycle time of the instrument is dedicated to the ion of interest rather than being used to scan a spectrum of other ions. The sensitivity of ion-traps is, however, limited by the capacity of the trap. When this is exceeded, space-charging occurs, in which the ions interact and no longer behave as would be predicted for the electrical field in the trap. Thus in many ion traps, SIM does not increase sensitivity, but allows faster scanning giving more measurements per second and a better measurement of peak shape in chromatography. SIM is unable to solve the problem of co-elution of near-isobaric ions although the correct product can be confirmed by spectral information. The UV absorption spectrum of the peak of interest can be compared with that of a standard, and peak-purity can be confirmed by ensuring that the UV spectrum remains constant throughout the peak. In GC-MS it is usual for the analyte to be fragmented during ionisation, so there will be a whole spectrum of ions, of which just one or two are typically chosen for quantification. However, GC-MS software allows the quantification of one or two extra ions as "qualifiers". If the intensity of these ions fails to fall within a certain window, the peak is rejected.

Where tandem mass spectrometry is available, even more security can be achieved by use of selected reaction monitoring (SRM) instead of SIM. Tandem mass spectrometry is the use of successive steps of MS to select or scan ions, create fragments from them, and scan the fragments. It can be carried out in time or space. Triple quadrupole systems are an example of tandem-in-space. The masses of parent ions are filtered and scanned by the first quadrupole, fragmentation occurs in the second quadrupole, and the fragments are scanned or filtered in the third. Ion traps are a typical example of tandem-in-time, in that all filtering and fragmentation occurs in the same place, in the trap. The electronic conditions of the trap are set up first to eject all ions apart from that chosen for MS2, then they are changed to encourage collision, and finally changed again to scan out the fragment ions. Both sorts of instrument are capable of SRM. In SRM, the parent ion of the desired product is selected and fragmented, and one of the fragments is quantified, possibly using other fragments as qualifier ions. For instance, to analyse benzoate-β-O-glucose our lab selected the parent ion 307 amu[1] (the sodium adduct), and monitored its MS2 fragment of 185 amu (which represents loss of benzoic acid; the fragment is a sodium adduct of the glucose residue). The benefit of this approach is that there can be very few analytes that have a mass of 307 amu, elute at 6.22 min (in this particular gradient), *and also* fragment by loss of 122 amu. Although SRM does not increase the signal (in fact it decreases the signal relative to SIM), it can nevertheless improve sensitivity because it removes background almost completely. Sensitivity is best measured not as raw signal-strength, but as signal:noise ratio, with SRM greatly reducing the noise. The method is very powerful if used correctly, but it can also be deceptive if the MS-MS fragment loss is badly chosen. For instance, losses of 18 amu (water) are usually possible in analytes that contain hydroxyl groups. This group is extremely common in plant natural products, so an SRM method using a loss of 18 adds only a little selectivity. SRM is only one application of tandem mass spectrometry; others will be discussed below.

Virtually always, detection and quantification rely on a standard. Firstly, the standard gives a reliable value for the retention time of the analyte. In HPLC, retention time varies with the dead volume and other characteristics of the HPLC, as well as the age and history of the column. Unless much development work has been specifically carried out to make a method portable between laboratories, it is always necessary to start by checking retention times and method validity in one's own lab. For many detectors it is also very difficult to calculate peak areas theoretically from data in literature. For methods such as ESI, it is impossible. Therefore a standard curve is run, measuring different quantities of standard, spanning the range of concentrations to be measured in the sample. There are two approaches to standard curves, external and internal. External curves are typical in UV-HPLC. Here the curve (standards containing various amounts of analyte; Fig. 3a) is run independently of the samples, before the samples begin, and possibly at intervals thereafter to account for gradual changes in system performance. This method fails if there is sample-dependent inefficiency in measuring. For instance, if ionisation efficiency in ESI is affected by a co-eluting (but undetected) ion, then an external calibration curve yields the wrong result as the samples are influenced by the co-suppressing ion, while the standards are not. The answer is to use an internal standard curve (Fig. 3b). The method of internal standards is described in detail below, because, while covered in standard analytical textbooks, it remains frequently unfamiliar to the non-analyst, and yet it is central to good measurement technique.

Again, standards are run before the start, but this time they include an internal standard (orange squares) not naturally found in the plant (the natural analyte is marked in blue).

[1] Amu = atomic mass unit. In this article amu is used for measured masses from instruments, in contrast to the known masses of molecules and ions, which are quoted in the more generally recognised Daltons. Instruments measure m/z, so 500 amu could, for example, correspond to 1,000 Da with charge (z) equal to 2. In this context, the correct unit is the Thomson (Th), but this is rarely used in biological articles, and is therefore avoided here.

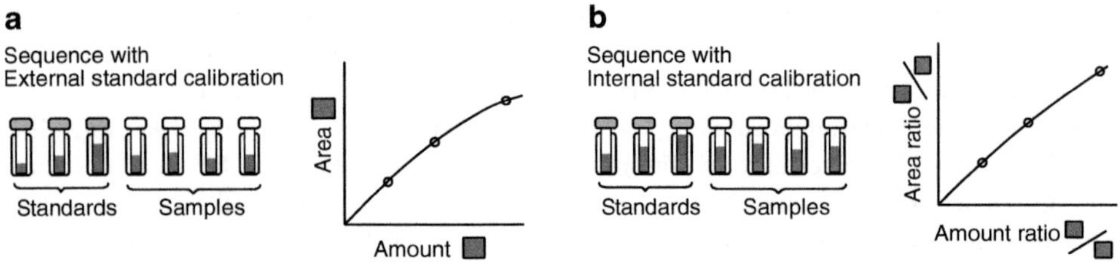

Fig. 3 Comparison of external and internal standardisation in LC-MS. In both cases a row of vials represents the sequence of samples run by the machine. The sequences begin with a standard curve (grey-topped vials) containing increasing amounts of the natural analyte (diagrammatically shown in blue). For external standard calibration (**a**) these are used to plot a calibration curve of signal versus amount of analyte. For internal calibration (**b**) all samples and standards also contain an internal standard that is different to the natural analyte, and shown in orange. Unlike the natural analyte, this is present in constant amount in all the standards. The calibration curve is now of a ratio of the analyte per internal standard. This ratio is not influenced by overall ionisation efficiency, or general losses during sample preparation

The internal standard is added at a constant level to all of the calibration standards, and also the samples. The instrument now measures the ratio of area of analyte (blue) per area of internal standard (orange) and constructs a calibration curve of changing area-ratio versus changing amount-ratio of natural analyte to internal standard. When samples are analysed, the instrument software uses this curve to convert a measured ratio of areas in the sample to the corresponding ratio of amounts of natural analyte and internal standard, and, using knowledge of the amount of internal standard added to that sample, converts the result back into total amount of natural analyte. The benefit of this is that if the efficiency of ionisation of the analyte is reduced, then the efficiency of ionisation of the internal standard is also reduced by the same amount, and the ratio remains unchanged. If the internal standard is added at the start of the extraction process, this method partially compensates for losses during extraction, assuming that the proportion of the internal standard that is lost is similar to that of the natural analyte. This requires that the internal standard behave exactly the same as the natural analyte. The best internal standards, therefore, are isotopically labelled versions of the natural analyte, which have different mass but very similar chemical and physical properties. Failing this, a different chemical may be chosen that is very similar to the analyte. For instance, in fatty acid analysis it is conventional to use fatty acids with odd numbers of carbons as standards since natural fatty acids all contain even numbers.

Unfortunately many plant natural products are extraordinarily hard to obtain as standards, even allowing some latitude about the exact structure, and ignoring isotopic labelling. For this reason, recent proposals for standards for reporting metabolomic data have allowed for partial identifications where the true identity cannot be confirmed (Sumner et al., 2007).

Finding Known Families of Compounds

This section examines the situation where a class of related analytes must be investigated, rather than one or two predefined individuals, which is often referred to as metabolite profiling. For example, there may be an interest in glucosinolates in a new brassica; the general nature of glucosinolates is known, but the actual examples present in the sample are unknown.

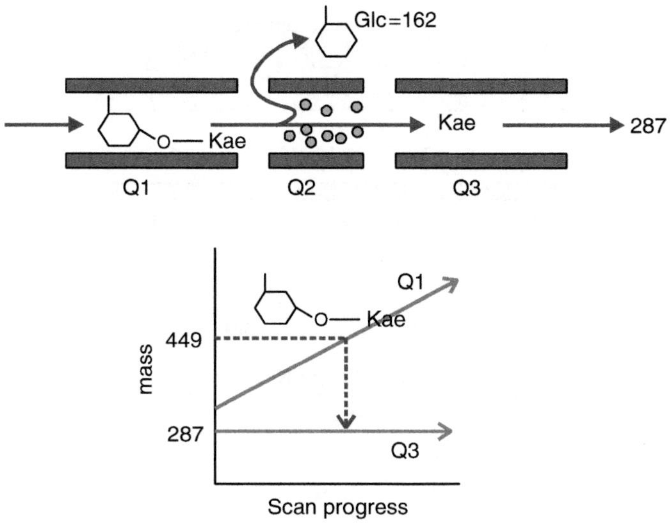

Fig. 4 Precursor ion scanning in a triple quadrupole instrument. Quadrupoles 1 and 3 behave as mass filters, scanning as shown in the lower part of the diagram. The instrument is analysing kaempferol-glucose, and the third quadrupole is set constantly to the mass of kaempferol. The first quadrupole scans. The spectrum resulting from this method shows any precursor ion that fragments to the mass of kaempferol

The method must therefore be chosen to detect all glucosinolates, but preferably to exclude as many irrelevant products as possible.

This is an opportunistic area, where the analyst exploits any feature of the analyte that happens to be generic. For instance, all glucosinolates have a sulphate group, which is not especially common amongst other plant products. There are no rules about how to take advantage of the feature. In the case of glucosinolates it is possible to exploit the sulphate group at a very early stage, during extraction, and produce an extract greatly enriched in glucosinolates (e.g. Cartea et al., 2008). The mixture is first bound to an ion exchange column by the negative charge of the sulphate group. Anything without a negative charge is washed off and rejected. The sulphate-mix is then eluted, and treated with an enzyme that removes sulphate, before putting it through the column again. Negative ions without sulphate will still bind, but the desulphated glucosinolates pass straight through.

Where it is not possible to bias the extract, it is often possible to bias the detection, especially with tandem MS-MS. Triple quadrupole systems offer a number of ways to look at families of chemicals; these methods can also be emulated in some ion traps. The first application is precursor ion scanning, demonstrated in Fig. 4 using kaempferol glycosides as an example. These nearly all fragment by loss of sugar units, and the smallest fragment is kaempferol itself, which forms a positive ion in ESI, of mass 287 amu. The triple quadrupole system can be set up so that the first quadrupole scans everything heavier than 287 amu, and the third quadrupole is fixed to 287 amu. As a result the MS will detect only those things that fragment to kaempferol, revealing their original mass as a precursor ion. This method is referred to as precursor ion scanning even though it is the product ion (287 amu) that is specified because it is, indeed, the precursor that is scanned.

Neutral loss scanning (Fig. 5) is another way to look for families of chemicals with common features. Again it works on the kaempferol glycosides. Many have glucose units amongst their glycosylation, and glucose is lost as a residue of

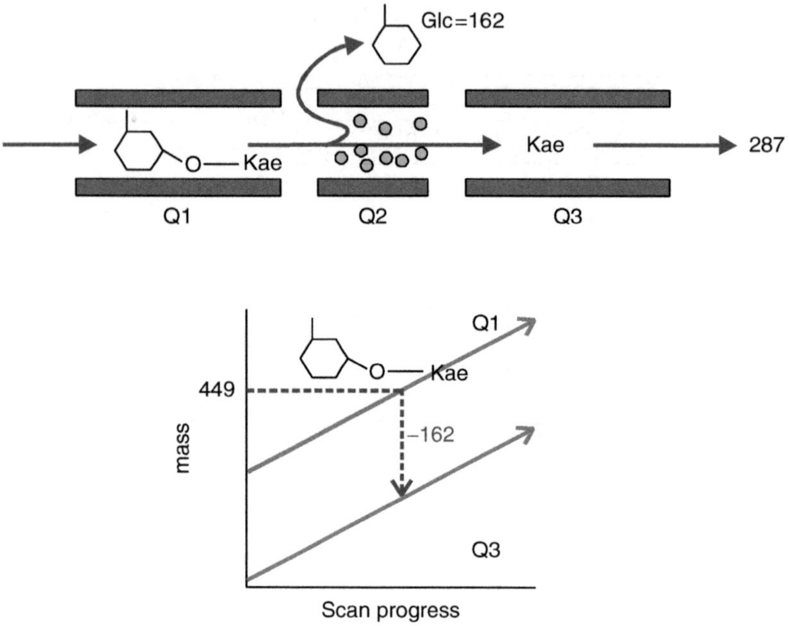

Fig. 5 Neutral loss scanning in a triple quadrupole instrument. Quadrupoles 1 and 3 now scan in parallel (contrast with Fig. 4), separated by the mass of the neutral group of interest. Here again we use kaempferol-glucose. The spectrum resulting from this method shows any precursor ion that can undergo a neutral loss of a glucose residue. Although Kaempferol-glucose will appear in both this method, and that of Fig. 4, here it appears because of the glucose, whereas in Fig. 4 it appeared because of the kaempferol

mass 162 amu. A triple quadrupole can be set up so that both the first and third quadrupoles scan, but scan simultaneously, in parallel, separated by 162 amu. The only ions that will get through (and be detected), are those that enter the first quadrupole, lose exactly 162 amu on collision, and are therefore just right to pass through the third. The method is now selective for glucose residues. There are two features of note in this method. Firstly, by slight changes in the MS, it is possible to select two different groups of chemicals from a common pool. The precursor scan method selects kaempferol glycosides, but would not see quercetin glycosides, because quercetin forms an ion of 303 amu, not 287. The neutral loss scan detects quercetin glycosides (and anything else with a glucose unit), but does not detect kaempferol-rhamnoside, because the rhamnose sugar has a residue mass of only 146 Da. Secondly, neither method is infallible. Glucose residues have the same mass as caffeate residues (162 Da), and rhamnose residues have the same mass as coumarate residues (146 Da, Fig. 6). Since coumarate and caffeate are themselves widespread and common, there is scope for confusion. This is an area where exact mass can be valuable. Currently no technology is available that can select an ion accurately (i.e. select a rhamnoside and reject a near-isobaric coumarate), but where the masses of the parent ion and fragment can be determined accurately, the loss can be calculated accurately, and fortunately, coumarate and rhamnose differ very slightly in mass.

Other simple characteristics can also be used in the search for features general to a whole

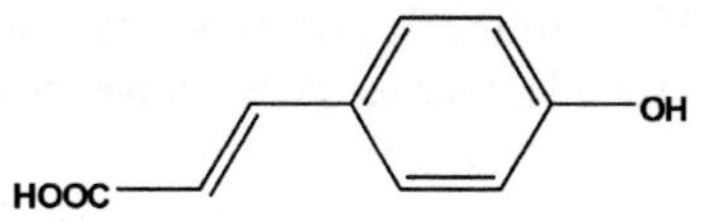

Coumarate

Mass = 164.0473Da

Exact mass of residue = 146.0368 (loss of water)

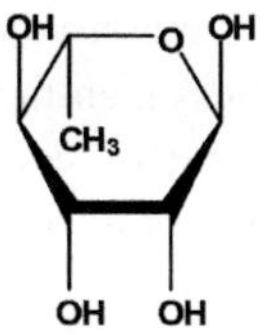

Rhamnose

Mass = 164.0685Da

Exact mass of residue = 146.0579 (loss of water)

Fig. 6 Exact masses of coumarate and rhamnose and their residues when incorporated into biological molecules. Both of these groups are common in plants, and they have the same nominal mass. They can be distinguished by accurate mass instruments

class of chemicals. For instance, anthocyanins (some of which are based on cyanidin, isobaric to kaempferol), can be located easily as single peaks in a UV chromatogram at 520 nm.

Identification of Unknown Compounds

Although metabolomics is not dealt with in this chapter in any depth, it can throw up the hardest of analytical problems: the structural characterisation of an unknown analyte that correlates with an interesting biological effect. This situation also arises where successive fractionation has pared down a biologically active extract potentially to a single, active ingredient. In this second case, it is important to note that purity is usually unmeasurable. Even a great Nobel laureate, Willstätter, concluded that enzymes were not proteins because it was possible to purify an extract to such an extent that no protein could be measured, and yet the activity remained (Deichmann, 2007). Of course this was a reflection on the sensitivity of the protein assay. His error should be a warning to the modern analyst, especially when working on samples that have been purified to the extent that even the target analyte is of abundance approaching the detection limit of the instrument.

Identification of unknowns is potentially very labour intensive. GC-MS instruments nearly all create fragments by electron ionisation, with a 70 eV beam giving similar spectra for which libraries are available (e.g. NIST; web ref). If the analyte has been found before, and matches a library entry, identification is complete. Many plant natural products, however, are too big for simple GC-MS analysis. Libraries in LC-MS are currently disjointed, technology-specific and frequently too small to be useful (Jiang et al., 2006). LC-MS typically uses collision induced fragmentation, where the ions are accelerated through a very low pressure of an inert collision gas. This is a low-energy method and the exact spectrum that is formed can vary greatly from one instrument to another. In some (for instance, triple quadrupoles and Q-ToFs), the fragments continue to be accelerated and can fragment further, yielding complicated, but information-rich spectra. In others (some ion traps) only the parent ion is accelerated; when it fragments, the process stops. These instruments can take a molecule apart, step by step, but often yield very simple (and less information-rich) MS2 spectra. For example, loss of 18 amu merely indicates a hydroxyl group, providing very little extra information to match up between an MS2 spectrum and a library. Ammonium adducts frequently fragment by loss of the adduct ammonia (17 amu), giving no structural information whatsoever about the parent molecule. Aware of this, one manufacturer (Thermo) has provided the option of extending the acceleration to include all masses down to slightly more than 18 amu less than the parent ion, although this yields spectra completely unlike either a simple ion trap or a triple quadrupole. Partly for this reason, those libraries that exist for LC-MS do not always take advantage of these technical advances (Halket et al., 2005). Thus there is no LC-MS equivalent of NIST. In fact, because of the nature of ion-trap LC-MS instruments, where the structure of the molecule is revealed spread over many MSn spectra, perhaps up to MS5, the very process of comparing measured spectra with library spectra becomes more complicated. It is necessary to compare a tree of spectra with another tree, possibly with a slightly different combination of branches. This is, indeed, now possible (e.g. in "Mass Frontier" software, HiChem, Slovakia).

Given that library searching is unlikely to solve all structural identification, it is necessary to turn to *de novo* data interpretation. The starting point is the mass of the ion. Taking into consideration charge-state (which can be found by looking at the spacing of the isotope peaks) and any adduct ions, the mass of the neutral molecule is calculated. Using accurate mass (to 2 ppm or better), it may be possible to narrow the molecule down to a handful of empirical formulae. Larger molecules have more possible formulae within a narrow mass range. For instance, the hydrogen ion adduct of kaempferol ($C_{15}H_{11}O_6^{(+)}$) has an exact mass of 287.05501 amu. There are only two empirical formulae of even-electron (non-radical) ions consisting of C, H, N and O within 0.001 amu of this mass. Addition of a single glucose to form kaempferol-Glc ($C_{21}H_{21}O_{11}^{(+)}$, exact mass 449.10784) increases this to 7 formulae. The situation is worse if other elements may be involved, or if it is uncertain whether the parent is a hydrogen or sodium adduct. Fragmentation and accurate mass are therefore complementary. Where the parent is so large that several alternative empirical formulae seem equally likely, it may be possible to get single formulae for some of the fragments. If one of the fragments is certain to contain, for example, four oxygens, all parent formulae that contain three or less can safely be rejected. Some manufacturers, appreciating that their accurate mass instruments lack the accuracy required to yield single empirical formulae over a biologically useful range, have attempted to supplement mass with isotope pattern matching. The common biological elements have only small proportions of heavy isotopes (H, C, N, O all of the order of 1% or less), so the differences in patterns between large near-isobaric molecules consisting of these elements are subtle. The method therefore requires utter certainty that the spectrum is not contaminated by near-isobaric ions. For instance, if the analyte is also present in a reduced form (+2H), even at very low abundance, the +2 peak will not be of sufficient quality to carry out reliable pattern matching. Paradoxically, high-resolution mass spectrometers can frequently resolve true isotope peaks from near-isobaric contaminants, but these instruments usually have sufficient mass accuracy that the isotope pattern is unnecessary. For most natural products in a biological matrix, however, commonly available ToF instruments will narrow down the possible formulae, but will not yield a single elemental composition.

Often it is necessary to deduce structure from the behaviour of the analyte. Its binding to anion or cation exchange columns, its solubility and the pH dependence of its retention time on reverse phase columns can all yield structural information. It is also helpful to check whether it forms both negative and positive ions as acidic properties and negative ions may suggest carboxylic acid groups whereas basic properties and an odd molecular mass may suggest an amino compound. (Biological products that contain no nitrogen have even masses, because nitrogen is the only common biological element with an even mass, but an odd valency. This is termed the "nitrogen rule.")

The major structural tool offered by mass spectrometry is fragmentation. Once again there are differences between fragmentation in GC-MS (EI) instruments, and LC-MS (typically ESI) instruments. EI creates free-radical ions for which various reactions are possible, and these have been studied in detail. Textbooks (e.g. Gross and Caprioli, 2005; see especially Chapter 7) are strongly biased towards EI-style ions because EI has been widely available for much longer than ESI. It is therefore generally harder to interpret LC-MS spectra *de novo* than GC-MS.

LC-MS ions as produced by ESI and APCI are very rarely free radicals. The positive ions are adducts with additional H^+, Na^+ or similar. The negative ions have lost H^+. Theoretical frameworks explaining the balance of products formed on fragmentation using kinetic and thermodynamic data do exist (e.g. "Quasi-equilibrium theory"; see Gross and Caprioli, 2005; De Hoffmann and Stroobant, 2002) but can rarely be applied rigorously by a biologist faced with a large natural product and its CID MS2 spectrum. Indeed, the theoretical difference between fragmentation of small products by EI, and larger products by CID, explains some of the difficulty in interpretation of the latter. The energy transmitted to a product will be distributed amongst its bonds, and fragmentation will occur when enough of the energy happens to be in one bond to exceed the bonding energy of that bond. This is statistically unlikely in a large molecule unless a large amount of energy has been transferred to the molecule. For this reason, huge biomolecules respond well to ion trap CID, where the molecule can be accelerated for a considerable period of time (ms instead of μs), creating numerous collisions, and giving a large amount of time for fragmentations to occur even when the rate of fragmentation is slow. Because the fragmentation of large molecules will tend to be slower for a given energy than that of small molecules (the energy being spread more thinly), there is more time for movement of the molecule (rotation around bonds) to form new geometries. Some of these geometries will favour intramolecular rearrangements and give rise to low-energy fragmentations. These are thus greatly favoured in CID, and they are very molecule-specific. The consequence of this will be illustrated below using glucosinolates as an example; clearly it is harder to evolve general fragmentation rules for ESI-CID than for EI.

It will usually be found in practice that the most abundant fragments are those formed by an easily derived mechanism, and where the products have reasonable stability. The bonds that break tend to be those that are already polarised (C–O and C–N bonds break more frequently than C–C bonds). The leaving groups seen in typical biological mass spectrometry tend to be stable neutral molecules (H_2O is a much more common loss than H_2), or ions where the charge is stabilised. Sometimes these rules can be helpful. For instance, Fig. 7 shows spectra of two glucosinolates previously isolated as described above. Candidate structures were available (drawn in Fig. 7 without regard to stereochemistry; CID is frequently inadequate to distinguish stereochemistry), but no standards. It was possible to assign tentative structures as indicated because in one case there is a loss of 31 amu. Although homolytic bond fusion giving rise to radicals is rare in CID (being energetically unfavourable), it is difficult to find any alternative explanation for this loss apart from loss of a methoxy radical. Loss of a radical is only possible with some strong stabilising feature, in this case delocalisation of the lone electron on the remaining ion, which is only possible in one of the two structures. It should be noted that this is not de novo structural identification, but assignment from a number of expected structures.

It is usually possible to explain in retrospect why an ion behaves as it does, but much harder to guess in advance how an ion will behave. Figure 8 shows an ion with a strong loss of 200. The parent ion was a sodium adduct, and it is almost invariable that sodium adducts produce fragments containing sodium (i.e. the sodium itself is rarely lost within a neutral fragment, which seems intuitively sensible since the

4

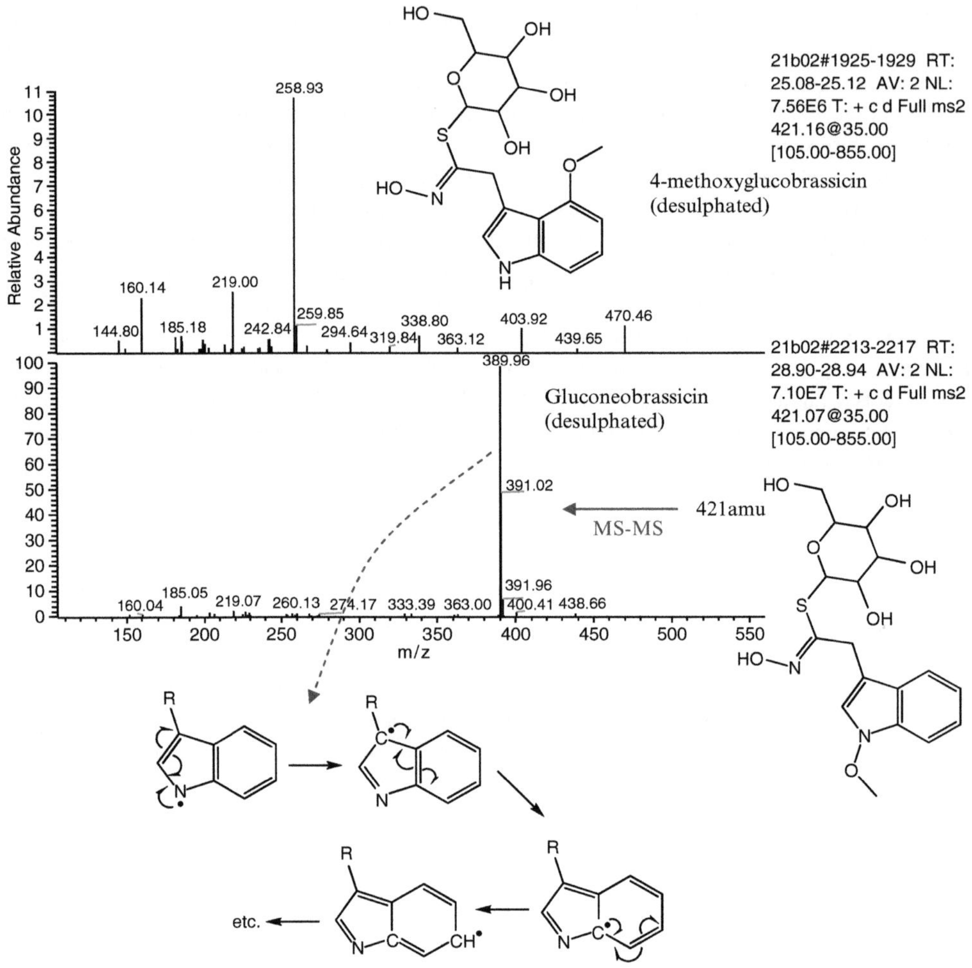

Fig. 7 MS2 spectra of two isomeric glucosinolates (desulphated). These products have the same parent mass, but can be distinguished by MS2. Standards were not available, but likely structures had been proposed in literature. Methoxy-glucobrassicin (*top*) has a methoxy group attached to a carbon on the ring system. Gluconeobrassicin has the methoxy group attached to nitrogen. The major fragment in the lower spectrum is clearly a loss of 31, for which a methoxy radical is the only reasonable explanation. This rare and energetically unlikely fragment is offset by a greatly stabilised radical fragment ion. The radical electron can be spread around the entire ring system (*bottom*). On this basis we can tentatively assign the two parents to the two MS2 spectra on theoretical grounds, though we would prefer additional evidence of their identities

sodium is the very means by which the parent ion became charged). Retrospectively it is possible to identify that the ion that is formed has the structure indicated – with no sodium – and that the neutral loss was the sugar part of the molecule *lost as a sodium salt.* It is likely that this occurs because the charge on the fragment is so highly stabilised by its delocalisation.

In *de novo* identification, accurate mass is very valuable as certain losses are diagnostic of

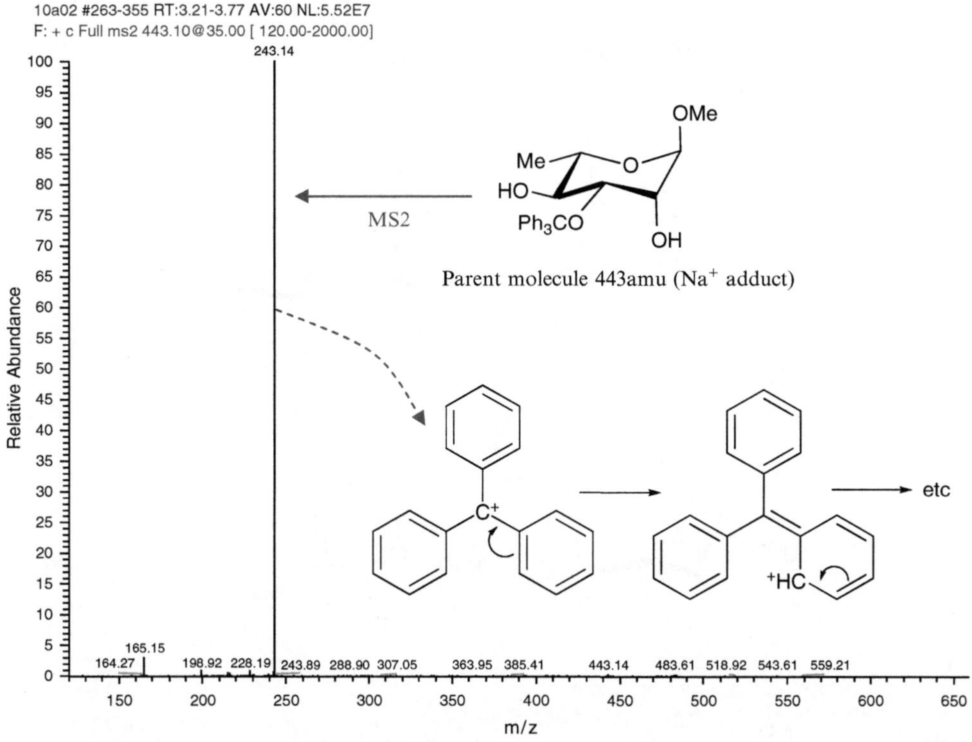

Fig. 8 Fragmentation of a sugar derivative. The parent molecule is seen as a sodium adduct, but forms a fragment in MS2 that cannot be explained as a sodium ion. In fact, the loss is a sodium salt of a glucose residue. This is an unusual loss, but is explained in this case by the high stability of the resulting sodium-free ion, where the charge can be delocalised over all three phenyl rings

certain groups. To those already mentioned above (18 for water and 31 amu for a methoxy radical) it is possible to add 15 amu for a methyl radical, and 44 amu (CO_2) lost from carboxylic acids. However, there are many groups that are semi-diagnostic. For instance, 28 amu is frequently CO from a carbonyl group, but can also be $H_2C=CH_2$ from an ethene side-chain. Given a sufficiently accurate instrument, these can be distinguished.

A combination of fragmentation and accurate mass, however, cannot solve a molecule neutral completely. Firstly, it very often happens that a large section of molecule is lost as a single loss. For example, if a parent ion of 1,007 amu fragments to 137 amu, it may be possible to identify the 137 amu with great certainty, but 870 Da of parent remains unknown. Sometimes this problem can be alleviated by trying a different ion; positive and negative ions of the same neutral molecule frequently fragment completely differently (Jiang et al., 2006). Different positive adducts also fragment differently. Secondly, although a fragment may be lost, it may be unclear from where it was lost. In a biological context, we may be aware of the building blocks involved in a complex anthocyanin (e.g. Fig. 9; Luo et al., 2007), but with MS-MS evidence alone, it would be impossible to state, for instance that the marked glucose residue is

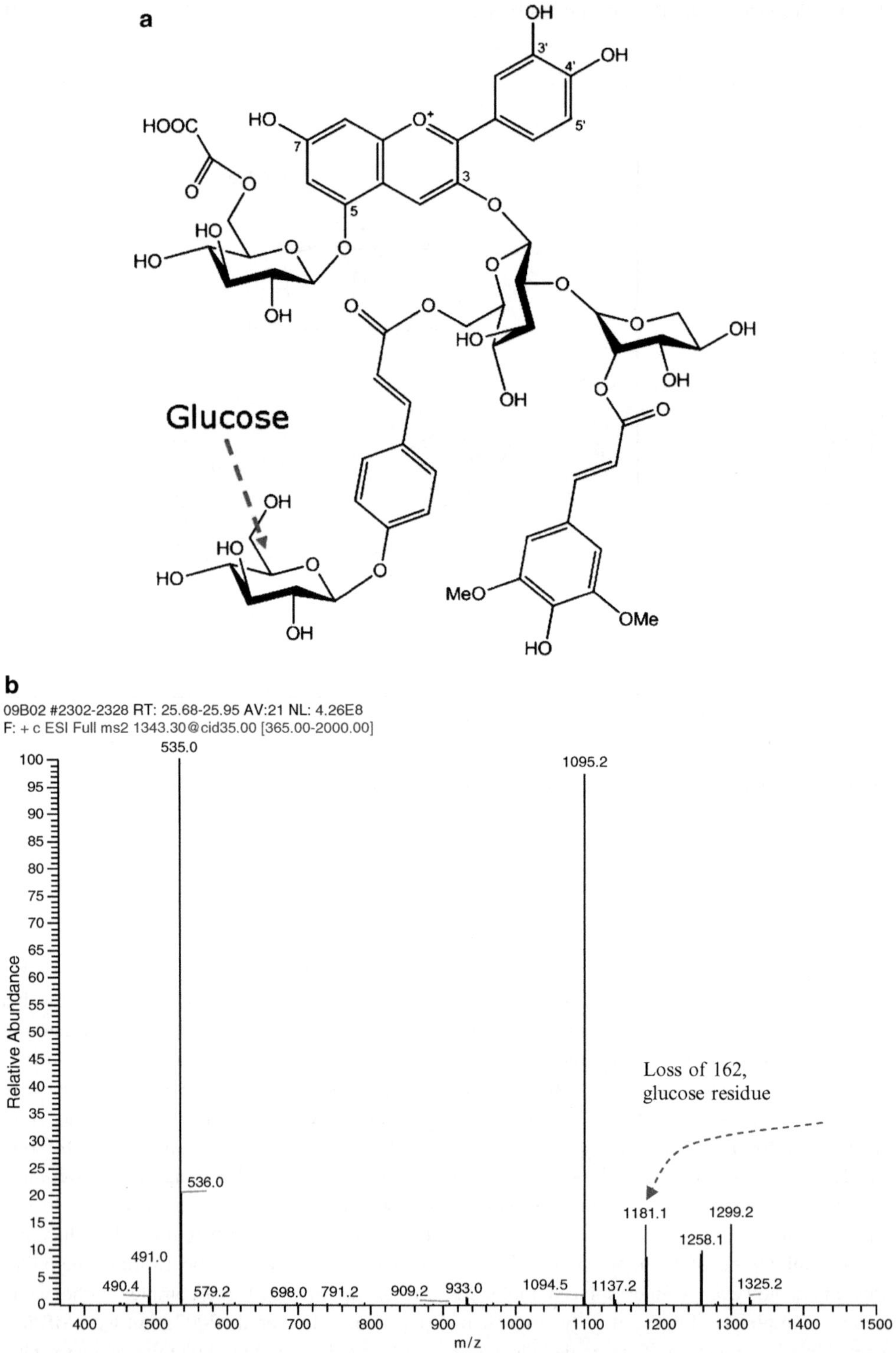

Fig. 9 An anthocyanin illustrating the challenges of structural elucidation by MS^n. (**a**) The structure of the anthocyanin. Note the terminal glucose, and the many free hydroxyl groups. (**b**) The MS2 spectrum. Note

attached to a coumaroyl residue rather than the sinapoyl residue or any of the free hydroxyl groups on cyanidin.

Thirdly, CID is complex (see above). In understanding CID, all the normal rules of organic chemistry apply, but molecules frequently undergo major rearrangements. Most obviously, nearly all CID fragmentations involve rearranging hydrogen to ensure that the products are not free radicals. But much more drastic and confusing rearrangements also occur. For instance, glucosinolates (undesulphated) produce a strong negative ion fragment of 259 amu, undoubtedly corresponding to glucose-sulphate, but the sulphate in a glucosinolate (Fig. 10) is not attached to the glucose. A further example is furnished by rearrangements of phenyl propanoid-derived spermidines. Spermidine is an asymmetric molecule with three nitrogens to which phenyl propanoids can be attached. There are therefore three positional isomers of a disubstituted spermidine, and yet these cannot be distinguished by MS-MS because of interactions between neighbouring groups during fragmentation (Bigler et al., 1996).

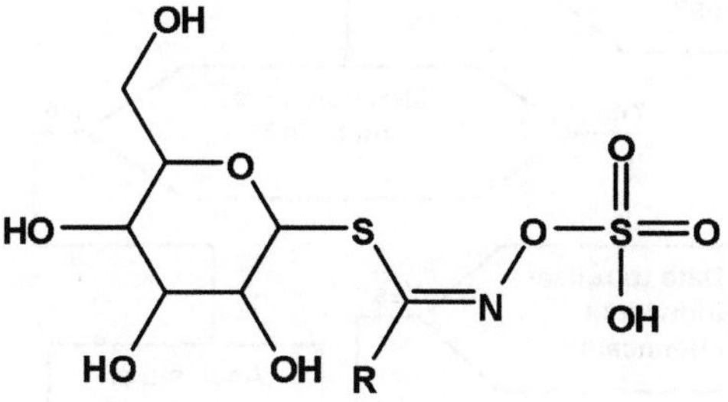

Fig. 10 General structure of a glucosinolate. Note that the sulphate group is not attached to the glucose group. In negative mode ESI-CID, however, an ion of 259 amu is frequently seen, which can be shown to be a glucose-sulphate. This indicates a rearrangement of the molecule. Such drastic rearrangements are common in ESI-CID, and can complicate the structural identification of natural products. At best they lead to inexplicable fragments. At worst they mislead the analyst into incorrect structures

In understanding the chemistry of fragments, it is essential to consider that they exist in the gas phase. The stabilisation of particular chemical groups expected from hydrogen bonding in aqueous solution does not necessarily apply.

Most publications using MS data for structural elucidation of plant natural products are either proof-of-concept papers where the authors begin with standards of known structure, or rely strongly on prior knowledge. The best structural work comes from groups who have been able to amass detailed knowledge of the behaviour of one or two specimen molecules, and have carried this over to the identification of a range of related products (triterpenoid saponins; Li et al., 2006). Hence, for the reasons cited so far in this section, the authors do not consider MS alone is adequate for *de novo* structural identification, but it is a powerful tool to distinguish between several biologically likely structures in conjunction with other sources of information.

For true identification of unknowns NMR is essential. Interpretation of NMR spectra is covered in any good degree-level organic chemistry textbook (e.g. Vollhardt, 1987) and has been

Fig. 9 (continued) that there is a loss of 162 corresponding to the loss of the terminal glucose. However, it cannot be determined where the glucose was attached. Nearly all fragments of this ion can be explained as breakages of single bonds between the constituent subunits, rather than breakages for instance across the rings of sugar units, as would be needed to differentiate between isomers varying only in sugar linkage. There is also an extremely abundant loss to of 248 amu, to 1095. This probably corresponds mostly to loss of the malonyl-glucose unit attached to position 5 of cyanidin; however, it is impossible to exclude the possibility of some loss of the malonyl unit together with the other, terminal glucose. Thus interpretation of MS3 spectra of 1343→1095→ is complicated by the possibility that there is more than one structure for the 1,095 amu parent

4

reviewed for secondary metabolites recently (Vogler and Setzer, 2006). Simple NMR spectra reflect one element only, though more complex NMR experiments monitor interactions of nuclei of different elements. For instance, proton NMR looks only at hydrogen atoms. Each atom gives a signal at a different point in the spectrum depending on its chemical surroundings. Particular groups fall in particular areas of the spectrum, so, for instance, a hydrogen in a benzene ring will fall in a different part of the spectrum to a methyl hydrogen. Indistinguishable hydrogens (for instance, the three hydrogens of a methyl group) give signals at the same point. Since the peak area is proportional to the number of atoms contributing to the peak, the hydrogens can be counted. Further, nuclei can interact with one another. This causes peaks to split into fine clusters of peaks. This fine structure can reveal how many neighbours an atom has. This, and more sophisticated experiments, gives NMR great power to determine structure, provided there is enough material to give a good signal.

It is no accident that NMR features last in this discussion, which is aimed at biologists. NMR is usually covered in a whole chapter of

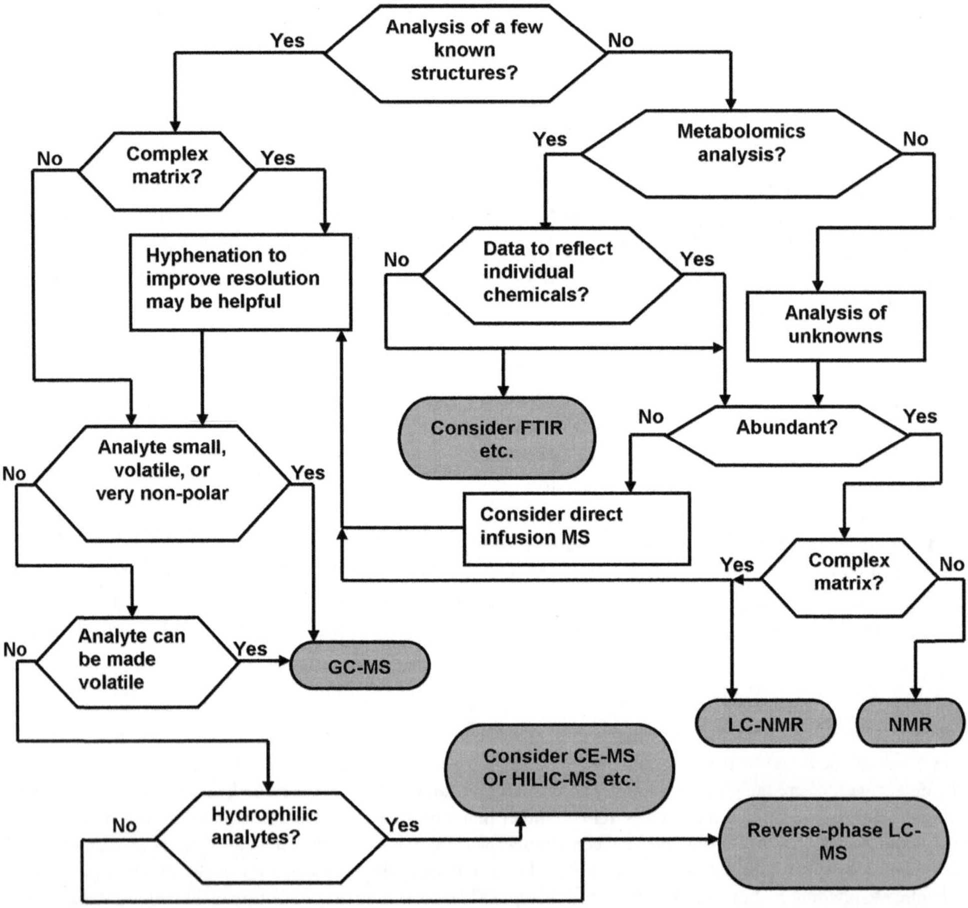

Fig. 11 A simple tree for the selection of analytical methods

chemistry texts, while mass spectrometry gets a few pages, much later in the book. It is perhaps true that biologists neglect NMR because it is "difficult", while chemists do not derive the full benefit from MS because NMR is enough to check their structure. In fact the two fields have much to offer one another. NMR can achieve structural identification only dreamt of by mass spectroscopists and biologists, but it can be misled by groups that contain few of the interesting nucleus. MS can check for these groups, reveal the overall empirical formula (which is a good starting point for the interpretation of NMR), handle certain strange elements that occasionally crop up, and rule out oligomers that can sometimes be hard to distinguish from the monomer by NMR alone. Hence the two technologies are complementary, not mutually exclusive. In summary, a simple decision tree (Fig. 11) is offered as aid to the novice.

The Future: Imaging?

An important biological question remains: "Where is this chemical found in the plant?" At the time of writing this chapter, functional imaging of chemicals is a growth area. It is also a no-man's land between microscopy and chemical analysis. The microscopist has histochemistry, enzyme systems coupled to light-producing reactions (Borisjuk et al., 2002), and fluorescent marker proteins that change their properties according to the amount of analyte they bind (Wouters et al., 2001). The chemical analysis lab has NMR imaging, and also more recently, MS imaging (McDonnell and Heeren, 2007). This is can be based on MALDI laser ionisation, where the laser can be scanned across a sample to yield an image. Secondary ion MS, where an ion-beam is "shot" at a sample to ionise it further, can also yield images when the incident beam is scanned. Currently these methods lack the resolution needed for intracellular localisation, and they are found in few laboratories. It is certain that they will improve and spread, and examples of their application to plant metabolites are accumulating, both in primary metabolism, and in secondary metabolites including cuticular waxes and flavonoids (Burrell et al., 2007; Cha et al., 2008).

References

Alali FQ, Tahboub YR, Ibrahim ES, Qandil AM, Tawaha K, Burgess JP, Sy A, Nakanishi Y, Kroll DJ, and Oberlies NH (2008) Pyrrolizidine alkaloids from Echium glomeratum (Boraginaceae). *Phytochemistry* 69:2341–2346

Allwood JW, Ellis DI, and Goodacre R (2008) Metabolomic technologies and their application to the study of plants and plant-host- interactions. *Physiol. Plant.* 132:117–135

Antonio C, Larson T, Gilday A, Graham I, Bergström E, and Thomas-Oates J (2007) Quantification of sugars and sugar phosphates in Arabidopsis thaliana tissues using porous graphitic carbon liquid chromatography-electrospray ionization mass spectrometry. *J. Chromatogr. A* 1172:170–178

Antonio C, Pinheiro C, Chaves MM, Ricardo CP, Ortuño MF, and Thomas-Oates J (2008) Analysis of carbohydrates in Lupinus albus stems on imposition of water deficit, using porous graphitic carbon liquid chromatography-electrospray ionization mass spectrometry. *J. Chromatogr. A* 1187:111–118

ap Rees T (1980) Integration of pathways of synthesis and degradation of hexose phosphates. In: Preiss J (ed) The biochemistry of plants: a comprehensive treatise, Vol 3. Academic, New York

Barkawi LS, Tam Y-Y, Tillman JA, Pederson B, Calio J, Al-Amier H, Emerick M, Normanly J, and Cohen JD (2008) A high-throughput method for the quantitative analysis of indole-3-acetic acid and other auxins from plant tissue. *Anal. Biochem.* 372:177–188

Bigler L, Schnider CF, Hu W, and Hesse M (1996) Electrospray-ionization mass spectrometry.3. Acid-catalysed isomerisation of N,N'-bis[(E)-3-(4-hydroxyphenyl)prop-2-enoyl]spermidines by the zip reaction. *Helv. Chim. Acta* 79:2152–2163

Borisjuk LS, Walenta H, Rolletschek W, et al. (2002) Spatial analysis of plant development: sucrose imaging within Vicia faba cotyledons reveals specific developmental patterns. *Plant J.* 29:521–530

4

Burrell MM, Earnshaw CJ, and Clench MR (2007) Imaging matrix assisted laser desorption ionization mass spectrometry: a technique to map plant metabolites within tissues at high spatial resolution. *J. Exp. Bot.* 58:757–763

Cartea ME, Rodriguez VM, de Haro A, et al. (2008) Variation of glucosinolates and nutritional value in nabicol (*Brassica napus pabularia* group). *Euphytica* 159:111–122

Cha S, Zhang H, Ilarslan HI, Wurtele ES, Brachova L, Nikolau BJ, and Yeung ES (2008) Direct profiling and imaging of plant metabolites in intact tissues by using colloidal graphite-assisted laser desorption ionization mass spectrometry. *Plant J.* 55:348–360

Chien S-C, Chen C-C, Chiu H-L, Chang C-I, Tseng M-H, and Kuo Y-H (2008) 18-nor-Podocarpanes and podocarpanes from the bark of Taiwania cryptomerioides. *Phytochemistry* 69:2336–2340

Davidovich-Rikanati R, Lewinsohn E, Bar E, Iijima Y, Pichersky E, and Sitrit Y (2008) Overexpression of the lemon basil -zingiberene synthase gene increases both mono- and sesquiterpene contents in tomato fruit. *Plant J.* 56:228–238

De Hoffmann E and Stroobant V (2002) Mass spectrometry principles and applications. Wiley, Chichester

De Person M, Chaimbault P, and Elfakir C (2008) Analysis of native amino acids by liquid chromatography/electrospray ionization mass spectrometry: comparative study between two sources and interfaces. *J. Mass Spectrom.* 43:204–215

Deichmann U (2007) "Molecular" versus "Colloidal": controversies in biology and biochemistry, 1900–1940. *Bull. Hist. Chem.* 32:105–118

Doehlemann G, Wahl R, Horst RJ, Voll LM, Usadel B, Poree F, Stitt M, Pons-Kühnemann J, Sonnewald U, Kahmann R, and Kämper J (2008) Reprogramming a maize plant: transcriptional and metabolic changes induced by the fungal biotroph Ustilago maydis. *Plant J.* 56:181–195

Dunn WB, Overy S, and Quick WP (2005) Evaluation of automated electrospray-TOF mass spectrometry for metabolic fingerprinting of the plant metabolome. *Metabolomics* 1:137–148

Edwards WR, Hall JA, Rowlan AR, Schneider-Barfield T, Sun TJ, Patil MA, Pierce ML, Fulcher RG Bell AA, and Essenberg M (2008) Light filtering by epidermal flavonoids during the resistant response of cotton *Xanthomonas* protects leaf tissue from light-dependent phytoalexin toxicity. *Phytochemistry* 69:2320–2328

Fiehn O (2002) Metabolomics – the link between genotypes and phenotypes. *Plant Mol. Biol.* 48:155–171

Fiehn O et al. (2007) The metabolomics standards initiative. *Metabolomics* 3:175–178

Giannoccaro E, Wang Y-J, and Chen P (2008) Comparison of two HPLC systems and an enzymatic method for quantification of soybean sugars. *Food Chem.* 106:324–330

Gibon Y, Vigeolas H, Tiessen A, Geigenberger P, and Stitt M (2002) Sensitive and high throughput metabolite assays for inorganic pyrophosphate, ADPGlc, nucleotide phosphates, and glycolytic intermediates based on a novel enzymic cycling system. *Plant J.* 30:221–235

Glauser G, Grata E, Dubugnon L, Rudaz S, Farmer EE, and Wolfender J-L (2008) Spatial and temporal dynamics of jasmonate synthesis and accumulation in Arabidopsis in response to wounding. *J. Biol. Chem.* 283:16400–16407

Goodacre R (2005) Making sense of the metabolome using evolutionary computation: seeing the wood with the trees. *J. Exp. Bot.* 56:245–254

Goossens A, Häkkinen ST, Laakso I, Seppänen-Laakso T, Biondi S, De Sutter V, Lammertyn F, Nuutila AM, Söderlund H, Zabeau M, Inzé D, and Oksman-Caldentey K-M (2003) A functional genomics approach toward the understanding of secondary metabolism in plant cells. *Proc. Natl. Acad. Sci. USA* 100:8595–8600

Gross ML and Caprioli RM (eds) (2005) In: Nibbering NMM (vol. ed) The encyclopedia of mass spectrometry, Vol 4. Fundamentals of and applications to organic (and organometallic) compounds. Elsevier, Amsterdam

Guy C, Kopka J, and Moritz T (2008) Plant metabolomics coming of age. *Physiol. Plant.* 132:113–116

Halket JM, Waterman D, Przyborowska AM, Patel RKP, Fraser PD, and Bramley PM (2005) Chemical derivatization and mass spectral libraries in metabolic profiling by GC/MS and LC/MS/MS. *J. Exp. Bot.* 56:219–243

Hall R, Beale M, Fiehn O, et al. (2002) Plant metabolomics: the missing link in functional genomic strategies. *Plant Cell* 14:1437–1440

Issaq HJ (1999) Capillary electrophoresis of natural products - II. *Electrophoresis* 20:3190–3202

Jellito T, Sonnewald U, Willmitzer L, Hajirezaei MR, and Stitt M (1992) Inorganic pyrophosphate content and metabolites in leaves and tubers of potato and tobacco plants expressing E. coli pyrophosphatase in their cytosol. *Planta* 188:238–244

Jenkins H, Hardy N, Beckmann M, Draper J, Smith AR, Taylor J, Fiehn O, Goodacre R, Bino R, Hall R, Kopka J, Lane GA, Lange BM, Liu JR, Mendes P, Nikolau BJ, Oliver SJ, Paton NW, Roessner-Tunali U, Saito K, Smedsgaard J, Sumner LW, Wang T, Walsh S, Wurtele ES, and Kell DB (2004)

A proposed framework for the description of plant metabolomics experiments and their results. *Nat. Biotechnol.* 22:1601–1606

Jiang H, Somogyi A, Timmermann BN, and Gang DR (2006) Instrument dependence of electrospray ionization and tandem mass spectrometric fragmentation of the gingerols. *Rapid Commun. Mass Spectrom.* 20:3089–3100

Jones WP and Klinghorn AD (2005) Extraction of plant secondary metabolites. In: Sarker SD, Latif Z, and Gray AI (eds) Natural products isolation, methods in biotechnology, Vol 20. Humana Press: Totowa, NJ

Kai K, Mizutani M, Kawamura N, Yamamoto R, Tamai M, Yamaguchi H, Sakata K, and Shimizu B-I (2008) Scopoletin is biosynthesized via orthohydroxylation of feruloyl CoA by a 2-oxoglutarate-dependent dioxygenase in Arabidopsis thaliana. *Plant J.* 55:989–999

Kishimoto K, Matsui K, Ozawa R, and Takabayashi J (2008) Direct fungicidal activities of C6-aldehydes are important constituents for defense responses in Arabidopsis against *Botrytis cinerea. Phytochemistry* 69:2127–2132

Klejdus B, Vacek J, Lojková L, Benešová L, and Kubáň V (2008) Ultrahigh-pressure liquid chromatography of isoflavones and phenolic acids on different stationary phases. *J. Chromatogr. A* 1195:52–59

Kobayashi N and DellaPenna D (2008) Tocopherol metabolism, oxidation and recycling under high light stress in Arabidopsis. *Plant J.* 55:607–618

Krishnan P, Kruger NJ, and Ratcliffe RG (2005) Metabolite fingerprinting and profiling in plants using NMR. *J. Exp. Bot.* 56:255–265

Kwon H-J, Jeong J-S, Lee Y-M, and Hong S-P (2008) A reversed-phase high-performance liquid chromatography method with pulsed amperometric detection for the determination of glycosides. *J. Chromatogr. A* 1185:251–257

Larkov O, Zaks A, Bar E, Lewinsohn E, Dudai N, Mayer AM, and Ravid U (2008) Enantioselective monoterpene alcohol acetylation in Origanum, Mentha and Salvia species. *Phytochemistry* 69:2565–2571

Lätti AK, Rihinen KR, and Kainulainen PS (2008) Analysis of anthocyanin variation in wild populations of bilberry (*Vaccinium myrtillus* L.) in Finland. *J. Agric. Food Chem.* 56:190–196

Laurens JB, Bekker LC, Steenkamp V, and Stewart MJ (2001) Gas chromatographic-mass spectrometric confirmation of atractyloside in a patient poisoned with *Callilepsis laureola. J. Chromatogr. B,* 765:127–133

Lee S-S, Lin H-C, and Chen C-K (2008) Acylated flavonol monorhamnosides, α-glucosidase inhibitors, from *Machilus philippinensis. Phytochemistry* 69:2347–2353

Lesney MS (2004) An apple a day. Today's Chemist at Work, Feb 2004, pp. 32–36

Li B, Abliz Z, Tang M, Fu G, and Yu S (2006) Rapid structural characterization of triterpenoid saponins in crude extract from *Symplocos chinensis* using liquid chromatography combined with electrospray ioniszation tandem mass spectrometry. *J. Chromatogr. A,* 1101:53–62

Lin CY, Wu HF, Tieerdema RS, and Viant MR (2007) Evaluation of metabolite extraction strategies from tissue samples using NMR metabolomics. *Metabolomics* 3:55–67

Linhardt RJ (1994) Capillary electrophoresis of oligosaccharides. *Methods Enzymol.* 230:265–280

Lísa M and Holčapek M (2008) Triacylglycerols profiling in plant oils important in food industry, dietetics and cosmetics using high-performance liquid chromatography–atmospheric pressure chemical ionization mass spectrometry. *J. Chromatogr. A* 1198-1199:115–130

Lísa M, Lynen F, Holčapek M, and Sandra P (2007) Quantitation of triacylglycerols from plant oils using charged aerosol detection with gradient compensation. *J. Chromatogr. A* 1176:135–142

Liu S, Tian N, Liu Z, Huang J, Li J, and Ferreira JFS (2008) Affordable and sensitive determination of artemisinin in *Artemisia annua* L. by gas chromatography with electron-capture detection. *J. Chromatogr. A* 1190:302–306

Luo J, Nishiyama Y, Fuell C, Taguchi G, Elliott K, Hill L, Tanaka Y, Kitayama M, Yamazaki M, Bailey P, Parr A, Michael AJ, Saito K, and Martin C (2007) Convergent evolution in the BAHD family of acyl transferases: identification and characterization of anthocyanin acyl transferases from *Arabidopsis thaliana. Plant J.* 50:678–695

Ma C, Wang H, Lu X, Xu G, and Liu B (2008) Metabolic fingerprinting investigation of *Artemisia annua* L. in different stages of development by gas chromatography and gas chromatography–mass spectrometry. *J. Chromatogr. A* 1186:412–419

Matsubara N and Terabe S (1996) Micellar electrokinetic chromatography. *Methods Enzymol.* 270:319–341

McDonnell LA and Heeren RMA (2007) Imaging mass spectrometry. *Mass Spectrom. Rev.* 26: 606–643

Mellon FA, Bennett RN, Holst B, and Williamson G (2002) Intact glucosinolate analysis in plant

4

extracts by programmed cone voltage electrospray LC/MS: performance and comparison with LC/MS/MS methods. *Anal. Biochem.* 306:83–91

Novák O, Hauserová E, Amakorová P, Doležal K, and Strnad M (2008) Cytokinin profiling in plant tissues using ultra-performance liquid chromatography-electrospray tandem mass spectrometry. *Phytochemistry* 69:2214–2224

Otieno DO and Shah NP (2007) A comparison of changes in the transformation of isoflavones in soymilk using varying concentrations of exogenous and probiotic-derived endogenous β-glucosidases. *J. Appl. Microbiol.* 103:601–612

Ouattara B, Angenot L, Guissou P, Fondu P, Dubois J, Frédérich M, Jansen O, van Heugen J-C, Wauters J-N, and Tits M (2004) LC/MS/NMR analysis of isomeric divanilloylquinic acids from the root bark of *Fagara zanthoxyloides* Lam. *Phytochemistry* 65:1145–1151

Pan J, Zhang S, Yan L, Tai J, Xiao Q, Zou K, Zhou Y, and Wu J (2008) Separation of flavanone enantiomers and flavanone glucoside diastereomers from *Balanophora involucrata* Hook by capillary electrophoresis and reversed-phase high-performance liquid chromatography on a C18 column. *J. Chromatogr. A* 1185:117–129

Papadopoulou K, Melton RE, Leggett M, Daniels MJ, and Osbourn AE (1999) Compromised disease resistance in saponin-deficient plants. *PNAS* 96:12923–12928

Pauli GF, Jaki BU, and Lankin DC (2005) Quantitative ^{1}H NMR: development and potential of a method for natural products analysis. *J Nat. Prod.* 68:133–149

Reichelt M, Brown PD, Schneider B, Oldham NJ, Stauber E, Tokuhisa J, Kliebenstein DJ, Mitchell-Olds T, and Gershenzon J (2002) Benzoic acid glucosinolate esters and other glucosinolates from *Arabidopsis thaliana*. *Phytochemistry* 59:663–671

Rellán-álvarez R, Hernández LE, Abadía J, and álvarez-Fernández A (2006) Direct and simultaneous determination of reduced and oxidized glutathione and homoglutathione by liquid chromatography–electrospray/mass spectrometry in plant tissue extracts. *Anal. Biochem.* 356:254–264

Řezanka T, Nedbalová L, and Sigler K (2008) Identification of very-long-chain polyunsaturated fatty acids from *Amphidinium carterae* by atmospheric pressure chemical ionization liquid chromatography-mass spectroscopy. *Phytochemistry* 69:2391–2399

Rupasinghe HPV, Jackson C-JC, Poysa V, Di Berardo C, Bewley JD, and Jenkinson J (2003) Soyasapogenol A and B distribution in soybean (Glycine max L. Merr.) in relation to seed physiology, genetic variability, and growing location. *J. Agric. Food Chem.* 51:5888–5894

Sadek PC (2002) The HPLC solvent guide, 2nd ed. Wiley, New York

Sana TR, Waddell K, and Fischer SM (2008) A sample extraction and chromatographic strategy for increasing LC/MS detection coverage of the erythrocyte metabolome. *J. Chromatogr. B.* 871:314–321

Scott RPW www.chromatography-online.org

Schulze B, Lauchli R, Sonwa MM, Schmidt A, and Boland W (2006) Profiling of structurally labile oxylipins in plants by in situ derivatization with pentafluorobenzyl hydroxylamine. *Anal. Biochem.* 348:269–283

Shimoda K, Sato N, Kobayashi T, Hamada H, and Hamada H (2008) Glycosylation of daidzein by the Eucalyptus cell cultures. *Phytochemistry* 69:2303–2306

Skorupinska-Tudek K, Poznanski J, Wojcik J, Bienkowski T, Szostkiewicz I, Zelman-Femiak M, Bajda A, Chojnacki T, Olszowska O, Grunler J, Meyer O, Rohmer M, Danikiewicz W, and Swiezewska E (2008) Contribution of the mevalonate and methylerythritol phosphate pathways to the biosynthesis of dolichols in plants. *J. Biol. Chem.* 283:21024–21035

Snyder LR, Kirkland JJ, and Glajch JL (1997) Practical HPLC method development, 2nd ed. Wiley, New York

Sumner LW, Amberg A, Barrett D, et al. (2007) Proposed minimum reporting standards for chemical analysis. *Metabolomics* 3:211–221

van der Klift EJC, Vivó-Truyols G, Claassen FW, van Holthoon FL, and van Beek TA (2008) Comprehensive two-dimensional liquid chromatography with ultraviolet, evaporative light scattering and mass spectrometric detection of triacylglycerols in corn oil. *J. Chromatogr. A* 1178:43–55

Vogler B and Setzer WN (2006) Characterisation of natural products. In: Cseke LJ, Kirakosyan A, Kaufman PB, Warber S, Duke JA, and Brielmann HL (eds) Natural products from plants. CRC Press, Boca Raton, FL, pp. 319–389

Vollhardt KPC (1987) Organic chemistry. WH Freeman, New York

Wade KL, Garrard IJ, and Fahey JW (2007) Improved hydrophilic interaction chromatography method for the identification and quantification of glucosinolates. *J. Chromatogr. A* 1154:469–472

Wang XK, He YZ, and Qian LL (2007) Determination of polyphenol components in herbal medicine by micellar electrokinetic capillary chromatography with Tween 20. *Talanta* 74:1–6

Ward JL, Harris C, Lewis J, Beale MH (2003) Assessment of ^{1}H NMR spectroscopy and multivariate analysis as a technique for metabolite fingerprinting of *Arabidopsis*. Phytochemistry 62:949–957

Ward JL, Baker JM, and Beale MH (2007) Recent applications of NMR spectroscopy in plant metabolomics. *FEBS J.* 274:1126–1131

Weckwerth W (2007) Metabolomics: methods and protocols. Humana press, Totowa, NJ

Weckwerth W (2008) Integration of metabolomics and proteomics in molecular plant physiology - coping with the complexity by data-dimensionality reduction. *Physiol. Plant.* 132:176–189

Weckwerth W, Wenzel K, and Fiehn O (2004) Process for the integrated extraction, identification and quantification of metabolites, proteins and RNA to reveal their co-regulation in biochemical networks. *Proteomics* 4:78–83

Wouters FS, Verveer PJ, and Bastiaens PIH (2001) Imaging biochemistry inside cells. *Trends Cell Biol.* 11:203–211

Yonekura-Sakakibara K, Tohge T, Matsuda F, Nakabayashi R, Takayama H, Niida R, Watanabe-Takahashi A, Inoue E, and Saito K (2008) Comprehensive flavonol profiling and transcriptome coexpression analysis leading to decoding gene-metabolite correlations in Arabidopsis. *Plant Cell* 20:2160–2176

Zhang Q, Wang G, Du Y, Zhu L, and Jiye A (2007) GC/MS analysis of the rat urine for metabonomic research. *J. Chromatogr. B* 854:20–25

Zhang H, Xie X, Kim M-S, Kornyeyev DA, Holaday S, and Paré PW (2008) Soil bacteria augment Arabidopsis photosynthesis by decreasing glucose sensing and abscisic acid levels in planta. *Plant J.* 56:264–273

Zhou Y, Han Q-B, Song J-Z, Qiao C-F, and Xu H-X (2008) Characterization of polyprenylated xanthones in *Garcinia xipshuanbannaensis* using liquid chromatography coupled with electrospray ionization quadrupole time-of-flight tandem mass spectrometry. *J. Chromatogr. A* 1206:131–139

Opportunities and Challenges for Ethnobotany at the Start of the Twenty-First Century 5

Monique S.J. Simmonds

Abstract Habitat destruction and urbanisation of many parts of the world have contributed to a decrease in the diversity of species being used in agriculture and for economic uses as well as our knowledge about the traditional uses of our biodiversity. The work undertaken by ethnobotanists can assist help access the impact biodiversity loss has on the well being of communities. However, scientists wishing to research biodiversity are faced with having to obtain the correct permits to collect species and any associated traditional knowledge about the uses of these species. This paper provides a summary of the types of agreements that are needed for ethnobotanical studies and the importance of having "prior informed consent" and "Access and Benefit-Sharing" clauses in research agreements, especially when dealing with traditional knowledge.

Introduction

Biodiversity is being threatened by many factors and the loss of plant diversity can impact negatively on the livelihoods of many communities. The research undertaken by ethnobotanists provides insights into how plants are used by different peoples throughout the world. This information can assist evaluate the social and economic importance of biodiversity and what impact biodiversity loss might have on the communities and can also support conservation strategies such as the Global Strategy on Plant Conservation. However, changes in legislation associated with ownership and access to biodiversity (Convention on Biological Diversity) now require ethnobotanists to obtain agreements before they undertake some of their studies. This paper provides a practical overview of the types of agreements that ethnobotanists now require before they start their studies and where some of the information can be obtained.

Observing and documenting the uses made of plants by different communities throughout the world has fascinated people for centuries. Today scientists undertaking this type of work are often called ethnobotanists and need to have a basic knowledge of anthropology, botany, pharmacology, linguistics, international politics, economics, history, communications, statistics, law and conservation – or have the input from those that have these skills as members of their team [1, 2]. Ethnobotany is without doubt a multidisciplinary subject. The growing interest in ethnobotany can be observed through the increased number of journals such as *Journal of Ethnobiology*, *Journal of Ethnopharmacology*, *Ethnobotany*, *Ethnobotany Research and Applications* and the range of university courses that

M.S.J. Simmonds (✉)
Royal Botanic Gardens, Kew,
Richmond Surrey UK, TW9 3AB
e-mail: m.simmonds@kew.org

A.E. Osbourn and V. Lanzotti (eds.), *Plant-derived Natural Products*,
DOI 10.1007/978-0-387-85498-4_5, © Springer Science+Business Media, LLC 2009

contain modules associated with different aspects of ethnobotany, especially the medicinal uses of plants.

Ethnobotanical studies can provide information about which species of plants people use in their day-to-day activities and how the species used for each activity can vary from country-to-country. These data illustrate the interdependency between people and their local flora and the factors that influence the selection of plants as food, medicine, fuel, etc. Information that links people with their plants is being lost in many parts of the world, especially as more communities become urbanised and adopt the trappings of a western life style. However, plants are still central to the lives of many people, especially in poor rural areas and their use is not static. For example, some species are used to treat specific medical conditions in one community but have other uses by different communities. A review of the uses of the genus *Plectranthus* (Lamiaceae) showed that one of the species, *P. barbatus* Andr., is used by communities in both the Old and New World to treat digestive, skin and infections, whereas another species, *P. amboinicus* (Lour.) Spreng, has similar uses to *P. barbatus* in Africa but in South America it is used to treat epilepsy and convulsions [3]. These species were taken from the Old to the New World and although there are data on the old traditional and modern uses no information could be found that would suggest why the use of one species changed but not the other.

Knowledge about the role of plants in the social and economic makeup of communities can provide an insight into what might happen if either biodiversity or cultural dilution occurs. For example, loss of medicinal plants through deforestation and the associated migration of traditional healers from villages to towns can result in villages not only loosing access to their plants but also the knowledge about the medicinal value of the plants to their communities. Thus ethnobotanical projects that evaluate the cultural importance of plants in an area can play an important role in plant conservation as these projects can provide qualitative and quantitative data about the diversity of species used in a community. These data along with data obtained from vegetation and animal surveys are also important for environment assessments. They can be used to assess the social importance of the genetic resources in specific habitats and thus the impact on communities if changes occur in that habitat and species are lost.

Information about the uses of specific species of plants in a community is usually obtained via interviews of individuals or groups of people and the scope of this information can vary depending on the methods used [1, 4]. Interviewing groups of people about the uses of plants often results in a greater diversity of species being identified, but more information about specific species, such as how it is prepared and when it should or should not be used is obtained from one-to-one interviews. The interviewing methods can also vary depending on the scope of the project. However, there are ethical, economic and legal issues associated with the collection and dissemination of these data, especially if "primary" information or traditional knowledge has been obtained in the course of the interviews [5]. Some communities want to restrict the dissemination of this information, whereas others want their data shared. Therefore it is critical to find out at the start of a project what use can be made of any data gathered, especially traditional knowledge. Once the data have been published it is difficult to monitor ownership. Data gathered for a study undertaken for purely social and conservation reasons can result in unforeseen consequences. For example, a survey of the medicinal plants in a village might results in data being gathered about the distribution of endangered medicinal plants of high economic value. This information if published could be used by plant hunters to target species. Similarly a guide to the medicinal plants in an area might provide information that increases wild harvesting of these species. In contrast, the data could empower a community to prioritise which species they need and thus the importance of developing methods to cultivate the threatened species. Therefore it is very important that the ethnobotanist identifies key people in the communities they work in and identifies and agrees with them exactly what is going to happen to any data gathered.

Conflicts about the ownership and rights of the information being exchanged about plant uses are not new. However, the rapid dissemination of information that can now occur via the internet highlights the challenges associated with trying to link "ownership of indigenous knowledge about a use" with the "academic research and then potential commercial development of that use". In most parts of the world scientists want to have a free right to exchange information and with the development of web-based interactive data exchange system it is going to be more and more difficult to keep track of who-owns what data. In many cases the exchange of "use" data are not the issue with indigenous communities it is usually how the data are used, especially if data are used to develop a commercial product without their involvement. This could happen indirectly. For example, an ethnobotanist publishes information about the plants used traditionally by a community in Brazil to treat diabetes, another author writes a review on what is known about plants used to treat diabetes and cites the work of the ethnobotanist in Brazil. Then a commercial company uses the review to select species for study. In this scenario the link between the original ethnobotanical study and the commercial study has been broken. However, if the company wants to justify that the species they have selected has been traditionally used to treat diabetes as part of patent claims or as part of their marketing package then they need to cite the community. To do this they might have to re-establish the link to the original community involved in the ethnobotanical study. In these cases the company could then negotiate some form of licence or revenues sharing agreement with the community or government appointed representatives. However, it might not be as simple as this, especially if other communities can show they used the same species for diabetes and thus feel that their traditional knowledge about the plant is being exploited. The development of *Hoodia* as an appetite suppressant product illustrates many of the challenges that need to be addressed if ethnobotanical data are to be used in drug development in the twenty-first century [6, 7].

Ethnobotany and the Convention on Biological Diversity

The importance of conserving biodiversity for sustainable use was a topic of the 1992 United Nations Conference on Environment and Development (Rio Summit) and resulted in a series of international action plans including the development of the Convention on Biological Diversity [8]. This Convention highlights not only the importance of biodiversity but also the importance of the links between biodiversity and the economic development of human societies, especially in areas of high biodiversity [9].

The Convention on Biological Diversity (CBD) came into force on 29 December 1993 and has now been ratified by over 90% of countries [8]. It was a landmark Convention as it was the first international Convention to state that a country has 'sovereign rights' over its genetic resources and to acknowledge the importance of sharing benefits that arise from using genetic resources (Box 1).

Many of the countries that have ratified the CBD are now developing legislation to implement the CBD, especially legislation associated with the access to genetic resources and the benefits that should be shared through the use of these resources [10]. Thus anyone undertaking a project that involves genetic resources, especially if the genetic resources are to be moved from country-to-country, should make themselves aware of the requirements of the CBD, of any existing national legislation relating to access to genetic resources and benefit sharing, and some

5

Box 1 Convention on Biological Diversity

Sections of the 42 Articles of the CBD that are relevant to ethnobotany

Article 1 Objective

The objectives of this convention … "are the conservation of biological diversity, the sustainable use of its components and the fair and equitable sharing of the benefits arising out of the utilisation of genetic resources, including by appropriate access to genetic resources."

Article 8 In-situ Conservation

(i) Endeavour to provide the conditions needed for compatibility between present uses and the conservation of biological diversity and the sustainable use of its components.
(j) Subject to its national legislation, respect, preserve and maintain knowledge, innovations and practices of indigenous and local communities embodying traditional lifestyles relevant for the conservation and sustainable use of biological diversity and promote their wider application with the approval and involvement of the holders of such knowledge, innovation and practices ad encourage the equitable sharing of the benefits arising from the utilization of such knowledge, innovation and practices;

Article 10 Sustainable use of components of biological diversity

(c) Protect and encourage customary use of biological resources in accordance with traditional cultural practices that are compatible with conservation or sustainable use requirements;
(e) Encourage cooperation between its governmental authorities and its private sector in developing methods for sustainable use o biological resources.

Article 15 Access to Genetic Resources

(1) Recognizing the sovereign rights of States over their natural resources, the authority to determine access to genetic resources rests with the national governments and is subject to national legislation.
(4) Access, where, granted shall be on mutually agreed terms and subject to the provisions of this Article.
(5) Access to genetic resources shall be subject to prior informed consent of the Contracting Party providing such resources, unless otherwise determined by that Party.
(7) Each Contracting Party shall take legislative, administrative or policy measures, as appropriate, and in accordance with Articles 16 and 19 and where necessary, through the financial mechanism established by Articles 20 and 21 with the aim of sharing in a fair and equitable way the results of research and development and the benefits arising from the commercial and other utilisation of genetic resources with the Contracting Party providing such resources. Such sharing shall be upon mutually agreed terms.

Article 16 Access to and Transfer of Technology

(1) Each Contracting Party, recognising that technology includes biotechnology, and that both access to and transfer of technology among Contracting Parties are essential elements for the attainment of the objectives of this Convention, undertakes subject to the provisions of this

(continued)

Box 1 (continued)

Article to provide and/or facilitate access for and transfer to other Contracting Parties of technologies that are relevant to the conservation and sustainable use of biological diversity or make use of genetic resources and do not cause significant damage to the environment.

Article 17 Exchange of information

(1) The Contracting Parties shall facilitate the exchange of information, from all public available sources, relevant to the conservation and sustainable use of biological diversity, taking into account the special needs of developing countries.

(2) Such exchange of information shall include exchange of results of technical, scientific and socio-economic research, as well as information on training and surveying programmes, specialized knowledge, indigenous and traditional knowledge as such and in combination with the technologies referred to in Article 16 (1) where feasible, include repatriation of information.

Article 18 Technical and Scientific co-operation

(4) The Contracting Parties shall, in accordance with national legislation and policies, encourage and develop methods of cooperation for the development and use of technologies, including indigenous and traditional technologies, in pursuance of the objectives of the Convention. For this purpose, the Contracting parties shall also promote cooperation in the training of personnel and exchange of experts.

The CBD web site provides more details about the Articles [8]

of the developing "best practise" agreements that deal with the transfer of these resources.

It is thus recommended that ethnobotanists familiarise themselves with how the country they propose to work in is implementing the CBD. This is because in many counties "Access and Benefit Sharing" (ABS) legislation covers not only what needs to be done to gain access to the genetic resources but can also include clauses that cover what happens to the genetic resources, including products derived from the genetic resources. For example, an ethnobotanist investigating the plants used in Africa for the treatment of malaria might take some plant material back to their research organisation to be verified by a botanist. The botanist talks about the uses of the species to a pharmacologist who is interested in anti-malarial plants and takes some of the verified material to the laboratory. They test the material for anti-malarial activity, isolate the active compound and pass the compound on to their colleagues in a chemistry department. The chemists then modify the compound to improve its efficacy and develop a method to synthesise the new derivative. Such a sequence of events on a small amount of material is fanciful. Nevertheless, if the ethnobotanist is going to include a clause in their ABS agreement saying they will monitor the use of the plants they are collecting and any derivatives from the plants, then they need to think whether they will be able to keep track of the movement of the material from laboratory-to–laboratory and to the compound (s) the chemists have synthesised that are derived from compound(s) isolated from the original plant material they collected. A

5

challenge that could only be met if very good tracking systems are in place that last beyond the time-scale of the ethnobotanical project.

Because of some of these challenges the CBD has had a huge impact on many aspects of ethnobotany projects, especially those that involve pharmacology and natural product research. However, this type of research can be facilitated by appropriate agreements being in place and clarity as to what is going to happen to material collected. The new bioprospecting legislation in South Africa "National Environmental Management: Biodiversity Act 2004, Regulations on Bio-Prospecting, Access and Benefit Sharing 2008" came into force on 1 April 2008 [11] and provides some guidelines for projects that could result in a commercial product being developed from genetic resources. This legislation requires researchers to obtain a "permit" if they wish to study the uses of the genetic diversity of South Africa, and this includes researchers within South Africa, especially if the use of the resources could result in the commercial development of the material being studied.

Some ethnobotanical projects report on the cultural uses of wild relatives of crop plants that have evolved over generations. The improved "crop" traits of these wild relatives could be considered as the Intellectual Property Rights (IPR) of the communities that have selected and grown the crops, yet many of these communities would not recognise their inputs as IP or giving rise to IPRs. The western concept of IPR is not appropriate or relevant for many indigenous communities as they have different concepts of property as illustrated by the work of Posey [12]. Lawyers working on the CBD have been trying to develop a uniform approach as to how agreements with these communities should deal with IP in ways that are compatible with the requirements of international policies such as the CBD, the International Undertaking on Plant Genetic Resources, the World Trade Agreements on Trade Related Aspects of Intellectual Property Rights (TRIPS) [13], and also respect the customs and lifestyles of indigenous communities.

Article 27 of the TRIPS agreement excludes plants from being patented but Article 27.3 allows for the protection of plant varieties which could be a link between the concepts of the CBD (namely obtaining prior informed consent for use of information or genetic material, and sharing benefits fairly and equitably with the original holder) with those in TRIPS (to promote effective and adequate protection of IPR, which could be plant varieties being selected for specific properties using the knowledge of local communities). However, the issues associated with IPR and Traditional Knowledge (TK) are diverse and are not resolved. It is clear that they will continue to be the topic of future CBD meetings. For instance the Working Group on CBD Article 8j is, amongst other things tasked with developing elements of a *sui generis* system to protect TK by 2010 [14].

Researchers working within a western context of IPR are used to dealing with issues associated with the rights to data generated during a project funded through their employment. When these researchers work with other organisations, be they commercial or academic, agreements are usually in place to deal with joint ownership of the resulting work. Surely it is just an extension of this practice to discuss and agree the terms with the indigenous communities and other stakeholders about how data collated about their resources are going to be used? However, it is clear that this has not always happened.

Within many research organisations the procedures are usually in place to facilitate the drawing up of the different IP agreements, whereas these procedures are often difficult to facilitate with village communities that sometimes do not recognise IP, or do not have knowledge about IPR nor easy access to lawyers to assist them. It can also take time for a researcher to develop a relationship within a community and identify whom within that community has the authority to give permission to disseminate information about the traditional knowledge about the uses of their plants. Because communities have expressed concerns about past injustices when their biodiversity has been exploited without their consent many are reluctant to get involved in ethnobotanical projects. The resentment and distrust is also fuelled because traditional information about the uses of the plants has been used in the past for the development of commercial products with no "prior informed consent" or "benefit sharing". However, the proportion of ethnobotanical projects that result directly in the development of a commercial product that does not involve the community are small, and usually input from many others is required before a commercial product is developed. Nevertheless those that have shared their traditional knowledge in the early stages of a project should be recognised. In most cases if the term "commercial" was replaced with "economic", then many ethnobotany projects do assist assess the economic value of plants to local communities. In these cases the communities are often involved in scoping the project and do not feel as threatened as it is their "economic" uses that are being evaluated. A project currently being undertaken in Panama is looking closely at the links among traditional uses of the local flora, their conservation and sustainable development [15].

Exchange of Plants "Material Transfer Agreements"

Plants being studied as part of an ethnobotanical project need to be correctly identified with help from a botanist. In many cases, especially if the results are to be published, a sample of each species needs to be deposited as a voucher in a national herbarium and the voucher number given to each specimen quoted in publications. If plant specimens are to be transferred between countries then a Material Transfer Agreement (MTA) is needed, such agreements are sometimes needed by a herbarium even when specimens are being transferred within a country. If specimens are to be taken out of a country then an agreement should be in place that deals with the "fair and equitable benefit-sharing" of benefits that could arise from the use of the material. CBD Article 15 states that access to genetic resources should be subject to national legislation (15.1), on mutually agreed terms (15.4), with Prior Informed Consent (15.5), and benefit sharing (15.7). The "Bonn Guidelines on Access and Benefit-Sharing" which were adopted by the CBD Conference of the Parties in April 2002 suggest that this permission should be in a written form, and Appendix 1 of the Guidelines provides a list of suggested elements for MTAs [16]. Another important aspect of the MTA, and a requirement of Article 15, is to show that access to genetic resources has been with the "prior informed consent" (PIC) of the Party providing the resource, and on Mutually Agreed Terms. This includes an explanation of why the material is being exchanged and what the plants are going to be used for, especially when someone is requesting to have plants taken from one country to another.

In the case of material being exchanged for identification then the agreement should indicate that the exchange is for taxonomic study and not for commercial use. In any case the sender can place restriction within the terms of the agreement as to what the material is to be used for as well as what it should not be used for. The sender might give permission for the use of the specimen in taxonomic research but restrict the use of the information they have supplied about the distribution and uses of the plants.

Only after an agreement is in place should there be any exchange of material. It can take time to develop these agreements and unless an organisation is used to working within the context of the CBD it can be difficult for an individual scientist to identify who in a foreign country needs to be involved in drawing up these agreements let alone dealing with the complexity of what communities and governments might define as "fair and equitable benefit-sharing".

An issue that is often encountered when starting to negotiate these agreements is who within each country has the authority to develop the terms for the agreement and who signs on behalf of a country. These are issues that can take time to resolve as in some countries government departments might need to be part of the agreement, whereas in other countries government departments do not need to be part of a formal agreement but might require a letter to support the aims of the proposed projects and terms outlined in the agreement. In recent years the government departments or organisations dealing with the CBD for most country have become available on the CBD web site [8].

One of the many challenges for the scientists wishing to comply with the CBD is that currently there are very few master agreements that can be downloaded from websites. In fact, agreements usually have to be developed country-by-country and project-by-project to reflect the needs and interest of the countries and the organisations involved in the research. However, there are agreements being used by different organisations that can serve as templates and the CBD web site does provide guidelines as to what should be in the agreements, up-to-date information from the Conference of the Parties that are debating "PIC" and "ABS" [8] and books such as Biodiversity and Traditional Knowledge [17] provide an overview of cases studies. The Swiss 'ABS-Management Tool: Best Practice Standard and Handbook for Implementing Genetic Resource Access and Benefit-Sharing Activities' has a very useful model MTA [18].

Botanic Gardens and Herbaria

When a scientist wants to have access to a species of plant for their research projects they often request material from a botanic garden. Nowadays although most botanic gardens continue to assist with these requests they might have restrictions placed on what they can do with the plants they have in the gardens. For example, unless the botanic garden was granted the rights to pass the material onto third parties in their "PIC" agreement with the country they obtained the material from, they are not free to pass the plants onto third parties. In these cases the third party needs to apply directly to the source county. Thus some botanic garden place restrictions, such as not to use a plant for commercial gain, on the supply of plants to third parties on plants they have obtained from other countries prior to 1992 and ensure that conditions in the specific MTAs associated with each plant are adhered to when dealing with material acquired since 1992, that is material obtained by the gardens after the CBD was adopted. These procedures reflect the impact the CBD has had on the exchange of non-indigenous plant material, especially on botanical gardens involved in different aspects of plant science, including ethnobotany.

Ethnobotanists are encouraged to contact botanical gardens in the country they are going to work in at the start of a project, as many gardens are very keen to increase their knowledge and displays of locally used species. These gardens might be able to not only provide support to the ethnobotanist but also to the communities involved in the project, especially if plants are being over-exploited and need to be conserved.

Most herbaria obtain specimens from different countries with the prime purpose of using them for taxonomic purposes, thus if this is agreed in the original MTA, the specimens can be exchanged with third parties. But as stated earlier the herbarium will require the recipient of the loan to sign a further MTA that requires them to state what they are going to use the plant material for before it is dispatched and that it is not going to be used commercially.

The Royal Botanic Gardens, Kew was one of the first botanical organisations to adopt an institutional policy on CBD and to use Material Acquisition Agreements (MAA) and MTAs to clarify on what terms material comes to Kew or is lent or given to others by Kew. The procedures adopted by Kew were work-shopped with 28 botanic gardens from 21 countries and resulted in the "Principles on Access to Genetic Resources and Benefit Sharing" [19]. This best practice model provides some clarity and guidance for staff dealing with requests to send material to or from Kew. For example, if an ethnobotanist wanted to have some specimens identified that could not be identified in the source country, it is likely that the ethnobotanists would be asked to either send their samples via their national herbarium or directly to Kew with a MTA or ABS agreements depending on the relationship that Kew has with the organisations involved in the project.

An overview of the terms that should be considered when drafting ABS agreement are provided in Appendix 1 of the Bonn Guidelines on Access and Benefit-Sharing [16]. Details about how the Royal Botanic Gardens, Kew developed their "Access and Benefit-Sharing Agreements (ABSA)" is illustrated by Cheyne [20]. She highlights the different stages of negotiations that can take place in developing an ASBA and the importance of having clarity in the terms used. As a result of these negotiations agreements have been successfully negotiated with different organisations and governments around the world taking part in the Millennium Seed Bank Project. The agreements cover access to the genetic resources in the source country, which in this case are seeds, and the scope of research that Kew can undertake on the seeds at Kew and the terms by which Kew can pass the seeds or any information about the seeds to third parties. These negotiations take time but when an ABSA has been signed then the exchange of material and associated research can take place. To date very few of these negotiations involve the sharing of ethnobotanical data about the uses of these plants. However, interest in this area is increasing and if seeds are going to be conserved then it is important to conserve information about their traditional uses.

Traditional Knowledge

If indigenous and/or traditional knowledge about the plants is going to be collected as part of an ethnobotanical study then it is important that this is covered in the original MTA and also that some thought has been given to how this information is going to be used and shared with others. In some cases the term indigenous traditional knowledge (ITK) is used to cover the current as well as historical uses of plants in a community. Within a community ITK is not static. Some ITK can be lost due to individuals with the knowledge leaving or it is dynamic and evolves to meet the needs of supporting the livelihoods of those in the community. The ethnobotanist should make themselves aware of the complexity of the cultural [5, 12] as well as legal issues associated with protecting ITK as provided by the discussions intergovernmental committee discussions [21] One issue that is very difficult to determine at the start of an ethnobotanical project is whether "unique" ITK is going to be identified and whether the information obtained from talking with one community is common to other communities. The recent introduction of terms such as "traditional resource rights" recognise the importance of ITK and the importance that communities retain or even regain control over their resources [22]. Thus ethnobotanists need to be sure they address ITK issues in their ABSA. The challenges are that many of these agreements are being developed at a national level as required by CBD Article 8j, but it is the community and individuals in that community that owns/controls the ITK. Thus the ethnobotanist often needs to establish contacts at a national and community level before starting the work. Notwithstanding the desire of most ethnobotanists to comply with the spirit of the CBD it can be very difficult for them to find out how they should obtain permission from all the appropriate stakeholders to gather ITK. There is a lack of clarity and in most cases no formal legally approved procedure that enables the ethnobotanist to know whether they have the appropriate approval from community, regional and national authorities. An IUCN report noted that "a conclusion of our initial research is that there is no framework in national and international law that is currently able to address the legal rights relating to genetic resources … it is not possible for countries to depend on normal contractual processes, documents and provisions, to protect their rights under ABS Agreements" [23]. Thus with debates still going on among those involved in drafting the legislation how do ethnobotanists make progress in furthering their studies on the uses of biodiversity? However, it is not an easy task to develop a regime that would provide an effective means to promote ABS in a fair and equitable way, that provides clarity to those involved, is legally binding so that it provides rights and protection to those involved in the agreement. Once ITK has been collated and left the community it is difficult, if not impossible, for the village elders to control what use is made of it. Advances in information exchange are now so rapid that it is very difficult to predict how best to control knowledge transfer, other than by empowering indigenous people with ways that they can be fully involved in the dissemination of their ITK by ensuring it is first published by them. More and more genetically rich countries will be closing their door to the exchange of genetic resources and ITK by outside researchers unless ABS agreements are in place and the procedures in the ABS agreements adhered to. Some case studies are available that illustrate the different types of benefits that can be considered as part of an academic project [17].

Terms in Agreements (Box 2 with Definitions)

When considering the scope of CBD related agreements it is important to agree on the definition of terms. Most ethnobotanical projects are undertaken as academic research projects with no interest in the commercial exploitation of the plants being studied. Others have a marginal commercial or "economic" interest, for examples, surveys of medicinal plants and surveys of plant species being traded in an area. However, some of the data and plants gathered during these projects could have commercial interest and it is thus important that the term "commercial" is defined. Many benefit-sharing and "PIC" clauses will allow academic research to be undertaken on the genetic resource but as soon as a commercial potential is realised then the parties to the agreement need to get together again to discuss the next stages and perhaps negotiate a new agreement to cover the commercial work. This is important if the parties want to avoid being labelled as "biopirates" [24]. However, compliance with all aspects of the CBD is a challenge. At the moment there is no international regime to ensure clarity in the procedures that need to be in place and implemented to monitor compliance. In some cases enforcement takes place through national laws and in China and Madagascar there have been cases of scientific researchers being imprisoned for not getting the correct permits. In these cases they were caught undertaking their research in country. Currently the enforcement regime is ad hoc and varies from country-to-country. Most ethnobotanists and those involved in this type of research want to ensure they and their collaborators adopt what they consider best practices for ethnobotanical research to avoid reputational risks through bad press and being labelled a biopirate. Therefore agreements should clearly define what should happen if an ethnobotanical project moves from being academic research to commercial. A clear robust definition of commercialisation should be agreed and a clause should indicate what those involved in the project should do is this happens. This does not have to mean that a project with commercial potential needs to be classed as commercial when it starts. Often a great deal of basic research need to be undertaken before this stage is reached and if all projects with commercial potential were classed as commercial from the start, then this could hinder many of these projects ever starting. Since the publication of the Bonn Guidelines [16] there has been increased international activity in debating different aspects of benefit sharing and this has resulted in some

Box 2 Some of the definitions used in agreements about the uses of genetic resources

Biopiracy: Utilisation and/or appropriation of genetic resources that is not based on the necessary access permits or does not fulfil the agreed conditions and therefore is illicit [18].

"**Commercialisation**" is not defined within the CBD [8].

Commercialise and Commercialisation: Filing a patent application, obtaining, or transferring intellectual property rights or other tangible or intangible rights by sale or licence or in any other manner; commencement of product development; conduct market research and seeking pre-market approval and/or the sale of any resulting product [20].

Genetic Resources: Genetic material of actual or potential value (as defined in the CBD [8].

Genetic Resources: Any biological material of plant, animal, microbial, fungal or other origin of actual or potential value containing functional units of heredity transferred "add name of institute" under this agreement and its progeny and derivatives, including extracts and compounds obtained from genetic resources and analogies of those compounds [20].

Material Transfer Agreement (MTA): Standardized contract or binding legal agreement between the owner of genetic material and the recipient of the material [18].

Prior Informed Consent (PIC): Prior Informed Consent is the consent of the relevant competent national authority/authorities in the provider country granted for the research and utilization of genetic resources. The consent of relevant stakeholders, such as indigenous and local communities, should also be obtained, as required by individual situations and subject to domestic law [18].

Stakeholders: All institutions, agencies, organizations, communities and individuals that may be involved in the ABS procedure in accordance with national law or based on case by case decisions: i.e. government agencies, regional and local governments and representatives of indigenous and local communities, local organizations [18].

Based on definitions in the Convention of Biodiversity [8], Cheyne [20] and the Swiss academy of Science Access and Benefit Sharing website [18].

national bodies having raised expectations that any request to gain access to genetic resources will result in commercial exploitation.

Ethnobotany and the Global Strategy for Plant Conservation

The overall objective of the CBD Global Strategy for Plant Conservation (GSPC) strategy is 'to halt the current and continuing loss of plant diversity'. The GSPC has five objectives covered by 16 targets to be reached by 2010 [25], although it is clear that countries will be asking for the timeframe to be extended. Of these 16 targets, five (Targets 3, 9, 11, 12 and 13) are particularly relevant to ethnobotany (Box 3). Much of the data being obtained in ethnobotanical projects support these international targets, although a trawl through the literature being published in the main ethnobotanical journals shows that very few publications make reference to the GSPC. Although organisations such

5

Box 3 Global Strategy for Plant Conservation

Objective 1: Understanding and documenting plant diversity

Target 1. A widely accessible working list of known plant species as a step towards a complete world flora.
Target 2. A preliminary assessment of the conservation status of all known plant species at national, regional and international levels.
Target 3. Development of models with protocols for plant conservation and sustainable use, based on research and practical experience.

Objective 2: Conserving plant diversity

Target 4. At least 10% of each of the world's ecological regions effectively conserved.
Target 5. Protection of 50% of the most important areas of plant diversity ensured.
Target 6. At least 30% of production lands managed consistent with the conservation of plant diversity.
Target 7. 60% of the world's threatened species conserved in situ.
Target 8. 60% of threatened plant species in accessible ex situ collections, preferably in the country of origin and 10% of them included in recovery and restoration programmes.
Target 9. 70% of the genetic diversity of crops and other major socio-economically valuable plant species conserved, and associated indigenous and local knowledge maintained.
Target 10. Management plans in place for at least 100 major alien species that threaten plant, plant communities ad associated habitats and ecosystems.

Objective 3: Using plant diversity sustainably

Target 11. No species of wild flora endangered by international trade.
Target 12. 30% of plant-based products derived from sources that are sustainable managed.
Target 13. The decline of plant resources, and associated indigenous and local knowledge, innovations and practices that support sustainable livelihoods, local food security and health care, halted.

Objective 4: Promoting education and awareness about plant diversity

Target 14. The importance of plant diversity and the need for its conservation incorporated within communication, educational and public-awareness programmes.

Objective 5: Building capacity for the conservation of plant diversity

Target 15. The number of trained people working with appropriate facilities in plant conservation increased, according to national needs, in order to achieve the targets of this strategy.
Target 16. Networks for plant conservation activities established or strengthened at national, regional and international levels.

More details about the GSPC available on http://www.UNEP/CBD/COP/6/IND/21/Add [25].

as the Food Agriculture Organisation (FAO) are providing mechanisms to support the implementation of some targets such as Target 12 through forest certification schemes and Non-Wood Forest Products programmes [26].

Those ethnobotanical projects collating data about the uses and sustainable harvesting of plants could be providing the much needed base-line data to support the monitoring of these GSPC targets [25]. As habitats are being threatened by changes in land use and climate change there is a real need for more ethnobotanical projects that can provide empirical data about the economic and social importance of plants with people, especially those living in rural communities throughout the world. Planta Europa [27] provides a very good overview of the progress being made towards meeting the GSPC targets in Europe and the many challenges that still need to be addressed. It is hoped that in the future more ethnobotanists will consider contacting the organisations monitoring the GSPC targets in the country they propose to work in, to see if their data could contribute to a GSPC target.

Questions to Ask Before Starting an Ethnobotanical Project

At the start of an ethnobotanical project there are many practical as well as scientific points that need to be addressed. Notwithstanding the challenges that CBD legislation can pose at the start of a project it is important that they are addressed. This is especially true if the data are to be published and used to support the sustainable utilisation of biodiversity for future generations. The Swiss Academy of Sciences web site [18] provides a simple step-by-step guide to what an academic should consider with respect to the implication of the CBD before undertaking an ethnobotanic study. Other texts by Carlson and Maffi [9] and Hamilton and Hamilton [28] provide many examples of the different types of interviewing techniques that can be used to obtain information and the equipment needed for field work. Many socio-cultural aspects of ethnobotany that need to be considered in planning a project are provided by Maffi [29] and best practice guidelines by Martin [1].

There is no doubt that ethnobotanical studies can be inspirational as well as physically and mentally hard work. There is so much to learn about how the peoples of the world use and have used our often threatened flora. Nowadays you need the stamina to make sure you have dealt with the ethical challenges of ethnobotanical research [30], but once these issues are addressed the opportunities offered by ethnobotany are well worth the effort.

Acknowledgements I would like to thank the many scientists and lawyers that have influenced how I have developed plant-based research projects that include traditional knowledge, especially Darrel Posey, Ghillean Prance, Alan Hamilton, Phyllida Middlemiss, Kerry ten Kate and China Williams as well as the people I have worked with in communities from different parts of Europe, Africa, South America and Asia.

References

1. Martin, G.J. (1995) Ethnobotany. Chapman & Hall, London
2. Cotton, C.M. (1997) Ethnobotany: Principles and Application. Wiley, Chichester
3. Lukhoba, C.W. et al. (2006) *Plectranthus*: a review of ethnobotanical uses. *J. Ethnopharm.* 103, 1–24
4. Asase, A. et al. (2006) Ethnobotanical study of some Ghanaian anti-malarial plants. *J. Ethnopharm.* 99, 273–279
5. Alexiades, M.N. (2004) Ethnobiology and globalization: science and ethics at the turn of the century. In *Ethnobotany and Conservation of Biocultural Diversity* (Carlson, T.J.S. and Maffi, L., eds.), pp. 283–305. *Adv. Econ. Bot.*

5

15, The New York Botanical Garden Press, Bronx, NY
6. Simmonds, M.S.J. (2003) Novel drugs from botanical sources. *Drug Discov.* Today 8, 721–722
7. Wynberg, R. (2004) Rhetoric, realism and benefit-sharing: use of traditional knowledge of *Hoodia* species in the development of an appetite suppressant. *J. World Intell. Prop.* 4, 856
8. Secretariat of the Convention on Biological Diversity (2008) Convention on Biological Diversity (www.cbd.int)
9. Carlson, T.J.S. and Maffi, L. (2004) Advances in Economic Botany, vol 15, The New York Botanical Garden Press, Bronx, NY
10. Secretariat of the Convention on Biological Diversity (2006) Database on ABS Measures (www.cbd.int/abs/measures.shtml)
11. Government Gazette, South Africa (2008) National Environmental Management: Biodiversity Act 2004, Regulations on Bio-Prospecting, Access and Benefit Sharing 2008 (www.cbd.int/doc/measures/abs/msr-abs-za1-en.pdf)
12. Posey, D.A. (2002) Commodification of the sacred through Intellectual Property Rights. *J. Ethnopharm.* 83, 3–12
13. World Trade Agreements (1994) Trade Related Aspects of Intellectual Property Rights (TRIPS) (www.uspto.gov/web/offices/com/doc/uruguay/finalact.html)
14. Secretariat of the Convention on Biological Diversity (2007) Working Group on Article 8(j) (www.cbd.int/convention/wg8j.shtml)
15. Kursar, T.A. et al. (2007) Linking bioprospecting with sustainable development and conservation: the Panama case. *Biodiv. Conserv.* 16, 2789–2800
16. Secretariat of the Convention on Biological Diversity (2002) Bonn Guidelines on Access to Genetic Resources and Fair and Equitable Sharing of the Benefits Arising out of their Utilization. Montreal (www.cbd.int/doc/publications/cbd-bonn-gdls-en.pdf)
17. Laird, S. (2002) Biodiversity and Traditional Knowledge. Earthscan, London
18. Swiss Academy of Science (2006) Benefit sharing (http://abs.scnat.ch/)
19. Royal Botanic Gardens Kew (2001) *Principles on Access to Genetic Resources and Benefit-sharing* (www.kew.org/conservation/principles.html)
20. Cheyne, P. (2004) Access and Benefit-Sharing Agreements: bridging the gap between scientific partnerships and the Convention on Biological Diversity. *In Seed Conservation: Turning Science into Practice* (Smith, R.D. et al., eds.), Royal Botanic Gardens, Kew
21. World Intellectual Property Organisations (WIPO) (2008) Traditional Knowledge, Genetic Resources and Traditional Cultural Expressions/Folklore (www.wipo.int/tk/)
22. Maffi, L. (2004) Maintaining and restoring biocultural diversity: the evolution of a role for ethnobotany. In *Ethnobotancy and Conservation of Biocultural* Diversity (Carlson, T.J.S. and Maffi, L., eds.), pp. 9–35. *Adv. Econ. Bot.* 15, The New York Botanical Garden Press, Bronx, NY
23. Carrizosa, S. et al. (2004) Accessing biodiversity and sharing the benefits: lessons from implementing the Convention on biological diversity. IUCN, Environmental Law Centre; IUCN, Environmental Law Programme; University of California, Genetic Resources Conservation Program; Germany, Ministry for Economic Co-operation and Development –IUCN, Gland
24. Shiva, V. (1997) Biopiracy: The Plunder of Nature and Knowledge. South End Press, Boston, MA
25. Secretariat of the Convention on Biological Diversity (2008) Global Strategy on Plant Conservation (www.cbd.int/gspc/)
26. Food and Agriculture Organization of the United Nations (2008) (www.fao.org/biodiversity/genetic resources/forests/en/)
27. Planta Europa (2008) *A sustainable future for Europe and European Strategy for Plant Conservation* 2008–2014. Plantlife International (Salisbury, UK) and the Council of Europe (Strasbourg, France)
28. Hamilton, A. and Hamilton, P. (2006) Plant Conservation: An Ecosystem Approach. Earthscan, London
29. Maffi, L. (2001) On Biocultural Diversity: Linking Language, Knowledge and the Environment. Smithsonian Institute Press, Washington, DC/London
30. Eshbaugh, W.H. (2008) A dilemma: economic/ethnobotanical research in the twenty-first century. *Econ. Bot.* 62, 3–11

Part II

Secondary Metabolite Biosynthesis

Introduction to the Different Classes of Biosynthetic Enzymes

6

Luzia V. Modolo, Angelika I. Reichert, and Richard A. Dixon

Abstract Plant natural products are intimately associated with traits such as quality, yield, disease resistance, stress tolerance, color, and fragrance, in addition to being important dietary components and phytomedicines. In spite of the apparent complexity of natural product biosynthesis, much of the rich chemical diversity of the plant kingdom arises from a limited number of chemical scaffold types, modified by specific chemical substitutions such as hydroxylation, glycosylation, acylation, prenylation, and *O*-methylation. The molecular genetic basis underlying plant natural product chemistry has recently been the subject of concerted genomic and genetic approaches, facilitated by the fact that many of the key enzymatic steps in scaffold formation and substitution are catalyzed by proteins originating from recognizable gene families (e.g. polyketide synthase, glucosyltransferase) that have undergone significant expansion throughout plant evolution. This overview summarizes the types of enzymatic reactions involved in plant secondary metabolism from a pathway organization perspective that highlights the entry points from primary metabolism, general scaffold formation and scaffold modification (Box 1).

L.V. Modolo, A.I. Reichert, and R.A. Dixon (✉)
Plant Biology Division, Samuel Roberts Noble Foundation, Ardmore, OK 73401, USA
e-mails: radixon@noble.org;
aireichert@noble.org;
lvmodolo@gmail.com

The Primary–Secondary Metabolism Interface

Phenylpropanoids

The aromatic amino acid L-phenylalanine (primary metabolite) is directed into the phenylpropanoid pathway leading to hydroxycinnamic acids, lignin and flavonoids by the activity of L-phenylalanine ammonia-lyase (PAL), which brings about its nonoxidative deamination yielding ammonia and *trans*-cinnamic acid (Fig. 1). PAL is one of the most studied plant enzymes, and its crystal structure has recently been solved [2]. PAL is related to the histidine and tyrosine ammonia-lyases of amino acid catabolism. A class of bifunctional PALs found in monocotyledonous plants and yeast can also deaminate tyrosine [3]. A single His residue is responsible for this switch in substrate preference [3, 4]. All three enzymes share a unique MIO (4-methylidene-imidazole-5-one) prosthetic group at the active site. This is formed autocatalytically from the tripeptide Ala-Ser-Gly by cyclization and dehydration during a late

A.E. Osbourn and V. Lanzotti (eds.), *Plant-derived Natural Products*,
DOI 10.1007/978-0-387-85498-4_6, © Springer Science+Business Media, LLC 2009

Box 1 A functional classification of natural product biosynthetic enzymes.

The immense variety of plant natural products is generated by an equally large and at first sight confusing array of enzymes. However, these enzymes catalyze a relatively limited number of reaction types [1]. For the purpose of this survey, the enzymes will be divided into three major groups based on their positions in the overall scheme of secondary metabolite biosynthesis:

- Enzymes at the interface between primary and secondary metabolism.
  - A small number of key enzymes catalyze the first committed steps which direct compounds from primary into secondary metabolism (Fig. 1). They control flux into natural products without depleting pools of primary metabolites, and must therefore act quickly and efficiently in response to developmental or environmental cues. For this reason they are often regulated both transcriptionally and post-transcriptionally.
- Enzymes forming plant secondary metabolite scaffolds
  - This group of enzymes directs flux into the major classes of plant natural products: e.g. polyketides (including flavonoids), alkaloids and terpenes. The initially formed scaffold molecules then enter different branch pathways as precursors for further downstream modifications.
- Enzymes for modification of scaffold structures
  - Modification reactions create the enormous diversity of plant natural products, providing new molecules with different biological activities from the basic scaffolds outlined above. The plant kingdom contains a large number of enzymes that catalyze hydroxylation, epoxidation, aryl migration, glycosylation, methylation, sulfation, acylation, prenylation, and reduction of secondary metabolite skeletons. Figure 6 shows how a single molecule (the isoflavone genistein) can be converted to a range of different products by such enzymes.

stage of chain folding [5]. Two reaction mechanisms have been proposed for the elimination of ammonia, with a Friedel-Crafts-like acylation mechanism being most favored [3].

Many plants utilize different PAL isoforms for stress responses or for biosynthesis of structural components, and these different PALs exhibit differential expression in distinct tissues. Metabolic channeling may help control the flux of phenylalanine through PAL into the different phenylpropanoid branch pathways [6, 7].

Polyketides

Many polyketide-derived plant natural products originate in part from acetyl CoA via malonyl CoA (Fig. 1). For example, the key reaction in flavonoid biosynthesis, catalyzed by chalcone synthase (CHS) (Fig. 2), combines a phenylpropanoid-derived moiety, 4-coumaroyl CoA, with three molecules of malonyl CoA. Although acetyl CoA carboxylase, the enzyme forming malonyl CoA, is essentially an enzyme of primary metabolism (Fig. 1), it is often co-regulated with the enzymes of plant polyketide biosynthesis [8].

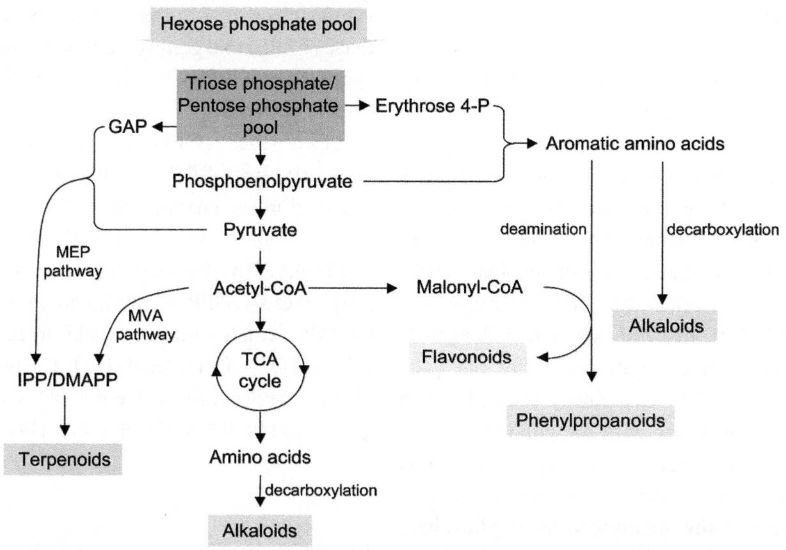

Fig. 1 The interface between primary and secondary metabolism in plants. GAP, glyceraldehyde 3-phosphate; IPP, isopentenyl diphosphate; DMAPP, dimethylallyl diphosphate; MEP, methylerythritol phosphate; MVA, mevalonic acid; TCA, tricarboxylic

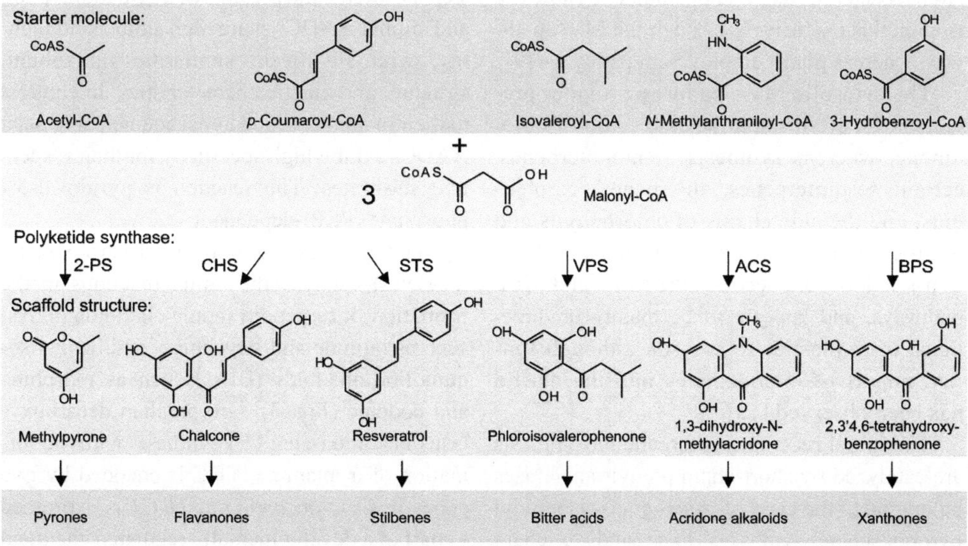

Fig. 2 Reaction types catalyzed by plant type III polyketide synthases. PS, pyrone synthase; CHS, chalcone synthase; STS, stilbene synthase; VPS, valerophenone synthase: ACS, acridone synthase; BPS, benzophenone synthase

Terpenoids

The interface between primary and secondary metabolism is less easy to define in the case of terpene (isoprenoid-derived) metabolites, because terpene units are also found in many compounds associated with primary metabolism, such as hormones and vitamins. The precursors of all isoprenoids, isopentenyl pyrophosphate (IPP) and its allylic isomer dimethylallyl pyrophosphate (DMAPP), are synthesized in higher plants by two independent pathways (Fig. 1). The mevalonic acid (MVA) pathway is localized in the cytosol and starts with the condensation of three molecules of acetyl-CoA. In plastids, IPP is formed from pyruvate and glyceraldehyde-3-phosphate via the methylerythritol phosphate (MEP) pathway [9]. MEP pathway enzymes are nuclear-encoded and imported into plastids [10]. In flowers of snapdragon (*Antirrhinum majus*), plant volatiles are assembled from IPP units originating from the MEP pathway, and flux through this pathway in controlled by a circadian clock following a diurnal rhythm [11]. Archaebacteria, fungi and animals synthesize isoprenoids exclusively through the MVA pathway, whereas plants employ both pathways [9].

The cytosolic MVA pathway provides precursors for sterols and the side chain of ubiquinone, whereas synthesis of monoterpenes, certain sesquiterpenes, diterpenes, carotenoids, and the side chains of chlorophylls and plastoquinone is carried out in plastids [12]. Cross-talk occurs between the MVA and MEP pathways, and appears to be mainly unidirectional from plastids to cytosol, although limited import of intermediates into the plastid has been observed [13].

The initial reactions of terpene biosynthesis are catalyzed by short-chain prenyltransferases belonging to the class of *trans*- or *cis*- isoprenyl pyrophosphate synthases that catalyze chain elongation of allylic pyrophosphate substrates with IPP to generate linear polymers with defined chain length. Geranyl pyrophosphate synthase (GPPS), farnesyl pyrophosphate synthase (FPPS) and geranylgeranyl pyrophosphate synthase (GGPPS) catalyze formation of the linear precursors of monoterpenes (C10), sesquiterpenes (C15) and diterpenes (C20), respectively. GPPS is a plastidic, homo- or hetero-dimeric enzyme that catalyzes the head-to-tail condensation of one IPP molecule and one DMAPP molecule to form *trans*-GPP (C10) [14]. *Trans*-GGPP is produced in plastids by GGPPS, which sequentially adds three IPP molecules to a DMAPP molecule [15]. The biosynthesis of *trans*-FPP occurs in the cytosol, where FPPS combines two IPP molecules with DMAPP.

Alkaloids

Most alkaloids are derived from amino acids (Fig. 3) and the first reaction in the otherwise independent pathways is the decarboxylation of the respective amino acid by an amino acid decarboxylase (AADC) (Figs. 1 and 3); this step is often under complex regulation. Plant and animal AADCs share high amino acid identity, with significant similarities in subunit structure and kinetic characteristics. In contrast to their mammalian and insect counterparts, plant AADCs exhibit high specificity for their respective substrates. The reaction is pyridoxal-5'-phosphate (PLP)-dependent.

Plant aromatic L-amino acid decarboxylases (AADCs) catalyze the initial reactions in the formation of terpenoid indole alkaloids (TIAs) such as quinine and strychnine, and benzylisoquinoline alkaloids (BIAs) such as morphine and codeine (Fig. 3). L-tryptophan decarboxylase (TDC) initiates TIA synthesis with the formation of tryptamine. TDC is encoded by two genes in *Cola accuminata*; *TDC1* is expressed as part of a developmentally regulated chemical defense system, whereas *TDC2* is induced after elicitation with yeast extract or methyl jasmonate (MJ).

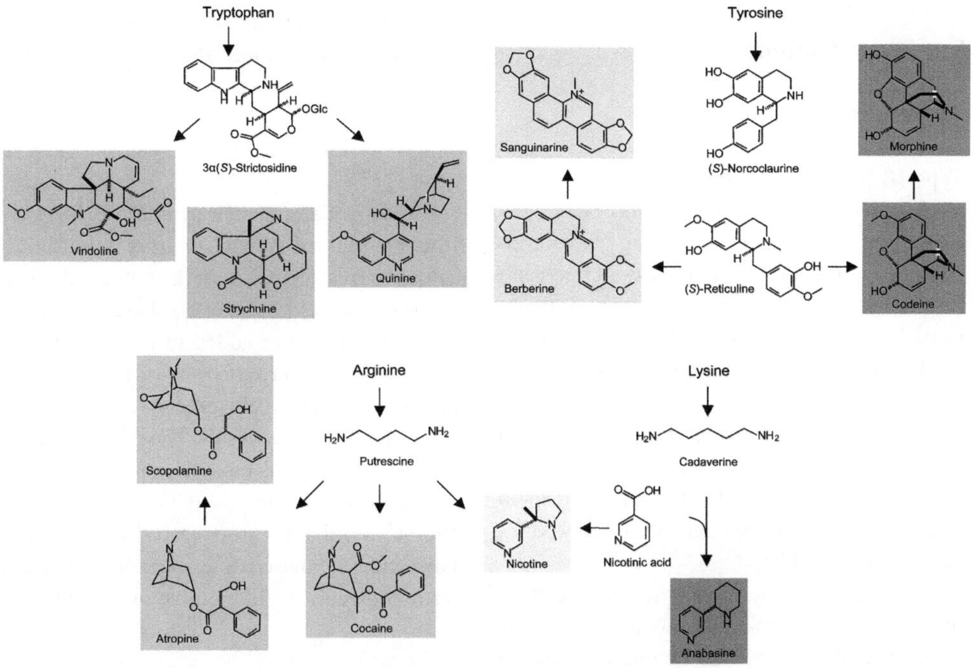

Fig. 3 Primary precursor–end product relationships in alkaloid biosynthesis. Color codes: blue, terpene indole alkaloids; green, benzophenanthridine alkaloids; gray, benzylisoquinoline alkaloids; red, quinolizidine alkaloids; yellow, pyridine alkaloids; pink, tropane alkaloids

L-tyrosine decarboxylase (TYDC) converts tyrosine and L-DOPA to tyramine and dopamine, respectively, the precursors for BIA biosynthesis [16]. TYDC is encoded by a single-copy gene in *Arabidopsis* (which does not produce BIAs), but a *TYDC* gene family of about 15 members divided into two subgroups is present in opium poppy (*Papaver somniferum*) [16]. The Arabidopsis *TYDC* gene contains 12 introns, whereas all other plant *AADC* genes described to date lack introns. TYDC is induced during plant defense responses, where it is involved in the synthesis of cell wall-bound hydroxycinnamic amides that provide a physical barrier against pathogens.

Lysine decarboxylase (LDC) catalyzes the formation of cadaverine by decarboxylation of lysine (Fig. 3). As an enzyme participating in polyamine biosynthesis, LDC links primary metabolism with biosynthesis of quinolizidine alkaloids like lupinine, which occurs in mesophyll chloroplasts of legumes. LDC is assumed to be the rate-limiting step in the biosynthesis of anabasine (Fig. 3), a pyridine alkaloid produced by tobacco species [17].

Ornithine decarboxylase (ODC) and arginine decarboxylase (ADC) are the first enzymes involved in the formation of tropane alkaloids (TPAs) such as atropine and cocaine (Fig. 3). Decarboxylation of ornithine yields putrescine, whereas arginine is converted to agmatine, which is metabolized to putrescine via a second route. ADC is assumed to play the primary role in TPA synthesis [18].

6

Scaffold Formation

Polyketides

Type III polyketide synthases (PKS) form the primary scaffolds for the synthesis of a range of secondary metabolites including flavonoids, stilbenes, bibenzyls, xanthones and pyrones [19] (Fig. 2). They catalyze a reaction similar to their fatty acid synthase (FAS) ancestors by facilitating the sequential head-to-tail addition of two-carbon acetate units to a growing polyketide chain. Type III PKSs differ from their type I and type II relatives by a simpler structure and the use of a CoA thioester substrate instead of an acyl carrier protein (ACP)-linked substrate [19].

CHS (Fig. 2), the most studied member of the type III PKS family, is a ubiquitous enzyme in plants that catalyzes the first committed step in flavonoid biosynthesis, the elongation of the starter molecule 4-coumaroyl-CoA by addition of three acetate units derived from three molecules of malonyl-CoA [19]. After binding of the 4-coumaroyl moiety to the active site Cys164, sequential polyketide chain elongation is initiated by the decarboxylation of malonyl-CoA to form an acetyl-CoA carbanion, followed by an intramolecular Claisen condensation step and subsequent cyclization and aromatization, yielding chalcone [20].

Other CHS-like PKSs accept different starter molecules, vary the length of the polyketide chain, or achieve molecular diversity through alteration of cyclization regiospecificity (Fig. 2). For example, stilbene synthase (STS) catalyzes a reaction similar to that of CHS in the initial stages, but the tetraketide intermediate undergoes a different cyclization reaction involving an intramolecular aldol condensation, hydrolysis from Cys164 and an additional decarboxylation step during formation of resveratrol. The structural bases for the differences in starter molecule, control of chain length, and overall cyclization mechanism (i.e. CHS- vs STS-type reactions) are now understood, and product formation has been altered rationally by point mutation to convert chalcone synthase to either stilbene synthase or pyrone synthase (Fig. 2) [21–23].

Phlorisovalerophenone synthase (VPS) from flower cones of hop (*Humulus lupulus* L.) utilizes isovaleryl-CoA or isobutyryl-CoA as starter molecules [24] (Fig. 2). Three molecules of malonyl-CoA are added to these starters to form phlorisovalerophenone or phorisobutyrophenone, respectively, precursors for the biosynthesis of hop bitter acids. The first committed step of cannabinoid biosynthesis in glandular trichomes of *Cannabis sativa* is catalyzed by a stilbene synthase carboxylate-like (STCSL) polyketide synthase using n-hexanoyl-CoA as starter molecule, yielding olivetolic acid [25].

Terpenoids

One group of terpene synthases (TPSs) use GPP, FPP or GGPP as substrates to form monoterpene (C10), sesquiterpene (C15) or diterpene (C20) scaffold molecules, respectively, which then undergo a variety of secondary modifications. This family of synthases is structurally distinct from triterpene (C30) or tetraterpene (C40) synthases [26]. The scaffolds produced are themselves highly divergent due to different folding patterns prior to cyclization (Fig. 4). TPSs fall into two major groups with regard to their modes of cyclization. Type A cyclization begins with the ionization of the polyprenyl pyrophosphate molecule, whereas type B cyclizations start with protonation at the terminal double bond. Both types of reactions are followed by cyclization and rearrangement ending with deprotonation of the final carbocation [27].

The sequence of reactions catalyzed by the type A limonene synthase (monoterpene cyclase) (Fig. 4) is initiated by ionization-isomerization

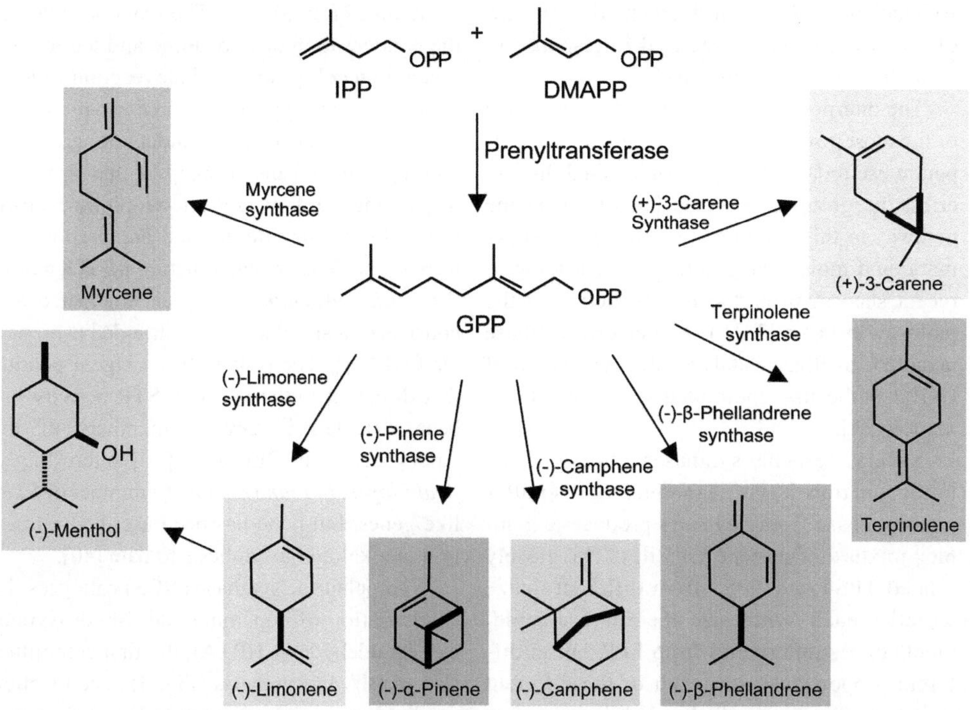

Fig. 4 Selected reactions catalyzed by monoterpene cyclases, illustrating the diversity of products that can be formed from a single precursor molecule (GPP, geranyl pyrophosphate)

of GPP to form the intermediate linalyl pyrophosphate in order to overcome the *trans*-geometry of the C2-C3 double bond, which prevents direct cyclization. This intermediate then undergoes an ionization-cyclization step resulting in a (*4R*)- or (*4S*)-α-terpinyl carbocation • pyrophosphate anion pair and subsequent termination reactions [28]. Many sesquiterpene and diterpene synthases employ type A cyclization [26].

5-*epi*-Aristolochene synthase (EAS) is one of the most studied sesquiterpene synthases. It catalyzes the conversion of FPP to 5-*epi*-aristolochene, an intermediate in the formation of sesquiterpene phytoalexins in tobacco, having (+) germacrene A and eudesmyl carbocation as intermediates [29]. EAS was the first terpene synthase for which a reaction mechanism could be confirmed by determination of the enzyme's crystal structure [30].

The diterpene synthase copalyl diphosphate synthase (CDP) is one of the most studied type B synthases. It forms the bicyclic intermediate (-)-copalyl pyrophosphate ((-)-CPP) or its diastereomer (+)-copalyl pyrophosphate ((+)-CPP) from GGPP. Several CDPs are expressed in rice; OsCyc1 forms (+)-CPP, while OsCyc2 and OsCPS1 form (-)-CPPs. OsCPS1 is believed to be involved in gibberellin biosynthesis and OsCyc2 in diterpene phytoalexin biosynthesis [31].

The abietadiene synthase from grand fir (*Abies grandis*) catalyzes two cyclization reactions at separate but interdependent active sites. The enzyme first converts GGPP in a type-B cyclization at one active site to (+)-CPP, which then undergoes a type A cyclization and additional reactions at the second active site to form a mixture of abietadiene isomers [26, 32].

6

Abietadiene is the main diterpenoid resin acid of oleoresin, which is secreted in response to wounding and herbivore attack.

The diterpene taxol from Pacific yew is one of the most powerful anticancer agents in therapeutic use today. In a series of seminal discoveries, the group of Croteau has dissected the pathway to this complex molecule at the enzymatic and molecular genetic levels, leading to recent success in partial reconstruction of the pathway in yeast [33]. The diterpene synthase taxadiene synthase catalyzes the cyclization of GGPP as the first committed step in taxol formation [34].

Several terpene synthases appear to be multi-functional. For example, a sesquiterpene synthase from *Zea mays* produces a complex mixture of terpene volatiles. The closely related TPS4 and TPS5 from different maize varieties each synthesize the same complements of sesquiterpenes from FPP, but in different proportions as a result of the different ratios of (*S*)- versus (*R*)-bisabolyl cation formation. This difference in stereoselectivity is determined by four amino acid residues in the active site [35].

Oxidosqualene cyclases (OSCs) or triterpene synthases convert oxidosqualene to one or more cyclic triterpene alcohols which are the precursors of sterols, steroids and saponins [36]. The reaction catalyzed is mechanistically similar to those of monoterpene, sesquiterpene and diterpene synthases, but the enzymes are phylogenetically distinct [37].

Alkaloids

A key reaction in assembly of scaffolds for several classes of alkaloid is the coupling of the amine derived by decarboxylation of an amino acid with a second molecule, the product of which then serves as the precursor for secondary modifications. Strictosidine synthase (STR) catalyzes the formation of strictosidine, the precursor for monoterpenoid indole alkaloids such as quinine (Fig. 3) [38]. The reaction involves the condensation of tryptamine and the monoterpenoid secologanin, a unique reaction called a Pictet-Spengler-type reaction. Formation of a Schiff base between the aldehyde group of secologanin and the primary amine group of tryptamine is followed by electrophilic cyclization between the iminium ion and carbon 2 of tryptamine. The crystal structure of STR1 from *Rauvolfia serpentina* has been elucidated and consists of a six-bladed four-stranded β-propeller fold [38]. The protein has a signal peptide that directs it to the vacuole. STR activity has been detected in several members of the Apocynaceae and Rubiaceae [39]. Interestingly, *Arabidopsis thaliana* contains a number of *STR*-like genes, but none has been ascribed a function in alkaloid biosynthesis to date [40].

Norcoclaurine synthase (NCS) catalyzes the condensation of dopamine and 4-hydroxyphenylacetaldehyde (4-HPAA), the first committed step in BIA biosynthesis (Fig. 3). The reaction mechanism is an asymmetric Pictet-Spengler reaction, similar to that of STR although no sequence homology exists between the two proteins. Deacetylipecoside synthase is the third enzyme known to utilize this reaction type, facilitating the condensation of dopamine and secologanin. It has been purified from *Alangium lamarchii* [41]. The homodimeric NCS from meadow rue (*Thalictrum flavum*) exhibits positive cooperativity towards dopamine but not 4-HPAA [42]. NCS belongs to the PR10 (pathogenesis-related) and Betv1 protein family [43]. *Coptis japonica* contains both a PR10-like NCS activity and an additional enzyme, with amino acid sequence similarity to 2-oxoglutarate-dependent dioxygenases (see below) (but lacking the 2-oxoglutarate binding domain), that likewise catalyzes formation of norcoclaurine from dopamine and 4-HPAA [44].

(*S*)-reticuline is a central branch-point metabolite in BIA biosynthesis (Fig. 3). It serves as a precursor for sanguinarine after conversion to (*S*)-scoulerine by the berberine bridge enzyme (BBE), originally purified from *Berberis beaniana*

[45]. The unique reaction catalyzed by BBE comprises the conversion of the *N*-methyl moiety of (*S*)-reticuline into the berberine bridge carbon of (*S*)-scoulerine through oxidative cyclization with a methylene iminium ion as reaction intermediate [45]. The enzyme uses FAD as cofactor, bi-covalently bound to the protein via a histidine and a cysteine residue. An N-terminal signal peptide targets the protein to the ER, and an adjacent vacuolar sorting determinant then directs BBE into the vacuole, where sanguinarine (Fig. 3) accumulates after elicitation with fungal elicitor in opium poppy. The low vacuolar pH suggests that alkaloid synthesis is completed before the ER-derived vesicles fuse with the vacuole [46].

Modifications of Secondary Metabolite Scaffolds

Modification reactions create the enormous diversity of plant natural products, providing new molecules with different biological activities from the basic scaffolds outlined above. The plant kingdom contains a large number of enzymes that catalyze hydroxylation, epoxidation, aryl migration, glycosylation, methylation, sulfation, acylation, prenylation, oxidation and reduction of secondary metabolite skeletons, examples of which are reviewed below, and illustrated, for phenylpropanoid and flavonoid biosynthesis, in Fig. 5. Figure 6

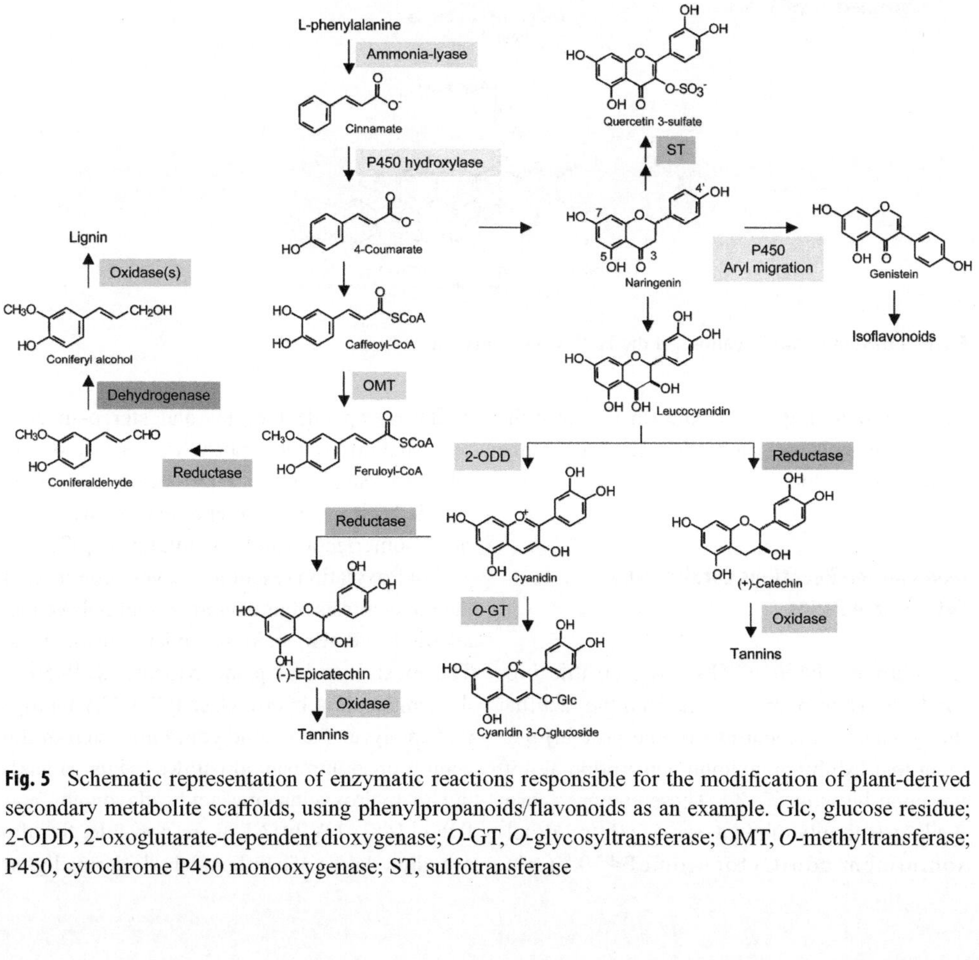

Fig. 5 Schematic representation of enzymatic reactions responsible for the modification of plant-derived secondary metabolite scaffolds, using phenylpropanoids/flavonoids as an example. Glc, glucose residue; 2-ODD, 2-oxoglutarate-dependent dioxygenase; *O*-GT, *O*-glycosyltransferase; OMT, *O*-methyltransferase; P450, cytochrome P450 monooxygenase; ST, sulfotransferase

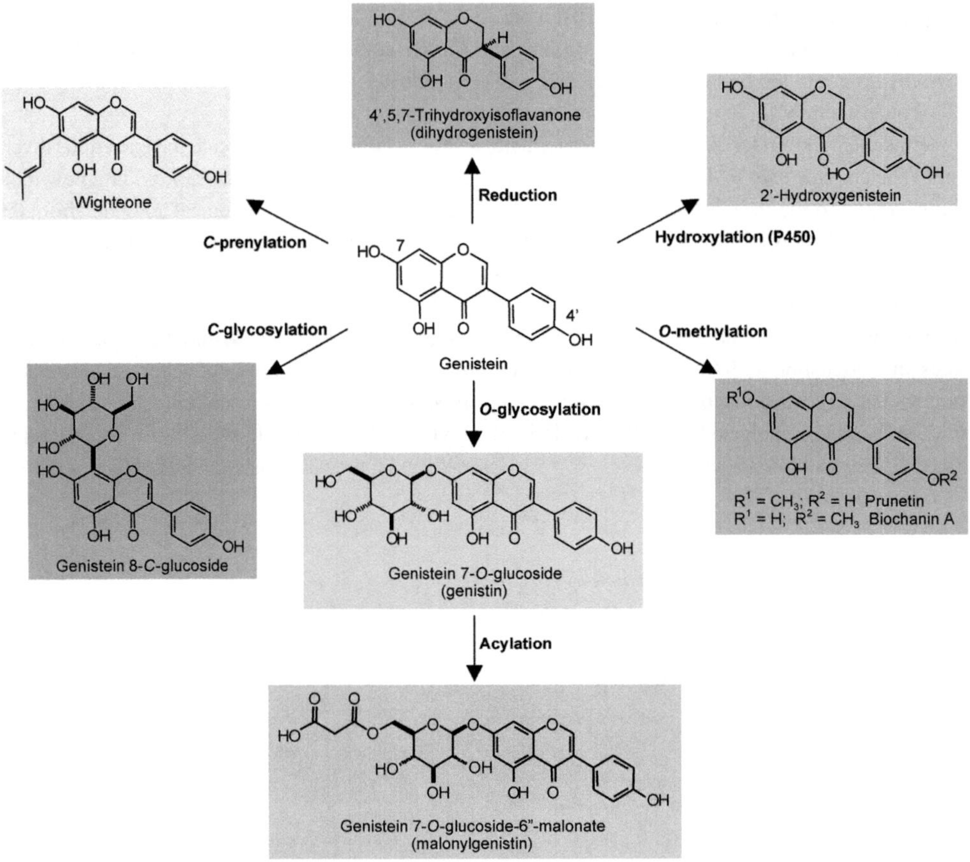

Fig. 6 Enzymatic modifications of the isoflavone genistein

shows how a single molecule (in this case the isoflavone genistein) can be converted to a range of different products.

Oxygenation Reactions Catalyzed by Cytochrome P450s

Cytochromes P450 (CYPs) are versatile biocatalysts. Their name derives from the fact that they contain a cytochrome (heme-protein) pigment that exhibits maximum absorption at 450 nm upon binding of CO. These enzymes form the largest family of plant proteins (http://drnelson.utmem.edu/CytochromeP450.html). NADPH-dependent regio- and stereo-specific oxygenations of lipids, phenolics, terpenoids and alkaloids catalyzed by P450 enzymes include simple hydroxylation or epoxidation, dealkylation, isomerization and aryl migration [47].

Hydroxylation reactions are very common in the biosynthesis of phenylpropanoid/polyketide, alkaloid and terpenoid secondary metabolites. The most abundant plant cytochrome P450 is the cinnamate 4-hydroxylase (CYP73A family) that catalyzes the second committed step of the central phenylpropanoid pathway (Fig. 5) leading to hydroxycinnamic acids, flavonoids and lignin [48]. CYP84A1 is responsible for the 5-hydroxylation of coniferaldehyde during lignin biosynthesis, and the *Arabidopsis thaliana* EMS mutant *fah 1* defective in CYP84A1 expression has altered lignin composition and suppressed sinapoyl malate accumulation [49]. CYP75A1 is a flavonoid 3',5'-hydroxylase involved in the biosynthesis of anthocyanin pigments [50] (Fig. 5). Two mutant alleles of this gene in Petunia (*Hf1* and *Hf2*) exhibit altered flower color (from blue to pink) as a result of loss of delphinidin and residual pigmentation from cyanidin or pelargonidin [50].

Geraniol 10-hydroxylase (CYP76B6) from *Catharanthus roseus* is a key regulatory enzyme in the synthesis of the terpene indole alkaloid vindoline [51]. Alkaloids such as vincristine and vinblastine (derived from vindoline) are used in modern medicine as anti-neoplastic agents (Fig. 3).

In recent years, a range of cytochrome P450 enzymes involved in cyclic terpene hydroxylation has been characterized at the molecular level. These include monoterpene hydroxylases involved in the biosynthesis of essential oils in commercial mint [52], a sesquiterpene hydroxylase performing successive hydroxylations of 5-*epi*-aristolochene in the biosynthesis of the tobacco phytoalexin capsidiol [53], diterpene hydroxylases involved in the formation of taxol [54, 55], and triterpene hydroxylases involved in the biosynthesis of saponins [56] and brassinosteroids [57].

Lutein, the most abundant carotenoid in photosynthetic tissues, is produced via hydroxylation of the ε-ring of zeionoxanthin catalyzed by CYP97C1 [58]. CYP707A1 and CYP707A2 are critical for the control of seed dormancy and germination in Arabidopsis by hydroxylating abscisic acid at the C-8' position for catabolism of this carotenoid-related phytohormone [59].

DIMBOA (2,4-dihydroxy-7-methoxy-1,4-benzoxazin-3-one) is a cyclic hydroxamic acid produced by members of the Gramineae as a defense against herbivores or microbial pathogens [60]. Maize CYP71C1 catalyzes the hydroxylation of indolin-2-one at the 3-position to form the precursor molecule for DIMBOA biosynthesis [61]. Transposon-tagged mutations (*Bx3::Mu*) in *CYP71C1* suppress the production of DIMBOA and make maize plants more susceptible to pathogens [61].

Epoxidation Reactions of P450 Monooxygenases

Oxylipins are biologically active signaling compounds of structural diversity generated by the coordinated action of lipases, lipoxygenases and P450s specialized in the metabolism of hydroperoxy fatty acids. CYP74A, CYP74B, and CYP74C catalyze epoxidation of the 9- and/or 13-hydroperoxides of linoleic and linolenic acid [62, 63]. CYP74C displays specificity toward 9-hydroperoxy fatty acid derivatives [62] while CYP74A and CYP74B are more selective for 13-hydroperoxy derivatives leading to the synthesis of the wound signal jasmonic acid (JA) [63]. Knock-out mutation in the *CYP74A* gene of Arabidopsis causes male sterility and impaired wound signal transduction due to suppression of JA production [64].

Hairy root cultures of *Catharanthus roseus* can synthesize tabersonine, a precursor of vindoline [65]. Methyl jasmonate induces the tabersonine 6,7-epoxidase P450 that converts tabersonine into lochnericine [65].

CYP714D1 (a catabolic enzyme) catalyzes 16α,17-epoxidation of non-13-hydroxylated gibberellins, and over-expression of this gene in *Oryza sativa* leads to a dwarf phenotype due to gibberellin deficiency [66].

Non-oxygenation Reactions Catalyzed by P450s

In isoquinoline alkaloid biosyntheses, several unique P450 reactions have been reported, such as methylenedioxy bridge formation, intramolecular C–C phenol-coupling and intermolecular C–O phenol-coupling reactions. Salutaridine synthase, involved in the conversion of (*R*)-reticuline to salutaridine during the formation of morphine (Fig. 3), is a cytochrome P450 that catalyzes an intramolecular phenol-coupling reaction that does not involve incorporation of oxygen into the substrate [67]. Recently, heterologous expression in yeast has indicated that CYP80G2 from *Coptis japonica* likewise exhibits intramolecular C–C phenol-coupling activity to produce (*S*)-corytuberine (aporphine-type alkaloid) from (*S*)-reticuline (benzylisoquinoline type) [68]. Methylenedioxy-bridge forming cytochrome P450s have been described in alkaloid [69] and isoflavonoid [70] biosynthesis.

Hydroxylation/Aryl Migration Catalyzed by P450 Monooxygenases

Isoflavonoids are widely produced by legumes and function in defense against pathogen attack and attraction of symbiotic microbes. An unusual aryl migration reaction constitutes the first committed step in their biosynthesis [71]. Isoflavone synthase (IFS, more correctly known as 2-hydroxyisoflavanone synthase) is a member of the CYP93C family that catalyzes migration of the B-ring of the flavanones liquiritigenin and naringenin from C-2 to C-3 leading to the isoflavones daidzein and genistein (Fig. 5) respectively [71, 72]. The reaction proceeds by abstraction of a hydrogen at C-3 followed by B-ring migration and subsequent hydroxylation of the resulting C-2 radical; dehydration with loss of the C-2 hydroxyl to yield isoflavone occurs enzymatically *in vivo* and non-enzymatically *in vitro*. The enzyme is stereo-selective and recognizes only 2*S*-flavanones as substrates. Down-regulation of IFS in soybean hairy roots compromises defense against *Fusarium solani* f. sp. *glycinea* by suppressing accumulation of the glyceollins, daidzein-derived phytoalexins [73].

Isoflavones are not limited to the Leguminosae. For example, sugarbeet (Chenopodiaceae) produces isoflavones in response to pathogen attack. Two IFS cDNAs from sugarbeet share higher than 95% similarity to IFS1 from soybean [74].

Phenylphenalenones are polycyclic natural products of the Haemodoraceae, Musaceae and Strelitziaceae. The biosynthetic pathway for 8-phenylphenalenone production in the water hyacinth *Eichhornia crassipes* may occur via the formation of diarylheptanoid and 9-phenylphenalenone as intermediates, and includes a 1,2-aryl migration reaction that may follow a mechanism similar to that of IFS [75]. However, no P450 enzyme able to catalyze this reaction has been reported to date.

2-Oxoglutarate-Dependent Dioxygenases (2-ODDs)

Plant 2-oxoglutarate-dependent dioxygenases (2-ODDs) are cytosolic, non-heme iron-containing enzymes that utilize an oxoacid to oxidize a target metabolite. Ascorbate is required *in vitro,* probably to maintain the iron moiety in the reduced form (Fe^{2+}). 2-ODD enzymes are involved in the biosynthesis of some amino acids, hormones, signaling molecules, and a large number of secondary metabolites [76].

Flavanone 3β-hydroxylase (F3H) converts the basic flavonoid skeleton flavanone to dihydroflavonol through hydroxylation at the 3-position [77], a critical early step in anthocyanin flower pigment biosynthesis. Flavonol synthase (FLS) is a 2-ODD that catalyzes the formation of a double bond between C-2 and C-3 in dihydroflavonols [78]. Flavonol 6-hydroxylase (F6H) is also a 2-ODD [79].

Anthocyanidin synthase (ANS), the key enzyme in the biosynthesis of anthocyanins, catalyzes oxidation of leucoanthocyanidin (flavan-3,4-diol) to a 2-flaven-3,4-diol that spontaneously isomerizes to 3-flaven-2,3-diol (anthocyanidin) (Fig. 5). This is subsequently glycosylated at C-3, transported to the vacuole, and finally converted to the colored flavilium cation at the acidic pH of the vacuole [77]. The first biochemical evidence for ANS being a 2-ODD came from studies with the recombinant enzyme from *Perilla* [80]. A catalytic mechanism has been proposed in which ANS promotes hydroxylation at C-3 as the initial step in anthocyanidin biosynthesis, and this is supported by structural data for ANS from Arabidopsis [81].

Hyoscyamine 6β-hydroxylase (H6βH) is a 2-ODD enzyme that catalyzes a critical step in the biosynthesis of the tropane alkaloid scopolamine (Fig. 3) in members of the Solanaceae [82].

Reductases

Several NADPH-dependent reductase enzymes play key roles in the biosynthesis of flavonoids. Isoflavone reductase (IFR), which is a member of the Reductase-Epimerase-Dehydrogenase (RED) family of proteins, catalyzes the stereospecific reduction of isoflavone to the corresponding (3*R*)-isoflavanone (Fig. 6), a key reaction in the biosynthesis of antimicrobial pterocarpan phytoalexins such as medicarpin in legumes. IFR from alfalfa catalyzes formation of (3*R*)-vestitone from 2'-hydroxyformononetin for the biosynthesis of medicarpin [83]. The IFR from soybean has activity toward 2'-hydroxydaidzein, 2'-hydroxyformononetin, and 2'-hydroxygenistein [84] while the IFRs from pea and chickpea recognize 7,2'-dihydroxy-4',5'-methylenedioxyisoflavone and 2'-hydroxyformononetin, respectively [85, 86].

IFRs belong to a large protein family that includes IFR-like proteins from non-legume plants. IFR-like proteins have high sequence identity to legume IFRs, but the functions of many are still unclear. Some clearly catalyze reduction reactions in pathways other than isoflavonoid biosynthesis. For example, pinoresinol-lariciresinol reductase and phenylcoumaran benzylic ether reductase are IFR-like proteins involved in the formation of lignans in *Forsythia intermedia* and *Pinus taeda*, respectively [87, 88], and it is thought that pinoresinol reductases may represent the progenitors of the IFRs.

Dihydroflavonol 4-reductase (DFR) is involved in the biosynthesis of anthocyanins and proanthocyanidins (PAs). DFRs catalyze the stereospecific reduction of (2*R*,3*R*)-dihydroflavonols to (2*R*,3*R*,4*S*)-leucoanthocyanidins [77] (Fig. 5). Petunia possesses three different *DFR* genes (*dfrA-C*), but only *dfrA* is transcribed in floral tissues. DFR-A does not accept dihydrokaempferol, the precursor for the synthesis of pelar-gonidin-type anthocyanins. Consequently, no orange-colored petunia flowers are found in nature [89]. Dihydroquercetin and dihydromyricetin are also substrates for DFRs and provide leucocyanidin and leucodelphinidin, respectively.

Leucoanthocyanidin reductase (LAR), which is related to the isoflavone reductase group of plant enzymes, catalyzes the reduction of leucoanthocyanidins to (+)-afzelechin, (+)-catechin, and (+)-gallocatechin, building blocks for PA biosynthesis [77] (Fig. 5). The above catechin series of flavan-3-ols possess 2,3-*trans* stereochemistry. The corresponding (*epi*)-catechin series with 2,3-*cis* stereochemistry is formed by a different mechanism involving an unrelated reductase (ANR) that acts at the level of anthocyanidin [90] (Fig. 5).

Reduction of a coenzyme a ester to the corresponding aldehyde, catalyzed by cinnamoyl CoA reductases (CCRs), is an important reaction in lignin biosynthesis (Fig. 5). Plants contain small *CCR* gene families [91] encoding enzymes with differences in overall specificity for monolignol precursors with different aromatic ring substitution patterns.

Glycosyltransferases

Among the reactions for the modification of secondary metabolite scaffolds, glycosylation plays a particularly important role in plants, contributing to the biosynthesis and storage of secondary metabolites, regulation of hormone homeostasis, detoxification of xenobiotics, enhancement of a molecules' solubility, and plant defense [92]. Glycosyltransferases of small molecules (UGTs) catalyze the transfer of sugar residues from uridine diphosphate sugars to an acceptor. The sugar moiety can be transferred to oxygen, nitrogen, or sulfur atoms of different classes of natural products [92]. UGTs comprise a superfamily of enzymes ubiquitous in living organisms (http://afmb.cnrsmrs.fr/CAZY/fam/acc_GT.html), the number of which in the plant kingdom is likely comparable to that of the cytochrome P450 enzymes.

The most common reaction catalyzed by UGTs is the transfer of sugar residues to oxygen atoms. Several plant UGT crystal structures have been reported in the literature, all for *O*-glucosyltransferases [93–95]. Glucose is the commonest sugar attached to plant secondary metabolites, although galactose, glucuronic acid and other "common" monosaccharides are also found; this contrasts to the situation in prokaryotes, where more "exotic" sugars are often found attached to secondary metabolites (e.g. in antibiotics).

Eight recombinant *Medicago truncatula* UGTs display *O*-glycosyltransferase activity toward (iso)flavonoids at different positions [96]. Recombinant UGT85H2 prefers flavonols whereas the substrate specificity of UGT78G1 is higher for isoflavones [95, 96]. UGT78G1 is regioselective, catalyzing *O*-glycosylation at C-7 unless the flavonoid has a hydroxyl group at the C-3 position (see labeling of naringenin in Fig. 5), in which case this position is preferred. UGT78G1 is also able to operate in the reverse direction, deglycosylating (iso)flavonoid glucoderivatives in the presence of uridine diphosphate [96]. Similar to grapevine *Vv*GT1 [94], UGT78G1 recognizes anthocyanidins as substrates [96]. Snapdragon 4'CGT converts chalcone to its 4'-*O*-glucoside derivative, an intermediate in the synthesis of yellow aurone flower pigments [97].

Three glucoyltransferases involved in the biosynthesis of the sweet diterpene glucosides of *Stevia rebaudiana* [98] and two involved in formation of triterpene saponins in *Medicago truncatula* [99], have been identified through genomic approaches. Whereas glycosylation often inactivates or targets plant natural products for storage, it is important for biological activity in the case of the triterpene saponins.

Arabidopsis UGT84B1 catalyzes the *O*-glycosylation of indole-3-acetic acid (IAA), and over-expression of *UGT84B1* leads to altered root phenotypes as a consequence of free IAA depletion [100]. UGT73C5 from Arabidopsis catalyzes the 23-*O*-glucosylation of brassinolide and castasterone, two plant steroid hormones. Brassinosteroid accumulation is dramatically reduced in transgenic plants over-expressing UGT73C5, and the phenotype is consistent with the deprivation of free steroid hormones [101]. UGT76C1 and UGT76C2 from *Arabidopsis thaliana* both exhibit activity toward cytokinins [102]. Plants over-expressing *UGT76C1* show increased accumulation of *trans*-zeatin 7-*N*-glucoside when supplemented with *trans*-zeatin [102]. Structural studies have recently revealed mechanisms controlling *N*- as compared to *O*-glucosylation in UGT72B1, a bifunctional N-/O-glucosyltransferase from Arabidopsis active in xenobiotic detoxification [103].

An *S*-GT from *Brassica napus* was the first thiohydroximate *S*-glycosyltransferase of the glucosinolate pathway to be partially characterized *in vitro* [104]. T-DNA insertions in the corresponding gene in Arabidopsis (UGT74B1) cause low levels of glucosinolates, leaf vein chlorosis, and impaired auxin metabolism [105].

Despite interest in the anti-microbial and anti-insect activities of flavonoid *C*-glycosides (e.g. from maize), and the potential therapeutic value of isoflavone *C*-glycosides such as genistein 8-*C*-glucoside (Fig. 6) or puerarin from *Pueraria lobata*, no plant gene encoding a *C*-glycosyltransferase has yet been cloned.

Methyltransferases

O-Methyltransferases (OMTs) catalyze the transfer of a methyl group from S-adenosyl-L-methionine (SAM) to oxygen atoms of hydroxyl groups on an acceptor molecule to yield the methyl ether derivative (Figs. 5 and 6). OMTs are regio- and stereo-selective and can collectively mono- or poly-methylate a great number of plant natural products [106]. Methylation of flavonoids alters their solubility and intracellular compartmentalization, and can increase their antimicrobial activity [107].

Small molecule OMT enzymes from plants are classified in two distinct groups. Group I OMTs have molecular weights in the range of 38–43 KDa and target many acceptors such as flavonoids, phenylpropanoids, alkaloids, and coumarins. Group II OMTs are of lower molecular weight (23–27 KDa) and are dependent on Mg^{2+} for activity [108].

O-Methylation at C-3 of the isoflavonoid derivative 6a-hydroxymaackiain by PsHMM (group I) is the final step in the production of the phytoalexin pisatin in *Pisum sativum*; the non-methylated precursor lacks antimicrobial activity [109]. Caffeic acid 3-*O*-methyltransferase (group I) and caffeoyl CoA 3-*O*-methyltransferase (group II) (Fig. 5) play important roles in lignin biosynthesis [110]. Group II OMTs from cell suspension cultures of meadow rue are involved in the biosynthesis of the isoquinoline alkaloid beberine (Fig. 3). The recombinant enzymes OMT II 1.1, OMT II 2.2, OMT II 3.3, and OMT II 4.4 also recognize a range of phenylpropanoids and catechols with different specificities [111]. OMT II 1.1, but not OMT II 4.4 is active toward the isoquinoline alkaloid (*R,S*)-norcoclaurine, and these enzymes only differ from each other at amino acid residue 21 (Tyr in the former and Cys in the latter) [111].

Putrescine *N*-methyltransferase (PMT) catalyzes the first committed step in TPA and nicotine biosynthesis, the SAM-dependent methylation of putrescine (Fig. 3). The enzyme has a high similarity to mammalian and plant spermidine synthase (SPDS), from which it is assumed to have evolved. SPDS uses a slightly different co-substrate, decarboxylated SAM (dcSAM), but can also accept putrescine as substrate [18]. PMT is expressed primarily in roots, but has also been detected in young potato tuber sprouts and wounded leaves of tobacco [112]. SAM-dependent *N*-methyltransferases play a key role in caffeine biosynthesis, where three steps of methylation of nitrogen atoms take place.

Sulfo- and Aromatic Prenyl-Transferases

Sulfate transfer to flavonoids and glucosinolate precursors is catalyzed by a small family of soluble sulfotransferases (STs) that use 3'-phosphoadenosine 5'-phosphosulfate (PAPS) as sulfate donor (Fig. 5). Sulfated flavonols may play a role in the transport of auxins [113]. Four position-specific flavonol STs are found in plants of the genus *Flaveria,* with preferences for the 3-position of the flavonoid aglycone, the 3' and 4'-positions of 3-sulfate derivatives, and the 7-position of 3,3'- or 3,4'-disulfate derivatives [113] (Fig. 5).

Aromatic prenyltransferases (PTs) utilize allyl diphosphate for the prenylation of aromatic compounds in a process dependent on divalent cations (Mg^{2+} or Mn^{2+}). The UbiA family includes PTs that catalyze the prenylation of 4-hydroxybenzoate as a key step in the formation of ubiquinone (UQ), an electron carrier in the respiratory chain [114]. UbiA PTs exhibit a broad substrate specificity accepting prenyl pyrophosphates of different chain lengths originating from UQ6 in *Saccharomyces cerevisae* to UQ10 in tobacco. The *AtPP1* gene from Arabidopsis encodes a 4-hydroxybenzoate polyprenyl diphosphate transferase (4-HPT) and its expression in a yeast mutant lacking 4-HPT activity restores ubiquinone synthesis as well as the cells' respiratory ability [114]. Disruption of the *AtPP1* gene through T-DNA insertion compromises the early stage of Arabidopsis embryo development [114]. PTs from soil bacteria also target small aromatic molecules catalyzing the production of antibiotic compounds. However, these bacterial PTs lack the prenyl diphosphate binding motif (N/D)DXXD characteristic of aromatic PTs from the UBiA family [115].

Prenylated (iso)flavonoids are relatively common antimicrobial compounds in plants (e.g. in fruits of Osage orange [*Maclura pomifera*] and roots of white lupin) (see wighteone in Fig. 6). The first characterization of a plant flavonoid PT at the molecular level was recently reported; naringenin 8-prenyltransferase from *Sophora flavescens* is related to the homogentisate PTs involved in the formation of tocopherols and tocotrienols [116].

Acyltransferases

Acylation of oxygen or nitrogen atoms to generate esters and amides, respectively, is a common reaction for natural product scaffold modification. Members of the large BAHD family of acyltransferases utilize CoA thioesters as a source of the acyl group. The term BAHD comes from the initials of the first four plant acyltransferases biochemically characterized (BEAT, AHCT, HCBT, and DAT) [117]. Benzylalcohol *O*-acetyltransferase (BEAT) is responsible for the production of the floral volatile benzylacetate in *Clarkia breweri*, whereas deacetylvindoline 4-*O*-acetyltransferase (DAT) participates in the last step in the biosynthesis of vindoline in *Catharanthus roseus*. *N*-hydroxycinnamoyl/benzoyltransferase (HCBT) acts in the production of anthramide phytoalexins in *Dianthus caryophyllus*, and anthocyanin *O*-hydroxycinnamoyltransferase (AHCT) is responsible for the 5-*O*-acylation of anthocyanins in *Gentiana triflora* [117]. Other BAHD acyltransferases include CmAAT4 from *Charentais melon* that catalyzes the formation of volatile medium-chain aliphatic esters, HMT/HLT, a tigloyltransferase crucial for the biosynthesis of quinolizidine alkaloids in *Lupinus albus* [117], and MtMat1–3, which catalyze the malonylation of the sugar residue on isoflavone 7-*O*-glucosides in *Medicago truncatula* [118] (Fig. 6).

Plants also possess various serine carboxypeptidase-like (SCPL) enzymes that function as acyltransferases [119]. In contrast to BAHD family acyltransferases, SCPL enzymes use 1-*O*-β-acyl acetals (most frequently the 1-*O*-β-ester of glucose) as the acyl donor. True serine carboxypeptidases are exclusively hydrolytic, and the discovery of serine carboxypeptidase-like enzymes with acyltransferase features brought a new perspective to gene annotation in plant secondary metabolism [120, 121]. Isolation of a cDNA encoding an SCPL protein responsible for the synthesis of glucose polyesters has been reported [120]. These compounds are produced in trichomes of *Lycopersicon pennellii* and *Solanum berthaultii* as a defense against insect attack. An SCPL enzyme responsible for the formation of UV-protecting sinapoyl malate in leaves of the Brassicaceae has also been described [121].

The Challenge of Predicting Enzyme Function in Plant Secondary Metabolism

The functional annotation of members of gene families involved in modification of secondary metabolite scaffolds is often challenging, and amino acid sequence identity by itself may be misleading for prediction of enzyme function. For example, although *Medicago truncatula* UGT71G1 clusters phylogenetically with UGT71C1 or UGT71C4 from *Arabidopsis thaliana* (enzymes known to glycosylate benzoic acid derivatives), recombinant UGT71G1 does not display activity toward benzoic acid [96]. Despite the high activity of recombinant UGT71G1 against quercetin, with all hydroxyl groups on the molecule being glycosylated, and much lower *in vitro* activity toward triterpenes, UGT71C1 is believed to catalyze the glycosylation of triterpenes *in vivo* based on correlated transcript and metabolite induction patterns, which seem to rule out an *in vivo* role in quercetin glycosylation [96, 99]. Difficulties associated with gene annotation have been widely discussed for methyltransferases, and the overlapping substrate specificities that are co-expressed in the same cell types becomes an additional issue [111, 122]. The existence of serine carboxypeptidases with acyl transfer but not hydrolytic properties, isoflavone reductase-like proteins in plants that do not synthesize isoflavonoids, and strictisodine synthase-like genes in Arabidopsis, clearly demonstrate that similarities in amino acid sequence *per se* are not enough for determination of protein function. Studies integrating spatially and temporally resolved metabolome and transcriptome analysis, together with loss of function genetic analysis using insertion/deletion mutants or transgenic plants (antisense or RNAi lines), will be crucial for elucidating the individual roles of these enzymes *in vivo*. Gain of function analyses alone may be confusing for enzymes with promiscuous *in vitro* substrate preferences.

References

1. Pichersky, E. and Gang, D.R. (2000) Genetics and biochemistry of secondary metabolites in plants: an evolutionary perspective. Trends Plant Sci 5, 439–445
2. Ritter, H. and Schulz, G.E. (2004) Structural basis for the entrance into the phenylpropanoid metabolism catalyzed by phenylalanine ammonia lyase. Plant Cell 16, 3426–3436
3. Watts, K.T. et al. (2006) Discovery of a substrate selectivity switch in tyrosine ammonia-lyase, a member of the aromatic amino acid lyase family. Chem. Biol. 13, 1317–1326
4. Louie, G.V. et al. (2006) Structural determinants and modulation of substrate specificity in phenylalanine-tyrosine ammonia-lyases. Chem. Biol. 13, 1327–1338
5. Baedeker, M. and Schulz, G.E. (2002) Autocatalytic peptide cyclization during chain folding of histidine ammonia-lyase. Structure 10, 61–67
6. Winkel, B.S.J. (2004) Metabolic channeling in plants. Annu. Rev. Plant Biol. 55, 85–107
7. Rasmussen, S. and Dixon, R.A. (1999) Transgene-mediated and elicitor-induced perturbation of metabolic channeling at the entry point into the phenylpropanoid pathway. Plant Cell 11, 1537–1552
8. Shorrosh, B.S. et al. (1994) Molecular cloning, characterization and elicitation of acetyl CoA carboxylase from alfalfa. Proc. Natl. Acad. Sci. USA 91, 4323–4327
9. Rodriguez-Concepcion, M. and Boronat, A. (2002) Elucidation of the methylerythritol phosphate pathway for isoprenoid biosynthesis in bacteria and plastids. A metabolic milestone achieved through genomics. Plant Physiol. 130, 1079–1089
10. Sauret-Gueto, S. et al. (2006) Plastid cues post-transcriptionally regulate the accumulation of key enzymes of the methylerythritol phosphate pathway in Arabidopsis. Plant Physiol. 141, 75–84
11. Dudareva, N. et al. (2005) The nonmevalonate pathway supports both monoterpene and sesquiterpene formation in snapdragon flowers. Proc. Natl. Acad. Sci. USA 102, 933–938
12. Laule, O. et al. (2003) Crosstalk between cytosolic and plastidial pathways of isoprenoid biosynthesis in Arabidopsis thaliana. Proc. Natl. Acad. Sci. USA 100, 6866–6871
13. Lichtenthaler, H.K. (1999) The 1-deoxy-D-xylulose-5-phosphate pathway of isoprenoid biosynthesis in plants. Annu. Rev. Plant Physiol. Plant Mol. Biol. 50, 47–65
14. Burke, C. et al. (1999) Geranyl diphosphate synthase: cloning, expression, and characterization of this prenyltransferase as a heterodimer. Proc. Natl. Acad. Sci. USA 96, 13062–13067
15. Cheng, A.-X. et al. (2007) Plant terpenoids: biosynthesis and ecological functions. J. Integrative Plant Biol. 49, 179–186
16. Facchini, P.J. et al. (2000) Plant aromatic -amino acid decarboxylases: evolution, biochemistry, regulation, and metabolic engineering applications. Phytochemistry 54, 121–138

17. Fecker, L.F. et al. (1993) Increased production of cadaverine and anabasine in hairy root cultures of *Nicotiana tabacum* expressing a bacterial lysine decarboxylase gene. Plant Mol. Biol 23, 11–21
18. Facchini, P.J. (2001) Alkaloid biosynthesis in plants: Biochemistry, cell biology, molecular regulation, and metabolic engineering applications. Annu. Rev. Plant Physiol. Plant Mol. Biol. 52, 29–66
19. Austin, M.B. and Noel, J.P. (2003) The chalcone synthase superfamily of type III polyketide synthases. Nat. Prod. Rep. 20, 79–110
20. Jez, J.M. et al. (2001) Structure and mechanism of chalcone synthase-like polyketide synthases. J. Indus. Microbiol. Biotechnol. 27, 393
21. Jez, J.M. et al. (2000) Structural control of polyketide formation in plant-specific polyketide synthesis. Chem. Biol. 7, 919–930
22. Jez, J.M. et al. (2002) Expanding the biosynthetic repertoire of plant type III polyketide synthases by altering starter molecule specificity. Proc. Natl. Acad. Sci. USA 99, 5319–5324
23. Austin, M.B. et al. (2004) An aldol switch discovered in stilbene synthases mediates cyclization specificity of type III polyketide synthases. Chem. Biol. 11, 1179–1194
24. Paniego, N.B. et al. (1999) Phlorisovalerophenone synthase, a novel polyketide synthase from hop (*Humulus lupulus* L.) cones. Eur. J. Biochem. 262, 612–616
25. Raharjo, T.J. et al. (2004) Comparative proteomics of *Cannabis sativa* plant tissues. J. Biomol. Tech. 15, 97–106
26. Tholl, D. (2006) Terpene synthases and the regulation, diversity and biological roles of terpene metabolism. Curr. Opin. Plant Biol. 9, 297–304
27. Ikeda, C. et al. (2007) Functional analysis of eubacterial *ent*-copalyl diphosphate synthase and pimara-9(11),15-diene synthase with unique primary sequences. J. Biochem. 141, 37–45
28. Schwab, W. et al. (2001) Mechanism of monoterpene cyclization: Stereochemical aspects of the transformation of noncyclizable substrate analogs by recombinant (-)-limonene synthase, (+)-bornyl diphosphate synthase, and (-)-pinene synthase. Arch. Biochem. Biophys. 392, 123–136
29. O'Maille, P.E. et al. (2006) Biosynthetic potential of sesquiterpene synthases: alternative products of tobacco 5-epi-aristolochene synthase. Arch. Biochem. Biophys. 448, 73–82
30. Starks, C.M. et al. (1997) Structural basis for cyclic terpene biosynthesis by tobacco 5-epi-aristolochene synthase. Science 277, 1815–1820
31. Otomo, K. et al. (2004) Biological functions of *ent*- and *syn*-copalyl diphosphate synthases in rice: key enzymes for the branch point of gibberellin and phytoalexin biosynthesis. Plant J. 39, 886–893
32. Peters, R.J. et al. (2003) Bifunctional abietadiene synthase: mutual structural dependence of the active sites for protonation-initiated and ionization-initiated cyclizations. Biochemistry 42, 2700–2707
33. Dejong, J.M. et al. (2006) Genetic engineering of taxol biosynthetic genes in *Saccharomyces cerevisiae*. Biotechnol. Bioeng. 93, 212–214
34. Wildung, M.R. and Croteau, R. (1996) A cDNA clone for taxadiene synthase, the diterpene cyclase that catalyzes the committed step of taxol biosynthesis. J. Biol. Chem. 271, 9201–9204
35. Kollner, T.G. et al. (2004) The variability of sesquiterpenes emitted from two *Zea mays* cultivars is controlled by allelic variation of two terpene synthase genes encoding stereoselective multiple product enzymes. Plant Cell 16, 1115–1131
36. Phillips, D.R. et al. (2006) Biosynthetic diversity in plant triterpene cyclization. Curr. Opin. Plant Biol. 9, 305–314
37. Xu, R. et al. (2004) On the origins of triterpenoid skeletal diversity. Phytochemistry 65, 261–291
38. Ma, X.Y. et al. (2004) Crystallization and preliminary X-ray crystallographic analysis of strictosidine synthase from *Rauvolfia*- the first member of a novel enzyme family. Biochim. Biophys. Acta 1702, 121–124
39. Kutchan, T.M. (1993) Strictosidine: from alkaloid to enzyme to gene. Phytochemistry 32, 493–506
40. Facchini, P.J. et al. (2004) Can Arabidopsis make complex alkaloids. Trends Plant Sci. 9, 116–122
41. De-Eknamkul, W. et al. (2000) Purification and characterization of deacetylipecoside synthase from *Alangium lamarckii* Thw. Phytochemistry 55, 177–181
42. Luk, L.Y.P. et al. (2007) Mechanistic studies on norcoclaurine synthase of benzylisoquinoline alkaloid biosynthesis: an enzymatic Pictet-Spengler reaction. Biochemistry 46, 10153–10161
43. Samanani, N. et al. (2004) Molecular cloning and characterization of norcoclaurine synthase, an enzyme catalyzing the first committed step in benzylisoquinoline alkaloid biosynthesis. Plant J. 40, 302–313

44. Minami, H. et al. (2007) Functional analysis of norcoclaurine synthase in *Coptis japonica*. J. Biol. Chem. 282, 6274–6282
45. Kutchan, T.M. and Dittrich, H. (1995) Characterization and mechanism of the berberine bridge enzyme, a covalently flavinylated oxidase of benzophenanthridine alkaloid biosynthesis in plants. J. Biol. Chem. 270, 24475–24481
46. Facchini, P.J. and St-Pierre, B. (2005) Synthesis and trafficking of alkaloid biosynthetic enzymes. Curr. Opin. Plant Biol. 8, 657–666
47. Bolwell, G.P. et al. (1994) Plant cytochrome P450. Phytochemistry 37, 1491–1506
48. Weisshaar, B. and Jenkins, G.I. (1998) Phenylpropanoid biosynthesis and its regulation. Curr. Opin. Plant Biol. 1, 251–257
49. Chapple, C.C. et al. (1992) An Arabidopsis mutant defective in the general phenylpropanoid pathway. Plant Cell 4, 1413–1424
50. Holton, T.A. et al. (1993) Cloning and expression of cytochrome P450 genes controlling flower colour. Nature 366, 276–279
51. Collu, G. et al. (2001) Geraniol 10-hydroxylase, a cytochrome P450 enzyme involved in terpenoid indole alkaloid biosynthesis. FEBS Lett 508, 215–220
52. Haudenschild, C. et al. (2000) Functional expression of regiospecific cytochrome P450 limonene hydroxylases from mint (*Mentha spp.*) in *Escherichia coli* and *Saccharomyces cerevisiae*. Arch. Biochem. Biophys. 379, 127–136
53. Takahashi, S. et al. (2005) Kinetic and molecular analysis of 5-epiaristolochene-1,3-dihydroxylase, a cytochrome P450 enzyme catalyzing successive hydroxylations of sesquiterpenes. J. Biol. Chem 280, 3686–3696
54. Schoendorf, A. et al. (2001) Molecular cloning of a cytochrome P450 taxane 10 beta-hydroxylase cDNA from Taxus and functional expression in yeast. Proc. Natl. Acad. Sci. USA 98, 1501–1506
55. Jennewein, S. et al. (2001) Taxol biosynthesis: taxane 13 alpha-hydroxylase is a cytochrome P450-dependent monooxygenase. Proc. Natl. Acad. Sci. USA 98, 13595–13600
56. Shibuya, M. et al. (2006) Identification of β-amyrin and sophoradiol 24-hydroxylase by expressed sequence tag mining and functional expression assay. FEBS J. 273, 948–959
57. Ohnishi, T. et al. (2006) C-23 hydroxylation by Arabidopsis CYP90C1 and CYP90D1 reveals a novel shortcut in brassinosteroid biosynthesis. Plant Cell 18, 3275–3288
58. Tian, L. et al. (2004) The Arabidopsis LUT1 locus encodes a member of the cytochrome P450 family that is required for carotenoid epsilon-ring hydroxylation activity. Proc. Natl. Acad. Sci. USA 101, 402–407
59. Okamoto, M. et al. (2006) CYP707A1 and CYP707A2, which encode abscisic acid 8'-hydroxylases, are indispensable for proper control of seed dormancy and germination in Arabidopsis. Plant Physiol. 141, 97–107
60. Niemeyer, H.M. (1998) Hydroxamic acids (4-hydroxy-1,4-benzoxazin-3-ones), defence chemicals in the gramineae. Phytochemistry 27, 3349–3358
61. Frey, M. et al. (1997) Analysis of a chemical plant defense mechanism in grasses. Science 277, 696–699
62. Stumpe, M. et al. (2006) Identification of an allene oxide synthase (CYP74C) that leads to formation of alpha-ketols from 9-hydroperoxides of linoleic and linolenic acid in below-ground organs of potato. Plant J. 47, 883–896
63. Howe, G.A. and Schilmiller, A.L. (2002) Oxylipin metabolism in response to stress. Curr. Opin. Plant Biol. 5, 230–236
64. Park, J.H. et al. (2002) A knock-out mutation in allene oxide synthase results in male sterility and defective wound signal transduction in Arabidopsis due to a block in jasmonic acid biosynthesis. Plant J. 31, 1–12
65. Rodriguez, S. et al. (2003) Jasmonate-induced epoxidation of tabersonine by a cytochrome P-450 in hairy root cultures of *Catharanthus roseus*. Phytochemistry 64, 401–409
66. Zhu, Y. et al. (2006) Elongated uppermost internode encodes a cytochrome P450 monooxygenase that epoxidizes gibberellins in a novel deactivation reaction in rice. Plant Cell 18, 442–456
67. Novak, B.H. et al. (2000) Morphine synthesis and biosynthesis- an update. Curr. Org. Chem. 4, 343–362
68. Ikezawa, N. et al. (2008) Molecular cloning and characterization of CYP80G2, a cytochrome P450 that catalyzes an intramolecular C–C phenol coupling of (S)-reticuline in magnoflorine biosynthesis, from cultured *Coptis japonica* cells. J. Biol. Chem. 283, 8810–8821
69. Bauer, W. and Zenk, M.H. (1991) Two methylenedioxy bridge forming cytochrome P-450 dependent enzymes are involved in (S)-stylopine biosynthesis. Phytochemistry 30, 2953–2961
70. Clemens, S. and Barz, W. (1996) Cytochrome P450-dependent methylenedioxy bridge

formation in *Cicer arietinum*. Phytochemistry 41, 457–460
71. Steele, C.L. et al. (1999) Molecular characterization of the enzyme catalyzing the aryl migration reaction of isoflavonoid biosynthesis in soybean. Arch. Biochem. Biophys. 367, 146–150
72. Dixon, R.A. and Ferreira, D. (2002) Genistein. Phytochemistry 60, 205–211
73. Lozovaya, V.V. et al. (2007) Modification of phenolic metabolism in soybean hairy roots through down regulation of chalcone synthase or isoflavone synthase. Planta 225, 665–679
74. Jung, W. et al. (2000) Identification and expression of isoflavone synthase, the key enzyme for biosynthesis of isoflavones in legumes. Nat. Biotechnol. 18, 208–212
75. Holscher, D. and Schneider, B. (2005) The biosynthesis of 8-phenylphenalenones from *Eichhornia crassipes* involves a putative aryl migration step. Phytochemistry 66, 59–64
76. Prescott, A.G. and Lloyd, M.D. (2000) The iron(II) and 2-oxoacid-dependent dioxygenases and their role in metabolism. Nat. Prod. Rep. 17, 367–383
77. Springob, K. et al. (2003) Recent advances in the biosynthesis and accumulation of anthocyanins. Nat. Prod. Rep. 20, 288–303
78. Holton, T.A. and Cornish, E.C. (1995) Genetics and biochemistry of anthocyanin biosynthesis. Plant Cell 7, 1071–1083
79. Anzellotti, D. and Ibrahim, R.K. (2004) Molecular characterization and functional expression of flavonol 6-hydroxylase. BMC Plant Biol. 4, 20
80. Saito, K. et al. (1999) Direct evidence for anthocyanidin synthase as a 2-oxoglutarate-dependent oxygenase: molecular cloning and functional expression of cDNA from a red forma of *Perilla frutescens*. Plant J. 17, 181–189
81. Wilmouth, R.C. et al. (2002) Structure and mechanism of anthocyanidin synthase from *Arabidopsis thaliana*. Structure 10, 93–103
82. Hashimoto, T. and Yamada, Y. (1986) Hyoscyamine 6-beta-hydroxylase, a 2-oxoglutarate-dependent dioxygenase, in alkaloid-producing root cultures. Plant Physiol. 81, 619–625
83. Paiva, N.L. et al. (1991) Stress responses in alfalfa (*Medicago sativa* L.) 11. Molecular cloning and expression of alfalfa isoflavone reductase, a key enzyme of isoflavonoid phytoalexin biosynthesis. Plant Mol. Biol. 17, 653–667
84. Fischer, D. et al. (1990) Phytoalexin synthesis in soybean: purification and characterization of NADPH:2'-hydroxydaidzein oxidoreductase from elicitor-challenged soybean cell cultures. Arch. Biochem. Biophys. 276, 390–395
85. Sun, Y.J. et al. (1991) Stereoisomerism in plant disease resistance: induction and isolation of the 7,2'-dihydroxy-4',5'-methylenedioxyisoflavone oxidoreductase, an enzyme introducing chirality during synthesis of isoflavonoid phytoalexins in pea (*Pisum sativum* L). Arch. Biochem. Biophys. 284, 167–173
86. Schlieper, D. et al. (1990) Stereospecificity of hydrogen transfer by fungal and plant NADPH: isoflavone oxidoreductases. Phytochemistry 29, 1519–1524
87. Dinkova-Kostova, A.T. et al. (1996) (+)-Pinoresinol/(+)-lariciresinol reductase from *Forsythia intermedia*. Protein purification, cDNA cloning, heterologous expression and comparison to isoflavone reductase. J. Biol. Chem. 271, 29473–29482
88. Gang, D.R. et al. (1999) Evolution of plant defense mechanisms. Relationships of phenylcoumaran benzylic ether reductases to pinoresinol-lariciresinol and isoflavone reductases. J. Biol. Chem. 274, 7516–7527
89. Gerats, A.G.M. et al. (1982) Genetic control of the conversion of dihydroflavonols into flavonols and anthocyanins in flowers of *Petunia hybrida*. Planta 155, 364–368
90. Xie, D.Y. et al. (2003) Role of anthocyanidin reductase, encoded by BANYULS in plant flavonoid biosynthesis. Science 299, 396–399
91. Lacombe, E. et al. (1997) Cinnamoyl-CoA reductase, the first committed enzyme of lignin branch biosynthetic pathway: cloning, expression and phylogenetic relationships. Plant J. 11, 429–441
92. Bowles, D. et al. (2006) Glycosyltransferases of lipophilic small molecules. Annu. Rev. Plant Biol. 57, 567–597
93. Shao, H. et al. (2005) Crystal structures of a multifunctional triterpene/flavonoid glycosyltransferase from *Medicago truncatula*. Plant Cell 17, 3141–3154
94. Offen, W. et al. (2006) Structure of a flavonoid glucosyltransferase reveals the basis for plant natural product modification. EMBO J 25, 1396–1405
95. Li, L. et al. (2007) Crystal structure of *Medicago truncatula* UGT85H2 – insights into the structural basis of a multifunctional (iso)flavonoid glycosyltransferase. J Mol. Biol. 370, 951–963
96. Modolo, L.V. et al. (2007) A functional genomics approach to (iso)flavonoid glycosylation in the model legume *Medicago truncatula*. Plant Mol. Biol. 64, 499–518

97. Ono, E. et al. (2006) Yellow flowers generated by expression of the aurone biosynthetic pathway. Proc. Natl. Acad. Sci. USA 103, 11075–11080
98. Richman, A. et al. (2005) Functional genomics uncovers three glucosyltransferases involved in the synthesis of the major sweet glucosides of *Stevia rebaudiana*. Plant J. 41, 56–67
99. Achnine, L. et al. (2005) Genomics-based selection and functional characterization of triterpene glycosyltransferases from the model legume *Medicago truncatula*. Plant J. 41, 875–887
100. Jackson, R.G. et al. (2002) Over-expression of an Arabidopsis gene encoding a glucosyltransferase of indole-3-acetic acid: phenotypic characterisation of transgenic lines. Plant J. 32, 573–583
101. Poppenberger, B. et al. (2005) The UGT73C5 of Arabidopsis thaliana glucosylates brassinosteroids. Proc. Natl. Acad. Sci. USA 102, 15253–15258
102. Hou, B. et al. (2004) *N*-glucosylation of cytokinins by glycosyltransferases of *Arabidopsis thaliana*. J. Biol. Chem. 279, 47822–47832
103. Brazier-Hicks, M. et al. (2007) Characterization and engineering of the bifunctional *N*- and *O*-glucosyltransferase involved in xenobiotic metabolism in plants. Proc. Natl. Acad. Sci. USA 104, 20238–20243
104. Marillia, E.F. et al. (2001) Molecular cloning of a Brassica napus thiohydroximate *S*-glucosyltransferase gene and its expression in *Escherichia coli*. Physiol. Plant. 113, 176–184
105. Grubb, C.D. et al. (2004) Arabidopsis glucosyltransferase UGT74B1 functions in glucosinolate biosynthesis and auxin homeostasis. Plant J. 40, 893–908
106. Ibrahim, R.K. et al. (1998) Plant *O*-methyltransferases: molecular analysis, common signature and classification. Plant Mol. Biol. 36, 1–10
107. Ibrahim, R.K. et al. (1987) Enzymology and compartmentation of polymethylated flavonol glucosides in *Chrysosplenium americanum*. Phytochemistry 26, 1237–1245
108. Noel, J.P. et al. (2003) Structural, functional, and evolutionary basis for methylation of plant small molecules. Rec. Adv. Phytochem. 37, 37–58
109. Preisig, C.L. et al. (1989) Purification and characterization of S-adenosyl-L-methionine: 6a-hydroxymaackiain 3-*O*-methyltransferase from *Pisum sativum*. Plant Physiol. 91, 559–566
110. Guo, D. et al. (2001) Downregulation of caffeic acid 3-*O*-methyltransferase and caffeoyl CoA 3-*O*-methyltransferase in transgenic alfalfa. Impacts on lignin structure and implications for the biosynthesis of G and S lignin. Plant Cell 13, 73–88
111. Frick, S. and Kutchan, T.M. (1999) Molecular cloning and functional expression of *O*-methyltransferases common to isoquinoline alkaloid and phenylpropanoid biosynthesis. Plant J. 17, 329–339
112. Teuber, M. et al. (2007) Putrescine *N*-methyltransferases – a structure-function analysis. Plant Mol. Biol. 63, 787–801
113. Varin, L. et al. (1987) Enzymatic synthesis of sulphated flavonoids in *Flaveria* spp. Phytochemistry 26, 135–138
114. Okada, K. et al. (2004) The AtPPT1 gene encoding 4-hydroxybenzoate polyprenyl diphosphate transferase in ubiquinone biosynthesis is required for embryo development in *Arabidopsis thaliana*. Plant Mol. Biol. 55, 567–577
115. Tian, L. et al. (2008) Biosynthesis and genetic engineering of health-promoting plant natural products. Phytochem. Rev. 7, 445–465
116. Sasaki, K. et al. (2008) Cloning and characterization of naringenin 8-prenyltransferase, a flavonoid-specific prenyltransferase of *Sophora flavescens*. Plant Physiol. 146, 1075–1084
117. D'Auria, J.C. (2006) Acyltransferases in plants: a good time to be BAHD. Curr. Opin. Plant Biol. 9, 331–340
118. Yu, X.-H. et al. (2008) Nucleocytoplasmic-localized acyltransferases catalyze the malonylation of 7-*O*-glycosidic (iso)flavones in Medicago truncatula. Plant J. 55, 382–396
119. Milkowski, C. and Strack, D. (2004) Serine carboxypeptidase-like acyltransferases. Phytochemistry 65, 517–524
120. Li, A.X. and Steffens, J.C. (2000) An acyltransferase catalyzing the formation of diacylglucose is a serine carboxypeptidase-like protein. Proc. Natl. Acad. Sci. USA 97, 6902–6907
121. Lehfeldt, C. et al. (2000) Cloning of the *SNG1* gene of Arabidopsis reveals a role for a serine carboxypeptidase-like protein as an acyltransferase in secondary metabolism. Plant Cell 12, 1295–1306
122. Deavours, B.E. et al. (2006) Functional analysis of members of the isoflavone and isoflavanone *O*-methyltransferase enzyme families from the model legume Medicago truncatula. Plant Mol. Biol. 62, 715–733

7 Methods for Molecular Identification of Biosynthetic Enzymes in Plants

Sarah O'Connor

Abstract This chapter highlights the major strategies used to identify the genes of plant natural product biosynthetic pathways. One or two key examples that illustrate these strategies are provided. A wide range of enzyme classes have been identified using the approaches described in this review including P450 enzymes, terpene systhases, and glycosyltransferases. The advantages and disadvantages of each approach is discussed.

Introduction

Plants produce some of the most important natural products used in medicine (Fig. 1). Plant-derived compounds have been used as powerful pharmaceuticals throughout the course of human history. For example, opium poppy has been used since neolithic times [1]; aspirin, synthesized by acetylation of salicyclic acid from willow bark, was discovered in the late nineteenth century [2]; and the powerful chemotherapy agent taxol, used in the treatment of advanced breast cancer, was discovered from the yew tree in the mid-twentieth century [3].

Sarah O'Connor (✉)
Massachusetts Institute of Technology,
Department of Chemistry,
18-592, 77 Massachusetts Ave,
Cambridge MA 02139, USA
e-mail: soc@mit.edu

The study of microbe derived natural products underwent a revolution in the 1980s as genetic strategies allowed the rapid discovery of the genes that encode natural product biosynthetic pathways [4]. This genetic information allows a detailed understanding of the chemical and biochemical mechanisms that Nature uses to construct complex molecules. Microbial metabolic pathways can now be reengineered at the genetic level to produce higher levels of natural products, or novel natural product derivatives with improved medicinal properties [5].

Analogously, if the biosynthetic enzymes leading to plant derived natural products are cloned, we can begin to understand how plants construct complex molecules [6]. Furthermore, genetic information is crucial for metabolic engineering efforts to increase the production levels of these compounds. For example, expression of plant biosynthetic pathways in fast growing organisms such as *Escherichia coli* or *Saccharomyces cerevisiae* could be explored, as it has been successfully achieved for several plant derived flavonoid [7] and terpenoid natural products [8]. Additionally, many plant biosynthetic pathways are highly branched, leading to the formation of "side products" that lack the desired bioactivity of the target natural product. If the enzymes at the branch points responsible for side product biosynthesis are cloned, they could be downregulated and production of the desired products in plants could be improved [9]. Alternatively, genes encoding

A.E. Osbourn and V. Lanzotti (eds.), *Plant-derived Natural Products*,
DOI 10.1007/978-0-387-85498-4_7, © Springer Science+Business Media, LLC 2009

7

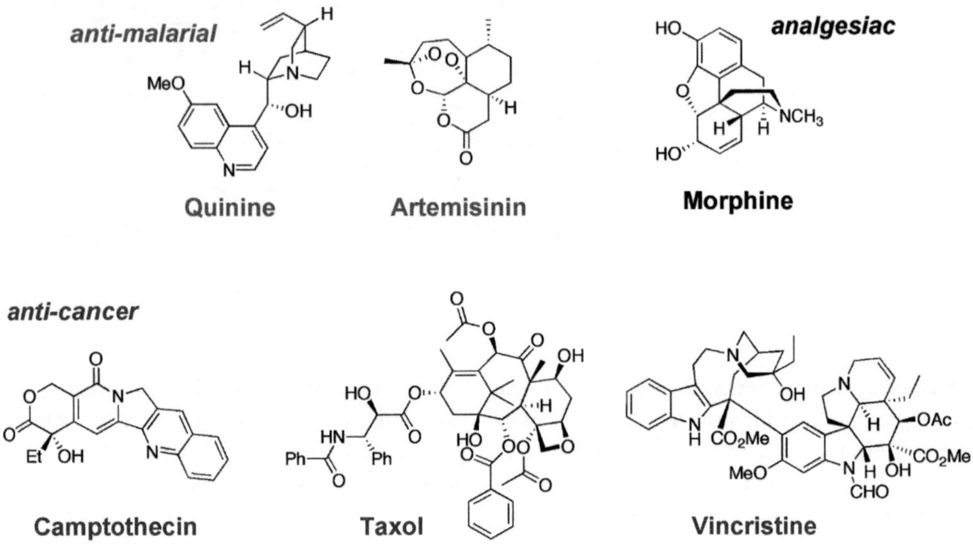

Fig. 1 Plant natural products. A selection of representative plant derived natural products that have important pharmaceutical activities

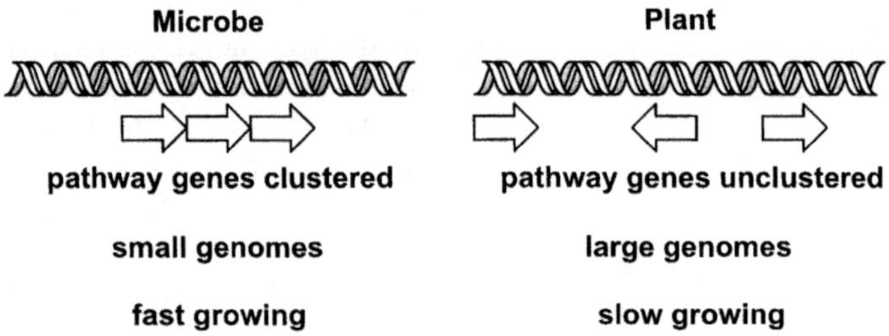

Fig. 2 Challenges of plants. Several challenges are encountered in plant genetics that have made pathway elucidation more difficult than pathway elucidation in microbes

biosynthesis of the desired product could be overexpressed [10]. Additionally, isolation of compounds from natural sources provides limited opportunities to modify their chemical and biological properties. Cloning the enzymes that catalyze natural product biosynthesis may allow reprogramming of the biosynthetic pathway to produce new unnatural products with improved pharmaceutical activities [11]. Finally, the expression levels of biosynthetic genes, and the levels of the corresponding natural products, are controlled by transcription factor proteins, which interact with the promoter regions of target genes and modulate the rate of mRNA production. If the transcription factors that control expression of biosynthetic genes are identified, then these proteins could be used to upregulate biosynthetic gene expression and natural product production levels [12].

Although plants produce a large fraction of known natural products, the number of plant pathways that have been elucidated pales in comparison to the number of well-characterized microbial pathways [13]. Elucidating the genes of a plant pathway is a task fraught with significant challenges, and the genetic tools that have unlocked the biosynthesis of countless bacterial and fungal natural products are not easily applied to plants. In contrast to microbes, the genes of plant pathways – with few exceptions - [14] are not clustered on the genome, so each gene of a plant pathway must be discovered individually (Fig. 2). Additionally, the genome sizes of medicinal plants are much larger (>1,000 Mbp) than the typical natural product producing bacteria (~8 Mbp), which makes finding and screening putative biosynthetic genes a daunting undertaking [15]. Direct purification of plant enzymes is complicated by low levels of protein expression, the presence of phenolic compounds and proteases that readily inactivate enzyme activity and the frequent requirement of accessory proteins or cofactors for catalysis. Finally, plants are slow growing, and the process of genetic transformation and subsequent selection is time-consuming. Although spectacular successes have been achieved in elucidating plant pathways, the challenges of plant biology have hindered the study of plant secondary metabolism (Fig. 2).

This review provides an overview of the major strategies used to identify the genes that comprise plant natural product biosynthetic pathways. One or two key examples that illustrate these strategies are provided though many more examples are available for most of the techniques outlined. A range of enzyme classes been identified using the approaches described in this review including P450 enzymes [16], terpene synthases [17], and glycosyltransferases [18]. Notably, many of the strategies for gene identification described below require prior knowledge of the biochemical transformation that the corresponding gene product catalyzes. This chemical knowledge can allow prediction of the enzyme class so that the gene can be identified in homology based cloning strategies. Additionally, the biochemical reaction must be known for design of enzymatic assay for *in vitro* enzymology. The biochemistry is typically elucidated by feeding isotopically labeled precursors to the plant or plant culture and mapping the placement of the isotopes within the final natural product. Additionally, isolation of biosynthetic intermediates as well as biomimetic synthetic chemistry provides insights into the chemical transformation that occur.

Methods of Pathway Elucidation in Plants

Traditional Biochemical Approaches

Countless plant biosynthetic enzymes have been identified using a classical approach (Fig. 3). The desired enzyme is purified from a crude plant or

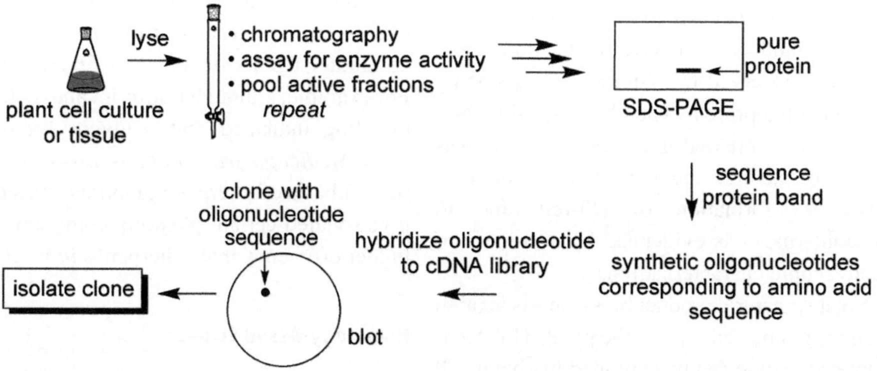

Fig. 3 Classical purification approaches. In this strategy, the desired enzyme is purified by traditional biochemical chromatography, and protein sequence information is used to isolate the desired gene from plant cDNA

plant cell culture lysate by traditional protein chromatography. The enzyme is typically monitored throughout the purification procedure by an *in vitro* enzymatic assay. Once a homogenous preparation of enzyme is prepared, amino acid sequences are obtained by subjecting the protein, or tryptic fragments of the protein, to N-terminal sequencingormassspectrometry.Oligonucleotide primers that correspond to the identified regions of protein sequence are designed, and these primers are used to identify the gene encoding the desired enzyme in a cDNA library. Usually, the identified gene is then expressed in a heterologous expression system and assayed *in vitro* to confirm that the gene encodes the enzyme having the expected function.

Example

The berberine bridge enzyme utilized in tetrahydroisoquinoline alkaloid pathways was cloned by this method [19]. This enzyme was purified from elicited *Eschscholzia californica* cell suspension cultures in six to seven steps, where approximately 10 μg of pure protein was obtained from 700 mg of crude protein. The purified protein was digested with trypsin and the resulting peptides were subjected to N-terminal sequencing. An oligonucleotide primer based on one of these amino acid sequences was shown to hybridize to a clone from an *E. californica* cDNA library. This clone was sequenced and shown to encode the amino acid sequences observed by N-terminal sequencing of the purified protein. The clone was then heterologously expressed in *S. cerevisea* and was shown to catalyze the expected biochemical activity – formation of (*S*)-reticuline to (S)-scoulerine – as evidenced by a number of spectroscopic characterizations.

Notably, natural product biosynthesis is often localized to one cell type of the plant. This compartmentalization can be exploited to obtain cell lysates that are highly enriched in the cell type harboring the enzyme of interest. For example, many of the enzymes involved in vinblastine biosynthesis are localized to the epidermal layer of *Catharanthus roseus* [20]. By selectively harves-ting the epidermal cells of *C. roseus* leaves using an abrasion technique, a highly enriched fraction of 16-hydroxytabersonine-*O*-methyltransferase could be purified in just three chromatographic steps [21].

Scope and Limitations

This classical cloning strategy has proven to be successful in many cases, but it is time-consuming and laborious. Purification of a native enzyme to homogeneity is difficult, and any undesired contaminant protein can be mistakenly subjected to sequencing. Cloning efforts from plant material expressing low levels of the desired enzyme are often unsuccessful. Plant tissues that are known to be enriched in the desired activity provide the best results. Furthermore, the enzymatic activity must be maintained *in vitro* as the desired enzyme is diluted and purified. If accessory proteins or cofactors are required for stability or functional activity, the purified protein will become inactive as it is isolated from the crude cell extract. A robust *in vitro* assay with correct substrates is also required, and this may not be possible if the substrates are unavailable or if the enzymatic products are unstable.

An additional limitation applies to *in vitro* enzyme assays described here and in all subsequent sections. Results of *in vitro* assays often contradict results obtained in *in vivo* studies. For example, although transcript and metabolite profiling indicated that a glycosyltransferase from *Medicago truncatula* is involved in the biosynthesis of triterpene saponins, this enzyme glycosylated certain phenolic compounds with higher efficiency than triterpenes *in vitro* [22].

Homology-Based Screens

Enzymes within a given class often have highly conserved regions in the protein sequence.

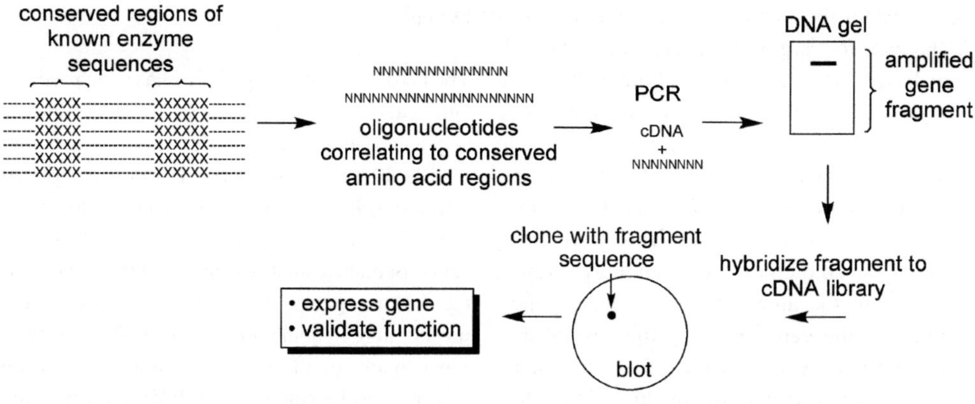

Fig. 4 Homology based cloning. Conserved regions of known enzyme classes are used to design oligonucleotide primers, which are used to isolate the desired gene from plant cDNA

Oligonucleotide primers complementary to these consensus sequences can be designed. These primers can be used to amplify the genes having the corresponding consensus sequence from cDNA libraries by polymerase chain reaction (PCR) (Fig. 4). This method, known as homology based cloning, has been widely used to elucidate genes of biosynthetic pathways. P450 enzymes, glycosyltransferases, acetyltransferases, and prenyl transferases have all been successfully cloned from plant genetic material using primers designed to recognize the known consensus sequences of these enzyme classes.

Example

A cDNA clone encoding a taxane 2a-*O*-benzoyltransferase that catalyzes a late-stage acylation step in the taxol biosynthetic pathway was isolated from *Taxus cuspidata* by employing a homology-based PCR cloning strategy [23]. After generating oligonucleotide primers specific for acyltransferases, several gene fragments were amplified by PCR and these fragments were used to screen a cDNA library constructed from mRNA isolated from induced *Taxus* cells. Several full-length acyltransferases were obtained and each was individually expressed in *E. coli* and assayed for function *in vitro*. These assays indicated that one of these genes encoded an enzyme having taxane 2a-*O*-benzoyltransferase activity.

P450 enzymes, which catalyze a wide range of oxidative transformations in many plant biosynthetic pathways, contain highly conserved regions that can be used to identify the corresponding genes from a library of clones [24]. For example, the P450 enzyme (*S*)-*N*-methylcoclaurine-3'-hydroxylase of tetrahydroisoquinoline alkaloid biosynthesis was cloned from *E. californica* cDNA using a primer based on a P450 consensus sequence [25]. Notably, membrane bound P450 enzymes are usually present in plant cell cultures in very low quantities. Consequently, purified preparation of these enzymes are not readily obtained, so biochemical purification approaches (Section 2.1) are often not possible with enzymes of this class [26]. Homology based cloning strategies, which circumvent the native purification process, have therefore been particularly crucial for discovery of new P450 enzymes involved in plant biosynthetic pathways.

Scope and Limitations

Homology-based cloning is limited to enzymes with a known conserved sequence. If no consensus sequence for the enzyme has been reported – as is the case for rare or novel enzymes – homology based cloning strategies cannot be used. Moreover, the biochemistry of the biosynthetic pathway must be sufficiently well understood to hypothesize which classes of enzyme are involved in the biosynthetic transformations. Additionally, certain enzyme classes, such as P450s, are ubiquitous and homology cloning can lead to the amplification of many genes. Therefore, the gene encoding the desired biosynthetic activity must be identified out of many that are cloned. Again, the quality of the plant material from which the mRNA is harvested is critical to ensure that the desired enzyme is present in high quantities in the library. The enzyme function and specific substrate transformation also need to be validated after isolation of the gene, either by *in vitro* enzyme assay, or by knockdown of the gene in the producing plant (see Section 2.6).

Functional Genomic Approach

As DNA sequencing technology has improved, numerous cDNA libraries can be sequenced within a relatively short time frame. The sequence information collected from the plant derived cDNA can then be compared to a gene database. If homologous genes have been functionally characterized, then a prediction about the biochemical function of the newly sequenced plant gene can be made. Often only short fragments of DNA, termed expressed sequence tags (EST), are sequenced. The EST, typically about 500 base pairs, provides enough sequence information for a protein function to be predicted by homology. The fragments that appear to encode enzymes involved in the natural product biosynthetic pathway are then used to clone the full length genes from a cDNA library. The open reading frame of an isolated cDNA can then be overexpressed in a heterologous organism such as *E. coli* and the recombinant protein can be assayed for function.

Example

The taxol biosynthetic pathway has been partially elucidated by a highly successful random library sequencing effort [27]. In this study, sequencing of 10,000 transcripts derived from an induced *Taxus cuspita* cell suspension culture led to the identification of approximately 3,500 unique ESTs. Sequence analysis indicated that several of the known genes of taxol biosynthesis were present in high abundance in this EST collection. Since many of the taxol biosynthetic steps are predicted to be catalyzed by P450 enzymes, the EST collection was searched for clones having homology to known P450 sequences. Nearly 100 unique P450s were identified by this homology search, out of which 19 displayed homology to previously identified enzymes involved in the hydroxylation steps of the taxol biosynthetic pathway. Ten of these hydroxylase-like P450 genes had never been observed in previous attempts to discover new taxol biosynthetic genes. Each of the ten full length genes was heterologously expressed in yeast. Subsequent *in vitro* enzyme assays suggested that one of these genes encoded a novel taxoid 10-hydroxylase while another encoded a taxoid 2-hydroxylase enzyme.

Importantly, these genomic libraries can be enriched for the appropriate gene transcripts if the tissue or cell type from which the genetic material is harvested corresponds to the site of natural product production. For example, 25% of the clones in cDNA obtained exclusively from the oil gland secretory cell of peppermint (*Mentha* × *piperita*) appear to be involved in oil metabolism [28]. In another example, mRNA for construction of a *C. roseus* cDNA library obtained from epidermal cells resulted in a collection enriched in genes involved in alkaloid biosynthesis [21].

Scope and Limitations

Until recently, a major disadvantage of this strategy was the expense and time incurred in large scale sequencing efforts, but modern high-throughput facilities can rapidly sequence thousands of clones per day. However, after the genetic information is obtained, strategies to identify the relatively small number of biosynthetic genes out of the large amount of sequenced clones must be developed. As with homology based cloning (Section 2.2), prediction of gene function by homology is not a useful way to identify enzymes that catalyze novel biochemical transformations; for example, in the taxol study, since little is known about the biochemical mechanism of a ring expansion that occurs late in taxol biosynthesis, analysis of the ESTs failed to provide any insight for this intriguing step. As with the other methods outlined, plant material with high levels of secondary metabolite enzyme expression provide the best results.

Analysis of Metabolic, Proteomic and Genomic Networks

The information that is obtained from large scale library sequencing as described in Section 2.3 can be further refined by clustering gene expression levels with metabolite production. By comparing the appearance of metabolites with gene expression levels (as indicated by the abundance of mRNA), a metabolic network that correlates genes with natural product biosynthesis can be obtained. This correlation network introduces an additional layer of refinement when analyzing the ESTs for genes involved in natural product production; transcripts that correlate with the appearance of natural products are the ones most likely to be part of the biosynthetic pathway [29].

Genes are randomly sequenced from the desired plant genetic material, and gene functions are predicted by homology when possible. The desired natural product, or, as is the case with many plant product pathways, a mixture of biosynthetically related natural products, is simultaneously monitored by mass spectrometry coupled with liquid chromatography. Genes that are upregulated with the advent of metabolite appearance are presumed to be involved in the biosynthesis or regulation of the natural product production (Fig. 5).

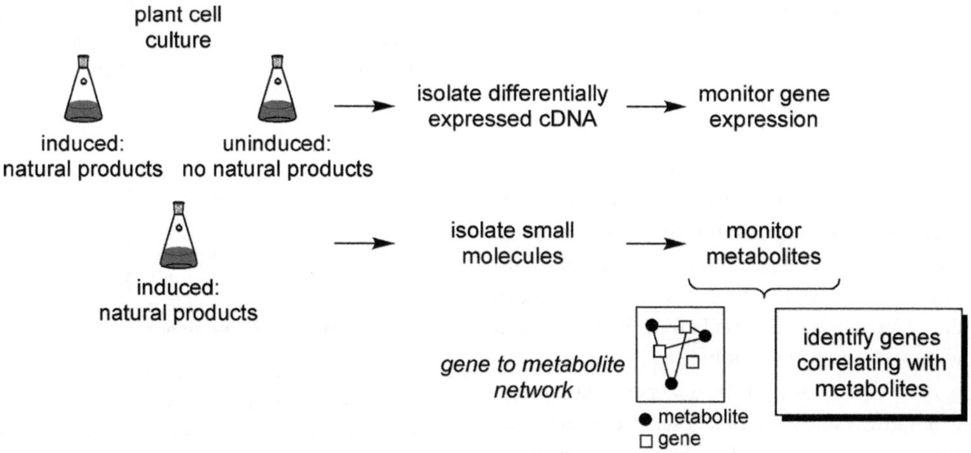

Fig. 5 Metabolic networks. Gene expression levels are correlated with the appearance of natural product production. This correlation provides clues as to which genes of the plant are involved in natural product biosynthesis

Example

A recent example illustrating the power of the genomics and metabolomics combination was recently demonstrated for *C. roseus*, which produces approximately 100 biosynthetically related alkaloid natural products [30]. A total of 178 metabolites were observed by mass spectrometry and nine of these metabolites were positively identified as alkaloid natural products by comparison with authentic standards. In the genomic analysis, approximately 400 unique genes were found from *C. roseus* mRNA. Almost 40% of these ESTs had no sequence homology to any genes in public databases, suggesting that novel enzymes are found within the *C. roseus* plant. Gene to metabolite networks were constructed from this information (as depicted in Fig. 5), making it clear which gene sequences were correlated with the appearance of natural products. Gene to gene networks provided information about groups of genes that had correlated expression levels. Although no conclusions regarding the mechanism of alkaloid biosynthesis were provided at the time of this report, this study provided a wealth of information that can be used to identify new genes involved in this biosynthetic pathway.

Scope and Limitations

As in Section 2.3, major disadvantages of this strategy includes the expense and time incurred in all large scale sequencing efforts. Additionally, plant material with different production levels of the desired metabolites must be available, so that gene expression levels from plant material with differential metabolite levels can be compared. Prediction of gene function by homology is not always possible, as described in previous sections, and the specific function of any putative biosynthetic enzyme candidates must be validated experimentally.

Subtracted cDNA Libraries

Genetic material from two related types of tissue can be "subtracted" from one another (Fig. 6). The genes that both sets of tissue have in common are identified, and the genes unique to each tissue type are readily obtained [31]. Subtractive hybridization techniques can be used to identify genes involved in plant secondary metabolism, provided that the appropriate plant tissue is available. For example, the genes in induced cell cultures – which produce natural products at high levels – can be compared with gene expression levels in uninduced cell cultures that produce low levels of natural products. Genes unique to the induced cell cultures will be likely to play a role in natural product biosynthesis catalysis or regulation. Additionally, different tissue types, such as root or leaf, often have different natural product production levels. Comparison of gene expression profiles from

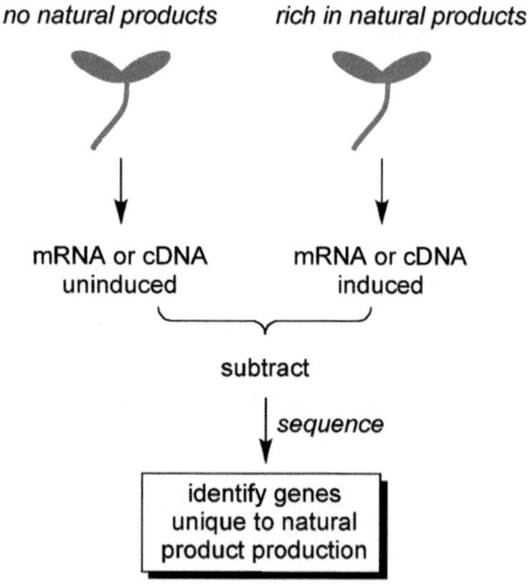

Fig. 6 Subtracted cDNA libraries. The gene expression levels of two plant tissue types, in which one produces natural products while the second does not, are compared. Genes found exclusively in the natural product producing tissue are likely to be involved in natural product production

two types of tissues from the same plant can also be used to predict which genes might play important roles in the biosynthesis of a particular natural product.

Example

A set of 60 cDNAs were isolated by subtractive hybridization of *Nicotiana tabacum* roots before and after removal of the flowers and young leaves, a process known as "topping" [32]. Topping is known to increase the amount of nicotine and other related alkaloids in the plant. A number of the subtracted cDNAs that were unique to the topped plants corresponded to known alkaloid biosynthetic enzymes. The functions of additional cDNAs observed in the subtracted library were not clear and remain under investigation. Further study of these new clones may provide insights into alkaloid natural product biosynthesis in *N. tabacum*.

A second example of subtractive hybridization took advantage of the variation of natural product production in two types of *C. roseus* tissue [33]. In *C. roseus*, roots and leaves produce a distinct spectrum of alkaloids. The leaf and root gene expression profiles were comparatively analyzed using subtractive hybridization. A total of 155 ESTs were subjected to homology-based classification and 16 EST sequences that had never been previously observed were obtained. Again, although new insights into the mechanism of the alkaloid pathway were not reported, functional analysis of these novel ESTs may provide new information into the mechanism of this alkaloid biosynthesis in *C. roseus*.

Scope and Limitations

This approach depends on the availability of tissues that have differential natural product expression. If natural products cannot be induced by an elicitor, or if various tissue types of the plant do not have different levels of natural product expression, then subtractive hybridization strategies cannot be applied. Additionally, subtractive hybridization, as with homology based cloning (Section 2.2) and large scale screening efforts (Sections 2.3 and 2.4), only provides putative biosynthetic genes that must be subjected to additional study. Validating the function of candidate genes must be addressed before the genes can be assigned a defined role in the biosynthetic pathway.

Forward Genetics Approach: Gene Suppression

Methods to suppress gene expression in plants have become widely available. For example, RNA interference (RNAi) can readily downregulate gene expression in plants. Large scale RNAi screens can be used to rapidly assess the function of genes, provided that a fast screen or selection is available to interpret the phenotypes of the transformed lines (Fig. 7) [34]. Typically, cDNA fragments of approximately 500 bp in size are cloned into the appropriate vector and then transformed into the desired plant. These vector constructs can then suppress any plant gene that contains sequence corresponding to the fragment. The transformed plant material with a suppressed gene is then monitored for changes in phenotype. For example, cell cultures that exhibit decreased levels of natural products are likely to have been transformed with a fragment that suppresses a gene involved in natural product biosynthesis. If an efficient way to assess the natural product production of the transformed lines is available, large scale RNAi screens can be used to identify secondary metabolite genes.

Example

To investigate a late step in tropane alkaloid biosynthesis in *Hyoscyamus niger*, a subtracted

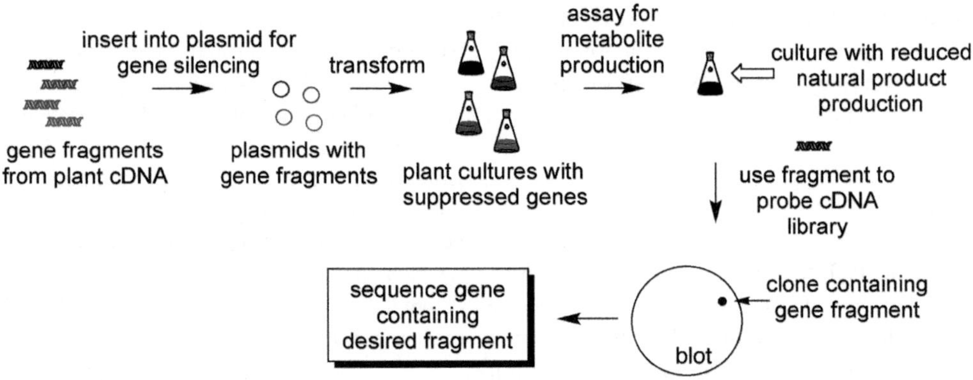

Fig. 7 Gene suppression. Gene fragments are used to silence genes in plant cell culture. If a fragment from a biosynthetic gene is used for silencing, then natural product production will be suppressed. The fragment can be used to identify the full length clone in a cDNA library

cDNA library of leaf and root was constructed to enrich the library in genes specifically expressed in the root, the site of tropane alkaloid biosynthesis [35]. The biosynthetic transformation of interest was predicted to be catalyzed by a P450 enzyme. A total of 25 ESTs from this cDNA library showed sequence homology to P450s, and the genes corresponding to these 25 ESTs were suppressed in *H. niger* using virus induced gene silencing. Virus-induced gene silencing exploits the RNA silencing pathway directed against invading viruses to silence host genes. Production of alkaloid products and alkaloid biosynthetic intermediates was then monitored by mass spectrometry in the suppressed cell lines. In one transformed line, formation of the final alkaloid product appeared to be inhibited, and the penultimate biosynthetic intermediate, littorine, was observed in large quantities. Gene function was validated by heterologous expression of littorine synthase in yeast and tobacco followed by an *in vitro* assay that validated enzymatic conversion of (*R*)-littorine to (*S*)-hyoscyamine.

Gene silencing can also be used to validate the function of a specific gene. For example, the cDNA CaMXMT1 encodes an *N*-methyltransferase involved in caffeine biosynthesis [36]. After suppression of this gene with RNAi in *Coffea Arabica*, caffeine biosynthesis was halted and the biosynthetic intermediate theobromine accumulated instead. This strongly suggested that the enzyme that CaMXMT1 encoded the *N*-methyltransferase enzyme that methylates theobromine to yield caffeine.

Scope and Limitations

Gene silencing requires that the plant be amenable to efficient transformation. Since a number of medicinal plants cannot be stably transformed by either *Agrobacterium* or particle bombardment technology, this strategy is unfortunately not universally applicable to the elucidation of all biosynthetic pathways of natural products. Additionally, suppression of a large number of genes requires a selection or screen that can rapidly assess the natural product profiles of the transformed cell lines.

Forward Genetics Approach: Expression Cloning

In expression cloning, a high quality cDNA library is constructed from the species of interest [37]. Plasmid DNA from this library is transformed into a fast growing organism such as *E. coli* or *S. cerevisiae* for heterologous expression of the clones contained within the cDNA library. Transformed *E. coli* or *S. cerevisiae* cultures are then cultivated on solid media containing appropriate enzyme substrates to allow the detection of enzyme activity. If a culture that appears to catalyze the desired enzymatic reaction is detected, then the corresponding clone can be further analyzed by DNA sequence analysis and more thorough *in vitro* enzymatic assays (Fig. 8). Expression cloning can be performed in the absence of any knowledge of the enzyme to be cloned, and many different enzymes can be screened simultaneously, provided that efficient screening assays are available [38].

Example

Expression cloning has had widespread success in discovery of prokaryotic enzymes [32], but relatively few examples of expression cloning for elucidation of plant metabolism have been reported. In one example, a cDNA library constructed from genetic material of pumpkin seedlings (*Cucurbita maxima* L.) was functionally expressed in bacteriophage [39]. The plaques were screened for hybridization to a polyclonal antibody for GA 20-oxidase, an enzyme involved in the gibberellin biosynthetic pathway. This antibody was raised using enzyme that had been purified from plant material by traditional biochemical chromatography. Plaques that hybridized to the antibody also catalyzed the expected enzymatic activity *in vitro*.

In a second example, cDNA from *C. roseus* was functionally expressed in yeast in an attempt to clone secologanin synthase, an enzyme that catalyzes the conversion of loganin to secologanin [40]. Secologanin is a key precursor for the indole alkaloid natural products produced in *C. roseus*. The plant cDNA was transformed into a strain of yeast that also overexpressed two alkaloid biosynthetic enzymes that convert secologanin into a bright, readily detectable yellow pigment. Yeast transformed with the cDNA and these two known enzymes were incubated with loganin and monitored for the appearance of a yellow color. Although no putative secologanin synthase candidate gene was obtained, the study demonstrated how a well-designed colorimetric assay can be used as a high throughput screen for large numbers of clones.

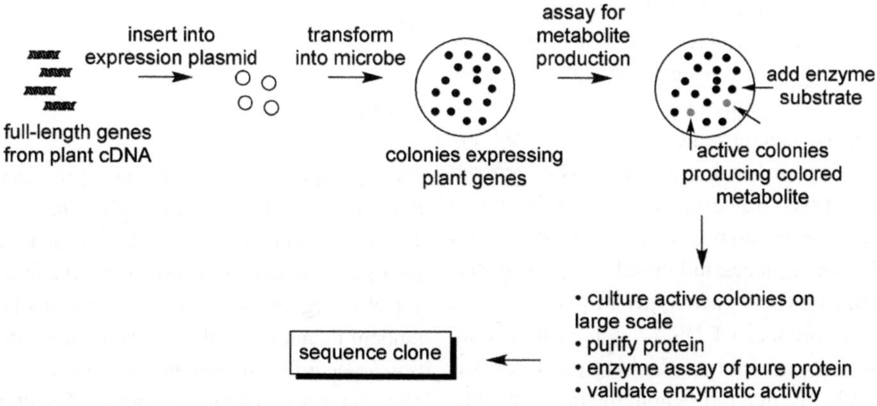

Fig. 8 Expression cloning. Plant cDNA is placed into an expression plasmid and is transformed into yeast or bacteria. Enzyme function can be directly assayed after transformation

7

Scope and Limitations

The greatest limitation of functional cloning is the design of effective enzyme assays that can rapidly assess the function of thousands of clones. Unfortunately, relatively few metabolic enzymes yield a colored product that can be easily detected. Furthermore, many biosynthetic enzymes act on unavailable, unknown or unstable biosynthetic intermediates that are not suitable for an *in vitro* enzyme assay. The plant enzymes must also be expressed in active form in *E. coli* or yeast for accurate functional assessment, and some plant proteins can only be functionally expressed in insect or plant cells, which are not amenable for this functional expression strategy. Finally, expression cloning requires exceptionally high quality cDNA that consists of full-length clones.

Forward Genetics Approach: T DNA Tagging

In the study of microbial natural product biosynthesis, analysis of mutants that express a desired phenotype is a powerful method to assess gene function. Once the desired phenotype – such as a decrease in natural product production – is observed, the corresponding genotype is analyzed. The mutated gene is presumed to be important in modulating the phenotype – i.e. natural product production - of the organism. This strategy is not widely applicable to plants since recessive mutations, in which the phenotype of the mutant is not easily visualized, are usually obtained with most methods of generating mutations. However, activation T-DNA tagging enables the dominant upregulation of genes and has been successfully used to rapidly generate and assess the phenotypes of plant mutants. In this technique, a T-DNA tag that causes transcriptional activation of flanking plant genes is randomly inserted throughout the plant genome. These transformed cells are then screened for changes in phenotype that result from the upregulation (Fig. 9). Provided that a screen or selection to detect changes in natural product production can be developed, this method provides a powerful approach to discovering the genes of secondary metabolic enzymes.

Example

Using an elegant selection approach, the ORCA3 transcriptional factor that regulates the expression of several enzymes of an alkaloid pathway in *C. roseus* was discovered [41, 42]. T-DNA tags were transformed into *C. roseus* cell suspension cultures and the T-DNA was randomly incorporated throughout the genome. The selection assessed overexpression of tryptophan decarboxylase, a known alkaloid biosynthetic enzyme that converts tryptophan into tryptamine. Cells were cultured in the presence of 4-methyltryptophan, which is toxic to *C. roseus*. Cells that grew in the presence of this compound were shown to have upregulated expression levels of tryptophan decarboxylase, which converts 4-methyltryptophan to non-toxic 4-methyltryptamine. Using this selection in combination with T-DNA tagging, the transcriptional factor ORCA3, which regulates expression of tryptophan decarboxylase as well as several other alkaloid biosynthetic enzymes, was discovered.

Scope and Limitations

The limitations that apply to gene silencing described in Section 2.6 also apply with activation T-DNA tagging. First, the plant must be amenable to efficient transformation. Additionally, T-DNA tagging results in the generation of many random mutants, and the natural product production levels of these mutants must be assessed by selection or a high throughput screen. Selection for production of natural products – which are not essential for the viability of the producing

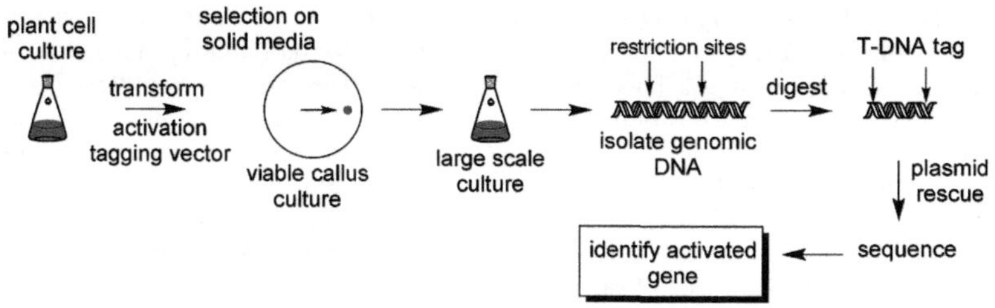

Fig. 9 T-DNA activation tagging. A T-DNA tag is randomly incorporated throughout the plant genome, where any gene adjacent to the tag is upregulated. If a biosynthetic gene is upregulated, then the natural product profile of the plant will be altered. The T-DNA tag can be recovered and the identity of the upregulated gene identified by DNA sequencing

organism – is not straightforward, and clever selection strategies such as the one described for tryptophan decarboxylase must be individually designed for each biosynthetic pathway.

Future Directions

Elucidation of metabolic pathways in plants pose significant challenges. Nevertheless, extraordinary advances have been made in the genetic elucidation of these complex metabolic pathways [6, 43]. As genomic technologies have improved, the genetic basis of the natural products chemistry catalyzed within medicinal plants has slowly – but surely – begun to emerge [44]. This genetic information allows detailed mechanistic explorations of plant pathways, and also enables metabolic engineering efforts that could improve production of medically important plant natural products [45]. This short review highlights a number of methods that have been used to elucidate the genes involved in plant biosynthetic pathways. Selected examples illustrate the successes and limitations encountered with each method. The advent of new technologies in proteomics, chemical biology and other disciplines ensure that novel strategies for elucidation of plant metabolism will continue to emerge. The increasing speed at which plant pathways are being elucidated bodes well for the future of plant derived natural products [46]. By using a variety of approaches in tandem to deconvolute plant metabolism, successes in plant natural products biosynthesis will become increasingly widespread [47].

Acknowledgments I gratefully acknowledge Dr. Xudong Qu for his helpful comments.

References

1. McCurdy, C. R. and Scully, S. S. (2005) Analgesic substances derived from natural products (natureceuticals). *Life Sci.* 78, 476–484.
2. Rainsford, K. D. History and development of the salicylates. (2004) In Aspirin and Related Drugs (K.D. Rainsford, ed.), Taylor & Francis, London/New York, pp. 1–23.
3. Altmann, K.-H. and Gertsch, J. (2007) Anticancer drugs from nature-natural products as a unique source of new microtubule-stabilizing agents. *Nat. Prod. Rep.* 24(2), 327–357.
4. Bode, H. B. and Mueller, R. (2005) The impact of bacterial genomics on natural product research. *Angew. Chem. Intl. Ed.* 44, 6828–6846.
5. Menzella, H. G. and Reeves, C. D. (2007) Combinatorial biosynthesis for drug development. *Curr. Opin. Microbiol.* 10, 238–245.
6. Dixon, R. A. (2005) Engineering of plant natural product pathways. *Curr. Opin. Plant Biol.* 8, 329–336.
7. Katsuyama, Y., Funa, N., Miyahisa, I., and Horinouchi, S. (2007) Synthesis of unnatural flavonoids and stilbenes by exploiting the plant

biosynthetic pathway in *Escherichia coli. Chem. Biol.* 14, 613–621.
8. Chang, M. C. Y., Eachus, R. A., Trieu, W., Ro, D. K., and Keasling, J. D. (2007) Engineering *Escherichia coli* for production of functionalized terpenoids using plant P450s. *Nat. Chem. Biol.* 3, 274–277.
9. Allen, R. S., Millgate, A. G., Chitty, J. A., Thisleton, J., Miller, J. A. C., Fist, A. J., Gerlach, W. L., and Larkin, P. J. (2004) RNAi-mediated replacement of morphine with the nonnarcotic alkaloid reticuline in opium poppy. *Nat. Biotechnol.* 22, 1559–1566.
10. Frick, S., Kramell, R., and Kutchan, T. M. (2007) Metabolic engineering with a morphine biosynthetic P450 in opium poppy surpasses breeding. *Met. Eng.* 9, 169–176.
11. Julsing, M. K., Koulman, A., Woerdenbag, H. J., Quax, W. J., and Kayser, O. (2006) Combinatorial biosynthesis of medicinal plant secondary metabolites. *Biomol. Eng.* 23, 265–279.
12. Gantet, P. and Memelink, J. (2002) Transcription factors: tools to engineer the production of pharmacologically active plant metabolites. *Trends Pharm. Sci.* 23, 563–569.
13. Hashimoto, T. and Yamada, Y. (2003) New genes in alkaloid metabolism and transport. *Curr. Opin. Biotech.* 14, 163–168.
14. Qi, X., Bakht, S., Leggett, M., Maxwell, C., Melton, R., and Osbourn, A. (2004) A gene cluster for secondary metabolism in oat: Implications for the evolution of metabolic diversity in plants. *Proc. Natl. Acad. Sci. USA.* 101, 8233–8238.
15. http://data.kew.org/cvalues/
16. Durst, F. and O'Keefe, D. P. (1995) Plant cytochromes P450: an overview. *Drug Metabol Drug Interact.* 12, 171–187.
17. Tholl, D. (2006) Terpene synthases and the regulation, diversity and biological roles of terpene metabolism. *Curr. Opin. Plant Biol.* 9, 297–304.
18. Gachon, C. M., Langlois-Meurinne, M., and Saindrenan, P. (2005) Plant secondary metabolism glycosyltransferases: the emerging functional analysis. *Trends Plant Sci.* 10, 542–549.
19. Dittrich, H. and Kutchan, T. M. (1991) Molecular cloning, expression, and induction of berberine bridge enzyme, an enzyme essential to the formation of benzophenanthridine alkaloids in the response of plants to pathogenic attack. *Proc. Natl. Acad. Sci. USA.* 88, 9969–9973.
20. Murata, J. and de Luca, V. (2005) Localization of tabersonine 16-hydroxylase and 16-OH tabersonine 16-O-methyl transferase to leaf epidermal cells defines them as a major site of precursor biosynthesis in the vindoline pathway in *Catharanthus roseus. Plant J.* 44, 581–594.
21. Levac, D., Murata, J., Kim, W. S., and De Luca, V. (2008) Application of carborundum abrasion for investigating the leaf epidermis: molecular cloning of *Catharanthus roseus* 16-hydroxytabersonine-16-O-methyltransferase. *Plant J.* 53, 225–236.
22. Achnine, L., Huhman, D. V., Farag, M. A., Sumner, L. W., Blount, J. W., and Dixon, R. A. (2005) Genomics-based selection and functional characterization of triterpene glycosyltransferases from the model legume *Medicago truncatula. Plant J.* 41, 875–887.
23. Walker, K. and Croteau, R. (2000) Taxol biosynthesis: Molecular cloning of a benzoyl-CoA:taxane 2a-O-benzoyltransferase cDNA from *Taxus* and functional expression in *Escherichia coli. Proc. Natl. Acad. Sci. USA.* 97, 13591–13596.
24. Kutchan, T. M. and Schroder, J. (2002) Selected cell cultures and induction methods for cloning and assaying cytochromes P450 in alkaloid pathways. *Meth. Enzymol.* 357, 370–381.
25. Pauli, H. H. and Kutchan, T. M. (1998) Molecular cloning and functional heterologous expression of two alleles encoding (S)-N- methylcoclaurine 3′-hydroxylase (CYP80B1), a new methyl jasmonate-inducible cytochrome P-450-dependent mono-oxygenase of benzylisoquinoline alkaloid biosynthesis. *Plant J.* 13, 793–801.
26. Chou, W. M. and Kutchan, T. M. (1998) Enzymatic oxidations of alkaloids. *Plant J.* 15, 289–300.
27. Jennewein, S., Wildung, M. R., Chau, M., Walker, K., and Croteau, R. (2004) Random sequencing of an induced *Taxus* cell cDNA library for identification of clones involved in Taxol biosynthesis. *Proc. Natl. Acad. Sci. USA.* 101, 9149–9154.
28. Lange, B. M., Wildung, M. R., Stauber, E. J., Sanchez, C., Pouchnik, D., and Croteau, R. (2000) Probing essential oil biosynthesis and secretion by functional evaluation of expressed sequence tags from mint glandular trichomes. *Proc. Natl. Acad. Sci. USA.* 97, 2934–2939.
29. Hirai, M. Y., Tohge, T., and Saito, K. (2006) Systems based analysis of plant metabolism by integration of metabolomics with transcriptomics. *Biotech. Agric. Forestry.* 57, 199–209.
30. Rischer, H., Oresic, M., Seppanen-Laakso, T., Katajamaa, M., Lammertyn, F., Ardiles-Diaz, W.,

Van Montagu, M. C. E., Inze, D., Oksman-Caldentey, K. M., and Goossens, A. (2006) Gene-to-metabolite networks for terpenoid indole alkaloid biosynthesis in *Catharanthus roseus* cells. *Proc. Natl. Acad. Sci. USA*. 103, 5614–5619.
31. Martin, K. J. and Pardee, A. B. (2000) Identifying expressed genes. *Proc. Natl. Acad. Sci. USA*. 97, 3789–3791.
32. Wang, J., Sheehan, M., Brookman, H., and Timko, M. P. (2000) Characterization of cDNAs differentially expressed in roots of tobacco (*Nicotiana tabacum* cv Burley 21) during the early stages of alkaloid biosynthesis. *Plant Sci.* 158, 19–32.
33. Shukla, A. K., Shasany, A. K., Gupta, M. M., and Khanuja, S. P. S. (2006) Transcriptome analysis in *Catharanthus roseus* leaves and roots for comparative terpenoid indole alkaloid profiles. *J. Exp. Bot.* 57, 3921–3932.
34. McGinnis, K., Chandler, V., Cone, K., Kaeppler, H., Kaeppler, S., Kerschen, A., Pikaard, C., Richards, E., Sidorenko, L., Smith, T., Springer, N., and Wulan, T. (2005) Transgene-induced RNA interference as a tool for plant functional genomics. *Meth. Enzymol.* 392, 1–24.
35. Li, R., Reed, D. W., Liu, E., Nowak, J., Pelcher, L. E., Page, J. E., and Covello, P. S. (2006) Functional genomic analysis of alkaloid biosynthesis in *Hyoscyamus niger* reveals a cytochrome P450 involved in littorine rearrangement. *Chem. Biol.* 13, 513–520.
36. Ogita, S., Uefuji, H., Morimoto, M., and Sano, H. (2004) Application of RNAi to confirm theobromine as the major intermediate for caffeine biosynthesis in coffee plants with potential for construction of decaffeinated varieties. *Plant Mol. Biol.* 54, 931–941.
37. Seed, B. (1995) Developments in expression cloning. *Curr. Opin. Biotechnol.* 6, 567–573.
38. Dalboge, H. and Lange, L. (1998) Using molecular techniques to identify new microbial biocatalysts. *Trends Biotechnol.* 16, 265–272.
39. Lange, T., Hedden, P., and Graebe, J. E. (1994) Expression cloning of a gibberellin 20-oxidase, a multifunctional enzyme involved in gibberellin biosynthesis. *Proc. Natl. Acad. Sci. USA*. 91, 8552–8556.
40. Geerlings, A., Redondo, F. J., Memelink, J., Contin, A., van der Heijden, R., and Verpoorte, R. (1999) Screening method for cDNAs encoding putative enzymes converting loganin into secologanin by a transgenic yeast culture. *Biotechnol. Tech.* 13, 605–608.
41. van der Fits, L. and Memelink, J. (2000) ORCA3, a Jasmonate-responsive transcriptional regulator of plant primary and secondary metabolism. *Science*. 289, 295–297.
42. van der Fits, L., Hilliou, F., and Memelink, J. (2001) T-DNA activation tagging as a tool to isolate regulators of a metabolic pathway from a genetically non-tractable plant species. *Transgenic Res.* 10, 513–521.
43. Dixon, R. A. (1999) Plant natural products: the molecular genetic basis of biosynthetic diversity. *Curr. Opin. Biotechnol.* 10, 192–197.
44. Memelink, J. (2005) The use of genetics to dissect plant secondary pathways. *Curr. Opin. Plant Biol.* 8, 230–235.
45. Gomez-Galera, S., Pelacho, A. M., Gene, A., Capell, T., and Christou, P. (2007) The genetic manipulation of medicinal and aromatic plants. *Plant Cell Rep.* 26, 1689–1715.
46. Littleton, J. (2007) The future of plant drug discovery. *Expert Opin. Drug Discov.* 2, 673–683.
47. McChesney, J. D., Venkataraman, S. K., and Henri, J. T. (2007) Plant natural products: Back to the future or into extinction. *Phytochemistry*. 68, 2015–2022.

8 Regulation of Secondary Metabolism by Jasmonate Hormones

Hongtao Zhang and Johan Memelink

Abstract The biosynthesis of many different types of secondary metabolites that serve defensive functions in different plant species is regulated by hormones belonging to the group of jasmonate compounds. Regulation acts at the level of transcription of structural genes encoding biosynthetic enzymes. Here we review recent insights into the mechanisms of signal transduction initiated by jasmonates leading to the activation of transcription factors. We present models for jasmonate signal transduction regulating tobacco alkaloid biosynthesis and terpenoid indole alkaloid biosynthesis in *Catharanthus roseus*. The models propose that perception of a bioactive jasmonate derivative by the receptor CORONATINE INSENSITIVE1 (COI1) results in the degradation of Jasmonate ZIM-domain (JAZ) proteins. Since these JAZ proteins repress the activity of the basic-Helix-Loop-Helix transcription factor MYC2, MYC2 then activates the expression of genes encoding certain members of the APETALA2/Ethylene Response Factor (AP2/ERF) family of transcription factors, which in turn activate the expression of alkaloid biosynthesis genes.

H. Zhang and J. Memelink (✉)
Institute of Biology, Leiden University,
P.O. Box 9505, 2300 RA Leiden, The Netherlands
e-mail: j.memelink@biology.leidenuniv.nl

Introduction

Secondary metabolites play diverse roles in plants. Flowers for example synthesize and accumulate anthocyanin pigments to attract pollinating insects. The biosynthesis of anthocyanins is under tight developmental control (Grotewold, 2006). Other compounds, including alkaloids in many plant species (Facchini, 2001) and glucosinolates in the Brassicaceae (Halkier and Gershenzon, 2006), function in protection against microorganisms or herbivores.

Biosynthesis of defensive secondary metabolites is often induced in plants following attack by microorganisms and/or herbivores. Primary signals specifying attack by fungal or bacterial microorganisms which are recognized by the plant are called elicitors, pathogen-associated molecular patterns (PAMPs) or microbe-associated molecular patterns (MAMPs) (Jones and Dangl, 2006). MAMPS that elicit a defence response include certain molecules derived from the microbial cell wall or the bacterial flagella. Attack by herbivores is recognized among others by certain molecules released from damaged plant cell walls and by certain molecules present in insect oral secretions. Elicitors activate signal transduction pathways that generate secondary signals within plants (Zhao et al., 2005). Three major plant secondary signalling molecules are jasmonates (JAs; Turner et al., 2002; Wasternack,

A.E. Osbourn and V. Lanzotti (eds.), *Plant-derived Natural Products*,
DOI 10.1007/978-0-387-85498-4_8, © Springer Science+Business Media, LLC 2009

2007; Balbi and Devoto, 2008), ethylene (Wang et al., 2002) and salicylic acid (Shah, 2003). Production of these hormones leads to cascades of events responsible for the physiological adaptation of the plant to the external stress. The JAs, ethylene and salicylic acid signal transduction pathways act synergistically or antagonistically in a variety of responses, leading to fine-tuning of the complex defence response (Kunkel and Brooks, 2002).

Among these three key regulatory signals, by far the most important molecules for induction of secondary metabolism are the JAs. JAs have been found to induce the biosynthesis of a variety of secondary metabolites in different plant species, including alkaloids, terpenoids, glucosinolates and phenylpropanoids (Memelink et al., 2001; Zhao et al., 2005). This chapter will review recent advances in our understanding of the mechanism of action of JAs in induction of alkaloid metabolism in tobacco and in *Catharanthus roseus* (Madagascar periwinkle). These advances have been enabled by pioneering research on signal transduction of JAs using the model plant *Arabidopsis thaliana* (Katsir et al., 2008a).

Jasmonates Are Essential for Elicitor Signal Transduction

Addition of elicitors is a common method of enhancing secondary metabolism in plant cell cultures for metabolic, enzymatic or regulatory studies. In its broadest definition, an elicitor is any compound or mixture of compounds that induces a plant defence reaction. Most elicitors used in plant research originate from microorganisms but others are derived from the plant cell wall. In addition, a variety of abiotic elicitors has been used, such as heavy metals.

Elicitors are often applied in the form of crude mixtures, such as a fungal cell wall extract. In a few cases, elicitors have been purified to homogeneity. In alkaloid research, an extract from baker's yeast is commonly used as an elicitor. Yeast extract contains several components that can elicit plant defence responses, including chitin, *N*-acetylglucosamine oligomers, β-glucan, glycopeptides and ergosterol. In addition, a low molecular weight component, which is probably a small peptide, induces the expression of terpenoid indole alkaloid biosynthesis genes in cells of *C. roseus* (Menke et al., 1999b).

Biotic elicitors induce a defence reaction in plant cells because they are recognized as "non-self" in the case of microbial elicitors or as "abnormally modified self" in the case of plant cell wall fragments. Intensive research efforts, including pharmacological studies, have uncovered components of the signal transduction pathway connecting elicitor perception to induction of defence genes (Zhao et al., 2005). In several elicitor responses, including secondary metabolite production (methyl)jasmonic acid and some of its bioactive precursors and derivatives play key roles as intermediate signals. In different plant species, elicitors were shown to induce accumulation of endogenous jasmonic acid, and (Me) jasmonic acid itself increased secondary metabolite production (Menke et al., 1999b; Zhao et al., 2005). In addition, in several species it was shown that blocking jasmonate biosynthesis abolished elicitor-induced metabolite accumulation and the expression of biosynthesis genes (Menke et al., 1999b; Zhao et al., 2005). Elicitors induce many intracellular events, including an increase in cytoplasmic calcium concentration, ion transport, production of reactive oxygen species and protein phosphorylation. How these events are exactly coupled to induced jasmonate biosynthesis remains largely unknown. A detailed review of intracellular events triggered by elicitors and their possible role in signal transduction is presented by Zhao et al. (2005).

The control points that govern the synthesis and accumulation of JAs remain to be identified. Timing and control of jasmonate biosynthesis suggest several ways in which jasmonate signalling might be modulated during stress perception. One level of control in jasmonate biosynthesis and/or signalling might be the sequestration of biosynthetic enzymes and substrates inside the chloroplasts (Stenzel et al., 2003). In this way, jasmonate biosynthesis and signalling would only be activated by the availability of substrate upon cellular decompartmentalization during wounding or pathogen attack. However, wounding induces the expression of several jasmonate biosynthesis genes (Turner et al., 2002), suggesting that, at least partly, the wound-induced production of JAs is a result of the increased transcription of genes encoding the jasmonate biosynthesis pathway enzymes and their subsequent *de novo* protein synthesis. In addition, JAs themselves induce the expression of genes involved in jasmonate biosynthesis (Turner et al., 2002), indicating the existence of a positive feedback regulatory mechanism for jasmonate biosynthesis in which JAs stimulate their own production.

Jasmonate Biosynthesis

JAs, including jasmonic acid (JA) and several of its cyclic precursors and derivatives, constitute a family of bioactive oxylipins that regulate plant responses to environmental and developmental cues (Turner et al., 2002; Wasternack, 2007; Balbi and Devoto, 2008). These signalling molecules affect a variety of developmental processes including fruit ripening, production of viable pollen, root elongation, and tendril coiling. In addition and more importantly for this review, JAs regulate responses to wounding and abiotic stresses, and defence against insects and necrotrophic pathogens.

An important defence response is the induction of secondary metabolite accumulation, which depends on JAs as a regulatory signal. JAs are fatty acid derivatives which are synthesized via the octadecanoid pathway (Fig. 1). Most of the enzymes of this pathway leading to jasmonate biosynthesis have been identified by a combination of biochemical and genetic approaches (Wasternack, 2007). The enzymes leading to JA biosynthesis are located in two different subcellular compartments. The octadecanoid pathway starts in the chloroplasts with phospholipase-mediated release of α-linolenic acid from membrane lipids. The fatty acid α-linolenic acid is then converted to 12-oxo-phytodienoic acid (OPDA) by the sequential action of the plastid enzymes lipoxygenase (LOX), allene oxide synthase (AOS), and allene oxide cyclase (AOC). The second part of the pathway takes place in the peroxisomes. OPDA is transported from the chloroplasts to the peroxisomes where it is reduced by OPDA reductase (OPR3) to give 3-oxo-2(2'[Z]-pentenyl)-cyclopentane-1-octanoic acid (OPC:8), followed by three rounds of beta-oxidation involving three enzymes to yield (+)-7-*iso*-jasmonic acid which equilibrates to the more stable (-)-JA.

Subsequently, JA can be metabolized in the cytoplasm by at least seven different reactions. Well-characterized reactions include methylation to methyl-jasmonate (MeJA) by *S*-adenosyl-L-methionine:jasmonic acid carboxyl methyl transferase (JMT), conjugation to amino acids by JA amino acid synthase (JAR1) or hydroxylation to 12-hydroxyjasmonic acid (12-OH-JA). OPDA, JA, MeJA and JA-Ile are active signalling molecules, whereas 12-OH-JA is thought to be a biologically inactive derivative (Wasternack, 2007).

Jasmonate Perception and Signalling in Arabidopsis and Tomato

How JAs induce gene expression has been mainly unravelled in studies using Arabidopsis and tomato (*Solanum lycopersicum*) (Katsir et al., 2008a). To identify molecular components of jasmonate signal transduction, screenings for Arabidopsis mutants that are insensitive to (Me) JA or to coronatine (a bacterial toxin which is a

8

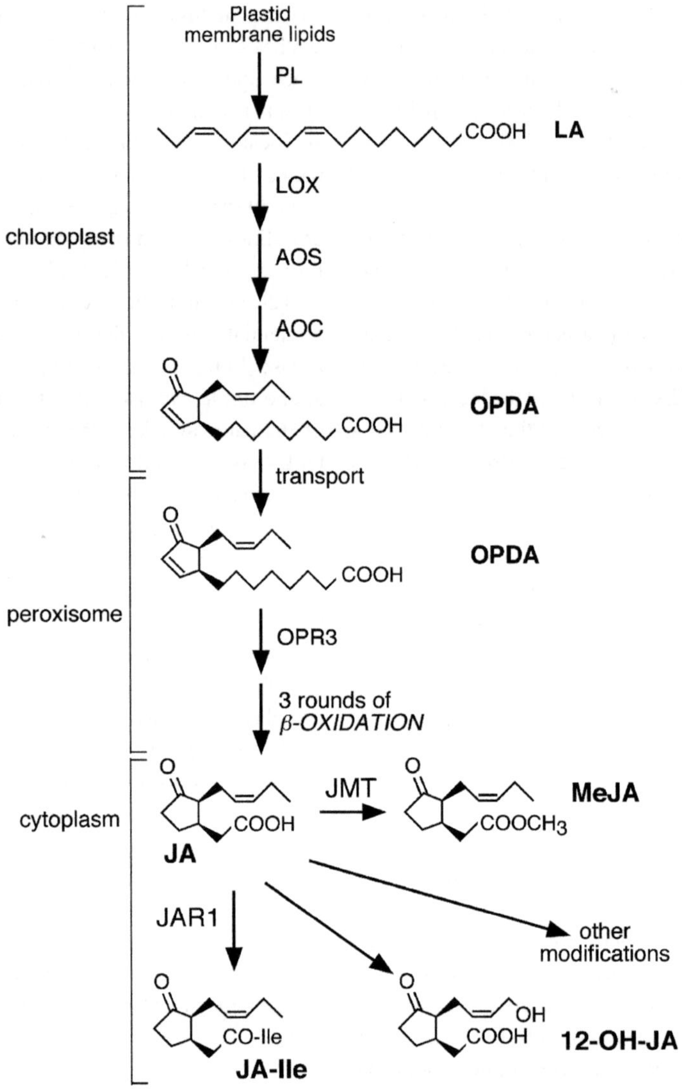

Fig. 1 Schematic representation of the octadecanoid pathway leading to jasmonic acid biosynthesis. 12-OH-JA, 12-hydroxy-jasmonic acid; AOC, allene oxide cyclase; AOS, allene oxide synthase; JA, jasmonic acid; JAR1, enzyme responsible for the conjugation of JA with isoleucine (JA-Ile); JMT, *S*-adenosyl-L-methionine:jasmonic acid carboxyl methyl transferase; LA, α-linolenic acid; LOX, lipoxygenase; MeJA, methyl jasmonate; OPDA, 12-oxo-phytodienoic acid; OPR3, OPDA reductase3; PL, phospholipase

structural and functional analogue of JA-Ile) or that show constitutive jasmonate responses have been performed (Lorenzo and Solano, 2005). A number of mutants have been characterized, and will be discussed below.

The *coronatine insensitive1* (*coi1*) mutant was isolated in a screen for Arabidopsis mutants insensitive to root growth inhibition by coronatine (Feys et al., 1994). The *coi1* mutant is also insensitive to JAs (Feys et al., 1994), is defective in resistance to certain insects and pathogens and fails to express jasmonate-regulated genes (Turner et al., 2002). The *COI1* gene encodes an F-box protein (Xie et al., 1998). F-box proteins associate with cullin, Skp1 and Rbx1 proteins to form an E3 ubiquitin ligase known as the SCF complex, where the F-box subunit functions as the specificity determinant targeting proteins for ubiquitin-mediated proteolysis by the 26S proteasome (del Pozo and Estelle, 2000). Co-immunoprecipitation experiments showed that COI1 associates *in vivo* with Skp1, cullin and Rbx1 proteins to form the SCF^{COI1} complex (Devoto et al., 2002; Xu et al., 2002). Therefore, the requirement for COI1 in jasmonate-dependent responses indicates that ubiquitin-mediated protein degradation is a crucial event in jasmonate signalling. Plants that are deficient in other components or regulators of SCF complexes, including AXR1, COP9 and SGT1b, also show impaired jasmonate responses (Lorenzo and Solano, 2005). The existence of a *COI1* function that is conserved in species other than Arabidopsis was demonstrated by the identification of *COI1* homologues in tomato (Li et al., 2004), tobacco (Shoji et al., 2008) and *Nicotiana attenuata* (Paschold et al., 2007). COI1 is a component that is specific to the JA pathway, whereas SGT1b and AXR1 are shared by other signalling pathways. Mutations in *AXR1* or *SGT1b* have pleiotropic effects that impair plant responses not only to JA but also to auxin and pathogens, suggesting that both SGT1b and AXR1 are regulators of SCF complexes and are involved in several different signalling pathways (Austin et al., 2002; Azevedo et al., 2002; Gray et al., 2003).

A particularly effective screen for jasmonate signalling mutants has been described by Lorenzo et al. (2004). Screening for mutants affected in JA-induced root growth inhibition in an *ethylene-insensitive*3 (*ein3*) background resulted in the identification of five loci called *JA-insensitive* (*JAI*) 1–5. The *JAI1* locus corresponds to the *AtMYC2* gene (Lorenzo et al., 2004), encoding a basic-Helix-Loop-Helix (bHLH) transcription factor which regulates a subset of jasmonate-responsive genes involved in wounding responses. The *JAI2* locus corresponds to the previously characterized *JAR1* gene (Staswick et al., 1992), encoding an enzyme that couples JA to amino acids with a preference for isoleucine (Staswick and Tiryaki, 2004). The *JAI4* locus corresponds to the *SGT1b* gene (Lorenzo and Solano, 2005). The *JAI5* locus corresponds to the *COI1* gene (Lorenzo et al., 2004).

Recently, the gene affected in the *jai3* mutant was identified. It encodes a protein with a zinc finger-like ZIM motif (Chini et al., 2007). There are several related genes in Arabidopsis forming a gene family called ZIM or TIFY (Vanholme et al., 2007). The members that are induced at the gene expression level by JAs are called Jasmonate ZIM domain (JAZ) proteins (Chini et al., 2007; Thines et al., 2007). They contain in addition to the highly conserved central ZIM motif a highly conserved C-terminal Jas motif and a less conserved N-terminal region. In the *jai3* mutant an aberrant protein is expressed with a deletion of the C-terminal domain including the Jas motif. The wild-type JAI3 (or JAZ3) protein is rapidly degraded in response to JA in a COI1-dependent manner, whereas the jai3 mutant protein is stable. The JAI3 protein was shown to interact in vitro and in yeast with AtMYC2. Based on these findings it was postulated that JAI3 is a repressor of AtMYC2 which is rapidly degraded in response to JA thereby activating AtMYC2 (Fig. 2; Chini et al., 2007).

In independent studies, members of the *JAZ* gene family in Arabidopsis were characterized as being predominant among genes induced in anthers after 30 min of JA treatment (Mandaokar et al., 2006). Subsequent study of the family member JAZ1 demonstrated that it is rapidly degraded in response to JA in a COI1-dependent manner (Thines et al., 2007). On the other hand a deletion derivative of JAZ1 lacking the C-terminal domain is stable.

Interestingly, these authors were able to detect interaction between JAZ1 and COI1 in

8

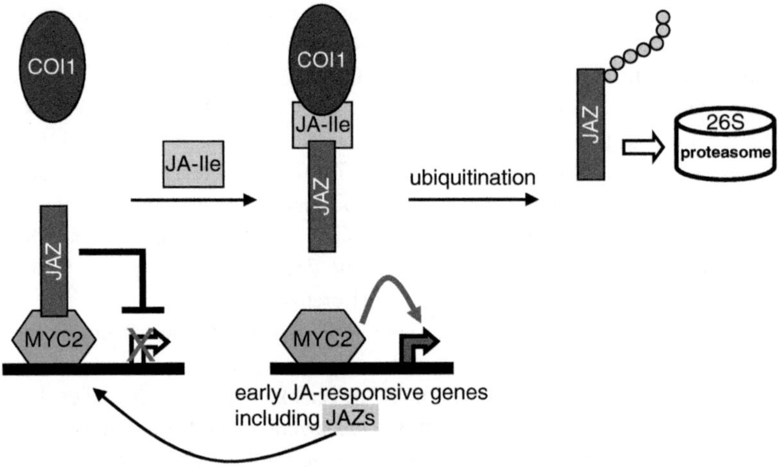

Fig. 2 Model for jasmonate signal transduction leading to expression of AtMYC2-regulated genes. Although depicted as a single protein, COI1 forms part of the E3 ubiquitin ligase SCF^{COI1}. In the absence of JA-Ile, JAZ repressors interact with AtMYC2 maintaining this transcription factor inactive. In the presence of JA-Ile, the F-box protein COI1 binds to JAZ proteins, which results in their ubiquitination by the SCF^{COI1} complex and their degradation by the 26S proteasome. AtMYC2 is liberated and activates transcription of target genes, including genes encoding JAZ proteins, resulting in a negative feedback loop

a yeast two-hybrid assay in the presence of JA conjugated to Ile (JA-Ile) in the yeast growth medium or in an in vitro pull-down assay in the presence of JA-Ile. No interaction was detected in the presence of OPDA, JA, MeJA or JA conjugated to Trp or Phe, whereas JA-Leu was about 50-fold less effective in promoting interaction between COI1 and JAZ1 than JA-Ile. JA-Ile and JA-Leu are products of the JAR1-mediated conjugation reaction (Staswick and Tiryaki, 2004). JA-Ile and coronatine also promote the interaction between JAZ3 and JAZ9 in a yeast two-hybrid assay, whereas JA or MeJA are ineffective (Melotto et al., 2008).

Using tomato SlCOI1 and SlJAZ1, it was shown that the complex binds radiolabeled coronatine (Katsir et al., 2008b). Binding can be displaced with unlabeled coronatine or JA-Ile. These experiments show that COI1 is the receptor for at least certain JAs including JA-Ile, as well as for the microbial JA-Ile mimic coronatine. The C-terminal domains containing the conserved Jas motif of tomato JAZ1 (Katsir et al., 2008b) and Arabidopsis JAZ1, JAZ3 and JAZ9 (Melotto et al, 2008) were shown to be necessary and sufficient for binding to COI1 in a JA-Ile or coronatine-dependent manner.

The expression of the JAZ genes in Arabidopsis is induced by JA (Mandaokar et al., 2006; Chini et al., 2007; Thines et al., 2007) and is controlled by AtMYC2 (Chini et al., 2007). AtMYC2 and JAZ proteins therefore form a jasmonate-responsive oscillator, where JAZ proteins negatively regulate AtMYC2 activity at the protein level, JAs cause JAZ degradation and AtMYC2 activation, and AtMYC2 switches on the expression of JAZ repressors at the gene level (Fig. 2).

The picture that emerges for jasmonate signal transduction is highly reminiscent of auxin signal transduction, which involves auxin-responsive degradation of AUX/IAA repressor proteins via the F-box protein TIR1 (Guilfoyle, 2007). TIR1 is the auxin receptor (Kepinski and Leyser, 2005; Dharmasiri et al., 2005) with auxin acting as the molecular glue between TIR1 and AUX/IAA proteins (Tan et al., 2007). COI1 is the closest relative to TIR1 that is not related to auxin perception among the about 700 members of the Arabidopsis F-box protein family (Gagne et al., 2002). JA-Ile enhances the interaction between COI1 and JAZ1, JAZ3, JAZ9 and possibly other JAZ family members (Fig. 2). It has been proposed that different biologically active JAs could promote the binding between COI1 and specific JAZ family members, and that these family members could act as repressors of specific downstream targets, presumably other transcription factors (Thines et al., 2007).

Challenges are to determine whether different JAs can indeed promote interaction of COI1 with specific JAZ family members, and to find out what are the specific targets of each member of the JAZ family of repressors. It is also conceivable that JA-Ile or other biologically active JAs enhance binding between COI1 and hitherto unidentified repressors distinct from the JAZ proteins.

Jasmonate Signalling in Terpenoid Indole Alkaloid Biosynthesis in *Catharanthus roseus*

Catharanthus roseus has the genetic potential to synthesize over a hundred terpenoid indole alkaloids (TIAs; van der Heijden et al., 2004). Several of the terpenoids that are known to be produced by *C. roseus* have pharmaceutical applications, including the monomeric alkaloids serpentine and ajmalicine, which are used as a tranquilizer and to reduce hypertension, respectively, and the dimeric alkaloids vincristine and vinblastine, which are potent antitumour drugs. TIAs consist of an indole moiety derived from tryptophan, and a terpenoid group derived from geraniol (Fig. 3). The condensation of the tryptophan derivative tryptamine with the terpenoid derivative secologanin is performed by the enzyme strictosidine synthase (STR) and results in the synthesis of 3α(*S*)-strictosidine. TIA biosynthesis has been shown to be induced by MeJA in developing seedlings and in cell cultures (Memelink and Gantet, 2007).

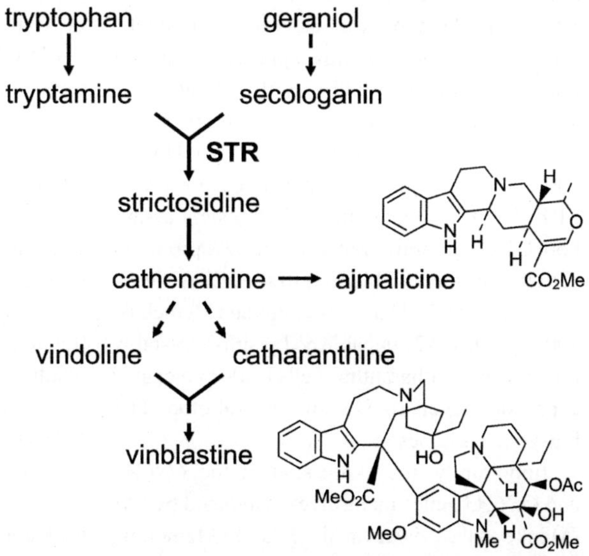

Fig. 3 Biosynthesis pathway for terpenoid indole alkaloids in *C. roseus*. Unbroken arrows indicate single enzymatic conversions and broken arrows indicate multiple enzymatic conversions. The structures of ajmalicine and vinblastine are shown. STR, strictosidine synthase

Over ten genes have now been cloned from the estimated total number of around 25 genes encoding enzymes involved in TIA biosynthesis (Memelink and Gantet, 2007). In addition, genes acting in primary precursor pathways leading to the formation of tryptophan (van der Fits and Memelink, 2000) and geraniol (Chahed et al., 2000; Veau et al., 2000) have been cloned. All TIA biosynthesis genes tested are induced by MeJA in Catharanthus suspension-cultured cells (van der Fits and Memelink, 2000). In addition, MeJA induces genes in primary metabolism leading to the formation of TIA precursors. This presents a good example of the profound effect of JAs on plant metabolism at the level of gene expression.

The promoter of the *STR* gene has been studied in detail to identify elicitor- and jasmonate-responsive sequences. A region close to the TATA box called jasmonate- and elicitor-responsive element (JERE) was found to dictate elicitor- and jasmonate-responsive reporter gene activation (Menke et al., 1999a). The JERE interacts with two transcription factors called Octadecanoid derivative-Responsive *Catharanthus* AP2-domain proteins (ORCAs). ORCA2 was isolated by yeast one-hybrid screening of a Catharanthus cDNA library with the JERE as bait (Menke et al., 1999a) and ORCA3 was isolated by a genetic T-DNA activation tagging approach (van der Fits and Memelink, 2000). Both belong to the AP2/ERF family of transcription factors, which are unique to plants and are characterized by the APETALA2/Ethylene Response Factor (AP2/ERF) DNA-binding domain. In transient assays both ORCA proteins transactivate *STR* promoter activity via specific binding to the JERE. Overexpression of ORCA3 (van der Fits and Memelink, 2000) or ORCA2 (unpublished results) in stably transformed Catharanthus cells leads to elevated expression levels of *STR* and several other TIA biosynthesis genes.

Importantly, the expression of the *ORCA2* and *ORCA3* genes themselves is induced by JAs (Menke et al., 1999a; van der Fits and Memelink, 2000, 2001). This suggests that JAs induce alkaloid metabolism by increasing the amount of the ORCA regulatory proteins. To study how the *ORCA3* gene is regulated, its promoter was used in loss- and gain-of- function experiments to identify a 74 bp D region containing a jasmonate-responsive element (JRE; Vom Endt et al., 2007). The JRE is composed of two important sequences, a quantitative sequence responsible for a high level of expression, and a qualitative sequence that acts as an on/off switch in response to MeJA. Using the JRE in yeast one-hybrid screening of Catharanthus cDNA libraries, several proteins belonging to the AT-hook class of DNA-binding proteins were isolated, which were found to interact specifically with the quantitative sequence within the JRE (Vom Endt et al., 2007).

The qualitative element consists of a G-box-like sequence (AACGTG). In a yeast one-hybrid screen of a Catharanthus cDNA library using a G-box as bait several genes encoding transcription factors were isolated, including five members of the bHLH family (Pré et al., 2000). One of these proteins called CrMYC2 is a close homologue of AtMYC2, and the corresponding gene was the only one among those five genes that was rapidly and strongly induced by MeJA in Catharanthus cells (unpublished results). CrMYC2 bound specifically to the qualitative sequence in the JRE of the *ORCA3* promoter in vitro, and it transactivated reporter gene expression specifically via interaction with the qualitative sequence in transient assays in Catharanthus cells (unpublished results). Silencing the expression of *CrMYC2* via RNAi strongly reduced the MeJA-responsive expression of the *ORCA2* and *ORCA3* genes in stably transformed Catharanthus cells (unpublished results). This demonstrates that MeJA-responsive *ORCA* expression is controlled by CrMYC2 (Fig. 4).

Transcript profiling of Catharanthus cells treated with MeJA identified two members of the Catharanthus JAZ family (Rischer et al., 2006).

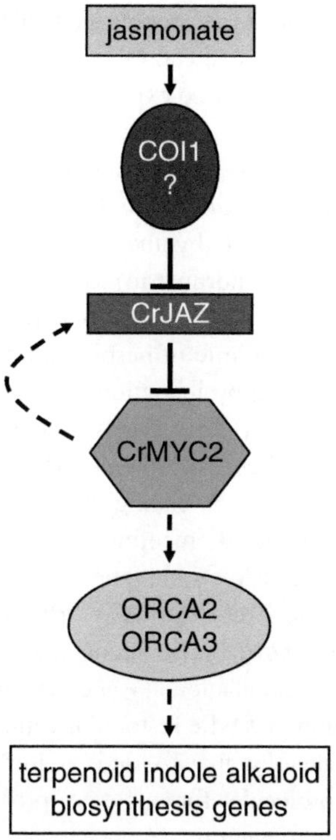

Fig. 4 Model for jasmonate signal transduction leading to the expression of terpenoid indole alkaloid biosynthesis genes in *Catharanthus roseus*. As depicted in Fig. 1, a bioactive jasmonate species enhances the interaction between CrCOI1 and CrJAZ, leading to degradation of the latter proteins. CrMYC2 then activates transcription of the genes encoding the AP2/ERF-domain transcription factors ORCA2 and ORCA3, which in turn activate the expression of terpenoid indole alkaloid biosynthesis genes. CrMYC2 also activates transcription of *CrJAZ* genes as part of a negative feedback loop. The position of CrCOI1 in this signal transduction pathway is hypothetical as indicated by the question mark. Solid lines indicate interactions between proteins and broken lines indicate interactions between proteins and genes

The corresponding genes were rapidly and strongly induced by MeJA in wildtype Catharanthus cells (Rischer et al., 2006; unpublished results). This response was strongly reduced in the transgenic cell line with silenced CrMYC2 expression, showing that MeJA-responsive *CrJAZ* gene expression is regulated by CrMYC2 (unpublished results). A CrJAZ1-GFP fusion protein was shown to be rapidly degraded in Catharanthus cells treated with MeJA, whereas a deletion derivative lacking the C-terminal Jas domain was much more stable. In transient reporter gene assays, the CrJAZ proteins abolished the activating activity of CrMYC2. This shows that the full-length CrJAZ proteins are indeed repressors that negatively regulate CrMYC2 activity. Yeast two-hybrid assays showed that both CrJAZ proteins were able to interact with CrMYC2 (unpublished results).

Our results prompted us to propose the model shown in Fig. 4 for signalling by JAs in *C. roseus* leading to alkaloid biosynthesis. Perception of a bioactive jasmonate derivative by CrCOI1 results in the degradation of CrJAZ proteins. CrMYC2 then activates the expression of the *ORCA* genes, which in turn activate the expression of a subset of TIA biosynthesis genes. Simultaneous activation of *JAZ* genes by CrMYC2 restores the uninduced situation by inhibition of CrMYC2 activity. The identity of the active jasmonate signalling molecule(s) and the involvement of CrCOI1 in this sequence of events still need to be experimentally confirmed.

Jasmonate Signalling in Tobacco Alkaloid Biosynthesis

The main alkaloid found in tobacco plants, nicotine, is composed of a pyrrolidine ring and a pyridine ring (Fig. 5). The pyrrolidine moiety is derived from *N*-methylputrescine, which is formed from putrescine by putrescine *N*-methyltransferase (PMT) (Katoh et al., 2005). The pyridine moiety of nicotine is derived from nicotinic acid. Nicotine is exclusively synthesized in the roots and is translocated to the leaves via the xylem. Multiple structural genes

8

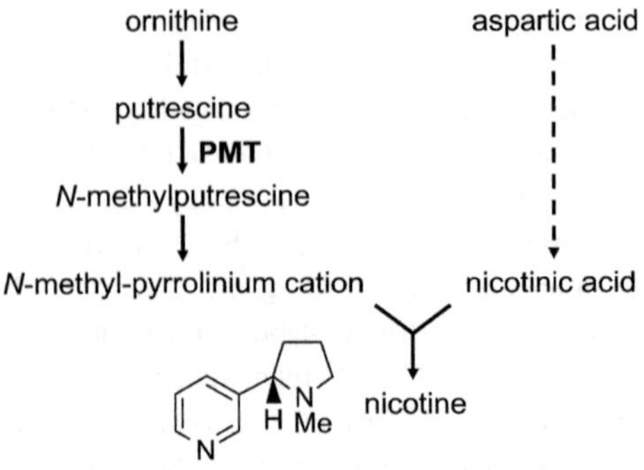

Fig. 5 Biosynthetic pathway of nicotine. Unbroken arrows indicate single enzymatic conversions and broken arrows indicate multiple enzymatic conversions. It is not known whether the *N*-methyl-pyrrolinium cation is coupled to nicotinic acid or a derivative of the latter. PMT; putrescine *N*-methyltransferase

for nicotine biosynthesis enzymes, including *PMT*, are transcriptionally activated by exogenous application of jasmonates in tobacco roots and in cultured tobacco cells (Goossens et al, 2003; Katoh et al., 2005). Promoter regions of ~250 base pairs from three *PMT* genes from *Nicotiana sylvestris* could confer jasmonate-responsive expression on a *GUS* reporter gene in transgenic hairy roots, showing that the jasmonate signal converges on relatively small promoter regions to confer transcriptional responses (Shoji et al., 2000). Two tobacco members of the AP2/ERF-domain transcription factor family called NtORC1 and NtJAP1 were shown to upregulate the activity of the tobacco *PMT* promoter in transient assays in tobacco protoplasts (De Sutter et al., 2005). Together the transcription factors caused a strong synergistic activation of the *PMT* promoter. NtORC1 is a close homologue of the Catharanthus AP2/ERF-domain transcription factor ORCA3. Both *NtORC1* and *NtJAP1* gene expression is induced by MeJA (Goossens et al., 2003).

Genetic studies using low-nicotine tobacco varieties demonstrated that the low-nicotine phenotype is caused by synergistic effects of two non-linked loci, called *nic1* and *nic2* (Katoh et al., 2005). The *nic1nic2* double mutant has highly reduced nicotine contents (about 5% of wild type) and strongly decreased expression levels of nicotine biosynthesis genes. The genes corresponding to the *nic* loci have not been cloned yet.

In *Nicotiana attenuata* JA accumulates in response to attack by the herbivore *Manduca sexta* (tobacco hornworm) or in response to wounding and application of caterpillar oral secretions (a mimic of herbivore attack). This is likely caused by induction of jasmonate biosynthesis genes, since the *NaLOX3* gene was induced by these treatments (Halitschke and Baldwin, 2003). Silencing the expression of *NaLOX3* reduced nicotine accumulation in response to JA. Simultaneous silencing of *NaJAR4* and *NaJAR6* (the *N. attenuata* orthologues of *AtJAR1*) also reduced JA-responsive nicotine accumulation (Wang et al., 2008). Application of JA-Ile restored nicotine accumulation, indicating that JA-Ile is an important signalling molecule for nicotine production in *Nicotiana attenuata*.

Silencing of the *COI1* gene in tobacco plants abolished the MeJA-responsive expression of nicotine biosynthesis genes including *PMT*, as well as MeJA-responsive nicotine accumulation (Shoji et al., 2008). The same report describes the isolation of three members of the tobacco *JAZ* gene family and their involvement in nicotine biosynthesis. The *NtJAZ*1–3 genes were induced by MeJA. For NtJAZ1 it was shown that the protein was rapidly degraded via the 26S proteasome in response to MeJA, whereas a derivative lacking the C-terminal Jas domain was stable. Overexpression of C-terminal deletion derivatives of NtJAZ1 or NtJAZ3 abolished MeJA-responsive *PMT* gene expression as well as nicotine accumulation.

These observations show that MeJA-responsive nicotine biosynthesis is controlled by the jasmonate receptor COI1 and depends on degradation of members of the JAZ repressor family (Fig. 6). There are no published data yet about the nature of the transcription factor(s) repressed

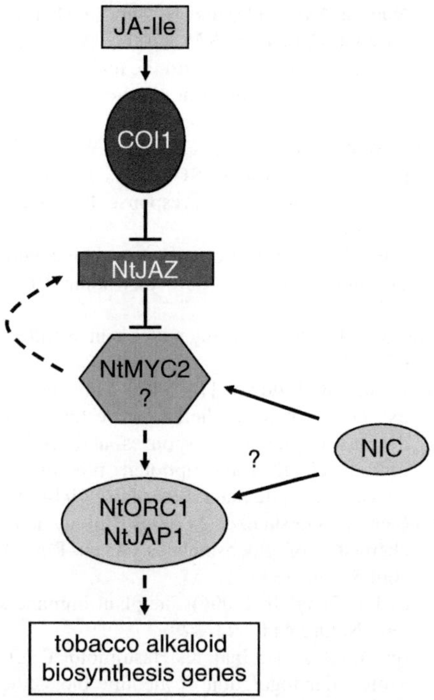

Fig. 6 Model for jasmonate signal transduction leading to the expression of tobacco alkaloid biosynthesis genes. As depicted in Fig. 1, JA-Ile enhances the interaction between NtCOI1 and NtJAZ, leading to degradation of the latter proteins. NtMYC2 then activates transcription of the genes encoding the AP2/ERF-domain transcription factors NtORC1 and NtJAP1, which in turn activate the expression of tobacco alkaloid biosynthesis genes. NtMYC2 also activates transcription of *NtJAZ* genes as part of a negative feedback loop. The elusive *NIC1* and *NIC2* genes may encode NtMYC2 and/or NtORC1 and NtJAP1 as indicated by the question mark. The position of NtMYC2 in this signal transduction pathway is hypothetical as indicated by the question mark. Solid lines indicate interactions between proteins and broken lines indicate interactions between proteins and genes

by the JAZ proteins in tobacco, but in Fig. 6 we speculate that it is the tobacco homologue of AtMYC2 and CrMYC2. We also hypothesize that this NtMYC2 transcription factor controls the MeJA-responsive expression of the *NtORC1* and *NtJAP1* genes, which in turn are hypothesized to control the MeJA-responsive expression of the nicotine biosynthesis genes.

Conclusion

We have given an overview of the current state of understanding of jasmonate signalling regulating alkaloid biosynthesis in tobacco and in *Catharanthus roseus*. For both species, certain elements in the models in Figs. 4 and 6 have not yet been experimentally confirmed. For the Catharanthus model, the involvement of the jasmonate receptor COI1 has not been experimentally confirmed. Given the conservation of COI1 as a jasmonate receptor in Arabidopsis, tomato, tobacco and *Nicotiana attenuata*, the position of COI1 in the jasmonate signal transduction pathway in Catharanthus seems highly probable. In tobacco, especially the identities and roles of transcription factors need more solid experimental confirmation. It will be interesting to see whether other secondary pathways regulated by JAs in different plant species are regulated in a similar manner.

Acknowledgements HZ was supported by the Research Council for Earth and Life Sciences (ALW) with financial aid from the Netherlands Organization for Scientific Research (NWO; grant # 812.06.002). The unpublished work on CrMYC2 was done in collaboration with Sabah Hedhili, Guillaume Chatel, Martial Pré and Pascal Gantet from the François Rabelais University in Tours, France. The unpublished work on the CrJAZ proteins was done in collaboration with Laurens Pauwels and Alain Goossens from the Department of Plant Systems Biology at Ghent University in Ghent, Belgium.

References

Austin MJ, Muskett P, Kahn K, Feys BJ, Jones JDG, Parker JE (2002) Regulatory role of SGT1 in early R gene-mediated plant defenses. Science 295: 2077–2080

8

Azevedo C, Sadanandom A, Kitagawa K, Freialdenhoven A, Shirasu K, Schulze-Lefert P (2002) The RAR1 interactor SGT1, an essential component of R gene-triggered disease resistance. Science 295: 2073–2076

Balbi V, Devoto A (2008) Jasmonate signalling network in *Arabidopsis thaliana*: crucial regulatory nodes and new physiological scenarios. New Phytol 177: 301–318

Chahed K, Oudin A, Guivarc'h N, Hamdi S, Chenieux JC, Rideau M, Clastre M (2000) 1-Deoxy-D-xylulose 5-phosphate synthase from perwinkle: cDNA identification and induced gene expression in terpenoid indole alkaloid-producing cells. Plant Physiol Biochem 38: 559–566

Chini A, Fonseca S, Fernández G, Adie B, Chico JM, Lorenzo O, García-Casado G, López-Vidriero I, Lozano FM, Ponce MR, Micol JL, Solano R (2007) The JAZ family of repressors is the missing link in jasmonate signalling. Nature 448: 666–671

De Sutter V, Vanderhaeghen R, Tilleman S, Lammertyn F, Vanhoutte I, Karimi M, Inzé D, Goossens A, Hilson P (2005) Exploration of jasmonate signalling via automated and standardized transient expression assays in tobacco cells. Plant J 44:1065–1076

del Pozo JC, Estelle M (2000) F-box proteins and protein degradation: an emerging theme in cellular regulation. Plant Mol Biol 44:123–128

Devoto A, Nieto-Rostro M, Xie D, Ellis C, Harmston R, Patrick E, Davis J, Sherratt L, Coleman M, Turner JG (2002) COI1 links jasmonate signalling and fertility to the SCF ubiquitin-ligase complex in Arabidopsis. Plant J 32: 457–466

Dharmasiri N, Dharmasiri S, Estelle M (2005) The F-box protein TIR1 is an auxin receptor. Nature 435: 441–445

Facchini PJ (2001) Alkaloid biosynthesis in plants: biochemistry, cell biology, molecular regulation, and metabolic engineering applications. Ann Rev Plant Physiol Plant Mol Biol 52: 29–66

Feys BF, Benedetti CE, Penfold CN, Turner JG (1994) Arabidopsis mutants selected for resistance to the phytotoxin coronatine are male sterile, insensitive to methyl jasmonate, and resistant to a bacterial pathogen. Plant Cell 6: 751–759

Gagne JM, Downes BP, Shiu SH, Durski AM, Vierstra RD (2002) The F-box subunit of the SCF E3 complex is encoded by a diverse superfamily of genes in Arabidopsis. Proc Natl Acad Sci USA 99: 11519–11524

Goossens A, Häkkinen ST, Laakso I, Seppänen-Laakso T, Biondi S, De Sutter V, Lammertyn F, Nuutila AM, Söderlund H, Zabeau M, Inzé D, Oksman-Caldentey KM (2003) A functional genomics approach toward the understanding of secondary metabolism in plant cells. Proc Natl Acad Sci USA 100: 8595–8600

Gray WM, Muskett PR, Chuang HW, Parker JE (2003) Arabidopsis SGT1b is required for SCF^{TIR1}-mediated auxin response. Plant Cell 15: 1310–1319

Grotewold E (2006) The genetics and biochemistry of floral pigments. Annu Rev Plant Biol 57: 761–780

Guilfoyle T (2007) Sticking with auxin. Nature 446: 621–622

Halitschke R, Baldwin IT (2003) Antisense LOX expression increases herbivore performance by decreasing defense responses and inhibiting growth-related transcriptional reorganization in Nicotiana attenuata. Plant J 36: 794–807

Halkier BA, Gershenzon J (2006) Biology and biochemistry of glucosinolates. Annu Rev Plant Biol 57: 303–333

Jones JD, Dangl JL (2006) The plant immune system. Nature 444: 323–329

Katoh A, Ohki H, Inai K, Hashimoto T (2005) Mole-cular regulation of nicotine biosynthesis. Plant Biotechnol 22: 389–392

Katsir L, Chung HS, Koo AJ, Howe GA (2008a) Jasmonate signaling: a conserved mechanism of hormone sensing. Curr Opin Plant Biol, doi:10.1016/j.physletb.2003.10.071

Katsir L, Schilmiller AL, Staswick PE, He SY, Howe GA (2008b) COI1 is a critical component of a receptor for jasmonate and the bacterial virulence factor coronatine. Proc Natl Acad Sci USA 105: 7100–7105

Kepinski S, Leyser O (2005) The Arabidopsis F-box protein TIR1 is an auxin receptor. Nature 435: 446–451

Kunkel BN, Brooks DM (2002) Cross talk between signaling pathways in pathogen defense. Curr Opin Plant Biol 5: 325–331

Li L, Zhao Y, McCaig BC, Wingerd BA, Wang J, Whalon ME, Pichersky E, Howe GA (2004) The tomato homolog of CORONATINE-INSENSITIVE1 is required for the maternal control of seed maturation, jasmonate-signaled defense responses, and glandular trichome development. Plant Cell 16: 126–143

Lorenzo O, Solano R (2005) Molecular players regulating the jasmonate signalling network. Curr Opin Plant Biol 8:532–540

Lorenzo O, Chico JM, Sanchez-Serrano JJ, Solano R (2004) JASMONATE-INSENSITIVE1 encodes a MYC transcription factor essential to discri-minate between different jasmonate-regulated defense responses in Arabidopsis. Plant Cell 16: 1938–1950

Mandaokar A, Thines B, Shin B, Lange BM, Choi G, Koo YJ, Yoo YJ, Choi YD, Choi G, Browse J (2006) Transcriptional regulators of stamen development in Arabidopsis identified by transcriptional profiling. Plant J 46: 984–1008

Melotto M, Mecey C, Niu Y, Chung HS, Katsir L, Yao J, Zeng W, Thines B, Staswick P, Browse J, Howe G, He SY (2008) A critical role of two positively charged amino acids in the Jas motif of Arabidopsis JAZ proteins in mediating coronatine- and jasmonoyl isoleucine-dependent interaction with the COI1 F-box protein. Plant J 55: 979–988

Memelink J, Gantet P (2007) Transcription factors involved in terpenoid indole alkaloid biosynthesis in Catharanthus roseus. Phytochem Rev 6:353–362

Memelink J, Verpoorte R, Kijne JW (2001) ORC Anization of jasmonate-responsive gene expression in alkaloid metabolism. Trends Plant Sci 6: 212–219

Menke FLH, Champion A, Kijne JW, Memelink J (1999a) A novel jasmonate- and elicitor-responsive element in the periwinkle secondary meta-bolite biosynthetic gene Str interacts with a jasmonate- and elicitor-inducible AP2-domain transcription factor, ORCA2. EMBO J 18: 4455–4463

Menke FLH, Parchmann S, Mueller MJ, Kijne JW, Memelink J (1999b) Involvement of the octadecanoid pathway and protein phosphorylation in fungal elicitor-induced expression of terpenoid indole alkaloid biosynthetic genes in *Catharanthus roseus*. Plant Physiol 119: 1289–1296

Paschold A, Halitschke R, Baldwin IT (2007) Co(i)-ordinating defenses: NaCOI1 mediates herbivore- induced resistance in *Nicotiana attenuata* and reveals the role of herbivore movement in avoiding defenses. Plant J 51: 79–91

Pré M, Sibéril Y, Memelink J, Champion A, Doireau P, Gantet P (2000) Isolation by the yeast one-hybrid system of cDNAs encoding transcription factors that bind to the G-box element of the strictosidine synthase gene promoter from Catharanthus roseus. Int J Bio-chrom 5: 229–244

Rischer H, Oresic M, Seppänen-Laakso T, Katajamaa M, Lammertyn F, Ardiles-Diaz W, Van Montagu MC, Inzé D, Oksman-Caldentey KM, Goossens A (2006) Gene-to-metabolite networks for terpenoid indole alkaloid biosynthesis in *Catharanthus roseus* cells. Proc Natl Acad Sci USA 103: 5614–5619

Shah J (2003) The salicylic acid loop in plant defense. Curr Opin Plant Biol 6: 365–371

Shoji T, Yamada Y, Hashimoto T (2000) Jasmonate induction of putrescine N-methyltransferase genes in the root of *Nicotiana sylvestris*. Plant Cell Physiol 41: 831–839

Shoji T, Ogawa T, Hashimoto T (2008) Jasmonate-induced nicotine formation in tobacco is mediated by tobacco *COI1* and *JAZ* genes. Plant Cell Physiol 49: 1003–1012

Staswick PE, Tiryaki I (2004) The oxylipin signal jasmonic acid is activated by an enzyme that conjugates it to isoleucine in *Arabidopsis*. Plant Cell 16: 2117–2127

Staswick PE, Su W, Howell SH (1992) Methyl jasmonate inhibition of root growth and induction of a leaf protein are decreased in an *Arabidopsis thaliana* mutant. Proc Natl Acad Sci USA 89: 6837–6840

Stenzel I, Hause B, Miersch O, Kurz T, Maucher H, Weichert H, Ziegler J, Feussner I, Wasternack C (2003) Jasmonate biosynthesis and the allene oxide cyclase family of *Arabidopsis thaliana*. Plant Mol Biol 51: 895–911

Tan X, Calderon-Villalobos LI, Sharon M, Zheng C, Robinson CV, Estelle M, Zheng N (2007) Mecha-nism of auxin perception by the TIR1 ubiquitin ligase. Nature 446: 640–645

Thines B, Katsir L, Melotto M, Niu Y, Mandaokar A, Liu G, Nomura K, He SY, Howe GA, Browse J (2007) JAZ repressor proteins are targets of the SCF^{COI1} complex during jasmonate signalling. Nature 448: 661–665

Turner JG, Ellis C, Devoto A (2002) The jasmonate signal pathway. Plant Cell 14 Suppl: S153–S164

Vanholme B, Grunewald W, Bateman A, Kohchi T, Gheysen G (2007) The tify family previously known as ZIM. Trends Plant Sci12: 239–244

van der Fits L, Memelink J (2000) ORCA3, a jasmonate-responsive transcriptional regulator of plant primary and secondary metabolism. Science 289: 295–297

van der Fits L, Memelink J (2001) The jasmonate-inducible AP2/ERF-domain transcription factor ORCA3 activates gene expression via interaction with a jasmonate-responsive promoter element. Plant J 25: 43–53

van der Heijden R, Jacobs DI, Snoeijer W, Hallard D, Verpoorte R (2004) The Catharanthus alkaloids: pharmacognosy and biotechnology. Curr Med Chem 11: 607–628

Veau B, Courtois M, Oudin A, Chénieux JC, Rideau M, Clastre M (2000) Cloning and expression of cDNAs encoding two enzymes of the MEP

8

pathway in Catharanthus roseus. Biochim Biophys Acta 1517: 159–163
Vom Endt D, Soares e Silva M, Kijne JW, Pasquali G, Memelink J (2007) Identification of a bipartite jasmonate-responsive promoter element in the *Catharanthus roseus ORCA3* transcription factor gene that interacts specifically with AT-hook DNA-binding proteins. Plant Physiol 144: 1680–1689
Wang L, Allmann S, Wu J, Baldwin IT (2008) Comparisons of LIPOXYGENASE3- and JASMONATE-RESISTANT4/6-silenced plants reveal that jasmonic acid and jasmonic acid-amino acid conjugates play different roles in herbivore resistance of *Nicotiana attenuata*. Plant Physiol 146: 904–915
Wang KL, Li H, Ecker JR (2002). Ethylene biosynthesis and signaling networks. Plant Cell 14 Suppl: S131–S151
Wasternack C (2007) Jasmonates: an update on biosynthesis, signal transduction and action in plant stress response, growth and development. Ann Bot 100: 681–697
Xie DX, Feys BF, James S, Nieto-Rostro M, Turner JG (1998) COI1: an Arabidopsis gene required for jasmonate-regulated defense and fertility. Science 280: 1091–1094
Xu L, Liu F, Lechner E, Genschik P, Crosby WL, Ma H, Peng W, Huang D, Xie D (2002) The SCF^{COI1} ubiquitin-ligase complexes are required for jasmonate response in Arabidopsis. Plant Cell 14: 1919–1935
Zhao J, Davis LC, Verpoorte R (2005) Elicitor signal transduction leading to production of plant secondary metabolites. Biotechnol Adv 23: 283–333

Metabolite Channeling and Multi-enzyme Complexes

9

Brenda S.J. Winkel

Abstract The assembly of cooperating enzymes into multicatalytic complexes, also known as "metabolons," has become a well-accepted concept in cellular metabolism, at least in principle. There are still relatively few examples where the existence of such systems is supported by solid experimental evidence and even fewer where there is evidence for "channeling" of metabolites through the complex. However, proteomic approaches are providing new evidence for the pervasiveness of this type of organization, while structural biology is offering insights into how these systems are constructed and regulated. New and improved technologies for analyzing protein interactions and assemblies, both *in vitro* and in intact cells, are opening the doors to exploring the intracellular organization of a growing number of metabolic complexes in plants and other organisms. There is also an increasing appreciation of the surprising scale of many protein interaction networks, the multiple functions of individual proteins, and the importance (and challenges) of compartmentalization. As a result, the concept of enzyme complexes is gaining wider acceptance and becoming an increasingly important consideration in efforts to engineer metabolism.

B.S.J. Winkel (✉)
Department of Biological Sciences, Virginia Tech, Blacksburg, VA 24061, USA
e-mail: winkel@vt.edu

Introduction to Metabolic Channeling and Enzyme Complexes

The concept of metabolic channeling of intermediates through large complexes of cooperating enzymes is quite appealing from the view point of cellular biochemistry. It offers the prospect of making sense of the incredible complexity of metabolism, providing means to enhance biochemical efficiency while orchestrating the production of specific products at defined times and locations within the cell. And yet the idea has stirred much debate, from both theoretical and experimental viewpoints, over the years (see discussion in [1]). New technologies for analyzing the structure, organization, and function of proteins and cells, as well as advances in computational biology, are doing a great deal to instill confidence that this phenomenon is, at a minimum, a critical aspect of several well-studied metabolic systems and perhaps a general characteristic of pathways in organisms across the evolutionary spectrum. For example, insights into the phenomenal complexity of the cell interior has come from the beautiful models drawn by David Goodsell over the past 20 years (Fig. 1a) and the remarkably similar images of actual cells generated more recently by cryo-electron tomography [2] (Fig. 1b). The protein "interactomes" established for model organisms such as *Helicobacter pylori*, yeast, *Drosophila*, *Caenorhabditis*, and humans [3–8],

A.E. Osbourn and V. Lanzotti (eds.), *Plant-derived Natural Products*,
DOI 10.1007/978-0-387-85498-4_9, © Springer Science+Business Media, LLC 2009

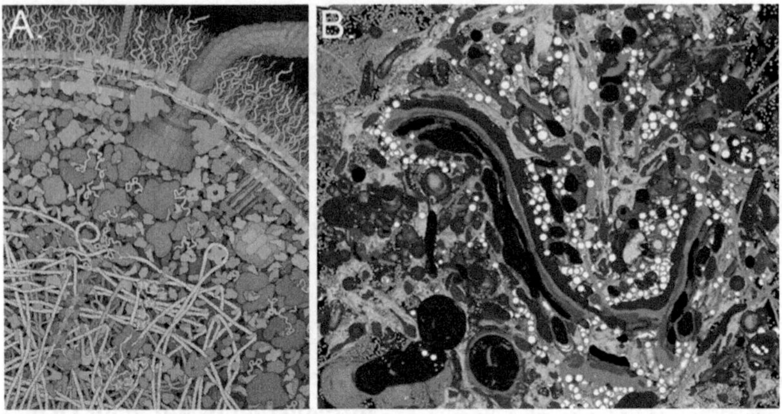

Fig. 1 Images of the cell interior. (**a**) Illustration of a cross-section of an *Escherichia coli* cell generated based on available biochemical, structural, and microscopic data. A flagellar motor is shown at the upper right (From The Machinery of Life, 2nd ed., scheduled for release in late 2008 through Springer-Verlag). (**b**) Three-dimensional model of pancreatic β cells derived by electron tomography from thin sections of plastic-embedded freeze-substituted cells. The Golgi complex with seven cisterna (C1–C7) is at the center (From [2])

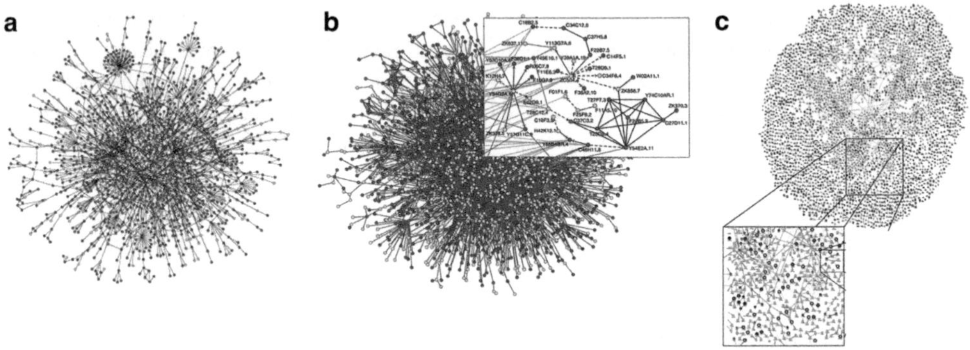

Fig. 2 Protein interaction networks for yeast (**a**), *Caenorhabditis* (**b**), and *Drosophila* (**c**) (From [4, 6, 96])

although still just a start to identifying biologically relevant interactions, point to surprising and intriguing connections among seemingly unrelated pathways (Fig. 2). New terms like "molecular sociology" are now being introduced to represent this type of thinking about the organization and interactive functioning of the cell [9]. However, it is important to bear in mind that the field has a long and distinguished history, as Paul Srere so eloquently reminded us in an article written shortly before his death [10]. Here we consider a number of recent advances in plants and other organisms that are helping to establish enzyme complexes as a central paradigm of cellular biochemistry.

Enzyme Complexes and Channeling

One of the inherent implications of organized metabolism is that it enhances the ability of cooperating enzymes to rapidly and efficiently hand off intermediates from one active site to the next, a concept known as "channeling." A number of recent reports provide new support for the occurrence of this phenomenon in a variety of systems. Examples include biochemical evidence for channeling between aminotransferase isozymes (BCAT) and the decarboxylase/dehydrogenase (E1) in branched chain amino acid metabolism in human mitochondria [11]. There have also been new insights into the structural basis for channeling of intermediates between active sites, notably in well-established supersystems like the fatty acid β-oxidation complex of *E. coli* [12] (Fig. 3a), the mammalian, fungal, and yeast fatty acid synthase (FAS) complexes [13, 14] (Fig. 3b), and the pyruvate dehydrogenase multienzyme complex of *Bacillus stearothermophilus* [15]. An excellent example on a smaller scale comes from studies of the shikimate pathway in fungi and plants. This is one of the first systems in which channeling was hypothesized to occur, by Norman Giles in 1978, based on studies of the pentafunctional AROM protein of *Aspergillus* [16]. Singh and Christendat [17] have now solved the quaternary structure of the Arabidopsis dehydroquinate dehydratase (DHQ)-shikimate dehydrogenase (SDH) bienzyme system (Fig. 3c). This complex, which is composed to two distinct polypeptides in plants, adopts a concave architecture with the two active sites face-to-face, thereby facilitating the transfer of the intermediate from one site to the next. Further evidence for channeling is coming from metabolic profiling experiments, which reinforce the fact that intermediates are often present in vanishingly small amounts, frequently far below stoichiometric levels with the corresponding enzymes, and must therefore be moving rapidly between active sites ([18, 19]; Daniel H. Kohl, personal communication). This is also the case, as discussed further below, in plant anthocyanin biosynthesis [20]. Conversely, there is evidence that channeling is not important in the CDP-ribitol synthase system [21] and, based on structural evidence, between dihydroneopterin aldolase/6-hydroxymethyl-7,8-dihydropterinin *Streptococcus pneumoniae* folate biosynthesis [22]. These findings suggest that channeling may not be a universal feature of multi-enzyme systems.

The Role of Channeling in Metabolic Evolution

Universal or not, channeling likely has an important bearing on the issue of enzyme evolution. Directed engineering efforts are reinforcing the idea that it is not possible to evolve activities that are not, at least in part, already present in an enzyme [23] – and the idea that early enzymes had very broad specificities that evolved into the highly specific enzymes that occur today (reviewed in [24]). This fits well with the accumulating evidence for the inherent

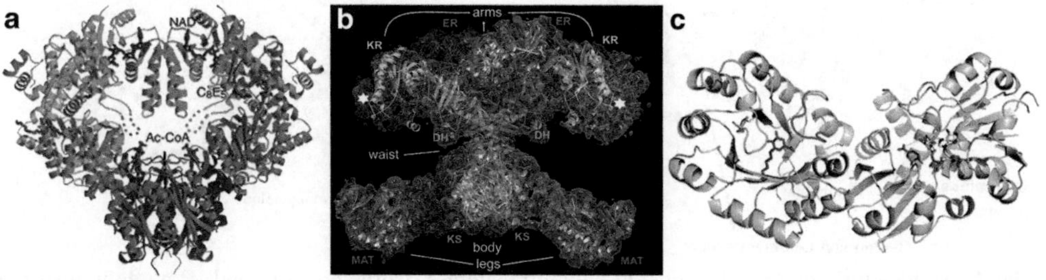

Fig. 3 Quaternary structures of the (**a**) fatty acid beta-oxidation complex of *E. coli* (From [12], (**b**) porcine fatty acid synthase (From [14]), and (**c**) active sites of the Arabidopsis shikimate pathway enzymes, DHQ (*left*) and SDH (*right*) (From [17])

promiscuity of enzymes. Originally believed to be a relatively rare feature when first reported in the 1960s, it now appears to be more the rule than the exception. Enzyme promiscuity mandates the existence of mechanisms to control access of relatively nonspecific enzymes to substrates, either through expression patterns, intracellular localization, and/or metabolic channeling. Some very nice examples in support of this concept have come from the pathways of plant secondary metabolism, which display enormous variation and have presumably been under strong evolutionary pressure since plants adopted terrestrial habitats some 40 million years ago. *O*-methyltransferases (OMTs) are a case in point. These enzymes catalyze terminal modifications in a wide variety of secondary pathways. OMTs from pine [25], meadow rue [26], and the freshwater weed, *Chrysosplenium americanum* [27], have been reported to have broad substrate specificities *in vitro*, yet appear to target very specific substrates *in vivo*. There is also evidence for differences in the *in vitro* and *in vivo* regiospecificity of isoflavone OMT from alfalfa [28] and it has been suggested that this enzyme associates with the ER-associated isoflavone synthase to ensure rapid conversion of the unstable product of this enzyme [29]. New studies indicate that promiscuous substrate utilization and regiospecificity characterize many members of this enzyme family in *Medicago* [30]. Structural studies are providing insights into the molecular basis of the substrate- and regiospecificity of these proteins and evidence that this plasticity also reflects the ability of these enzymes to rapidly evolve new activities in response to environmental pressures such as pathogens [31, 32] (Fig. 4). These findings echo what has been observed in enzymes from a wide variety of organisms. They also reinforce the idea that, rather than limiting the evolution of metabolic pathways, the assembly of cooperating enzymes into complexes that facilitate channeling may, in fact, allow enzymes to carry substantial evolutionary plasticity without compromising metabolic specificity.

There may also be other ways in which enzyme complexes contribute to the evolution of novel or enhanced enzyme function. One mechanism is suggested by recent examples from monocots in which multimers with altered substrate binding affinities or specificities have arisen through the interaction of closely-related isoforms. In the case of nitrilase in *Sorghum bicolor*, individual isoforms of NIT4 are catalytically inactive, but

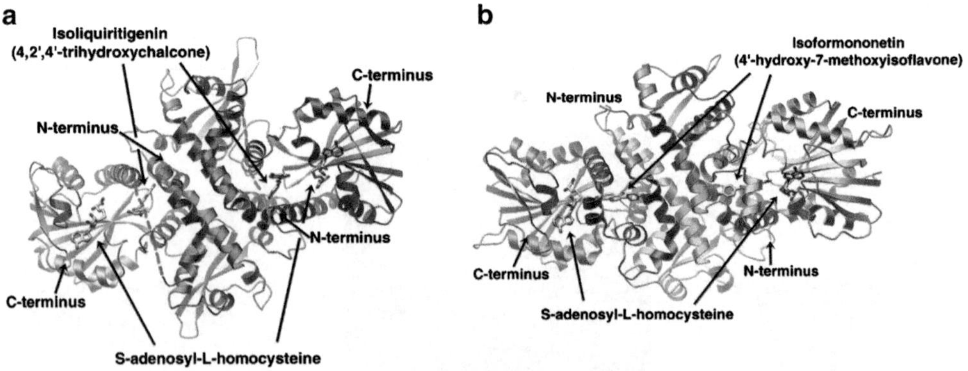

Fig. 4 Architecture of the chalcone *O*-methyltransferase isoflavone (**a**) and *O*-methyltransferase (IOMT) (**b**) active sites. Structural analysis shows that the tertiary structures of these enzymes are highly conserved; substrate selectivity is determined by variations in side chains on the active site surface that alter van der Walls interactions and hydrogen bonding patterns (From [32])

heterocomplexes containing various combinations of SbNIT4A, B, and C alternatively catalyze reactions in either auxin biosynthesis or cyanide detoxification [33]. A slightly different system operates in *Zea mays*, where ZmNIT2 functions in auxin metabolism by itself, but participates in cyanide detoxification by heterocomplexation with the highly-inefficient isomer, ZmNIT1 [34]. A second piece of evidence for the role of multimerization in the evolution of novel enzyme functions come from studies of β-glucosidase in *Avena sativa* [35]. In this case, the AsGlu1 isomer forms long fibrillar assemblies either by itself, or in complex with AsGlu2, which otherwise assembles as a dimer. This multimerization not only increases the affinity of AsGlu2 for its substrate, but likely also substantially enhances substrate specificity. Presumably, many more examples remain to be uncovered in which interactions among the members of multienzyme families modulate the functions of the individual isozymes. It is noteworthy that even isoforms with little or no apparent activity can thereby contribute to cellular biochemistry, particularly in view of the growing number of enzyme families that include members with little or no *in vitro* activity (recent examples in plants are described in [36–38]). Whether enzyme interactions have also contributed to the evolution of novel biochemical functions in the context of more complex multienzyme systems remains to be seen.

The Role of Microtubules in Organizing Enzyme Complexes

The cytoskeleton has long been proposed to play a key role in cellular metabolism, serving as a matrix for the assembly, and even regulation, of enzyme complexes (reviewed in [10]). A striking example comes from the cellulose synthase complex of plants, an enzyme system that is currently of particular interest with regard to enhancing plant-based biofuels (reviewed in [39–41]. This system is organized in vascular plants as a 30 nm diameter plasma membrane complex, known as the rosette TC, that was first described by Mueller and Brown more than 25 years ago [42]. The rosette is composed of an estimated 36 subunits comprising at least three types of cellulose synthase A proteins. All three components have been shown to be essential for correct assembly and enzymatic activity of the complex in Arabidopsis [43, 44]. The physical interaction of these three components has recently been demonstrated by bimolecular fluorescence complementation [45]. The system also offers a beautiful example of the functional aspects of this organization, as cortical microtubules guide the movement of cellulose synthase complexes through the plasma membrane during cell wall biosynthesis [46, 47].

Further evidence for the role of microtubules in metabolic channeling comes from an intriguing recent study aimed at identifying microtubule-associated proteins in plant cells [48]. In these experiments, tubulin affinity chromatography was used to purify 122 tubulin-binding proteins from Arabidopsis suspension cell extracts, identified by LC-MS/MS analysis of spots recovered from two-dimensional gels. In addition to known microtubule-associated proteins (6%) and proteins involved in RNA binding (21%), translation (19%), and signaling (12%), a full 21% were proteins known to function in metabolism. These included the glycolytic enzyme, glyceraldehyde 3-phosphate dehydrogenase, which has known MT-binding activity. The metabolic proteins also included five enzymes of the folate-dependent pathways of one-carbon metabolism, five that function in fatty acid metabolism in the peroxisome (note that there is previous evidence for interaction of peroxisomal enzymes with microtubules), three components of the pantothenate pathway, and two enzymes each of the pentose phosphate pathway and amino acid biosynthesis. Although the MT-binding activity of many of these proteins remains to be confirmed, the identification of multiple components from specific

9

pathways suggests that MTs could function to organize these systems as multi-enzyme complexes, similar to what has been described for the cellulose synthase complex. This type of affinity chromatography/mass spectrometry-based approach may well be a very effective new means for discovering metabolic complexes in model plants such as Arabidopsis, rice, tomato, and poplar, for which full or nearly-full genome information is available.

A Tethering Protein Involved in Complex Assembly at the ER

Membranes represent another critical matrix for the assembly of multi-enzyme complexes in cells and it has long been suggested that the membrane-associated enzymes that are part of many biochemical pathways could serve as anchor points for the rest of the pathway enzymes (reviewed in [10, 49]). Now there is evidence that non-catalytic proteins may serve this function in at least some cases. In 2000, in one of the early examples of gene discovery by whole-genome expression profiling, the yeast gene *YER044C* was identified as being highly correlated with known ergosterol biosynthetic genes [50]. Since designated *ERG28*, the gene was predicted to encode a 148 amino acid protein with two transmembrane domains. Subsequent experiments in the laboratory of Martin Bard have shown that *ERG28* deletion strains grow slowly, have ergosterol levels approximately one third that of wild type, and accumulate sterol precursors normally found in strains deficient in *ERG26* and *ERG27*, which encode key enzymes in the C-4 demethylation component of the sterol pathway (summarized in [51]). This group then showed that the Erg28p protein functions as a transmembrane scaffold to anchor the C-4 demethylation complex, as well as downstream enzymes, to the ER. To more fully characterize interactions among the various proteins involved in sterol biosynthesis, they recently used the split-ubiquitin membrane protein yeast two-hybrid system to investigate all 196 possible pairwise interactions [52] (Fig. 5). Interestingly, enzymes in the middle of the pathway appear to act as a "hub" for enzymes that function either earlier or later. However, enzymes such as Erg2p and Erg3p that catalyze sequential reactions do not interact with the same subset of partners in this assay and enzymes in the late portion of the pathway interact more strongly with a core of enzymes in the center of the pathway. These researchers therefore suggest that the physical organization of the sterol pathway explains the different products produced in mutant lines and in various organisms; because synthesis is not physically organized in a linear array, loss (by mutation or evolution) of one enzyme does not interrupt the biosynthetic process. In fact, a similar organization is proposed to exist for sterol biosynthesis in other organisms, including plants and mammals, in which homologs of *ERG28* have been found ([51, 53]; Joe Chappell, personal communication). It may be noted that in the flavonoid pathway of Arabidopsis, a similar

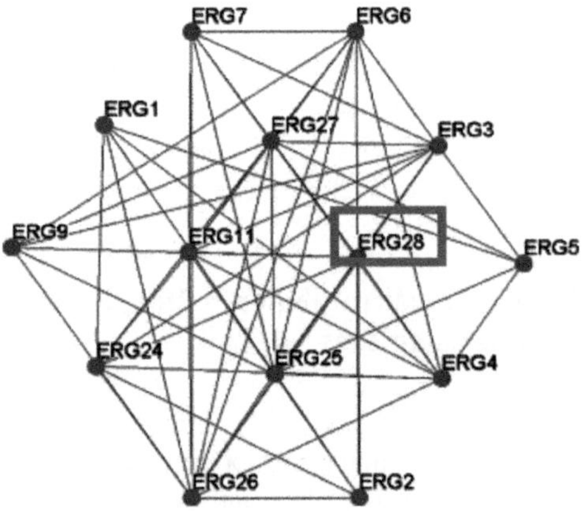

Fig. 5 Schematic of the yeast sterol protein interaction network based on biochemical data and results of split-ubiquitin membrane yeast two-hybrid assays. The tethering protein, ERG28, is highlighted in red (Adapted from [52])

non-linear protein interaction network has been observed, in this case with the first enzyme of the pathway serving as an apparent "hub" ([36, 54, 55]; Christopher Dana, Daniel Owens, Justin Runac, Jonathan Watkinson, and Brenda Winkel, unpublished data). However, Erg28p appears to represent the first instance where a non-catalytic protein has been shown to be an essential component of a biochemical pathway. It is possible that similar proteins will be identified in other systems as more extensive global expression and protein interaction networks are described.

Reversible Assembly of Enzyme Complexes as a Regulatory Mechanism

One of the most appealing features of enzyme complexes, one that has long been touted in the literature, is that they offer a means for extremely rapid responses to changes in the metabolic status of the cell, through reversible assembly and disassembly or all or some of the constituent parts. The cysteine synthase complex (CSC) in bacteria and plants is providing insights into just how this may occur. The CSC plays a pivotal role in sulfur assimilation and cysteine biosynthesis in these organisms and has been the subject of extensive study since it was first purified from *Salmonella typhimurium* in 1969 [56]. The hetero-oligomeric ($\alpha_6\beta_4$) complex consists of two enzymes, serine acetyltransferase (SAT) activity and *O*-acetylserine sulfhydrylase (OASS) (also known as *O*-acetylserine (thiol) lyase [OAS-TL]) that convert serine to cysteine [57] (Fig. 6). In plants, these enzymes are located in three cellular compartments, plastids, mitochondria, and the cytosol, with the added challenge of coordinating metabolite levels across multiple compartments. *In vitro* biochemical and physiological studies have given rise to a model in which intracellular sulfur levels regulate the assembly and disassembly of the complex and thereby the assimilation of reduced sulfur into cysteine (summarized in [57, 58]). When sulfur levels are low, *O*-acetylserine accumulates because OASS cannot efficiently convert it to cysteine (Fig. 6). This results in dissociation of the complex with a corresponding reduction in SAT activity and elevation of OASS activity. Increased *O*-acetylserine levels also cause activation of genes encoding SAT, OASS, and sulfur transporters. The system is therefore poised to react to subsequent increases in sulfur levels, when OASS can rapidly begin consuming accumulated *O*-acetylserine, leading to reassembly of the complex, activation of SAT, and resumption of cysteine biosynthesis. The first *in vivo* test of this model was described recently by Wirtz and Hell [59], who generated transgenic tobacco plants in which an Arabidopsis SAT gene was overexpressed in the cytosol in either a catalytically active or inactive form. Expression of the active form of the enzyme resulted in substantial increases in reaction intermediates and endproducts, consistent with evidence for a large excess of OASS activity in wild-type plants. Surprisingly, the inactive form of SAT also resulted in a large (30X) increase in cysteine levels. These researchers propose that, in this situation, the Arabidopsis SAT would out-compete endogenous SAT in the cytoplasm, shutting down cysteine biosynthesis in that

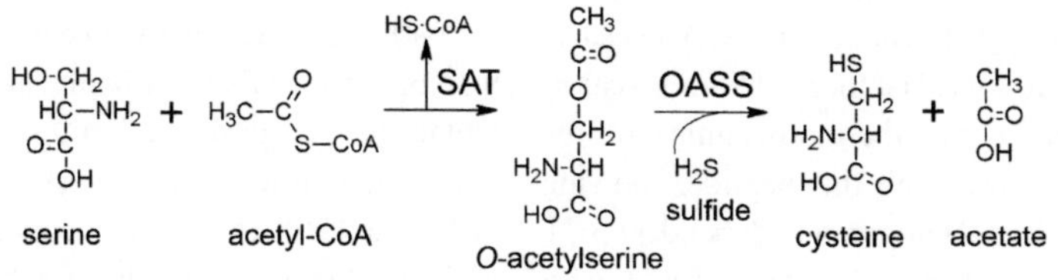

Fig. 6 Biochemical conversion of serine to cysteine by the enzymes of the cysteine synthase complex (Adapted from [57])

compartment. Cysteine synthesis in the mitochondria and/or plastids may overcompensate due to limited transfer of cysteine to the cytosol, despite efficient exchange of metabolites between these compartments. These experiments confirm the regulatory function of the CSC and provide new insights into the coordination of metabolism and sulfur homeostasis among different intracellular compartments.

The macromolecular CSC has been characterized by a variety of methods, including yeast two-hybrid analysis, gel filtration chromatography, analytical ultracentrifugation, and surface plasmon resonance refractometry (summarized in [57, 58]). These studies, together with biochemical analysis of the enzymes from *E. coli* and *Haemophilus influenza*, provided evidence that the C terminus binds to the OASS active site, although other parts of SAT are also critical for the interaction. Although the architecture of the $\alpha_6\beta_4$ CRC remains to be resolved, homology models of the plant enzymes, based on the corresponding bacterial proteins, were used in a preliminary effort that placed the OASS dimers at the distal ends of the SAT dimer of trimers [57]. Shortly thereafter, the crystal structure of Arabidopsis OASS was solved by Joe Jez's group and site-directed mutagenesis was used to identify a highly conserved surface loop near the OASS active site that appeared to be involved in the interaction with SAT [58]. This group has gone on to determine the crystal structure of AtOASS bound with a peptide corresponding to the C-terminal ten residues of Arabidopsis SAT [60]. This work showed that the peptide blocks the OASS active site through hydrogen bonding to key catalytic residues, thus demonstrating how complex formation downregulates OASS activity. They have also used isothermal titration calorimetry to explore the thermodynamics of the interaction of OASS with this peptide, showing that it is both salt and temperature dependent [61]. Intriguingly, the enhanced interaction of OASS SAT at lower temperatures is consistent with known differences in the metabolic demands of plant cells for sulfur and cysteine under elevated or chilling temperatures. Once the structure of a plant SAT is solved, it should be possible to explore the mechanism by which association with OASS upregulates SAT activity. It should then also be possible to develop a model of the full CSC system using approaches such as low-resolution X-ray diffraction or small angle X-ray or neutron scattering [13, 14, 62], leading to a molecular-level understanding of this key biochemical system, including regulation of its assembly and disassembly by *O*-acetylserine.

Enzyme Complexes in Plant Natural Product Biosynthesis

Numerous pathways of plant central metabolism are now known to function as multienzyme systems; in addition to those mentioned so far, notable examples include the pathways of glycolysis [63, 64] and the respiratory chain [65]. There are also growing numbers of pathways of plant natural product synthesis for which experimental evidence supports their organization as multienzyme complexes. Particularly well-studied in this regard are the pathways of cyanogenic glycoside, isoprenoid, alkaloid, and phenylpropanoid metabolism (reviewed in [1, 66, 67]). Experimental approaches used to study these systems in recent years have included co-localization, chromatography and co-immunoprecipitation, yeast two-hybrid assays, and the reconstitution of the pathways in heterologous systems. Several recent advances have relied on new technologies, such as the use of ^{13}C NMR spectroscopy to analyze stilbene and anthocyanin metabolism in grape cell cultures [20]. Consistent with a large body of work based on the use of TLC and HPLC, several intermediates remained undetectable even with this much more sensitive technology. This reinforces previous evidence for the existence of a multi-enzyme complex, and metabolic channeling, in this system. New efforts to use yeast two-hybrid assays to probe the protein-protein interactions of flavonoid enzymes have uncovered evidence of an interaction network that may extend beyond the phenylpropanoid pathway (Jonathan Watkinson and Brenda Winkel, unpublished data).

As a result of the mounting evidence for the importance of intracellular organization in metabolism, the issues of channeling and the assembly and localization of enzyme complexes are gaining increasing attention with regard to efforts to engineer the production of natural products in plants and other organisms (reviewed in [68, 69]). These include efforts to characterize isoflavonoid biosynthetic enzymes from soybean and licorice in a Lotus hairy root culture system [70]. In these experiments, analysis of gene expression and metabolite profiles suggested the possibility that 2-hydroxyisoflavanone dehydratase (HID) from soybean was able to assemble into a complex with the preceding enzymes in the Lotus pathway, while the endogenous enzyme was not. Channeling has also been cited as a possible explanation for the relative lack of success in engineering sesquiterpene biosynthesis, despite the presumed presence of substantial quantities of the precursor molecule, farnesyl diphosphate, in plants [71]. Similarly, Yan et al. [72] used enzymes from several different plant species in an effort to engineer anthocyanin production in *E. coli*, with marginal success compared to what they have achieved for other flavonoids. Developing a better understanding of the flavonoid enzyme complex was cited as one approach to optimizing this system.

Future Prospects

It has been nearly more than 50 years since Marko Zalokar [73, 74] and Ellis Kempner and Jay Miller [75] provided early insights into the extent to which enzymes are organized in cells, at membranes, the cytoskeleton, and in organelles (Fig. 7). These experiments, performed with Neurospora, Phyco-myces, and Euglena cells, used an emerging technology, ultracentrifugation (the first commercial instruments were introduced in 1947), to challenge the view that the cell was simply a "bag of enzymes." New technologies have continued to open doors to understanding the complexities of biochemistry in its native environment over the years, now perhaps more than ever. Among the most exciting advances are those in the area of structural biology, finally making it possible to probe the molecular basis for enzyme complex formation (Fig. 8). New global profiling technologies, for gene expression, protein interactions, metabolic output, and even metabolic flux are further underscoring these themes. The power of protein interaction mapping for uncovering previously unknown functional connections is just now becoming apparent, particularly in the study of human disease [76].

Technological advances offer further promise for detecting enzyme interactions, which are likely to be transient or unstable and may be particularly sensitive to the conditions of the native

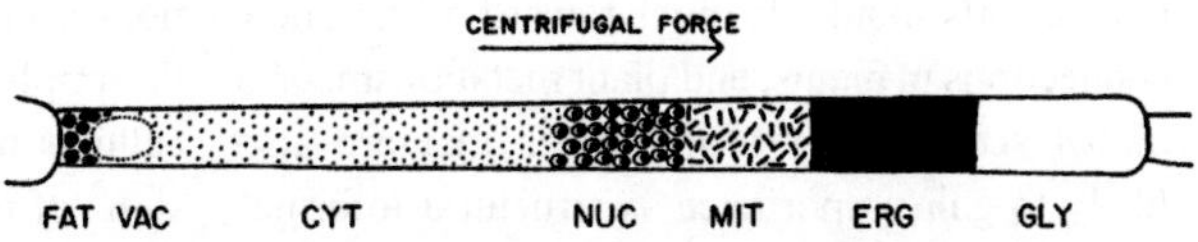

Fig. 7 Schematic of a *Neurospora* hypha centrifuged at 35,000 rpm. FAT, fat; VAC, vacuoles; CYT, enchylema; NUC, nuclei; MIT, mitochondria; ERG, ergastoplasm; GLY, glycogen (From [73])

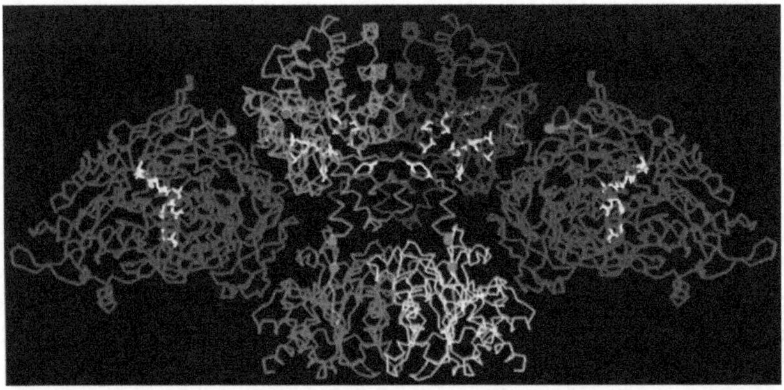

Fig. 8 One of the earliest efforts to model the quinary structure of an enzyme complex, showing the predicted docking orientation between two monomers of pig mitochondrial citrate synthase (red and orange, with the first four N-terminal residues in green and the last four C-terminal residues in pale blue), aconitase (blue, with the first four C-terminal residues in green), and malate dehydrogenase (bright and pale pink, with the first four N-terminal residues in pale blue) (From [97])

environment. An excellent example is the recent use of a protein fragment complementation assay based on murine dihydrofolate reductase, which makes it possible to assay protein interactions in their natural context. In yeast, this system identified 2,770 interactions among 1,124 proteins, many of which were previously unknown, including interactions involving a substantial number of components of cellular metabolism [8]. It seems that we are poised to see this type of approach take off in the area of plant metabolism. Large-scale efforts are being undertaken by Marc Vidal to map 12,000 proteins on the Arabidopsis interactome by high-throughput yeast two-hybrid analysis, by Wolf Frommer to map the potential interactome of Arabidopsis membrane proteins using the split-ubiquitin system, and by Michael Snyder, S.P. Dinesh-Kumar, and Mark Gerstein to produce a first-generation Arabidopsis protein chip. Although each approach has its own inherent limitations [77, 78], together these efforts should do much toward making new connections in plants, and plant metabolism, on a global scale. Other technologies that are also likely to gain importance as structural information about macromolecular systems becomes available are advanced microscopy methods for the analysis of whole cells such as cryo-electron tomography, particularly as these approach molecular resolution, making it possible to identify intact multi-enzyme systems *in situ* [79]. Other emerging microscopy methods, such as FLIM-FRET (see, for example, [80]), are likely to become powerful tools in the drive to understand protein interactions *in vivo*, even for the dynamic interactions that characterize enzyme systems (Kevin Crosby, Brenda Winkel, and Theodorus Gadella, unpublished data). An even within the field of structural biology, understanding the influence of protein interactions on structure is an area of emerging interest and importance, including the extreme example of intrinsically disordered proteins, which acquire a unique tertiary structure only upon interaction with their partners [81].

The growing imperative for "translational" migration from basic science to practical applications is bringing new urgency to reunderstanding the implications of subcellular organization for cellular metabolism. This is evident in discussions of the surprising failure of efforts to engineer key processes, such as the production of valuable plant oils, by the targeted introduction of transgenes [82]. There are still only a handful of examples where localization and function are clearly connected, such as evidence for assembly of glycolytic pathway enzymes on the cytosolic face of the outer mitochondrial membrane in *Arabidopsis* [64, 83]. This finding, which has now also been extended to yeast and humans [84], led to the intriguing suggestion that this organization facilitates transport of pyruvate directly to the mitochondrion for use in respiration. Adding to the challenge of defining enzyme function are confounding factors such as the increasingly common finding that specific enzymes have moonlighting functions (i.e., additional functional activities, often in distinct subcellular compartments) [85–87]. Plant enzymes with known or suspected moonlighting functions already include pea glyceraldehyde-3-phosphate dehydrogenase, phosphofructoaldolase and phosphoglycerate kinase isozymes [88–90], Arabidopsis flavonoid enzymes ([91]; Melissa Ramirez and Brenda Winkel, unpublished data), ferredoxin-dependent glutamate synthase [92], tomato cysteine protease [93], and Arabidopsis amidase I [94]. Moreover, we are still in the early stages of understanding how metabolic networks span compartments, in other words, how pathway intermediates and signal metabolites are transported (or not) across organellar boundaries (discussed in [95]). Many of the same questions that have plagued cellular biochemists for decades remain unanswered and, indeed, some seem to have even more complex answers than ever anticipated. Yet there is promise that we are entering a new era of understanding, as taking the intracellular organization of enzyme systems into account in descriptions of cellular biochemistry becomes the norm rather than the exception.

Acknowledgments The author acknowledges the insights of Joe Chappel (on the erg28p system), Danny Kohl (on the implications of recent metabolic profiling experiments in *E. coli*), and Joe Noel (on the issue of enzyme promiscuity supporting the existence of enzyme complexes), as well as the very helpful comments of three anonymous reviewers. She is grateful to the National Science Foundation for supporting the work in her laboratory on the flavonoid enzyme complex (currently grant number MCB 0445878). This article is dedicated to the memory of H. Olin Spivey, Professor Emeritus of Biochemistry and Molecular Biology at Oklahoma State University, who died on December 5, 2007. A pioneer in the field of metabolic organization, he will be remembered by many as a valued colleague who encouraged newcomers to join the network.

References

1. Jorgensen, K.et al. (2005) Metabolon formation and metabolic channeling in the biosynthesis of plant natural products. *Curr. Opin. Plant Biol.* 8, 280–291
2. Marsh, B.J.et al. (2001) Organellar relationships in the Golgi region of the pancreatic beta cell line, HIT-T15, visualized by high resolution electron tomography. *Proc. Natl. Acad. Sci. USA*. 98, 2399–2406
3. Uetz, P.et al. (2000) A comprehensive analysis of protein-protein interactions in *Saccharomyces cerevisiae*. *Nature* 403, 623–627
4. Giot, L.et al. (2003) A protein interaction map of *Drosophila melanogaster*. *Science* 302, 1727–1736
5. Rual, J.F.et al. (2005) Towards a proteome-scale map of the human protein-protein interaction network. *Nature* 437, 1173–1178
6. Li, S.et al. (2004) A map of the interactome network of the metazoan *C. elegans*. *Science* 303, 540–543
7. Han, J.D.et al. (2004) Evidence for dynamically organized modularity in the yeast protein-protein interaction network. *Nature* 430, 88–93
8. Tarassov, K.et al. (2008) An *in vivo* map of the yeast protein interactome. *Science* 320, 1465–1470
9. Robinson, C.V.et al. (2007) The molecular sociology of the cell. *Nature* 450, 973–982
10. Srere, P.A. (2000) Macromolecular interactions: tracing the roots. *Trends Biochem. Sci.* 25, 150–153
11. Islam, M.M.et al. (2007) A novel branched-chain amino acid metabolon – protein-protein interactions in a supramolecular complex. *J. Biol. Chem.* 282, 11893–11903

9

12. Ishikawa, M.et al. (2004) Structural basis for channelling mechanism of a fatty acid beta-oxidation multienzyme complex. *EMBO J.* 23, 2745–2754
13. Jenni, S. et al. (2006) Architecture of a fungal fatty acid synthase at 5 angstrom resolution. *Science* 311, 1263–1267
14. Maier, T. et al. (2006) Architecture of mammalian fatty acid synthase at 4.5 angstrom resolution. *Science* 311, 1258–1262
15. Fries, M.et al. (2007) Distinct modes of recognition of the lipoyl domain as substrate by the E1 and E3 components of the pyruvate dehydrogenase multienzyme complex. *J. Mol. Biol.* 366, 132–139
16. Giles, N.H. (1978) The organization, function, and evolution of gene clusters in eucaryotes. *Am. Naturalist* 112, 641–657
17. Singh, S.A. and Christendat, D. (2007) The DHQ-dehydroshikimate-SDH-shikimate-NADP(H) complex: insights into metabolite transfer in the shikimate pathway. *Cryst. Growth Des.* 7, 2153–2160
18. Luo, B. et al. (2007) Simultaneous determination of multiple intracellular metabolites in glycolysis, pentose phosphate pathway and tricarboxylic acid cycle by liquid chromatography-mass spectrometry. *J. Chromatogr. A* 1147, 153–164
19. Ishii, N.et al. (2007) Multiple high-throughput analyses monitor the response of *E. coli* to perturbations. *Science* 316, 593–597
20. Saigne-Soulard, C. et al. (2006) C-13 NMR analysis of polyphenol biosynthesis in grape cells: Impact of various inducing factors. *Anal. Chim. Acta* 563, 137–144
21. Pereira, M.P. and Brown, E.D. (2004) Bifunctional catalysis by CDP-ribitol synthase: Convergent recruitment of reductase and cytidylyltransferase activities in *Haemophilus influenzae* and *Staphylococcus aureus*. *Biochemistry (Mosc).* 43, 11802–11812
22. Garcon, A. et al. (2006) Crystal structure of the bifunctional dihydroneopterin aldolase/6-hydroxymethyl-7,8-dihydropterin pyrophosphokinase from *Streptococcus pneumoniae*. *J. Mol. Biol.* 360, 644–653
23. Taglieber, A. et al. (2007) Alternate-site enzyme promiscuity. *Angew. Chem. Int. Ed.* 46, 8597–8600
24. Peisajovich, S.G. and Tawfik, D.S. (2007) Protein engineers turned evolutionists. *Nature Methods* 4, 991–994
25. Chiron, H. et al. (2000) Molecular cloning and functional expression of a stress-induced multifunctional O-methyltransferase with pinosylvin methyltransferase activity from Scots pine (*Pinus sylvestris* L.). *Plant Mol. Biol.* 44, 733–745
26. Frick, S. and Kutchan, T.M. (1999) Molecular cloning and functional expression of *O*-methyltransferases common to isoquinoline alkaloid and phenylpropanoid biosynthesis. *Plant J.* 17, 329–339
27. Gauthier, A. et al. (1998) Characterization of two cDNA clones which encode *O*-methyltransferases for the methylation of both flavonoid and phenylpropanoid compounds. *Arch. Biochem. Biophys.* 351, 243–249
28. He, X.-Z. and Dixon, R.A. (2000) Genetic manipulation of isoflavone 7-*O*-methyltransferase enhances biosynthesis of 4'-*O*-methylated isoflavonoid phytoalexins and disease resistance in alfalfa. *Plant Cell* 12, 1689–1702
29. Liu, C.-J. and Dixon, R.A. (2001) Elicitor-induced association of isoflavone *O*-methyltransferase with endomembranes prevents the formation and 7-*O*-methylation of daizein during isoflavoniod phytoalexin biosynthesis. *Plant Cell* 13, 2643–2658
30. Deavours, B.E. et al. (2006) Functional analysis of members of the isoflavone and isoflavanone *O*-methyltransferase enzyme families from the model legume *Medicago truncatula*. *Plant Mol. Biol.* 62, 715–733
31. Liu, C.J. et al. (2006) Structural basis for dual functionality of isoflavonoid *O*-methyltransferases in the evolution of plant defense responses. *Plant Cell* 18, 3656–3669
32. Zubieta, C. et al. (2001) Structures of two natural product methyltransferases reveal the basis for substrate specificity in plant *O*-methyltransferases. *Nat. Struct. Biol.* 8, 271–279
33. Jenrich, R. et al. (2007) Evolution of heteromeric nitrilase complexes in Poaceae with new functions in nitrile metabolism. *Proc. Natl. Acad. Sci. USA* 104, 18848–18853
34. Kriechbaumer, V. et al. (2007) Maize nitrilases have a dual role in auxin homeostasis and β-cyanoalanine hydrolysis. *J. Exp. Bot.* 58, 4225–4233
35. Kim, S.Y.et al. (2005) Novel type of enzyme multimerization enhances substrate affinity of oat β-glucosidase. *J. Struct. Biol.* 150, 1–10
36. Owens, D.K. et al. (2008) Functional analysis of a predicted flavonol synthase gene family in Arabidopsis. *Plant Physiol.* 147, 1046–1061
37. Kim, S.J. et al. (2004) Functional reclassification of the putative cinnamyl alcohol dehydrogenase multigene family in *Arabidopsis*. *Proc. Natl. Acad. Sci. USA* 101, 1455–1460

38. Yamagami, T. et al. (2003) Biochemical diversity among the 1-amino-cyclopropane-1-carboxylate synthase isozymes encoded by the *Arabidopsis* gene family. *J. Biol. Chem.* 278, 49102–49112
39. Dhugga, K.S. (2007) Maize biomass yield and composition for biofuels. *Crop Sci.* 47, 2211–2227
40. Somerville, C. (2006) Cellulose synthesis in higher plants. *Annu. Rev. Cell Dev. Biol.* 22, 53–78
41. Saxena, I.M. and Brown, R.M. (2005) Cellulose biosynthesis: current views and evolving concepts. *Ann. Bot.* 96, 9–21
42. Mueller, S.C. and Brown, R.M. (1980) Evidence for an intramembrane component associated with a cellulose microfibril-synthesizing complex in higher plants. *J. Cell Biol.* 84, 315–326
43. Taylor, N.G. et al. (2003) Interactions among three distinct CesA proteins essential for cellulose synthesis. *Proc. Natl. Acad. Sci. USA* 100, 1450–1455
44. Persson, S. et al. (2007) Genetic evidence for three unique components in primary cell-wall cellulose synthase complexes in *Arabidopsis*. *Proc. Natl. Acad. Sci. USA* 104, 15566–15571
45. Desprez, T. et al. (2007) Organization of cellulose synthase complexes involved in primary cell wall synthesis in *Arabidopsis thaliana*. *Proc. Natl. Acad. Sci. USA* 104, 15572–15577
46. Paradez, A. et al. (2006) Microtubule cortical array organization and plant cell morphogenesis. *Curr. Opin. Plant Biol.* 9, 571–578
47. DeBolt, S. et al. (2007) Morlin, an inhibitor of cortical microtubule dynamics and cellulose synthase movement. *Proc. Natl. Acad. Sci. USA* 104, 5854–5859
48. Chuong, S.D.X. et al. (2004) Large-scale identification of tubulin-binding proteins provides insight on subcellular trafficking, metabolic channeling, and signaling in plant cells. *Mol. Cell. Proteomics* 3, 970–983
49. Hrazdina, G. and Wagner, G.J. (1985) Metabolic pathways as enzyme complexes: evidence for the synthesis of phenylpropanoids and flavonoids on membrane associated enzyme complexes. *Arch. Biochem. Biophys.* 237, 88–100
50. Hughes, T.R. et al. (2000) Functional discovery via a compendium of expression profiles. *Cell* 102, 109–126
51. Mo, C. and Bard, M. (2005) Erg28p is a key protein in the yeast sterol biosynthetic enzyme complex. *J. Lipid Res.* 46, 1991–1998
52. Mo, C. and Bard, M. (2005) A systematic study of yeast sterol biosynthetic protein-protein interactions using the split-ubiquitin system. *Biochim. Biophys. Acta Mol. Cell. Biol. Lipids* 1737, 152–160
53. Ottolenghi, C. et al. (2000) The genomic structure of C140rf1 is conserved across eukarya. *Mamm. Genome* 11, 786–788
54. Burbulis, I.E. and Winkel-Shirley, B. (1999) Interactions among enzymes of the *Arabidopsis* flavonoid biosynthetic pathway. *Proc. Natl. Acad. Sci. USA* 96, 12929–12934
55. Owens, D.K. et al. (2008) Biochemical and genetic characterization of Arabidopsis flavanone 3β-hydroxylase. *Plant Physiol. Biochem.* 46, 833–843
56. Kredich, N. et al. (1969) Purification and characterization of cysteine synthetase, a bifunctional protein complex, from *Salmonella typhimurium*. *J. Biol. Chem.* 244, 2428–2439
57. Wirtz, M. and Hell, R. (2006) Functional analysis of the cysteine synthase protein complex from plants: Structural, biochemical and regulatory properties. *J. Plant Physiol.* 163, 273–286
58. Bonner, E.R. et al. (2005) Molecular basis of cysteine biosynthesis in plants – structural and functional analysis of *O*-acetylserine sulfhydrylase from *Arabidopsis thaliana*. *J. Biol. Chem.* 280, 38803–38813
59. Wirtz, M. and Hell, R. (2007) Dominant-negative modification reveals the regulatory function of the multimeric cysteine synthase protein complex in transgenic tobacco. *Plant Cell* 19, 625–639
60. Francois, J.A. et al. (2006) Structural basis for interaction of *O*-acetylserine sulfhydrylase and serine acetyltransferase in the *Arabidopsis* cysteine synthase complex. *Plant Cell* 18, 3647–3655
61. Kumaran, S. and Jez, J.M. (2007) Thermodynamics of the interaction between *O*-acetylserine sulfhydrylase and the C-terminus of serine acetyltransferase. *Biochemistry (Mosc).* 46, 5586–5594
62. Petoukhov, M.V. and Svergun, D.I. (2007) Analysis of X-ray and neutron scattering from biomacromolecular solutions. *Curr. Opin. Struct. Biol.* 17, 562–571
63. Anderson, L.E. and Carol, A.A. (2005) Enzyme co-localization in the pea leaf cytosol: 3-P-glycerate kinase, glyceraldehyde-3-P dehydrogenase, triose-P isomerase and aldolase. *Plant Sci.* 169, 620–628
64. Graham, J.W.A. et al. (2007) Glycolytic enzymes associate dynamically with mitochondria in response to respiratory demand and support substrate channeling. *Plant Cell* 19, 3723–3738

9

65. Dudkina, N.V. et al. (2006) Respiratory chain supercomplexes in the plant mitochondrial membrane. *Trends Plant Sci.* 11, 232–240
66. Winkel, B. (2004) Metabolic channeling in plants. *Annu. Rev. Plant Biol.* 55, 85–107
67. Facchini, P.J. and St-Pierre, B. (2005) Synthesis and trafficking of alkaloid biosynthetic enzymes. *Curr. Opin. Plant Biol.* 8, 657–666
68. Gómez-Galera, S. et al. (2007) The genetic manipulation of medicinal and aromatic plants. *Plant Cell Rep.* 26, 1689–1715
69. Morant, A.V. et al. (2007) Lessons learned from metabolic engineering of cyanogenic glucosides. *Metabolomics* 3, 383–398
70. Shimamura, M. et al. (2007) 2-hydroxyisoflavanone dehydratase is a critical determinant of isoflavone productivity in hairy root cultures of *Lotus japonicus*. *Plant Cell Physiol.* 48, 1652–1657
71. Aharoni, A. et al. (2005) Volatile science? Metabolic engineering of terpenoids in plants. *Trends Plant Sci.* 10, 594–602
72. Yan, Y.J. et al. (2005) Metabolic engineering of anthocyanin biosynthesis in *Escherichia coli*. *Appl. Environ. Microbiol.* 71, 3617–3623
73. Zalokar, M. (1960) Cytochemistry of centrifuged hyphae of *Neurospora*. *Exp. Cell Res.* 19, 114–132
74. Zalokar, M. (1969) Intracellular centrifugal separation of organelles in *Phycomyces*. *J. Cell Biol.* 41, 494–509
75. Kempner, E.S. and Miller, J.H. (1968) The molecular biology of *Euglena gracilis*. *Exp. Cell Res.* 51, 141–149
76. Goehler, H. et al. (2005) A protein interaction network links GIT1, an enhancer of huntingtin aggregation, to Huntington's disease (vol 15, pp. 853, 2004). *Mol. Cell* 19, 287–287
77. Morsy, M. et al. (2008) Charting plant interactomes: possibilities and challenges. *Trends Plant Sci.* 13, 183–191
78. Lalonde, S. et al. (2008) Molecular and cellular approaches for the detection of protein-protein interactions: latest techniques and current limitations. *Plant J.* 53, 610–635
79. Best, C. et al. (2007) Localization of protein complexes by pattern recognition. *Cell. Electron Microsc.* 79, 615–638
80. Adjobo-Hermans, M.J.W. et al. (2006) Plant G protein heterotrimers require dual lipidation motifs of G alpha and G gamma and do not dissociate upon activation. *J. Cell Sci.* 119, 5087–5097
81. Serdyuk, I.N. (2007) Structured proteins and proteins with intrinsic disorder. *Mol. Biol.* 41, 262–277
82. Dyer, J.M. and Mullen, R.T. (2008) Engineering plant oils as high-value industrial feedstocks for biorefining: the need for underpinning cell biology research. *Physiol. Plant.* 132, 11–22
83. Giegé, P. et al. (2003) Enzymes of glycolysis are functionally associated with the mitochondrion in Arabidopsis cells. *Plant Cell* 15, 2140–2151
84. Brandina, I. et al. (2006) Enolase takes part in a macromolecular complex associated to mitochondria in yeast. *Biochim. Biophys. Acta* 1757, 1217–1228
85. Jeffery, C.J. (2005) Mass spectrometry and the search for moonlighting proteins. *Mass Spectrom. Rev.* 24, 772–782
86. Moore, B.D. (2004) Bifunctional and moonlighting enzymes: lighting the way to regulatory control. *Trends Plant Sci.* 9, 221–228
87. Srere, P.A. (1997) An exception that proves the rule. *Trends Biochem. Sci.* 22, 11–11
88. Anderson, L.E. et al. (2005) Both chloroplastic and cytosolic phosphofructoaldolase isozymes are present in the pea leaf nucleus. *Protoplasma* 225, 235–242
89. Anderson, L.E. et al. (2004) Both chloroplastic and cytosolic phosphoglycerate kinase isozymes are present in the pea leaf nucleus. *Protoplasma* 223, 103–110
90. Anderson, L.E. et al. (2004) Cytosolic glyceraldehyde-3-P dehydrogenase and the B subunit of the chloroplast enzyme are present in the pea leaf nucleus. *Protoplasma* 223, 33–43
91. Saslowsky, D. et al. (2005) Nuclear localization of flavonoid metabolism in *Arabidopsis thaliana*. *J. Biol. Chem.* 280, 23735–23740
92. Shimojima, M. et al. (2005) Ferredoxin-dependent glutamate synthase moonlights in plant sulfolipid biosynthesis by forming a complex with SQD1. *Arch. Biochem. Biophys.* 436, 206–214
93. Matarasso, N. et al. (2005) A novel plant cysteine protease has a dual function as a regulator of 1-aminocyclopropane-1-carboxylic acid synthase gene expression. *Plant Cell* 17, 1205–1216
94. Pollmann, S. et al. (2006) Subcellular localization and tissue specific expression of amidase 1 from *Arabidopsis thaliana*. *Planta* 224, 1241–1253
95. Lunn, J.E. (2007) Compartmentation in plant metabolism. *J. Exp. Bot.* 58, 35–47
96. Jeong, H. et al. (2001) Lethality and centrality in protein networks. *Nature* 411, 41–42
97. Vélot, C. et al. (1997) Model of a quinary structure between Krebs TCA cycle enzymes: a model for the metabolon. *Biochemistry (Mosc).* 36, 14271–14276

Glycosylation of Secondary Metabolites and Xenobiotics 10

Fabián E. Vaistij, Eng-Kiat Lim, Robert Edwards, and Dianna J. Bowles

Abstract Glycosyltransferases of small molecules are encoded by a large multigene family in the plant kingdom. These enzymes transfer sugars from nucleotide sugars to a wide range of acceptors, from hormones and secondary metabolites to biotic and abiotic chemicals, thereby alter the physical and chemical properties of the acceptors, their bioactivity and subcellular localisation. This chapter focuses on the current understanding in the biochemistry of these enzymes and their biological roles in the plant.

Multigene Families of Glycosyltransferases

Phytochemical studies throughout the twentieth century have demonstrated the number and complexity of small molecule glycosides in higher plants. Often, these molecules exist in numerous glycoforms, regioselectively glycosylated at different positions and carrying single or multiple sugars, linked as mono, di-, tri- or higher oligosaccharides. As an example, from a total of 5,000 different flavonoids, 300 different glycosides of a single flavonol, quercetin, have been identified [1].

In nature glycoconjugates are formed by glycosyltransferases (GTs), a class of enzymes catalysing the transfer of a glycosyl moiety from an activated donor to an acceptor molecule. Currently the Carbohydrate-Active Enzyme database (CAZy, http://www.cazy.org/) has collected over 12,000 sequences encoding GTs in different organisms. These sequences have been classified into 90 families based on the biochemical studies of their gene products or through sequence homology comparison of the genes encoding enzymes of known catalytic activity. Among the GTs contributing to the many different glycosyl transfer reactions, those glycosylating natural products and small lipophilic molecules belong to the GT1 family.

In this chapter, an overview of the GT1 enzymes with respect to their chemistry and biology will be provided together with the recent work reported from 2006 to 2007. For other earlier works and more details about this GT family, readers are referred to the reviews published in the last few years [2–4].

F.E. Vaistij, E.-K. Lim, and D.J. Bowles (✉)
Centre for Novel Agricultural Products,
Department of Biology, University of York,
York YO10 5DD, UK
e-mails: fv4@york.ac.uk;
ekl100@york.ac.uk;
djb32@york.ac.uk

R. Edwards
School of Biological & Biomedical Sciences,
Durham University, South Road, Durham,
DH1 3LE, UK
e-mail: robert.edwards@durham.ac.uk

A.E. Osbourn and V. Lanzotti (eds.), *Plant-derived Natural Products*,
DOI 10.1007/978-0-387-85498-4_10, © Springer Science+Business Media, LLC 2009

Gene Phylogeny of GT1 Members

Most plant GT1 members carry a signature motif of 44–46 residues in their amino acid sequences. This is reflected in the complete genome of *Arabidopsis*. 107 of the 120 *Arabidopsis* sequences classified into GT1 family contain this signature motif. These GTs are thought to be cytosolic proteins since they do not carry targeting peptides [5]. In contrast, three GTs without the motif are membrane-anchored proteins glycosylating sterols and lipids [6]. Comprehensive phylogenetic analysis of the 107 *Arabidopsis* GT1 sequences has revealed 14 well-defined evolutionary groups. As a result of unequal gene recombination and duplication events, many sequences from the same phylogenetic groups form clusters on individual chromosomes [5].

When a further 63 full-length *Medicago truncatula* GT sequences were included in the phylogenetic analysis, 59 of these sequences clustered within the evolutionary groups previously identified in *Arabidopsis*, with the other four forming distantly related groups [7]. Thus, it is likely that there are additional ancestral genes in higher plants not found in the *Arabidopsis* genome. The *M. truncatula* genome is anticipated to contain over 150 different GT sequences [7]. When all these sequences are completely annotated, their phylogenetic analysis will provide a wider view of the evolution of GT1 members in dicotyledonous plants.

GT1 is one of the biggest GT families in the CAZy collection, and is also the largest GT family in the entire *Arabidopsis* genome database. These high numbers of GTs glycosylating small molecular acceptors in plants reflect the diversity of plant secondary metabolites, such as those involved in protection against oxidative stress and those involved in interactions between plants and other organisms, as well as the enzymes that regulate their cellular homeostasis, that are derived under evolutionary pressures.

Three-Dimensional Structures of VvGT1, UGT72B1, UGT71G1 and UGT85H2

Prior to 2005, only three GT1 protein structures were available in the public domain [8-10]. All of these were bacterial GTs involved in antibiotic biosynthesis. The structures consisted of two Rossmann-folds, each constructed by a central sheet of several β-strands flanked on either side by α-helices. Results from the co-crystallisation of these proteins with their ligands indicated that residues in the N-terminal half of the protein were responsible for acceptor binding, whereas those in the C-terminal half were mainly involved in donor interactions. Due to the low sequence similarity between the bacterial and plant GTs, as well as the lack of a signature motif in the bacterial sequences, few studies used the published structures to model the respective plant enzymes [11, 12].

Between 2005 and 2007, four plant GT1 enzymes have been crystallised and their 3-D structures solved by two research groups. Two *Medicago truncatula* GTs, UGT71G1 and UGT85H2, were crystallised and their structures solved at 2.6 and 2.1 Å, respectively [13, 14]. The crystal structure of UGT71G1 was solved in the presence of the native sugar donor UDP-glucose (UDP-Glc); however, no acceptor molecules were co-crystallised with these *Medicago* GTs. In contrast, the *Vitis vinifera* VvGT1 and the *Arabidopsis* UGT72B1 were crystallized with the acceptor molecules kaempferol and trichlorophenol respectively and a non-hydrolysable donor, UDP-2-deoxy-2-fluoro-Glc; these structures were solved at 1.9 and 1.45 Å, respectively [15, 16]. Collectively, these results show that the plant proteins also contain two Rossmann-folds, with acceptors binding to residues in the N-terminal half, whereas activated donor sugars bind mainly to amino acids in the C-terminal region.

The structures of these GTs clearly illustrate the role of the signature motif and the residues that constitute the donor-sugar binding pocket.

Using the VvGT1 structure as an example, tryptophan 332 forms the hydrophobic platform that stacks with the uracil base of UDP-Glc, while glutamate 358 binds the two hydroxyls of ribose. Histidine 350 interacts with the O2 of the β-phosphate of UDP, and serine 355 interacts with the equivalent atom of the α-phosphate. Aspartate 374 and glutamine 375 play key roles in recognising the hydroxyl at the C2, C3 and C4 positions of the Glc moiety (Fig. 1). Equivalent residues can be found in the other three structures and are conserved in almost all of the plant GT1 sequences.

In contrast, the N-terminal domains of the GTs have greater variation in their sequences. This is conceivable since the N-domains are known to interact with the sugar acceptors, which range from natural products to xenobiotics. All four GT structures contain a hydrophobic core centre in the N-terminus which forms a pocket for interactions with lipophilic acceptors.

Within the N-domain, a conserved histidine residue, which resides on the loop before the first α-helix, is positioned to serve as a Brønsted base to deprotonate the OH group of *O*-acceptor molecules. The charge on the histidine from proton abstraction is balanced by an aspartate residue located some 100 residues away in terms of amino acid sequence [15]. Interestingly, the configuration of the $O_{acceptor}$–His–Asp residues is analogous to that seen in the Ser–His–Asp triad of serine hydrolases, and is found in the sequences of all GTs that catalyse *O*-glycosylation. However, this catalytic histidine is not found in the sequences of the two cytokinin *N*-glycosyltransferases in *Arabidopsis* [16], suggesting that a different reaction mechanism is involved in *N*-conjugation. Remarkably, in the crystallography study with the *Arabidopsis* UGT72B1, a GT which catalyses both *O*- and *N*-glycosylation *in vitro*, when the enzyme is presented with an aryl amide group for *N*-glucosylation, the catalytic histidine side-chain is rotated through 90° and does not interact directly with the Asp triad residue [16]. This non-canonical geometry indicates that *O*- and *N*-glycosylation proceed via distinct mechanisms.

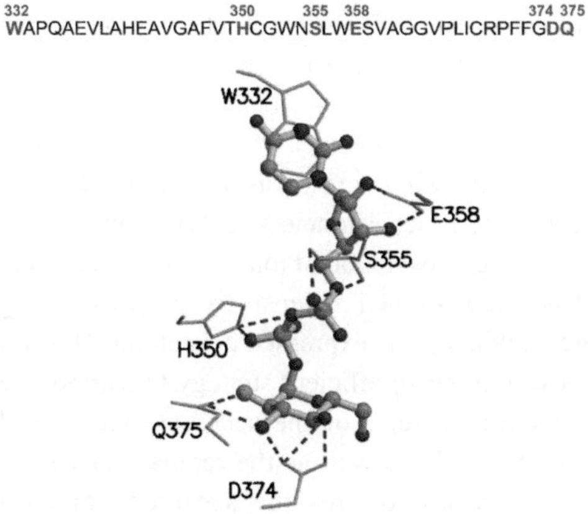

Fig. 1 Sequence of the signature motif and the 3-D structure of the sugar-binding pocket of VvGT1

Donors and Acceptors

Donors

The activated sugar donor of plant GT1 enzymes is typically UDP-Glc, although UDP-rhamnose (UDP-Rha), UDP-galactose (UDP-Gal), UDP-xylose (UDP-Xyl) and UDP-glucuronic acid (UDP-GlcUA) have also been identified as activated sugars in the transfer reaction [3]. These activated donors, except UDP-Rha, have their leaving group in the α-configuration. GT1 enzymes catalyse a simple S_N2 nucleophilic displacement reaction with these sugar donors and the nucleophile acceptor to form β-glycosides. Similarly, glycosylation involving UDP-β-Rha results in α-rhamnosylated products (Fig. 2).

Donor selectivity is an interesting feature of the GT1 enzymes. Although the four plant GT1 structures have determined how UDP-Glc is bound to the proteins, the features defining donor-selectivity are yet to be identified. It has been suggested that the last residue in the signature motif plays a crucial role in discriminating

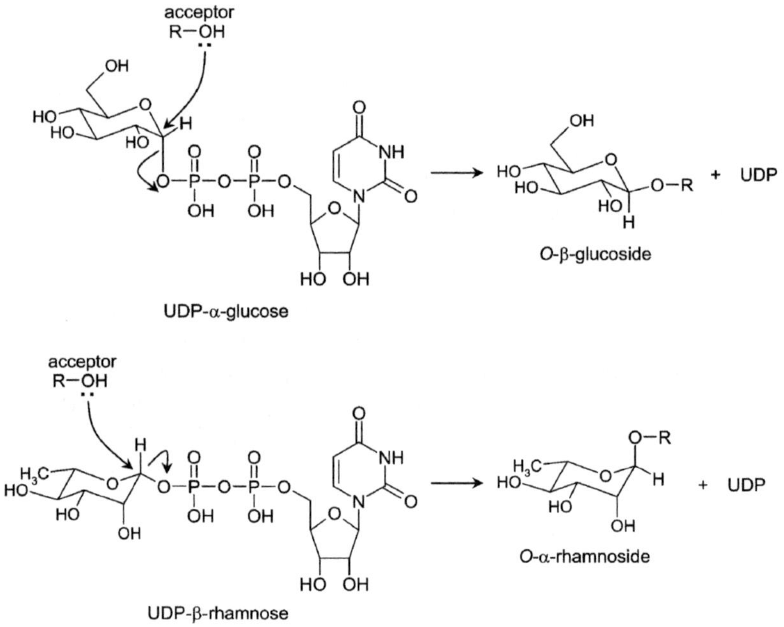

Fig. 2 The S_N2 reaction catalysed by GT1

Glc from Gal binding [17]. However, another study showed that the donor preference of a GT can be dependent on the acceptor molecule [7].

Acceptors

Natural products such as phytohormones, derivatives of shikimic acid (phenylpropanoids, benzoates, flavonoids) and mevalonate (terpenoids, steroids), alkaloids and many amino acid derivatives exist as glycosides in plants. Sugars can be transferred to the OH, SH, NH and carboxyl groups of these acceptors. To a lesser extent, certain acceptors can also be glycosylated at their CH group. In addition to natural products, man-made chemicals may also be glycosylated when these compounds are applied to plants.

Acceptors with more than one reactive group are often differentially glycosylated in different plants. Glycosides may also act as acceptors for further glycosylation. For example, quercetin, one of the most widely observed flavonoids in nature, exists predominantly as a 3-*O*-rhamnosyl (1→2) glucoside-7-*O*-rhamnoside in *Arabidopsis* and as 3,4'-di-*O*-glucoside in onion [18, 19] (Fig. 3a). The glycosylation profile of an acceptor can also be distinct between different species within a plant genus. A classical example is the rhamnosylation of naringenin-7-*O*-glucoside which occurs at the 2-OH of the glucose moiety in *Citrus paradisi* (grapefruit) leading to a bitter tasting glycoside, and at the 6-OH in other *Citrus* species including *Citrus reticulata* (mandarin orange) resulting in conjugate which is tasteless [20] (Fig. 3b).

The glycosylation of many of these acceptors has been studied extensively *in vitro* using recombinant GTs expressed in bacteria. This has proven to be an efficient strategy to explore the chemical features of the acceptors recognised by plant GTs as well as the regioselectivity of the enzymes towards the acceptors. *In vitro* biochemical studies have revealed that multiple GTs from the same plant can glycosylate the same acceptor, with individual GTs having the capacity to glycosylate multiple substrates.

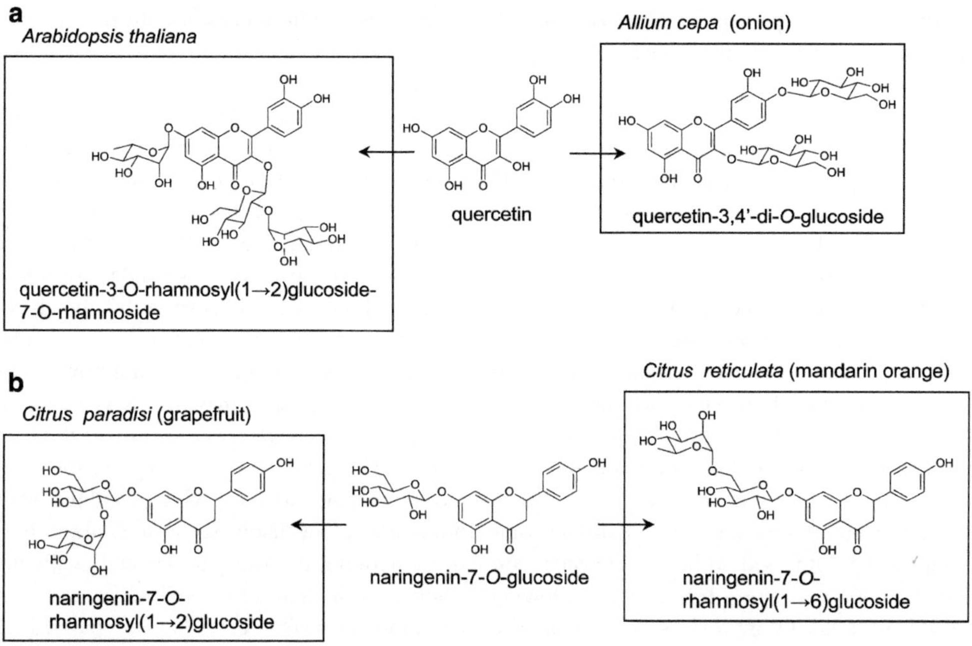

Fig. 3 (**a**) Glycosides of quercetin found in *Arabidopsis* and onion; (**b**) glycosides of naringenin found in two different *Citrus* species

A key question is the extent to which any of these *in vitro* biochemical studies are useful for in predicting the respective GT activity and physiological functions *in planta*. Thus, *in vivo* factors which can influence GT activity in plant cells, such as cofactors, protein-protein interactions and metabolite channelling are absent in the *in vitro* assays. It is thus essential to examine the functional activity of GTs by altering their expression *in planta*, when probing the roles of these enzymes in development, metabolic or stress/defense-related pathways.

The Biology of Glycosyltransferases

Plants have an immense capacity to perceive changes in their environment and can adapt rapidly to maximise opportunity and minimise risk. This plasticity in adaptive responses is dependent on the integration of growth, development and metabolism and the evolution of diverse mechanisms to regulate cellular homeostasis. There is increasing evidence to suggest that glycosylation is a key component of these mechanisms.

Addition of a sugar onto a lipophilic acceptor will make the compound more polar and prevent any further possibility of its free diffusion across lipid bilayers and between intra-cellular compartments. In *Arabidopsis,* with the exception of the three membrane associated GTs [21–23], it appears that the reminder of the family are cytosolic proteins [24]. This suggests the GTs function in the cytosol where they may locate within multi-protein complexes at the surface of membranes to enable efficient channelling of metabolites [25].

Whilst the GT transfer reactions are cytosolic, the conjugates serve as substrates for membrane-bound transporter systems that recognise the glycosyl residues. ATP-dependent binding cassette

(ABC) transporter proteins of glucosides of both endogenous metabolites and xenobiotics have been defined in tonoplast membranes [26, 27], with the corresponding accumulation of glycosylated compounds in the vacuolar compartment documented [28]. Glycosides and glycosidases of small lipophilic molecules have also been identified in the apoplast [29–31], although their exit route has not been defined. The 'clearance' of glycosylated acceptors from the cytosol underpins the view that GTs are involved in 'detoxification', since both the vacuolar and the apoplastic space can accumulate many different products which are only slowly turned over.

The above points highlight that glycosylation will alter the solubility of compounds and their movement within the cell. There is also evidence that glycosylation can stabilise compounds, as witnessed with the glucosinolates, which undergo rapid degradation following deglycosylation by myrosinase [32]. It is also known that the bioactivity of phytohormones and defence-related small molecules of plant and non-plant origin, is regulated by glycosylation/deglycosylation *in planta* [33].

Regulation of the Activities of Small Molecules

There are extensive studies on the GTs that recognise and conjugate endogenous or foreign acceptors thereby regulating their activity. Evidence exists that the key regulatory event can be either the addition of a sugar by a GT or its removal by a glycosidase, dependent on the nature of the acceptor. Here we will summarise current information on the roles of these reactions in regulating the biological activity of plant hormones, glucosinolates and cyanogenic glucosides as well as pathogen toxins.

Hormones

There are many compounds in a plant that influence developmental processes and enable rapid adaptation to environmental change. The homeostasis of hormones is critical and many mechanisms have evolved to control levels of different hormones within the cells and tissues of a plant. All of the classical hormones, with the exception of ethylene, have been found to occur as glycosides *in planta*. Many early biochemical papers described GT activities toward different hormones, as well as profiling the existence of hormone glycosides in plant extracts. There is good evidence that the glycosylation can be reversible or irreversible, dependent on the individual hormone. The glycosidic conjugates may also have different bioactivities from the 'free' forms of the hormones. The genetic tools to explore the relative importance of glycosylation and the other homeostatic mechanisms in regulating hormone action are starting to become available. Though most studies involve *in vitro* assays of recombinant GTs, some have expanded to investigate their potential roles *in planta*.

Auxins: In vegetative tissues of *Arabidopsis*, some 90% of indole-3-acetic-acid (IAA) (Scheme 1) is conjugated to biomolecules by amide linkages, with 10% ester linked and only 1% present as free IAA [34]. The first gene encoding a GT active toward IAA was the maize *IAGLU* [35]. Subsequently, *in vitro* screening of recombinant *Arabidopsis* GTs revealed that UGT84B1 synthesised the 1-*O*-indole acetyl glucose ester (IAAGlc), with its over-expression leading to a phenotype consistent with auxin deficiency [36, 37]. Thus, an altered root phenotype of the transgenic lines was not rescued by IAA, but was restored by treatment with 2,4-dichlorophenoxyacetic acid, an IAA functional analogue not glucosylated *in vitro* by UGT84B1. Metabolic profiling of over-expressor lines of UGT84B1 showed high levels of IAAGlc, but levels of free IAA were also increased, suggesting overproduction of an IAA-conjugating enzyme led to complex changes in auxin homeostasis [36].

Brassinosteroids: A wide range of glycosides of many different brassinosteroids (BRs) have been identified in plant extracts [38]. An

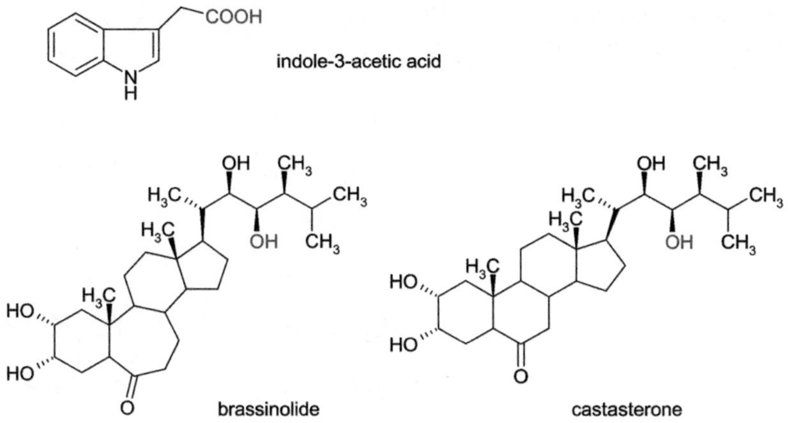

Scheme 1 Residues in red indicate position where the conjugation has been described

Arabidopsis GT capable of catalysing the 23-*O*-glucosylation of brassinolide (BL) (Scheme 1) and its biosynthetic precursor, castasterone (CS) has been identified [39]. Over-expression of this GT, UGT73C5, resulted in BR-deficient phenotypes, as well as reduced levels of active BRs in transgenic plants, thereby suggesting that glucosylation of BRs reduced their bioactivity. Significantly, silencing the GT led to the effective abolition BL-23-*O*-glucosylating activity in the seedlings used in BL feeding studies.

Cytokinins: Glycosylation of cytokinins can involve *O*-glucosylation, *O*-xylosylation and *N*-glucosylation [40]. Genes encoding GTs recognising cytokinins have been identified in several plant species [41–44]. When the expression of *ZOG1*, a *trans*-zeatin-*O*-glucosyltransferase from *Phaseolus lunatus*, was inducibly-regulated in transgenic tobacco leaf disc cultures, a tenfold higher zeatin level was required for the formation of shoots and callus compared to controls, suggesting that the conjugating activity resulted in an increased inactivation of the hormone [45]. The first two GTs with *N*-glucosyltransferase activity towards cytokinins, UGT76C1 and UGT76C2, have now also been identified in *Arabidopsis,* with an increased accumulation of *trans*-zeatin-7-*N*-glucoside detected when *trans*-zeatin (Scheme 2) was applied to transgenic plants constitutively over-expressing UGT76C1 [46].

Abscisic acid: Several glycosides of abscisic acid (ABA) (Scheme 2) have been identified in plants, with the glucose ester typically being the most abundant conjugate [47]. A recombinant GT from *Vigna angularis* converted 2-*trans*-(+)-ABA to its glucose ester *in vitro* [48]. Eight recombinant GTs from *Arabidopsis* where shown to recognise ABA *in vitro*, with only one, UGT71B6 showing enantioselective glucosylation with the naturally occurring substrate *cis*-S-(+)-ABA [49, 50]. Two plant studies have explored the role of glucosylation of ABA. Over-expression of UGT71B6 affected ABA metabolism, particularly under water-stress conditions [51]. Interestingly, deglycosylation of the hormone was found to have major impacts on the pool of free ABA [52], suggesting that stress-induced activation of glucosidase(s) plays a central role in ABA homeostasis.

Salicylic acid: Two glucosylated forms of salicylic acid (SA) (Scheme 2) have been identified in plant species, the glucose ester and the 2-*O*-glucoside [53]. Several publications have shown the elevation of SA during biotic or abiotic stress/pathogen responses and associated increased levels of the glucose ester and glucoside [54]. When SA was applied to plants, the

Scheme 2 Residues in red indicate position where the conjugation has been described

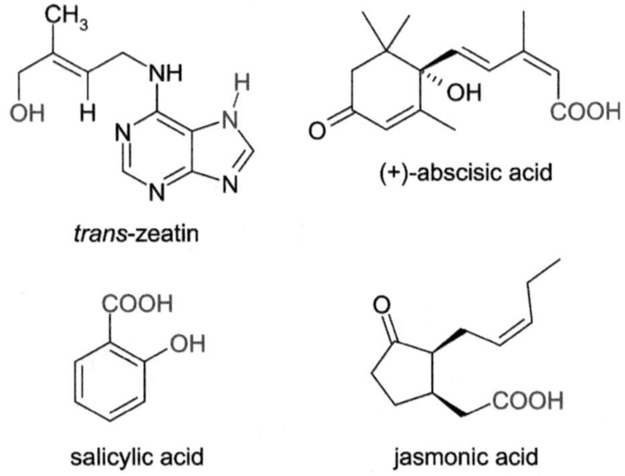

expression of a number of defence/stress-related genes was elevated, including those that encode GTs [32, 55–60]. However, when recombinant enzymes encoded by these genes were assayed *in vitro*, their substrate recognition was found to be broad [32, 57], and in one instance, SA was not an acceptor [58]. Screening *Arabidopsis* GTs with SA and other benzoates *in vitro* revealed that only UGT74F1 and UGT74F2 were active toward SA [61]. The expression of both genes was found to be up-regulated by SA application as well as by challenge with a bacterial pathogen [62].

Jasmonic acid: An *Arabidopsis* GT, AtJGT1, recognized jasmonic acid (JA) (Scheme 2) as a substrate *in vitro* whilst also recognizing a variety of other compounds, particularly indole-3-butyric acid [63]. However, *in planta* glucosylation of JA is not thought to be a major conjugation route of the hormone [64].

Glucosinolates, Cyanogenic Glucosides and Cyclic Hydroxamic Acids

The storage of glycosides in vacuoles, that on tissue damage come into contact with glycosidases leading to hydrolysis and production of bioactives compounds, is a well characterized chemical defence strategy in plants.

Glucosinolates (Scheme 3), found almost exclusively in the Brassicales, represent a diverse set of compounds defined by a common glucone comprising a sulfonated oxime and a β-thioglucose residue, with a variable side chain that is responsible for the high degree of glucosinolate chemical diversity [65]. The hydrolytic action of the plant thioglucosidase, myrosinase leads to unstable intermediates that spontaneously rearrange to form bioactive compounds including isothiocyanates, thiocyanates and nitriles. In contrast to the glucosinolates, cyanogenic glucosides (Scheme 3) are found much more widely in the plant kingdom [66]. On disruption of cell integrity, these glucosides also come into contact with glucosidases and are hydrolysed to release toxic hydrogen cyanide and an aldehyde or ketone [67]. Thus, whilst glucosylation regulates activity, in both instances the action of the glycosidases is the trigger to the production of bioactive molecules.

Genes encoding thiohydroximate *S*-glucosyltransferases have been cloned from *Brassica napus* and *Arabidopsis* [68, 69]. Knocking out the *Arabidopsis* UGT74B1 impacted on auxin metabolism, a feature also observed with other mutants impaired in glucosinolate synthesis [69]. Synthesis of the cyanogenic glucoside, dhurrin, has also been studied [70, 71] and its accumulation in transgenic *Arabidopsis* achieved

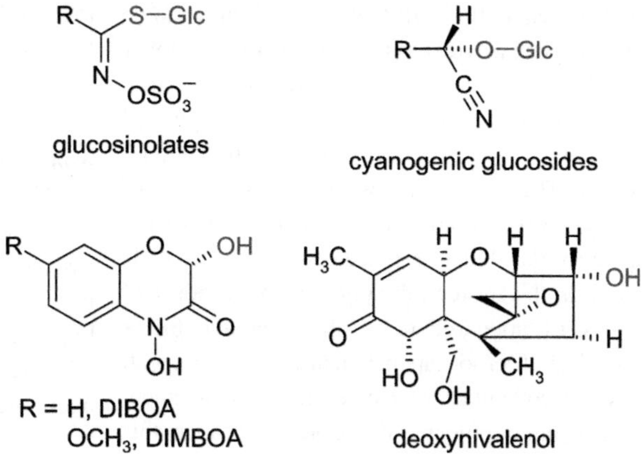

Scheme 3 Residues in red indicate position where the conjugation has been described

by transformation with different genes, including UGT85B1 and a cytochrome P450 [72, 73].

1,4-Benzoxazinones are cyclic hydroxamic acids (Scheme 3) occurring constitutively as glucosides in certain members of the Gramineae [74]. The glucosides are located in vacuoles and become available for hydrolysis by plastidic glycosidases when cell integrity is lost in response to tissue damage or pathogen invasion. The resulting aglycones, in rye, DIBOA (2, 4-dihydroxy-1, 4-benzoxazin-3-one), and in maize and wheat, DIMBOA (2, 4-dihydroxy-7-methoxy-1, 4-benzo-xazin-3-one) decompose rapidly with a liberation of formic acid. Thus, both the benzoxazinone aglygones and their degradation products exhibit multifunctional bioactivities, whereas their parent glucosides are inactive [75]. Two GTs have been purified from maize seedlings, with their over-expression in *Arabidopsis* reducing the toxic effects of DIBOA and DIMBOA application [76]. Since it is known that these compounds can also act as allelochemicals in the rhizosphere [77, 78], it was suggested that the GTs may detoxify exogenous benzoxazinones delivered in root exudates from neighbouring plants. There is some evidence for GTs down-regulating the impact of allelochemicals in field communities [79], and this is an interesting example where glycosylation events controlling bioactivity involve plant enzymes and substrates, originating from different plants.

Additional Plant and Pathogen Acceptors

The process of glycosylation and deglycosylation can regulate a range of activities, both of metabolites synthesized by the plants as well as those synthesized by pathogens. This section will briefly describe some examples of these activities where the GTs involved have been characterized.

Pisum sativum UGT1 (PsUGT1): Root caps, attached to root apices of many species play a number of roles, that include protection of the apical meristem, perception of gravity, and the production of 'border cells' that become detached and play a key role in influencing interactions of the plant within the rhizosphere [80]. Using pea as a model system, the removal of border cells was shown to trigger mitosis in the root cap meristem [81]. Since the system could be synchronised, genes whose expression was rapidly up-regulated at the onset of cell division in the root cap could be identified and one of these was found to be PsUGT1 [23]. The donor was initially identified as UDP-GlcUA, but subsequently UDP-Glc was also shown to be active, with *in vitro* assays suggesting a number of flavonoids could act as acceptors [23, 82]. When PsUGT1 expression was down-regulated in alfalfa, renewed border cell production did not occur. These data suggested that expression of the GT was required for the root cap meristem activity that led to replacement of border cells [23]. In this system, the glycosylated product of PsUGT1 appeared to be the bioactive factor necessary for root cap meristem cell division, with the aglycone inactive. However, the involvement and nature of potential flavonoid acceptors in this regulatory system remain to be elucidated.

***Arabidopsis thaliana* UGT73C5:** To date, only one plant GT capable of recognising a fungal toxin has been identified, its gene cloned and the ability of its recombinant gene product to detoxify the compound demonstrated [83]. The trichothecene deoxynivalenol (DON) (Scheme 3) is produced by fungi of the genus *Fusarium* that infect cereal species such as wheat, barley, rye and maize [84]. The mycotoxins produced are harmful when present in food or feed products and DON is also considered to act as a virulence factor in fungal pathogenesis [85]. Screening an *Arabidopsis* cDNA library expressed in yeast for an enhanced resistance to DON, identified UGT73C5. The recombinant GT catalysed the formation of DON-3-*O*-glucoside in yeast cells and when over-expressed in transgenic *Arabidopsis* conferred enhanced tolerance to DON [83]. Interestingly, as summarised in the hormone section, UGT73C5 also glucosylates BL, which suggests the GT may play a dual role in the plant, glucosylating endogenous and exogenous acceptors.

***Nicotiana tobaccum* TOGTs:** The expression of several GTs of tobacco can be up-regulated by SA, even though their *in vitro* acceptor recognition is broad and includes the hydroxycoumarins, such as scopoletin, as well as hydroxycinnamic acids [32, 55, 56, 58]. The consequences of changing the expression of *TOGT1* on the course of a pathogen response was analysed in transgenic tobacco [86-88]. The data were conflicting, but suggested a role for scopoletin and/or its glycosylated form in defence responses to viruses and other plant pathogens.

***Arabidopsis thaliana* UGT73B3 and UGT73B5:** T-DNA disrupted mutants of two *Arabidopsis* GTs, known to be up-regulated by SA, MeJA, hydrogen peroxide as well as by a fungal pathogen, displayed decreased resistance during the hypersensitive response. Whilst the acceptors of these GTs are unknown, a role in plant-pathogen interactions is likely [60].

Metabolic Homeostasis

As discussed in the introduction to this section, Family 1 GTs and their acceptors are generally located in the cytosol of the cell, whereas their products are found in the vacuolar compartment or apoplast. If the reactions are cytosolic, but the glycosylated products are transported out of the cytosol, it is possible that the removal of the product from the reaction mix will change the flux through the metabolic pathway in which the GT is involved. There is good evidence that most secondary metabolites can be found in glycosylated forms and therefore their glycosylation status may play a role in pathway regulation and homeostasis as well as in the formation of precursors of the final products.

Phenylpropanoids: The phenylpropanoid pathway leads from phenylalanine to a vast complexity of secondary metabolites with many different functions [89, 90]. In members of the Brassicaceae, 1-*O*-sinapoylglucose (SG) is an intermediate in the synthesis of sinapoylmalate (SM), a UV protectant found in foliar tissue, and sinapoylcholine (SC) (Scheme 4), made during seed development and degraded during germination to provide sinapic acid and choline for the developing seedling. SG is formed via the action of a GT on sinapic acid and genes encoding enzymes capable of this transfer reaction have been identified both in oil seed rape (*Brassica*

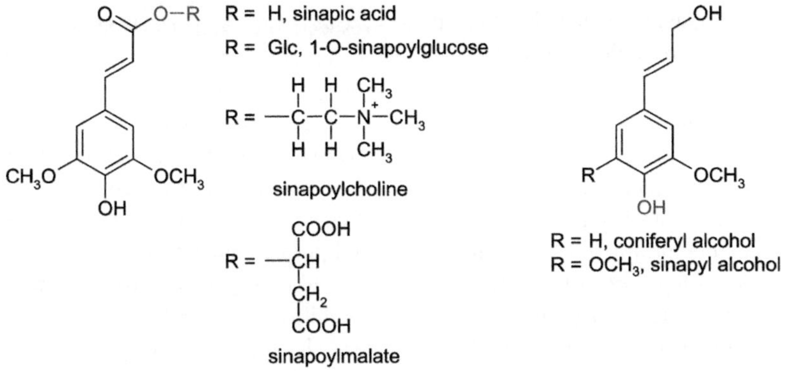

Scheme 4 Residues in red indicate position where the conjugation has been described

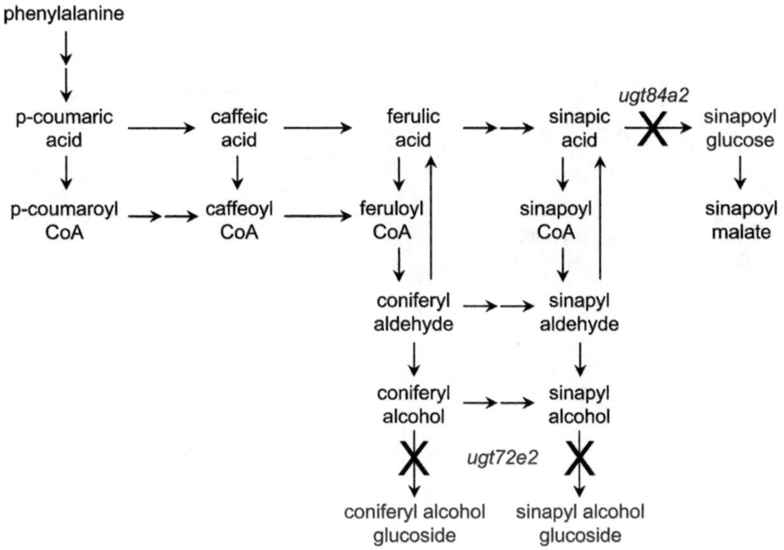

Fig. 4 The *Arabidopsis* GTs involved in the phenylpropanoid pathway. The UGT84A2- and UGT72E2 lack-of-function do not accumulate the highlighted glucosides

napus) and *Arabidopsis* [91–95]. Decreasing the expression of BnSGT in transgenic rape plants, led to a reduction in many different sinapate esters, including SG, SC and SM [96]. Similarly, in the *Arabidopsis brt1* mutant, deficient in UGT84A2, a reduction in the glucosylation of sinapic acid was observed [95] (Fig. 4).

The accumulation of glucosides of coniferyl alcohol and sinapyl alcohol (Scheme 4) was found to be light-dependent in roots of *Arabidopsis* [97]. UGT72E2 has been found to be responsible for the glucosylation of these monolignols in *Arabidopsis* roots [98] (Fig. 4). Overexpression of UGT72E2 led to a massive accumulation of the glucoside of coniferyl alcohol in roots. In leaves of those plants, the increase in glucoside was accompanied by a reduction in SM [98]. Since the overall phenylpropanoid pool was

10

unaffected, it was suggested that the overexpression of the GT led to a redirection of flux within the pathway.

Flavonoids – flavonols and anthocyanins: Flavonols are the widest spread metabolites in higher plants and predominantly accumulate as glycosides. *In vitro* biochemical analyses and *in planta* metabolite profiling have identified a number of interesting *Arabidopsis* GTs involved in flavonol glycoside biosynthesis, such as UGT78D1 and UGT89C1 which catalyze the 3-OH and 7-OH rhamnosylation of flavonols respectively, and UGT78D2 and UGT73C6 which glucosylate the 3-OH and 7-OH, respectively [19, 99, 100] (Fig. 5).

The pigmentation of flowers and fruits is due to the accumulation of flavonoids, carotenoids or betalains. In this context, anthocyanins (Scheme 5) predominantly determine red, blue and purple pigmentation. Glycosylation is considered to play a major role in defining both the solubility and stability of the pigments. The best-characterised GTs active in anthocyanin metabolism are those catalysing the primary

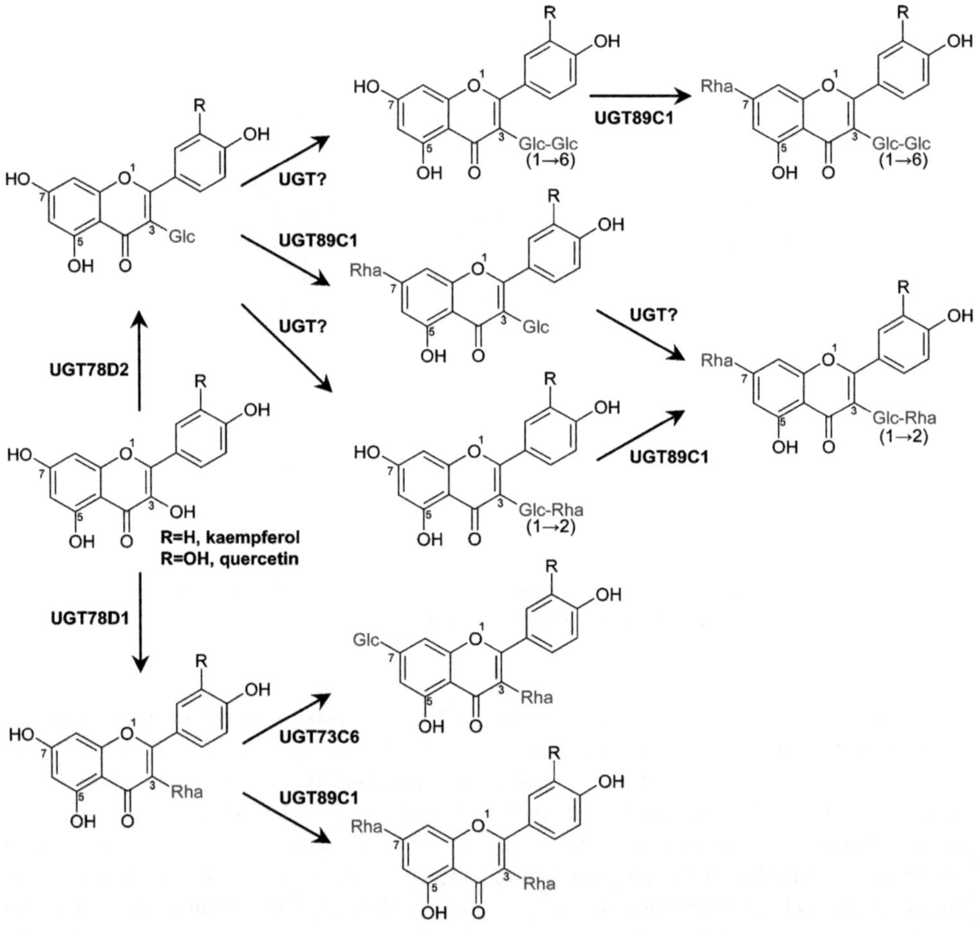

Fig. 5 Flavonol glycosylation in *Arabidopsis*

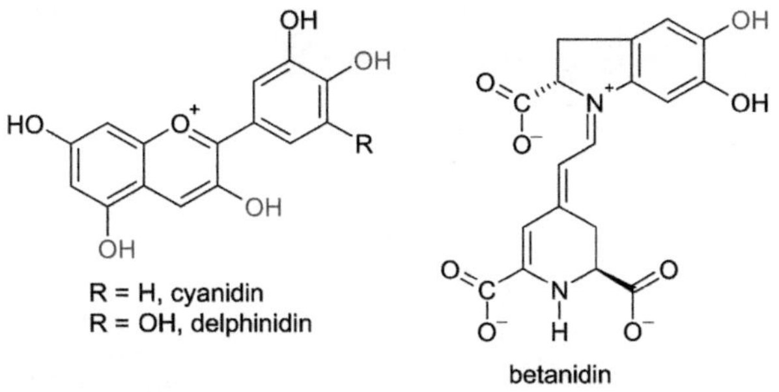

Scheme 5 Residues in red indicate position where the conjugation has been described

3-*O*-glucosylation event common to the biosynthetic pathway of all anthocyanins (for references see [38]). This step was thought to precede subsequent glycosylation events, but a single GT (RhGT1) was identified in *Rosa hybrida* that synthesised first cyanidin 5-*O*-glucoside, then the cyanidin 3, 5-di-*O*-glucoside [101]. Flowers of wild-type Japanese morning glory (*Ipomoea nil*) are red or blue due to the accumulation of anthocyanidin 3-*O*-sophorosides, whereas flowers of the *dusky* mutants are reddish-brown or purplish-gray as they only accumulate the 3-*O*-glucoside precursor [102]. This is due to the lack of a GT activity in the mutant background. Other identified GT activities involved in flower pigmentation include the enzymes responsible for the synthesis of gentiodelphin in *Gentiana triflora* [38] and cyanidin 3-*O*-(6"-*O*-malonyl-2"-*O*-glucuronyl)-glucoside in red daisy [103].

Betalains: In the Caryophyllales, the chromogenic anthocyanins are replaced as fruit and flower pigments by the betalains, comprising betacyanins (red and purple) and betaxanthins (yellow), with both compounds derived from the shikimic acid pathway. A number of studies have focussed on the synthesis of betacyanins and their glycosylation. Following the partial purification of two betanidin GTs (5-GT and 6-GT) from *Dorotheanthus bellidiformis* [104], the gene encoding the 5-GT was cloned and highest activity of the recombinant protein *in vitro* was shown towards betanidin (Scheme 5), with regio-specific transfer of Glc to the 5-OH position [105]. A GT from *Beta vulgaris* (red beet) was also identified and its activity towards betanidin investigated by assaying its expression during the accumulation of the respective pigmented glycoside in the leaves [106].

Terpenoids and steroids: Carotenoids are synthesized in the chloroplast, but modification of apocarotenoid cleavage products largely takes place outside of the plastidial compartment, prior to accumulation of water-soluble glycosylated metabolites in the vacuole [107]. For example, the final biosynthetic steps in the formation of the crocin yellow pigment of *Crocus sativus* includes glucosylation of the insoluble aglycone crocetin (Scheme 6) leading to its enhanced stability and water-solubility [22]. Another example of glycosylation affecting water solubility is seen in the metabolism of diterpenoid glycosides in *Stevia rebaudiana*, in which the extent and regioselectivity of glycosylation also influences the taste perception of sweetness. The most abundant of these compounds are the triglucoside, stevioside and the tetraglucoside, rebaudioside A. The initial steps of stevioside biosynthesis occur in plastids followed by further modifications in which glucosylation of the C4 carboxyl position of stevioside is thought to be critical for its transport into the vacuole. Three GTs (UGT74G1, UGT76G1 and UGT85C2) have been identified and their regioselective glucosylation of steviol (Scheme 6) confirmed [108].

10

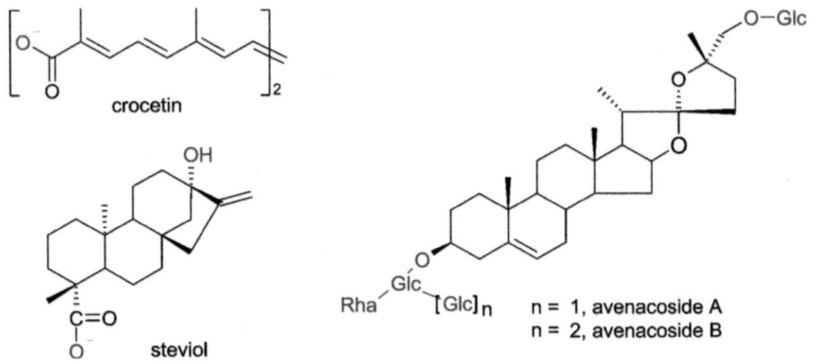

Scheme 6 Residues in red indicate position where the conjugation has been described

Saponins are glycosylated triterpenoids, steroids or steroidal alkaloids, found in many plant species [33, 109, 110]. These compounds are characterised by an oligosaccharide chain attached at the C3 position, consisting of up to five sugars, typically Glc, Gal, Ara, GlcUA, Xyl or Rha. Some saponins may also have a single Glc attached at C26 or C28. In oat leaves, the steroidal avenacosides A and B (Scheme 6) are biologically inactive but are converted into antifungal compounds on tissue damage or pathogen invasion by a plant glucosidase specific for the C26 position. This is an example in which bioactivity is revealed by deglycosylation, with both the substrate and the hydrolase made by the plant. The sugar chain attached to C3 is thought to be critical for both the membrane permeabilizing and antifungal properties of saponins, since removal of these sugar residues results in a loss in bioactivity. Interestingly, the C3 oligosaccharide is the target for hydrolases produced by invading fungal pathogens leading to saponin detoxification [111].

Detoxification

Typically, whether in the plant or animal kingdoms, detoxification can occur in three phases [112, 113]. When necessary, phase I oxidative or hydrolytic reactions activate chemicals for detoxification through the introduction of functional groups. Phase II enzymes conjugate hydrophilic molecules, such as through the addition of a glycosyl, malonyl or glutathione residues to those functional groups. In plants, these modifications lead to the elimination of the chemicals from the cytosol via membrane-bound transporters into the vacuolar or apoplastic space through the action of phase three transporter proteins. There is some evidence to suggest that the carriers for the transport of glycosylated and glutathionylated xenobiotics across the tonoplast membrane may be different from one another, the former using a carrier system energetically coupled to the transmembrane H^+ gradient, the latter transported by ABC transporter(s) directly energised by ATP [26].

There are many studies describing the partial purification of GT activities towards exogenous chemical compounds from plants [114–117], with xenobiotics included in the screening of recombinant GT activities [58, 118, 119]. *In planta,* the pollutant, 3,4-dichloroaniline (DCA) was rapidly detoxified by root cultures of *Arabi-dopsis* with the *N*-glucosylated-DCA exported into the medium [120]. The native *N*-glucosyltransferase responsible was purified and identified by proteomics as UGT72B1. Following the cloning of its gene, the *in vitro* activity of the recombinant

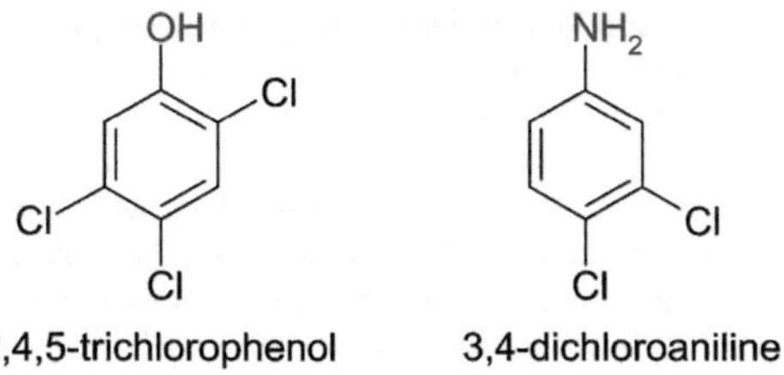

Scheme 7 Residues in red indicate position where the conjugation has been described

protein was determined towards DCA and other substrates. Interestingly the recombinant enzyme displayed both *N*-glucosylating activity towards DCA and *O*-glucosylating activity toward chlorinated phenols, notably 2,4,5-trichlorophenol (TCP) (Scheme 7). Cell extracts from the knock-out line of *UGT72B1* displayed reduced GT activity *in vitro* towards DCA and TCP [121]. However, feeding with radiolabelled TCP suggested that whilst UGT72B1 was the primary enzyme responsible for DCA conjugation, other GTs in *Arabidopsis* could compensate for TCP conjugation in the knock-out. Subsequent studies showed that many *Arabisopsis* UGTs were active toward this substrate [16]. Surprisingly, whilst glucosylation of DCA was severely compromised in the knock-out, the plants were more resistant to the compound, with the authors suggesting that the accumulation of non-extractable DCA residues in the cell walls of the knock-outs being a more effective detoxification route than glucosylation and vacuolar compartmentation of glycosides. Intriguingly, this is one example where a glucosylation reaction intended to detoxify a foreign compound may actually result in an unexpected potentiation of toxicity.

Conclusion

Phytochemical studies have provided detailed information on the existence, diversity, and complexity of small molecule glycosides. Studying the enzymes responsible for the synthesis of these glycosides highlights a common problem in post-genomic science – how to relate gene sequence and *in vitro* activity to cellular function. The multiplicity of GTs, both in numbers of enzymes and potential acceptors, is considerable, making *in vivo* studies complicated but essential to undertake. Through integrating biochemical, genetic and metabolic knowledge we are now beginning to understand the role of GTs in phytohormone and secondary metabolite homeostasis during biotic and abiotic challenges. In a wider context, the research of GTs has led to the development of these enzymes as biocatalysts to provide clean, efficient and regioselective synthesis of glycoconjugates of small molecules as a means to improve the chemical properties, such as solubility and stability, and to alter the biological activities of small molecule drugs. The coming years will continue to provide increasing insights into the role and utility of this fascinating family of enzymes.

References

1. Harborne, J.B. and Baxter, H. (1999) *The Handbook of Natural Flavonoids,* Vol. 1, pp. 297–549, Wiley, Chichester
2. Bowles, D. et al. (2005) Glycosyltransferases: managers of small molecules. *Curr. Opin. Plant Biol.* 8, 254–263
3. Bowles, D. et al. (2006) Glycosyltransferases of lipophilic small molecules. *Annu. Rev. Plant Biol.* 57, 567–597
4. Lim, E.K. and Bowles, D.J. (2004) A class of plant glycosyltransferases involved in cellular homeostasis. *EMBO J.* 23, 2915–2922
5. Ross, J. et al. (2001) Higher plant glycosyltransferases. Genome Biol. 2, 3004.3001–3004.3006
6. Paquette, S. et al. (2003) On the origin of family 1 plant glycosyltransferases. *Phytochemistry* 62, 399–413
7. Modolo, L.V. et al. (2007) A functional genomics approach to (iso)flavonoid glycosylation in the model legume Medicago truncatula. *Plant Mol. Biol.* 64, 499–518

8. Mulichak, A.M. et al. (2003) Structure of the TDP-*epi*-vancosaminyltransferase GtfA from the chloroeremomycin biosynthetic pathway. *Proc. Natl. Acad. Sci. USA* 100, 9238–9243
9. Mulichak, A.M. et al. (2001) Structure of the UDP-glucosyltransferase GtfB that modifies the heptapeptide aglycone in the biosynthesis of vancomycin group antibiotics. *Structure* 9, 547–557
10. Mulichak, A.M. et al. (2004) Crystal structure of vancosaminyltransferase GtfD from the vancomycin biosynthetic pathway: interactions with acceptor and nucleotide ligands. *Biochemistry* 43, 5170–180
11. Thorsoe, K.S. et al. (2005) Determination of catalytic key amino acids and UDP sugar donor specificity of the cyanohydrin glycosyltransferase UGT85B1 from Sorghum bicolor. Molecular modeling substantiated by site-specific mutagenesis and biochemical analyses. *Plant Physiol.* 139, 664–673
12. Hans, J. et al. (2004) Site-directed mutagenesis and protein 3D-homology modelling suggest a catalytic mechanism for UDP-glucose-dependent betanidin 5-O-glucosyltransferase from *Dorotheanthus bellidiformis. Plant J.* 39, 319–333
13. Shao, H. et al. (2005) Crystal Structures of a multifunctional triterpene/flavonoid glycosyltransferase from *Medicago truncatula. Plant Cell* 17, 3141–3154
14. Li, L. et al. (2007) Crystal structure of *Medicago truncatula* UGT85H2–insights into the structural basis of a multifunctional (iso)flavonoid glycosyltransferase. *J. Mol. Biol.* 370, 951–963
15. Offen, W. et al. (2006) Structure of a flavonoid glucosyltransferase reveals the basis for plant natural product modification. *EMBO J.* 25, 1396–1405
16. Brazier-Hicks, M. et al. (2007) Characterisation and engineering of the bifunctional N- and O-glucosyltransferase involved in xenobiotic metabolism in plants. *Proc. Natl. Acad. Sci. USA* 104, 20238–20243
17. Kubo, A. et al. (2004) Alteration of sugar donor specificities of plant glycosyltransferases by a single point mutation. *Arch. Biochem. Biophys.* 429, 198–203
18. Price, K.R. and Rhodes, M.J.C. (1997) Analysis of the major flavonol glycosides present in four varieties of onion (*Allium cepa*) and changes in composition resulting from autolysis. *J. Sci. Food Agric.* 74, 331–339
19. Yonekura-Sakakibara, K. et al. (2007) Identification of a flavonol 7-*O*-rhamnosyltransferase gene determining flavonoid pattern in Arabidopsis by transcriptome coexpression analysis and reverse genetics. *J. Biol. Chem.* 282, 14932–14941
20. Frydman, A. et al. (2004) Citrus fruit bitter flavors: isolation and functional characterization of the gene Cm1, 2RhaT encoding a 1, 2 rhamnosyltransferase, a key enzyme in the biosynthesis of the bitter flavonoids of citrus. *Plant J.* 40, 88–100
21. Miller, K.D. et al. (1999) Purification, cloning, and heterologous expression of a catalytically efficient flavonol 3-*O*-galactosyltransferase expressed in the male gametophyte of Petunia hybrida. *J. Biol. Chem.* 274, 34011–34019
22. Moraga, A.R. et al. (2004) Glucosylation of the saffron apocarotenoid crocetin by a glucosyltransferase isolated from *Crocus sativus* stigmas. *Planta* 219, 955–966
23. Woo, H.H. et al. (1999) Meristem-localized inducible expression of a UDP-glycosyltransferase gene is essential for growth and development in pea and alfalfa. *Plant Cell* 11, 2303–2315
24. Li, Y. et al. (2001) Phylogenetic analysis of the UDP-glycosyltransferase multigene family of *Arabidopsis thaliana. J. Biol. Chem.* 276, 4338–4343
25. Burbulis, I.E. and Winkel-Shirley, B. (1999) Interactions among enzymes of the *Arabidopsis* flavonoid biosynthetic pathway. *Proc. Natl. Acad. Sci. USA* 96, 12929–12934
26. Bartholomew, D.M. et al. (2002) Alternate energy-dependent pathways for the vacuolar uptake of glucose and glutathione conjugates. *Plant Physiol.* 130, 1562–1572
27. Rea, P.A. (1999) MRP subfamily ABC transporters from plants and yeast. *J. Exp. Bot.* 50, 895–913
28. Martinoia, E. et al. (2001) Vacuolar transport of secondary metabolites and xenobiotics. In *Vacuolar Compartments* (Robinson, D. and Rogers, J., eds), pp. 221–253, Sheffield Academic Press, Sheffield
29. Dietz, K.J. et al. (2000) Extracellular beta-glucosidase activity in barley involved in the hydrolysis of ABA glucose conjugate in leaves. *J. Exp. Bot.* 51, 937–944
30. Samuels, A.L. et al. (2002) Cellular machinery of wood production: differentiation of secondary xylem in Pinus contorta var. latifolia. *Planta* 216, 72–82
31. Bouarab, K. et al. (2002) A saponin-detoxifying enzyme mediates suppression of plant defences. *Nature* 418, 889–892
32. Fraissinet-Tachet, L. et al. (1998) Two tobacco genes induced by infection, elicitor and salicylic

acid encode glucosyltransferases acting on phenylpropanoids and benzoic acid derivatives, including salicylic acid. *FEBS Lett.* 437, 319–323
33. Morrissey, J.P. and Osbourn, A.E. (1999) Fungal resistance to plant antibiotics as a mechanism of pathogenesis. *Microbiol. Mol. Biol. Rev.* 63, 708–724
34. Woodward, A.W. and Bartel, B. (2005) Auxin: Regulation, action, and interaction. *Ann. Bot.* 95, 707–735
35. Szerszen, J.B. et al. (1994) *IAGLU*, a gene from *Zea mays* involved in conjugation of growth hormone indole-3-acetic acid. *Science* 265, 1699–1701
36. Jackson, R.G. et al. (2002) Over-expression of an *Arabidopsis* gene encoding a glucosyltransferase of indole-3-acetic acid: phenotypic characterisation of transgenic lines. *Plant J.* 32, 573–583
37. Jackson, R.G. et al. (2001) Identification and biochemical characterization of an *Arabidopsis* indole-3-acetic acid glucosyltransferase. *J. Biol. Chem.* 276, 4350–4356
38. Fukuchi-Mizutani, M. et al. (2003) Biochemical and molecular characterization of a novel UDP-glucose:anthocyanin 3'-*O*-glucosyltransferase, a key enzyme for blue anthocyanin biosynthesis, from gentian. *Plant Physiol.* 132, 1652–1663
39. Poppenberger, B. et al. (2005) The UGT73C5 of *Arabidopsis thaliana* glucosylates brassinosteroids. *Proc. Natl. Acad. Sci. USA* 102, 15253–15258
40. Mok, D.W.S. and Mok, M.C. (2001) Cytokinin metabolism and action. *Annu. Rev. Plant Physiol. Plant Mol. Biol.* 52, 89–118
41. Martin, R.C. et al. (1999) Isolation of a cytokinin gene, ZOG1, encoding zeatin *O*-glucosyltransferase from Phaseolus lunatus. *Proc. Natl. Acad. Sci. USA* 96, 284–289
42. Martin, R.C. et al. (1999) A gene encoding the cytokinin enzyme zeatin O-xylosyltransferase of Phaseolus vulgaris. *Plant Physiol.* 120, 553–558
43. Martin, R.C. et al. (2001) A maize cytokinin gene encoding an *O*-glucosyltransferase specific to cis-zeatin. *Proc. Natl. Acad. Sci. USA* 98, 5922–5926
44. Veach, Y.K. et al. (2003) *O-glucosylation* of *cis*-zeatin in maize. Characterization of genes, enzymes, and endogenous cytokinins. *Plant Physiol.* 131, 1374–1380
45. Martin, R.C. et al. (2001) Development of transgenic tobacco harbouring a zeatin *O*-glucosyl-transferase gene from Phaseolus. In Vitro Cell. Dev. Biol. Plant. 354-360
46 Hou, B. et al. (2004) N-glucosylation of cytokinins by glycosyltransferases of *Arabidopsis thaliana. J. Biol. Chem.* 279, 47822–47832
47. Schwartz, S.H. and Zeevaart, J.A.D. (2004) Abscisic acid biosynthesis and metabolism. In *Plant Hormones, Biosynthesis, Signal Transduction, Action!* (Davies, P.J., ed), pp. 137–155, Kluwer, Dordrecht
48. Xu, Z.J. et al. (2002) Cloning and characterization of the abscisic acid-specific glucosyltransferase gene from adzuki bean seedlings. *Plant Physiol.* 129, 1285–1295
49. Lim, E.K. et al. (2005) Resolution of (+)-abscisic acid using an Arabidopsis glycosyltransferase. *Tetrahedron: Asymmetry* 16, 143–147
50. Priest, D.M. et al. (2005) The use of abscisic acid analogues to analyse the substrate selectivity of UGT71B6, a UDP-glycosyltransferase of *Arabidopsis thaliana. FEBS Lett.* 579, 4454–4458
51. Priest, D.M. et al. (2006) Use of the glucosyltransferase UGT71B6 to disturb abscisic acid homeostasis in *Arabidopsis thaliana. Plant J.* 46, 492–502
52. Lee, K.H. et al. (2006) Activation of glucosidase via stress-induced polymerization rapidly increases active pools of abscisic acid. *Cell* 126, 1109–1120
53. Verberne, M.C. et al. (1999) Salicylic acid biosynthesis. In *Biochemistry and Molecular Biology of Plant Hormones* (Libbenga, K. et al. , eds), pp. 295–314, Elsevier, London
54. HammondKosack, K.E. et al. (1996) Race-specific elicitors of *Cladosporium fulvum* induce changes in cell morphology and the synthesis of ethylene and salicylic acid in tomato plants carrying the corresponding Cf disease resistance gene. *Plant Physiol.* 110, 1381–1394
55. Horvath, D.M. and Chua, N.H. (1996) Identification of an immediate-early salicylic acid-inducible tobacco gene and characterization of induction by other compounds. *Plant Mol. Biol.* 31, 1061–1072
56. Horvath, D.M. et al. (1998) Four classes of salicylate-induced tobacco genes. *Mol. Plant Microbe Interact.* 11, 895–905
57. Lee, H.I. and Raskin, I. (1999) Purification, cloning, and expression of a pathogen inducible UDP-glucose: salicylic acid glucosyltransferase from tobacco. *J. Biol. Chem.* 274, 36637–36642
58. Taguchi, G. et al. (2001) Molecular cloning and heterologous expression of novel glucosyltransferases from tobacco cultured cells that have broad substrate specificity and are induced by

salicylic acid and auxin. *Eur. J.Biochem.* 268, 4086–4094
59. O'Donnell, P.J. et al. (1998) A novel tomato gene that rapidly responds to wound- and pathogen-related signals. *Plant J.* 14, 137–142
60. Langlois-Meurinne, M. et al. (2005) Pathogen-responsive expression of glycosyltransferase genes UGT73B3 and UGT73B5 is necessary for resistance to Pseudomonas syringae pv tomato in *Arabidopsis*. *Plant Physiol.* 139, 1890–1901
61. Lim, E.K. et al. (2002) The activity of *Arabidopsis* glycosyltransferases toward salicylic acid, 4-hydroxybenzoic acid, and other benzoates. *J. Biol. Chem.* 277, 586–592
62. Song, J.T. (2006) Induction of a salicylic acid glucosyltransferase, AtSGT1, is an early disease response in *Arabidopsis thaliana*. *Mol. Cells* 22, 233–238
63. Song, J.T. (2005) Biochemical characterization of an *Arabidopsis* glucosyltransferase with high activity toward jasmonic acid. *J. Plant Biol.* 48, 422–428
64. Miersch, O. et al. (2008) Hydroxylated jasmonates are commonly occurring metabolites of jasmonic acid and contribute to a partial switch-off in jasmonate signaling. *New Phytol.* 177, 114–127
65. Fahey, J.W. et al. (2001) The chemical diversity and distribution of glucosinolates and isothiocyanates among plants. *Phytochemistry* 56, 5–51
66. Zagrobelny, M. et al. (2004) Cyanogenic glucosides and plant-insect interactions. *Phytochemistry* 65, 293–306
67. Poulton, J.E. (1988) Localization and catabolism of cyanogenic glycosides. *CIBA Found. Symp.* 140, 67–91
68. Marillia, E.F. et al. (2001) Molecular cloning of a *Brassica napus* thiohydroximate S-glucosyltransferase gene and its expression in Escherichia coli. *Physiol. Plant.* 113, 176–184
69. Grubb, C.D. et al. (2004) Arabidopsis glucosyltransferase UGT74B1 functions in glucosinolate biosynthesis and auxin homeostasis. *Plant J.* 40, 893–908
70. Hansen, K.S. et al. (2003) The *in vitro* substrate regiospecificity of recombinant UGT85B1, the cyanohydrin glucosyltransferase from Sorghum bicolor. *Phytochemistry* 64, 143–151
71. Jones, P.R. et al. (1999) The UDP-glucose: p-hydroxymandelonitrile-*O*-glucosyltransferase that catalyzes the last step in synthesis of the cyanogenic glucoside dhurrin in *Sorghum bicolor*. *J. Biol. Chem.* 274, 35483–35491
72. Tattersall, D.B. et al. (2001) Resistance to an herbivore through engineered cyanogenic glucoside synthesis. *Science* 293, 1826–1828
73. Kristensen, C., et al. (2005) Metabolic engineering of dhurrin in transgenic Arabidopsis plants with marginal inadvertent effects on the metabolome and transcriptome. *Proc. Natl. Acad. Sci. USA* 102, 1779–1784
74. Niemeyer, H.M. (1988) Hydroxamic acids (4-hydroxy-1, 4-benzoxazin-3-ones), defense chemicals in the Gramineae. *Phytochemistry* 27, 3349–3358
75. Sicker, D. et al. (2000) Role of natural benzoxazinones in the survival strategy of plants. In *International Review of Cytology – A Survey of Cell Biology*, Vol 198 (Jeon, K.W., ed), pp. 319–346, Academic, New York
76. von Rad, U. et al. (2001) Two glucosyltransferases are involved in detoxification of benzoxazinoids in maize. *Plant J.* 28, 633–642
77. Barnes, J.P. and Putnam, A.R. (1987) Role of benzoxazinones in allelopathy by rye (*Secale cereale* L). *J. Chem. Ecol.* 13, 889–906
78. Perez, F.J. and Ormenonunez, J. (1993) Weed growth interference from temperate cereals: the effect of a hydroxamic-acids-exuding Rye (*Secale cereale* L.) cultivar. *Weed Res.* 33, 115–119
79. Schulz, M. and Wieland, I. (1999) Variation in metabolism of BOA among species in various field communities – biochemical evidence for co-evolutionary processes in plant communities. *Chemoecology* 9, 133–141
80. Hawes, M.C. et al. (1998) Function of root border cells in plant health: pioneers in the rhizosphere. *Annu. Rev. Phytopathol.* 36, 311–327
81. Brigham, L.A. et al. (1998) Meristem-specific suppression of mitosis and a global switch in gene expression in the root cap of pea by endogenous signals. *Plant Physiol.* 118, 1223–1231
82. Woo, H.H. et al. (2007) Modifying expression of closely related UDP-glycosyltransferases from pea and *Arabidopsis* results in altered root development and function. *Physiol. Plant.* 130, 250–260
83. Poppenberger, B. et al. (2003) Detoxification of the *Fusarium* mycotoxin deoxynivalenol by a UDP-glucosyltransferase from Arabidopsis thaliana. *J. Biol. Chem.* 278, 47905–47914
84. Gale, L.R. (2003) Population structure of *Fusarium* species causing head blight of grain crops. In *Fusarium Head Blight of wheat and barley* (Leonard, K.J., and Bushnell, W.R., eds), pp. 120–143, American Phytopathological Society, St Paul, MN
85. McCormick, S.M. (2003) The Role of DON in Pathogenicity. In *Fusarium Head Blight of wheat and barley* (Leonard, K.J. and Bushnell, W.R., eds), pp. 165–183, American Phytopathological Society, St Paul, MN

86. Chong, J. et al. (2002) Downregulation of a pathogen-responsive tobacco UDP-Glc: phenylpropanoid glucosyltransferase reduces scopoletin glucoside accumulation, enhances oxidative stress, and weakens virus resistance. *Plant Cell* 14, 1093–1107
87. Gachon, C. et al. (2004) Over-expression of a scopoletin glucosyltransferase in *Nicotiana tabacum* leads to precocious lesion formation during the hypersensitive response to tobacco mosaic virus but does not affect virus resistance. *Plant Mol. Biol.* 54, 137–146
88. Matros, A. and Mock, H.P. (2004) Ectopic expression of a UDP-glucose:phenylpropanoid glucosyltransferase leads to increased resistance of transgenic tobacco plants against infection with potato virus Y. *Plant Cell Physiol.* 45, 1185–1193
89. Dixon, R.A. et al. (2002) The phenylpropanoid pathway and plant defence – a genomics perspective. *Mol. Plant Pathol.* 3, 371–390
90. Dixon, R.A. et al. (2001) The biosynthesis of monolignols: a "metabolic grid", or independent pathways to guaiacyl and syringyl units. *Phytochemistry* 57, 1069–1084
91. Lim, E.K. et al. (2001) Identification of glucosyltransferase genes involved in sinapate metabolism and lignin synthesis in Arabidopsis. *J. Biol. Chem.* 276, 4344–4349
92. Milkowski, C. et al. (2004) Molecular regulation of sinapate ester metabolism in Brassica napus: expression of genes, properties of the encoded proteins and correlation of enzyme activities with metabolite accumulation. *Plant J.* 38, 80–92
93. Milkowski, C. et al. (2000) Cloning and heterologous expression of a rape cDNA encoding UDP-glucose: sinapate glucosyltransferase. *Planta* 211, 883–886
94. Milkowski, C. et al. (2000) Identification of four *Arabidopsis* genes encoding hydroxycinnamate glucosyltransferases. *FEBS Lett.* 486, 183–184
95. Sinlapadech, T. et al. (2007) The hyper-fluorescent trichome phenotype of the *brt1* mutant of Arabidopsis is the result of a defect in a sinapic acid: UDPG glucosyltransferase. *Plant J.* 49, 655–668
96. Baumert, A. et al. (2005) Formation of a complex pattern of sinapate esters in Brassica napus seeds, catalyzed by enzymes of a serine carboxypeptidase-like acyltransferase family. *Phytochemistry* 66, 1334–1345
97. Hemm, M. et al. (2004) Light induces phenylpropanoid metabolism in Arabidopsis roots. *Plant J.* 38, 765–778
98. Lanot, A. et al. (2006) The glucosyltransferase UGT72E2 is responsible for monolignol 4-O-glucoside production in Arabidopsis thaliana. *Plant J.* 48, 286–295
99. Jones, P. et al. (2003) UGT73C6 and UGT78D1, glycosyltransferases involved in flavonol glycoside biosynthesis in *Arabidopsis thaliana*. *J. Biol. Chem.* 278, 43910–43918
100. Tohge, T. et al. (2005) Functional genomics by integrated analysis of metabolome and transcriptome of Arabidopsis plants over-expressing an MYB transcription factor. *Plant J.* 42, 218–235
101. Ogata, J. et al. (2005) Anthocyanin biosynthesis in roses. *Nature* 435, 757–758
102. Morita, Y. et al. (2005) Japanese morning glory *dusky* mutants displaying reddish-brown or purplish-gray flowers are deficient in a novel glycosylation enzyme for anthocyanin biosynthesis, UDP-glucose: anthocyanidin 3-*O*-glucoside-2"-*O*-glucosyltransferase, due to 4-bp insertions in the gene. *Plant J.* 42, 353–363
103. Sawada, S. et al. (2005) UDP-glucuronic acid: anthocyanin glucuronosyltransferase from red daisy (Bellis perennis) flowers: enzymology and phylogenetics of a novel glucuronosyltransferase involved in flower pigment biosynthesis. *J. Biol. Chem.* 280, 899–906
104. Vogt, T. et al. (1997) Are the characteristics of betanidin glucosyltransferases from cell-suspension cultures of Dorotheanthus bellidiformis indicative of their phylogenetic relationship with flavonoid glucosyltransferases. *Planta* 203, 349–361
105. Vogt, T. et al. (1999) Cloning and expression of a cDNA encoding betanidin 5-*O*-glucosyltransferase, a betanidin- and flavonoid-specific enzyme with high homology to inducible glucosyltransferases from the Solanaceae. *Plant J.* 19, 509–519
106. Sepulveda-Jimenez, G. et al. (2005) A red beet (*Beta vulgaris*) UDP-glucosyltransferase gene induced by wounding, bacterial infiltration and oxidative stress. *J. Exp. Bot.* 56, 605–611
107. Bouvier, F. et al. (2005) Oxidative tailoring of carotenoids: a prospect towards novel functions in plants. *Trends Plant Sci.* 10, 187–194
108. Richman, A. et al. (2005) Functional genomics uncovers three glucosyltransferases involved in the synthesis of the major sweet glucosides of Stevia rebaudiana. *Plant J.* 41, 56–67
109. Haralampidis, K. et al. (2002) Biosynthesis of triterpenoid saponins in plants. *Adv. Biochem. Eng. Biotechnol.* 75, 31–49
110. Hostettmann, K.A. and Marston, A. (1995) *Saponins*. Cambridge University Press, Cambridge

10

111. Sandrock, R.W. and VanEtten, H.D. (1998) Fungal sensitivity to and enzymatic degradation of the phytoanticipin α-tomatine. *Phytopathology* 88, 137–143
112. Cole, D.J. and Edwards, R. (2000) Secondary metabolism of agrochemicals in plants. In *Agrochemicals and Plant Protection* (Roberts, T.R., ed), pp. 107–154, Wiley, Chichester
113. Coleman, J.O.D. et al. (1997) Detoxification of xenobiotics by plants: chemical modification and vacuolar compartmentation. *Trends Plant Sci.* 2, 144–151
114 Brazier, M. et al. (2002) O-Glucosyltransferase activities toward phenolic natural products and xenobiotics in wheat and herbicide-resistant and herbicide-susceptible black-grass (*Alopecurus myosuroides*). *Phytochemistry* 59, 149–156
115 Brazier, M. et al. (2003) Partial purification and characterisation of a 2,4,5-trichlorophenol detoxifying *O*-glucosyltransferase from wheat. *Phytochemistry* 64, 419–424
116. Lao, S.H. et al. (2003) 3, 4-Dichloroaniline is detoxified and exported via different pathways in Arabidopsis and soybean. *Phytochemistry* 63, 653–661
117. Pflugmacher, S. and Sandermann, H. (1998) Taxonomic distribution of plant glucosyltransferases acting on xenobiotics. *Phytochemistry* 49, 507–511
118. Hefner, T. et al. (2002) Arbutin synthase, a novel member of the NRD1beta glycosyltransferase family, is a unique multifunctional enzyme converting various natural products and xenobiotics. *Bioorg. Med. Chem.* 10, 1731–1741
119. Landmann, C. et al. (2007) FaGT2: a multifunctional enzyme from strawberry (*Fragaria x ananassa*) fruits involved in the metabolism of natural and xenobiotic compounds. *Planta* 226, 417–428
120. Loutre, C. et al. (2003) Isolation of a glucosyltransferase from *Arabidopsis thaliana* active in the metabolism of the persistent pollutant 3, 4-dichloroaniline. *Plant J.* 34, 485–493
121. Brazier-Hicks, M. and Edwards, R. (2005) Functional importance of the family 1 glucosyltransferase UGT72B1 in the metabolism of xenobiotics in *Arabidopsis thaliana*. *Plant J.* 42, 556–566

Handling Dangerous Molecules: Transport and Compartmentation of Plant Natural Products*

11

Markus Klein† and Werner Roos

Abstract The plant cell faces a dilemma: secondary products provide a multitude of defence and signalling functions, but their biosynthesis poses a severe burden, as it competes for energy sources and building blocks and may generate toxic products. Thus, evolution of recent secondary metabolites is not only driven by their advantageous functions but also selects for strict control mechanisms including the integration of biosynthesis into the cellular ultrastructure. In order to minimize the risk of self-intoxication, secondary products are usually targeted into compartments of low metabolic activity, notably the vacuole and the extracellular space. This is most obvious for phenolic substances but also for alkaloids, the best studied plant toxins. Compartmentation on a cellular or subcellular level is also instrumental in plants synthesizing preformed defence substances such as cyanogenic glycosides in order to assure that the active toxins are only liberated in case of an attack. Biosynthetic pathways and regulatory elements are well-established at least for some natural compound classes such as the flavonoids. In contrast, our knowledge of transport steps behind the subcellular distribution of these substances is just scratching the surface.

This chapter provides an overview on transport processes involved in secondary metabolite compartmentation that is concentrated at the best known areas of flavonoid and alkaloid production. Starting from 'classical' data of secondary metabolite transport we characterize the actually known transporters – which mainly belong to the ATP-Binding Cassette (ABC) or Multidrug and Toxic Extrusion (MATE) superfamilies – and their specific functioning in cells and tissues as analyzed by modern experimental techniques. The 'transporter' hypothesis is confronted with 'vesicle transport' models of subcellular trafficking. Although it appears premature to find common ground between these alternative models, the discovery of novel cellular functions of secondary metabolites facilitates our understanding of an intimate interplay between biosynthetic steps, transmembrane

---

M. Klein
Zurich Basel Plant Science Center,
University of Zurich, Plant Biology,
Zollikerstrasse 107, CH-8008 Zurich, Switzerland
e-mail: markus.klein@botinst.uzh.ch

W. Roos (✉)
Martin-Luther University Halle-Wittenberg,
Pharmaceutical Biology, Molecular Cell Biology,
Kurt-Mothes-Str. 3, D-06120, Halle,
Germany
e-mail: werner.roos@pharmazie.uni-halle.de

*Dedication: This chapter is dedicated to Prof. N. Amrhein, ETH Zurich, on the occasion of his retirement.

† Present address: Philip Morris Products S.A.,
Quai Jeanrenaud 56, CH-2000 Neuchâtel,
Switzerland
e-mail: markus.klein@pmintl.com

A.E. Osbourn and V. Lanzotti (eds.), *Plant-derived Natural Products*,
DOI 10.1007/978-0-387-85498-4_11, © Springer Science+Business Media, LLC 2009

fluxes and metabolic channels, i.e. the plant's solution to the 'toxic dilemma'.

Introduction

The diversity of plant secondary metabolites with more than 100,000 known compounds and the bewildering variability of their chemical structures reflects the forces of metabolic evolution: the origin of new enzymes and regulatory proteins by mutagenesis, the natural selection of acquired biosynthetic abilities for optimum adaptation to environmental challenges, and the genetic drift, i.e. the survival of individual chemical patterns in small populations until a positive selection pressure is acquired. Advantageous functions of secondary products are detectable mainly in the defence against biotic (pathogens, herbivores) and abiotic (UV light) stress and in the communication between organisms of an ecosystem (e.g. plants and pollinators). To realize such benefits requires to cope with the burden inherent in any secondary biosynthesis: a drain of substrates and energy sources, metabolic perturbations and the potential toxicity of metabolites.

Plants escape this dilemma by a complex organization of secondary biosynthesis, especially by integrating it into the structural organization of the producing cell and into developmental programs of tissues and organs. In particular, two such elements create the need for channeling, membrane transport and compartmentation of secondary metabolites:

1. Enzymes of a secondary biosynthesis are usually localized in different cell types and different subcellular compartments. For instance, three cell types are involved in the biosynthesis of morphinane alkaloids in *Papaver*, a similar diversity of distribution holds true for the enzymes of terpenoid indole alkaloids in *Catharanthus* (Facchini and St-Pierre, 2005). Other examples for secondary biosynthesis linked to cell specialization are trichome-specific tobacco diterpenoids (Guo et al., 1994; Wang and Wagner, 2003) or triterpenoid avenacins in the rhizodermis layer of oat root tips (Osbourn, 2003). In rye, the flavone glucuronides made in mesophyll cells are structurally distinct from the flavonoid glucosides synthesized in the adjacent epidermal layer (Schulz and Weissenböck, 1986) and only the epidermal glucosides respond to UV stress while the mesophyll compounds exhibit a characteristic developmental pattern (Burchard et al., 2000; Klein et al., 2000).
2. Water-soluble secondary metabolites must be separated from the living cytoplasm. This is accomplished by their accumulation in vacuoles, functionally similar compartments or by their export into the extracellular space. Highly efficient transport mechanisms allow plants to solve an apparent contradiction: the need to accumulate secondary products to high concentrations in order to accomplish beneficial ecological functions and the need of protection from self-made toxins. This concept, early highlighted by P. Matile (1984), has been substantiated throughout all classes of secondary compounds (Roos and Luckner, 1986), including phytoalexins (Wink, 1997) and phytoanticipins (Osbourn, 1996).
3. At least some groups of secondary metabolites have acquired functions beyond the "classical" strategies as protectants or interorganismic signals. Mainly flavonoids appear to be deeply integrated in essential mechanisms of cell specialization and in this way influence a variety of cellular transport processes, not only of secondary metabolites. One such function substantiated by flavonoid mutants in Arabidopsis is the regulation of polar auxin transport by aglycone flavonols in a process which may include cycling of auxin transport proteins between

endosomal and plasma-membrane bound pools (Brown et al., 2001; Buer and Muday, 2004; Peer et al., 2004; Santelia et al., 2008)

Because of the diversity of cell types that express secondary biosynthetic enzymes and the different strategies of cellular and subcellular compartmentation of secondary metabolites followed by different plants, a unifying concept of transport in secondary metabolism is unlikely to emerge in the near future. In spite of growing 'omics' datasets, only a few transporters of secondary compounds are characterized at a biochemical level and only a handful of their genes has been cloned (reviewed by Yazaki, 2005, 2006; Kitamura, 2006). In fact, the transport puzzle contains more pieces than just transporters and includes potential metabolite-binding proteins such as glutathione S-transferases (GSTs) (Mueller et al., 2000), factors involved in vesicle trafficking (Collins et al., 2003), inclusions (Markham et al., 2000) or multivesicular bodies and must consider membrane-bound biosynthetic metabolons (Bock et al., 2002; Winkel, 2004; Jorgensen et al., 2005; Facchini and St-Pierre, 2005; Fig. 1). In this chapter, compartmentation, transport and channeling in secondary metabolism will be discussed in the context of its integration into the cellular and subcellular architecture. While this aspect of a "cell biology of plant natural products" is still in its infancy, most molecular details are actually available in the fields of flavonoid and alkaloid biosynthesis. Therefore we will focus on these two classes of secondary compounds. For a historic perspective on secondary metabolite compartmentation and 'early' transport experiments, we refer to Wink (1997) and Martinoia et al. (2000). Secondary metabolite transport has been recently reviewed (Yazaki, 2005, 2006; Yazaki et al., 2007). With regard to the multitude of functions of ATP-binding cassette (ABC) transporters in plants and their apparent expansion in plant genomes, a recent contribution suggested a novel nomenclature for the plant ABC transporter families and provides the interested reader also with further, ABC transporter-related references (Verrier et al., 2008).

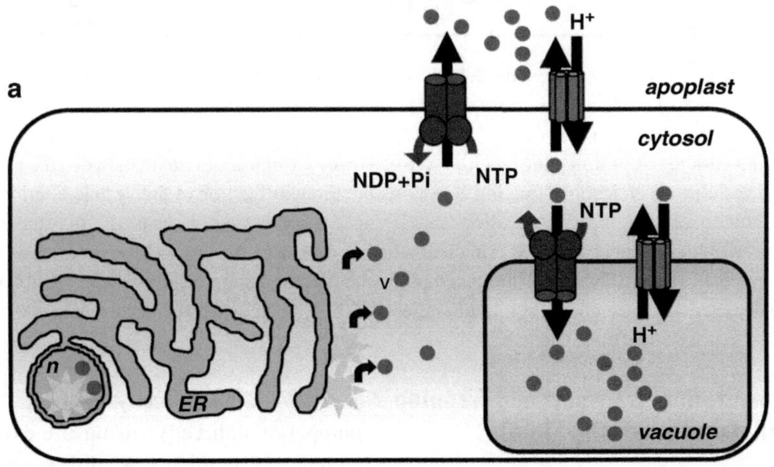

Fig. 1 Three models explaining aspects of subcellular transport of phenolic compounds. (**a**) Transporter model. Following biosynthesis at multienzyme complexes localized at the cytosolic face of ER membranes (yellow stars), flavonoids (red cicrcles) diffuse in the cytosol and are transported into the vacuole or into the apoplastic space by directly energized pumps (blue) or H^+-driven antiporters (violet)

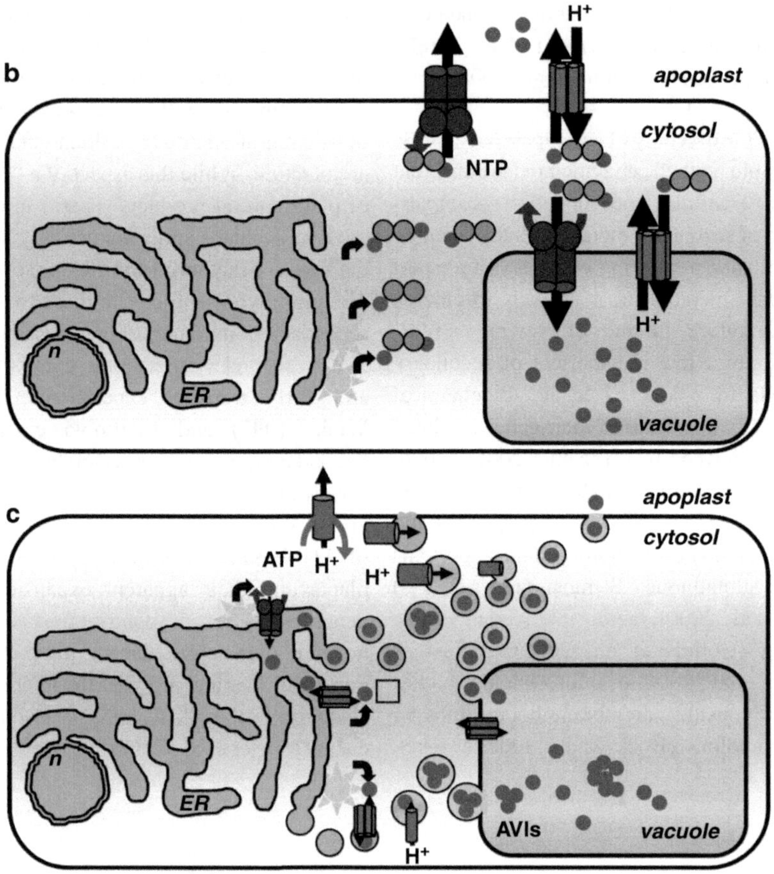

Fig. 1 (continued) (**b**) Binding protein model. Substrates such as anthocyanins are bound to proteins such as GSTs (green dimers) which deliver the substrates to the corresponding transporters. (**c**) Vesicle model. Transporters at the ER or in ER-derived vesicles transport the flavonoids into the ER lumen or into corresponding vesicles. A proton pump such as AHA10 (grey) which could travel between the plasmalemma and internal vesicles via endosomal cycling is responsible for acidification of the vesicle interior (color shift from light blue to yellow). Flavonoid-filled vesicles need to be sorted either to the plasmalemma or the tonoplast by an unknown mechanism. It is unclear whether the Golgi, the trans-Golgi network or prevacuolar compartments are involved in vesicle-mediated phytochemical transport. Anthocyanic vacuolar inclusions (AVIs) are depicted as droplets of several red circles. NT(D)P, nucleoside tri(di)phosphate

Compartmentation, Transport and Trapping of Phenolic Compounds (M. Klein)

Vacuolar Transport Coupled to Metabolic Reactions

Any accumulation of substances against a transmembrane concentration gradient needs to be energized either directly (via an ATP-driven 'pump') or indirectly (using pre-existing electrochemical gradients at this membrane, i.e. ΔpH or $\Delta\Psi$). When techniques to isolate vacuoles became available in the late 1970s/early 1980s, examples of vacuolar accumulation of secondary metabolites, energized by either way, were characterized. In some cases, however, neither ATP- nor ΔpH-dependence of the uptake step was observed, which indicated a passive or facilitated diffusion followed by the conversion of the transportant into a non-permeable, 'trapped' form thus preventing its efflux. Here, since any accumulation requires energy, this energy is provided by metabolic reactions that fuel the conversion inside the vacuole. Trapping mechanisms were first suggested as the well-known "*ion trap*" model (e.g. Matile, 1976). It implies that uncharged basic metabolites, preferentially alkaloids, diffuse across the tonoplast and get trapped by protonation in the acidic vacuolar lumen, as the membrane is impermeable to the protonated, positively charged acid (cf. Section Compartmentation and Transport in Alkaloid Biosynthesis). Notably, while H^+-antiport is driven by the ΔpH established by the vacuolar proton pumps and vacuolar accumulation thereby controlled by the electrochemical equilibrium, luminal protonation for trapping is limited by the absolute H^+ concentration in the vacuole which depends on the vacuolar buffer capacity.

The ion trapping model has been extended by Matern (1987) who suggested that vacuolar retention of some secondary metabolites can be explained by so-called *isomer traps*. Here, the tonoplast acts as a specific isomer filter being permeable for one but not the other isomer, that is formed by post-transport conformational changes in the acidic vacuolar lumen. Examples are *cis/trans*- or conformational isomerizations of *Melilotus alba* coumaryl-glucosides (Rataboul et al., 1985) and the parsley flavonoid apigenin 7-*O*-(6-*O*-malonylglucoside) (Matern et al., 1983, 1986), respectively, which according to this model are trapped in the vacuole. Clearly, while conformational changes and pH-dependent reactions are equilibrium-dependent and reversible, the *trans/cis*-isomerization of coumarylglucosides in the vacuole represents a chemical reaction that is practically irreversible in the absence of a vacuolar isomerase.

Lumenal 'biosynthetic' changes to vacuolar constituents have been unravelled for both primary (Frehner et al., 1984; Braun and Keller, 2000) and secondary metabolites, thus arguing against the historical view of the vacuole as a 'litter bin' for toxins and catabolites. In the field of secondary metabolism, sinapic acid ester metabolism in Brassicaceae is an important example for irreversible chemical conversions in the vacuolar lumen. The sinapic acid esters *O*-sinapoyl-L-malate and *O*-sinapoyl-choline, are formed from 1-*O*-sinapoyl-*β*-D-glucose in vegetative and regenerative tissues, respectively (Strack, 1982). Forward genetic screens allowed the isolation of the Arabidopsis *sinapoylglucose accumulator* mutants *sng1* and *sng2* which are defective in genes encoding serine carboxypeptidase-like proteins (SCPLs) that catalyze the transacylations of sinapoyl-glucose to sinapoyl-malate (SMT) and –choline (SCT), respectively, instead of acting as hydrolytic peptidases (Lehfeldt et al., 2000; Shirley et al., 2001; Shirley and Chapple, 2003). Antibodies raised against the Arabidopsis SMT localized this protein to the vacuolar space (Hause et al., 2002), confirming older biochemical fractionation data in radish seedlings (Sharma and Strack, 1985). Taken together, vacuolar formation of sinapoyl-malate seems likely although direct biochemical evidence in Arabidopsis is lacking. Even if vacuolar acylation may shift the chemical equilibrium towards a 'trapped' derivative, substances as sinapoyl-glucosides are too hydrophilic to be considered as freely diffusing, membrane-permeable compounds. Thus, a hitherto undiscovered membrane transporter has to be postulated for the transfer of the glucoside into the vacuole.

Another example arguing for vacuolar localization of distinct biosynthetic steps is the finding that the oxidative formation of the yellow aurones in snapdragon flowers from chalcones via aureusidin synthase, a polyphenol oxidase (PPO), occurs in the vacuole. This may represent a means to overcome the instability of the precursor chalcones in the cytosol and to concentrate aurones in the flower (Ono et al., 2006).

11

Any discussion of vacuolar metabolization of flavonoids would be incomplete without mentioning the copigmentation reactions that alter the color of flower anthocyanins (Brouillard and Dangles, 1994). Apart from color changes caused by the acidic vacuolar pH, complexations with metal ions and intra- or intermolecular reactions with hydroxy-cinnamic acids, flavones or flavonols strongly contribute to the spectral properties of the anthocyanins (for review, see Harborne and Williams, 2000). As a recent example, the crystal structure of the blue cornflower pigment was shown to be composed of a supermolecular complex consisting of four metal ions, six anthocyanin [cyanidin 3-*O*-(6-*O*-succinyl glucoside)-5-*O*-glucoside] and six flavone [apigenin 7-*O*-glucuronide-4'-*O*-(6-*O*-malonyl glucoside] molecules, resulting in a tetranuclear metal complex with a pseudo-three-fold axis (Shiono et al., 2005). If the final formation of such a 'superpigment' occurs within the vacuolar lumen, co-pigmentation and complex formation represent examples of a trapping mechanism according to Matern (1987). However, it is again likely that the components of such a complex use transporters to pass the tonoplast. Along this line, vacuolar uptake of the *Daucus carota* anthocyanin was time-dependent and protonophore-sensitive but not ATP-dependent (Hopp and Seitz, 1987). Although vacuolar uptake of this anthocyanin was independent of ATP, sensitivity towards the protonophore CCCP suggested an H^+-dependent uptake mechanism. Notably, absence or only low rates of ATP-stimulation of transport are not necessarily evidence for absence of energized transport (Box 1).

Box 1 Means to differentiate between ABC- and H^+-driven transporters

Pharmacological profiling allows to differentiate between direct, ATP-driven (ABC-type; blue transporters) or secondary energized transport (violet transporters).

Uptake into the vacuole: Depending on the pH and buffer composition, isolated vacuoles can still possess an electrochemical gradients across the tonoplast that is sufficient to energize secondary transport in the absence of ATP resulting in low levels of ATP-stimulation. In contrast, ABC transporters should exhibit zero activity in the absence of ATP. In the case of vesicle assays, both mechanisms, direct and secondary energization require ATP since the ΔpH is lost during vesicle preparation. The inside acidic ΔpH and cytosolic negative $\Delta\Psi$ used by secondary transporters is established by two vacuolar proton pumps, the V-type H^+-ATPase (orange) and the H^+-pyrophosphatase (brown). ABC transporters cannot hydrolyze PPi but accept several nucleoside triphosphates (NTPs) while the V-ATPase, which is strongly sensitive to the macrolide antibiotics bafilomycin A1 (baf; all inhibitors in red) and concanamycin A (conc; Drose and Altendorf, 1997), is rather specific for ATP. Therefore, ABC transporters are identified by energization by several NTPs and inhibition by vanadate acting as a phosphoryl transition state analog. In contrast, strong inhibition of ATP or PPi-energized transport by agents destroying either the ΔpH (e.g. NH_4Cl, protonophores such as carbonyl cyanide m-chlorophenylhydrazone [CCCP]) or the $\Delta\Psi$ (ionophores such as valinomycin [valin]) are a strong argument for secondary energization. An elegant way to demonstrate ΔH^+- or $\Delta\Psi$-dependence is to show substrate-dependent decrease of an imposed transmembrane gradient using fluorescence quenching of pH- or potential-sensing fluorescent dyes (e.g. Apse et al., 1999). For many secondary metabolites, this assay is complicated by overlapping spectral properties and direct fluorescence quenching by the substrates.

(continued)

Box 1 (continued)

Transport across the plasmalemma. Here, the situation is more simple and complex at the same time. On the one hand, only one type of H^+-pump, the P-type H^+-ATPase (grey), is responsible for the generation of the electrochemical gradient. On the other hand, since the P-type proton pump is also sensitive to vanadate, this inhibitor is not suitable to distinguish between an ABC-type pump or H^+-driven transport (symport for substrate uptake or antiport for substrate secretion). Thus, clear evidence comes only from inhibition by iono- and protonophores, by hyper-activation of the P-type H^+-ATPase by fuscicoccin (fus, green) or from experiments using inside-out plasmalemma vesicles with an artificially imposed pH gradient.

Only few compounds known to inhibit classical ABC transporters are specific for this type of pumps. Many (e.g. verapamil) also affect ion channels (Lee and Tsien, 1983), are not properly evaluated in plants and should therefore be used with caution. Recently, inhibition of long-distance transport of flavonoids by the sulfonylurea glibenclamide, which inhibits certain but not all MRP-type ABC transport processes on plant vacuoles (Forestier et al., 2003), was used an argument that MRP-type transporters are involved in this process (Buer et al., 2007). However, glibenclamide can also inhibit H^+-dependent transport processes (Martinoia and Klein, unpublished).

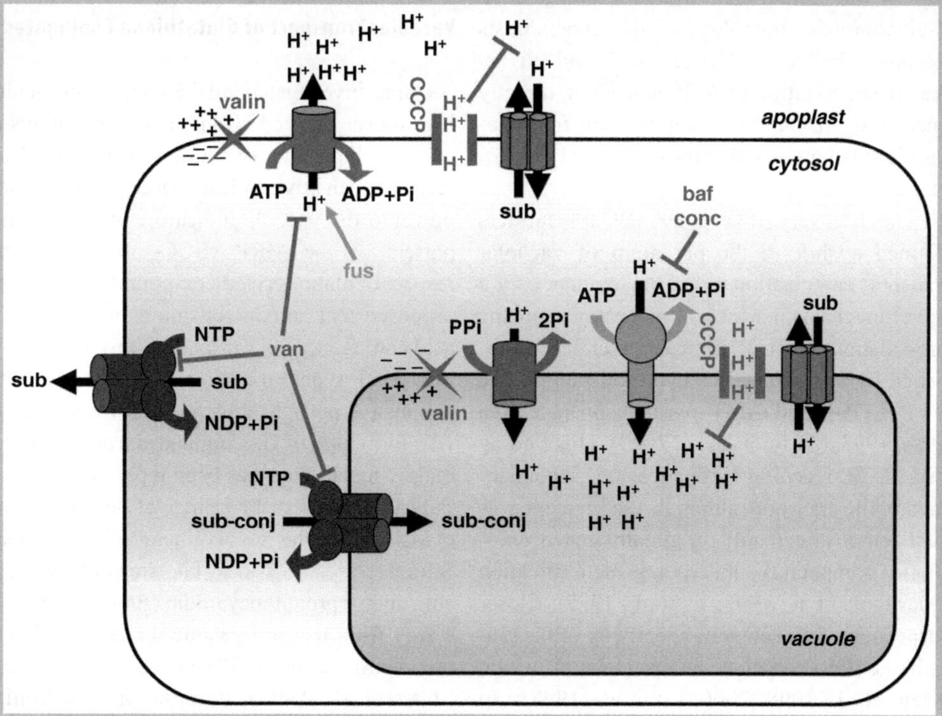

Early studies of vacuolar uptake repeatedly suggested a narrow species- and substrate specificity, including stereospecificity: Vacuoles isolated from plant species not producing the substrates tested or rather small changes in the chemical structure of the substrate (e.g. removal of attached sugars) resulted in loss of transport activity in most cases (Wink, 1997). It was therefore surprising that many of the transporters identified in the last 15 years represent so-called 'multidrug transporters' which display a low selectivity and accept structurally diverse substances. The existence of transporters with broad specificity was supported (i) by the discovery that multidrug transporters present in prokaryotic and eukaryotic cells, including the prominent family of mammalian P-glycoprotein are able to accept structurally diverse substances (Paulsen et al., 1996; Saier and Paulsen, 2001), (ii) by the identification and cloning of a P-glycoprotein homolog in the Arabidopsis genome (Dudler and Hertig, 1992) and (iii) by the characterization of ATP-dependent, directly energized, ABC-type vacuolar pumps for xenobiotic substances (Martinoia et al., 1993; Li et al., 1995).

The discovery of vacuolar ABC transporters defined a shift in the paradigm of vacuolar transport energization, providing evidence for a novel mechanism which is independent of the pre-existing electrochemical potential established by the two vacuolar proton pumps, the V-H^+-ATPase and the H^+-pyrophosphatase (Rea et al., 1998; Martinoia et al., 2000). This finding had a far-reaching influence on secondary metabolite transport although the concept was first demonstrated only for glutathionated xenobiotic compounds undergoing detoxification (Martinoia et al., 1993; Li et al., 1995). Closer inspection of the substrate specificity of the glutathione (GS)-conjugate pumps (Li et al., 1995; Klein et al., 1998, 2000; Lu et al., 1998; Liu et al., 2001) and the discovery of multidrug resistance associated-proteins forming the MRP (or ABCC) subfamily of ABC transporters, in mammals, plants and yeast (Cole et al., 1992; Leier et al., 1994; Li et al., 1996; Lu et al., 1997) suggested that on the one hand the substrate spectrum of MRP transporters is rather variable and mainly defined by the amphiphilic character of the substrates. On the other hand, it became evident that preferred MRP substrates are 'organic anions' that possess one or more negatively charged hydrophilic groups including GS, glucuronate or sulfate, apart from a lipophilic core. As a consequence, several studies addressed the question whether presence and nature of a hydrophilic tag attached to either natural or xenobiotic compounds were sufficient to destine them for a distinct transport mechanism. However, the type of conjugation may give some hint but not generally identify the transporter to be used by this substrate.

Vacuolar Transport of Glutathione Conjugates

As far as investigated, all GS conjugates including those derived from plant metabolism, undergo high-affinity, direct, ATP-dependent, vanadate-sensitive and $\Delta\mu_H^+$-independent transport into the vacuole indicative of ABC transporters of the MRP/ABCC-subfamily. With respect to plant-derived compounds, it has been proposed that certain flavonoids and cinnamic acid conjugates are transported into the vacuole by MRP-type transporters forming GS-conjugates prior to transport. However, the *in vivo* formation of GS conjugates from plant secondary metabolites has been reported only sporadically and is still a matter of debate. Strong evidence for the involvement of glutathione S-transferases (GSTs) in late steps of anthocyanin and/or proanthocyanidin (PA) biosynthesis comes from forward genetic screens that led to the identification of *GST* genes in maize (*Bz2*; Marrs et al., 1995), Petunia (*An9*; Alfenito et al., 1998), Arabidopsis (*TT19*; Kitamura et al., 2004) and carnation (*Fl3*; Larsen et al., 2003). Typically, mutants in those *GST* genes display a large reduction in anthocyanin pigmentation in flowers, maize kernels and other vegetative tissues or in the brown PA pigmentation of Arabidopsis seeds together with a weak, fuzzy coloration of the tissue. This phenotype was interpreted as a substantial block in vacuolar transport in the absence of GSTs resulting in reduced cytoplasmic accumulation of precursors (Marrs, 1996). In accordance with a role of GSTs in the formation of anthocyanin-GS conjugates, at least two of the fourteen Arabidopsis MRPs, *AtMRP1* and *AtMRP2*, exhibited in yeast transport of an *in vitro*-product of a GST-catalyzed reaction between the anthocyanin cyanidin 3-glucoside (C3G) and reduced glutathione (GSH) though with different capacities, while transport of unconjugated C3G was negligible (Lu et al., 1997, 1998). Although neither single mutants in all Arabidopsis *MRP* nor an *atmrp1 atmrp2* double mutant display a clear anthocyanin-related phenotype (Klein, Frelet, Kolukisaoglu, Martinoia, unpublished), the recent genetic characterization of the maize transporter *ZmMRP3* suggests that MRP-type ABC transporters may be involved in vacuolar anthocyanin transporter (Goodman et al., 2004). Similar to *bz2* mutants, *ZmMRP3* antisense lines accumulate lower levels of anthocyanins and display features of a pigment mislocalization phenotype. In accordance with a function in anthocyanin transport, ZmMRP3 was localized to the tonoplast (Goodman et al., 2004). Yet, biochemical evidence for the action of ZmMRP3 or the related ZmMRP4 as an anthocyanin transporter is missing.

In spite of this genetic evidence, implying an evolutionary conserved role of sequence-divergent GSTs in anthocyanin transport (Alfenito et al., 1998; Larsen et al., 2003), the biochemical function of anthocyanin-related GSTs remains obscure, most importantly because neither Bz2 nor An9 catalyze the formation of an anthocyanin-conjugate with GSH (Mueller et al., 2000). In contrast, *trans*-cinnamic acid and some derivatives had been shown to be capable of the formation of GS-conjugates when exposed to specific GSTs *in vitro* (e.g. Dean et al., 1995). The elicitation of chickpea cell cultures leads to massive *de novo* synthesis and secretion of isoflavonoids and pterocarpans. Under these conditions, externally added cinnamic acid was found to be metabolized to a GS-conjugate derivative (Barz and Mackenbrock, 1994) which may provide some support for a physiological significance of glutathionated phenolics. However, the attachment of GSH to unsaturated phenylpropanoids is not due to electrophilic Michael-addition occurring during a transferase reaction of GSTs. Instead, a GST-containing maize extract formed a conjugate via an intermediary thiyl radical indicating a peroxidase reaction mechanism (Dean and Devarenne, 1997). Strikingly and in accordance with general transport properties of GS conjugates via MRP-type transporters, the product of a horseradish peroxidase-mediated reaction between GSH and cinnamic acid is transported into *Beta vulgaris* tonoplast vesicles in a strict ATP-dependent, uncoupler-insensitive, vanadate-inhibitable manner (Walczak and Dean, 2000). In this context it is instructive to note that also the isoflavonoid medicarpin when reacted with GST-containing cellular extracts from maize forms a conjugate that is transported into mung bean tonoplast vesicles by a similar ATP-dependent mechanism and a four-fold higher velocity when compared to unconjugated medicarpin (Li et al., 1997). Taken together, the question arises whether under elicitation conditions where the flux of metabolites is redirected towards large-scale production of phytoalexins (Barz and Mackenbrock, 1994) together with drastic changes in the organization and channeling properties of the biosynthetic complexes (Rasmussen and Dixon, 1999) and a strong requirement for massive transmembrane movements of precursors and end-products, MRP-type transporters that support high accumulation rates (Kreuz et al., 1996), serve as modulating valves preventing overflow of cytosolic pools.

In view of the toxicity of phenolic compounds, the model that anthocyanins produced in large amounts in pigmented tissues travel from their site of biosynthesis at an ER-localized metabolon (Winkel, 2004; Jorgensen et al., 2005) towards the vacuole in a protected manner by binding to highly abundant proteins is highly attractive (Fig. 1). However, refined cell biological tools will be necessary to prove anthocyanin binding of GSTs *in vivo*. Only circumstantial evidence suggests that GSTs can act as flavonoid-binding proteins (Mueller et al., 2000; Fig. 1). Selected flavonoids were found to inhibit the GS-conjugating activity of An9 with the standard substrate 1-chloro 2,4-dinitrobenzene. An9, but also other GSTs not related to anthocyanin metabolism, could bind isoquercitrin and luteolin in equilibrium-dialysis experiments. Although the physiological relevance of such a process still requires more in-depth investigation, an interesting link to *in vitro* transport properties of MRP proteins exists because mammalian MRPs, such as human MRP1 (ABCC1) and MRP4 (ABCC4), have been reported to co-transport GSH together with conjugated or unconjugated substrates (Cole and Deeley, 2006; Deeley and Cole, 2006; Russel et al., 2008). Strikingly, several flavonoid aglycones inhibit or activate organic anion and GSH transport capacities, ATPase activities and drug-resistance properties of human MRP1 and MRP4 (Leslie et al., 2001, 2003; Wu et al., 2005) indicating that flavonoids are able to modulate transport efficiency of ABC transporters.

An intriguing hypothesis to test experimentally is whether GSTs bind flavonoids in addition to GSH and affect the MRP's transport properties through interaction while delivering one or both substances. One simple but unproven prediction of this model is that the GSTs need to interact physically with terminal enzymes of the metabolon to accept the metabolite load on the one side and with the transporter delivering the substrate on the other (Fig. 1).

Vacuolar Transport of Glucuronide Conjugates and Sulphonates

Glucuronidated compounds form a second prominent class of typical MRP/ABCC-type transporter substrates. Using vacuolar transport experiments, evidence for directly energized, ABC-type transport of β-estradiol 17-(β-D-glucuronide) ($E_2$17G), a mammalian metabolite formed during steroid degradation, was presented (Klein et al., 1998). Investigation of the transport properties of individual MRP-type transporters confirmed that apart from mammalian also individual plant MRPs catalyze $E_2$17G-transport (Liu et al., 2001; Klein et al., 2003). However, when compared to glylcosylated compounds, negatively charged glucuronates of flavonoids form only a minor group (Harborne, 1988). The best investigated example of vacuolar transport of glucuronated compounds are the flavone glucuronides luteolin 7-*O*-diglucuronyl-4'-*O*-glucuronide (R1) and luteolin 7-*O*-diglucuronide (R2) found in the mesophyll tissue of rye primary leaves (Schulz and Weissenböck, 1986). The metabolism of these flavone glucuronates starting from luteolin is sequentially catalyzed by three cytosolic UDP-glucuronate:flavone-glucuronosyltransferases (Anhalt and Weissenböck, 1992). Both, R1 and R2 are transported into rye vacuoles in a directly energized manner again suggesting that MRP-type transporters are responsible for this process (Klein et al., 2000). Interestingly, flavone glucuronide- and also $E_2$17G transport was not restricted to rye vacuoles but occurred with similar kinetic properties also when barley or broccoli vacuoles were tested which are not known to contain glucuronates (Klein et al., 2001). In addition, cross-competitive inhibition between R1, R2 and $E_2$17G suggests that the vacuolar glucuronate pump is also not specific for flavone glucuronides. Thus, the addition of a glucuronide tag to the molecule – like attachment of glutathione to form a GS conjugate – is apparently sufficient to direct a structurally diverse substrate to a MRP-type directly energized pump.

The interactions between typical plant MRP-transportants including glucuronides, GS conjugates and GSH at the tonoplast was found to be complex and could not be described by competitive inhibition (Hinder et al., 1996; Klein et al., 1998, 2000; Lu et al., 1998; Liu et al., 2001; see Section Vacuolar Transport of Glucuronide Conjugates and Sulphonates). A detailed kinetic study using heterologously expressed *AtMRP2* led to a multisite transport model with at least three transport pathways, cross-talk in between transport channels and additional allosteric binding sites (Liu et al., 2001). Interestingly, comparable complex substrate interactions have been described *in vitro* for several human MRPs (Cole and Deeley, 2006; Deeley and Cole, 2006; Russel et al., 2008) indicative of more general principles. Yet, it is complicated to speculate on physiological roles of such transport properties that also cannot yet be related to structural features of the transporter.

A third group of potential MRP substrates are sulfated, i.e. flavonoids which are common in Asteraceae. Although transport experiments with sulfated flavonoids have not yet been performed, indirect evidence suggests that a directly energized transporter is responsible for vacuolar uptake of sulfated compounds because directly energized uptake of the disulfonated fluorescent tracer Lucifer Yellow CH into rye mesophyll vacuoles was strongly inhibited by sulfated flavones (Klein et al., 1997).

Vacuolar Transport of Glucosylated Compounds: A Tag is not Enough!

In barley primary leaves, different types of flavone glucosides are synthesized and finally stored in the vacuole of mesophyll and epidermal cells. The major barley flavone, saponarin (apigenin 6-*C*-glucosyl-7-*O*-glucoside) is generated from isovitexin (apigenin 6-*C*-glucoside) by a terminal cytosolic glucosylation step (Blume et al., 1979). When the vacuolar uptake of isovitexin, the 'endogenous' glucoside, was compared to transport of hydroxyprimisulfuron-glucoside, a 'xenobiotic' glucoside derived from the herbicide primisulfuron, two mechanistically distinct uptake mechanisms were identified (Klein et al., 1996): a ΔpH-energized uptake of the flavonoid-glucoside could be distinguished from a directly energized, ABC-type mechanism for the abiotic glucoside. In addition, two independent mechanisms, a proton-antiport and an ABC-type mechanism, respectively, were found when saponarin uptake in barley vacuoles was compared to Arabidopsis vacuoles which are not known to contain simple flavone glucosides (Frangne et al., 2002). In spite of this, three studies provide evidence that the tonoplast does not act as a filter for 'self' and 'non-self' glucosides by using antiporters and ABC transporters, respectively. First, transport of glucosylated chlorsulfuron and of glucosides formed with the naturally occurring phenylpropanoids, *p*-hydroxycinnamic acid and *p*-hydroxybenzoic acid, into *Beta vulgaris* tonoplast vesicles both occurred via a H^+-antiport mechanism (Bartholomew et al., 2002). Second, salicylic acid 2-*O*-β-D-glucoside, a conjugate formed with the signal molecule salicylic acid, is transported into soybean tonoplast vesicles by an ABC-type mechanism while uptake into red beet vesicles was via H^+-antiport (Dean and Mills, 2004). Third, the coumaric acid glucoside esculin undergoes H^+-dependent uptake into barley vacuoles although this compound is not known to exist in barley (Werner and Matile, 1985).

It remains to be shown which ABC transporters facilitate directly energized glucoside transport. However, the recent identification of the first membrane protein related to the transport of phenolic compounds based on the analysis of the Arabidopsis *transparent testa (tt)12* mutant provides evidence that the phenolic glucoside/H^+-antiporters are likely to be members of the

11

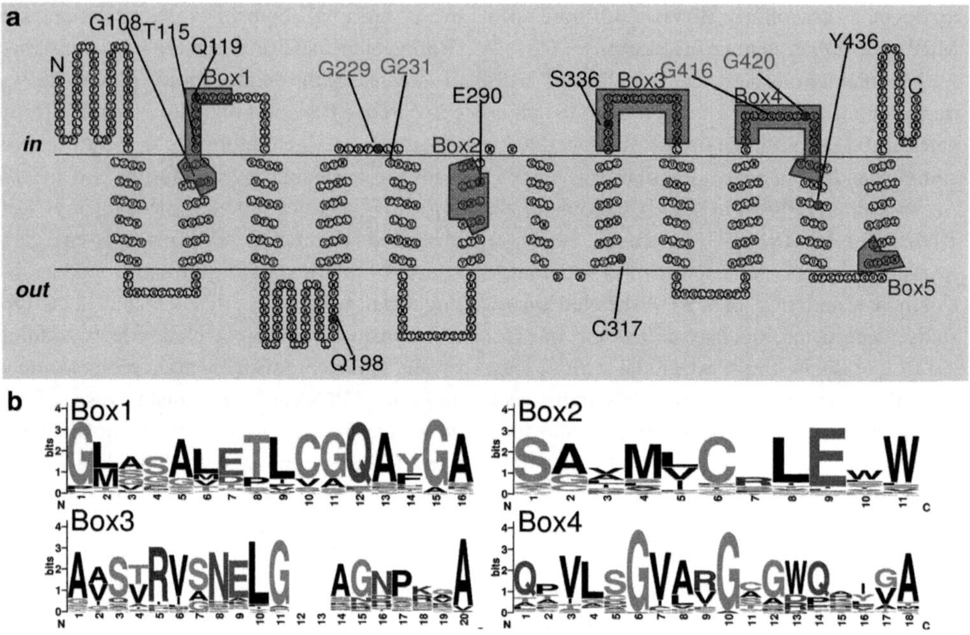

Fig. 2 Two-dimensional model of the Arabidopsis MATE transporter TT12. (**a**) The prediction of transmembrane helices was performed using ConPred2. Conserved sequence motifs identified by ClutsalW are highlighted by blue boxes and single amino acids conserved in eukaryotic and prokaryotic MATE proteins (black letters) or all Arabidopsis MATE proteins (red letters) are indicated. (**b**) Weblogo presentation of the conserved Boxes 1–4 using all annotated Arabidopsis MATE protein sequences

multidrug and toxic extrusion (MATE) transporter family (Debeaujon et al., 2001; Marinova et al., 2007; Fig. 2). The synthesis of PA polymers in Arabidopsis is restricted to the seed coat, mainly the innermost layer of the inner integument (endothelium) (Debeaujon et al., 2003; Lepiniec et al., 2006). PA precursors and polymers were strongly reduced or absent in *tt12* seeds suggesting that the vacuolar transfer of PA precursors in PA synthesizing cells is blocked (Debeaujon et al., 2001; Marinova et al., 2007). *TT12* expression occurs specifically in cells of the seed coat that accumulate PA polymers and the TT12 protein was localized to the tonoplast. Vesicles isolated from *TT12* expressing yeasts did not transport detectable quantities of the PA precursor flavan-3-ol epicatechin, PA dimers or glycosylated flavonols. ATP-dependent, uncoupler- and bafilomycin A-sensitive uptake was found when the anthocyanin C3G was tested as a substrate but not with the corresponding aglycone. Most importantly, TT12-dependent C3G/H^+-antiport could be inhibited in a dose-dependent manner by a glucosylated but not by an unconjugated flavan-3-ol. In consequence, TT12 is a vacuolar flavonoid/H^+-antiporter with a remarkable expression and substrate specificity which suggests that it is neither a general 'glucoside transporter' nor a strict 'anthocyanin transporter' although it needs a glycosylated 'anthocyanin-like' substrate.

Independent genetic evidence for H^+-dependent PA precursor transport was obtained by the characterization of the *aha10* Arabidopsis mutant (Baxter et al., 2005). AHA10 is a member of the Autoinhibitory H^+-ATPase family of P-type proton pumps crucial for the acidification of the apoplast and for establishing the electrochemiosmotic gradient across the plasmalemma (Sondergaard et al., 2004). *AHA10* is preferentially expressed in Arabidopsis seeds (Harper et al., 1994). *aha10* mutants display a metabolic phenotype comparable to *tt12* in the seeds with an absence of PAs (Baxter et al., 2005) which suggests that AHA10 may provide the proton gradient necessary for TT12-mediated flavan-3-ol glycoside/H^+ antiport (Fig. 1). However, the subcellular localization of AHA10 remains to be elucidated.

Seen together, the 'sweet' sugar tag of substrates is obviously not sufficient to specify transport mechanisms and transporters. This is in contrast to organic anion transport by MRP-type ABC transporters. At least two pharmacologically distinguishable transport mechanisms exist. With regard to anthocyanins and flavonols naturally occurring in Arabidopsis, it needs to be investigated whether paralogs of *TT12* expressed in vegetative organs are anthocyanin and flavonol glycoside transporters (in addition to *MRPs* as exemplified by the maize *ZmMRP3*?).

Non-vacuolar Transport

Compartmentation and enzyme localisation studies suggest that complex networks involving different organelles, the vacuole and the apoplastic space and several membranes exist with regard to phenolic compound transport. For example, hydroxycinnamic acids are classically separated into soluble, vacuolar and insoluble, cell wall-associated pools. Monolignol-glucosides, the glucosylated precursors of the lignin polymer, are supposed to represent a transport form being secreted to the apoplastic space but whether the glucosides undergo transient vacuolar storage prior to extracellular polymerisation is unclear (Lim et al., 2001; Ehlting et al., 2005; Kaneda et al., 2008). Likewise, the flux of the phytoalexin coumarines such as scopoletin and its glucoside scopolin secreted massively by tobacco plants and cells cultures in response to elicitation between cytosolic biosynthesis, transient vacuolar storage and signal-induced secretion is not understood (Taguchi et al., 2000; Chong et al., 2002; Costet et al., 2002; Matros and Mock, 2004). Histochemical data argues for the nuclear localisation of structurally diverse flavonoids in several plant species (e.g. Grandmaison and Ibrahim, 1996; Hutzler et al., 1998; Peer et al., 2001; Buer and Muday, 2004; Polster et al., 2006) which correlates with the immunological detection of the biosynthetic enzymes chalcone synthase (CHS) and chalcone isomerase (CHI) in the nucleus of Arabidopsis cells (Saslowsky et al., 2005). Thus, flavonoid biosynthesis is not restricted to the cytoplasm and a second biosynthetic pool of metabolites may exist in the nucleus raising the question whether and how metabolic exchange occurs between a cytosolic and a nuclear pool and whether there is a physiological function for nuclear flavonoids.

Many species that are exposed to high irradiance living in arid and xeric habitats accumulate flavonoids on the surface of leaves, flowers and other tissues. These 'surface' or 'exudate flavonoids' are in most cases lipophilic, predominantly aglycones that in many cases are strongly methylated (Valant-Vetschera and Wollenweber, 2001; Wollenweber et al., 2003; Onyilagha and Grotewold, 2004) which was defined as a characteristic modification necessary to achieve passing through the lipophilic waxes or lipids (Stafford, 1990). However, the polar or apolar nature of flavonoids cannot be strictly associated with their extrusion to the extracellular surface or vacuolar storage, respectively, because as well hydrophobic flavonoids such as quercetin 3-methyl ethers appearing in leaf vacuoles of *Vellozia streptophylla* as presence of polar flavonoid glucosides on the leaf surface of *Heteranthemis viscidehirta* or *Chrysanthemum segetum* have been reported (Harborne et al., 1994; Valant-Vetschera et al.,

2003). In many species though, vacuolar flavonoids are structurally different from leaf surface flavonoids and it will be interesting to establish the structural basis of differential compartmentation.

Direct transport data with regard to flavonoid exudation is scarce. Antibodies raised against the major polymethylated flavonol glucosides of *Chrysosplemium americanum* detected these compounds in the cell wall of epidermal cells, the plasmalemma, the periplasm and associated vesicles (Ibrahim et al., 1987; Ibrahim, 1990). In accordance with further evidence given below, it is tempting to speculate that the polymethylated flavonol glucosides are loaded into vesicles at their site of synthesis on a membrane-associated multienzyme complex from where they are secreted in flavonoid-containing vesicles towards the plasmalemma without passing the Golgi (Ibrahim, 1990; Latchinian-Sadek and Ibrahim, 1991). Metabolites found in root exudates actively orchestrate plant interactions with the rhizosphere including symbiotic, pathogenic and allelopathic interaction (Walker et al., 2003; Bais et al., 2004; Weisskopf et al., 2006) while the mechanisms involved in secretion of rhizosphere compounds are not well understood. In an attempt to use Arabidopsis as a model, 15 major secondary metabolites were found in the root exudate.

Pharmacological profiling, the investigation of mutants and the mining of the root-specific transcriptomics data set of Arabidopsis (Birnbaum et al., 2003) suggest that transporters are involved in root exudation processes, although at present none of the identified substances can be assigned to specific transporters (Loyola-Vargas et al., 2007; Sugiyama et al., 2007; Badri et al., 2008). Interestingly, *Glycine max* root plasmalemma vesicles exhibit time-dependent, saturable, uncoupler-sensitive transport of the isoflavonoid genistein, which serves as a signal molecule in the legume-*Rhizobium* symbiosis, suggesting that ABC transporters are likely involved in root flavonoid secretion (Sugiyama et al., 2007).

In contrast to the secretion of phenolic compounds where no transporters have been identified, two examples exist where specific ABC transporters likely catalyse the excretion of secondary metabolites. First, NpABC1, a Pleitropic Drug Resistance (PDR/ABCG)-type ABC transporter characterized in *Nicotiana plumbaginifolia*, is involved in the plasmalemma export of sclareol, an antifungal diterpene secreted by glandular trichomes on the leaf surface of *Nicotiana* species (Jasinski et al., 2001). Second, the isolation and characterization of Arabidopsis mutants in two so-called half-size ABC transporter genes of the *White-brown complex* (WBC /ABCG subfamily, termed *cer5/abcg12/wbc12* and *pel1/cof1/dos/abcg11/wbc11* suggests that a minimum of two WBC-type ABC transporters play an important role in providing components of the cuticular wax layer to extracellular lipid transfer proteins (Pighin et al., 2004; Bird et al., 2007; Luo et al., 2007; Panikashvili et al., 2007; Ukitsu et al., 2007). *wbc11* and *wbc12* mutants exhibit pleiotropic phenotypes including changes in the thickness of the cuticle, drastic alterations in the composition and amount of extracellular cutin monomers, most importantly alkanes and lipids, accumulation of intracellular alkanes which coincides with the appearance of cytoplasmatic inclusions in epidermal cells. Although biochemical evidence is lacking, the mutant phenotypes and the localization of fluorescent protein fusion of WBC11 and WBC12 to the plasmalemma are consistent with a role of both ABC transporters in the export of cutin monomers (Schulz and Frommer, 2004).

Beyond the ABC: Membrane Protein Families Involved in Secondary Metabolite Transport

Ignoring trapping mechanisms, it can be stated that all transmembrane transport processes presently known for secondary metabolites are either directly ATP-dependent, thus driven by

ABC transporters (Transporter classification [TC] number 3.A.1 according to the Transport Classification Database [www.tcdb.org] and TransportDB [www.membranetransport.org]), or H^+-dependent (Fig. 1). Up to now, no $\Delta\Psi$-dependent transport process has been described.

It is important to note that the model calculation of the maximal accumulation rate of a GS conjugate in vacuoles exploiting different energy sources clearly demonstrates kinetic perfection of ABC transporters: direct energisation maximally allows vacuolar accumulation by a factor of ~4 x 10^8 compared to a factor of 100 or 3.2 only for ΔH^+- or $\Delta\Psi$-dependent energisation, respectively (Kreuz et al., 1996). Thus, if potentially toxic molecules need to be removed from the cytosol against a steep concentration gradient, ABC transporters can be expected to be more efficient than antiporters.

In spite of this energetic 'advantage' of ABC transporters, at least four classes of ion-dependent multidrug transporters have been detected based on phylogenetic relationships, overall structure and number of transmembrane-spanning helices (TM) which are the Major Facilitator Superfamily (MFS, TC 2.A.1, minimum of 12 TM), the Small Multidrug Resistance Family (SMR, TC 2.A.7.1, 4 TM), the Resistance/Nodulation/Cell division family (RND, TC 2.A.6, 12 TM) and the MATE transporters (TC 2.A.66.1, 9–12 TM) which form the major subfamily of the Multidrug/Oligosaccharidyl-lipid/Polysaccharide (MOP) Flippase Superfamily (TC 2.A.66; Paulsen et al., 1996; Brown et al., 1999; Saier and Paulsen, 2001; Saidijam et al., 2006). The functional identification of most of these membrane proteins as secondary multidrug transporters still relies mainly on experiments performed with prokaryotic organisms. Moreover, while SMR transporter genes are absent in plant genomes, multidrug transport properties of plant MFS (90 and 149 genes in Arabidopsis and rice, respectively) and RND transporters (2 and 3 genes in Arabidopsis and rice, respectively) remain to be shown. Interestingly, all predicted *MOP* genes in the genomes of Arabidopsis, rice and *Chlamydomonas reinhardtii* (56 in Arabidopsis, about 50 in rice depending on annotation and 13 in Chlamydomonas) belong to the *MATE* subfamily.

The taxonomic coverage demonstrates that MATE proteins have mainly been identified in prokaryotes where they mediate resistance against fluoroquinolone antibiotics such as norfloxacin but also to structurally unrelated antibiotics (e.g. kanamycin, streptomycin) and small toxic molecules such as ethidium (Morita et al., 1998). Importantly, NorM acts as a Na^+-driven multidrug efflux transporter (Morita et al., 2000). When sequenced genomes are compared, it is evident that comparable to the situation of ABC proteins, MATE protein-coding genes are overrepresented in plants. In Arabidopsis, mutant characterisation provides evidence that MATE transporters are involved in n lateral root formation (ALF5; Diener et al., 2001), iron homeostasis (FRD3; Rogers and Guerinot, 2002) or salicylic acid-dependent signaling for disease resistance (EDS5; Nawrath et al., 2002) apart from flavonoid transport. Complementation of a hypersensitive *E. coli* strain with an Arabidopsis cDNA library led to functional cloning of the MATE transporter *AtDTX1* (Li et al., 2002) which catalyzes proton-coupled efflux of ethidium, antibiotics, the alkaloid berberine and mediates cadmium resistance in *E. coli.* Although the specific endogenous substrates of plant MATE transporters have not been established in most cases, the mutant phenotypes are consistent with the hypothesis that plant MATE proteins are involved in transmembrane transport of small organic molecules. Importantly, the substrate spectrum in plants includes primary metabolites as exemplified by Arabidopsis FRD3 and related transporters identified by positional cloning of the major loci conferring aluminium resistance in *Sorghum bicolor* and *Hordeum vulgare* which are proposed to act as a plasmalemma H^+-dependent citrate efflux system (Durrett et al., 2007; Magalhaes et al., 2007; Wang et al., 2007).

With respect to mammals, *HsMATE1* and *HsMATE2*, two homologous genes in humans as well as a rodent *MATE* gene were described to encode transporters mediating H^+-coupled cellular efflux of toxic organic cations such as the electroneutral exchange of tetraethyl ammonium. In spite of differences in tissue-specific expression and substrate specificities, the mammalian MATEs represent the polyspecific organic cation exporters that catalyze the terminal transfer of toxic organic cations into urine and bile (e.g. Otsuka et al., 2005; Masuda et al., 2006). A recent phylogenetic analysis groups all plant MATE-type transporters together into one clade containing no other fungal or mammalian MATEs (Omote et al., 2006). In contrast to ABC transporters which can be clearly identified based on the characteristic Walker A and B motifs as well as the ABC signature, MATE transporters lack obvious consensus sequences, although multiple sequence alignments of MATE transporters from all kingdoms (Omote et al., 2006) and all Arabidopsis MATEs (Fig. 2) indicate conserved regions in TMs 1 and 7 as well as in several cytosolic loops. In the Arabidopsis MATEs several cytosolic glycine residues are highly conserved. A glutamate residue in TM7 which is highly conserved has been shown to be essential for the transport function of *HsMATE1* (Otsuka et al., 2005).

On the Road Again: Transport Versus Trafficking and the Dawn of a Cell Biology of Natural Products

An unresolved issue is the question whether the transporters at their 'target' membranes such as the tonoplast or the plasmalemma use substrates that diffuse freely in the cytosol (*transporter model*; Fig. 1). As an alternative to diffusion, a 'protected delivery' or *binding protein model* suggests that flavonoids such as anthocyanins are associated with GSTs acting as binding proteins as described in Section Vacuolar Transport Coupled to Metabolic Reactions. A third model proposes trafficking of flavonoids and other phytochemicals in ER-derived vesicles towards target membranes where the vesicles would fuse and deliver their cargo into the vacuolar lumen or extracellular space (*vesicle model*). This model is based on different subcellular observations that can be interpreted as consequences of vesicle-mediated flavonoid transport.

ER-derived structures have been proposed to represent intermediates of vesicular flavonoid transport supporting a model of direct loading of flavonoids across the ER membrane into the lumen after or during synthesis by ER-localized metabolons facing the cytoplasm. Green and yellow autofluorescent bodies which presumably indicate the presence of flavonoids, morphological alterations in the ER and electron-dense structures that appear to originate from the ER traveling towards the plasma membrane were observed in Maize Black Mexican Sweet (BMS) cells ectopically expressing the MYB transcription factor *P1* (Grotewold et al., 1998). The green and yellow bodies are presumably targeted to the extracellular space and the vacuole, respectively, and the yellow bodies are not evenly dispersed within the vacuole but form inclusions (Grotewold et al., 1998; Lin et al., 2003). Brefeldin A (BFA) and monensin, two agents known to disrupt endocytosis, Golgi function and the Golgi-ER vesicle trafficking, do not inhibit accumulation of green autofluorescence in the cell wall suggesting that secretion of these compounds does not necessarily pass the Golgi apparatus (Lin et al., 2003). Likewise, a recent investigation of *Brassica* and Arabidopsis tapetum cells describes the formation of ER-derived structures termed tapetosomes which accumulate flavonols and alkanes destined to impregnate the pollen surface after secretion (Hsieh and Huang, 2007). These authors also report a profound reduction of flavonol-filled tapetosomes in tapetum cells of the *tt12* mutants which is puzzling because transport experiments with TT12, rather specific expression in the seed coat as well as subcellular localisation of TT12 to the tonoplast (Marinova et al., 2007) would argue against TT12 as a transporter for pollen flavonols.

Signal-based communication between the ER-localized biosynthetic complexes and the vacuolar transporters has been proposed because the mutant *ant310* barley line, defective in *CHI*, exhibited strongly reduced vacuolar uptake of flavone-glucosides which could be reactivated when protoplasts or leaf segments were chemically complemented with naringenin, the product of the CHI reaction (Frangne et al., 2002; Marinova et al., 2007a). Considering that biosynthesis takes place at the ER while transport is tested on isolated vacuoles, it has to be postulated that the flavonoid antiporter at the tonoplast gets activated by an unknown mechanism if flavonoid biosynthesis is functional.

Anthocyanins have been repeatedly found in inclusions in several plant species with no clear phylogenetic relation including lisianthus, sweet potato, red cabbage or grapevine (Pecket and Small, 1980; Small and Pecket, 1982; Nozue and Yasuda, 1985; Conn et al., 2003; Irani and Grotewold, 2005; Zhang et al., 2006). Interestingly, anthocyanin-filled inclusion occur in the cytoplasm as well as in the vacuole itself where the vacuolar bodies are generally referred to as 'anthocyanic vacuolar inclusions' (AVIs) (Markham et al., 2000). It is assumed that the cytoplasmic inclusions represent transport forms of anthocyanins on their way to the vacuole although direct evidence for this is lacking. In lisianthus flower petals, transmission electron microscopy (TEM) demonstrated the presence of electron-dense structures in vesicles adjacent to deformed ER structures suggesting that they originate from there (Zhang et al., 2006) which is in accordance with dilated ER near the autofluorescent bodies found in BMS cells (Grotewold et al., 1998). A prominent example for cytosolic inclusions are the red 3-deoxyanthocyanidin-containing structures formed as phytoalexins in *Sorghum* leaves in response to attack by the pathogen *Colletorichum sublineum* which causes Sorghum anthracnose (Snyder and Nicholson, 1990). These bodies move towards the site of infection in a polar manner where the phytoalexins are released. In spite of the attractive model suggesting controlled movement of phytoalexin compounds, the hypothesis that the observed deoxyanthocyanidin-containing structures are membrane-bound vesicles has been challenged and the formation of self-organizing protein-containing spheres lacking a surrounding lipid bilayer has been proposed (Nielsen et al., 2004).

Likewise, the vacuolar AVIs are not surrounded by membranes (Zhang et al., 2006) which leaves open the possibility that AVIs present in ER-derived vesicles are delivered to the vacuole by vesicle fusion at the tonoplast and released as a whole into the vacuolar lumen. A recent study in Arabidopsis describes novel *in vivo* and *in vitro* autofluorescent properties of anthocyanins. Co-localization studies with GFP reporters of the plant endomembrane system demonstrates that naringenin-induced anthocyanin autofluorescence in cotyledons is present in the ER as well as ER-derived vesicles and passes to protein storage vacuoles in a Golgi-independent manner where AVIs are formed in cotyledons (Poustka et al., 2007).

The vacuole is a highly dynamic organelle, but the mechanisms underlying the various fusion and budding processes are poorly understood. For example, fusion of vesicles and formation of vacuolar fragments derived from the central vacuole are important processes supporting the drastic volume changes of guard cells during stomatal opening or closure (Gao et al., 2005). Recent evidence suggests that in the PA-containing cells of the Arabidopsis seed coat, formation of the vacuole is linked to PA biosynthesis, which proposes a novel cellular function of flavonoids as a signal molecule.

External loading of immature seeds with the fluorescein-derivative carboxy-DCFDA, which accumulates in the acidic vacuole in vegetative tissues, labels central, cell-filling structures in endothelial cells presumably corresponding to the vacuole. In contrast, when seeds of mutants lacking PAs caused by the disruption of late steps of the PA pathway such as *tds4* (leuanthocyanidin dioxygenase mutant), *aha10* (P-type H^+-ATPase) and *tt12* (transporter mutant) were exposed to carboxy-DCFDA, fluorescence signals were obtained from multiple vesicular structures (Abrahams et al., 2003; Baxter et al., 2005; Fig. 3) TEM investigation of the PA-synthesizing endothelial cells verifies that in contrast to the wild type, *tds4* and *tt12* endothelial cells exhibit an accumulation of vesicular structures while the central, cell-filling vacuole is absent. The effects of PA absence on the endothelial vacuoles seems to be specific for this cell type since root or mesophyll cells of PA biosynthesis mutants possess a normal vacuolar ultrastructure (Abrahams et al., 2003; Ballmann, C.; Marinova, K. and Klein, M., unpublished). Interestingly, a functional link between PA synthesis and vacuolar biogenesis related to programs of cellular differentiation has also been reported for *Petunia* where mutants in the bHLH-type transcription factor Anthocyanin1 (AN1) which controls anthocyanin synthesis and the vacuolar pH in petal cells exhibit strongly altered morphology of the outer integumental layer of the seed coat where PAs are synthesized in this species (Spelt et al., 2002). However, disruption of PA biosynthesis could also result in pathway intermediates that exert toxic effects inhibiting vacuole formation. Since more general flavonoid Arabidopsis mutants such as the *chs* mutant *tt4* where all flavonoids are absent possess normal vacuole formation in vegetative cells, control of vacuolar biogenesis is obviously not a general feature of flavonoids. If only PA-synthesizing cells exhibit vacuolar alterations in the absence of PA biosynthesis, the question arises whether PA-containing vacuoles are 'functionally specialized' vacuoles in a way that is comparable to the classical differentiation between lytic and protein storage vacuoles. Thus, the determination of vacuolar identity in these cells is instrumental.

Disruption of the proper glycosylation of the major triterpenoid saponin avenacin A-1, which is an antimicrobial triterpene glycoside that accumulates in the vacuole of oat root epidermal cells (Haralampidis et al., 2001) leads to pleiotropic effects on root development including root hair formation, degeneration of epidermal cells and changes in the cell wall structure (Mylona et al., 2008). A careful mutant analysis demonstrates that missing glucosylation rather than absence of avenacin biosynthesis is responsible for these effects on root morphology which is in accordance with a model that explains the deleterious effects on root epidermal cells by the missing vacuolar deposition of toxic avenacins.

Consequences of the Vesicular Transport Model

To date, it seems complicated to find common ground between the *transporter* or *binding protein model* on the one side and the *vesicle model* for flavonoids on the other side. It is important to note, that from the transporter's point of view, the vesicular model raises the issue that the site of active flavonoid transport is not primarily at the tonoplast or the plasmalemma but in vicinity to the site of metabolon-mediated synthesis across the ER membrane. Vesicle loading at the ER is in disagreement with available data on the subcellular localization of MRP, PDR or MATE transporters. Since in most cases localization experiments used constitutive but not the endogenous promoters, the possibility that functional transporters traffick between the ER and the target membrane still exists. In addition, it is complicated to experimentally address the question whether a transporter which enters the endomembrane system and travels towards the target membrane is immediately active in the ER which could partially explain loading of

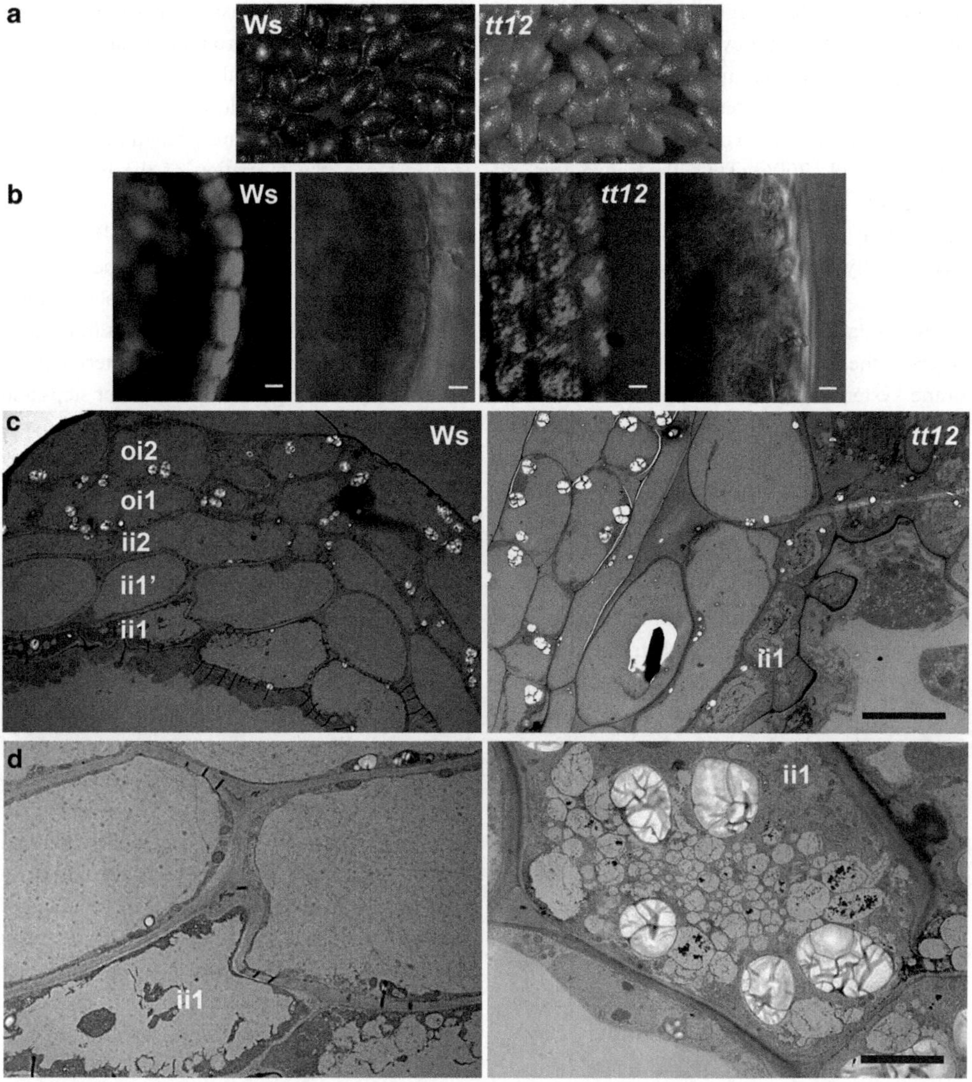

Fig. 3 Effects of the *tt12* mutation on the vacuolar structure in the endothelial layer of Arabidopsis seeds. (**a**) Staining of mature wild-type (Ws) or *tt12* mutant seeds with DMACA demonstrates absence of the black coloration indicative of PA polymers in *tt12* seeds. (**b**) Loading of immature seeds with the fluorescent dye carboxy-DCFDA shows changes in the fluorescence pattern in seed coat cells towards punctuate fluorescence in *tt12* seeds. Bar = 30 μm. (**c, d**) TEM images of immature seeds (5 days after pollination) after high-pressure freezing/freeze-substitution sample preparation. Depicted are the two outer (oi1, oi2) and three inner integuments (ii2, ii1', ii1). The appearance of multiple vesicular structures instead of a large vacuole is visible in the ii1 (endothelium) layer of *tt12* seeds. (**c**) Bar = 20 μm. (D) Bar = 5 μm

ER-derived vesicles. It must be postulated that flavonoid accumulation at the ER or in vesicles by H^+-dependent transporters such as TT12 is driven by ΔpH thus requiring an acidic ER lumen. V-ATPases have been found to be distributed throughout the endomembrane system, including the ER, Golgi, and provacuoles and are required for endocytic and secretory trafficking clearly indicating that acidification by the V-ATPase is not restricted to the vacuole and is a major player in defining vesicle identity (Dettmer et al., 2006). Thus, the transporter and the vesicle transport model may merge if we accept that also the flavonoid transporters are dynamically cycling within the endomembrane system.

A second challenge to the vesicle model refers to the 'sorting' capacity of the endomembrane system since it would predict that structurally related compounds destined to travel to the vacuole or the extracellular space need to be 'sorted' in the ER or downstream of this compartment in a manner comparable to vacuolar or secretory proteins. In consequence, either specific receptors recognizing the different phytochemicals must be postulated or the synthesis of metabolites that need to be distributed to distinct compartments takes place on spatially separated complexes. Thus, the "interface" between phenolic biosynthesis, including distinct metabolons, and the loading of end products into primary transport vesicles, most probably at the ER surface, requires and surely will attract much research interest in the future.

Compartmentation and Transport in Alkaloid Biosynthesis (W. Roos)

Principles

Alkaloids are amino acid -derived secondary compounds which often display a high pharmacological activity (Wink et al., 1998; Schmeller. et al., 1997; Xu et al., 2006; Faddeeva and Beliaeva, 1997) and include several established remedies of traditional and present medicine. Many of them selectively interfere with essential processes of cellular metabolism, gene expression or signaling, i.e. they hit molecular targets shared by most living cells. Their separation from the living protoplasm is therefore of critical importance for the producing plant.

Biosynthesis of alkaloids requires flow of metabolites through multi-enzyme sequences in branched pathways that are often rather elaborate (for instance, the formation of the benzophenanthridine alkaloid macarpine from tyrosine requires 19 enzymatic steps, cf. Fig. 4). The biosynthetic enzymes are typically localized in multiple cell types and different subcellular compartments and separated from the sites of product accumulation. Therefore, pathway intermediates and products undergo multiple transfer steps and are channelled between precursor pools, biosynthetic and storage sites. As with other secondary biosynthesis, the expression of biosynthetic enzymes, transporters and proteins related to channeling and storage of alkaloids is coordinated within programs of cell differentiation and tissue development, the architecture of these programs provides a growing challenge for research. It appears that evolution has brought about no organelles with an exclusive destination to alkaloid metabolism but rather adapted elements of primary metabolism and of its basic compartmentation to new functions. As an example, homologs of two committed enzymes of alkaloid biosynthesis, berberine bridge enzyme and strictosidine synthase (both localized in vacuoles or related vesicles, as shown below) are present in the vacuoles of Arabidopsis, a plant that is not known to synthesize corresponding alkaloids. This suggests an evolutionary adaptation of preexisting vacuolar enzymes to new functions within a newly evolved alkaloid biosynthesis (De Luca and St-Pierre, 2000).

At present, only a few membrane compartments, transporters and channels are known to have a defined role in alkaloid transport and compartmentation. One major challenge for research is the high individuality in the cellular organization that rests on the species-specific localization of biosynthetic enzymes and the opportunistic use of transporters, metabolite pools, storage sites and accumulation mechanisms even between taxonomically related genera.

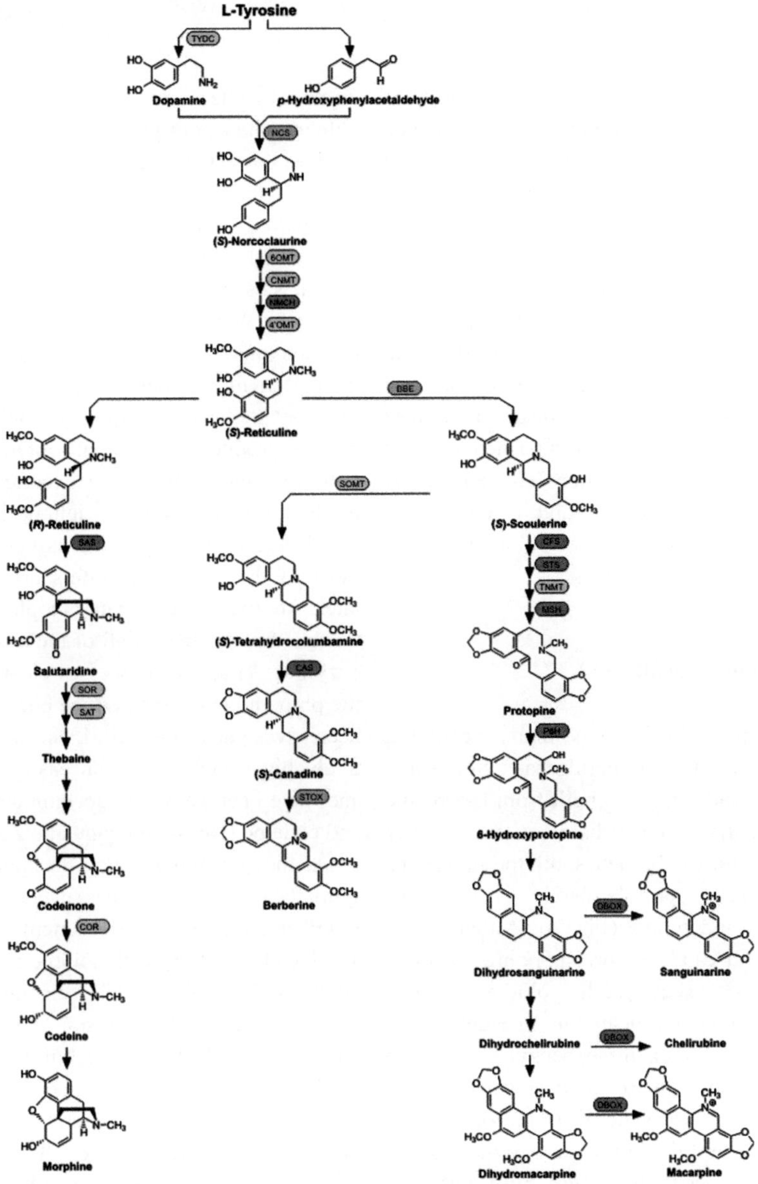

Fig. 4 Pathways of benzylisoquinoline biosynthesis. A selection of biosynthetic enzymes is notified with their localization in the cytosol (yellow), the ER membrane (red) or in the lumen of cytosolic vesicles (blue). Informations are taken mainly from Facchini and St-Pierre (2005), Bock et al. (2002), own experiments and other references cited in the text. BBE, berberine bridge enzyme; CFS, cheilanthifoline synthase; CNMT, coclaurine N-methyltransferase; COR, codeinone reductase; DBOX, dihydrobenzophenanthridine oxidase; MSH, N-methylstylopine 14-hydroxylase; NCS, norcoclaurine synthase; NMCH, N-methylcoclaurine 3'-hydroxylase; 4'OMT, 3'-hydroxy-N-methylcoclaurine 4'-O-methyltransferase; 6OMT, norcoclaurine 6-O-methyltransferase; 7OMT, reticuline 7-O-methyltransferase; P6H, protopine 6-hydroxylase; SAT, salutaridinol-7-O-acetyltransferase; SOR, salutaridine:NADPH 7-oxidoreductase; STS, stylopine synthase; SAS, salutaridine synthase; TNMT, tetrahydroprotoberberine *cis*-N-methyltransferase; TYDC, tyrosine decarboxylase; CAS, canadine synthase; SOMT, scoulerine 9-O-methyltransferase; STOX, (S)-tetrahydroprotoberberine oxidase

The biochemistry and molecular biology of alkaloid transporters has actually been reviewed (Shitan and Yazaki, 2007; Yazaki, 2006) as well as the subcellular localization of enzymes of major alkaloid families (Wink and Roberts, 1998; Facchini, 2001; Facchini and St-Pierre, 2005; Ziegler and Facchini, 2008). The following overview rests on data of the best investigated alkaloid biosynthetic pathways and illustrates the requirement for transport and channeling as predicted from the distribution of metabolic enzymes, pools of intermediates and sites of product accumulation. This enzymatic and ultrastructural background is confronted with the actual knowledge of alkaloid transporters and detoxifying enzymes.

Benzylisoquinoline Alkaloids

Benzylisoquinolines are typically occurring in Papaveraceae, Berberidaceae and some Ranunculaceae. They are built from L-tyrosine via the central intermediate reticuline, after which the pathway branches into the subgroups of morphinanes, protoberberines, benzophenanthridines and others (Fig. 4). Although the enzymes that catalyze corresponding reactions show extensive sequence homologies between the species, they are located in different tissues and cell types. For example, benzylisoquinoline biosynthesis in roots of *Papaver* was localized to the vascular tissues, but in *Thalictrum* occurs in the root's and rhizom's non-vascular tissues (endodermis, pericycle, cortex, pith), although the first enzymes of either pathway, acting between tyrosine and (S)- reticuline, are nearly identical. This might indicate that a primordial benzylisoquinoline biosynthetic pathway migrated into diverging cell types during evolution (Samanani et al., 2005).

The **morphinan branch** is best investigated (but not exclusively occuring) in *Papaver*, where six of the biosynthetic enzymes acting between (S)-norcoclaurine and codeine were immunolocalized to sieve elements of the phloem adjacent or proximal to laticifers. Most of the corresponding gene transcripts were found in the companion cells, i.e. parenchyma cells paired with each sieve element, suggesting post translational cell-to-cell transfer of the enzymes (Samanani et al., 2006). Earlier experiments by Weid et al. (2004) had favoured such parenchymatic cells to harbour also the biosynthetic enzymes. Whatever the precise location of biosynthetic sites within the phloem might be, the products and /or late intermediates of morphine biosynthesis are excreted from the producing cells into the neighbouring lacticifers, that accompany vascular tissues throughout most of the plant. Lacticifers are filled with latex, i.e. the multinucleate cytoplasm of fused vessels, but are free of all or most biosynthetic enzymes. Transport processes clearly occur at the phloem/lacticifer interface but have not yet been investgated in detail. Latex contains relatively big vesicles that can easily be isolated and have been shown to accumulate morphine and codeine (Deus-Neumann and Zenk, 1984).

At the subcellular level, the morphine biosynthetic pathway combines several cytochrome P-450 enzymes, bound to ER membranes (Chou and Kutchan, 1998) with soluble enzymes that reside in ER derived vesicles, e.g. norcoclaurine synthase, or in the cytosol, e.g. codeinone reductase (Zenk, 1994; Facchini and St-Pierre, 2005). The diverse localization of the biosynthetic enzymes requires several transport steps of the intermediates between cytosol, vesicles and vacuoles. Only one of them has been characterized by pioneering experiments (Deus-Neumann and Zenk, 1986): vacuolar vesicles prepared from *Fumaria capreolata* accumulate (S)-reticulin or (S)-scoulerin via highly specific transporters, that discriminate between (S)- and (R) stereoisomer and exclude other benzylisoquinolines (sanguinarine, protopine, morphine) and alkaloids of unrelated families (e.g. indoles or tropanes). Uptake is energized by the pH gradient across the tonoplast. Accumulated alkaloids completely efflux after dissipation of ΔpH, indicating an H^+/alkaloid – antiport (Km < 1 μM) and the absence of stable intravacuolar alkaloid complexes. It is not known whether the protonation of the accumulated alkaloids is required for maintaining their vacuolar pool.

Enzymes of the morphine branch display a coordinate regulation as seen after the suppression of codeinone reductase (COR) via RNAi-based gene silencing. Cells deprived of COR did not accumulate codeine, the substrate of this enzyme, but only (S)-reticuline, the first committed substrate of the morphinane branch (Allen et al., 2004). Silencing of salutaridinol O-acetyltransferase (SAT), an enzyme acting in the same branch (cf. Fig. 4), caused no accumulation of its substrate salutaridinol but of salutaridine, the substrate of the previous enzyme (Allen et al., 2008; Kempe, 2008). Overexpression of SAT increased the yield of morphine indicating a rate limiting function of this enzyme (Allen et al., 2008). The data support the existence in the morphine branch of one or more metabolons, i.e. multi-enzyme complexes that allow the internal hand-over of pathway intermediates and are disrupted by the removal of one enzyme. If this interpretation can be finally confirmed (the alternate explanation, i.e. co-silencing of other enzymes downstream of (S)-reticuline is not yet excluded) it would meet expectations derived from metabolic channels acting in different areas of plant metabolism (Winkel, 2004) and serve the channeling of intermediates independent of their cytoplasmic pools (phenylpropanoids: Achnine et al., 2004; polyamines: Panicot et al., 2004; cf. Section Compartmentation, Transport and Trapping of Phenolic Compounds).

The existence of a metabolic channel would be in line with the regulatory independence of the enzymes of the morphinane branch from the general benzylisoquinoline biosynthesis. A well known expression of this phenomenon is the intimate link between lacticifer development and biosynthesis of morphine: dedifferentiated, cultured cells of opium poppy do not produce codeine and morphine (Facchini and Bird, 1998) although COR and other late biosynthetic enzymes are present (e.g. Gerardy and Zenk, 1993; Lenz and Zenk, 1995). At the other hand, it remains challenging to explore whether and how a metabolon is compatible with the ER-based localization of several biosynthetic enzymes and the export of the end products into the adjacent lacticifers.

The first committed enzyme of the **protoberberine and benzophenanthridine branch** (cf. Fig. 4) is the berberine bridge enzyme (BBE), a flavoproteinated oxidase. Together with most enzymes of benzylisoquinoline biosynthesis it resides in sieve elements of root and hypocotyl (cf. above, Samanani et al., 2006). Subcellularly, visualization by ELMI or fluorescence tracing of GFP-conjugates congruently show that BBE is a lumenal protein within a distinct class of vesicles that have a higher density as ER and the tonoplast (in *Berberis*: Amann et al., 1986; in *Eschscholzia* und *Papaver*: Bock et al., 2002). These vesicles are derived from the ER and may fuse with the vacuolar membrane (Bock et al., 2002; Alcantara et al., 2005). The morphological data are supported by the amino acid sequence of the nascent BBE protein that contains a functional ER-targeting signal peptide and an adjacent vacuolar-sorting determinant (Bird and Facchini, 2001). Due to its slightly alkaline pH optimum, BBE is likely inactive at the acidic vacuolar pH, suggesting that the catalyzed step (formation of scuolerine) occurs before the enzyme crosses the tonoplast (Bird and Facchini, 2001). It is assumed that other soluble biosynthetic enzymes reside in vesicles of similar type as shown with enzymes of berberine formation (STOX, CDO, COMT: Amann et al., 1986; Galneder et al., 1988; Rueffer et al., 1986). While in non-latex tissues such "alkaloid synthesizing vesicles" likely harbor intermediate steps of biosynthesis, the end products of the berberine and the benzophenanthridine biosynthetic branch are handled differently.

The biosynthesis and compartmentation of **Berberine** is an impressive example of variability and adaptation to different antimicrobial defense strategies.

In *Coptis japonica* the alkaloid is produced in the roots as judged from the specific expression of biosynthetic enzymes, but accumulates mainly in the rhizomes (Ikezawa et al., 2003). This implicates post-synthetic unloading of berberine from root cells followed by transfer to and uptake into rhizome cells. An essential component of this process is an ABC- transporter in the plasma membrane of rhizome cells that acts as an ATP driven influx pump and allows the efficient absorption of external berberine (Sakai et al., 2002). The cDNA of the transport protein (CjMDR1) was identified from a berberine accumulating cell line and shown to be expressed in xylematic cells of the rhizome (Yazaki et al., 2001; Shitan et al., 2003). The essential role of *Cjmdr1* in the accumulation of berberine is supported by co-suppression of the transporter protein (Shitan et al., 2005). CjMDR1 is a member of the currently expanding family of plant ABC-transporters (Yazaki, 2006) and the subfamily of multidrug-resistance transporters (MDR, cf. Section Compartmentation, Transport and Trapping of Phenolic Compounds). Its specificity, measured after heterologous expression in oocytes, is surprisingly broad and includes reticuline, sanguinarine, quinine and even quercetine with only small differences in transport efficiency.

Subsequent to its CjMDR1-catalyzed uptake into the cytoplasm, the final accumulation of berberine in *Coptis* rhizome cells occurs in the vacuole by a different mode of transport. The tonoplast harbours a H^+/berberine antiporter that operates with a Km of about 45 μM for berberine (Fig. 5). Competition experiments with berberine analogs and non-related alkaloids revealed that this transporter is fairly specific, although not exclusive, for berberine (Otani et al., 2005).

A different strategy of berberine compartmentation is realized in *Thalictrum minus* cells where most of the produced alkaloid is excreted into the apoplast and the outer medium, as observed in cell suspension cultures (Nakagawa et al., 1984). The export step is likely catalyzed by an ABC transporter of the plasma membrane, that displays remarkable sequence similarity to CjMDR1 from *Coptis*, although it operates here in the opposite (outward) direction (Terasaka et al., 2003; Yazaki, 2006).

Seen together, the transport of berberine in roots and rhizomes of *Coptis* and *Thalictrum* is dominated by the activities of transporters at the plasm membrane and the tonoplast. These mechanisms are in contrasting with observations from *Berberis*, where the transfer of the alkaloid to the vacuole most probably occurs in ER derived vesicles that contain the final enzyme of berberine biosynthesis (Bock et al., 2002). It appears that vesicular transfer is employed in plants that synthesize and store berberine in the same cells, whereas the above berberine carriers serve the intercellular transfer of this alkaloid in plants that established a spatial separation of alkaloid formation and storage (Shitan and Yazaki, 2007).

Berberine, as a hydrophobic cation, is a preferred substrate of multidrug resistance pumps (MDR's) which do not only operate in animals and plants but also in bacteria. In staphylococci, the rapid extrusion by MDR transporters is a main determinant of resistance against the antimicrobial effect of berberine (Stermitz et al., 2000). Berberine producing species of *Berberis* also produce efficient inhibitors of the bacterial MDR pump NorA: the flavonolignan 5'-methoxy-hydnocarpin and the porphyrin pheophorbide. These compounds strongly potentiate the growth inhibiting effect of berberine at *Staphylococcus aureus* (Guz et al., 2001). Similar synergistic effects are known from other medicinal plants. They might indicate a coordinated evolution of antimicrobial alkaloids and phenolic inhibitors of pathogen resistance mechanisms.

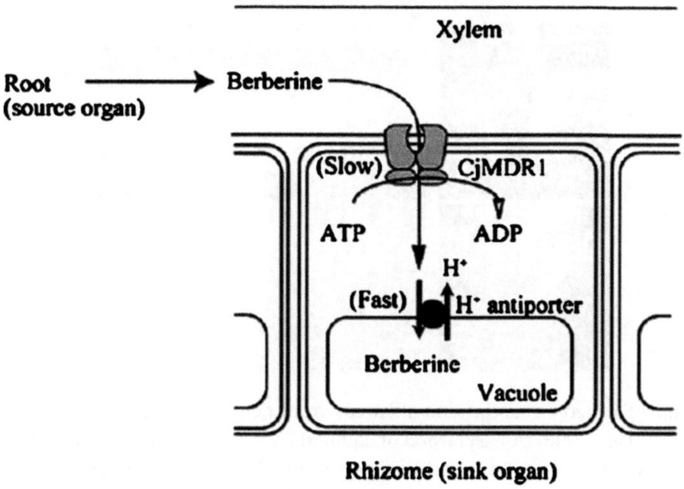

Fig. 5 The two step accumulation of berberine from the xylem soap into the vacuole of rhizome cells of *Coptis japonica*. (Taken with permission from Shitan and Yazaki, 2007

In this respect, it is noteworthy that plants not producing berberine contain a detoxifying berberine efflux transporter: the finding that the Arabidopsis MATE transporter AtDTX1 catalyzes the efflux of various toxins, including berberine, across the plasmalemma energized by the pH gradient (Li et al., 2002) supports the idea that berberine is an important constituent of soil microenvironments.

Benzophenanthridines (cf. Fig. 4) are mainly biosynthesized and stored in idioblasts of the root cortex parenchyma that are not connected to the lacticifer system (with the exception of *Chelidonium*, cf. below). Five of the seven enzymes between (S)-reticulin and sanguinarine, the first benzophenanthridine alkaloid, have been localized to a membrane fraction with a specific density of ± D 1.14 g/mL, that appears similar to the "alkaloid producing vesicles" isolated from *Berberis* and *Papaver* (cf. above). Except BBE, these enzymes are P450-dependent monooxygenases, i.e. integral proteins of ER membranes (Zenk, 1994; Facchini, 2001). The final step of benzophenanthridine formation is catalyzed by dihydrobenzophenanthridine oxidase (Schumacher and Zenk, 1988; Arakawa et al., 1992), which is not P450-dependent, but a soluble enzyme with a pH optimum >7, located in the lumen of ER-derived vesicles, where its products temporarily accumulate (M. Heinze and W. Roos, 2007, unpublished).

In contrast to some other benzylisoquinolines, benzophenanthridines are not finally stored in vacuoles but are excreted into the apoplast. Accumulation of the cationic alkaloids in the cell wall region has been visualized in elicited, cultured cells of *Eschscholzia* and is facilitated by their binding to cellulosic cell wall material (Färber et al., 2003; Viehweger et al., 2006; Roos et al., 2006; Fig. 6a). Earlier data from elicited, cultured cells of *Papaver* likewise aligned the orange-red fluorescence of newly made benzophenanthridines to the cell surface and the external medium (Cline and Coscia, 1989; Mahady and Beecher, 1994). About 85% of the excreted alkaloid were bound to osmiophilic particles. They may be derived from the rough ER as indicated by their high RNA content and their apparent similarity with ELMI-visualized particles that were repeatedly

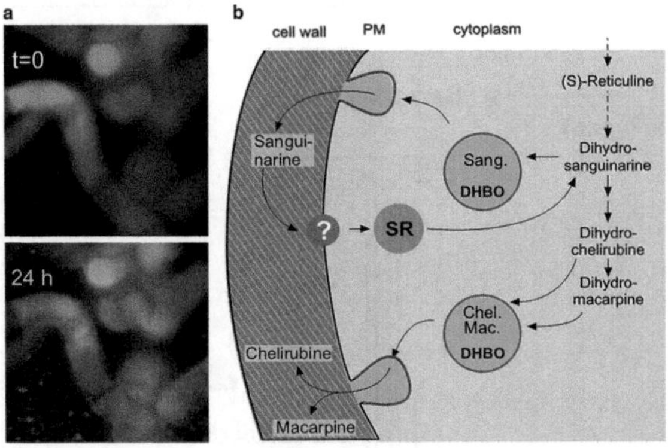

Fig. 6 a Excretion of benzophenanthridine alkaloids from cells of Eschscholzia after pathogen contact (according to Roos et al., 2006). A suspension of cultured cells of Eschscholzia californica received a yeast glycoprotein elicitor at t=0. After 24 h, cells were stained with the vitality indicator 5-carboxy-fluorescein (green flourescence indicates vital cells) and assayed by fluorescece microscopy. Benzophenanthridines, indicated by their red fluorescence, almost exclusively accumulate in the cell wall region and outer medium. **b**. Detoxication and recycling of benzophenanthridine alkaloids via sanguinarine reductase (according to Weiss et al., 2006). SR, sanguinarine reductase; DHBO, dihyrobenzophenanthridine oxidase; PM, plasma membrane, Sang., sanguinarine, Chel., chelerythrine., Mac., macarpine

visualized in the cytoplasm (Cline and Coscia, 1989; Bock et al., 2002; Alcantara et al., 2005). Membrane vesicles that contain sanguinarine have also been visualized in vivo and isolated from producing cells (in *Papaver*: Alcantara et al., 2005, in *Eschscholzia*: W. Roos, K. Viehweger and M. Heinze, 2007, unpublished). Their involvement in the biosynthesis and export (exocytosis) of benzophenanthridines is supported by the fact that newly made alkaloids escape the reduction by the highly active, cytoplasmic enzyme sanguinarine reductase (Fig. 6b, cf. below).

Despite several attempts, a vacuolar accumulation of sanguinarine or other benzophenanthridines was never demonstrated. The abovementioned ELMI studies (Bock et al., 2002; Alcantara et al., 2005) suggest that ER-derived, alkaloid-containing vesicles may fuse with the vacuole, but it is not known which cargo they would bring in. A yellow-green fluorescence (benzophenanthridines emit yellow-to-red fluorescence) observed in a minority of vacuoles from *Papaver* cells (Alcantara et al. 2005) could well arise from intermediates like dopamine or others (e.g. Kutchan et al., 1986). As mentioned already, sanguinarine **formation** does not occur in the vacuole, due to the alkaline pH optimum of the BBE (Bird and Facchini, 2001) and of the finalizing enzyme, dihydrobenzophenanthridine oxidase (Arakawa et al., 1992; Weiss et al., 2006).

The export of benzophenanthridines into the apoplast appears advantageous as the high antimicrobial potency of these root-made alkaloids protects the plant from pathogenic soil microbes. At the same time, the cytotoxic alkaloids are separated from the cellular interior thus avoiding self-intoxication by DNA intercalation or enzyme inhibition (see below). Nevertheless, the excreted alkaloids do not escape the control of the producing cell. Cultured cells of *Eschscholzia* rapidly take up external benzophenanthridines, preferentially sanguinarine, and reduce it to dihydrosanguinarine, its biosynthetic precursor, that can be further substituted according to the biosynthetic route. The resulting dihydroalkaloids, e.g. dihydrochelirubine, are oxidized and reappear in the outer medium (Weiss et al., 2006). Thus, the rapid reduction of incoming sanguinarine drives a recycling process that allows to present the toxic phytoalexin at the cellular surface and at the same time controls its concentration (Fig. 6b). It is catalyzed by sanguinarine reductase, a soluble cytoplasmic, NADPH-dependent enzyme with high substrate specificity: sanguinarine is preferred over other benzophenanthridines, berberine and other isoquinolines are excluded (Weiss et al., 2006). The mode of uptake of benzophenanthridines across the plasma membrane that is coupled to their subsequent reduction is currently under investigation.

Sanguinarine reductase has now been cloned and appears to be present in many if not all benzophenanthridine producing Papaveraceae (M. Vogel and W. Roos, 2008, in preparation). Its preferred substrate, sanguinarine, is the most cytotoxic benzophenanthridine and likely the most effective antimicrobial alkaloid made in plants (Schmeller et al. 1997). Its toxicity is due to intercalation into dsDNA, inhibition of ion pumps and several SH- dependent proteins, as well as interaction with cytoskeletal components (Faddeeva and Beliaeva, 1997; Bajaj et al., 1990; Wolff and Knipling, 1993; Slaninova et al., 2001; Bartak et al., 2003). Many plant cell cultures that do not produce benzophenanthridines and contain no sanguinarine reductase are intoxicated by low micromolar concentrations of sanguinarine (Weiss et al., 2006). Most probably, the risk of self-intoxication established a selection pressure towards the evolution of a detoxication mechanism based upon sanguinarine reductase and coupled transport steps.

A basically different strategy in the compartmentation of benzophenanthridines is realized by their sequestration in the latex of *Chelidonium*. The latex vesicles of this plant have been used for early transport experiments that supported an ion-trap mechanism of these alkaloids. Sanguinarine and chelerythrine that are cationic under acidic pH accumulate to a higher degree than the non-charged (and less toxic) berberine and dihydrocoptisine (Jans, 1974; Matile, 1976). The high vesicular content of alkaloids is further maintained by complexing them with chelidonic acid and similar vacuolar phenols. Transfer across the vesicular membrane is facilitated by a passive carrier mechanism of low specificity that accepts also structurally unrelated alkaloids, as vinblastin, strychnin oder dihydroergocryptin (Hauser and Wink, 1990).

Monoterpenoid Indole Alkaloids

Terpenoid indole alkaloids, the largest group of known alkaloids, are mainly produced in Loganiaceae, Apocynaceae and Rubiaceae. Their biosynthesis, which leads e.g. to vindoline, catharanthine and their dimers vinblastine and vincristine (medically important antineoplastic drugs) is another example of embedding complex enzymic pathways into the tissue - and subcellular morphology. The most thorough analysis has been done with leaves of *Catharanthus roseus*, but alkaloid biosynthesis is also detectable in roots and root cultures (St-Pierre et al., 1999; De Luca and St-Pierre, 2000; Burlat et al., 2004; Murata and De Luca, 2005; Ziegler and Facchini, 2008).

The formation of the first committed intermediate, strictosidine, requires to coordinate two sources of precursor biosynthesis: the MEP pathway which provides the monoterpene secologanine, and the tryptophane decarboxylase reaction which yields tryptamine (Fig. 7).

The MEP pathway is localized in the phloem parenchyma cells, as shown by mRNA distribution analysis (Burlat et al., 2004). More recent data imply that this pathway is also expressed in epidermal cells, where strictosidine synthase,

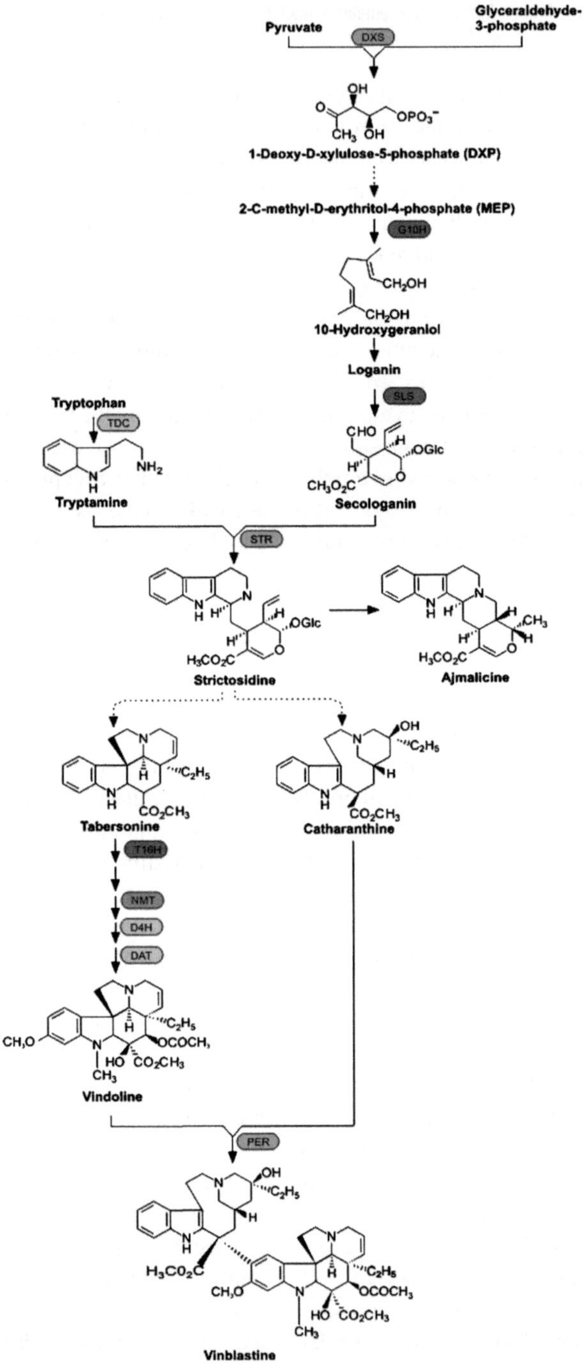

Fig. 7 Biosynthetic pathways leading to monoterpenoid indole alkaloids. A selection of biosynthetic enzymes is notified with their localization in the cytosol (yellow), the ER membrane (red), the chloroplast (green) or in the vacuole (blue). Informations are taken mainly from Facchini and St-Pierre (2005), Murata and De Luca (2005), De Luca and St-Pierre (2000) and other references cited in the text. DAT, deacetylvindoline 4-O-acetyltransferase; D4H, desacetoxyvindoline 4-hydroxylase; G10H, geraniol 10-hydroxylase; OMT, O-methyltransferase; NMT, N-methyltransferase; PER, peroxidase; SLS, secologanin synthase; STR, strictosidine synthase; TDC, tryptophan decarboxylase; T16H, tabersonine 16-hydroxylase

tryptophane decarboxylase and several of the alkaloid biosynthetic enzymes are located. Thus, it appears that the precursor secologanin is provided from cells of the same tissue and does not necessarily need to be imported from vascular cells (Murata and De Luca, 2005). Late genes, as those coding for desacetoxyvindoline 4-hydroxylase and deacetylvindoline 4-*O*-acetyltransferase, were found expressed only in specialized lacticifers or idioblasts of leaves and shoots (St-Pierre et al., 1999). This might ensure that the toxic antineoplastic dimers made from vindoline remain sequestered to non-living cells. The expression of some finalizing enzymes is tightly coupled to the development of lacticifers: vindoline, in contrast to catharanthine, is not produced in cultured, dedifferentiated cells of *Catharanthus* (De Luca, 1993) but the ability to produce this alkaloid reappears after regeneration of shoots and in shoot cultures (De Luca and St-Pierre, 2000). This analogy to morphine formation in opium poppy (cf. above) indicates that the expression of the vindoline branch of alkaloid biosynthesis is likewise embedded into the program of lacticifer differentiation of Catharanthus.

The distribution of the biosynthetic enzymes implicates that pathway intermediates must travel from epidermal cells to the lacticifer-idioblast system. For instance, as in the root no late biosynthetic enzymes were found, root-synthesized intermediates like catharanthine and tabersonine might be transported to laticifers via xylem and laticifer-associated tracheids.

At the subcellular level, a bewildering variety of compartments is involved in the course of biosynthesis (De Luca and St-Pierre, 2000; Facchini, 2001; Facchini and St-Pierre, 2005). Tryptamin that is formed from tryptophane in the cytosol must be transported into the vacuole (McKnight et al., 1991; Stevens et al., 1993). The biosynthesis of the other precursor, secologanine begins with the MEP pathway of the chloroplasts and the product is formed at provacuolar, ER-derived membranes (Madyastha et al., 1977). These data, together with the localization of strictosidine synthetase in the vacuole (McKnight et al., 1991), suggest that strictosidine is synthesized inside this organell. Several enzymes that act downstream of strictosidine are associated wit the ER (St-Pierre and De Luca, 1995) or thylakoid membranes (De Luca and Cutler, 1987). The last two steps of vindoline biosynthesis, catalyzed by D4H and DAT, likely occur in the cytoplasm (De Luca and Cutler, 1987). For the condensation with catharanthine, vindoline must reenter the vacuole where the peroxidases catalyzing this step are residing.

The channeling of intermediates that is enforced by the diverse distribution of enzymes is not yet understood. The first transport step investigated in some detail was the pH dependent accumulation of strictosidine, vindoline and ajmalicine (a side product made from strictosidine) in isolated vacuoles of *Catharanthus roseus*. The transporter selectively accepts endogenous alkaloids of *Catharanthus* and excludes others not produced in this species (Deus-Neumann and Zenk, 1984). Later, more detailed experiments with ajmalicine confirmed that alkaloid uptake was energized by the pH gradient across the tonoplast but not by the membrane potential, indicating an H^+/alkaloid antiport. The high accumulation ratio is assumed to be maintained by an ion-trap mechanism based upon the conversion of ajmalicine to the charged molecule serpentine by a vacuolar peroxidase (Blom et al., 1991).

The presented data, taken from two large alkaloid families, demonstrate that most plant organelles can harbor elements of alkaloid biosynthesis, i.e. cytosol, ER membranes, ER-derived vesicles, vacuoles, provacuolar vesicles and chloroplasts. In rare cases, even an actively metabolizing organell as the mitochondrion can be involved in alkaloid biosynthesis: the lysine-derived quinolizidine alkaloids epilupinine and multiflorine are synthesized in the chloroplasts of mesophyll cells in several legumes. The finalizing step is an acylation of these molecules, which occurs in the cytoplasm, except in the case of tigloyloxy-multiflorine: this compound is made in the mitochondria, which provide the acyl donor tigloyl-CoA (Saito et al., 1992; Suzuki et al., 1996).

There is little doubt that the few examples of alkaloid transporters known today will be greatly expanded in the next future as the perfection of molecular and cellular techniques should allow to characterize suspected transport mechanism and proteins and that hitherto escaped the experimental access. Candidates will be found not only among the steps of intracellular accumulation and transport as postulated above, but also within the long known inter-organ transfers, e.g. of nicotine from the root to leave vacuoles in Nicotiana (Hashimoto and Yamada, 1994) or of senecionine-N-oxide from the roots to vacuoles of inflorescences in *Senecio* (Hartmann, 1999).

Conclusions

Seen together, our present knowledge of transport and compartmentation of secondary metabolites evades any tight and unifying concept, even if we realize that much elucidation has yet to be done. Obviously, natural selection, acting as a "blind clockmaker", has combined transporters, energy sources, channels and compartments according to actual metabolic needs and structural opportunities of the plant that invented a new biosynthetic path. Although different in individual details, the examples from alkaloid and phenolic biosynthesis compiled in this chapter consistently indicate the opportunistic use of a few modules to serve common principles:

1. To combine biosynthetic channeling (e.g. in metabolons) with intracellular transfer mechanisms in order to avoid high local accumulation of intermediates. The sequestration of metabolites and enzymes in vesicles derived from the ER is an ubiquitious strategy in this respect.
2. To drain off end products from the cytoplasm with a final destination to either apoplast or vacuole and to stabilize this separation by complex formation or subsequent metabolic reactions,
3. To transport precursors, intermediates and products across the tonoplast, vesicular membranes or the plasma membrane by a variety of either ATP-driven pumps or ΔpH - energized transporters. The specificity of transport proteins ranges from very high to very broad and sometimes includes xenobiotics, heavy metals and primary metabolites.
4. To edit programs of morphological differentiation that include the expression of biosynthetic enzymes and proteins involved in transport, channeling, binding and detoxication of secondary products. These programs may coordinate multiple secondary pathways to allow for complex biosynthesis or for the potentiation of biological effects of the final products.

Acknowledgements Work in the former lab of MK was supported by the Swiss National Science Foundation. MK wishes to thank his former team, especially E. Martinoia, B. Weder, K. Marinova, C. Ballmann, K. Schmid, A. Polinceusz, H. Grob, B. Burla and T. Kretschmar (all Univ. of Zurich). MK acknowledges Daniel Studer, University of Bern, for his experimental help to obtain Fig. 3. Work in the lab of WR was supported by the Deutsche Forschungsgemeinschaft and the Excellence Cluster Sachsen-Anhalt. WR gratefully acknowledges contributions of D. Weiss, K. Färber, K. Viehweger, M. Heinze, M. Hieke, G. Danders and K. Thomasch.

References

Abrahams S, Lee E, Walker AR, Tanner GJ, et al. (2003) The Arabidopsis *TDS4* gene encodes leucoanthocyanidin dioxygenase (LDOX) and is essential for proanthocyanidin synthesis and vacuole development. Plant J 35:624–636

Achnine L, Blancaflor EB, Rasmussen S, et al. (2004) Colocalization of L-phenylalanine ammonia-lyase and cinnamate 4-hydroxylase for metabolic channeling in phenylpropanoid biosynthesis. Plant Cell 16:3098–3109

Alcantara J, Bird DA, Franceschi VR, et al. (2005) Sanguinarine biosynthesis is associated with the endoplasmic reticulum in cultured opium poppy cells after elicitor treatment. Plant Physiol 138:173–183

Alfenito MR, Souer E, Goodman CD, et al. (1998) Functional complementation of anthocyanin sequestration in the vacuole by widely divergent glutathione S-transferases. Plant Cell 10: 1135–1149

Allen RS, Miller JAC, Chitty JA, et al. (2008) Metabolic engineering of morphinan alkaloids by over-expression and RNAi suppression of salutaridinol7-O-acetyltransferase in opium poppy. Plant Biotech J 6:22–30

Allen RS, Millgate AG, Chitty JA, et al. (2004) RNAi-mediated replacement of morphine with the nonnarcotic alkaloid reticuline in opium poppy. Nature Biotech 22:1559–1568

Amann M, Wanner G, Zenk MH (1986) Intracellular compartmentation of two enzymes of berberine biosynthesis in plant cell cultures. Planta 167: 310–320

Anhalt S, Weissenböck G (1992) Subcellular localization of luteolin glucuronides and related enzymes in rye mesophyll. Planta 187:83–88

Apse MP, Aharon GS, Snedden WA, et al. (1999) Salt tolerance conferred by overexpression of a vacuolar Na^+/H^+ antiport in Arabidopsis. Science 285:1256–1258

Arakawa H, Clark WG, Psenak M, et al. (1992) Purification and characterization of dihydrobenzophenanthridine oxidase from elicited *Sanguinaria canadensis* cell cultures. Arch Biochem Biophys 299:1–7

Badri DV, Loyola-Vargas VM, Broeckling CD, et al. (2008) Altered profile of secondary metabolites in the root exudates of Arabidopsis ATP-Binding Cassette (ABC) transporter mutants. Plant Physiol 146:762–771

Bais HP, Park SW, Weir TL, et al. (2004) How plants communicate using the underground information superhighway. Trends Plant Sci 9:26–32

Bajaj NP, McLean MJ, Waring MJ, et al. (1990) Sequence-selective, pH-dependent binding to DNA of benzophenanthridine alkaloids. J Mol Recognit 3:48–54

Bartak P, Simanek V, Vlckova M, et al. (2003) Interactions of sanguinarine and chelerythrine with molecules containing a mercapto group. J Phys Org Chem 16:803–810

Bartholomew DM, Van Dyk DE, Lau S-MC, et al. (2002) Alternate energy-dependent pathways for the vacuolar uptake of glucose and glutathione conjugates. Plant Physiol 130:1562–1572

Barz W, Mackenbrock U (1994) Constitutive and elicitation induced metabolism of isoflavones and pterocarpans in chickpea (*Cicer arietinum*) cell suspension cultures. Plant Cell, Tissue Organ Culture 38:199–211

Baxter IR, Young JC, Armstrong G, et al. (2005) A plasma membrane H^+-ATPase is required for the formation of proanthocyanidins in the seed coat endothelium of *Arabidopsis thaliana*. Proc Natl Acad Sci USA 102:2649–2654

Bird DA, Facchini PJ (2001) Berberine bridge enzyme, a key branch-point enzyme in benzylisoquinoline alkaloid biosynthesis, contains a vacuolar sorting determinant. Planta 213:888–897

Bird D, Beisson F, Brigham A, et al. (2007) Characterization of Arabidopsis ABCG11/WBC11, an ATP binding cassette (ABC) transporter that is required for cuticular lipid secretion. Plant J 52:485–498

Birnbaum K, Shasha DE, Wang JY, et al. (2003) A gene expression map of the Arabidopsis root. Science 302:1956–1960

Blom TJM, Sierra M, van Vliet TB, et al. (1991) Uptake and accumulation of ajmalicine into isolated vacuoles of cultured cells of Catharanthus roseus (L.) G. Don. and its conversion into serpentine. Planta 183:170–177

Blume DE, Jaworski JG, McClure JW (1979) Uridinediphosphate-glucose: Isovitexin 7-O-glucosyltransferase from barley protoplasts: Subcellular localization. Planta 146:199–202

Bock A, Wanner G, Zenk MH (2002) Immunocytological localization of two enzymes involved in berberine biosynthesis. Planta 216:57–63

Braun R, Keller F (2000) Vacuolar chain elongation of raffinose oligosaccharides in Ajuga reptans. Funct Plant Biol 27:743–746

Brouillard R, Dangles O (1994) Flavonoids and flower colour. In: H Harbord, JB Harborne, eds, The flavonoids: advances in research since 1986. CRC Press, Boca Raton, FL, pp 565–588

Brown MH, Paulsen IT, Skurray RA (1999) The multidrug efflux protein NorM is a prototype of a new family of transporters. Mol Microbiol 31:394–395

Brown DE, Rashotte AM, Murphy AS, et al. (2001) Flavonoids act as negative regulators of auxin transport in vivo in Arabidopsis. Plant Physiol 126:524–535

Buer CS, Muday GK, Djordjevic MA (2007) Flavonoids are differentially taken up and transported long distances in Arabidopsis. Plant Physiol 145:478–490

Buer CS, Muday GK (2004) The *transparent testa4* mutation prevents flavonoid synthesis and alters auxin transport and the response of Arabidopsis roots to gravity and light. Plant Cell 16:1191–1205

Burchard P, Bilger W, Weissenböck G (2000) Contribution of hydroxycinnamates and flavonoids to epidermal shielding of UV-A and UV-B radiation in developing rye primary leaves as assessed by ultraviolet-induced chlorophyll fluorescence measurements. Plant Cell Environ 23:1373–1380

Burlat V, Oudin A, Courtois M, et al. (2004) Co-expression of three MEP pathway genes and geraniol 10-hydroxylase in internal phloem parenchyma of *Catharanthus roseus* implicates multicellular translocation of intermediates during the biosynthesis of monoterpene indole alkaloids and isoprenoid-derived primary metabolites. Plant J 38:131–141

Chong J, Baltz R, Schmitt C, et al. (2002) Downregulation of a pathogen-responsive tobacco UDP-Glc:phenylpropanoid glucosyltransferase reduces scopoletin glucoside accumulation, enhances oxidative stress, and weakens virus resistance. Plant Cell 14:1093–1107

Chou WM., Kutchan TM (1998) Enzymatic oxidations in the biosynthesis of complex alkaloids. Plant J 15:289–300

Cline SD, Coscia CJ (1989) Ultrastructural changes associated with the accumulation and secretion of sanguinarine in *Papaver bracteatum* suspension cultures treated with fungal elicitor. Planta 178:303–314

Cole SP, Bhardwaj G, Gerlach JH, et al. (1992) Overexpression of a transporter gene in a multidrug-resistant human lung cancer cell line. Science 258:1650–1654

Cole SPC, Deeley RG (2006) Transport of glutathione and glutathione conjugates by MRP1. Trends Pharmacol Sci 27:438–446

Collins NC, Thordal-Christensen H, Lipka V, et al. (2003) SNARE-protein-mediated disease resistance at the plant cell wall. Nature 425: 973–977

Conn S, Zhang W, Franco C (2003) Anthocyanic vacuolar inclusions (AVIs) selectively bind acylated anthocyanins in *Vitis vinifera* L. (grapevine) suspension culture. Biotechnol Lett 25:835–839

Costet L, Fritig B, Kauffmann S (2002) Scopoletin expression in elicitor-treated and tobacco mosaic virus-infected tobacco plants. Physiol Plant 115:228–235

De Luca, V. (1993) Indole alkaloid biosynthesis. In: Lea, P., ed, Methods in plant biochemistry. Enzymes of secondary metabolism, Vol. 9, pp 345–368, Academic Press

De Luca V, St-Pierre B (2000) The cell and developmental biology of alkaloid biosynthesis. Trends Plant Sci 4:168–173

De Luca V, Cutler AJ (1987) Subcellular localization of enzymes involved in indole alkaloid biosynthesis in *Catharanthus roseus*. Plant Physiol 85:1099–1102

Dean JV, Devarenne TP (1997) Peroxidase-mediated conjugation of glutathione to unsaturated phenylpropanoids. Evidence against glutathione S-transferase involvement. Physiol Plant 99: 271–278

Dean JV, Devarenne TP, Lee IS, et al. (1995) Properties of a maize glutathione S-transferase that conjugates coumaric acid and other phenylpropanoids. Plant Physiol 108:985–994

Dean JV, Mills JD (2004) Uptake of salicylic acid 2-O-ß-D-glucose into soybean tonoplast vesicles by an ATP-binding cassette transporter-type mechanism. Physiol Plant 120:603–612

Debeaujon I, Peeters AJ, Leon-Kloosterziel KM, et al. (2001) The TRANSPARENT TESTA12 gene of Arabidopsis encodes a multidrug secondary transporter-like protein required for flavonoid sequestration in vacuoles of the seed coat endothelium. Plant Cell 13:853–871

Debeaujon I, Nesi N, Perez P, et al. (2003) Proanthocyanidin-accumulating cells in Arabidopsis testa: regulation of differentiation and role in seed development. Plant Cell 15:2514–2531

Deeley RG, Cole SPC (2006) Substrate recognition and transport by multidrug resistance protein 1 (ABCC1). FEBS Lett 580:1103–1111

Dettmer J, Hong-Hermesdorf A, Stierhof Y-D, et al. (2006) Vacuolar H^+-ATPase activity is required for endocytic and secretory trafficking in Arabidopsis. Plant Cell 18:715–730

Deus-Neumann B, Zenk MH (1984) A highly selective alkaloid uptake system in vacuoles of higher plants. Planta 162:250–260

Deus-Neumann B, Zenk MH (1986) Accumulation of alkaloids in plant vacuoles does not involve an ion-trap mechanism. Planta 167:44–53

Diener AC, Gaxiola RA, Fink GR (2001) Arabidopsis *ALF5*, a multidrug efflux transporter gene family member, confers resistance to toxins. Plant Cell 13:1625–1638

Drose S, Altendorf K (1997) Bafilomycins and concanamycins as inhibitors of V-ATPases and P-ATPases. J Exp Biol 200:1–8

Dudler R, Hertig C (1992) Structure of an Mdr-like gene from *Arabidopsis thaliana* - Evolutionary implications. J Biol Chem 267:5882–5888

Durrett TP, Gassmann W, Rogers EE (2007) The FRD3-mediated efflux of citrate into the root vasculature is necessary for efficient iron translocation. Plant Physiol 144:197–205

Ehlting J, Mattheus N, Aeschliman DS, et al. (2005) Global transcript profiling of primary stems from *Arabidopsis thaliana* identifies candidate genes for missing links in lignin biosynthesis and transcriptional regulators of fiber differentiation. Plant J 42:618–640

Facchini PJ, Bird DA (1998) Developmental regulation of benzylisoquinoline alkaloid biosynthesis in opium poppy plants and tissue cultures. In Vitro Cellular and Developmental Biology Plant 34:69–79.

Facchini PJ, St-Pierre B (2005) Synthesis and trafficking of alkaloid biosynthetic enzymes. Curr Opin Plant Biol 8:657–666

Facchini PF (2001) Alkaloid biosynthesis in plants: biochemistry, cell biology, molecular regulation, and metabolic engineering applications. Annu Rev Plant Physiol Plant Mol Biol 52:29–66

Faddeeva MD, Beliaeva TN (1997) Sanguinarine and ellipticine cytotoxic alkaloids isolated from well-known antitumor plants. Intracellular targets of their action. Tsitologiia 39:181–208

Färber K, Schumann B, Miersch O, et al. (2003) Selective desensitization of jasmonate- and pH-dependent signaling in the induction of benzophenanthridine biosynthesis in cells of *Eschscholzia californica*. Phytochemistry 62:491–500

Forestier C, Frangne N, Eggmann T, et al. (2003) Differential sensitivity of plant and yeast MRP (ABCC)-mediated organic anion transport processes towards sulfonylureas. FEBS Lett 554:23–29

Frangne N, Eggmann T, Koblischke C, et al. (2002) Flavone glucoside uptake into barley mesophyll and Arabidopsis cell culture vacuoles. Energization occurs by H^+-antiport and ATP-binding cassette-type mechanisms. Plant Physiol 128:726–733

Frehner M, Keller F, Wiemken A (1984) Localization of fructan metabolism in the vacuoles isolated from protoplasts of Jerusalem artichoke tubers (*Helianthus tuberosus L.*). J Plant Physiol. 116:197–208

Galneder E, Rueffer M, Wanner G, et al. (1988) Alternative final steps in berberine biosynthesis in *Coptis japonica* cell cultures. Plant Cell Rep 7:1–4

Gao XQ, Li CG, Wei PC, et al. (2005) The dynamic changes of tonoplasts in guard cells are important for stomatal movement in *Vicia faba*. Plant Physiol 139:1207–1216

Gerardy R, Zenk MH (1993) Formation of salutaridine from (R)-Reticuline by a membrane bound cytochrome P-450 enzyme from '*Papaver somniferum*'. Phytochemistry 32, Nr. 1:79–86

Goodman CD, Casati P, Walbot V (2004) A multidrug resistance-associated protein involved in anthocyanin transport in *Zea mays*. Plant Cell 16:1812

Grandmaison J, Ibrahim RK (1996) Evidence for nuclear protein binding of flavonol sulfate esters in *Flaveria chloraefolia*. J Plant Physiol 147:653–660

Grotewold E, Chamberlin M, Snook M, et al. (1998) Engineering secondary metabolism in maize cells by ectopic expression of transcription factors. Plant Cell 10:721–740

Guo Z, Severson RF, Wagner GJ (1994) Biosynthesis of the diterpene cis-abienol in cell-free extracts of tobacco trichomes. Arch Biochem Biophys 308:103–108

Guz NR, Stermitz FR, Johnson JB, et al. (2001) Flavonolignan and flavone inhibitors of a *Staphylococcus aureus* multidrug resistance pump: structure – activity relationships. J Med Chem 44:261–268

Haralampidis K, Bryan G, Qi X, et al. (2001) A new class of oxidosqualene cyclases directs synthesis of antimicrobial phytoprotectants in monocots. Proc Natl Acad Sci USA 98: 13431–13436

Harborne JB, Williams CA (2000) Advances in flavonoid research since 1992. Phytochemistry 55:481–504

Harborne JB, Williams CA, Greenham J, et al. (1994) Variations in the lipophilic and vacuolar flavonoids of the genus Vellozia. Phytochemistry 35:1475–1480

Harborne JB (1988) Flavonoids in the environment: structure-activity relationships. Prog Clin Biol Res 280:17–27

Harper JF, Manney L, Sussman MR (1994) The plasma membrane H^+ ATPase gene family in Arabidopsis: Genomic sequence of *AHA10* which is expressed primarily in developing seeds. Mol Gen Genet 244:572–587

11

Hartmann T. (1999) Chemical ecology of pyrrolizidine alkaloids. Planta, 207, 483–495
Hashimoto T. Yamada Y. (1994) Alkaloid biogenesis: molecular aspects. Annu Rev Plant Physiol Plant Mol Biol, 45, 257–285.
Hause B, Meyer K, Viitanen PV, et al. (2002) Immunolocalization of 1-O-sinapoylglucose: malate sinapoyltransferase in *Arabidopsis thaliana*. Planta 215:26–32
Hauser M-T, Wink M (1990) Uptake of alkaloids by latex vesicles and isolated mesophyll vacuoles of Chelidonium majus (Papaverceae). Z. Naturforsch 45c, 949–995
Hinder B, Schellenberg M, Rodon S, et al. (1996) How plants dispose of chlorophyll catabolites – Directly energized uptake of tetrapyrrolic breakdown products into isolated vacuoles. J Biol Chem 271:27233–27236
Hopp W, Seitz HU (1987) The uptake of acylated anthocyanin into isolated vacuoles from a cell suspension culture of *Daucus carota*. Planta 170:74–85
Hsieh K, Huang AHC (2007) Tapetosomes in *Brassica* tapetum accumulate endoplasmic reticulum-derived flavonoids and alkanes for delivery to the pollen surface. Plant Cell 19:582–596
Hutzler P, Fischbach R, Heller W, et al. (1998) Tissue localization of phenolic compounds in plants by confocal laser scanning microscopy. J Exp Bot 49:953–965
Ibrahim RK, De Luca V, Khouri H, et al. (1987) Enzymology and compartmentation of polymethylated flavonol glucosides in *Chrysosplenium americanum*. Phytochemistry 26:1237–1245
Ibrahim R (1990) Immunocytochemical localization of plant secondary metabolites and enzymes involved in their biosynthesis. Phytochem Anal 1:49–59
Ikezawa N, Tanaka M, Nagayoshi M, et al. (2003) Molecular cloning and characterization of CYP719, a methylenedioxy bridge-forming enzyme that belongs to a novel P450 family, from cultured *Coptis japonica* cells. J Biol Chem 278:38557–38565
Irani NG, Grotewold E (2005) Light-induced morphological alteration in anthocyanin-accumulating vacuoles of maize cells. BMC Plant Biol 5:7
Jans B (1974) Untersuchungen am Milchsaft des Schöllkrauts (*Chelidonium majus*) Ber. Schweiz Bot Ges 83:306–344
Jasinski M, Stukkens Y, Degand H, et al. (2001) A plant plasma membrane ATP binding cassette-type transporter is involved in antifungal terpenoid secretion. Plant Cell 13:1095–1107
Jorgensen K, Rasmussen AV, Morant M, et al. (2005) Metabolon formation and metabolic channeling in the biosynthesis of plant natural products. Curr Opin Plant Biol 8:280–291
Kaneda M, Rensing KH, Wong JCT, et al. (2008) Tracking monolignols during wood development in lodgepole pine. Plant Physiol. 147:1750–1760
Kempel(2008)AnalysederBenzylisochinolinalkaloid-Biosynthese des *Papaver somniferum* L. mit Hilfe der RNAi-Technik. Dissertation (Ph.D.) Martin-Luther-Universität Halle, 2008.
Kitamura S (2006) Transport of flavonoids. In: E Grotewold, ed, The science of flavonoids. Springer Science and Business Media, New York, pp 123–146
Kitamura S, Shikazono N, Tanaka A (2004) *TRANSPARENT TESTA 19* is involved in the accumulation of both anthocyanins and proanthocyanidins in Arabidopsis. Plant J 37: 104–114
Klein M, Martinoia E, Hoffmann-Thoma G, et al. (2000) A membrane-potential dependent ABC-like transporter mediates the vacuolar uptake of rye flavone glucuronides: regulation of glucuronide uptake by glutathione and its conjugates. Plant J 21:289–304
Klein M, Martinoia E, Hoffmann-Thoma G, et al. (2001) The ABC-like vacuolar transporter for rye mesophyll flavone glucuronides is not species-specific. Phytochemistry 56:153–159
Klein M, Martinoia E, Weissenböck G (1997) Transport of lucifer yellow CH into plant vacuoles – evidence for direct energization of a sulphonated substance and implications for the design of new molecular probes. FEBS Lett 420:86–92
Klein M, Martinoia E, Weissenböck G (1998) Directly energized uptake of ß-estradiol 17-(ß-D-glucuronide) in plant vacuoles is strongly stimulated by glutathione conjugates. J Biol Chem 273:262–270
Klein M, Perfus-Barbeoch L, Frelet A, et al. (2003) The plant multidrug resistance ABC transporter AtMRP5 is involved in guard cell hormonal signalling and water use. Plant J 33:119–129
Klein M, Weissenböck G, Dufaud A, et al. (1996) Different energization mechanisms drive the vacuolar uptake of a flavonoid glucoside and a herbicide glucoside. J Biol Chem 271: 29666–29671
Kreuz K, Tommasini R, Martinoia E (1996) Old enzymes for a new job - Herbicide detoxification in plants. Plant Physiol 111:349–353

Kutchan TM, Rush M, Coscia CJ (1986) Subcellular localization of alkaloids and dopamine in different vacuolar compartments of *Papaver bracteatum*. Plant Physiol 81:61–166

Larsen ES, Alfenito MR, Briggs WR, et al. (2003) A carnation anthocyanin mutant is complemented by the glutathione S-transferases encoded by maize *Bz2* and petunia *An9*. Plant Cell Rep 21:900–904

Latchinian-Sadek L, Ibrahim RK (1991) Flavonol ring B-specific *O*-glucosyltransferases: purification, production of polyclonal antibodies, and immunolocalization. Arch Biochem Biophys 289:230–236

Lee KS, Tsien RW (1983) Mechanism of calcium channel blockade by verapamil, D600, diltiazem and nitrendipine in single dialysed heart cells. Nature 302:790–794

Lehfeldt C, Shirley AM, Meyer K, et al. (2000) Cloning of the *SNG1* gene of Arabidopsis reveals a role for a serine carboxypeptidase-like protein as an acyltransferase in secondary metabolism. Plant Cell 12:1295–1306

Leier I, Jedlitschky G, Buchholz U, et al. (1994) The MRP gene encodes an ATP-dependent export pump for leukotriene C4 and structurally related conjugates. J Biol Chem 269:27807–27810

Lenz R, Zenk MH (1995) Purification and properties of codeinone reductase (NADPH) from *Papaver somniferum* cell cultures and differentiated plants. Eur J Biochem 233:132–139

Lepiniec L, Debeaujon I, Routaboul JM, et al. (2006) Genetics and biochemistry of seed flavonoids. Annu Rev Plant Biol 57:405–430

Leslie EM, Deeley RG, Cole SPC (2003) Bioflavonoid stimulation of glutathione transport by the 190-kDa multidrug resistance protein 1 (MRP1). Drug Metab Dispos 31:11–15

Leslie EM, Mao Q, Oleschuk CJ, et al. (2001) Modulation of multidrug resistance protein 1 (MRP1/ABCC1) transport and ATPase activities by interaction with dietary flavonoids. Mol Pharmacol 59:1171–1180

Li ZS, Alfenito M, Rea PA, et al. (1997) Vacuolar uptake of the phytoalexin medicarpin by the glutathione conjugate pump. Phytochemistry 45:689–693

Li L, He Z, Pandey GK, et al. (2002) Functional cloning and characterization of a plant efflux carrier for multidrug and heavy metal detoxification. J Biol Chem 277:5360–5368

Li ZS, Szczypka M, Lu YP, et al. (1996) The yeast cadmium factor protein (YCF1) is a vacuolar glutathione S-conjugate pump. J Biol Chem 271:6509–6517

Li ZS, Zhao Y, Rea PA (1995) Magnesium adenosine 5'-triphosphate-energized transport of glutathione-S-conjugates by plant vacuolar membrane vesicles. Plant Physiol 107:1257–1268

Lim EK, Li Y, Parr A, et al. (2001) Identification of glucosyltransferase genes involved in sinapate metabolism and lignin synthesis in Arabidopsis. J Biol Chem 276: 4344–4349

Lin Y, Irani NG, Grotewold E (2003) Sub-cellular trafficking of phytochemicals explored using auto-fluorescent compounds in maize cells. BMC Plant Biol 3:10

Liu G, Sanchez-Fernandez R, Li ZS, et al. (2001) Enhanced multispecificity of Arabidopsis vacuolar multidrug resistance-associated protein-type ATP-binding cassette transporter, AtMRP2. J Biol Chem 276:8648–8656

Loyola-Vargas VM, Broeckling CD, Badri D, et al. (2007) Effect of transporters on the secretion of phytochemicals by the roots of *Arabidopsis thaliana*. Planta 225:301–310

Lu YP, Li ZS, Rea PA (1997) AtMRP1 gene of Arabidopsis encodes a glutathione S-conjugate pump: isolation and functional definition of a plant ATP-binding cassette transporter gene. Proc Natl Acad Sci USA 94:8243–8248

Lu YP, Li ZS, Drozdowicz YM, et al. (1998) AtMRP2, an Arabidopsis ATP binding cassette transporter able to transport glutathione S-conjugates and chlorophyll catabolites: Functional comparisons with AtMRP1. Plant Cell 10:267–282

Luo B, Xue XY, Hu WL, et al. (2007) An ABC transporter gene of *Arabidopsis thaliana*, *AtWBC11*, is involved in cuticle development and prevention of organ fusion. Plant Cell Physiol 48:1790–1802

Madyastha KM, Ridgway JE, Dwyer JG, et al. (1977) Subcellular localization of a cytochrome P-450-dependent monooxygenase in vesicles of the higher plant *Catharanthus roseus*. J Cell Biol 72:303–313

Magalhaes JV, Liu J, Guimares CT, et al. (2007) A gene in the multidrug and toxic compound extrusion (MATE) family confers aluminum tolerance in sorghum. Nature Genetics 39:1156–1161

Mahady GB, Beecher CW (1994) Quercetin-induced benzophenanthridine alkaloid production in suspension cell cultures of *Sanguinaria canadensis*. Planta Med 60:553–557

Marinova K, Kleinschmidt K, Weissenböck G, et al. (2007a) Flavonoid biosynthesis in barley primary leaves requires the presence of the vacuole and controls the activity of vacuolar flavonoid transport. Plant Physiol 144:432–444

11

Marinova K, Pourcel L, Weder B, et al. (2007) The Arabidopsis MATE transporter TT12 acts as a vacuolar flavonoid/H^+-antiporter active in proanthocyanidin-accumulating cells of the seed coat. Plant Cell 19:2023–2038

Markham KR, Gould KS, Winefield CS, et al. (2000) Anthocyanic vacuolar inclusions - their nature and significance in flower colouration. Phytochemistry 55:327–336

Marrs KA (1996) The functions and regulation of glutathione S-transferases in plants. Annu Rev Plant Physiol Plant Mol Biol 47:127–158

Marrs KA, Alfenito MR, Lloyd AM, et al. (1995) A glutathione S-transferase involved in vacuolar transfer encoded by the maize gene *Bronze-2*. Nature 375:397

Martinoia E, Grill E, Tommasini R, et al. (1993) ATP-dependent glutathione S-conjugate export pump in the vacuolar membrane of plants. Nature 364:247–249

Martinoia E, Klein M, Sanchez-Fernandez R, et al. (2000) Vacuolar uptake of secondary metabolites and xenobiotics. In: DG Robinson, JC Rogers, eds, Vacuolar compartments, Vol 5. Sheffield Academic Press, Sheffield, pp 221–253

Masuda S, Terada T, Yonezawa A, et al. (2006) Identification and functional characterization of a new human kidney-specific H^+/organic cation antiporter, kidney-specific Multidrug and Toxin Extrusion 2. J Am Soc Nephrol 17:2127–2135

Matern U (1987) Die Isomerenfall für Sekundärmetabolite, eine Alternative zum Ionenfallen-Modell. Biologie in unserer Zeit 17:148–152

Matern U, Heller W, Himmelspach K (1983) Conformational-changes of apigenin 7-*O*-(6-*O*-malonylglucoside), a vacuolar pigment from parsley, with solvent composition and proton concentration. Eur J Biochem 133:439–448

Matern U, Reichenbach C, Heller W (1986) Efficient uptake of flavonoids into parsley (*Petroselinum hortense*) vacuoles requires acylated glycosides. Planta 167:183–189

Matile P (1976) Lokalization of alkaloids and mechanism of their accumulation in vacuoles of *Chelidonium majus* lacticifers. Nova Acta Leopoldina, Suppl. Nr 7, Symposium Secondary Metabolism and Coevolution, 139–156

Matile P (1984) Das toxische Kompartiment der Pflanzenzelle. Naturwiss. 71:18–24

Matros A, Mock H-P (2004) Ectopic expression of a UDP-glucose:phenylpropanoid glucosyltransferase leads to increased resistance of transgenic tobacco plants against infection with Potato Virus Y. Plant Cell Physiol 45:1185–1193

McKnight TD, Bergey DR, Burnett RJ, et al. (1991) Expression of enzymatically active and correctly targeted strictosidine synthase in transgenic tobacco plants. Planta 185:148–152

Morita Y, Kodama K, Shiota S, et al. (1998) NorM, a putative multidrug efflux protein, of *Vibrio parahaemolyticus* and its homolog in *Escherichia coli*. Antimicrob Agents Chemother 42:1778–1782

Morita Y, Kataoka A, Shiota S, et al. (2000) NorM of *Vibrio parahaemolyticus* is an Na^+-driven multidrug efflux pump. J Bacteriol 182:6694–6697

Mueller LA, Goodman CD, Silady RA, et al. (2000) AN9, a petunia glutathione S-transferase required for anthocyanin sequestration, is a flavonoid-binding protein. Plant Physiol 123:1561–1570

Murata J, De Luca V (2005) Localization of tabersonine 16-hydroxylase and 16-OH tabersonine-16-O-methyltransferase to leaf epidermal cells defines them as a major site of precursor biosynthesis in the vindoline pathway in Catharanthus roseus. Plant J 44: 581–594

Mylona P, Owatworakit A, Papadopoulou K, et al. (2008) *Sad3* and *Sad4* are required for saponin biosynthesis and root development in oat. Plant Cell 20:201–212

Nakagawa K, Konagai A, Fukui H, et al. (1984) Release and crystallization of berberine in the liquid medium of *Thalictrum minus* cell suspension cultures. Plant Cell Reports 3:254–257

Nawrath C, Heck S, Parinthawong N, et al. (2002) EDS5, an essential component of salicylic acid-dependent signaling for disease resistance in Arabidopsis, is a member of the MATE transporter family. Plant Cell 14:275–286

Nielsen KA, Gotfredsen CH, Buch-Pedersen MJ, et al. (2004) Inclusions of flavonoid 3-deoxyanthocyanidins in *Sorghum bicolor* self-organize into spherical structures. Physiol Molec Plant Pathol 65:187–196

Nozue M, Yasuda H (1985) Occurrence of anthocynoplasts in cell suspension culture of sweet potato. Plant Cell Reports 4:252–255

Omote H, Hiasa M, Matsumoto T, et al. (2006) The MATE proteins as fundamental transporters of metabolic and xenobiotic organic cations. Trends Pharmacol Sci 27:587–593

Ono E, Hatayama M, Isono Y, et al. (2006) Localization of a flavonoid biosynthetic polyphenol oxidase in vacuoles. Plant J 45:133–143

Onyilagha JC, Grotewold E (2004) The biology and structural distribution of surface flavonoids Recent Res Devel Plant Sci 2:1–18

Osbourn AE (1996) Preformed antimicrobial compounds and plant defense against fungal attack. Plant Cell 8:1821–1831

OsbournAE(2003)Saponinsincereals.Phytochemistry 62:1–4

Otani M, Shitan N, Sakai K, et al. (2005) Characterization of vacuolar transport of the endogenous alkaloid berberine in *Coptis japonica*. Plant Physiol 138:1939–1946

Otsuka M, Matsumoto T, Morimoto R, et al. (2005) A human transporter protein that mediates the final excretion step for toxic organic cations. Proc Natl Acad Sci USA 102:17923–17928

Panicot M, Minguet EG, Ferrando A, et al. (2004) A polyamine metabolon involving aminopropyl transferase complexes in Arabidopsis. Plant Cell 14:2539–2551

Panikashvili D, Savaldi-Goldstein S, Mandel T, et al. (2007) The Arabidopsis DESPERADO/AtWBC11 transporter is required for cutin and wax secretion. Plant Physiol 145:1345–1360

Paulsen IT, Brown MH, Skurray RA (1996) Proton-dependent multidrug efflux systems. Microb Mol Biol Rev 60:575–608

Pecket RC, Small CJ (1980) Occurrence, location and development of anthocyanoplasts. Phytochemistry 19:2571–2576

Peer WA, Brown DE, Tague BW, et al. (2001) Flavonoid accumulation patterns of *transparent testa* mutants of Arabidopsis. Plant Physiol 126:536–548

Peer WA, Bandyopadhyay A, Blakeslee JJ, et al. (2004) Variation in expression and protein localization of the PIN family of auxin efflux facilitator proteins in flavonoid mutants with altered auxin transport in *Arabidopsis thaliana*. Plant Cell 16:1898–1911

Pighin JA, Zheng H, Balakshin LJ, et al. (2004) Plant cuticular lipid export requires an ABC transporter. Science 306:702–704

Polster J, Dithmar H, Burgemeister R, et al. W (2006) Flavonoids in plant nuclei: detection by laser microdissection and pressure catapulting (LMPC), in vivo staining, and UV-visible spectroscopic titration. Physiol Plant 128: 163–174

Poustka F, Irani NG, Feller A, et al. (2007) A trafficking pathway for anthocyanins overlaps with the endoplasmic reticulum-to-vacuole protein-sorting route in Arabidopsis and contributes to the formation of vacuolar inclusions. Plant Physiol 145:1323–1335

Rasmussen S, Dixon RA (1999) Transgene-mediated and elicitor-induced perturbation of metabolic channeling at the entry point into the phenylpropanoid pathway. Plant Cell 11:1537–1552

Rataboul P, Alibert G, Boller T, et al. (1985) Intracellular transport and vacuolar accumulation of *o*-coumaric acid glucoside in *Melilotus alba* mesophyll cell protoplasts. Biochim Biophys Acta 816:25–36

Rea PA, Li ZS, Lu YP, et al. (1998) From vacuolar GS-X pumps to multispecific ABC transporters. Annu Rev Plant Physiol Plant Mol Biol 49:727–760

Rogers EE, Guerinot ML (2002) FRD3, a member of the multidrug and toxin efflux family, controls iron deficiency responses in Arabidopsis. Plant Cell 14:1787–1799

Roos, W., Luckner, M. (1986) The spatial organization of secondary metabolism in microbial and plant cells. In: T. A. V. Subramanian, ed, Cell metabolism: growth and environment, Vol. 1. CRC Press, Boca Raton, FL, pp 45–73

Rueffer M, Amman M, Zenk MH (1986) S-Adenosyl-L-methionine: columbamine-O-methyl transferase, a compartmentalized enzyme in protoberberine biosynthesis. Plant Cell Rep 3: 182–185

Russel FGM, Koenderink JB, Masereeuw R (2008) Multidrug resistance protein 4 (MRP4/ABCC4): a versatile efflux transporter for drugs and signalling molecules. Trends Pharmacol Sci 29: 200–207

Saier MH, Jr., Paulsen IT (2001) Phylogeny of multidrug transporters. Semin Cell Dev Biol 12:205–213

Saito K, Suzuki H, Takamatsu S, Murakoshi I, (1992) Acyltransferases for lupin alkaloids in *Lupinus hirsutus*. Phytochemistry 32: 87–91

Sakai K, Shitan N, Sato F, et al. (2002) Characterization of berberine transport into *Coptis japonica* cells and the involvement of ABC protein. J Exp Bot 53:1879–1886

Samanani N, Alcantara J, Bourgault R, et al. (2006) The role of phloem sieve elements and laticifers in the biosynthesis and accumulation of alkaloids in opium poppy. Plant J 47:547–563

Samanani N, Park SU, Facchini PJ (2005) Cell type-specific localization of transcripts encoding nine consecutive enzymes involved in protoberberine alkaloid biosynthesis. Plant Cell 17:915–926

Santelia D, Henrichs S, Vincenzetti V, et al. (2008). Flavonoids redirect PIN-mediated polar auxin fluxes during root gravitropic responses. J. Biol. Chem. 283:31218–31226.

Saslowsky DE, Warek U, Winkel BS (2005) Nuclear localization of flavonoid enzymes in Arabidopsis. J Biol Chem 280:23735–23740

Schmeller T, Latz-Brüning B, Wink M (1997) Biochemical activities of berberine, palmatine and sanguinarine mediating chemical defence against microorganisms and herbivores. Phytochemistry 44:257–266

11

Schulz B, Frommer WB (2004) A plant ABC transporter takes the lotus seat. Science 306:622–625

Schulz M, Weissenböck G (1986) Isolation and separation of epidermal and mesophyll protoplasts from rye primary leaves – tissue specific characteristics of secondary phenolic product accumulation. Z Naturforsch [C] 41:22–27

Schumacher HM, Zenk MH (1988) Partial purification and characterization of dihydrobenzophenanthridine oxidase from *Eschscholtzia californica* cell suspension cultures. Plant Cell Rep 7:43–46

Sharma V, Strack D (1985) Vacuolar localization of 1-sinapolglucose: l-malate sinapoyltransferase in protoplasts from cotyledons of *Raphanus sativus*. Planta 163:563–568

Shiono M, Matsugaki N, Takeda K (2005) Phytochemistry: structure of the blue cornflower pigment. Nature 436:791

Shirley AM, McMichael CM, Chapple C (2001) The *sng2* mutant of Arabidopsis is defective in the gene encoding the serine carboxypeptidase-like protein sinapoylglucose:choline sinapoyltransferase. Plant J 28:83–94

Shirley AM, Chapple C (2003) Biochemical characterization of sinapoylglucose:choline sinapoyltransferase, a serine carboxypeptidase-like protein that functions as an acyltransferase in plant secondary metabolism. J Biol Chem 278:19870–19877

Shitan N, Bazin I, Dan K, et al. (2003) Involvement of CjMDR1, a plant multidrug-resistance-type ATP-binding cassette protein, in alkaloid transport in *Coptis japonica*. Proc Natl Acad Sci USA 100:751–756

Shitan N, Kiuchi F, Sato F, Yazaki K, Yoshimatsu K. (2005) Establishment of Rhizobium-mediated transformation of *Coptis japonica* and molecular analyses of transgenic plants. Plant Biotechnol., 22,:113–118.

Shitan N, Yazaki K (2007) Accumulation and membrane transport of plant alkaloids. Current Pharmaceutical Biotechnology 8:244–252

Slaninova I., Taborska E, Bochorakova H, et al. (2001) Interaction of benzo[c]phenanthridine and protoberberine alkaloids with animal and yeast cells. Cell Biol Toxicol 17:51–63

Small CJ, Pecket RC (1982) The ultrastructure of anthocyanoplasts in red cabbage. Planta 154:97–99

Snyder BA, Nicholson RL (1990) Synthesis of phytoalexins in sorghum as a site-specific response to fungal ingress. Science 248:1637–1639

Sondergaard TE, Schulz A, Palmgren MG (2004) Energization of transport processes in plants. Roles of the plasma membrane H^+-ATPase. Plant Physiol 136:2475–2482

Spelt C, Quattrocchio F, Mol J, et al. (2002) *ANTHOCYANIN1* of petunia controls pigment synthesis, vacuolar pH, and seed coat development by genetically distinct mechanisms. Plant Cell 14:2121–2135

St-Pierre B, Vazquez-Flota FA, De Luca V (1999) Multicellular compartmentation of *Catharanthus roseus* alkaloid biosynthesis predicts intercellular translocation of a pathway intermediate. Plant Cell 11:887–900

St-Pierre B, De Luca V (1995) A cytochrome P-450 monooxygenase catalyzes the first step in the conversion of tabersonine to vindoline in *Catharanthus roseus*. Plant Physiol 109:131–139

Stafford HA (1990) Flavonoid metabolism. CRC Press, Boca Raton, FL

Stermitz FR, Lorenz P, Tawara JN, et al. (2000) Synergy in a medicinal plant: antimicrobial action of berberine potentiated by 5'-methoxyhydnocarpin, a multidrug pump inhibitor. PROC NATL ACAD SCI USA 97:1433–1437

Stevens LH, Blom TJM, Verpoorte R (1993) Subcellular localization of tryptophan decarboxylase, strictosidine synthase and strictosidine glucosidase in suspension cultured cells of *Catharanthus roseus* and *Tabernaemontana divaricata*. Plant Cell Rep 12:563–576

Strack D (1982) Development of 1-*O*-sinapoyl-b-D-glucose: L-malate sinapoyltransferase activity in cotyledons of red radish (*Raphanus sativus* L. var. sativus). Planta 155:31–36

Sugiyama A, Shitan N, Yazaki K (2007) Involvement of a soybean ATP-binding cassette-type transporter in the secretion of genistein, a signal flavonoid in legume-rhizobium symbiosis. Plant Physiol 144:2000–2008

Suzuki H, Koike Y, Murakoshi I, et al. (1996) Subcellular localization of acyltransferases for quinolizidine alkaloid biosynthesis in *Lupinus*. Phytochemistry 42:1557–1562

Taguchi G, Fujikawa S, Yazawa T, et al. (2000) Scopoletin uptake from culture medium and accumulation in the vacuoles after conversion to scopolin in 2,4-D-treated tobacco cells. Plant Sci 151:153–161

Terasaka K, Sakai K, Sato F, et al. (2003) *Thalictrum minus* cell cultures and ABC-like transporter. Phytochemistry 62:483–489

Ukitsu H, Kuromori T, Toyooka K, et al. (2007) Cytological and biochemical analysis of COF1, an Arabidopsis mutant of an ABC transporter gene. Plant Cell Physiol 48:1524–1533

Valant-Vetschera KM, Wollenweber E (2001) Exudate flavonoid aglycones in the alpine species of *Achillea sect. Ptarmica*: Chemosystematics of A. moschata and related species (Compositae-Anthemideae). Biochem Syst Ecol 29:149–159

Valant-Vetschera KM, Wollenweber E, Faure R, et al. (2003) New exudate flavonoids of species from the Chrysanthemum complex (Asteraceae-Anthemideae). Biochem Syst Ecol 31:545–548

Verrier PJ, Bird D, Burla B, et al. (2008) Plant ABC proteins – a unified nomenclature and updated inventory. Trends Plant Sci 13:151–159

Viehweger K, Lein W, Schumann B, Roos W, et al. (2006). A G protein controls a pH dependent signal path to the induction of phytoalexin biosynthesis in *Eschscholzia californica*. Plant Cell 18:1510–1523

Walczak HA, Dean JV (2000) Vacuolar transport of the glutathione conjugate of trans-cinnamic acid. Phytochemistry 53:441–446

Walker TS, Bais HP, Grotewold E, et al. (2003) Root exudation and rhizosphere biology. Plant Physiol 132:44–51

Wang E, Wagner GJ (2003) Elucidation of the functions of genes central to diterpene metabolism in tobacco trichomes using posttranscriptional gene silencing. Planta 216:686–691

Wang J, Raman H, Zhou M, et al. (2007) High-resolution mapping of the *Alp* locus and identification of a candidate gene *HvMATE* controlling aluminium tolerance in barley (*Hordeum vulgare* L.). Theor Appl Genet 115:265–276

Weid M, Ziegler J, Kutchan TM (2004) The roles of latex and the vascular bundle in morphine biosynthesis in the opium poppy, *Papaver somniferum*. Proc Natl Acad Sci USA 101:13957–13962

Weiss D, Baumert A, Vogel M, Roos W (2006) Sanguinarine reductase, a key enzyme of benzophenanthridine detoxification. Plant Cell Environ 29:291–302

Weisskopf L, Abou-Mansour E, Fromin N, et al. (2006) White lupin has developed a complex strategy to limit microbial degradation of secreted citrate required for phosphate acquisition. Plant Cell Environ 29:919–927

Werner C, Matile P (1985) Accumulation of coumarylglucosides in vacuoles of barley mesophyll protoplasts. J Plant Physiol 118:237–249

Wink M (1997) Compartmentation of secondary metabolites and xenobiotics in plant vacuoles. In: RA Leigh, D Sanders, JA Callow, eds, The plant vacuole. Advances in Botanical Reserach, Vol 25. Academic Press, London/New York, pp 141–169

Wink M, Roberts MF (1998) Compartmentation of alkaloid biosynthesis, transport and storage. In: Roberts MF, Wink M eds, Alkaloids: biochemistry, ecology and medicinal applications, Plenum Press, New York, pp 239–262

Wink M, Schmeller T, Latz-Bruning B (1998) Modes of action of allelochemical alkaloids: Interaction with neuroreceptors, DNA, and other molecular targets. J Chem Ecol 24: 1881–1937

Winkel BS (2004) Metabolic channeling in plants. Annu Rev Plant Biol 55:85–107

Wolff J, Knipling L (1993) Antimicrotubule properties of benzophenanthridine alkaloids. Biochem 32:13334–13339

Wollenweber E, Dorr M, Rivera D, et al. (2003) Externally accumulated flavonoids in three Mediterranean *Ononis* species. Z Naturforsch [C] 58:771–775

Wu CP, Calcagno AM, Hladky SB, Ambudkar SV, Barrand MA (2005) Modulatory effects of plant phenols on human multidrug-resistance proteins 1, 4 and 5 (ABCC1, 4 and 5). FEBS J. 272: 4725–4740

Xu LF, Chu WJ, Qing XY, et al. (2006) Protopine inhibits serotonin transporter and noradrenaline transporter and has the antidepressant-like effect in mice models. Neuropharmacology 50:934–940

Yazaki K (2005) Transporters of secondary metabolites. Curr Opin Plant Biol 8:301–307

Yazaki K (2006) ABC transporters involved in the transport of plant secondary metabolites. FEBS Lett 580:1183–1191

Yazaki K, Shitan N, Takamatsu H, et al. (2001) A novel *Coptis japonica* multidrug resistance protein preferentially expressed in the alkaloid-accumulating rhizome. J Exp Bot 52: 877–879

Yazaki K, Sugiyama A, Morita M, et al. (2007) Secondary transport as an efficient membrane transport mechanism for plant secondary metabolites. Phytochem Rev. DOI 10.1007/s11101-007-9079-8

Zenk MH (1994) The formation of benzophenanthridine alkaloids. Pure Appl Chem 66:2023–2028

Zhang H, Wang L, Deroles S, et al. (2006) New insight into the structures and formation of anthocyanic vacuolar inclusions in flower petals. BMC Plant Biology 6:29

Ziegler J, Facchini PJ (2008) Alkaloid Biosynthesis: Metabolism and Trafficking. Annu Rev Plant Biol 59:735–769

Participation of Phytochemicals in Plant Development and Growth 12

Lucille Pourcel and Erich Grotewold

Abstract Phytochemicals, also known as natural products and specialized compounds, display well known functions in plants providing varying levels of protection to biotic and abiotic stress conditions. The biosynthesis of phytochemicals is tightly spatio-temporally regulated, often restricted to specialized cells, yet their transport within plants allow them to interact with, and modulate, other signalling networks. In this chapter, we describe how phytochemicals participate in plant development and growth, further blurring the boundaries between primary and secondary metabolism, and between hormones and phytochemicals.

Introduction

As part of their adaptation to the environment, plants accumulate a large number of chemical compounds or phytochemicals, often also described as secondary metabolites. The number of phytochemicals has been estimated in the hundreds of thousands [1]. These compounds can be classified into large families including, among others, flavonoids, terpenoids, glucosinolates and alkaloids. Many of these compounds function as phytoalexins or as phytoanticipins [2, 3], increasing tolerance of plants to herbivore or pathogen attack. Others participate in shielding plants from unfavourable environmental conditions, such as high UV [4] or drought. Beyond these classical functions of phytochemicals in the response to biotic or abiotic stress conditions, an increasing number of studies show an involvement of particular types of phytochemicals in developmental processes, implicating them as integral components of the plant signalling machinery. This raises the question of whether phytochemicals can act as phytohormones by displaying primary functions in plants, as it has been previously suggested [5]. This chapter will summarize these findings, with a special focus on the emerging involvement of the flavonoid and glucosinolate families of phytochemicals (Fig. 1) in plant development and growth. In the last section of the chapter, we describe how phytochemicals can be part of a signalling network. Two examples are used to exemplify these functions. First, we describe the movement of phytochemicals within plants, allowing them to act like signal molecules and interact with other plant networks (highlighted by the case of allelopathy). Second, and exemplified by the concept of xenohormesis, we describe how chemicals produced by plants can interact with signalling networks in other organisms, such as for example humans.

L. Pourcel and E. Grotewold (✉)
Department of Plant Cellular and Molecular Biology, The Ohio State University, Columbus, Ohio, USA
e-mails: grotewold.1@osu.edu; pourcel.1@osu.edu

A.E. Osbourn and V. Lanzotti (eds.), *Plant-derived Natural Products*,
DOI 10.1007/978-0-387-85498-4_12, © Springer Science+Business Media, LLC 2009

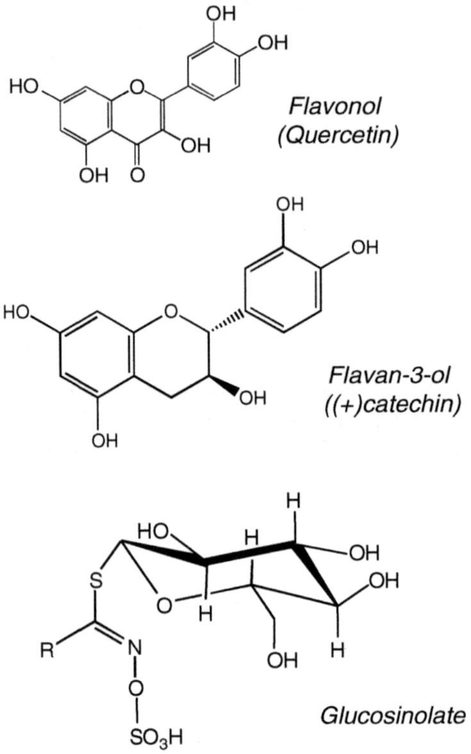

Fig. 1 Chemical structure of some phytochemicals involved in plant growth and development. One example is provided for each class of phytochemicals

Temporal and Spatial Accumulation of Phytochemicals

The accumulation of phytochemicals can be regulated by endogenous signals (e.g., hormones), or induced by exogenous/environmental conditions. The specific type of compounds, their level of accumulation and the spatio-temporal distribution of phytochemicals varies between different plant species. For example, during development and growth, plants accumulate monoterpenes and sesquiterpenes. These volatile compounds are primarily released from aerial plant parts and play a major role in the interaction between plants and the environment (e.g., defense against pathogens) [6]. Arabidopsis glucosinolates accumulate mainly in young leaves and reproductive tissues (siliques and seeds). Intermediate amounts of these compounds are present in roots, stems and leaves, while lower levels are associated with senescing rosette leaves [7]. Flavonoids, one large class of phenolic compounds, can be found in most plant tissues, although individual flavonoid sub-classes display a much more specialized distribution. For example, Arabidopsis accumulates three main sub-classes of flavonoids: (i) anthocyanins, which are present in the vegetative tissues as well as in the embryo, (ii) flavonols, which are present in vegetative and reproductive tissues, and (iii) proanthocyanidins (polymers of flavan-3-ol, PAs), which are present only in the endothelium of developing seed coats. Flavan-4-ol 3-deoxyflavonoids apiferol and luteoferol, which are precursors of the phlobaphene brick-red polymers, are present in the floral tissues (e.g., pericarp and cob glumes) of maize and other monocots [8]. Isoflavones are colorless compounds that accumulate in the embryo and in the seed coat of soybean. Highlighting the taxon-specific distribution of some phytochemicals, neither the phlobaphenes nor the isoflavones are present in Arabidopsis. In contrast, glucosinolates, for example, appear to be almost exclusive to *Brassicaceas* family, in which Arabidopsis belongs [9].

Plants often synthesize and store active phytochemicals in specific organs. For example, glandular trichomes on the surface of leaves and stems in a variety of plants provide a major site for phytochemical synthesis, storage and secretion [10]. These differentiated organs express genes of various secondary metabolic pathways, as well as proteins associated with defence responses (e.g., superoxide dismutase in tobacco leaves] [10] or sulfur metabolism in *Arabidopsis* [11]). The biosynthesis of various benzylisoquinoline-derived alkaloids in opium poppy occurs in specific cells localized in sieve elements, whereas the final products accumulate in specialized cell type, the laticifers, that accompany the vascular tissues throughout the plant [12, 13].

Once synthesized, metabolites may be immediately utilized by the plant. For example, when photosynthetic tissues receive high energy-quanta, photoinhibition results, and chloroplasts generate reactive oxygen species (ROS) that will damage the cell at multiple levels. Anthocyanins have been showed to accumulate in higher plant in response to light [14, 15]. Leaf anthocyanins can then protect the photosynthetic apparatus by reducing photoinhibition and scavenging ROS [16], but other phytochemicals are likely to have similar functions because anthocyanins are often not present. Alternatively, metabolites can be stored inside the cell, often in specialized compartments until being "activated". For example, Arabidopsis seed coat proanthocyanidins accumulate in the storage vacuole after being synthesized in the cytoplasm [17]. They are released by the vacuole to exercise their protective functions after seed coat collapsing (which involves programmed cell death) or wounding [18]. Products of glucosinolate metabolism (such as isothiocyanates, oxazolidine-2-thiones, nitrile, epithionitriles and thiocyanates), which are responsible for most of the biological activity displayed by members of this family of phytochemicals, are released after myrosinase-catalyzed hydrolysis of the respective glucosinolate [9]. In intact plants, myrosinase and glucosinolates accumulate in distinct subcellular compartments or even in distinct cells and the glucosinolate and corresponding metabolizing enzymes only come into contact following cell damage [19].

Phytochemicals and their Influence on Plant Development

So far, we have described how the biosynthesis of phytochemicals, their accumulation and often their activation are spatio-temporally regulated during plant development and growth. What is however the effect of phytochemicals on plant developmental processes? Two possibilities should be considered. First, phytochemicals can play structural roles with important consequences for plant development. Obvious examples include cellulose and lignin, yet more subtle effects such as the role of starch in root gravitropic responses [20] should not be ignored. Second, phytochemicals can directly serve as signalling molecules, or interfere with the signalling activity of other molecules. In this section, we will summarize findings describing how phytochemicals act at several steps during plant development, from the fertilization of the flowers to the allelopathy displayed by some invasive plant species (Fig. 2).

Phytochemicals as Signal Molecules

Flavonols and Pollen Germination

Pollen grains undergo maturation and are stored in the anther until being released, where they germinate on the surface of the stigma and fertilize the ovule. Flavonols are stored in the tapetosomes, organelles found in tapetum cells, which correspond to the innermost anther coat layer. During pollen development, these metabolites are released from the tapetosomes and interact with the pollen coat, licensing pollen tube growth germination [21]. This phenomenon has been well characterized in petunia [22]. Indeed, pollen-specific gene products induced by the flavonol kaempferol have been characterized in this plant, including the leucine-rich repeat protein SHY, which is implicated in signalling [23]. Recent studies have also shown that the rate of pollen germination obtained from plants compromised in flavonol biosynthesis (e.g., *tt4* mutant) [24] or transport (*tt12* and *tt19* mutants) [25, 26] is lower when subjected to UV-B irradiation damage, compared to wild-type pollen

12

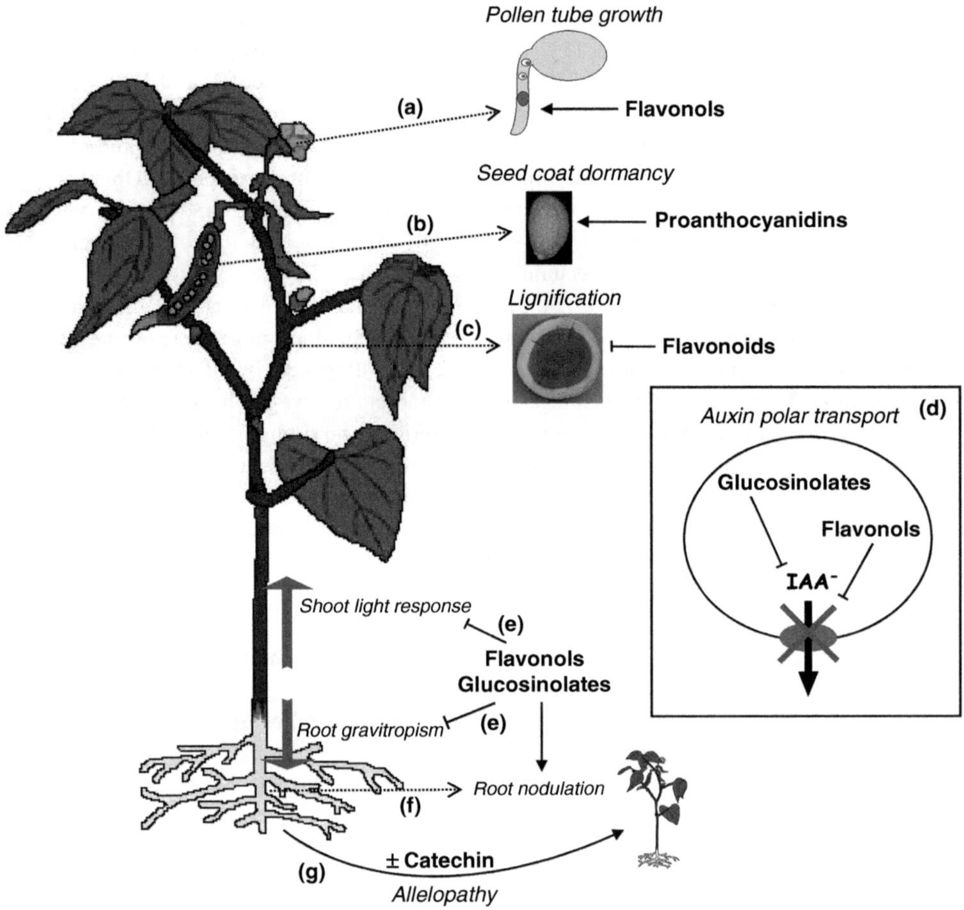

Fig. 2 Functions of phytochemicals in plant growth and development. Phytochemicals play multiple functions in plant growth and development, including pollen tube growth elongation (**a**), seed coat imposed dormancy reinforcement (**b**), modulation of lignin biosynthesis (**c**), regulation of auxin polar transport (**d**), shoot/root gravitropic responses (**e**), root nodulation (**f**) and allelopathy (**g**)

[27]. Cellular targets of flavonol in the pollen coat have not yet been characterized. Although the mechanisms of action of flavonols remain to be determined, it has been suggested that these compounds may act as signal molecules to control pollen germination, rather than as structural components of the pollen tube or pollen coat [22]. Indeed, bioactive aglycone flavonols induced a fast germinative response in *in vitro* assays [28], a finding that is more consistent with a signalling response.

Flavonols and Auxin Polar Transport

Auxin (indole-3-acetic acid, IAA) is a phytohormone that plays an essential role in plant growth, affecting cell division and expansion. Auxin promotes the development of organs such as fruits, controls apical dominance through the inhibition of the formation of lateral buds, and regulates root growth. It is also involved in several tropic responses, including gravitropism and phototropism. Auxin is not synthesized in all plant tissues but can move from cell to cell by directional polar transport, mediated by auxin uptake and efflux carrier proteins localized on the plasma membrane. Three families of carriers, PIN, MDR and AUX proteins, have been identified through the characterization of the corresponding mutants [29, 30]. Flavonols bind to the plasma membrane 1-naphthylphthalmic (NPA) binding protein (NBP) associated with multidrug resistance transporter MDR/P-glycoproteins (MDR/Pgp) [31]. MDR/Pgp acts as an ATP-dependent auxin transporter, whose interaction with PIN proteins confers directionally and substrate specificities to the auxin efflux machinery [29]. Flavonols bind to the MDR/Pgp and function as negative regulators of auxin efflux. The Arabidopsis flavonoid mutant *tt4* displays an elevated auxin transport in young seedlings, roots and inflorescences, consistent with an alteration in auxin export, resulting in the inhibition of the light and root gravitropic responses [31, 32]. Gravitropic alterations in *tt4* can be complemented with the flavonol precursor naringenin [33]. Despite these findings, the biological significance of the role of flavonoids in auxin transport needs yet to be reconciled with the modest developmental effects of flavonoid mutants on development, and with the vacuolar sequestration of flavonoids (including flavonols). Flavonols would therefore need to be actively exported out of the vacuole and transported from cell to cell in order to regulate auxin transport and nodulation. Such phenomena would suggest the involvement of transporters and stabilizers of vesicles that would direct flavonols to their site of action. Alternatively, a small quantity of flavonols, undetectable by the available histochemical stainings (such as diphenylboric acid-2-aminoethyl ester, DPBA, [34]) may still remain in the cytoplasm and may participate in these signalling functions. An attractive possibility that needs to be further explored is whether proteins (e.g., glutathione *S*-transferases [35]) might stabilize cytoplasmic pools of flavonols, and hence participate in the modulation of auxin transport. However, many of the "actors" participating in flavonol export and stabilization remain to be discovered, and there remains significant controversy with regards to the biological relevance of flavonols in auxin signalling [36]. Perhaps, there has been a co-evolution of auxin and flavonoids, and the appearance of the angiosperm may have led flavonols to display new functions for plant survival, as auxin transport regulators.

Isoflavonoid/flavonoids and Nodulation

Flavonoids and isoflavonoids play multiples functions in nodulation, the process by which Rhizobium species colonize the roots of some plants and induce the host plant to form nitrogen fixing nodules. First, they act as signal molecules secreted by root exudates, so providing the first level of communication between the two symbiotic partners. Flavonoids bind to the bacteria Nodulating D protein (Nod D), which in turn activates the transcription of the *Nod* genes [37]. Then, the formation of the nodule primordium is initiated by the inhibition of the auxin polar transport by flavonoids. Highlighting the significance of flavonols in this step, Wasson et al. [38] showed that Medicago truncatula flavonoid-deficient roots displayed increased auxin transport and inhibition of nodule formation, compared to wild-type roots. Finally, flavonoids may act as secondary inducers of Nod factor inside the plant root, in order to increase the specificity of the host [39].

Glucosinolate and Auxin Biosynthesis

The indole-3-acetaldoxime (IAOx) is an intermediate in the biosynthesis of indole glucosinolate and also a precursor of IAA [9]. Therefore, glucosinolate and IAA share common steps in their biosynthesis. Several Arabidopsis mutants [e.g., *sur1* (C-S lyase mutant), *sur2* and *ugt74B1* (glucosyltransferase)] displaying a high-auxin phenotype were shown to be also affected in glucosinolate biosynthesis [9]. While a possible direct role of glucosinolates on auxin biosynthesis cannot be yet formally ruled out, the observed phenomenon can most likely be explained by a re-directioning of the glucosinolate intermediates to the auxin pathway.

Phytochemicals and Plant Development

Proanthocyanidins and Seed-Coat Imposed Dormancy

Dormancy is a natural process where the mature seed cannot germinate, awaiting environmental conditions favourable for the initiation of germination and seedling development. Two major types of dormancy mechanisms exist: (a) embryo dormancy where the agents inhibiting germination are inherent to the embryo, and (b) coat-imposed dormancy, where the inhibition is conferred by the seed envelope [40].

PAs, present in the Arabidopsis endothelium cell layer of the seed coat, reinforce seed coat imposed dormancy [41]. In fact, flavan-3-ol polymers are likely to furnish a barrier for important processes to activate embryo germination. Such processes include water uptake, radicle protrusion, gas exchange (particularly molecular oxygen and carbon dioxide) and light filtration. Moreover, flavonoid oxidation produces semiquinones and quinones, highly reactive species that can further react with phenolic compounds, proteins, or scavenge free radicals. These oxidative products may protect the plant during oxidative stresses and reinforce the testa structure by crosslinking with protein and carbohydrates of the cell wall [18]. Arabidopsis *transparent testa* mutant affected in the accumulation of PAs are less dormant than wild type seeds [41]. Flavonoids can also act as direct and indirect inhibitors of germination [40]. For example, (+/-)-catechin synthesized in the seed coat has been shown to inhibit giberellic acid (GA) metabolism in the embryo. The mechanisms by which catechin is transported from the seed coat to the embryo remain unknown [40]. In addition, flavonoids may also decrease the oxygen available for the embryo to develop, by fixing it through enzymatic reactions catalyzed by phenol oxydoreductases [40]. It is possible that the phlobaphenes, present in some monocot grasses such as maize, sorghum and wheat, display a similar function as PAs in increasing seed dormancy. For example, white wheat seeds that do not accumulate red pigments are less dormant than the red seeds [42].

Flavonoids and Lignification

Lignin is a polymer of monolignol subunits, present in the cell wall of plant fibres and tracheids. The lignification process is crucial for the structural rigidity and water impermeability of plant tissues. Similar to flavonoids, lignins are derived from the phenylpropanoid pathway, and hence share common early steps of the pathway with flavonoids (from phenylalanine to *p*-coumarate). Therefore, it is expected that alterations of either flavonoid or lignin biosynthesis would lead to indirect effects on the other branch. Recent studies showed that mutants repressed in lignin biosynthesis exhibited an increase in flavonoid (flavonols in particular) content, in addition to showing significant growth reductions [43]. The study concluded that the growth defects were a consequence of alterations in auxin transport resulting from the altered flavonoid levels, rather than by the decreased lignin levels. Similarly as described earlier for the glucosinolate and auxin pathways, the effects of the reduction of lignin on flavonoid accumulation are likely a consequence of the re-direction of flux between these two branches of phenolics biosynthesis. The biological and evolutionary significance of maintaining a balance between lignin and flavonoid production remains to be explored.

The Plant Chemical Information Network

Transport of Phytochemicals Between Cells

The term phytohormone refers to a discrete group of chemical compounds that affect diverse developmental processes, that are produced throughout the plant, and that are able to act over distances and in low concentrations [44] (see glossary). However, over the past few years evidence has accumulated indicating that phytochemicals can move between cells. For example, several lines of evidence suggest that glucosinolates are transported over long distances *via* the phloem [9]. Desulfoglucosinolate, a glusosinolate precursor, is likely to be the substrate for phloem transport [19]. Supporting this model, glucosinolates are found in seeds, where none of the biosynthetic enzyme is expressed. Flavonoids have recently been shown to be transported over long distances in *Arabidopsis*, suggesting a mechanism by which green tissue-synthesized light-induced flavonoids might continue to influence root-specific functions, such as lateral root development and gravitropic responses [33]. Similarly, intermediates in alkaloid biosynthesis move between cells [12] and many compounds, including root-synthesized nicotine [45], are transported to leaf tissues or from the leaves to the roots, as found for linamarin, a cassava cyanogenic glucoside [46].

Phytochemicals and Allelopathy

Phytochemicals often provide a selective advantage to plants, for example by eliminating the competition by other species, in a process that is generally included within the broader context of allelopathy [47, 48]. Allelochemicals released to the rizosphere, for example as root exudates or by decaying leaf tissues, or accumulating in the aerial space as volatile emissions, have been proposed to play an important function in invasive plant species. Recent data for the invasive plant Centaurea maculosa (spotted knapweed) are suggestive of such a function [49], although these data have subsequently been challenged by others [50]. Allelochemicals belong to various families of phytochemicals. For example, sorgeoleone, a sorghum root exudate, is a quinone, and juglone, present in the leaves and bark of black walnut and other *Junglandaceae* family members, is a naphtoquinone. In contrast, (-)-catechin, secreted by the roots of C. maculosa is a flavonoid, and two benzoxazinones, DIBOA (2,4-dihydroxy-1,4-benzoxazin-3-one) and DIMBOA (dihydroxy-7-methoxy-1,4-benzoxazin-3-one), are secreted by the roots of various members of the *Poaceae* [51].

The mechanisms by which allelochemicals exert their phytotoxic activity also vary, although the direct target for any of these phytochemicals are yet to be identified. Sorgoleone, for example, interferes with a number of essential processes such as nutrient uptake and electron transfer [52]. In contrast, (-)-catechin induces in the roots of susceptible species a wave of ROS (reactive oxygen species), which, through a Ca^{2+} signal transduction cascade, culminates in the death of the root system [49].

It is therefore conceivable that allelochemicals function by interfering ("jamming") with normal signaling processes, pretty much in the way in which a personal cell phone signal blocker device interferes with proximal cell phone communications.

Phytochemicals and Signaling Across Kingdoms: The Concept of Xenohormesis

The field of chemical ecology is rapidly expanding, and an increasing number of phytochemicals that function as signals to other plants (e.g., allelochemical), to microorganisms or to animals (e.g., in tritrophic interactions [53]) is being identified. This idea of communication across kingdoms has recently been developed into the concept of xenohormesis, which we briefly describe here.

Calorie restriction results in lifespan increases in every organism tested. While several theories have been proposed to explain how calorie restriction impacts lifespan, a provocative idea is that mild stress (such as induced by calorie restriction) is beneficial by augmenting the organism's defense response (hormesis) [54]. Central to the calorie restriction-dependent lifespan increase are the silent information regulator (SIR) proteins, encoding protein deacetylases and ADP-ribosyl transferases [55]. Many phytochemicals activate sirtuins, resulting in extended lifespan, at least as demonstrated in S. cerevisiae [56]. Yet, these phytochemicals accumulate in plants often in response to stress (see above) suggesting that somehow stress in the plant is transmitted to the animals (and humans) that feed on them. This, together with a number of related studies, provided the premises for xenohormesis, which refers to the ability of an organism (e.g., animal or human) to respond to an ecological or environmental stress signal provided by another organisms (xeno), for example a plant, and is ingrained in the diet [57, 58] (Fig. 3). Indeed, it has even been proposed that preferences for taste may have evolved as a consequence of the need to sense stress [59].[1] Inherent to the concept of xenohormesis is the idea that plants and animals share cellular factors that respond to phytochemicals in similar ways. A good example is provided by the conservation of the role played by jasmonic acid in plants and prostaglandins in animals, both functioning as signal molecules in response to wounding. Salicylic acid, which is induced by stress conditions in plants and which functions as an inhibitor of JA biosynthesis, has the ability to inhibit prostaglandin biosynthesis in animals, highlighting a conserved function of this plant-specific chemical across kingdoms [60].

[1]The xenohormesis theory assumes that longevity provides an evolutionary advantage, yet the consequences of mild stress on reproductive output, which is under strong selective constrains, remains unknown

Conclusions

Phytochemicals serve important functions in plant development and growth by acting as structural components and/or by interacting with plant signaling networks. Indeed, this last observation is supported by the fact that the number of phytochemicals that appear to participate in short- or long-distance signaling is increasing. Currently, only a handful of plant compounds are formally considered phytohormones. It is conceivable that as new cellular functions of phytochemicals are uncovered, particularly in regards to signaling and plant development, new members will be added to the select club of the phytohormones. While the controversy persists on whether the field of plant neurobiology has a rational foundation [61, 62], it is undisputable that plants have developed unique opportunities for complex chemical signaling.

Glossary

Phytohormones

Phytohormones are substances that, at low concentration, function to coordinate plant growth and development. The compounds that have been considered as plant hormones include indole-3-acetic acid (auxin), cytokinins, gibberellins (GA), ethylene and abscisic acid (ABA). In addition, brassinosteroids, jasmonic acid (JA) and salicylic acid (SA) have been shown to display important growth regulating activities and are also considered to function as phytohormones.

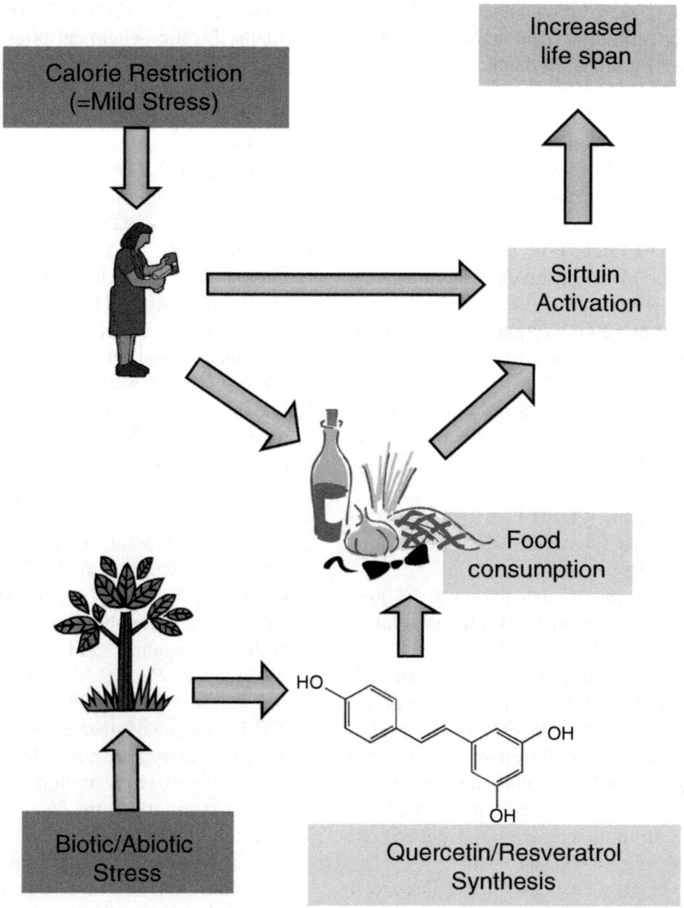

Fig. 3 Proposal effect of flavonoids on human lifespan and the xenohormesis theory. The theory states that phytochemicals (e.g., flavonoids) can "communicate" a stress signal present in the plant to different organisms, when they are consumed as food. This signal leads to the activation of sirtuins involved in lifespan extension

Xenohormesis

Interspecies communication of stress signals. This term as been proposed by Howitz and Sinclair [56] to explain the ability of animals and fungi to "sense" and being activated by molecules that are not produced in these organisms, such as phytochemicals.

References

1. Wink, M. (1999) Introduction. In *Functions of Plant Secondary Metabolites and their Exploitation in Biotechnology* (Wink, M., ed), 1–16, CRC Press LLC, Boca Raton, FL
2. Hammerschmidt, R. (1999) PHYTOALEXINS: What have we learned after 60 years? *Annu Rev Phytopathol* 37, 285–306

12

3. Curir, P. et al. (2005) A phytoalexin-like flavonol involved in the carnation (Dianthus caryophyllus)-Fusarium oxysporum f. Sp dianthi pathosystem. *J. Phytobiol* 153, 65–67
4. Bieza, K., and Lois, R. (2001) An Arabidopsis mutant tolerant to lethal ultraviolet-B levels shows constitutively elevated accumulation of flavonoids and other phenolics. *Plant Physiol* 126, 1105–1115
5. Seigler, D., and Price, P.W. (1976) Secondary compounds in plants: primary functions. *Am Nat* 110, 101–105
6. Aharoni, A. et al. (2003) Terpenoid metabolism in wild-type and transgenic Arabidopsis plants. *Plant Cell* 15, 2866–2884
7. Brown, P.D. et al. (2003) Variation of glucosinolate accumulation among different organs and developmental stages of Arabidopsis thaliana. *Phytochemistry* 62, 471–481
8. Lepiniec, L. et al. (2006) Genetics and biochemistry of seed flavonoids. *Annu Rev Plant Biol* 57, 405–430
9. Halkier, B.A., and Gershenzon, J. (2006) Biology and biochemistry of glucosinolates. *Annu Rev Plant Biol* 57, 303–333
10. Amme, S. et al. (2005) A proteome approach defines protective functions of tobacco leaf trichomes. *Proteomics* 5, 2508–2518
11. Wienkoop, S. et al. (2004) Cell-specific protein profiling in Arabidopsis thaliana trichomes: identification of trichome-located proteins involved in sulfur metabolism and detoxification. *Phytochemistry* 65, 1641–1649
12. Bird, D.A., et al. (2003) A tale of three cell types: Alkaloid biosynthesis is localized to sieve elements in opium poppy. *Plant Cell* 15, 2626–2635
13. Weid, M. et al. (2004) The roles of latex and the vascular bundle in morphine biosynthesis in the opium poppy, Papaver somniferum. *Proc Natl Acad Sci USA* 101, 13957–13962
14. Chalker-Scott, L. (1999) Environmental significance of anthocyanins in plant stress responses. *Photochem Photobiol* 70, 1–9
15. Winkel-Shirley, B. (2002) Biosynthesis of flavonoids and effects of stress. *Curr Opin Plant Biol* 5, 218–223
16. Gould, K.S. (2004) Nature's Swiss army knife: the diverse protective roles of Anthocyanins in leaves. *J Biomed Biotechnol* 2004, 314–320
17. Debeaujon, I. et al. (2003) Proanthocyanidin-accumulating cells in Arabidopsis testa: regulation of differentiation and role in seed development. *Plant Cell* 15, 2514–2531
18. Pourcel, L. et al. (2007) Flavonoid oxidation in plants: from biochemical properties to physiological functions. *Trends Plant Sci* 12, 29–36
19. Grubb, C.D., and Abel, S. (2006) Glucosinolate metabolism and its control. *Trends Plant Sci* 11, 89–100
20. Kiss, J.Z. et al. (1996) Gravitropism in roots of intermediate-starch mutants of Arabidopsis. *Physiol Plant* 97, 237–244
21. Taylor, L.P., and Grotewold, E. (2005) Flavonoids as developmental regulators. *Curr Opin Plant Biol* 8, 317–323
22. Taylor, L.P., and Hepler, P.K. (1997) Pollen germination and tube growth. *Annu Rev Plant Physiol Plant Mol Biol* 48, 461–491
23. Guyon, V. et al. (2004) Antisense phenotypes reveal a role for SHY, a pollen-specific leucine-rich repeat protein, in pollen tube growth. *Plant J* 39, 643–654
24. Shirley, B.W. et al. (1995) Analysis of Arabidopsis mutants deficient in flavonoid biosynthesis. *Plant J.* 8, 659–671
25. Kitamura, S. et al. (2004) TRANSPARENT TESTA 19 is involved in the accumulation of both anthocyanins and proanthocyanidins in Arabidopsis. *Plant J* 37, 104–114
26. Debeaujon, I. et al. (2001) The *TRANSPARENT TESTA12* gene of *Arabidopsis* encodes a multidrug secondary transporter-like protein required for flavonoid sequestration in vacuoles of the seed coat endothelium. *Plant Cell* 13, 853–871
27. Hsieh, K., and Huang, A.H. (2007) Tapetosomes in Brassica tapetum accumulate endoplasmic reticulum-derived flavonoids and alkanes for delivery to the pollen surface. *Plant Cell* 19, 582–596
28. Mo, Y.Y. et al. (1992) Biochemical complementation of chalcone synthase mutants defines a role for flavonols in functional pollen. *Proc Natl Acad Sci USA* 89, 7213–7217
29. Blakeslee, J.J. et al. (2005) Auxin transport. *Curr Opin Plant Biol* 8, 494–500
30. Zazimalova, E. et al. (2007) Polar transport of the plant hormone auxin - the role of PIN-FORMED (PIN) proteins. *Cell Mol Life Sci* 64, 1621–1637
31. Murphy, A. et al. (2000) Regulation of auxin transport by aminopeptidases and endogenous flavonoids. *Planta* 211, 315–324
32. Buer, C.S., and Muday, G.K. (2004) The transparent testa4 mutation prevents flavonoid synthesis and alters auxin transport and the response of Arabidopsis roots to gravity and light. *Plant Cell* 16, 1191–1205

33. Buer, C.S. et al. (2006) Ethylene modulates flavonoid accumulation and gravitropic responses in roots of Arabidopsis. *Plant Physiol* 140, 1384–1396
34. Sheahan, J.J., and Rechnitz, G.A. (1992) Flavonoid-specific staining of Arabidopsis thaliana. *Biotechniques* 13, 880–883
35. Mueller, L.A. et al. (2000) AN9, a petunia glutathione S-transferase required for anthocyanin sequestration, is a flavonoid-binding protein. *Plant Physiol.* 123, 1561–1570
36. Bennett, T. et al. (2006) The Arabidopsis MAX pathway controls shoot branching by regulating auxin transport. *Current Biology* 16, 553–563
37. Kobayashi, H. et al. (2004) Flavanoids induce temporal shifts in gene-expression of nod-box controlled loci in Rhizobium sp. NGR234. *Mol Microbiol* 51, 335–347
38. Wasson, A.P. et al. (2006) Silencing the flavonoid pathway in Medicago truncatula inhibits root nodule formation and prevents auxin transport regulation by rhizobia. *Plant Cell* 18, 1617–1629
39. Subramanian, S. et al. (2007) Distinct, crucial roles of flavonoids during legume nodulation. *Trends Plant Sci* 12, 282–285
40. Debeaujon, I. et al. (2007) Seed coat development and dormancy. In Seed development, dormancy and germination (Bradford, K. and Nonogaki, H., (editors) Oxford, UK Blackwell Publishing), 25–49
41. Debeaujon, I. et al. (2000) Influence of the testa on seed dormancy, germination, and longevity in Arabidopsis. *Plant Physiol* 122, 403–413
42. Mares, D. et al. (2005) A QTL located on chromosome 4A associated with dormancy in white- and red-grained wheats of diverse origin. *Theor Appl Genet* 111, 1357–1364
43. Besseau, S. et al. (2007) Flavonoid accumulation in Arabidopsis repressed in lignin synthesis affects auxin transport and plant growth. *Plant Cell* 19, 148–162
44. Bishopp, A. et al. (2006) Signs of change: hormone receptors that regulate plant development. *Development* 133, 1857–1869
45. Nugroho, L.H., and Verpoorte, R. (2002) Secondary metabolism in tobacco. *Plant Cell Tissue Organ Cult* 68, 105–125
46. Siritunga, D. et al. (2004) Over-expression of hydroxynitrile lyase in transgenic cassava roots accelerates cyanogenesis and food detoxification. *Plant Biotech J* 2, 37–43
47. Inderjit, and Duke, S.O. (2003) Ecophysiological aspects of allelopathy. *Planta* 217, 529–539
48. Weir, T.L. et al. (2004) Biochemical and physiological mechanisms mediated by allelochemicals. *Curr Opin Plant Biol* 7, 472–479
49. Bais, H.P. et al. (2003) Allelopathy and exotic plant invasion: from molecules and genes to species interactions. *Science* 301, 1377–1380
50. Blair, A.C. et al. (2006) A lack of evidence for an ecological role of the putative allelochemical (+/-)-catechin in spotted knapweed invasion success. *J Chem Ecol* 32, 2327–2331
51. Niemeyer, H.M. (1988) Hydroxamic acids (4-Hydroxy-1,4-Benzoxazin-3-Ones), defense chemicals in the Gramineae. *Phytochemistry* 27, 3349–3358
52. Hejl, A.M., and Koster, K.L. (2004) The allelochemical sorgoleone inhibits root H+-ATPase and water uptake. *J Chem Ecology* 30, 2181–2191
53. Marvier, M.A. (1996) Parasitic plant-host interactions: plant performance and indirect effects on parasite-feeding herbivores. *Ecology* 77, 1398–1409
54. Anderson, R.M. et al. (2003) Nicotinamide and PNC1 govern lifespan extension by calorie restriction in Saccharomyces cerevisiae. *Nature* 423, 181–185
55. Haigis, M.C., and Guarente, L.P. (2006) Mammalian sirtuins - emerging roles in physiology, aging, and calorie restriction. *Genes Dev* 20, 2913–2921
56. Howitz, K.T., et al. (2003) Small molecule activators of sirtuins extend Saccharomyces cerevisiae lifespan. *Nature* 425, 191–196
57. Lamming, D.W. et al. (2004) Small molecules that regulate lifespan: evidence for xenohormesis. *Mol Microb* 53, 1003–1009
58. Howitz, K.T., and Sinclair, D.A. (2008) Xenohormesis: Sensing the chemical cues of other species. *Cell* 133, 387–391
59. Yun, A.J., and Doux, J.D. (2007) Unhappy meal: how our need to detect stress may have shaped our preferences for taste. *Medical Hypotheses* 69, 746–751
60. Schultz, J.C. (2002) Shared signals and the potential for phylogenetic espionage between plants and animals. *Int Comp Biology* 42, 454–462
61. Alpi, A. et al. (2007) Plant neurobiology: no brain, no gain? *Trends Plant Sci* 12, 135–136
62. Brenner, E.D. et al. (2006) Plant neurobiology: an integrated view of plant signaling. *Trends Plant Sci* 11, 413–419

Part III

Biological Activity

Biological Activity of Defence-Related Plant Secondary Metabolites

13

John P. Morrissey

Abstract Althought is accepted that secondary metabolites and natural products in plants are involved in diverse activities, the function of most of the thousands of phenolics, quinones, terpenes, flavonoids and other low molecular weight metabolites remains unknown. The best understood secondary metabolites are implicated in defence against pathogens, with the mode of action of some of these established. Interestingly, to date, a relatively small number of processes have been shown to be the targets of plant metabolites and these include electron transport chains, mitochondrial function and membrane integrity. It is now emerging, however, that other specific enzymes and processes may also be the targets of particular metabolites. There is a general hope that modern genomic approaches will identify new targets and modes of action of plant metabolites. Molecules, especially triterpenoids, that trigger apoptosis or autophagy in tumour cells are of particular interest in this regard. When considering the approaches taken in plant science hitherto, and the strategies that have yielded success in the biomedical sector, we discuss whether there is a case to be made for carrying out initial studies on mode of action in a genetically tractable system like the yeast *Saccharomyces cerevisiae*, before moving to specific studies in plant or human cells.

J.P. Morrissey (✉)
Microbiology Department University College
Cork, Ireland
e-mail: j.morrissey@ucc.ie

Introduction

Secondary Metabolites in Plants

Secondary metabolism is generally used to descibe metabolic pathways that produce molecules or metabolites that appear to be dispensable for normal growth, or are required only under particular conditions. In contrast, primary metabolism traditionally describes core housekeeping functions such as energy generation or production of essential metabolites and macromolecules. These distinctions may be somewhat misleading, however, as it is now known that secondary metabolites play very important roles in the biology of many different organisms. Indeed, it is clear that evolution would not have selected and maintained the complex pathways that give rise to secondary metabolites were it not for competitive benefits that must accrue to the producing organism. This logic, coupled with the fact that the biological function of the majority of plant and microbial secondary metabolites is poorly understood, has led to an alternative description of plant metabolites as "natural products" (Field et al., 2006), though that description also carries some limitations. Nature produces a tremendous array of secondary metabolites/natural products, with the most diversity seen in micro-organisms and plants (Wink, 2003; Konig et al., 2006). This is a hugh resource for mankind, and there are many

A.E. Osbourn and V. Lanzotti (eds.), *Plant-derived Natural Products*,
DOI 10.1007/978-0-387-85498-4_13, © Springer Science+Business Media, LLC 2009

examples of microbial or plant metabolites that are exploited by man, for example as antibiotics and pharmaceutical drugs. It is commonly accepted, however, that we have only scratched the surface and there are vast numbers of different natural metabolites that may have applications, particularly in the biomedical sector. This is the basis of many natural product discovery programmes, for example the efforts of apply metagenomics to explore marine microbial diversity (Kennedy et al., 2007; Singh and Pelaez, 2008). In contrast to these efforts to explore metabolic diversity in new niches, plant metabolic diversity has been exploited by man throughout history, initially through the use of plant extracts, and more recently by the application of the scientific method to identify metabolites with particular functions and to subsequently exploit these products directly or as leads for therapeutic compounds (Balunas and Kinghorn, 2005; Baker et al., 2007; Schmidt et al., 2007).

Secondary Metabolites and Defence Against Pathogens

Exploitation of natural products requires knowedge of how these molecules exert their effects and often follows from an understanding of the role of the metabolite in the producing organism. In plants, the best understood secondary metabolites are implicated in pathogen defence or sensing and signalling. With respect to pathogen defence, the major threat to plant health is fungal disease, with estimates of > 13,000 phytopathogenic fungal species in the US alone. It is unsurprising, therefore, that plants have evolved elaborate protective mechanisms against fungal pathogens, with chemical defence one of the key weapons in the plant arsenal (Madden and Wheelis, 2003; Maor and Shirasu, 2005; Field et al., 2006). Although thousands of different molecular entities are believed to play a role in plant defence against bacterial and fungal pathogens, the mode of action of relatively few has been been the subject of detailed study. Plant defence molecules can be either preformed in plant tissues (Fig. 1a) or synthesised in reponse to pathogen attack, distinctions that lead to the terms phytoanticipans and phytoalexins, respectively (VanEtten et al., 1994; Morrissey and Osbourn, 1999). This differentiation does not provide any particular information on the chemical structure or mode of action of the metabolite and is sometimes misleading as some defence molecules that are preformed but accumulate in high concentrations at the site of infection could reasonably be considered either phytoanticipins or phytoalexins. In practice, when studying the range of potential biological activities that a metabolite may have, the chemical structure of the natural product is more pertinent than the precise timing of production in the plant.

Role of Secondary Metabolites in Plant-Microbe Signalling

In terms of signalling, the best understand metabolites are probably the flavonoids that are involved in the symbiotic legume-rhizobia interaction that leads to formation of nitrogen fixing nodules in root tissue (Peters et al., 1986; Perret et al., 2000). Collectively, plants produce more that 5,000 different flavonoids, with only a small subset involved in specific interactions with rhizobia. This interaction commences with secretion of signal flavonoids in root exudates, followed by perception of the signal by the bacteria and direct interaction with the bacterial nodD transcriptional activator. This triggers a series of events that culminates in a faciltiated rhizobial infection of the plant root and formation of nitrogen fixing nodules. The other major beneficial plant-microbe interaction that occurs in nature is the formation of mycorrhizal roots. Once again, there is a facilitated infection of plant roots, this time by arbuscular mycorrhiza fungi, which develop specialised structures called arbuscles within the root for nutrient exchange between the plant and fungus. Although a role for signalling has long been postulated, it is only in recent years that the first

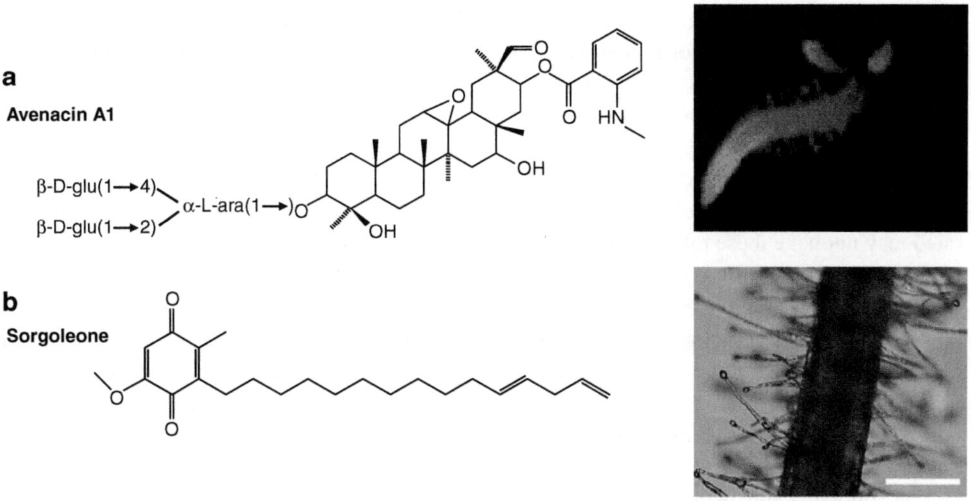

Fig. 1 Natural products may be localised within plant tissue or secreted externally. Panel A shows avenacin A1, a triterpenoid saponin (see Section 2.4.3), that is produced in the roots of oats. The molecule is autfluorescent and can be visualised accumulating in the tips of actively growing roots. Avenacin A1 is antifungal and provides the plant with protection against phytopathogenic fungi. Panel B shows sorgoleone, a quinone that is produced by sorghum roots. In this case, the molecule is secreted into the rhizosphere, and in this image sorgoleone can be seen as little droplets coming from the tips of root hairs. Sorgoelone has phytotoxic properties and has been implicated in allelopathic interactions with other plants (see Sections 1.4; 2.2 and 2.3.1). The avenacin image is reproduced from Osbourn, 2003, and the sorgoleone image from Field et al., 2006, with permission

experimental evidence demonstrating a role for a plant chemical has been obtained, with work from Hayashi and colleagues showing that a particular class of sesquiterpene, the strigolactones, can induce hyphal branching, an important step in the symbiosis (Akiyama et al., 2005; Parniske, 2005; Akiyama and Hayashi, 2006). In an added twist, several studies have now reported that some stigolactones are actually plant hormones and play a role in regulating shoot branching in the plant, thus regulating processes above and below ground (Gomez-Roldan et al., 2008; Klee, 2008; Umehara et al., 2008).

Allelopathic Interactions

Allelopathy is typically defined as the inhibition of growth in one species by chemicals produced by another species and although it has been a somewhat controversial topic in the scientific literature, the concept has gained general acceptance in recent years (Bais et al., 2006; Field et al., 2006; Macias et al., 2007). This definition is considerably narrower than the original use of the term, which would have encompassed positive and negative interactions, but it is a reflection of the perceived importance of allelopathy in the interaction between indigenous and introduced species of plants, in particular when the introduced species is invasive and can displace native plants. There is considerable interest in exploiting allelopathy in agriculture by engineering plants, especially cereals, to control weeds in their own environs (Field et al., 2006). The underlying basis of allelopathy is that plants secrete phytotoxic metabolites into their surroundings (primarily the rhizosphere), and these metabolites inhibit the growth of susceptible plants. This process could reasonably be characterised as

13

either defence or signalling, and, in fact, molecules such as the strigolactones may have dual roles. There is believed to be an evolutionary dimension to allelopathy, and thus plants that have co-existed for long periods are co-adapted and have evolved tolerance mechanisms, whereas plants that have been ecologically separated may not have these tolerance or resistance mechanisms. Many different molecules that are present in root secretions are known to have phytoxic properties at biologically relevant concentrations (Fig. 1b). The majority are either phenolics, including simple phenolics, flavonoids, and quinones; terpenes, including monoterpenoids, sesquiterpene lactones and diterpenes; benzoxazinoids or glucosinolates. When considering plant defence against microbial or insect pathogens, signalling, and allelopathy, one of the striking features is that the same overall classes of molecules are implicated in these phenomena. Our capacity to determine whether particular metabolites originally evolved as signal molecules, for defence against pathogens, or as phytotoxic agents to improve competitive ability may be lost in evolutionary history. In terms of exploiting these natural products as (lead) herbicides, plant protection products, or pharmaceuticals, however, the important quest now is to understand their mode of action on both target and non-target species.

Mode of Action of Plant Natural Products

Identifying Targets of Plant Metabolites

Despite the vast number range of biological structures and biological reactions in cells, a relatively small number are exploited by man. For example, the 270 herbicides in commercial use target only 17 different processes, and clinical and agricultural antifungal drugs target only six different processes (Cole et al., 2000; Odds et al., 2003; Francois et al., 2005; Thevissen et al., 2007). Since the repertoire of natural products in plants runs to many thousands of different molecules, it is likely that many new inhibitors of cellular functions can be identified. This belief drives much of the research on plant natural products and their mode of action. Numerous plant metabolites have been described chemically, and although many have been ascribed roles in signalling, defence and allelopathy, the precise mode of action of relatively few has been determined in any detail. In cases where efforts have been made to determine how the chemicals exert their effects, interpretation of the results is often complicated by multiple targets, difficulty in separating primary and secondary effects, and problems in ascertaining whether data obtained from *in vitro* studies have any *in vivo* relevance. To a large extent, these difficulties are symptomatic of the limitations with the experimental systems used, and there is certainly a case for carrying out studies in genetically tractable systems such as yeast (see Section 3 for a more detailed discussion). Nevertheless, it has been possible to identify key processes that are commonly targetted by plant metabolites, as well as particular enzymes that may be inhibited by specific metabolites.

Inhibition of Specific Enzymes

Plant secondary metabolites may inhibit specific enzymes, either in plants or in other species such as fungi or animals. In some cases, this appears to be the sole mode of action of the metabolite, whereas in others, enzyme inhibition forms part of a suite of effects. It must be noted, however, that the specificity and *in vivo* biological relevance of some findings remains questionable. An example of this is the inhibition of a variety of enzymatic reactions, including plant hormone biosynthetic enzymes, catalase, maltase and phosphatase by phenolics and phenolic acids (Macias et al., 2007).

Sesquiterpenes are one of the largest families of plant natural products and there are many general effects associated with this class of molecule. It is believed that some sesquiterpenes inhibit the activity of sufhydryl-containing enzymes (e.g. phosphofructokinase), and this may be responsible for the general apoptopic effects of plant sesquiterpenes on animal cells, but more detailed investigations in this area are required. In contrast to those generic effects, the quinone sorgoleone (Fig. 1b) specifically inhibits the enzyme 4-hydroxyphenylpyruvate dioxygenase (HPPD) (Meazza et al., 2002). HPPD activity is required for plastoquinone and ultimately chloroplast synthesis and is also the target of sulcotrione and other herbicides (Cole et al., 2000). Other quinones, for example juglone produced by the walnut tree, may also inhibit HPPD activity. Another example is the steroidal alkaloid tomatidine, which specifically inhibits the C24 sterol methyltransferase reaction that is required for synthesis of the essential fungal membrane sterol, ergosterol (Simons et al., 2006). This anti-fungal metabolite is synthesised in tomatos as a glycosylated form called α-tomatine and is coverted to the steroidal alkaloid tomatidine by fungal enzymes during plant infection (Fig. 2). Studies with the yeast *Saccharomyces cerevsiae* demonstrated that α-tomatine and tomatidine have distinct modes of action with tomatidine up to fifty times more potent than α-tomatine. Interestingly, the importance of C24 sterol methyltransferase for ergosterol biosynthesis had already been recognised and commercial fungicides such as fenpropimorph target the same enzyme. The fact that in both the case of sorgoleone/HPPD and tomatidine/C24 sterol methyltransferase the enzymes in question had already been identified and used as drug targets validates the strategy of identifying new enzyme targets of plant natural products as leads for intervention drugs/chemicals. Some novel natural products/enzymes are under investigation in this regard. For example, 1,4-cineole (a monoterpene) has been found to inhibit asparagine synthesis (Romagni et al., 2000) and quassinoids (diterpenes) are believed to inhibit membrane NADH oxidase (Morre et al., 1998).

Inhibition of Electron Transport Systems

Targetting of Photosynthesis and Respiration

Photosynthesis is centrally important to plant health, therefore it is an obvious target for natural and synthetic inhibitory molecules. At least fifty nine different herbicides target photosystem II (PSII), primarily by interfering with electron transport (Cole et al., 2000). PSII was also found to be the major target of the quinone sorgoleone, the same metabolite that inhibits the HPPD enzyme (above). Sorgoleone is believed to compete with plastoquinone for binding to the D1 proteins at PSII (Rimando et al., 1998) and is secreted in droplets from the root hairs, accumulating at levels from 10 to 100 μM in the soil around the plant root. The imbalance between the number of herbicides and natural metabolites that inhibit photosynthesis is surprising and suggests that there may be many more natural inhibitors of photosynthesis yet to be identified. Respiration is the other key process in the cell that is based on electron transport chains and this is also a target of inhibitory molecules. The clearest example is probably the cyanogenic glycosides that are produced by over 200 different species of plants. These are synthesised by the conversion of amino acid precursors to oximes, which are then glycosylated. Hydrolysis of cyanogenic glycosides in response to tissue damage generates hydrogen cyanide (HCN), a potent respiratory toxin (Morrissey and Osbourn, 1999). Glucosinolates are evolutionarily related molecules that are synthesised by only a subset of species, mainly within the order Capparales, including the agriculturally important Brassicaceae family (Halkier and Gershenzon, 2006). Hydrolysis of glucosinolates yields isothiocyanates, thiocyanates and nitriles and although the antifungal mode of action of

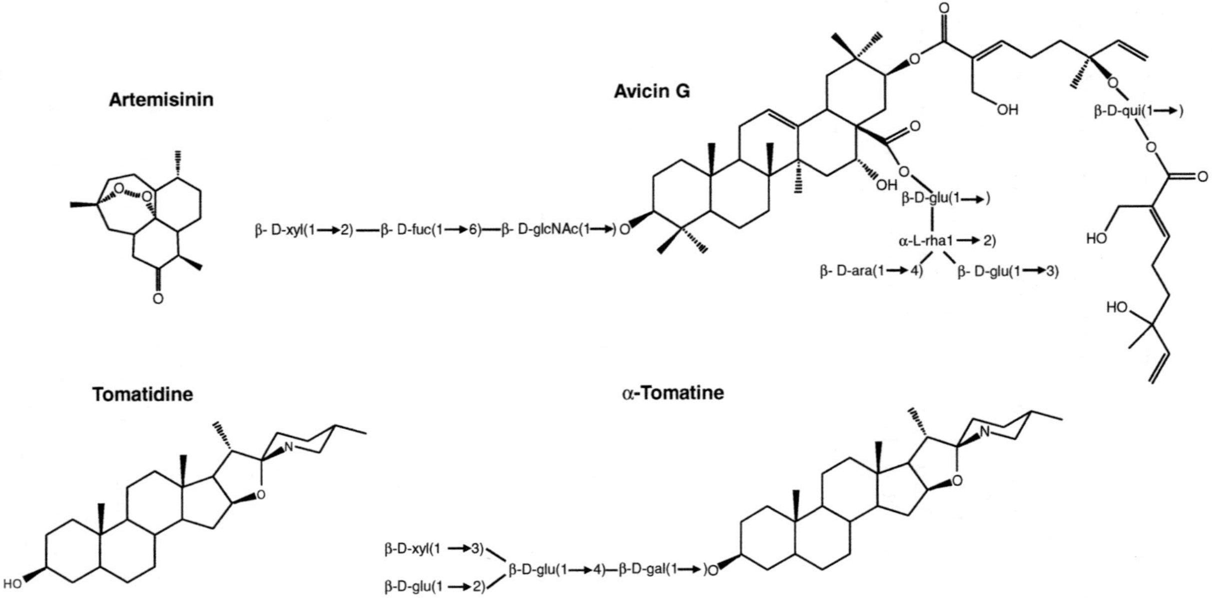

Fig. 2 Structures of some plant secondary metabolites. The structures of the key natural products discussed in the text are shown. Artemisinin is a sesquiterpene lactone produced by the Chinese sweet wormwood plant. It is a front-line anti-malaria drug that may have several targets in the cell including calcium homestasis and mitochondrial function (Section 2.3.2). Avicin G and α-tomatine are saponins produced by the legume *Acacia victoria* and the tomato plant, respectively. Avicins are triterpenoid molecules and are under investigation because of anti-tumour properties against cells in tissue culture (Section 2.3). α-tomatine is a steroidal glycoalkaloid that has anti-fungal activity and is believed to work by permeabilising cell membranes (Section 2.2.3). Tomatidine is the aglycone of α-tomatine and has potent anti-fungal activity. Its main effect is to inhibit an enzyme required for the synthesis of the fungal membrane sterol, ergosterol (Section 2.2)

these metabolites has not been demonstrated, the cyanide moiety suggests that respiration is the target of at least some of these metabolites. It is also believed that other low molecular weight plant natural products target respiration but in many cases this has proved difficult to establish definitively and studies with isolated mitochondria sometimes have produced conflicting results. Thus, although some phenolic acids inhibit oygen uptake by mitochondria, the concentrations of phenolics required seem unrealistically high. And while there are suggestions that phenolics may block electron transport in the b/c1 cytochrome complex, there are also data that phenolics may actually stimulate respiration in certain cases (Macias et al., 2007).

Biological Activity of the Antimalarial Drug Artemisinin

The sesquiterpene lactone artermisinin is reported to exert a number of diverse physiological effects on target cells, including disruption of mitochondrial function (Golenser et al., 2006; Krishna et al., 2006). Artemisinin (*qingaosu*) is a natural product synthesised by the chinese plant *Artemesia annua* (sweet wormwood or *qinghao*) and its importance is that this molecule and its derivatives are now part of front-line anti-malarial therapy. The effects of artemisinin on plant cells is not known, but many studies have attempted to determine why this metabolite is toxic to the malaria parasite *Plasmodium falciparum* and other protozoa. Althought a sesquiterpene, artemisinin is unusual in possessing an endoperoxide moiety that is essential for its biological activity (Fig. 2). It was shown that artemisinin inhibits oxygen uptake by *P. falciparum* indicating that the target may be the respiratory chain (Krungkrai et al., 1999; Krungkrai, 2004). In a novel strategy, Li and colleagues used a yeast model to search for the mode of action of artemisinin and, by applying yeast genetics, produced evidence that artimisinin disrupted mitochondrial membrane potential (Li et al., 2005a). Their work indicated that the electron transport chain actually activated the mitochondrial depolarising activity of artemisinin. In contrast, Nagamune and colleagues found that artemisinin failed to affect the mitochondrial membrane potential of another protozoan parasite, *Toxoplasma gondii*, suggesting that the mitochondrion is not a primary target in *T. gondii* (Nagamune et al., 2007b). In fact, there is now strong evidence that artemisinin affects calcium homeostasis in target species. Eukaryotic cells use Ca^{2+} as a second messenger and typically maintain very low cytoplasmic concentrations of Ca^{2+} by partitioning Ca^{2+} in organelles and compartments such as the endoplasmic reticulum. One of the key enzymes in this process is the Sarcoplasmic/Endoplasmic Reticulum Ca^{2+}-ATPase (SERCA). Initial work using a heterologous host, *Xenopus laevis*, demonstrated that artemisinin inhibited *P. falciparum* SERCA activity (Eckstein-Ludwig et al., 2003), and more recent experiments expressing the *T. gondii* SERCA gene in *S. cerevisiae* demonstrated that the *T. gondii* SERCA enzyme is also inhibited by artemisinin (Nagamune et al., 2007a). Physiological experiments in several protozoa are consistent with effects of artemisinin on calcium homeostasis, suggesting that this is likely to be responsible for a considerable portion of the biological activity of this metabolite (Golenser et al., 2006; Krishna et al., 2006; Nagamune et al., 2007a, b). Obviously, there are conflicting opinions on the biological activities of artemisinin with further studies required to resolve whether artemisinin does affect mitochondria, either as a primary or a secondary target, in the malarial parasite *P. falciparum*, and indeed in plant cells. Artemisinin also appears to exert non-specific effects such as production of free radicals and immune stimulation and it is completely plausible that, like sorgoleone, artemisinin has more than one target/effect. The links between mitochondrial function, calcium signalling and

13

apoptosis should not be overlooked and it is possible that effects that appear pleiotropic may in fact turn out to be part of the same process.

Disruption of Plamsa Membrane Integrity

Importance of Fungal Membrane as a Target

As highlighted earlier, one of the major roles in plants for secondary metabolites is defence against fungal pathogens. This is also a key area of interest in modern agriculture and medicine with the fungal cell membrane being the most common target of clinical and agricultural anti-fungal drugs. The fungal membrane possesses unique features, most notably the sterol ergosterol rather than cholesterol or stigmasterol, which are present in animal and plant membranes, respectively. Other differences include the presence of specific lipids in the outer leaf of the membrane. Common anti-fungal compounds include amphotericin B, which binds ergsoterol leading to pore formation, and azoles and morpholines, which inhibit ergosterol biosynthesis. Evolution has not overlooked this vulnerability in fungi, and plants produce different classes of anti-fungal defence metabolites that target the membrane of phytopathogenic fungi. The best understood of these are the defensins and the saponins.

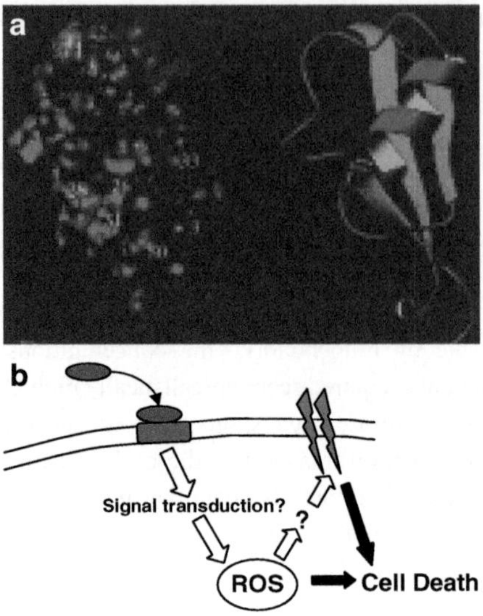

Fig. 3 The antifungal activity of plant defensins is specific and involves receptors and signal transduction pathways. Panel A shows the three dimensional structure of the radish defensin *Rs*-AFP2.The ribbon diagram on the right shows the classical defensin folds, and the image on the left depicts the hydrophobic surface area. Panel B shows one model for how *Rs*-AFP2 may exert its antifungal activity. The defensin (red) binds to specific sphingolipids, in this case glucosylceramides, in the membrane and transmits a signal internally in the cell. This leads to formation of reative oxygen species (ROS), loss of membrane integrity and cell death. The signal transduction components and the links between the different effects are not yet determined, but this mode of action may prove to be a paradigm for how plant metabolites exert effect within target cells. Panel A is reproduced with permission from Landan et al., 2000, and panel B is a modified version of a model proposed by Thevissen et al., 2007. See Section 2.4.2 for further details

Plant Defensins have Specific Binding Sites on Fungal Membranes

Defensins are highly basic, cysteine-rich peptides, typically 40–45 amino acids in length, that are produced as anti-microbial defence molecules by plants, insects and other animals, including humans (Fig. 3a) (Wong et al., 2007). Molecular phylogenetic analysis suggests that while the evolutionary root of these molecules is probably in plants, through evolution there has been significant functional divergence in the defensin family of cationic anti-microbial peptides (cAMPs) (Thomma et al., 2002). Many different defensins have been reported to have antiviral, antibacterial and antifungal activity, with limited information on mode of action. The prevailing view is that the positive charge of the peptides mediates non-specific binding to phospholipids, leading to pore formation and loss of membrane integrity (Verma et al., 2007). Although that may be a general feature of cAMPs, in recent years, it has emerged that specific interactions play a part in the activity of some cAMPs. For example, it was shown that human α–defensins inactivate adenovirus through direct binding to the virus protein (Buck, 2008; Smith and Nemerow, 2008), and intracellular targets for cAMPS have been identified in some bacteria (Hale and Hancock, 2007). It was already known that plant defensins, and also some insect defensins, have a specific binding target and mode of action (Thevissen et al., 2007). This discovery initially came from work using *S. cerevisiae* as a model to study the anti-fungal activity of plant defensins. It was found that a yeast strain mutated in a gene required for synthesis of sphingolipids had altered sensitivity to DmAMP1, a plant defensin (Thevissen et al., 2005). Sphingolipids are commonly found in the outer leaf of eukaryotic membranes and resemble phospholipids except that the backbone is ceramide rather than diacylglycerol (Thevissen et al., 2003a). Many variations of sphingolipid are found in eukaryotic membranes with some unique structures present in different fungi. Follow-on studies from Thevissen and colleagues established that (certain) plant and insect defensins bind to different fungal-specific sphingolipids, or even different motifs within the same sphingolipid (Thevissen et al., 2003b, 2004). Following binding, membrane permeabilisation does occur, but it is as yet unknown whether this is a consequence of a signalling cascade or a biophysical effect. It is clear, however, that plant defensins do not non-specifically insert into fungal membranes, forming pores and destroying membrane integrity. Interestingly, in *Candida albicans*, the anti-fungal activity of one plant defensin RsAFP2, which binds glucosylceramide, was found to involve the production of reactive oxygen species (ROS) suggesting that binding to the membrane ligand initiated a signal transduction cascade that culminated in the production of ROS and membrane permeabilisation (Fig. 3b) (Aerts et al., 2007; Thevissen et al., 2007). Despite the progress that has been made in the study of plant defensins, some serious questions and challenges remain to be addressed. First, most of the detailed work has been performed with a small number of specific defensins and it must be determined whether this is the only mode of action. Second, it is not known what signal transduction pathways are activated in response to defensins. Third, it is still unclear whether defensins are internalised, either following binding to sphingolipids, or independently. Work with human cAMPS suggests that at least some are internalised by bacteria (Hale and Hancock, 2007), and it was reported that a pea defensin is internalised by fungal cells (Cabral et al., 2003).

Lysis of Fungal Membranes by Saponins

Saponins are a structurally diverse class of secondary metabolite that are found in many different plants. For example, one survey lists over two hundred different plants from which saponins were isolated between the years 1998 and 2003 (Sparg et al., 2004). The basic structure of all saponins comprises a non-polar core and a polar glycosyl group or groups, which give the molecules amphipathic characteristics. Conventionally, saponins are classified as either triterpenoid or steroidal, with a subclass of steroidal alkaloids (steroidal glycoalkaloids), depending on the structure of the hydrophobic core. Some authors, however, treat the steroidal glycoalkaloids as a distinct type of natural product (Friedman, 2002; Korpan et al., 2004; Friedman, 2006) and recently, a new saponin classification into eleven different families depending on the structure of the backbone was proposed (Vincken et al., 2007). Saponins are present in appreciable concentrations in many traditional plant medicines, including common products such as ginseng, with diverse beneficial activities often ascribed to the saponin component. Within plants, saponins are believed to have evolved to provide defence against phytopathogenic fungi as they have potent antifungal activity, are generally localised to epidermal layers of plant tissues, and have been demonstrated to have a defence role in several pathogenic interactions (Morrissey and

13

Osbourn, 1999). The amphipathic nature of saponins suggests a mode of action, and it was demonstrated that saponins can permeabilise fungal membranes. The proposed mode of action is that the hydrophobic core inserts into the outer membrane, forming a complex with ergosterol. Subsequent interaction between the polar glycosidic sidechains leads to aggregation, pore formation and loss of membrane integrity (Armah et al., 1999). The ability to permeabilise membranes has been demonstrated *in vitro* on model membranes and *in vivo* in a study that used *S. cerevisiae* to explore the anti-fungal activity of the steroidal glycoalkaloid saponin, α-tomatine (Simons et al., 2006). That study also showed, however, that tomatidine, the algycone of α-tomatine, did not permeabilise membranes, is more potent than α-tomatine, and in fact inhibits ergosterol biosynthesis. Furthermore, a number of studies have proposed additional activities for α-tomatine and its derivatives. It was found that β_2-tomatine (formed by removal of a single sugar from α-tomatine) is capable of supressing the plant defence response (Bouarab et al., 2002), and it was reported that α-tomatine is capable of inducing programmed cell death in the fungus *Fusarium oxysporum* in the absence of membrane permeabilisation (Ito et al., 2007). Finally, studies with the potato steroidal glycoalkaloids (saponins), α-chaconine and α-solanine, have identified a range of toxic effects in animal systems that are distinct from membrane permeabilising activity (Friedman, 2006). In conclusion, therefore, it can be stated that although membrane permeabilising activity is one characteristic of saponins that contributes to their anti-fungal activity, saponins are likely to possess other biological properties, including those that may have beneficial roles on human health.

Anti-tumour Activity of Plant Natural Products

Of the various properties putatively associated with saponins, the capacity of some saponin preparations to inhibit the growth of tumour cells *in vitro* attracts the most attention. Indeed, numerous saponins have been reported to possess such activity, raising the possibility of developing novel saponin-based anti-cancer therapeutic drugs (Rao and Gurfinkel, 2000). Some commentators have questioned the relevance of these *in vitro* data and confirmation of their significance will require demonstration that the effects involve specific intracellular targets and are not associated with membrane permeabilisation. Particular progress in this direction is being made by researchers working on two different groups of legume triterpenoid saponins, the avicins, which are synthesised by *Acacia victoria* (Fig. 2), and the soyasaponins, which are synthesised by soybean plants. This work was stimulated by reports that avicins possess pro-apoptotic activity against human tumour cells (Haridas et al., 2001; Mujoo et al., 2001). Several studies have established that this is mediated by perturbation of the mitochondria, with data indicating that the effects are twofold – disruption of the outer membrane potential, and induced closure of the voltage dependent anion channel (VDAC) in the mitochondrial membrane (Haridas et al., 2001; Mujoo et al., 2001; Li et al., 2005b; Lemeshko et al., 2006; Haridas et al., 2007). The link between saponin-induced mitochondrial dysfunction and apoptosis was given further support by a recent report showing that treatment of HeLa cells with a soyasaponin preparation led to apoptosis via the mitochondrial pathway (Xiao et al., 2007). Other intracellular targets for specific avicins have also been reported, however, indicating that pro-apoptotic effects may involve multiple targets, or that different avicins have specific target processes. In a yeast model, evidence for modulation/inhibition of rho-dependent signalling and the cAMP/PKA signal transduction pathways by avicin G was obtained (Gutterman et al., 2005). A more direct link to apoptosis/autophagy was obtained from studies with avicin D, where it was shown that avicin D activates the AMP-Activated Kinase (AMPK), thereby inhibiting mTORC1 and downstream targets. It is woth noting here that although many studies of plant natural products have reported "pro-apoptotic" activity, there is generally not much differentiation in the plant literature between apoptosis and autophagy. In fact, although the end result is the same, the pathways and processes involved are quite different, and it is a topic that needs closer attention in the future (Codogno and Meijer, 2005). This link to autophagy in given greater significance by the finding that a preparation of B-group soysasaponins also reduced mTORC activity, this time apparently by activating another kinase, Akt (Ellington et al., 2006). These data are exciting but outstanding questions remain: for example, how do glycosylated molecules enter the cell? And is the glycone group required/retained for activity in the cell? Alternatively, are there lessons to be learned from the mode of actions of at least some defensins that act via a membrane receptor and signal transduction pathway? The general importance of apoptotic pathways as a target of plant natural metabolites is also illustrated by data that other non-saponin metabolites, for example the sesquiterpenoid helenalin, which may inhibit telomerase, also have pro-apototic effects on cultured mammalian cells (Huang et al., 2005). Some of the more than over 5,000 different flavonoids that occur naturally in plants also have effects that may be linked to apoptosis in the future. The knotweed flavonoid, (-)catechin, for example, inhibits seed germination and causes cell death in sensitive species (Bais et al., 2002, 2003b, 2006; Field et al., 2006). The effect appears to involve the generation of reactive oxygen species, and may also be linked to calcium signalling or homeostasis (Bais et al., 2003a; Iwashina, 2003; Friedman, 2007). Once again, it is worth bearing in mind the connections between ROS, calcium homeostasis, mitochondrial function, autophagy and apoptosis.

Conclusions

Current Status

Understandably, in the field of natural product biology, the focus of research is on metabolites that are secreted or are present in appreciable concentrations. It is salient, however, that most plant secondary metabolites are neither of those, and of the thousands of plant natural products, we know the biological activities of a tiny minority. And of those, our interest is concentrated on metabolites involved in processes that concern us – especially signalling and defence, so we may be overlooking vast swathes of functionality. In mitigation, perhaps, one could also suggest that metabolites designed to communicate between species are most likely to have biological activities of relevance to agricultural or biotechnological applications. Considering the body of knowledge as a whole, it is possible to draw some general conclusions about the biological activities of plant secondary metabolites. First, different metabolites with a defence role, whether against pathogens or other plants, often target a small number of processes. Thus, membrane integrity, photosynthesis and mitochondrial function are disrupted by metabolites that have very different chemical structures. Second, although many secondary metabolites have anti-cellular, phytotoxic, or apoptopic activity that appear quite generic and non-specific, more detailed studies, for example as with the avicins, usually reveals that interference with specific targets underpins these activities. The problem is that identification of these targets is difficult. Third, metabolites may have more that one target, which is not the same as being "non-specific". Sorgoleone, for example, inhibits both the enzyme HPPD and PSII, either of which alone would be enough to prevent growth. Some of the conflicting data on artemisinin may also ultimately be resolved by similar findings. Fourth, the activities that we uncover for metabolites, may or may not reflect the normal function of that molecule. Thus, a sesquiterpene or quinone produced by a plant as a signal molecule to interact with microbes in the soil could also induce apoptosis in human tumour cells. Although this would be a peculiarity of the chemical structure rather than a "natural" role for the metabolite, it is very relevant to exploitation of plants or their natural products.

Limitations with Current Approaches

Although some estimates put the number of plant species that are used in traditional medicines at between 25,000 and 50,000 (Kartal, 2007) relatively few plant natural products are used in purified form in pharmaceuticals. There are examples, of course, that show the potential benefits of plant secondary metabolites – for instance, the use of the saponin disgenin as a precursor for production of the hormone progesterone. Furthermore, the success of the sesquiterpene lactone artemisinin as an anti-malarial drug highlights possible applications in the pharmaceutical sector. Nonetheless, there is a clear dichotomy between the chemical diversity in plants and the number of bioactive leads for the pharmaceutical or agri-biotech industries. Two of the reasons for this were the decisions in the 1990s by many biotechnology companies to replace natural product screening programmes with discovery approaches employing libraries of synthetic chemical compounds, and the transfer of emphasis to target-based, rather than cell-based, assays for biological activity (Baker et al., 2007). Combinatorial chemistry has yielded, few if any major new drugs, and there is now a realisation that the diversity of natural products, in plants and microbes, may represent the best prospects for future lead molecules. This, notwithstanding, new approaches are required if significant progress is to be made. Some of the limitations are fragmentation among scientists – plant chemists and biologists often work within very narrow parameters and do not engage sufficiently in multidisciplinary research; reliance on traditional approaches that are not amenable to high-throughput biology; and limitations in the choice of model systems for determining mode of action.

Exploiting the Power of Genomics

One of the justifications for genome sequencing programmes has always been that they will advance drug discovery and lead to new products that benefit society. Although unquestionable true, it is really only in recent years that this is starting to become a reality. With respect to plant natural products, there are at least two areas where genomics can make a significant contribution. First, is the plant genomes themselves. As it becomes possible to inexpensively sequence plant genomes, we gain access to the genes that encode the biosynthetic enzymes for plant secondary metabolites. For example, it is estimated that 25% of the genes in *Arabidopsis* are involved in synthesis of natural products (D'Auria and Gershenzon, 2005). The biosynthetic pathways are often complex, but work by Osbourn and colleagues illustrates how these pathways can be dissected and the coding genes identified (Osbourn, 2003). Knowing the genes will facilitate production of molecules of interest in heterologous systems for mode of action studies, and also opens up possibilies of combining genes to generate pathways that create new variant metabolites for testing. This approach has been used for some time in microbial systems and is currently popularised by the term "synthetic biology". Engineering pathways to produce new functional variants is an exciting prospect for scientists and biotechnologists alike. Second, genomic tools can be applied to understand how metabolites exert their biological effects. One way of doing this is to perform global analysis (transcriptome/proteome) of the effects of treating a cell/organism with a metabolite of interest. In terms of looking at the effects of metabolites on plants, studies of benzoxazolinone (BOA) is one of the few attempts to carry out this type of investigation with plants (Baerson et al., 2005). The other major application of genomic tools is the use of model systems. As genomics and molecular biology advance with plant and mammalian systems, these will become useful models for high throughput (HT) strategies, but arguably, yeast is currently the optimum system for a number of reasons. These include ease of manipulation, libraries of knockout mutants, complete sets of complementing clones, heterozygous libraries for gene dosage experiments, detailed knowedge of biochemical pathways and extensive databases on gene expression (Fig. 4). Combining large-scale screens of chemicals or natural products with libraries of mutants facilitates an approach to drug discovery termed chemical genomics, an approach now widely used in yeast (Baerson et al., 2005). In general, plant biologists have been slower to employ yeast models than human cell biologists, but some of the cases discussed earlier where most progress towards discovering mode of action was made applied a yeast model. There are limitations of course, in particular in target-based screens where an organelle like the *Plasmodium* apicoplast, or a process like photosynthesis is targetted. Yeast are ideal, however, for cell-based appraoches, which are regaining popularity because of the unbiased nature of the assay and the requirement for uptake of the test molecule into the cell. Even in a case of artemisinin where a yeast screen generated some controversy, a yeast model was used to heterologously express the *T. gondii* SERCA gene and show that activity was inhibited by artemisinin.

Concluding Remarks

Plant natural products, in particular those involved in defence against pathogens, offer great potential as leads for biotechnological applications. There is a pressing need, however, to move beyond phytochemical cataloging and generic studies, to specific screens and experiments to identify mode of action. This should take two forms, identification of the biological

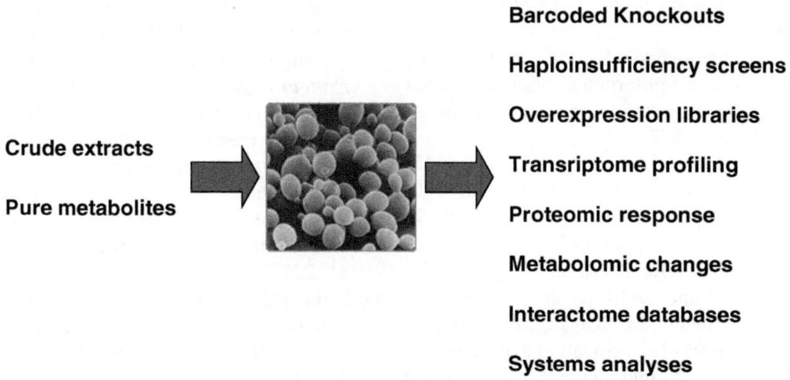

Fig. 4 Chemical genomics in yeast to determine mode of action of natural products. *S. cerevisiae* provides a molecular resource for chemical genomics and has been extensively developed for drug discovery applications in the pharmaceutical sector. Some of the key tools and strategies are listed and these are discussed further in the text (Section 3.3 and references therein). Although yeast has some very obvious differences to higher eukaryotes, it is notable that the key data on mode of action of some of the better understood metabolites like defensins, artemsisin, avicins and tomatidine were obtained in yeast (see relevant sections for more information). (Landon et al., 2000)

13

role of the metabolite in the plant, and determination of the effects of the metabolite on other organisms. For the latter, there is a compelling argument to use unbiased genomic/proteomic methods and cell-based assays to avoid confusion with non-specific targets. Ultimately, once candidate targets are identified, it will be necessary to undertake detailed structural and functional studies of the interaction in authentic hosts. For preliminary screens and analyses, however, plant natural product scientists should consider emulating their biomedical colleagues by applying chemical genomics and molecular techniques in the model eukaryote, *S. cerevisiae*.

References

Aerts, A.M., Francois, I.E., Meert, E.M., Li, Q.T., Cammue, B.P., and Thevissen, K. (2007) The antifungal activity of RsAFP2, a plant defensin from raphanus sativus, involves the induction of reactive oxygen species in Candida albicans. *J Mol Microbiol Biotechnol* **13**: 243–247.

Akiyama, K., and Hayashi, H. (2006) Strigolactones: chemical signals for fungal symbionts and parasitic weeds in plant roots. *Ann Bot (Lond)* **97**: 925–931.

Akiyama, K., Matsuzaki, K., and Hayashi, H. (2005) Plant sesquiterpenes induce hyphal branching in arbuscular mycorrhizal fungi. *Nature* **435**: 824–827.

Armah, C.N., Mackie, A.R., Roy, C., Price, K., Osbourn, A.E., Bowyer, P., and Ladha, S. (1999) The membrane-permeabilizing effect of avenacin A-1 involves the reorganization of bilayer cholesterol. *Biophys J* **76**: 281–290.

Baerson, S.R., Sanchez-Moreiras, A., Pedrol-Bonjoch, N., Schulz, M., Kagan, I.A., Agarwal, A.K., et al. (2005) Detoxification and transcriptome response in Arabidopsis seedlings exposed to the allelochemical benzoxazolin-2(3H)-one. *J Biol Chem* **280**: 21867–21881.

Bais, H.P., Walker, T.S., Stermitz, F.R., Hufbauer, R.A., and Vivanco, J.M. (2002) Enantiomeric-dependent phytotoxic and antimicrobial activity of (+/-)-catechin. A rhizosecreted racemic mixture from spotted knapweed. *Plant Physiol* **128**: 1173–1179.

Bais, H.P., Walker, T.S., Kennan, A.J., Stermitz, F.R., and Vivanco, J.M. (2003a) Structure-dependent phytotoxicity of catechins and other flavonoids: flavonoid conversions by cell-free protein extracts of Centaurea maculosa (spotted knapweed) roots. *J Agric Food Chem* **51**: 897–901.

Bais, H.P., Vepachedu, R., Gilroy, S., Callaway, R.M., and Vivanco, J.M. (2003b) Allelopathy and exotic plant invasion: from molecules and genes to species interactions. *Science* **301**: 1377–1380.

Bais, H.P., Weir, T.L., Perry, L.G., Gilroy, S., and Vivanco, J.M. (2006) The role of root exudates in rhizosphere interactions with plants and other organisms. *Annu Rev Plant Biol* **57**: 233–266.

Baker, D.D., Chu, M., Oza, U., and Rajgarhia, V. (2007) The value of natural products to future pharmaceutical discovery. *Nat Prod Rep* **24**: 1225–1244.

Balunas, M.J., and Kinghorn, A.D. (2005) Drug discovery from medicinal plants. *Life Sci* **78**: 431–441.

Bouarab, K., Melton, R., Peart, J., Baulcombe, D., and Osbourn, A. (2002) A saponin-detoxifying enzyme mediates suppression of plant defences. *Nature* **418**: 889–892.

Buck, C.B. (2008) Defensins' offensive play: exploiting a viral achilles' heel. *Cell Host Microbe* **3**: 3–4.

Cabral, K.M., Almeida, M.S., Valente, A.P., Almeida, F.C., and Kurtenbach, E. (2003) Production of the active antifungal Pisum sativum defensin 1 (Psd1) in Pichia pastoris: overcoming the inefficiency of the STE13 protease. *Protein Expr Purif* **31**: 115–122.

Codogno, P., and Meijer, A.J. (2005) Autophagy and signaling: their role in cell survival and cell death. *Cell Death Differ* **12 Suppl 2**: 1509–1518.

Cole, D., Pallet, K., and Rodgers, M. (2000) Discovering new modes of action for herbicides and the impact of genomics. *Pesticide Outlook* **11**: 223–229.

D'Auria, J.C., and Gershenzon, J. (2005) The secondary metabolism of Arabidopsis thaliana: growing like a weed. *Curr Opin Plant Biol* **8**: 308–316.

Eckstein-Ludwig, U., Webb, R.J., Van Goethem, I.D., East, J.M., Lee, A.G., Kimura, M., et al. (2003) Artemisinins target the SERCA of Plasmodium falciparum. *Nature* **424**: 957–961.

Ellington, A.A., Berhow, M.A., and Singletary, K.W. (2006) Inhibition of Akt signaling and enhanced ERK1/2 activity are involved in induction of macroautophagy by triterpenoid

B-group soyasaponins in colon cancer cells. *Carcinogenesis* **27**: 298–306.

Field, B., Jordan, F., and Osbourn, A. (2006) First encounters - deployment of defence-related natural products by plants. *New Phytol* **172**: 193–207.

Francois, I.E., Aerts, A.M., Cammue, B.P., and Thevissen, K. (2005) Currently used antimycotics: spectrum, mode of action and resistance occurrence. *Curr Drug Targets* **6**: 895–907.

Friedman, M. (2002) Tomato glycoalkaloids: role in the plant and in the diet. *J Agric Food Chem* **50**: 5751–5780.

Friedman, M. (2006) Potato glycoalkaloids and metabolites: roles in the plant and in the diet. *J Agric Food Chem* **54**: 8655–8681.

Friedman, M. (2007) Overview of antibacterial, antitoxin, antiviral, and antifungal activities of tea flavonoids and teas. *Mol Nutr Food Res* **51**: 116–134.

Golenser, J., Waknine, J.H., Krugliak, M., Hunt, N.H., and Grau, G.E. (2006) Current perspectives on the mechanism of action of artemisinins. *Int J Parasitol* **36**: 1427–1441.

Gomez-Roldan, V., Fermas, S., Brewer, P.B., Puech-Pages, V., Dun, E.A., Pillot, J.P., et al. (2008) Strigolactone inhibition of shoot branching. *Nature* **455**: 189–194.

Gutterman, J.U., Lai, H.T., Yang, P., Haridas, V., Gaikwad, A., and Marcus, S. (2005) Effects of the tumor inhibitory triterpenoid avicin G on cell integrity, cytokinesis, and protein ubiquitination in fission yeast. *Proc Natl Acad Sci U S A* **102**: 12771–12776.

Hale, J.D., and Hancock, R.E. (2007) Alternative mechanisms of action of cationic antimicrobial peptides on bacteria. *Expert Rev Anti Infect Ther* **5**: 951–959.

Halkier, B.A., and Gershenzon, J. (2006) Biology and biochemistry of glucosinolates. *Annu Rev Plant Biol* **57**: 303–333.

Haridas, V., Li, X., Mizumachi, T., Higuchi, M., Lemeshko, V.V., Colombini, M., and Gutterman, J.U. (2007) Avicins, a novel plant-derived metabolite lowers energy metabolism in tumor cells by targeting the outer mitochondrial membrane. *Mitochondrion* **7**: 234–240.

Haridas, V., Higuchi, M., Jayatilake, G.S., Bailey, D., Mujoo, K., Blake, M.E., et al. (2001) Avicins: triterpenoid saponins from Acacia victoriae (Bentham) induce apoptosis by mitochondrial perturbation. *Proc Natl Acad Sci U S A* **98**: 5821–5826.

Huang, P.R., Yeh, Y.M., and Wang, T.C. (2005) Potent inhibition of human telomerase by helenalin. *Cancer Lett* **227**: 169–174.

Ito, S., Ihara, T., Tamura, H., Tanaka, S., Ikeda, T., Kajihara, H., et al. (2007) alpha-Tomatine, the major saponin in tomato, induces programmed cell death mediated by reactive oxygen species in the fungal pathogen Fusarium oxysporum. *FEBS Lett* **581**: 3217–3222.

Iwashina, T. (2003) Flavonoid function and activity to plants and other organisms. *Biol Sci Space* **17**: 24–44.

Kartal, M. (2007) Intellectual property protection in the natural product drug discovery, traditional herbal medicine and herbal medicinal products. *Phytother Res* **21**: 113–119.

Kennedy, J., Marchesi, J.R., and Dobson, A.D. (2007) Metagenomic approaches to exploit the biotechnological potential of the microbial consortia of marine sponges. *Appl Microbiol Biotechnol* **75**: 11–20.

Klee, H. (2008) Plant biology: Hormones branch out. *Nature* **455**: 176–177.

Konig, G.M., Kehraus, S., Seibert, S.F., Abdel-Lateff, A., and Muller, D. (2006) Natural products from marine organisms and their associated microbes. *Chembiochem* **7**: 229–238.

Korpan, Y.I., Nazarenko, E.A., Skryshevskaya, I.V., Martelet, C., Jaffrezic-Renault, N., and El'skaya, A.V. (2004) Potato glycoalkaloids: true safety or false sense of security? *Trends Biotechnol* **22**: 147–151.

Krishna, S., Woodrow, C.J., Staines, H.M., Haynes, R.K., and Mercereau-Puijalon, O. (2006) Re-evaluation of how artemisinins work in light of emerging evidence of in vitro resistance. *Trends Mol Med* **12**: 200–205.

Krungkrai, J. (2004) The multiple roles of the mitochondrion of the malarial parasite. *Parasitology* **129**: 511–524.

Krungkrai, J., Burat, D., Kudan, S., Krungkrai, S., and Prapunwattana, P. (1999) Mitochondrial oxygen consumption in asexual and sexual blood stages of the human malarial parasite, Plasmodium falciparum. *Southeast Asian J Trop Med Public Health* **30**: 636–642.

Landon, C., Pajon, A., Vovelle, F., and Sodano, P. (2000) The active site of drosomycin, a small insect antifungal protein, delineated by comparison with the modeled structure of Rs-AFP2, a plant antifungal protein. *J Pept Res* **56**: 231–238.

Lemeshko, V.V., Haridas, V., Quijano Perez, J.C., and Gutterman, J.U. (2006) Avicins, natural anticancer saponins, permeabilize mitochondrial membranes. *Arch Biochem Biophys* **454**: 114–122.

13

Li, W., Mo, W., Shen, D., Sun, L., Wang, J., Lu, S., et al. (2005a) Yeast model uncovers dual roles of mitochondria in action of artemisinin. *PLoS Genet* 1: e36.

Li, X.X., Davis, B., Haridas, V., Gutterman, J.U., and Colombini, M. (2005b) Proapoptotic triterpene electrophiles (avicins) form channels in membranes: cholesterol dependence. *Biophys J* **88**: 2577–2584.

Macias, F.A., Molinillo, J.M., Varela, R.M., and Galindo, J.C. (2007) Allelopathy - a natural alternative for weed control. *Pest Manag Sci* **63**: 327–348.

Madden, L.V., and Wheelis, M. (2003) The threat of plant pathogens as weapons against U.S. crops. *Annu Rev Phytopathol* **41**: 155–176.

Maor, R., and Shirasu, K. (2005) The arms race continues: battle strategies between plants and fungal pathogens. *Curr Opin Microbiol* **8**: 399–404.

Meazza, G., Scheffler, B.E., Tellez, M.R., Rimando, A.M., Romagni, J.G., Duke, S.O., et al. (2002) The inhibitory activity of natural products on plant p-hydroxyphenylpyruvate dioxygenase. *Phytochemistry* **60**: 281–288.

Morre, D.J., Grieco, P.A., and Morre, D.M. (1998) Mode of action of the anticancer quassinoids - inhibition of the plasma membrane NADH oxidase. *Life Sci* **63**: 595–604.

Morrissey, J.P., and Osbourn, A.E. (1999) Fungal resistance to plant antibiotics as a mechanism of pathogenesis. *Microbiol Mol Biol Rev* **63**: 708–724.

Mujoo, K., Haridas, V., Hoffmann, J.J., Wachter, G.A., Hutter, L.K., Lu, Y., et al. (2001) Triterpenoid saponins from Acacia victoriae (Bentham) decrease tumor cell proliferation and induce apoptosis. *Cancer Res* **61**: 5486–5490.

Nagamune, K., Beatty, W.L., and Sibley, L.D. (2007a) Artemisinin induces calcium-dependent protein secretion in the protozoan parasite Toxoplasma gondii. *Eukaryot Cell* **6**: 2147–2156.

Nagamune, K., Moreno, S.N., and Sibley, L.D. (2007b) Artemisinin-resistant mutants of Toxoplasma gondii have altered calcium homeostasis. *Antimicrob Agents Chemother* **51**: 3816–3823.

Odds, F.C., Brown, A.J., and Gow, N.A. (2003) Antifungal agents: mechanisms of action. *Trends Microbiol* **11**: 272–279.

Osbourn, A.E. (2003) Saponins in cereals. *Phytochemistry* **62**: 1–4.

Parniske, M. (2005) Plant-fungal associations: cue for the branching connection. *Nature* **435**: 750–751.

Perret, X., Staehelin, C., and Broughton, W.J. (2000) Molecular basis of symbiotic promiscuity. *Microbiol Mol Biol Rev* **64**: 180–201.

Peters, N.K., Frost, J.W., and Long, S.R. (1986) A plant flavone, luteolin, induces expression of Rhizobium meliloti nodulation genes. *Science* **233**: 977–980.

Rao, A.V., and Gurfinkel, D.M. (2000) The bioactivity of saponins: triterpenoid and steroidal glycosides. *Drug Metabol Drug Interact* **17**: 211–235.

Rimando, A.M., Dayan, F.E., Czarnota, M.A., Weston, L.A., and Duke, S.O. (1998) A new photosystem II electron transfer inhibitor from sorghum bicolor. *J Nat Prod* **61**: 1456.

Romagni, J.G., Duke, S.O., and Dayan, F.E. (2000) Inhibition of plant asparagine synthetase by monoterpene cineoles. *Plant Physiol* **123**: 725–732.

Schmidt, B.M., Ribnicky, D.M., Lipsky, P.E., and Raskin, I. (2007) Revisiting the ancient concept of botanical therapeutics. *Nat Chem Biol* **3**: 360–366.

Simons, V., Morrissey, J.P., Latijnhouwers, M., Csukai, M., Cleaver, A., Yarrow, C., and Osbourn, A. (2006) Dual effects of plant steroidal alkaloids on Saccharomyces cerevisiae. *Antimicrob Agents Chemother* **50**: 2732–2740.

Singh, S.B., and Pelaez, F. (2008) Biodiversity, chemical diversity and drug discovery.*Prog Drug Res* **65**: 141, 143–174.

Smith, J.G., and Nemerow, G.R. (2008) Mechanism of adenovirus neutralization by Human alpha-defensins. *Cell Host Microbe* **3**: 11–19.

Sparg, S.G., Light, M.E., and van Staden, J. (2004) Biological activities and distribution of plant saponins. *J Ethnopharmacol* **94**: 219–243.

Thevissen, K., Ferket, K.K., Francois, I.E., and Cammue, B.P. (2003a) Interactions of antifungal plant defensins with fungal membrane components. *Peptides* **24**: 1705–1712.

Thevissen, K., Kristensen, H.H., Thomma, B.P., Cammue, B.P., and Francois, I.E. (2007) Therapeutic potential of antifungal plant and insect defensins. *Drug Discov Today* **12**: 966–971.

Thevissen, K., Francois, I.E., Takemoto, J.Y., Ferket, K.K., Meert, E.M., and Cammue, B.P. (2003b) DmAMP1, an antifungal plant defensin from dahlia (Dahlia merckii), interacts with sphingolipids from Saccharomyces cerevisiae. *FEMS Microbiol Lett* **226**: 169–173.

Thevissen, K., Warnecke, D.C., Francois, I.E., Leipelt, M., Heinz, E., Ott, C., et al. (2004)

Defensins from insects and plants interact with fungal glucosylceramides. *J Biol Chem* **279**: 3900–3905.

Thevissen, K., Idkowiak-Baldys, J., Im, Y.J., Takemoto, J., Francois, I.E., Ferket, K.K., et al. (2005) SKN1, a novel plant defensin-sensitivity gene in Saccharomyces cerevisiae, is implicated in sphingolipid biosynthesis. *FEBS Lett* **579**: 1973–1977.

Thomma, B.P., Cammue, B.P., and Thevissen, K. (2002) Plant defensins. *Planta* **216**: 193–202.

Umehara, M., Hanada, A., Yoshida, S., Akiyama, K., Arite, T., Takeda-Kamiya, N., et al. (2008) Inhibition of shoot branching by new terpenoid plant hormones. *Nature* **455**: 195–200.

VanEtten, H.D., Mansfield, J.W., Bailey, J.A., and Farmer, E.E. (1994) Two Classes of Plant Antibiotics: Phytoalexins versus "Phytoanticipins". *Plant Cell* **6**: 1191–1192.

Verma, C., Seebah, S., Low, S.M., Zhou, L., Liu, S.P., Li, J., and Beuerman, R.W. (2007) Defensins: antimicrobial peptides for therapeutic development. *Biotechnol J* **2**: 1353–1359.

Vincken, J.P., Heng, L., de Groot, A., and Gruppen, H. (2007) Saponins, classification and occurrence in the plant kingdom. *Phytochemistry* **68**: 275–297.

Wink, M. (2003) Evolution of secondary metabolites from an ecological and molecular phylogenetic perspective. *Phytochemistry* **64**: 3–19.

Wong, J.H., Xia, L., and Ng, T.B. (2007) A review of defensins of diverse origins. *Curr Protein Pept Sci* **8**: 446–459.

Xiao, J.X., Huang, G.Q., and Zhang, S.H. (2007) Soyasaponins inhibit the proliferation of Hela cells by inducing apoptosis. *Exp Toxicol Pathol* **59**: 35–42.

The Role of Natural Products in Plant-Microbe Interactions

14

Giuliano Bonanomi, Francesco Vinale, and Felice Scala

Abstract Plants during their life cycles interact with a large diversity of microbial species. To simultaneously manage symbiotic, competitive and pathogenic interactions, plants rely on their chemical compounds which are capable to recognize, coordinate and regulate the exchange of resources and information with the myriads of potentially interacting microbes. This chapter deals with plant-microbe interactions mediated by natural products and focuses on their role, chemical nature and ecological significance. Particular attention is devoted to role of natural compounds in the cross-talk between plants and beneficial microbes like vesicular-arbuscular mycorrhizal fungi, bacteria inducing legume nodulation and plant-growth promoting rhizobacteria. The role of plant natural compounds with antimicrobial activity against pathogenic bacteria, fungi and oomycetes is also reviewed. Finally, we discuss the potential applications of natural products and future directions of studies including the use of new experimental approaches and techniques such as metagenomics, proteomics, microscopy, isotope labeling and NMR microspectrophotometry.

G. Bonanomi, F. Vinale, and F. Scala (✉)
Dipartimento di Arboricoltura, Botanica e Patologia Vegetable, University of Naples Federico II, Via Università 100, 80055 Portici, Napoli, Italy
e-mails: giuliano.bonanomi@unina.it; frvinale@unina.it; felice.scala@unina.it

Introduction

Plants during their life cycles interact with a vast range of different microbial species. The effect of these interactions can be positive (as in symbiotic), neutral (as in amensalic and commensalic), and negative (as in competitive and pathogenic). Plant-microbe interactions are of fundamental importance in determining plant fitness and, as a consequence, biogeochemical cycles and ecosystem functions. The ways by which plants recognize, coordinate and regulate the exchange of resources and information with the myriads of potentially interacting microbes are still not yet completely understood.

Higher plants produce a great variety of chemical compounds (Wink, 1999; Dixon, 2001; Hartmann, 2008), traditionally know as secondary metabolites because it was thought that plant performances were not negatively influenced by their absence, at least under controlled conditions. However, since it has been shown that plant secondary metabolites have important and different functions in the natural environment that are commonly named as “natural products” (Field et al., 2006). Natural compounds are largely involved in plant-microbe interactions although microbes can also use thigmotrophic and electrical signals (van West et al., 2002) to target the host plants. Much progress in understanding the role of specific chemicals has been made by studying very

A.E. Osbourn and V. Lanzotti (eds.), *Plant-derived Natural Products*,
DOI 10.1007/978-0-387-85498-4_14, © Springer Science+Business Media, LLC 2009

simple plant-microbe systems. Studies performed under more realistic ecological conditions are still a challenge, but advances in techniques such as genomics, proteomics, microscopy, isotope labelling, NMR, microspectrophotometry, etc. may produce a significant increase in our current knowledge. A very large and continuously expanding body of literature is available on plant chemical products. It is beyond the purpose of this paper to extensively review the findings on this topic. Here, we focus our attention on the role of some plant derived compounds in interactions with selected microbial groups with positive and negative impacts on plant performances.

Origin and Site of Action of Natural Compounds

Plants show a remarkable variability in their relationships with microbes, ranging from interactions with endophyte non-pathogenic organisms hosted within specialized plant tissues, to those

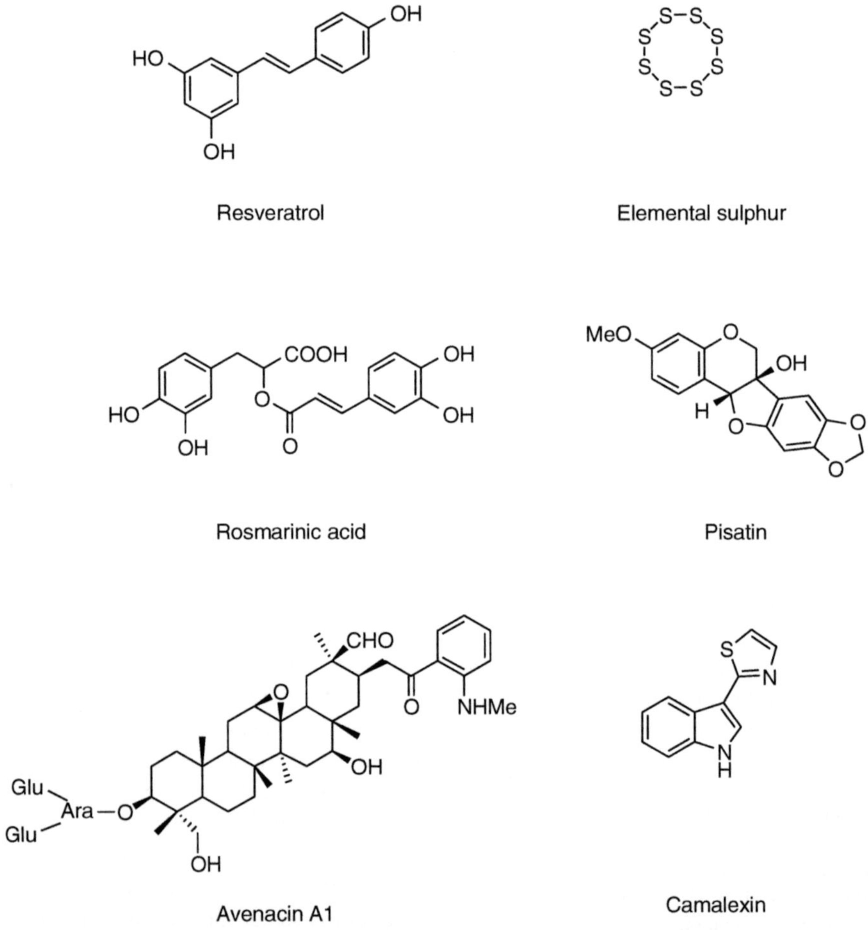

Fig. 1 Examples of plant natural products with negative effects on microbes (see text for details)

with extra-cellular microorganisms inhabiting plant surfaces, such as the phylloplane and the rhizoplane (Gray and Smith, 2005). In addition, plants continuously modify their surrounding habitat by exuding chemicals that select and regulate the microbial populations by, for instance, enhancing beneficial microbes and deterring pathogens.

In order to successfully manage all these different types of interactions with microbes, plants have evolved different strategies for the use of their natural products. First, chemicals can be locally produced in response to microbial infection, which occurs in many plant-pathogen interactions (Fig. 1; Par. 4.2). Plant chemicals can also be constitutively produced and then stored in specific plant cells. For example, the epidermal cells of the root of several *Avena* species accumulate the saponin avenacin A-1 at a concentration capable of antifungal activity against the soilborne pathogen *Gaeumannomyces graminis* var. *tritici* (Fig. 1; Osbourn et al., 1994). Plants can also synthesize chemicals in some cells and then move the products to the site of action (Field et al., 2006). To this purpose, Snyder and Nicholson (1990) demonstrate that the antimicrobial flavonoids 3-deoxyanthocyanidins are produced in vesicles that form in the cytoplasm of the plant cell and then move towards and accumulate at the site of pathogen penetration. Finally, chemicals can be released outside the plant body by germinating seeds (Nelson, 2004), growing roots (Bais et al., 2006), root border cells (Hawes et al., 2000), and also leaves (Baldwin et al., 2006).

Plant Natural Compounds and Beneficial Microbes

The rhizosphere, the area immediately adjacent to the root surface, is an extremely complex environment. This depends on the diversity of inhabiting microbial species (Rappé and Giovannoni, 2003), in association with the heterogeneity of abiotic factors, such as pH, redox potential, ion concentration and organic carbon quality and quantity. The key feature of the rhizosphere environment is the rhizodeposition (Jones et al., 2004), that is, the release of organic carbon compounds from roots with the consequent proliferation of microbial populations. Rhizodeposition is characterized by a great number of different low molecular weight compounds including sugars, amino acids, organic acids, phenols and many other natural products. Polysaccharides, proteins, and also detached, still living, root border cells (Hawes et al., 2000), account mostly for the mass of exudates. Rhizodeposition represents a significant energetic cost for the plant, which may reach up to 40% of the dry photosynthesized matter (Lynch and Whipps, 1990). The enhanced diversity and availability of organic compounds into the rhizosphere sustain and allow the coexistence of potentially beneficial microbes (Denison et al., 2003). Unraveling the chemical nature of the cross-talk between plant and microbes is thus important for understanding and successfully managing the rhizosphere (Raaijmakers et al., 2008). In the following sections, we discuss how plant derived compounds control the relationships between plant and vesicular-arbuscular mycorrhizas, nitrogen fixing nodulating rhizobacteria and extracellular beneficial microbes (Figs. 2 and 3).

Plant Natural Compounds Controlling Vesicular-Arbuscular Mycorrhizal Symbiosis

Approximately 80% of terrestrial higher plants are involved in mutualistic relationship with mycorrhizal fungi (Smith and Read, 1997; Parniske, 2008). These interactions, which played a crucial role for land colonization by plants (Brundrett, 2001), are essential for plant mineral nutrition (Smith et al., 2003). Some evidence indicates that mycorrhizas also have a relevant role in water up-take (Ebel et al., 1994), soil organic matter exploitation (Hodge et al.,

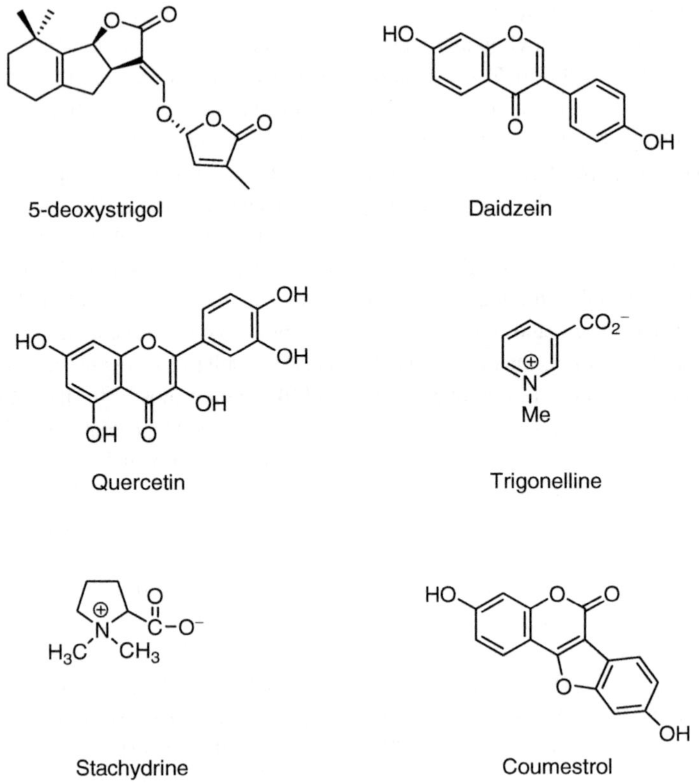

Fig. 2 Examples of plant natural products with positive effects on microbes (see text for details)

2001; Read and Perez-Moreno, 2003) and plant protection from soilborne pathogens (Graham, 2001; Vierheilig et al., 2008). Several types of mycorrhizas are known, comprising orchids, ericoids, ectomycorrhizas (ECM) and arbuscular mycorrhizas fungi (AM) (Smith and Read, 1997). Here we focus mainly on AM (recently placed in the phylum Glomeromycota) because they are the most widespread underground symbiont of angiosperm plants (Brundrett, 2001). AM fungi are obligate biotrophic organisms and completely rely on the host plant for survival through a mutual exchange of organic carbon and mineral nutrients. AM fungi are usually non-specific and the ~200 species that have been described (Harrison, 2005) form associations with 60% of the plant species (Smith and Read, 1997). In contrast, for ECM an exceptionally high fungal diversity has been reported on a limited number of host plants (Dickie, 2007).

The establishment and maintenance of symbiosis involves different plant natural compounds. In spite of the great diversity of both fungi and plants involved (Klironomos, 2003), a remarkable similarity is observed in the developmental stages of symbiosis, including spore germination into the soil, hyphal growth and branching, appressoria formation and penetration, and intercellular and arbuscule development (a branched hypha that develops within the plant cell). Each of these phases is affected by chemical compounds between plant and fungi (Fig. 4).

AM spores are usually large with significant amounts of reserve in the form of lipids, glyco-

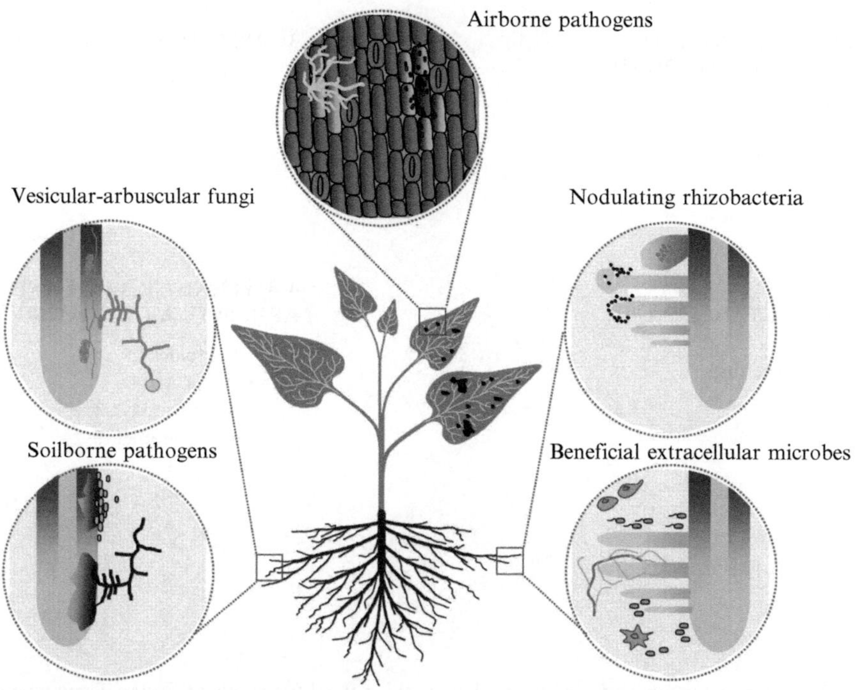

Fig. 3 Schematic representation of the major types of microbe–plant interactions

gen and sugars. Although spore germination can occur in pure water, this process is stimulated by plant root exudates of host plants (Gianninazzi-Person et al., 1989; Phillips and Tsai, 1992), chemical volatiles and high CO_2 concentrations (Smith and Read, 1997). Several phenolic compounds, especially flavonoids such as kaempferol, myricetin and quercetin present in root exudates, stimulated subsequent mycelial growth (Gianninazzi-Person et al., 1989; Smith and Read, 1997; Buee et al., 2000).

Following germination, AM fungi use spore reserves to sustain the germ tube growth searching for host roots. This process lasts from a few days to several weeks in relation to the fungal species, but ceases before the complete consumption of reserves in absence of the host. This allows several attempts at germination to reach the rhizosphere of the host plants. Afterwards, the hyphae profusely branch, suggesting that plant chemical compounds influence this stage. The branching phase greatly increases the probability that AM will come into contact with the host roots for the subsequent infection process (Fig. 4). Although hyphal branching has been observed in several studies as a response to root exudates (Bécard et al., 1995; Smith and Read, 1997; Buee et al., 2000) chemical identification of the involved compounds has remained obscure until recently. Some studies have proposed flavonoids from root exudates as branching factors (Phillips and Tsai, 1992). However, the result that flavonoid-deficient maize mutants stimulated branching at a level comparable to that of wild-type maize root exudates does not support this hypothesis (Bécard et al., 1995). Akiyama et al. (2005) identified plant strigolactones (5-deoxystrigol, released from the roots of *Lotus japonicus*, strigol from *Menispermum dauricum* and sorgolactone from sorghum) that induce hyphal branching in AM *Gigaspora margarita* (Fig. 4). These compounds act at very low

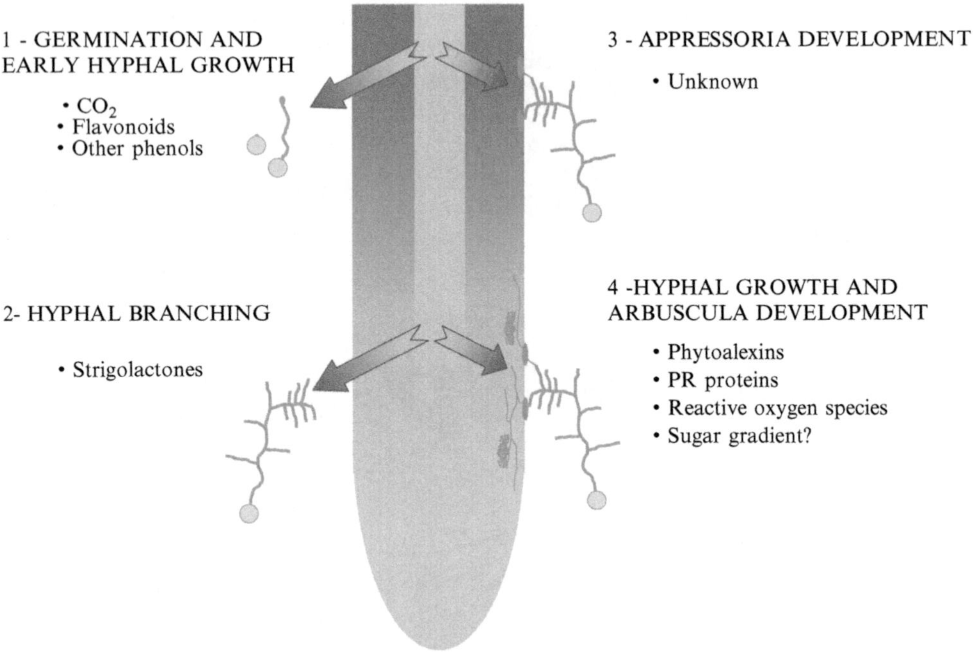

Fig. 4 Plant chemical compounds and their relative role in the different phases of the development of the plant–arbuscular mycorrhizal symbiosis. Arrows represent chemical compounds released by the plant. Note that different phases represent a temporal sequence of the events and do not relate to different spatial locations on the root system

concentrations and are relatively instable. This has an ecological relevance because it does not allow accumulation of these compounds in the soil which, otherwise, would give poor information about the position of suitable host roots. Recently, López-Ráez et al. (2008) demonstrated that phosphate starvation increases the production of several strigolactones in root exudates of tomato plants, thus enhancing the hyphal branching of AM fungi. Considering the low species-specificity of AM fungi, and at the same time, the wide number of plants capable of forming a symbiosis with them, it is necessary to understand if different stimulatory compounds are present or if plants produce similar branching factors. Interestingly, strigolactones are produced by a wide number of higher plants (Matusova et al., 2005; Humphrey and Beale, 2006; Rani et al., 2008) and recent evidences indicate that such compounds might also act as a plant hormone that control shoot branching (Gomez-Roldan et al., 2008; Umehara et al., 2008).

The next developmental stage is appressorium formation and hyphal penetration into the roots (Fig. 4). The stimuli that trigger appressorium formation are unknown, but probably they are of chemical nature and not thigmotrophic, because synthetic surfaces do not stimulate this process (Smith and Read, 1997). In compatible interactions, the fungi penetrate and extensively invade both intra- and intercellularly the root cortex and produce the typical arbuscules. It seems very likely that the plant host exerts a strict control over the fungal development which is limited to cortical parenchyma (Brundrett, 2001). Plant chemicals involved in plant defense response (see

Par. 4.1) are thought to play a significant role in this function. Several studies report a significant induction of plant defense responses with the production, during AM penetration and early stage development, of phytoalexins (Harrison and Dixon, 1993), pathogenesis-related proteins such as chitinases and β-1,3-glucanases (Lambais and Mehdy, 1996), and reactive oxygen species (Lambais et al., 2003). However, this activation of plant defense responses has been proved to be weak, spatially limited and/or temporally transitory, compared to that observed for plant-pathogen interactions (Volpin et al., 1995; Gianinazzi-Pearson et al., 1996).

When endosymbiosis is established in the inner cortical cells, it is thought that both intercellular hyphae and arbuscules are involved in carbohydrate transfer from plant to fungi and mineral nutrients go in the opposite direction (Smith and Read, 1997). The molecules that control arbuscule development, maintenance and senescence are unknown. However, because of the proximity with vascular tissue (the source of sugars for roots), it has been proposed that a carbon gradient may be involved as a signal (Blee and Anderson, 1998).

In many plant species the interaction with AM fungi determines the activation of the carotenoid biosynthesis and their oxidation, leading to the accumulation in roots of several apocarotenoids such as cycloexenone and mycorradicin (Strack and Fester, 2006). Mycorradicin, that together with a mixture of unknown compounds confers a typical yellow colour to the infected roots, has been isolated from several plant species (Klingner et al., 1995; Fester et al., 2002a). Although the role of these compounds in the interaction plant-AM fungi is still unravelled, growing evidence indicates that they can be involved in promoting symbiosis (Fester et al., 2002b), controlling root colonization by fungi (Park et al., 2004), and protecting from oxidative stresses (Strack and Fester, 2006).

The development of AM symbiosis involves several steps that require both specific and non-specific plant chemical compounds. These act either positively (germination stimuli, branching factor, appressoria formation, carbohydrate feeding) and negatively (phytoalexins, PR proteins, reactive oxygen species, etc.) on the fungus. The balance of these effects allows for the establishment, progress and maintenance of symbiosis (Fig. 4).

Plant Natural Compounds Inducing Legume Nodulation

Nitrogen fixation is widely distributed among species of bacteria either free living or in association with higher plants. Several bacteria of the family of Rhizobiaceae (comprising *Rhizobium*, *Bradyrhizobium*, *Mesorhizobium*, *Sinorhizobium* and *Azorhizobium* genera) are able to establish a mutualistic nitrogen-fixing association with the roots of legumes. Effective N_2 fixation is ecologically relevant and gives a competitive advantage where nitrogen is a limiting factor (Walker, 1993). For this reason it is especially important in mixed cropping systems in the developing world.

Symbiosis is a result of a complex chemical and molecular interplay between plant and bacteria (review in Perret et al., 2000; Brencic and Winans, 2005). Here we limit our analysis to the early developmental stages of this interaction. It is well-known that the process of nodule formation requires specific chemical compounds released by plants during seed and root exudation. Flavonoids and isoflavonoids are the most important chemical inducers for the establishment of mutualistic association (Peters et al., 1986). Daidzein, genistein, coumestrol, naringenin, luteolin and 4,4'-dihydroxy-2'methoxychalcone (Fig. 2) are only some of the most representative flavonoids released by legumes that induce Rhizobia nodulation. Rhizobia, when in close proximity with host plant roots that secrete specific flavonoids, in turn respond by activating the expression of the nodulation genes, whose products direct the synthesis of the Nodulation (Nod) factors. These Nod factors are lipochitooligosaccharide that promote the curling and elongation of the meristematic root cells that will host bacteria. By this mechanism, bacteria penetrate the root and a globular structure (the nodule) develops. Inside the nodule, in a low-oxygen environment, bacterioide cells actively fix molecular nitrogen. In addition to flavonoids, other compounds such as alkaloids (trigonelline) and betaines (stachydrine) (Phillips et al., 1992), and aldonic acids (Gagnon and Ibrahim, 1998; Dakora, 2003) have been proved to elicit the activation of the Nod factor, but often at a higher concentration compared to flavonoids.

Flavonoid production is not restricted to legumes. Several thousands of different flavonoids have been identified in the plant kingdom (Harborne, 1988; Harborne and Williams, 2000) with a great variety of critical functions (see Par. 4.1). This point is relevant for understanding the relationship between the diversity of chemical inducers produced by legumes and the species-specificity of the symbiosis. The Rhizobia-legume interaction is more specific compared to AM-plant association, although a large variability is observed (Perret et al., 2000). For instance, *S. meliloti* forms associations with only three genera of legumes (*Medicago*, *Trigonella* and *Melilotus*), while several *Bradyrhizobium* and *Rhizobium* species can be much more promiscuous (Perret et al., 2000). The species-specificity of these associations may be controlled by both plant chemical inducers and Nod factors. For instance, *R. etli* and *R. loti* have distinct host ranges (*Phaseolus* spp. and *Lotus* spp., respectively), but secrete Nod factors with the same structure. The specificity of the two bacteria is determined by elicitors produced by each host plant that differentially activate the transcriptional genes of the Nod factor (Albrecht et al., 1999; Radutoiu et al., 2007).

Chemical Basis of the Cooperative Interaction Between Plant and Extracellular Beneficial Microbes

Exudates released by germinating seeds and growing roots contain mainly simple sugars such as glucose, fructose, cellobiose, arabinose, xylose, mannitol, lactose, maltose, sucrose, etc. (Nelson, 2004; Jones et al., 2004). The continual release of carbon from the roots is only partially under plant control (Jones et al., 2004). The cost of this rhizodeposition may appear excessive for the plant, especially taking into account that this process triggers an enhancement of the microbial populations that compete with plants for mineral nutrients (Hodge et al., 2000). Furthermore, in many cases, root exudates attract harmful soil-borne pathogens (Schroth and Cook, 1964; van Dijk and Nelson, 2000), and positive correlations between the amount of exudates and disease incidence have been observed (Nelson, 2004). For instance, zoospores of oomycete pathogens are attracted by sugars, amino acids (Nelson, 2004), long-chain fatty acids such as linoleic acid (Ruttledge and Nelson, 1997), and isoflavones (Morris et al., 1998) present in seed and root exudates of host plants.

To explain the cost of rhizodeposition, Cook et al. (1995) proposed that plants could have the ability to support and selectively control the populations of rhizosphere beneficial microorganisms antagonistic to their pathogens. The rhizosphere environment is inhabited by a wide array of beneficial microbes, such as plant growth promoting rhizobacteria (PGPR) and antagonistic fungi (*Trichoderma* spp., *Gliocladium* spp., etc.). PGPR comprise bacteria (*Azospirillum* spp., *Pseudomonas* spp., *Serratia* spp., *Enterobacter* spp., *Bacillus* spp., *Pantoea* spp., *Agrobacterium* spp., etc.) and actinomycetes (*Actinobacter* spp., *Streptomyces* spp., etc). Beneficial microbes improve plant mineral nutrition (Kloepper et al., 1980; Vessey, 2003), release stimulatory compounds such as phytormones (Srinivasan et al., 1996), and act as biocontrol agents towards soilborne pathogens by using a number of different mechanisms including competition for plant exudates (van Dijk and Nelson, 2000), production of antibiotics (Fravel, 1988) and/or induction of plant resistance (Pieterse and van Loon, 1999).

Chemical compounds produced by beneficial microbes to interact with other rhizosphere microorganisms and plants have been extensively studied (Lugtenberg et al., 2001; Harman et al., 2004; Gray and Smith, 2005), while those originating from plants have received much less attention. In contrast to the intimate association between Rhizobia and AM with plants where the symbiosis is driven by specific compounds, beneficial microbe colonization of rhizosphere and rhizoplane seems related to a general increased availability of organic carbon and nutrients around the roots (Gray and Smith, 2005). Many microbes such as *T. harzianum*, *P. fluorescens* and *A. brasiliese*, etc. have a very low host specificity and are commonly recovered from the rhizosphere of a wide array of host plants. However, some evidence suggests that the levels of specificity of these microorganisms may be higher than what has been previously thought. Root exudates stimulate PGPR chemotaxis towards root surface by determining the rhizosphere competence of these microbes (Lugtenberg et al., 2001). de Weert et al. (2002) reported that *P. fluorescens* responds chemotactically to organic acids (malic and citric), amino acids (especially isoleucine), but not sugars from root exudates of tomato. Indirect evidence of some degree of beneficial microbe specificity comes from studies on "suppressive soils", where the indigenous microflora effectively protect plants against soilborne pathogens (Weller et al., 2002). Suppressive soils can be induced by the cultivation of determined plant species that select specific microbial communities (Menzies, 1959; Raaijmakers and Weller, 1998; Mazzola and Gu, 2000). Fox example, the cultivation of wheat cultivar susceptible to the pathogen *Gaeumannomyces graminis* var. *tritici* enhances the population of *Pseudomonas* strains that produce the antibiotic 2,4 – diacetylphloroglucinol. The mechanisms of these plant-microbe interactions, through which plants selectively favour specific rhizosphere beneficial microbes, are still obscure. However, it could be hypothesized that natural products play a role in determining the specificity level of these interactions.

A further explanation for the excessive cost of rhizodeposition was given more than 20 years ago by Clarholm (1985). In Clarholm's model, plants indirectly exchange organic carbon for mineral nitrogen, the most limiting nutrient in terrestrial ecosystems (Vitousek, 1982), through a complex multitrophic interaction. The sugar exudates stimulate a large proliferation of microbial populations that are no longer constrained by carbon availability. However, microbial populations require for growth also mineral nutrients and especially nitrogen. When this nutrient is present at a low concentration in the soil, it will be scavenged by bacteria through the release of exoenzymes that degrade soil organic matter. Higher plants are commonly inefficient in using nitrogen in organic forms (Hodge, 2004), although remarkable exceptions are present in environments where N mineralization is extremely slow (Chapin et al., 1993; Persson and Näsholm, 2001). Increased microbial populations attract natural predators such as nematode, amoebae and protozoa grazers that actively forage on microbial populations (Fig. 3). Afterwards, because of stechiometric constraints (bacteria biomass has a very low C/N ratio compared to root exudates), the grazers release as metabolic waste large amounts of ammonia that will be available for plant uptake. Recent evidence seems to partially support this model (Bonkowski and Brandt, 2002; Bonkowski, 2004).

Plant Natural Compounds and Pathogens

Plants cope with thousands of infectious diseases caused by pathogens such as fungi, bacteria and viruses. However, only a relatively small amount of these pathogens successfully cause disease because plants have evolved a complex array of defense mechanisms. These include mechanical reinforcement of the cell wall through deposition of callose and lignin (Develey-Rivière and Galiana, 2007), production of reactive oxygen species (Apel and Hirt, 2004), pathogen-related proteins (Maleck et al., 2000) and preformed and inducible antimicrobial compounds. In the following paragraphs we limited our discussion to some cases in which antimicrobial natural products play a relevant role for plant defense.

Antimicrobial Activity of Natural Compounds Towards Pathogenic Bacteria, Fungi and Oomycetes

The antimicrobial activity of plant natural compounds has been reviewed a number of times. Different aspects such as their phytochemical diversity (Senevirante and Harborne, 1992), involvement in mechanisms of resistance (Morrisey and Osbourn, 1999; Hammerschmidt, 1999), constitutive (Wittstock and Gershenzon, 2002) and inducible defense (Smith, 1996; Kuć, 1995), trafficking and site of action (Field et al., 2006), fitness cost of resistance (Heil, 2002) and their potential exploitation to improve plant disease resistance (Dixon, 2001) have been extensively analyzed.

Thousands of diverse natural products are produced by plants and many of these are involved in plant defense. The phytochemical diversity of antimicrobial compounds include terpenoids, saponins, phenolics and phenylpropanoids, pterocarpans, stilbenes, alkaloids, glucosinolates, hydrogen cyanide, terpenoids, indole and also elemental sulphur, the sole inorganic compound (Cooper et al., 1996). Investigation of chemotaxonomic relationships provided evidence that plant families may be good proxy for the chemical composition of antimicrobial compounds: for instance, Fabaceae produce primarily isoflavonoids and Solanaceae sesquiterpenes. However, many exceptions exist and some plant families produce various antimicrobial compounds (Smith, 1996; Hammerschmidt, 1999). Poaceae produce stilbenes, diterpenes and deoxyanthocyanidins (Grayer and Harborne, 1994) and the species *Theobroma cacao*, when challenged with the vascular fungus *Verticillium dahliae*, accumulates pentacyclic triterpene, arjunolic acid, hydroxylated acetophenones and elemental sulphur (Dixon, 2001). Generally, the analysis of the chemical compound classes gives little information about their role in protection against pathogens. Very similar compounds could have different functions: flavonoid-derived compounds may act in plants as a photoprotectant from UV radiation (Reuber et al., 1996; Ryan et al., 2002), attractors of pollinators (Simmonds, 2003), signal molecules for beneficial microbes (Par. 3.2) and antimicrobial factors (Grayer and Harborne, 1994).

Commonly, plant antimicrobial compounds are classified as constitutive (phytoanticipins) when already present in the plant tissue or inducible (phytoalexins) if produced after pathogen attack. Benzoxazinoides, glucosinolates, glycoalkaloids, hydrogen cyanide and saponins are only some of the most common phytoanticipins. The chemistry of isolated metabolites is not very useful for unraveling if they are phytoanticipins or phytoalexins, because, for instance, the same compound could be constitutive in some plant organs, but inducible in others (Hart, 1981). The isoflavonoid maackiain is constitutive in the roots of the legume *Trifolium pratense* (McMurchy and Higgins, 1984), but can be also synthesized de novo in other plant organs following microbial infection (Higgins and Smith, 1972). Phytoanticipins have several different mechanisms of action. Hydrogen cyanide released from cyanogenic glycosides inhibits cellular respiration (Jones et al., 2000), while saponins display their antimicrobial activity through the ability to form complexes with the sterols present in the membrane of microorganisms (Osbourn, 1996). The limited toxicity of saponins to pathogens such as *Pythium* and *Phytophthora* (Arneson and Durbin, 1968; Barile et al., 2007) has been related to the lack of sterols in the membranes of oomycetes (Morrisey and Osbourn, 1999). Phytoanticipins can be very effective in protecting plant tissue because they are already present before the pathogen attack. However, biosynthesis and maintenance of these compounds, even in absence of pests, is energy consuming for the plant (Heil, 2002). Further, to avoid self-toxicity, several phytoanticipins (glucosinolates, isoflavonoid and some saponins) are stored as inactive precursors and separated from the enzymes that convert them in antimicrobial molecules, or as active compounds kept in specific plant cells or cell compartments (Hallahan, 2000). Producing antimicrobial compounds only when necessary (i.e. during pathogen attack) may limit these problems, although this alternative way is risky because the attack can be too rapid to allow an effective response by the plant. Optimal defense strategies are a delicate balance between costs and benefits of the production of constitutive or inducible antimicrobial compounds (McKey, 1974; Zangerl and Rutledge, 1996).

Phytoalexins are inducible low-molecular weight antimicrobial compounds produced and accumulated in plants after exposure to pathogens. Phytoalexins belong to very diverse chemical families such as stilbenes (Jeandet et al., 1995), polyketides (Kurosaki, 1996), sesquiterpenes (Kuć, 1992) and flavonoids (Treutter, 2006). The isoflavonoids present in legumes such as pisatin, maackiain and kievitone are among the best characterized (Smith et al., 1982; Zhang and Smith, 1983; VanEtten et al., 1989). Phytoalexins usually are not particularly potent and, because of their diverse chemical nature, the mechanisms of action are variable and often poorly understood (Smith, 1996; Morrisey and Osbourn, 1999). Most phytoalexins are not site-specific inhibitors, but exert their toxic effect because of a disruption of the membrane integrity thus controlling their permeability (Smith, 1996). However, as consequences of membrane alteration, other cellular processes such as respiration and synthesis of nucleic acid and protein (Amin et al., 1998) can be negatively affected.

Although higher plants can produce an astonishing diversity of natural compounds that have *in vitro* antimicrobial activity, their role in plant resistance has been clearly elucidated only in a few cases (Morrisey and Osbourn, 1999). Identification of the sites and mechanisms of action is necessary for understanding if antimicrobial compounds are just a response to infection or if they are determinant for plant resistance (Hammerschmidt, 1999). First, the assessment that antimicrobial compounds are present or accumulate to inhibitory concentrations at the site of infection provides a correlative support. Snyder and Nicholson (1990) demonstrate that the flavonoid phytoalexin 3-deoxyanthocyanidin accumulates in the cells of the sorghum plant attacked by *Colletotrichum graminicola* at a concentration higher than that required to inhibit the pathogen in *in vitro* experiments. Other studies that localized and quantified the amount of antimicrobial compounds include avenalumin in *Avena sativa* (Mayama and Tani, 1982), glyocellin in *Glycine soja* (Hahn et al., 1985), avenacin A-1 in *Avena* spp. (Osbourn et al., 1994), elemental sulfur in *T. cacao* (Cooper et al., 1996), and several sesquiterpenoid phytoalexins in *Gossypium hirsutum* (Pierce et al., 1996). Further evidence demonstrates that pathogens may have the enzymatic capability to degrade into non-toxic molecules the antimicrobial compounds of the host plant (VanEtten et al., 1989), and/or tolerate them. For example, one mechanism of tolerance may involve the activation of efflux membrane pumps, such as ABC transporters, to avoid the intracellular accumulation of toxic compounds (Schoonbeek et al., 2001; Duffy et al., 2003). The fungus *G. graminis* var. *avenae* is able to infect oat plants because it produces avenacinase, an enzyme capable of detoxifying avenacins, the oat saponins, by removing the terminal D-glucose molecules from the sugar chains (Crombie et al., 1986; Bowyer et al., 1995). Furthermore, several studies found that fungi specifically impaired in their detoxifying capability lose or reduce their pathogenicity. Examples include mutants of *G. graminis* var. *avenae* which do not produce avenacinase (Bowyer et al., 1995), *Septoria lycopersici* unable to detoxify the α–tomatine (Martin-Hernandez et al., 2000), and *Nectria haematococca* incapable of detoxifying phytoalexin pisatin (Wasmann and VanEtten, 1996). The pathogen capability of detoxifying host antimicrobial compounds may also determine the species-specificity of the plant-pathogen interaction, because several antimicrobial natural products have a relatively wide spectrum of activity (Dixon, 2001). On the host plant side, Papadopoulou et al. (1999) demonstrate that *Avena strigosa* mutants defective for the production of the saponin avenacin A-1, are more susceptible to several fungal soilborne pathogens. Genetic approaches that manipulate, either by increasing or decreasing, the host production of antimicrobial compounds (Dixon et al., 1996), may be very useful for supporting their role in plant defense. Several phytoalexin-deficient mutants have been selected for *Arabidopsis thaliana*, a plant which produces only one phytoalexin, the indole-derived camalexin (Tsuji et al., 1992; Fig. 1). The results show that disease resistance to both bacterial (*P. syringae*) and fungal (*Cochliobolus carbonum* and *Peronospora parasitica*) pathogens depends on the amounts of camalexin produced, the pathogen involved (Glazebrook and Ausubel, 1994; Glazebrook et al., 1997) and the activation of other defense mechanisms as the pathogenesis-related proteins. However, the difficulties in identifying the specific genes that need to be manipulated represent still a limitation to the diffusion of this approach (Slater et al., 2003).

In addition to antimicrobial compounds produced and transported within the plant, growing evidence demonstrates that plants also release antimicrobial compounds in the surrounding environment during seed and root exudation (Schenk et al., 1991). Bais et al. (2005) showed that *A. thaliana* root exudates contain a broad range of compounds with antimicrobial activity (e.g. butanoic, *trans*-cinnamic, *o*-coumaric, *p*-coumaric, ferulic, vanillic and syringic acid, etc.) that confer resistance to several pathovars of the bacterium *P. syringae*. The application of activated charcoal, which adsorbs selectively organic compounds, markedly increased *A. thaliana* disease symptoms and plant mortality. The constitutive production of antimicrobial compounds was enhanced when the plant was challenged with microbes. Another study (Walker et al., 2004) showed that rosmarinic acid (Fig. 1) present in root exudates of sweet basil (*Ocimum basilicum*) has significant antimicrobial activity against free living cells of the pathogen *P. aeruginosa*, but not on cells aggregated in biofilm. This type of bacterial organization causes a strong resistance increase to antibiotic treatments (Davey and O'Toole, 2000). Plants can also produce chemicals that by negatively affecting biofilm formation (see Par. 4.2), act as defence factors against pathogens (Prithiviraj et al., 2005). In addition to low-molecular weight compounds, plants can also release into the rizosphere molecules such as cystein-rich peptides including defensins (Broekaert et al., 1995), proteins with proven antimicrobial activity such as ribosome-inactivating proteins (Park et al., 2002) and other pathogen-related proteins such as β-1,3-glucanase, chitinase and protease (Linthorst, 1991; Park et al., 2002). Turrini et al. (2004) transformed an aubergine plant (*Solanum melongena*) that releases into root exudates the antimicrobial defensin Dm-AMP1 with negative effects on the pathogens *V. albo-atrum*. It must be pointed out that Dm-AMP1 had no effect on the beneficial AM *Glomus mosseae*.

There is growing evidence that roots consistently secrete compounds with broad antimicrobial activity (Bais et al., 2005). This seems to be in contradiction with the obvious proliferation of microbial populations in the rhizosphere following the rhizodeposition of organic carbon (Jones et al., 2004). However, only a few microbial species exhibit compatible interaction with plant roots, suggesting that rhizosphere competence is the net effect of the balance between stimulating and inhibitory chemicals released simultaneously in root exudates.

Plant Natural Compounds Involved in Quorum Sensing

Quorum sensing (QS) is a mechanism that allows bacteria to coordinate group activities by sensing their own population density. This mechanism is commonly based, for many gram-negative bacteria, on acyl-homoserine lactone compounds that act as signals (Zhang and Dong, 2004; Williams, 2007), although other signal molecules are known (Holden et al., 1999; Chen et al., 2002). QS controls many virulence factors such as the transfer of Ti plasmid in agrobacteria, motility, production of extracellular enzymes, antibiotics and toxins, capability of biofilm formation, and epiphytic fitness (von Bodman et al., 2003). Consequently, great attention has been paid to searching for specific QS mimics as tools for the control of either human and plant pathogens or biofilm formation.

Increasing evidence indicates that plants produce and release metabolites named QS mimics because they interfere with bacteria signals. The marine algae *Delisea pulchra* produces halogenated furanones that by binding to the receptors of acyl-homoserine lactones inhibit the expression of QS mediated genes (Manefield et al., 1999), thus preventing biofilm formation (Givskov et al., 1996). Following studies report that root exudates of different higher plants such as pea, rice, tomato, soybean and *Medicago trunculata* are able to either inhibit or stimulate QS-regulated microbial functions (Teplitski et al., 2000; Gao et al., 2003), although chemicals involved still escape identification. There is no evidence for plant-derived compounds capable of either inhibiting the synthesis or degrading bacterial QS signals, as observed for several bacteria species (Dong et al., 2000; Leadbetter, 2001; Bauer and Mathesius, 2004).

Conclusion

Plant growth and survival rely on the ability of roots to communicate with microbes *via* chemical compounds. In the last few years significant advances in understanding plant-microbe interactions have been achieved. However, the prevailing approach in these studies uses over-simplified systems in which only one or a few microbial species interact with the plant. The relevance of these results in real field conditions remains to be explored. In fact, a single gram of rhizospheric soil is inhabited by billions of microbes, and the application of molecular tools (Muyzer et al., 1993; Rappé and Giovannoni, 2003) demonstrates the enormous diversity and complexity of microbial communities. Over the root system of a single plant, a great diversity of both eukaryotes and prokaryotes coexist (Fig. 3), and all of these send and receive chemicals and other types of signals. In this context, future research will be oriented to identify and understand how the myriads of signals are received, categorized and prioritized to produce plant and microbe responses. The contemporaneous presence of multiple, partially overlapping signals may suggest that both plants and microbes use and integrate different signals to perceive their surrounding environments. The study of van West et al. (2002) demonstrates how oomycete pathogens use and integrate chemical and electrical signals to target host roots. In other cases, the same plant-derived chemical compounds represent a signal for a range of different microbes. For instance, the isoflavones daidzein and genistein present in the root exudates of *Glycine soja* are involved in the development of the mutualistic relationship with *B. japonicum*, but also attract the oomycete pathogen *P. sojae* (Morris et al.,

1998). Examples of similar "side effects", where the plant chemical signal aimed to reach beneficial microbes can also be used by deleterious microbes or parasitic weeds (Parniske, 2008), are increasing. This indicates how important it is to study the chemical activity of natural compounds in a more realistic ecological context. Significant progresses in understanding the role of plant-derived chemicals will be obtained by optimizing the trade-off between the need to control experiment factors and create realistic experimental conditions.

A further fast moving research area is the exploitation of natural compounds for increasing plant resistance to microbial diseases. This goal has been addressed, for example, by engineering crop plants to produce novel antimicrobial compounds (Dixon et al., 1996; Dixon, 2001). In fact, by constitutively expressing in *Medicago sativa* plants the metabolic pathways for the production of the grapevine phytoalexin resveratrol (Hipskind and Paiva, 2000), an enhanced resistance to the fungus *Phoma medicaginis* resulted. Similar results have been obtained by overexpressing the gene coding the isoflavone *O*-methyltransferase for a more rapid accumulation of the isoflavone medicarpin (He and Dixon, 2000). Shadle et al. (2003) obtained tobacco plants over-expressing l-phenylalanine ammonia-lyase and producing high levels of chlorogenic acid that showed increased resistance to the fungus *Cercospora nicotianae*. These finding are promising, although additional studies are needed to overcome the difficulties of introducing a large number of coordinated new genes in plants and the capability of pathogens to rapidly modify their detoxification systems (Dixon, 2001).

References

Akiyama, K., Matsuzaki, K. and Hayashi, H. (2005) Plant sesquiterpenes induce hyphal branching in arbuscular mycorrhizal fungi. *Nature* 435, 824–827

Albrecht, C., Geurts, R. and Bisseling, T. (1999) Legume nodulation and mycorrhizae formation; two extremes in host specificity meet. *EMBO J.* 18, 281–288

Amin, M., Kurosaki, F. and Nishi, A. (1998) Carrot phytoalexin alters the membrane permeability of *Candida albicans* and multilamellar liposomes. *J. Gen. Microbiol.* 134, 241–246

Apel, K. and Hirt, H. (2004) Reactive oxygen species: metabolism, oxidative stress and signal transduction. *Ann. Rev. Plant Biol.* 55, 373–399

Arneson, P.A. and Durbin, R.D. (1968) The sensitivity of fungi to α-tomatine. *Phytopathology* 58, 536–537

Bais, H.P., Prithiviraj, B., Jha, A.K., Ausubel, F.M. and Vivanco, J.M. (2005) Mediation of pathogen resistance by exudation of antimicrobials from roots. *Nature* 434, 217–221

Bais, H.P., Weir, T.L., Perry, L.G., Gilroy, S. and Vivanco, J.M. (2006) The role of root exudates in rhizosphere interactions with plants and other organisms. *Ann. Rev. Plant Biol.* 57, 233–266

Baldwin, I.T., Halitschke, R., Paschold, A., von Dahl, C.C. and Preston, C.A. (2006) Volatile signalling in plant–plant interactions: 'Talking trees' in the genomics era. *Science* 311, 812–815

Barile, E., Bonanomi, G., Antignani, V., Zolfaghari, B., Sajjadi, S.E., Scala, F. and Lanzotti, V. (2007) Saponins from *Allium minutiflorum* with antifungal activity. *Phytochemistry* 68, 596–603

Bauer, W.D. and Mathesius, U. (2004) Plant responses to bacterial quorum sensing signals. *Curr. Opin. Plant Biol.* 7, 429–433

Bécard, G., Taylor, L.P., Douds, D.D., Pfeffer, P.E. and Doner, L.W. (1995) Flavonoids are not necessary plant signal compounds in arbuscular mycorrhizal symbiosis. *Mol. Plant-Microbe Interact.* 8, 252–258

Blee, K.A. and Anderson, A.J. (1998) Regulation of arbuscule formation by carbon in the plant. *Plant J.* 16, 523–530

Bonkowski, M. (2004) Protozoa and plant growth: the microbial loop in soil revisited. *New Phytol.* 162, 617–631

Bonkowski, M. and Brandt, F. (2002) Do soil protozoa enhance plant growth by hormonal effects? *Soil Biol. Biochem.* 34, 1709–1715

Bowyer, P., Clarke, B.R., Lunness, P., Daniels, M.J. and Osbourn, A.E. (1995) Host range of a plant pathogenic fungus determined by a saponin detoxifying enzyme. *Science* 267, 371–374

Brencic, A. and Winans, C. (2005) Detection of and response to signals involved in host-micobe interactions by plant-associated bacteria. *Microbiol Mol. Biol. R.* 69, 155–194

Broekaert, W., Terras, F.R.G., Cammue, B.P.A. and Osborn, R.W. (1995) Plant defensins: novel antimicrobial peptides as components of the host defense system. *Plant Physiol.* 108, 1353–1358

Brundrett, M.C. (2001) Coevolution of roots and mycorrhizas of land plants. *New Phytol.* 154, 275–304

Buee, M., Rossignol, M., Jauneau, A., Ranjeva, R. and Bécard, G. (2000) The pre-symbiotic growth of arbuscular mycorrhizal fungi is induced by a branching factor partially purified from plant root exudates. *Mol. Plant-Microbe Interact.* 13, 693–698

Chapin, F.S.III, Moilainen, L. and Kielland, K. (1993) Preferential use of organic nitrogen by a non-mycorrhizal arctic sedge. *Nature* 361, 150–153

Chen, X., Schauder, S., Potier, N., van Drosselaer, A., Pelczer, I., Bassler, B.L. and Hughson, F.M. (2002) Structural identification of a bacterial quorum-sensing signal containing boron. *Nature* 415, 545–549

Clarholm, M. (1985) Interactions of bacteria, protozoa and plants leading to mineralization of soil-nitrogen. *Soil Biol. Biochem.* 17, 181–187

Cook, R.J., Thomashow, L.S., Weller, D.M., Fujimoto, D. and Mazzola, M. (1995) Molecular mechanisms of defense by rhizobacteria against root disease. *Proc. Natl. Acad. Sci. USA* 92, 4197–4201

Cooper, R.M., Resende, M.L.V., Flood, J., Rowan, M.G., Beale, M.H. and Potter, U. (1996) Detection and cellular localization of elemental sulphur in disease-resistant genotypes of *Theobroma cacao*. *Nature*, 379, 159–162

Crombie, W.M.L., Crombie, L., Green, J.B. and Lucas, J.A. (1986) Pathogenicity of the take-all fungus to oats: its relationship to the concentration and detoxification of the four avenacins. *Phytochemistry* 25, 2075–2083

Dakora, F.D. (2003) Defining new roles for plant and rhizobial molecules in sole and mixed plant cultures involving symbiotic legumes. *New Phytol.* 158, 39–49

Davey, M.E. and O'Toole, G.A. (2000) Microbial biofilms: from ecology to molecular genetics. *Microbiol. Mol. Biol. R.*, 64, 847–867

de Weert, S., Vermeiren, H., Mulders, I.H.M., Kuiper, I., Hendrickx, Bloemberg, G.V., Vanderleyden, J., De Mot, R. and Lugtenberg, B.J.J. (2002) Flagella-driven chemotaxis towards exudate components is an important trait for tomato root colonization by *Pseudomonas fluorescens*. *Mol. Plant-Microbe Interact.* 15, 1173–1180

Denison, R.F., Bledsoe, C., Kahn, M., O'Gara, F., Simms, E.L. and Thomashow, L.S. (2003) Cooperation in the rhizosphere and the "free rider" problem. *Ecology* 84, 838–845

Develey-Rivière, M. and Galiana, E. (2007) Resistance to pathogens and host developmental stage: a multifaceted relationship within the plant kingdom. *New Phytol.* 175, 405–416

Dickie, I.A. (2007) Host preference, niche and fungal diversity. *New Phytol.* 174, 228–230

Dixon, R.A. (2001) Natural products and plant disease resistance. *Nature* 411, 843–847

Dixon, R.A., Lamb, C.J., Masoud, S., Sewalt, V.J.H. and Paiva, N.L. (1996) Metabolic engineering: prospects for crop improvement through the genetic manipulation of phenylpropanoid biosynthesis and defense responses – a review. *Gene* 179, 61–71

Dong, Y.H., Xu, J.L., Li, X.Z. and Zhang, L.H. (2000) AiiA, an enzyme that inactivates the acylhomoserine lactone quorum-sensing signal and attenuates the virulence of *Erwinia carotovora*. *Proc. Natl. Acad. Sci. USA* 97, 3526–3531

Duffy, B., Schouten, A. and Raaijmakers, J.M. (2003) Pathogen self-defense: mechanisms to counteract microbial antagonism. *Ann. Rev. Phytopathol.* 41, 501–538

Ebel, R.C., Stodola, A.J.W., Duan, X. and Auge, R.M. (1994) Non hydraulic root-to-shoot signaling in mycorrhizal and non-mycorrhizal sorghum exposed to partial soil drying or root severing. *New Phytol.* 127, 495–505

Fester, T., Hause, B., Schmidt, D., Halfmann, K., Schmidt, J., Wray, V., Hause, G. and Strack, D. (2002a). Occurrence and localization of apocarotenoids in arbuscular mycorrhizal plant roots. *Plant Cell Physiol.* 43, 256–265

Fester, T., Schmidt, D., Lohse, S., Walter, M.H., Giuliano, G., Bramley, P.M., Fraser, P.D., Hause, B. and Strack, D. (2002b) Stimulation of carotenoid metabolism in arbuscular mycorrhizal roots. *Planta* 216, 148–154

Field, B., Jordán, F. and Osbourn, A. (2006) First encounters - deployment of defence-related natural products by plants. *New Phytol.* 172, 193–207

Fravel, D.R. (1988) Role of antibiosis in the biocontrol of plant diseases. *Ann. Rev. Phytopathol.* 26, 75–91

Gagnon, H. and Ibrahim, R.K. (1998) Aldonic acids: a novel family of nod gene inducers of *Mesorhizobium loti*, *Rhizobium lupini*, and

14

Sinorhizobium meliloti. *Mol. Plant-Microbe Interact.* 11, 988–998

Gao, M., Teplitski, M., Robinson, J.B. and Bauer, W.D. (2003) Production of substances by *Medicago truncatula* that affect bacterial quorum sensing. *Mol. Plant-Microbe Interact.* 16, 827–834

Gianinazzi-Pearson, V., Dumas-Gaudot, E., Gollotte, A., Tahiri-Alaoui, A. and Gianinazzi, S. (1996) Cellular and molecular defence-related root responses to invasion by arbuscular mycorrhizal fungi. *New Phytol.* 133, 45–57

Gianninazzi-Person, V., Branzanti, B. and Gianninazzi, S. (1989) In vitro enhancement of spore germination and early hyphal growth of a vesicular-arbuscular mycorrhizal fungus by host root exudates and plant flavonoids. *Symbiosis* 7, 243–255

Givskov, M., Nys, R.D., Manefield, M., Gram, L., Maximilien, R., Eberl, L., Molin, S., Steinberg, P.D. and Kyelleberg, S. (1996) Eukaryotic interference with homoserine lactone-mediated prokaryotic signalling. *J. Bacteriol.* 178, 6618–6622

Glazebrook, J. and Ausubel, F.M. (1994) Isolation of phytoalexin-deficient mutants of *Arabidopsis thaliana* and characterization of their interactions with bacterial pathogens. *Proc. Natl. Acad. Sci. USA* 91, 8955–8959

Glazebrook, J., Zook, M., Mert, F., Kagan, I., Rogers, E.E., Crute, I.R., Holub, E.B., Hammerschmidt, R. and Ausubel, F.M. (1997) Phytoalexin-deficient mutants of *Arabidopsis* reveal that PAD4 encodes a regulatory factor and that four PAD genes contribute to downy mildew resistance. *Genetics* 146, 381–392

Gomez-Roldan, V., Fermas, S., Brewer, P.B., Puech-Pages, V., Dun, E.A., Pillot, J., Letisse, F., Matusova, R., Danoun, S., Portais, J., Bouwmeester, H., Bécard, G., Beveridge, C.A. Rameau, C. and Rochange, S.F. (2008) Strigolactone inhibition of shoot branching. *Nature* 455, 189–195

Graham, J.H. (2001) What do root pathogens see in mycorrhizas? *New Phytol.* 149, 357–359

Gray, E.J. and Smith, D.L. (2005) Intracellular and extracellular PGPR: commonalities and distinctions in the plant-bacterium signaling processes. *Soil Biol. Biochem.*, 37, 395–412

Grayer, R.J. and Harborne, J.B. (1994) A survey of antifungal compounds from higher plants, 1982–1993. *Phytochemistry* 37, 19–42

Hahn, M.G., Bonhoff, A. and Grisebach, H. (1985) Quantitative localization of the phytoalexin glyceollin I in relation to fungal hyphae in soybean roots infected with *Phytophthora megasperma f.sp. glycinea*. *Plant Physiol.* 77, 591–601

Hallahan, D.L. (2000) Monoterpenoid biosynthesis in glandular trichomes of Labiate plants. *Adv. Bot. Res.* 31, 121–151

Hammerschmidt, R. (1999) Phytoalexins: what have we learned after 60 years? *Ann. Rev. Phytopathol.* 37, 285–306

Harborne, J.B. (1988) Flavonoids: advances in research since 1980. Volume 2. Chapman & Hall, London, UK

Harborne, J.B. and Williams, C.A. (2000) Advances in flavonoid research since 1992. *Phytochemistry* 55, 481–504

Harman, E., Howell, R., Viterbo, A., Chet, I. and Lorito, M. (2004) *Trichoderma* species – opportunistic, avirulent plant symbionts. *Nature Rev.* 2, 43–56

Harrison, M.J. (2005) Signaling in the arbuscular mycorrhizal symbiosis. *Ann. Rev. Microbiol.* 59, 19–42

Harrison, M.J. and Dixon, R.A. (1993) Isoflavonoid accumulation and expression of defense gene transcripts during the establishment of vesicular-arbuscular mycorrhizal associations in roots of *Medicago truncatula*. *Mol. Plant-Microbe Interact.* 6, 643–654

Hart, J.H. (1981) Role of phytostilbenes in decay and disease resistance. *Ann. Rev. Phytopathol.* 19, 437–458

Hartmann, T. (2008) The lost of origin of chemical ecology in the late 19th century. *Proc. Natl. Acad. Sci. USA* 105, 4541–4546

Hawes, M.C., Gunawardena, U., Miyasaka, S. and Zhao, X. (2000) The role of root border cells in plant defense. *Trends Plant Sci.* 5, 128–133

He, X.Z. and Dixon, R.A. (2000) Genetic manipulation of isoflavone 7-O-methyltransferase enhances the biosynthesis of 48-O-methylated isoflavonoid phytoalexins and disease resistance in alfalfa. *Plant Cell* 12, 1689–1702

Heil, M. (2002) Ecological costs of induced resistance. *Curr. Opin. Plant Biol.* 5, 1–6

Higgins, V.J. and Smith, D.G. (1972) Separation and identification of two pterocarpanoid phytoalexins produced by red clover leaves. *Phytopathology* 62, 235–238

Hipskind, J.D. and Paiva, N.L. (2000) Constitutive accumulation of a resveratrol-glucoside in transgenic alfalfa increases resistance to *Phoma medicaginis*. *Mol. Plant-Microbe Interact.* 13, 551–562

Hodge, A., Campbell, C.D. and Fitter, A.H. (2001) An arbuscular mycorrhizal fungus accelerates decomposition and acquires nitrogen directly from organic material. *Nature* 413, 297–299

Hodge, A., Robinson, D. and Fitter, A.H. (2000) Are microorganisms more effective than plants at competing for nitrogen? *Trends Plant Sci.* 5, 304–308

Hodge, H. (2004) The plastic plant: root responses to heterogeneous supplies of nutrients. *New Phytol.* 162, 9–24

Holden, M.T.G., Chhabra, S.R., de Nys, R., Stead, P., Bainton, N.J., Hill, P.J., Manefield, M., Kumar, N., Labatte, M., England, D., Rice, S., Givskov, M., Salmond, G.P.C., Stewart, G.S.A.B., Bycroft, B.W., Kjelleberg, S. and Williams, P. (1999) Quorum-sensing cross-talk: isolation and chemical characterization of cyclic dipeptides from *Pseudomonas aeruginosa* and other gram-negative bacteria. *Mol. Microbiol.* 33, 1254–1266

Humphrey, A.J. and Beale, M.H. (2006) Strigol: biogenesis and physiological activity. *Phytochemistry* 67, 636–640

Jeandet, P., Bessis, R., Sbaghi, M. and Meunier, P. (1995) Production of the phytoalexin resveratrol by grapes as a response to *Botrytis* attack under natural conditions. *J. Phytopathol.* 143, 135–139

Jones, D.L., Hodge, A. and Kuzyakov, Y. (2004) Plant and mycorrhizal regulation of rhizodeposition. *New Phytol.* 163, 459–480

Jones, P., Andersen, M.D., Nielsen, J.S., Høj, P.B. and Møller, B.L. (2000) The biosynthesis, degradation, transport and possible function of cyanogenic glucosides. In: Romeo JT, Ibrahim R, Varin L, Pergamon (ed) Recent advances in phytochemistry, vol. 34. Pergamon, Amsterdam

Klingner, A., Bothe, H., Wray, V. and Marner, F-J. (1995). Identification of a yellow pigment formed in maize roots upon mycorrhizal colonization. *Phytochemistry* 38, 53–55

Klironomos, J.N. (2003) Variation in plant response to native and exotic abuscular mycorrhizal fungi. *Ecology*, 84, 2292–2301

Kloepper, J.W., Leong, J., Teintze, M. and Schroth, M.N. (1980) Enhanced plant growth by siderophores produced by plant growth-promoting rhizobacteria. *Nature* 286, 885–886

Kuć, J. (1992) Antifungal compounds in plants. In: Nigg HN, Seigler D (ed) Phytochemical resources for medicine and agriculture. Plenum Press, New York

Kuć, J. (1995) Phytoalexins, stress metabolism and disease resistance in plants. *Ann. Rev. Phytopathol.*, 33, 275–397

Kurosaki, F. (1996) Effect of NADPH associated keto-reducing domain on substrate entry into 6-hydroxymellein synthase, a multifunctional polyketide synthetic enzyme involved in phytoalexin biosynthesis in carrot. *Arch. Biochem. Biophys.* 328, 213–217

Lambais, M.R. and Mehdy, M.C. (1996) Soybean roots infected by *Glomus intraradices* strains differing in infectivity exhibit differential chitinase and β-1,3-glucanase expression. *New Phytol.* 134, 531–538

Lambais, M.R., Ríos-Ruiz, W.F. and Andrade, R.M. (2003) Antioxidant responses in bean (*Phaseolus vulgaris*) roots colonized by arbuscular mycorrhizal fungi. *New Phytol.* 160, 421–428

Leadbetter, J.R. (2001) Plant microbiology. Quieting the raucous crowd. *Nature* 411, 748–749

Linthorst, H.J.M. (1991) Pathogenesis-related proteins of plants. *CRC Crit. Rev. Plant Sci.* 10, 305–308

López-Ráez, J.A., Charnikhova, T., Gómez-Roldán, V., Matusova, R., Kohlen, W., De Vos, R., Verstappen, F., Puech-Pages, V., Bécard, G., Mulder, P. and Bouwmeester, H.J. (2008) Tomato strigolactones are derived from carotenoids and their biosynthesis is promoted by phosphate starvation. *New Phytol.* 178, 863–874

Lugtenberg, B.J.J., Dekkers, L. and Bloemberg, G.V. (2001) Molecular determinants of rhizosphere colonization by *Pseudomonas*. *Ann. Rev. Phytopathol.* 39, 461–490

Lynch, J.M. and Whipps, J.M. (1990) Substrate flow in the rhizosphere. *Plant Soil* 129, 1–10

Maleck, K., Levine, A., Eulgem, T., Morgan, A., Schmidn, J., Lawton, K.A., Dangl, J. and Dietrich, R.A. (2000) The transcriptome of *Arabidopsis thaliana* during systemic acquired resistance. *Nat. Genet.* 26:403–410

Manefield, M., de Nys, R., Kumar, N., Read, R., Givskov, M., Steinberg, P. and Kjelleberg, S. (1999) Evidence that halogenated furanones from *Delisea pulchra* inhibit acylated homoserine lactone (AHL)-mediated gene expresson by displacing the AHL signal from its receptor protein. *Microbiology* 145, 283–291

Martin-Hernandez, A.M., Dufresne, M., Hugouvieux, V., Melto, R. and Osbourn, A. (2000) Effects of targeted replacement of the tomatinase gene on the interaction of *Septoria lycopersici* with tomato plants. *Mol. Plant-Microbe Interact.* 13, 1301–1311

Matusova, R., Rani, K., Verstappen, F.W.A., Franssen, M.C.R., Beale, M.H. and Bouwmeester, H.J. (2005) The strigolactone germination stimulants of the plant-parasitic *Striga* and *Orobanche* spp. are derived from the carotenoid pathway. *Plant Physiol.* 139, 920–934

Mayama, S. and Tani, T. (1982) Microspectrophotometric analysis of the location of avenalumin accumulation in oat leaves in response to fungal infection with Puccinia coronata. *Physiol. Plant Pathol.* 21, 141–149

Mazzola, M. and Gu, Y.H. (2000) Impact of wheat cultivation on microbial communities from replant soils and apple growth in greenhouse trials. *Phytopathology* 90, 114–119

McKey, D. (1974) Adaptive patterns in alkaloid physiology. *Am. Nat.* 108, 305–320

McMurchy, R.A. and Higgins, V.J. (1984) Trifolirhizin and maackiain in red clover: changes in *Fusarium roseum* 'Avenaceum"- infected roots and *in vitro* effects on the pathogen. *Physiol. Plant Pathol.* 25, 229–238

Menzies, J.D. (1959) Occurrence and transfer of a biological factor in soil that suppresses potato scab. *Phytopathology* 49, 648–652

Morris, P.F., Bone, E. and Tyler, B.M. (1998) Chemotropic and contact responses of *Phytophthora sojae* hyphae to soybean isoflavonoids and artificial substrates. *Plant Physiol.* 117, 1171–1178

Morrisey, J.P. and Osbourn, A.E. (1999) Fungal resistance to plant antibiotics as a mechanism of pathogenesis. *Microbiol. Mol. Biol. R.* 63, 708–724

Muyzer, G., de Waal, E.C. and Uitterlinden, A.G. (1993) Profiling of complex microbial populations by denaturing gradient gel electrophoresis analysis of polymerase chain reaction-amplified genes encoding for 16S rRNA. *Appl. Environ. Microb.* 59, 695–700

Nelson, E.B. (2004) Microbial dynamics and interactions in the spermosphere. *Ann. Rev. Phytopathol.* 42, 271–309

Osbourn, A. (1996) Saponins and plant defence – a soap story. *Trends Plant Sci.* 1, 4–9

Osbourn, A.E., Clarke, B.R., Lunness, P., Scott, P.R. and Daniels, M.J. (1994) An oat species lacking avenacin is susceptible to infection by *Gaeumannomyces graminis* var. tritici. *Physiol. Mol. Plant Pathol.* 45, 457–467

Papadopoulou, K., Melton, R.E., Leggett, M., Daniels, M.J. and Osbourn, A. (1999) Compromised disease resistance in saponin-deficient plants. *Proc. Natl. Acad. Sci. USA* 96, 12923–12928

Park, S., Takano, Y., Matsuura, H. and Yoshihara, T. (2004) Antifungal compounds from the root and root exudates of *Zea mays*. *Biosci. Biotechnol. Biochem.* 68, 1366–1368.

Park, S.W., Lawrence, C.B., Linden, J.C. and Vivanco, J.M. (2002) Isolation and characterization of a novel ribosome-inactivating protein from root cultures of pokeweed and its mechanism of secretion from roots. *Plant Physiol.* 130, 164–178

Parniske, M. (2005) Cue for the branching connection. *Nature*, 435, 750–751

Parniske, M. (2008) Arbuscular mycorrhiza: the mother of plant root endosymbioses. *Nature Review* 6, 763–775

Perret, X., Staehelin, C., William, J.B. (2000) Molecular basis of symbiotic promiscuity. *Microbiol. Mol. Biol. R.* 64, 180–201

Persson, J. and Näsholm, T. (2001) Amino acid uptake: a widespread ability among boreal forest plants. *Ecol. Lett.* 4, 434–438

Peters, N.K., Frost, J.W. and Long, S.R. (1986) A plant flavone, luteolin, induces expression of *Rhizobium meliloti* nodulation genes. *Science* 233, 977–980

Phillips, D.A., Joseph, C.M. and Maxwell, C.A. (1992) Trigonelline and stachydrine released from alfalfa seeds activate NodD2 protein in *Rhizobium meliloti*. *Plant Physiol.* 99, 1526–1531

Phillips, D.A. and Tsai, S.M. (1992) Flavonoids as plant signals to rhizosphere microbes. *Mycorrhiza* 1, 55–58

Pierce, M.L., Cover, E.C., Richardson, P.E., Scholes, V.E. and Essenberg, M. (1996) Adequacy of cellular phytoalexin concentrations in hypersensitively responding cotton leaves. *Physiol. Mol. Plant Pathol.* 48, 305–324

Pieterse, C.M.J., van Loon, L.C. (1999) Salicylic acid independent plant defense pathways. *Trends Plant Sci.*, 4, 52–58

Prithiviraj, B., Bais, H.P., Weir, T., Suresh, B., Najarro, E.H., Dayakar, B.V., Schweizer, H.P., Vivanco, J.M. (2005) Down regulation of virulence factors of *Pseudomonas aeruginosa* by salicylic acid attenuates its virulence on *Arabidopsis thaliana* and *Caenorhabditis elegans*. *Infect. Immun.* 73, 5319–5328

Raaijmakers, J.M., Paulitz, T.C., Steinberg, C., Alabouvette, C. and Moënne-Loccoz, Y. (2008) The rhizosphere: a playground and battlefield for soilborne pathogens and beneficial microorganisms. *Plant Soil.* DOI 10.1007/s11104-008-9568-6

Raaijmakers, J.M. and Weller, D.M. (1998) Natural plant protection by 2,4-diacetylphloroglucinol-producing *Pseudomonas* spp. in take-all decline soils. *Mol. Plant-Microbe Interact.* 11, 144–152

Radutoiu, S., Madsen, L.H. Madsen, E.B., Jurkiewicz, A., Fukai, E., Quistgaard, E.M.H., Albrektsen, A.S., James, E.K., Thirup, S. and Stougaard, J. (2007) LysM domains mediate lipochitin–oligosaccharide recognition and *Nfr* genes extend the symbiotic host range. *EMBO J.* 26, 3923–3935

Rani, K., Zwanenburg, B., Sugimoto, Y., Yoneyama, K. and Bouwmeester, H.J. (2008) Biosynthetic considerations could assist the structure elucidation of host plant produced rhizosphere signalling compounds (strigolactones) for arbuscular mycorrhizal fungi and parasitic plants. *Plant Physiol. Biochem.* DOI 10.1016/j.plaphy.2008.04.012.

Rappé MS, Giovannoni SJ (2003) The uncultured microbial majority. *Annu Rev Microbiol* 57:369–394

Read, D.J. and Perez-Moreno, J. (2003) Mycorrhizas and nutrient cycling in ecosystems - a journey towards relevance? *New Phytol.* 157, 475–492

Reuber, S., Bornman, J.F. and Weissenböck, G. (1996) Phenylpropanoid compounds in primary leaf tissues of rye (*Secale cereale*). Light response of their metabolism and the possible role in UV-B protection. *Physiol. Plant* 97, 160–168

Ruttledge, T.R. and Nelson, E.B. (1997) Extracted fatty acids from *Gossypium hirsutum* stimulatory to the seed-rotting fungus, *Pythium ultimum*. *Phytochemistry* 46, 77–82

Ryan, K.G., Swinny, E.E., Markham, K.R. and Winefield, C. (2002) Flavonoid gene expression and UV photoprotection in transgenic and mutant *Petunia leaves*. *Phytochemistry* 59, 23–32

Schenk, S., Lambein, F. and Werner, D. (1991) Broad antifungal activity of beta-isoxazolinonyl-alanine, a nonprotein amino-acid from roots of pea (*Pisum sativum* L.) seedlings. *Biol. Fert. Soils*, 11, 203–209

Schoonbeek, H., Del Sorbo, G. and De Waard, M.A. (2001) The ABC transporter BcatrB affects the sensitivity of *Botrytis cinerea* to the phytoalexin resveratrol and the fungicide fenpiclonil. *Mol. Plant-Microbe Interact.* 14, 562–571

Schroth, M.N. and Cook, R.J. (1964) Seed exudation and its influence on pre-emergence damping-off of bean. *Phytopathology* 54, 670–673

Senevirante, G.I. and Harborne, J.B. (1992) Constitutive flavonoids and induced isoflavonoids as taxonomic markers in the genus *Vigna*. *Biochem Syst. Ecol.* 20, 459–467

Shadle, G.L., Wesley, S.V., Korth, K.L., Chen, F., Lamb, C. and Dixon, R.A. (2003) Phenylpropanoid compounds and disease resistance in transgenic tobacco with altered expression of l-phenylalanine ammonia-lyase. *Phytochemistry* 64, 153–161

Simmonds, M.S.J. (2003) Flavonoid–insect interactions: recent advances in our knowledge. *Phytochemistry* 64, 21–30

Slater, A., Scott, N. and Fowler, M. (2003) Plant Biotechnology: the genetic manipulation of plants. Oxford University Press, Oxford

Smith, C.J. (1996) Accumulation of phytoalexins: defence mechanism and stimulus response system. *New Phytol.* 132, 1–45

Smith, D.A., Harrer, J.M. and Cleveland, T.E. (1982) Relation between extracellular production of kievitone hydratase by isolates of *Fusarium* and their pathogenicity on *Phaseolus vulgaris*. *Phytopathology* 72, 1319–1323

Smith, S.E. and Read, D.J. (1997) Mycorrhizal symbiosis. Second Edition, Academic Press, London

Smith, S.E., Smith, F.A. and Jakobsen, I. (2003) Mycorrhizal fungi can dominate phosphate supply to plants irrespective of growth responses. *Plant Physiol.* 133, 16–20

Snyder, B.A. and Nicholson, R.L. (1990) Synthesis of phytoalexins in sorghum as a site-specific response to fungal ingress. *Science* 248, 1637–1639

Srinivasan, M., Peterson, D.J. and Holl, F.B. (1996) Influence of IAA producing *Bacillus* isolates on the nodulation of *Phaseolus vulgaris* by *Rhizobium etli*. Can. *J. Microbiol.*, 42, 1006–1014

Strack, D. and Fester, T. (2006) Isoprenoid metabolism and plastid reorganization in arbuscular mycorrhizal roots. *New Phytol.* 172, 22–34

Teplitski, M., Robinson, J.B. and Bauer, W.D. (2000) Plants secrete substances that mimic bacterial N-acyl homoserine lactone signal activities and affect population density-dependent behaviors in associated bacteria. *Mol. Plant-Microbe Interact.* 13, 637–648

Treutter, D. (2006) Significance of flavonoids in plant resistance: a review. *Environ. Chem. Lett.* 4, 147–157

Tsuji, J., Jackson, E.P., Gage, D.A., Hammerschmidt, R. and Somerville, S.C. (1992) Phytoalexin accumulation in *Arabidopsis thaliana* during the hypersensitive reaction to *Pseudomonas syringae* pv. *syringae*. *Plant Physiol.* 98, 1304–1309

Turrini, A., Sbrana, C., Pitto, L., Ruffini Castiglione, M., Giorgetti, L., Briganti, R., Bracci, T., Evangelista, M., Nuti, M.P. and Giovannetti, M. (2004) The antifungal Dm-AMP1 protein from *Dahlia merckii* expressed in *Solanum melongena* is released in root exudates and differentially affects pathogenic fungi and mycorrhizal symbiosis. *New Phytol.* 163, 393–403

Umehara, M., Hanada, A., Yoshida, S., Akiyama, K., Arite, T., Takeda-Kamiya, N., Magome, H., Kamiya, Y., Shirasu, K., Yoneyama, K., Kyozuka, J., and Yamaguchi, S. (2008) Inhibition of shoot branching by new terpenoid plant hormones. *Nature* 455, 195–200

van Dijk, K. and Nelson, E.B. (2000) Fatty acid competition as a mechanism by which *Enterobacter cloacae* suppresses *Pythium ultimum* sporangium germination and damping-off. *Appl. Environ. Microbiol.* 66, 5340–5347

van West, P., Morris, B.M., Reid, B., Appiah, A.A., Osborne, M.C., Campbell, T.A., Shepherd, S.J. and Gow, N.A.R. (2002) Oomycete plant pathogens use electric fields to target roots. *Mol. Plant-Microbe Interact.* 15, 790–798

VanEtten, H.D., Matthews, D.E. and Matthews, P.S. (1989) Phytoalexin detoxification: importance for pathogenicity and practical implications. *Ann. Rev. Phytopathol.* 27, 143–164

Vessey, K.V. (2003) Plant growth promoting rhizobacteria as biofertlizers. *Plant Soil* 255, 571–586

Vierheilig, H., Steinkellner, S., Khaosaad, T. and Garcia-Garrido, J.M. (2008) The biocontrol effect of mycorrhization on soilborne fungal pathogens and the autoregolation of the AM symbiosis: one mechanism, two effect? In: Varma A, (ed), *Mycorrhiza*. Springer, Berlin

Vitousek, P. (1982) Nutrient cycling and nutrient use efficiency. *Am. Nat.* 119, 553–572

Volpin, H., Phillips, D.A., Okon, Y. and Kapulnik, Y. (1995) Suppression of an isoflavonoid phytoalexin defense response in mycorrhizal alfalfa roots. *Plant Physiol.* 108, 1449–1454

von Bodman, S.B., Bauer, W.D., Coplin, D.L. (2003) Quorum-sensing in plant pathogenic bacteria. *Ann. Rev. Phytopathol.* 41, 455–482

Walker, L.R. (1993) Nitrogen fixers and species replacements in primary succession. In: Miles J, Walton DWH (ed), Primary succession on land. Blackwell Scientific, Oxford

Walker, T.S., Bais, H.P., Deziel, E., Schweizer, H.P., Rahme, L.G., Fall, R., Vivanco, J.M. (2004) *Pseudomonas aeruginosa*-plant root interactions. Pathogenicity, biofilm formation, and root exudation. *Plant Physiol.* 134, 320–331

Wasmann, C.C. and VanEtten, H. (1996) Transformation-mediated chromosome loss and disruption of a gene for pisatin demethylase decreases the virulence of *Nectria haematococca* on pea. *Mol. Plant-Microbe Interact.* 9, 793–803

Weller, D.M., Raaijmakers, J.M., Gardener, B.B.M., Thomashow, L.S. (2002) Microbial populations responsible for specific soil suppressiveness to plant pathogens. *Ann. Rev. Phytopathol.* 40, 309–348

Williams, P. (2007) Quorum sensing, communication and cross-kingdom signalling in the bacterial world. *Microbiology* 153, 3923–3938

Wink, M. (1999) Functions of plant secondary metabolites and their exploitation in biotechnology. Academic Press, Sheffield, UK

Wittstock, U. and Gershenzon, J. (2002) Constitutive plant toxins and their role in defense against herbivores and pathogens. *Curr. Opin. Plant Biol.* 5, 1–8

Zangerl, A.R. and Rutledge, C.E. (1996) The probability of attack and patterns of constitutive and induced defense: a test of optimal defense theory. *Am. Nat.* 147, 559–608

Zhang, L. and Dong, Y. (2004) Quorum sensing and signal interference: diverse implications. *Mol. Microbiol.* 53, 1563–1571

Zhang, Y. and Smith, D.A. (1983) Concurrent metabolism of the phytoalexins phaseollin, kievitone and phaseollinisoflavan by *Fusarium solani* f.sp. *phaseoli*. *Physiol. Plant Pathol.* 23, 89–100

Role of Natural Products in Nature: Plant–Insect Interactions

15

Ruth Gordon-Weeks and John A. Pickett

Abstract The chapter discusses the interactions of relatively low molecular weight and largely lipophilic secondary plant metabolites with insects and some other invertebrates. This includes compounds stored within plant tissues that are toxic to insects by a range of mechanisms. It covers metabolites that are both constitutively produced and those that are induced in response to defence signalling stimuli including insect attack. Also included are volatile compounds released from plants that act as signals (semiochemicals) detected by herbivorous insects and those that interact with insects at higher trophic levels. Finally, plant to plant communication is described.

Metabolites with Toxic Modes of Action

General Properties of Plant Toxicants

The production of chemicals that are capable of deterring insect pests by toxic activity is an important survival strategy for plants. Structurally, such toxicants are usually compounds that are non-volatile, due to their molecular weight or hydrophilicity. If they accumulate in the tissue of healthy plants prior to insect attack, they are considered to be constitutive (Stamp, 2003; Wittstock and Gershenzon, 2002). Alternatively, they may only be present, or present in much higher concentrations, after plants have encountered attack, or after exposure to natural plant or insect derived defence activators. In this case they are considered to be induced toxicants (Walling, 2000). Induction provides economic advantage to the plant as metabolic energy has to be diverted from primary metabolism for toxin production. In addition, insect herbivores are less likely to develop resistance to induced defence products as they will be subjected to less exposure to them. However, the balance between these two strategies may depend on the likelihood of the plant coming under attack. Those that encounter more frequent colonisation by pests may be forced to rely more heavily on constitutive rather than induced defence, despite its greater energetic cost to the plant (McKey, 1979). Also, defence metabolites are often restricted in their distribution, both spatially and temporally and plant organs associated with survival or reproduction tend to contain the highest concentrations of constitutive defence metabolites (Wittstock and Gershenzon, 2002). They may be developmentally regulated, being present at highest concentrations when the plant is young and less able to protect itself against predators, or they may be concentrated around the region of contact with the invader.

R. Gordon-Weeks (✉) and J.A. Pickett
Department of Biological Chemistry,
Centre for Pest and Disease Management,
Rothamsted Research, West Common,
Harpenden, Hertfordshire, AL5 2JQ
e-mails: ruth.gordon-weeks@bbsrc.ac.uk;
John.pickett@bbsrc.ac.uk

A.E. Osbourn and V. Lanzotti (eds.), *Plant-derived Natural Products*,
DOI 10.1007/978-0-387-85498-4_15, © Springer Science+Business Media, LLC 2009

Compounds may be stored in specialised subcellular compartments from which they are released, or normally present in inactive forms, ensuring the most economically efficient means of maximum contact with the invader, whilst protecting the host from its own toxic substances (discussed in detail in the following section) (Koroleva et al., 2000; Hallahan, 2000; Grubb and Abel, 2006; Sicker et al., 2000; Neal et al., 1990).

Plant defence may rely on strategies relating to high levels of apparency, as with trees, where weakly acting toxicants are present in high concentrations, as is the case with phenolic compounds. Non-apparent defence, for example, in annuals, relies on highly active compounds that are present in low concentrations (Feeny, 1976). Alkaloids, the largest and most diverse family of plant defence metabolites, are frequently involved in non-apparent defence (Schuler, 1996; De Luca and St Pierre, 2000; Steppuhn et al., 2004; Velozo et al., 1999). These compounds are present in many plant families, including cereals and solanaceous plants, most notably of the genus Nicotiana.

Insecticidal activities of plant metabolites can be demonstrated by their effect on insects when added to artificial diets (De Boer and Hanson, 1987; Broadway and Duffey, 1988; Snook et al., 1997; Jassbi et al., 2006). However, comparative quantification between the toxicity of metabolites applied artificially and amounts measured in whole plant extracts are potentially misleading because of the lower intrinsic acceptability of artificial diets. Compounds within plants may have additional effects that deter insects indirectly, and these defence mechanisms are discussed later in this chapter (see Section "Higher Trophic Level Interactions").

Plant toxicants may interact with specific molecular targets within the herbivore, including proteins, such as enzymes (Feeny, 1976; Rhoades and Cates, 1976; Downs et al., 2003), structural proteins (Morimoto et al., 2001), ion-channels and receptors (Wittstock et al., 1997;Wink, 2000), and a toxicant may derive its potency by mimicking the structure of endogenous ligands or other plant compounds, such as amino acids (Wink, 2003). Alkaloids derive their bioactivity from their ability to affect neurotransmitter activity (Zenk and Juenger, 2007) and nicotine, for example, is a nicotinic acetylcholine receptor agonist (Itier and Bertrand, 2001). Due to their bioactivity alkaloids have been developed for use as powerful drugs and halucegens. Non specific compounds may also disrupt biomembranes and thereby destroy the activity of ion channels or receptors, that reside within them (Osbourn, 1996). The action of phenolics tends also to be non-specific and frequently involves inactivation of a range of proteins by the formation of hydrogen bonded complexes via their multiple hydroxyl groups (Bennet and Wallsgrove, 1994; Nickolson and Hammerschmidt, 1992).

Storage and Release of Plant Toxicants

The potential advantages to the plant of storing toxicants in discreet compartments as inactive precursors have been described. In this section the biological properties, regulation and distribution of some well characterised examples of such metabolites will be discussed.

A family of plant toxicants that clearly falls into this category are the cyanogenic glucosides (Fig. 1a) (Bennet and Wallsgrove, 1994; Seigler, 1991). These amino acid derived compounds are stored as glucosides within the vacuole and are present in representatives from all vascular plants orders. Although they are constitutively produced throughout the life of the plant, young or developing tissues (including seeds or nuts) contain the highest concentrations (Conn, 1981). Cyanide is released when the glucoside is hydrolysed by a glucosidase (Seigler, 1998), which, in addition to its role as a respiratory poison, may also have antifeedant activities.

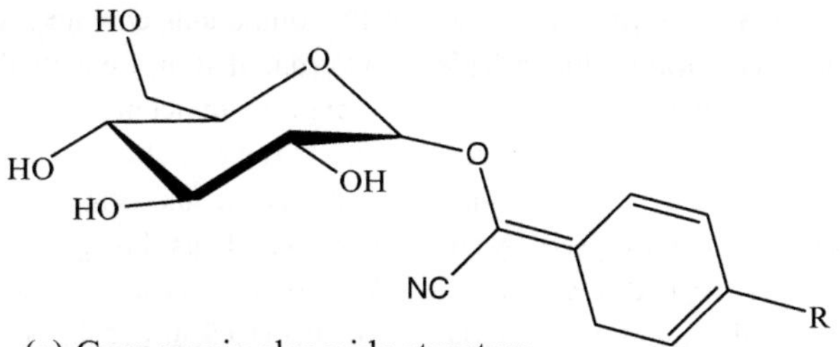

(a) Cyanogenic glucoside structure

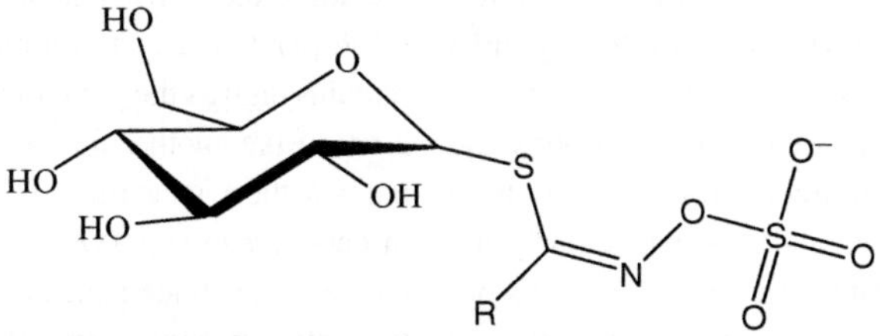

(b) Glucosinolate structure

Fig. 1 The general structure of cyanogenic glucosides (**a**) and glucosinolates (**b**) structure. In (**b**), the R group is derived from amino acids and is highly variable. It may be aliphatic, aromatic or heterocyclic

Furthermore, consumption of cyanide producing compounds can be an advantage to certain herbivores (see Section "Sequestration of Toxins by Herbivores Against Predators") as they defend insects from their own predators (Spencer, 1988). The associated carbonyl compounds have also been shown to be directly toxic to insects.

The glucosinolates are a diverse group of nitrogen and sulphur-containing defence compounds, also derived from amino acids (Fig. 1b). They are found primarily in species of the order Brassicales (for reviews, see Grubb and Abel, 2006; Fahey et al., 2001; Halkier and Gershenzon, 2006) and are thought to have evolved from the cyanogenic glycosides. They are also stored as inactive parent molecules in the vacuole and specialised cells (Koroleva et al., 2000). They are derived from eight different amino acids, and in thale cress, *Arabidopsis thaliana*, the most common precursors are methionine, tryptophan and phenylalanine. Biosynthesis involves three distinct stages: construction or elongation of side chains added to a primary protein amino acid (Textor et al., 2004), core structure elaboration and secondary side chain modification (Grubb and Abel, 2006). These stages, which have been largely confirmed by both biochemical and molecular genetic studies, are responsible for the diversity of this family of metabolites and more than 120 different structural forms have been identified. In the healthy plant gene expression and reporter gene fusion analysis has indicated that the pathway is expressed most highly in reproductive and young developing tissues and in sulphur rich cells in the immediate vicinity of the vascular tissues (Koroleva et al., 2000). Their biosynthesis is up regulated by herbivory (Kim and Jander, 2007; Agrawal and Kurashige, 2003), or plant defence activators (Doughty et al., 1995) and down regulated during sulphur depletion (Bones and Rossiter, 1996).

The toxicity derived from of the glucosinolates is produced in response to cell damage when the glycosidic bond is hydrolysed, by specific β-thioglucosidases termed myrosinases. These enzymes are contained within neighbouring idioblasts or myrosin cells and they release unstable thiohydroximate-O-sulfate products (Grubb and Abel, 2006). These products yield a range of biologically active isothiocyanates, thiocyanates, nitriles oxazolidine-2-thiones and epithioalkanes. Both glucosinolates and to a greater extent their catabolites, particularly the organic isothiocyanates, are toxic to insects but some specialist herbivores are able to detoxify them (Ratzka et al., 2002) (see Section "Sequestration of Toxins by Herbivores Against Predators"). Isothiocyanates have also been shown to attract specialist parasitoids when released from plants (Pope et al., 2008) (see Section on "Synomones").

Another group of plant defence secondary metabolites are the hydroxamic acids or benzoxazinones. These compounds are chiefly produced in cereals and they are involved in defence against a broad range of insect pests, pathogens and weeds. (Kluge et al., 1997; Niemeyer, 1988; Wu et al., 2001). Their role in plant defence was first fully recognised in 1969 when the correlation between susceptibility to the European corn borer, *Ostrinia nubilalis*, and benzoxazinone concentration in maize, *Zea mays*, plants was established (Klun and Robinson, 1969). Similarly, a correlation between elevated tissue concentration in wheat, *Triticum aestivum*, and reduced growth rate of a number of aphid species including the rose-grain aphid, *Metopolophium dirhodum*, the grain aphid, *Sitobion avenae* and the bird cherry oat aphid, *Rhopalosiphum padi* has been found (Argandona et al., 1980; Thackaray et al., 1990). Their toxicity to aphids has been demonstrated in artificial diets (Argandona et al., 1980). The compounds contain a very reactive carbonyl group that may enable them to interact with a range of molecular targets (Niemeyer et al., 1989; Friebe et al., 1997).

In cereals, the benzoxazinones accumulate in young seedlings throughout the plant tissues but decrease in concentration during the later stages of growth (Thackaray et al., 1990). They are also produced in other plant species (Sicker et al., 2000) including dicots (Schullehner et al., 2008), and in some plants the concentration of the compounds remains high as the plant matures (Baumeler et al., 2000). Like another family of plant defence compounds, the avenacins, Fig. 2 (1), that are produced in oats, *Avena* spp. (Osbourn, 1996) the genes in the benzoxazinone pathway are clustered together in a single genomic locus (Frey et al., 1997; Gierl and Frey, 2001). The most generally toxic of the compounds in *Z. mays* and hexaploid wheat is 2,4-dihydroxy-7-methoxy-1,4-benzoxazin-3-one, Fig. 2 (2),whereas in some diploid wheat and rye, *Secale cereale*, its non methoxylated precursor, 2,4-dihydroxy-1,4-benzoxazin-3-one is the major component. These compounds are biosynthesised on the endoplasmic reticulum and stored in the vacuole as inactive glucosides (Sicker et al., 2000). Upon tissue damage by the pest or pathogen the glucoside is released from the vacuole, although active release via a vacuolar membrane transporter may occur. In both foliar and root tissue cytoplasmic glucosidases then catabolise the glucoside to release the active compound (Esen, 1992; Sue et al., 2006). This process in roots may precede rapid release of the compounds into the rhizosphere where they deter soil borne pathogens.

Trichomes are glandular structures situated on regions of the plant surface, including the leaves, where the biosynthesis and storage of a range important secondary metabolites, including terpenes, flavones and phenolics occurs (Bisio et al., 1999; Combrinck et al., 2007; Hallahan, 2000). After insect attack, or in response to pant defence activators (see Section Regulation of Plant Toxicants by Natural Activators) these stored compounds are released

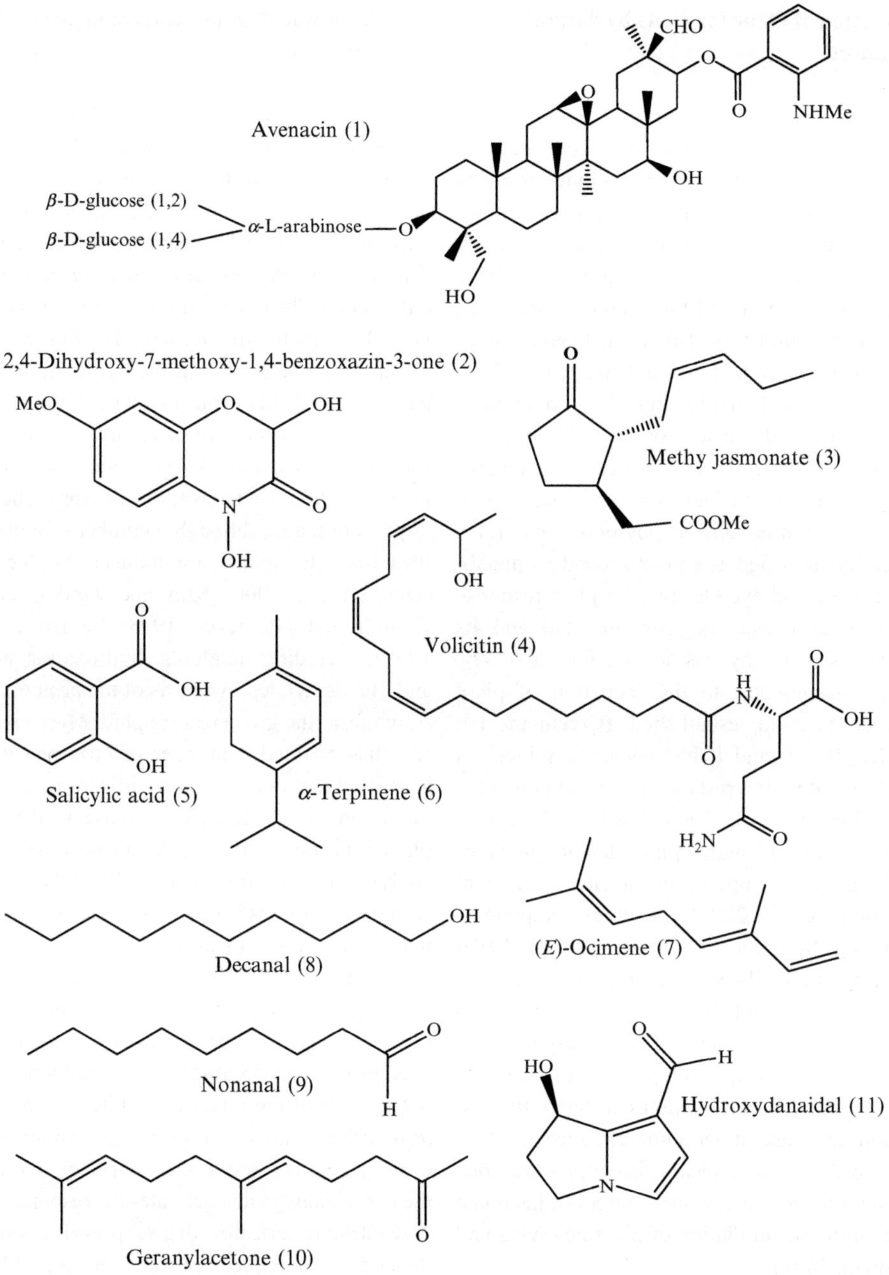

Fig. 2 Chemical structures of compounds 1–11

to protect the plant against herbivory (Jassbi et al., 2008; Schie et al., 2007). Cotton, *Gossypium* spp., for example stores terpenoids in glands at the leaf surface and these are released when the glands are ruptured upon attack (Elzen et al., 1985; Loughrin et al., 1994).

15

Regulation of Plant Toxicants by Natural Activators

The vast majority of toxicants display some level of induction or increase in concentration after insect attack or exposure to phytohormones that act as disease resistance elicitors. Examples of toxicants that are increased in concentration due to release from non toxic precursors have already been described but accumulation of an active compound may also be due to an up regulation of expression of biosynthetic genes. These changes are mediated by signalling compounds released from the insect or from the plant in response to the insect to enhance the defensive capacity of the attacked host. The major plant signalling compound or phytohormone (also generally classified as a plant secondary metabolite) that is believed to mediate plant response to chewing insects is jasmonic acid and its methyl ester, methyl jasmonate, Fig. 2(3). The role of jasmonates in the induction of plant defence is well established (Creelman and Mullet, 1997), and it has been the subject of much scientific interest (see reviews Leon et al., 2001; Farmer et al., 2003; Wasternack, 2007). However, not all major plant defence metabolites increase upon jasmonate treatment (Keinänen et al., 2001) and other compounds, including abscisic acid (Peña-Cortés et al., 1989) and ethylene (O'Donnell et al., 1996) are also released after insect attack and these compounds have been shown to act synergistically in insect defence (Leon et al., 2001). Conversely, insect induced ethylene production can block the jasmonate response under some circumstances. In tobacco, *Nicotiana attenuate*, leaves, for example, ethylene reduces the positive effect of jasmonic acid on the accumulation of nicotine (Winz and Baldwin, 2001).

Although these phytohormones are released after mechanical wounding, the effects of herbivory are distinct, involving additional responses that can be triggered by chemical elicitors in insect oral secretions. These elicitors include fatty acid-amino acid conjugates, which themselves induce the production jasmonic acid and ethylene.

The most extensively studied of these is volicitin, Fig. 2 (4), an N-linolenoylamino acid conjugate isolated from regurgitate from the beet armyworm caterpillar, *Spodoptera exigua* (Alborn et al., 1997; Halitschke et al., 2001). This compound, and further related molecules have been shown to induce the release of volatiles, including terpenoids and indole, which may compete with the hydroxamic acid pathway, and their role in plant defence will be discussed later (see Section on "Negative Interactions"). Little is known about compounds in aphid saliva that are responsible for induction of plant defence, although a number of metabolites toxic to aphids are induced by feeding (Ding et al., 2000; Kim and Jander, 2007; Gianoli and Niemeyer, 1998; Velozo et al., 1999), including alkaloids, hydroxamic acids and glucosinylates. Analysis of the proteome of saliva from the green peach aphid, *Myzus persicae*, has enabled a number of enzymes to be identified that may play a significant role in the induction of the defence response in the host plant (Harmel et al., 2008). Gene expression analysis suggests that in *A. thaliana* the plant's response to aphids is different from its response to chewing insects (Walling, 2000; de Vos et al., 2007; Moran et al., 2007), and pathways regulated by another phytohormone, salicylic acid, Fig. 2 (5), appear to be primarily involved. This is even more noticeable with the phloem feeding nymph of the silverleaf white fly, *Bemisia argentifolii*. It appears that this insect may either evade jasmonate-induced defences by avoiding the tissue damage that activates the response or it may introduce effectors that suppress jasmonate-dependent defences (Zarate et al., 2007). However, there is evidence that jasmonate, ethylene, abscisic acid and gibberellic acid also play significant role in aphid defence (Thompson and Goggin, 2006). Moreover, it is generally accepted that although chewing insects that cause wounding preferentially trigger jasmonate regulated responses and phloem-feeding insects and biotrophic pathogens trigger salicylate pathways there is considerable overlap, trade off and cross talk between the two signalling pathways (Bostock, 2005; Beckers and Spoel, 2006).

Behaviourally and Developmentally Active Components

Host Recognition

Chemicals, known as semiochemicals, are compounds produced by plants that play a pivotal role in their selection by insects as hosts due to their ability to exert behavioural or developmental effects. They may be released into the atmosphere and used as cues by herbivorous insects to locate their hosts or they may be involved in the modulation of insect feeding. Alternatively, they may protect the plant by attracting insects that prey on or parasitize herbivorous pests.

Aerial Plant Attractants (Kairomones) and the Potentiation of Pheromones

The first stage in host recognition by herbivorous insects is the perception of volatile semiochemicals released from the plant. Kairomones, are chemical substances produced and released by a living organism that benefit the receiver and disadvantage the emitting organism. Hence the definition applies to plant volatiles that are used by herbivores to locate their plant hosts. (Visser, 1988; Pickett et al., 1998; Bengtsson et al., 2005). Insects are able to detect general plant volatiles but just as they possess neurones that enable them to taste individual plant compounds (see Section on "Contact Action Including Phagostimulants"), so they possess sensory neurons tuned to host volatile compounds that enable them to recognise specific plant species. Using electroantenography and single olfactory neuron recordings coupled to gas chromatography it is possible to identify volatile compounds released by plants that are detected by insect olfactory neurons (Pickett, 1990). The activity of these compounds can then be further tested in bioassays to confirm their behavioural role. A comprehensive list of the common volatiles released by plants and the range of different insect species that respond to them have been recently reviewed (Bruce et al., 2005).

In some cases insect neurons are able to distinguish between different structural types of the same family of compounds, for example, selective detection of aromatic isothiocynates (Pickett et al., 1998). This phenomenon has now been found to extend to other insect plant systems, and several examples of this degree of selectivity are described in Pickett et al. (1999). However, there appears to be insufficient structural range in the compounds released by plants to enable insects to locate narrow taxonomic groups of plants. Coupled to this, insects possess neurons that detect a range of compounds present in the volatile emissions of many plants. It has been observed that olfactory cells that detect specific plant volatiles are paired with those that detect other identified plant volatiles. This has led to the hypothesis that insects can detect their hosts on the basis of the ratios of certain volatiles emitted, as opposed to specific compounds. It has now been demonstrated that blends of compounds attract phytophagous insects (Bruce et al., 2005) and if the ratios are changed, attraction fails to occur. In a recent study a mixture of compounds (α-terpene, Fig. 2 (6), *(E)*-ocimene, (7), decanal, (8), nonanal, (9), geranylacetone, (10), benzyl alcohol, phenylacetaldehyde and methyl salicylate assembled in the same ratio as was released by tomato plants, *Lycopersicon esculentum*, attracted the tobacco hornworm, *Manduca sexta* (Fraser et al., 2003). However, although

15

the wheat midge, *Sitodiplosis mosellana*, an oligophagous insect, is attracted by the correct blend of *T. aestivum*, volatiles, small changes in the relative composition removes the effect in wind tunnel bioassays (Birkett et al., 2004).

The ability of plant volatile emissions to affect insects' responses to their own pheromones (Dawson et al., 1987), in particular sex pheromones, has long been recognised (Blight et al., 1984). It is appropriate for insects to encounter their mates in the vicinity of their food source and if the attractivity of a pheromone is enhanced by interaction with a plant volatile the chance of encounters will be increased. There are now several examples of the synergistic affects of volatile plant secondary metabolites and insect sex pheromones (Landholt and Phillips, 1997; Sadek and Anderson, 2007), but the effect also extends to the potentiation of aggregation or dispersal pheromones (Dawson et al., 1987; Witzgall et al., 2008). These observation are being used to develop improved methods of control of pest populations. For example, serricornins, which are insect derived sex pheromones, more effective as traps for the beetles *Stegobium paniceum* and *Lasioderma serricorne* when applied in the presence of dried red chilli, *Capsicum frutescens* L, volatiles, which enhanced the effect of the sex pheromone (Mahroof and Phillips, 2008).

There is also evidence that insects are capable of sequestering plant compounds, including alkaloids, and using them both in defence and to attract members of the opposite sex. Male *Estigmene acrea* moths consume plant derived pyrollizidine alkaloids to produce a sex pheromone, hydroxydanaidal, Fig. 2 (11) (Jordan et al., 2005; Edgar et al., 2007) and certain beetles species also sequester plant compounds for similar purpose. Fruit flies, *Drosophila melanogaster*, utilise methyl eugenol, Fig. 3 (12), as a sexual attractant (Raghu and Clarke, 2003), and euglossine bees collect plant compounds to use as aggregates (Lunau, 1992). However, there remains controversy over the degree to which insects sequester compounds and to which they are synthesise *de novo* within their own tissues.

External Attraction Due to Stress Perception

Some insects have evolved to be attracted to volatiles released from stressed or dead plants taking advantage of the weakened defense mechanisms but sustained nutritional quality of the plant tissue (Ginzel and Hanks, 2005; Miller, 2006; Pureswaran et al., 2004; Kalberer et al., 2001; Kendrick and Raffa, 2006). The release of a combination of ethylene and α-pinene, Fig. 3 (13), has been shown to attract longhorn beetle, *Xylotrechus longitarsis*, to its host (Morewood et al., 2002). Damage of a plant by conspecific or heterospecific insects also influences host-plant selection by herbivorous insects (Schindek and Hilker, 1996; Fernandez and Hilker, 2007). The willow beetle, *Phratora vulgatissima*, preferentailly selects willow, *Salix* spp., plants in plantations (Peacock et al., 1999) possibly due to the effect of plant cues induced by feeding and attractants released by the herbivores themselves. Artificiality damaged plants attract herbivores, due to the release of increased amounts of green leaf volatiles (Peacock et al., 2001; Karban and Baldwin, 1997) which are a mixture of volatiles biosynthesised from 18 carbon fatty acid precursors. They include (*E*)-2-hexenal, Fig. 3 (14), (*Z*)-3-hexenal, (15), hexanal, (*Z*)-3-hexenol and hexanol, (16), (Hatanaka, 1993) and the ratio of these components may play a role in host–plant selection (Bruce et al., 2005).

Contact Action Including Phagostimulants

Compounds that elicit feeding by insects via sensory rather than nutritional mechanisms fall into this category. Phagostimulants are detected

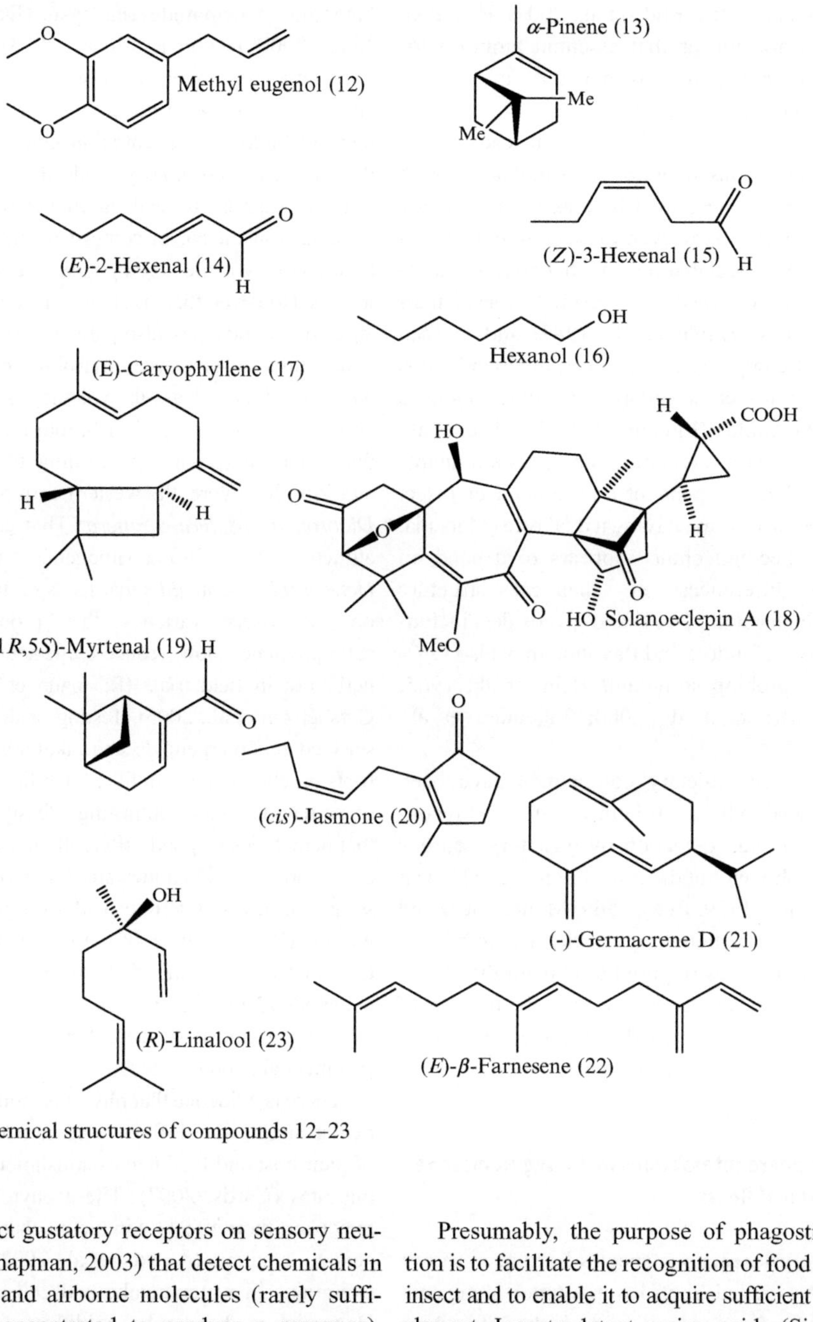

Fig. 3 Chemical structures of compounds 12–23

by contact gustatory receptors on sensory neurones (Chapman, 2003) that detect chemicals in solution and airborne molecules (rarely sufficiently concentrated to produce a response). The process is termed "gustation" or "contact chemoreception" as opposed to taste (Wieczorek and Koppl, 1978).

Presumably, the purpose of phagostimulation is to facilitate the recognition of food by the insect and to enable it to acquire sufficient nourishment. Insects detect amino acids (Simpson et al., 1990) and sugars (Schoonhoven and van Loon, 2002; Glendinning et al., 2000) but their response may be modulated by the presence of other metabolites (Endo et al., 2004). However, it does not appear that essential amino acids have a greater phagostimulatory effect than those that are nonessential.

Non dietary metabolites that can act as phagostimulants include glucosinolates, which have already described here as toxic to many insect species. However, these compounds enable certain specialist feeders that possess gustatory neurons sensitive to them to locate their host plants (Blight et al., 1989) and to feed without competition. However, the interactions between insect herbivores and glucosinolates can be complex (Li et al., 2000; Lambrix et al., 2001). Certain flavonoids can act as stimulants or repellents, dependent on the plant or insect species, with some having a dual role (Matsuda, 1978). Feeding choice appears to depend on small differences in chemical structure (Onyilagha et al., 2004) and glycosides, including those of indole and flavonol, may also function as probing stimulants (Kim et al., 1985; Adjei-Afriyie et al., 2000; Takemura et al., 2002).

Other non dietary compounds have been implicated to have a role in specific host recognition through detection by gustatory neurons (Schoonhoven and van Loon, 2002; van Drongelen, 1979; Rees, 1969; Montgomery and Arn, 1972), although, given the degree of host recognition cues required to explain the attraction of insects to a specific host, the number of gustatory cells identified that can recognise host chemicals remains insufficient.

Rhizosphere Interactions Including Nematode Location of Roots

Many cyst nematode species have been shown to move towards roots by orientation along concentration gradients of root exudates to which they are attracted, and root diffusates have been shown to contain factors that induce phagostimulation (Rolfe et al., 2000), movement or hatching of nematode egg cysts (Devine and Jones, 2000; Devine and Jones, 2003).

Compounds involved in such long distance attraction must be sufficiently stable to enable them to build up concentration gradients within the soil (Perry, 2005). They would also be expected to be hydrophilic, to facilitate movement through aqueous soil media, in contrast to the air borne hydrophobicvolatile compounds that attract insects. However, there is some evidence that volatile compounds may also play an important role in nematode host localisation (Robinson, 2002). *Z. mays*, seedlings release the volatile sesquiterpene (*E*)-caryophyllene, Fig. 3 (17), into the soil from their roots in response to feeding by the root-attacking herbivore, the western corn root worm, *Diabrotica virgifera virgifera*. This compounds attracts the entomopathogenic nematode, *Heterorhabditis megidis* that feeds on the herbivore. *Z. mays* varieties that produce (*E*)-caryophyllene show reduced infestation by the herbivore in field trials (Rasmann et al., 2005; Gershenzon et al., 2005). Testing of the effect of selected monoterpenoids and alkaloids found in roots on chemotaxis, motility, viability and hatch of nematodes with contrasting life styles shows that nematodes respond differently to the individual compounds. Phenolics and flavonols repel to some species and ferulic acid strongly inhibits motility (Wuyts et al., 2006). A hydrophobic fructose related compound, that attracts the beet cyst nematode *Heterodera schachtii* has been purified from white mustard seed, *Sinapis alba*, roots (Rühm et al., 2003).

There is evidence that phytohormones, in particular auxin, may influence nematode invasion of their host and facilitate establishment of feeding sites (Curtis, 2007). These phytohormones may exert their affect by inducing changes in the nematode cuticle (Akhkha et al., 2002), causing alterations in water uptake and cell signalling. Dopamine and salicylic acid have also been shown to attract nematodes (Wuyts et al., 2006).

Solanoeclepin A, Fig. 3 (18), is a complex compound derived from a triterpenoid steroid structure with nine chemical steriocentres. The compound is released from roots of young potato, *Solanum tuberosum*. Biological activity tests demonstrate that synthetic derivatives of the compound induce premature hatching of potato cyst nematode, *Globodera pallida*, juveniles and therefore may have the potential to be used as an environmentally compatible strategy for pest control (Netherlands Organization For Scientific Research, 2001).

Non-host Avoidance and Stress Perception in Hosts

Initial detection by insects and other invertebrates of unsuitable hosts can involve the perception of deterrent volatile compounds released from their leaves. Such emissions benefit the survival of the emitter and are hence known as allomones. The second stage of non host avoidance involves the encountering of unfavourable secondary defence metabolites within plant tissues.

Negative Interactions (e.g. Allomones)

Initially it was assumed that plant derived volatiles acted only as insect attractants but now it is clear that they can also have a repellent effect (Pickett et al., 1999) due to stimulation of specific neurones that detect non-host semiochemicals (Hori, 1996; Hori and Kamatsu, 1997; Guerrero et al., 1997). Highly specific neurones for isothiocyanates that act as repellents have been found in insects for which brassica are non-hosts, for example, the damson-hop aphid, *Phorodon humuli*, and the black bean aphid, *Aphis fabae* (Nottingham et al., 1991). *A. fabae* olfactory neurones respond to the repellent compound (1*R*,5*S*)-myrtenal, Fig. 3 (19), released from the nonhost herb savory, *Satureja montana* (Hardie et al., 1994).

Where an insect requires more than one host to complete its life cycle, its migration from one host to another may be orchestrated by the release of repellent volatiles from the primary host (Pickett and Glinwood, 2007). In aphid species these alterations in host preference may be accompanied by changes in winged or wingless forms that may differ in their response to certain plant semiochemicals. The spring migrant of *R. padi*, is repelled by volatiles from its winter host, the bird-cherry, *Prunus padus*, and here methyl salicylate may be the active component. The winter host for *A. fabae*, the spindle tree, *Euonymus europeaus*, repels the insect species in the spring, but attracts it in the autumn for the sexual colonization stage (Pettersson et al., 1994). The life cycle of the lettuce aphid, *Nasonovia ribis-nigri*, has led to the identification of a further deterrent compound, *cis*-jasmone, Fig. 3 (20). In spring the aphid is repelled by its winter host, the black currant, *Ribes nigrum*, when the volatiles released from the plant contain high concentrations of the compound. Although *cis*-jasmone is a metabolite of methyl jasmonate, it stimulates a specific olfactory neurone and is also repellent to other aphid species (Birkett et al., 2000). There is also evidence that *cis*-jasmone can act as an inducer of plant defence and when applied to *T. aestivum* plants it causes an increase in the concentration phenolic and hydroxamic acids (Moraes et al., 2008).

Stressed or herbivore attacked plants have altered volatile emission profiles and they may then be perceived as non hosts by approaching insects that would normally colonise them (Pickett et al., 2003). This includes the emission of green leaf volatiles, the first stage of the plants response to herbivory (Turlings et al., 1995), frequently followed by the production of terpenoids (Takabayashi and Dicke, 1996; Halitschke et al., 2001), including monoterpenes ((*E*)- ocimene) and sesqueterpenes ((-)-germacrene D), Fig. 3 (21). The plant's response is partly dependent on the nature of the herbivore and although interactions between plants and chewing insects are well documented (Karban and Baldwin, 1997;

Walling, 2000), less is known about plant responses to sucking or piercing insects (Du et al., 1998; Williams III et al., 2005) (see Section on "Regulation of Plant Toxicants by Natural Activators"). A recent study of the interactions between different types of herbivorous insects and plant defence volatiles showed that tobacco, *Nicotiana tabacum*, releases terpenoids and nicotine after caterpillar regurgitate treatment. Colonisation by western flower thrips, *Frankliniella occidentalis*, was reduced on the regurgitate treated plants, possibly due to the production of nicotine (Delphia et al., 2007).

The aphid alarm pheromone, (*E*)-β-farnesene, Fig. 3 (22), is released from aphids in response to predator attack, but it is also produced by some plants in response to herbivory, although its effect may be masked by other volatile emissions. The wild potato, *Solanum berthaultii*, releases sufficiently high and pure levels of the compound from the trichomes to induce avoidance behaviour in *M. persicae* (Gibson and Pickett, 1983). Transgenic *A. thaliana* plants, expressing an (*E*)-β-farnesene synthase gene are also unattractive to these aphids (Beale et al., 2006). Transformation of *A. thaliana* plants with a dual linalool/nerolidol synthase produced plants constitutively containing 40- to 60-fold higher levels of linalool, Fig. 3 (23), and its glycosylated and hydroxylated derivatives and lower levels of nerolidol, Fig. 4 (24). The high expressing transgenic plant significantly repelled *M. persicae* in dual-choice assays (Aharoni et al., 2003).

Contact Action Including Antifeedants

Antifeedants, as the converse of phagostimulants, stimulate deterrent neurones within the insect taste receptor system (Chapman, 2003) and some compounds possess both activities (Onyilagha et al., 2004 and references therein). Plant tissues contain both stimulatory and deterrent compounds and insects select their hosts on the basis of the balance between the two, with the deterrent molecules carrying the most weight. (Chapman, 2003).

Metabolites present in the phloem of coniferous trees (Scots pine, *Pinus sylvestris*, and Norway spruce, *Picea abies*,) can affect the feeding behaviour of the conifer bark beetle, *Ips typographus* (Faccoli et al., 2005). This has been attributed to phenolic compounds and host plant phenolics, including catchetin, Fig. 4 (25), taxifolin, (26), and resveratrol, (27), have been compared to a nonhost compound, *E*-conophthonin, Fig. 4 (28), in *in vitro* feeding experiments. The antifeedant effects of the phenolics was greater in males than in females and the nonhost compound had the greatest antifeedant properties. The authors suggest that females may have developed resistance to the antifeedants by extended feeding on resistant trees at a stage where the tree's hypersensitive response, and hence antifeedant production, is most active. Males, on the other hand, are more sensitive. This may be because, as the pioneers, they have to select the suitable host for colonisation and are therefore more sensitive to repellents. In addition, they spend less time feeding on living trees gaining less opportunity to develop resistance than females (Faccoli and Schlyter, 2007).

The bean aphid, *Megoura crassicauda*, discriminates between its host, *Vicia faba*, and a non-host, the vetch, *Vicia hirsute*, by tasting specific chemicals during stylet penetration. (*E*)-2-methyl-2-butene-1,4-diol 4-*O*-β-D-glucopyranoside, Fig. 4 (29), in the vetch plant is responsible for the observed non-host avoidance by the aphid, and it is present in sufficiently high concentrations to overcome the effect of feeding stimulants (Ohta et al., 2006). Some terpenoids have been shown to be detected by and to deter insects. Using an *in vitro* assay system, Omar et al. (2007) have identified five terpenoid compounds from the bark of a tropical plant, *Lansium domesticum*, that act as antifeedants against the rice weevil, *Sitophilus oryzae*, at very low concentrations.

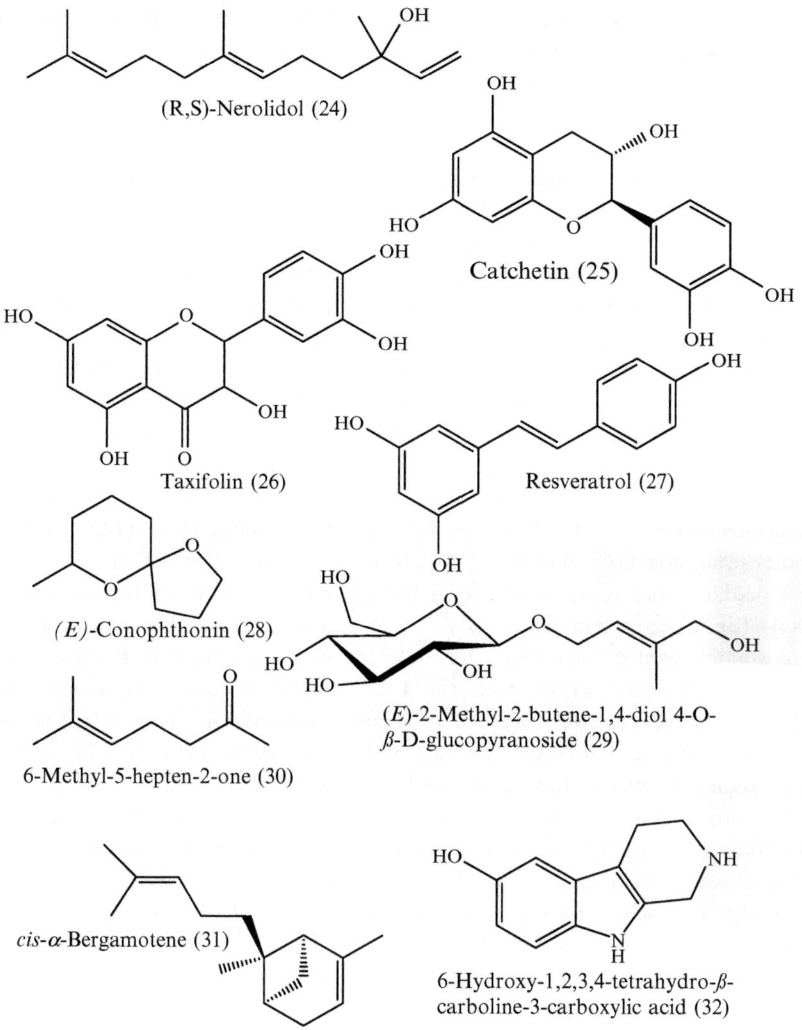

Fig. 4 Chemical structures of compounds 24–32

Higher Trophic Level Interactions

Sequestration of Toxins by Herbivores Against Predators

Some herbivorous insects consume and sequester plant toxins within their tissues to provide protection from predators, either instead of, or in addition to, their own defence mechanisms (Pickett et al., 1999; Macias et al., 2007). This sophisticated strategy frequently leads to highly species specific interactions between host plant and sequestering herbivore. A considerable degree of co-evolutionary development can occur, rendering this phenomenon ideal for the study of the processes underlying natural selection. Pioneering work by Miriam Rothschild

(1972) has led to significant discoveries in more recent years, and it is clear that the sequestration of toxins has been effectively exploited by Lepidoptera and is often associated with aposematic coloration. This can lead to learnt avoidance behaviour by predators or mimicry by other herbivores (Brower, 1969). The chemical structures of the compounds sequestered by moth and butterfly species is posted on the Annual reviews of Entomology web site (www.annualreviews.org) and a review (Nishida, 2002) provides a comprehensive description of many of the intriguing ways that these insects make use of plant compounds to enhance their survival and reproductive capacity.

Some pyrrolizidine alkaloids stored for defence (Bernays et al., 1977), are used to biosynthesise hormones (Nishida et al., 1996), or are converted to other toxic substances (Brown, 1984). The monarch butterfly, *Danaus piexippus*, exploits cardiac glycosides for defence against vertebrate predators (Roeske et al., 1975) and glucosinolates are also favoured by some insects for this type of defence (Aliabadi et al., 2002; Aplin et al., 1975; Müller et al., 2001).

The production of substances that attract herbivores is not of obvious evolutionary benefit to plants. Nonetheless, it is clear that there is a complex interplay between the insects and their hosts and plant induced defence can be converted into a benefit for specialist herbivores by a process of adaptation to toxins that were originally defensive.

Induced Stress Volatiles for Increased Predation and Parasitism (Synomones)

Synomones are semiochemicals that benefit both the emitting and receiving organism. It has been recognised for some time that many plants release volatiles in response to herbivore attack that are capable of attracting insect predators, or parasitoids (parasitic insects that kill their hosts) and this process represents an efficient form of indirect defence for the host plant (Du et al., 1996; Dicke et al., 2003; Turlings and Wäckers, 2004). Wind tunnel bioassays that demonstrate that *A. ervi*, is equally deterred by *V. faba* plants infested by the pea aphid herbivore, *Acyrthosiphon pisum*, as by those that have been infested but have had the aphids removed, demonstrate existence of emissions that function in this way (Du et al., 1996). Many plants exploit this form of indirect defence, including legumes, brassicas and a range of cereal species (Dicke et al., 1998; Takabayashi and Dicke, 1996; Heil, 2008). Several species of spiders, caterpillars and aphids have been shown to induce their host plants to attract their predators or parasitoids (Turlings et al., 1995; Pickett and Glinwood, 2007; De Moraes et al., 1998).

Green leaf volatiles and terpenoids are released from healthy and mechanically damaged plants (Mattiacci et al., 1994; Shiojiri et al., 2001) but their emissions increase upon herbivory and these compounds are one of the herbivore-induced synomones that attract carnivores (Takabayashi and Dicke, 1996; Takabayashi et al., 1994). An intriguing aspect of this tritrophic phenomenon is that the predatory insects can identify the precise nature of the herbivore feeding upon the same host. Female *A. ervi* are more attracted to *V. faba* plants previously infested with host than non host aphids (Du et al., 1996). *Cardiochiles nigriceps*, a parasitic wasp, can distinguish *N. tabacum*, cotton, *Gossypium* spp, and *Z. mays* plants infestated by its host aphid, *Heliothis virescens*, from plants infested by the closely related non-host aphid, *Helicoverpa zea*. (De Moraes et al., 1998). The diamond back moth, *Cotesia plutellae*, discriminates between cabbage plants, *Brassica oleracea*, infested with *Plutella xylostella* (host) and plants infested with *Pieris rapae* (nonhost) aphids (Shiojiri et al., 2000). Analysis of the volatiles released showed both qualitative and quantitative differences in the compounds produced in response to the two aphids (Agelopoulos and Keller,

1994; Shiojiri et al., 2001). Compounds released from these plants in differing amounts included (*Z*)-3-hexen-1-ol, linalool, *cis*-jasmone, humulene, (*E,E*)-α-farnesene, (*E*)-ocimene and (*E*)-caryophyllene. Similar compounds, but in different ratios, are released by apple plants, *Malus* spp, infested by two different spider mites, *Panonychus ulmi* and *Tetranychus urticae* (Takabayashi and Dicke, 1996) enabling the predatory mite, *Phytoseiulus persimilis*, to distinguish between the two species. *V. faba*, plants produce higher levels of similar compounds in response to the aphid host of *A. ervi* than its non host, but one compound, 6-methyl-5-hepten-2-one, Fig. 4 (30), appears to have a particularly profound effect on the wasp's behavior (Guerrieri et al., 1999; Pickett and Glinwood, 2007)

The release of volatiles from plant leaves that attract third trophic level insects can be systemic (Dicke et al., 1990). The upper leaves of *V. faba* plants infested with *A. pisum*, become attractive to the parasitoid, *A. ervi*, at least 2 days later than infested leaves (Guerrieri et al., 1999; Pickett and Glinwood, 2007). This suggests that a mobile signal, possibly transported through the vascular system is responsible although more rapid volatile signalling may also occur (See Fig. 5 and the following section).

The production by phylogenetically diverse plant species and the exploitation by parasitoids

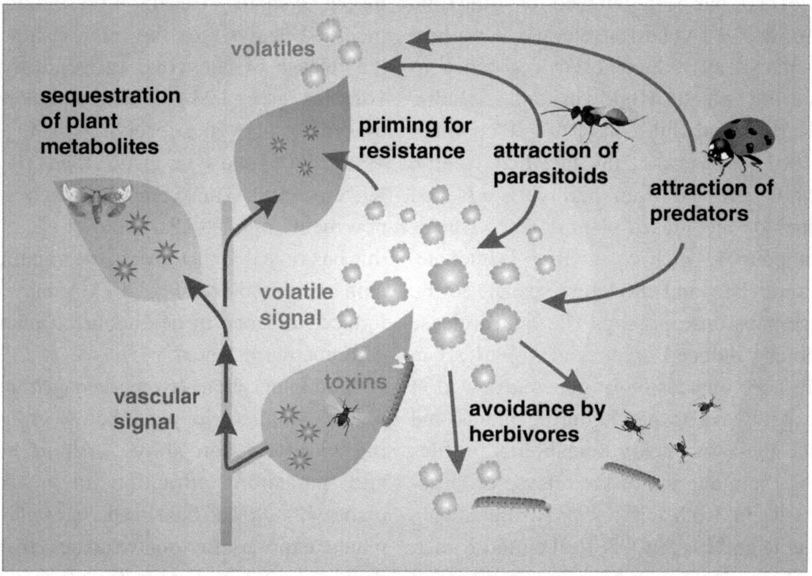

Fig. 5 A diagrammatical representation of the above ground interactions between plants and insects. Feeding by both sucking and chewing insects induces the accumulation of compounds, including those that are directly toxic (basal leaf, orange stars), to defend themselves against the herbivorous insects. The release of volatile emissions is also induced and these cause approaching herbivores to avoid the plant (*grey arrows*) but attract parasitoids and predators (*green arrows*). Volatile emissions also prime neighbouring parts of the plant (upper leaf) for resistance so that when these leaves are challenged herbivores they respond more rapidly to attack. Signals are released into the plants vascular system from insect infested leave that induce the production of defence compounds, including volatiles, from neighbouring, uninfected leaves (*red arrows*). However, some insects consume these compounds and store them within their tissues to deter their own predators (second leaf)

of highly specific chemical signals, keyed to individual herbivore species, indicate that the interaction between plants and the natural enemies of the herbivores that attack them is a highly sophisticated process.

Plant: Plant Interactions

Induction of Direct and Indirect Defence (Parasitoid or Predator Attraction) by External Signals, Both Aerially and Through the Rhizosphere

In addition to their role in defence against herbivores, both directly and involving the third trophic level, it has been believed for some time that plant derived volatiles released upon herbivore attack can induce insect resistance in neighbouring plants (Baldwin and Schultz, 1983; Karban and Baldwin, 1997; Tscharntke et al., 2001). In view of the fact that volatile compounds released from herbivore attacked plants include recognised plant defence signalling compounds, such as methyl jasmonate, methyl salicylate, and ethylene (Farmer, 2001), this is perhaps unsurprising. The rapid release of herbivory induced green leaf volatiles can also affect the expression of genes involved in defence pathways in neighbouring plants and can alter their secondary metabolite profiles including the induction of the release of terpenoids (Bate and Rothstein, 1998; Arimura et al., 2002; Farag and Pare, 2002; Ruther and Kleier, 2005). These effects can be between plants of the same or different species (Karban et al., 2003) and can involve both deterrence of feeding by herbivores and attraction of their predators or parasitoids (see Fig. 5).

The concept has been received with some scepticism as on occasions the observed effect has required unrealistically high levels of compounds, which are often synthetic or purified rather than applied in the ratios released from the plant. Frequently, methods have tested effects in sealed containers (bell jars) which may cause a depletion of CO_2 resulting in non-specific stress responses. The responses in the receiver plants are consider too weak to be of any physiological significance, sufficient replication or reproducibility is often lacking and evidence of the robustness of the phenomenon under field conditions is limited. Furthermore, receiver plants located more than 20 cm away from the emitter plant cease to respond (Karban et al., 2003). The relative merits and short comings of the experimental approaches and the conclusions drawn from them have been the subject of a comprehensive review (Dicke and Bruin, 2001).

The phenomenon continues to be the focus of scientific interest (Baldwin et al., 2006) and has involved the study of a wide range of plant/insect systems. Recent investigations have included analysis of the relationship between the release of converted exogenously applied volatiles (green leaf volatiles) and their *de novo* biosynthesis by the receiving plant (Yan and Wang, 2006) and synergistic effects (Ruther and Kleier, 2005). Transgenic approaches are also now being applied (Paschold et al., 2006) and this has revealed that green leaf volatiles or *cis*-alpha-bergamotene, Fig. 4 (31), may suppress induced defence in neighbouring plants under certain circumstances.

Communication between damaged and healthy plants can also take place below ground, leading to affects on above ground herbivory and parasitoid attraction in neighbouring plants. *V. faba*, and lima bean, *Phaseolus lunatus*, plants exposed to root exudates from plants damaged by *A. pisum*, or *T. urticae*, were more attractive to carnivorous enemies of the herbivores (Dicke and Dijkman, 2001; Guerrieri et al., 2002). Also, root exudates from *V. faba* plants infested by *A. pisum*, cause intact neighbouring plants to become more attractive to parasitoids (Chamberlain et al., 2002). Plant-plant communication affecting herbivore resistance has also been demonstrated between healthy plants. Barley, *Hordeum vulgare*, cultivars experience less aphid settling when exposed to volatiles from other barley cultivars (Pettersson et al., 1999). Root exudate collected from couch grass, *Elytrigia repens*, a pernicious weed known to release a number of biologically active compounds from its roots (see references in Glinwood et al., 2003) induces resistance to *R. padi* when added to *H. vulgare*, roots. The most active ingredient in the exudate was found to be 6-hydroxy-1,2,3,4-tetrahydro-β-carboline-3-carboxylic acid, Fig. 4 (32). Root colonisation of *Z. mays*, plants by the parasitic weed, *Striga hermatothica*, reduces susceptibility of the maize plant to infestation by the stem borer, *Chilo partellus* (Khan et al., 2007).

Because neighbouring plants compete for resources, it is not clear how the emitting plant can benefit from this form of plant-plant communication. Possibly, the volatile signals are really intended for other leaves on the same plant, rather than leaves on neighbouring plants. The transmission of an air borne signal would be faster than one transmitted through the plant's vascular system and the latter route would be too slow to provide a defence response in time to protect adjacent leaves effectively from attack by chewing insects (Heil and Ton, 2008). Further more, if the release of volatiles by damaged leaves was intended for leaves of the same plant, this would address the anomaly that the plant to plant effects require the plants to be very close (Karban et al., 2003). Leaves of hybrid poplar, *Populus deltoides* × *nigra*, damaged by the gypsy moth, *Lymantria dispar*, larvae have been shown to release volatiles that enhance defense against the larvae in neighbouring leaves. This has confirmed that damaged leaves can transmit resistance to leaves of the same plant (Frost et al., 2007).

Priming Effects

The benefits to the plant of inducible as opposed to constitutive defence have already been discussed in this chapter with respect to conservation of the plant's energy resources and the reduction of the likelihood of the development of resistance within the insect population. The ability of damaged or herbivore infested plants to induce direct or indirect defence in neighbouring plants has been described above. However, a further novel aspect of plant defence involves the ability of plants that have received prior exposure to certain signals to respond more rapidly to a subsequent encounter by a insect pest (Conrath et al., 2006). The existence of this phenomenon, known as "priming", would indicate that plants have the capacity for memory, better described as "stress imprinting" as the latter definition would not imply that plants are cognisant (Bruce et al., 2007). The advantage to the plant of such a mechanism would be obvious as resources would not be expended but the plant would be on guard to respond more effectively. The initial priming event would be related to the subsequent attacking agent as it would be volatile blends released from damaged plants that would indicate a likelihood of damage in the near future to the receiving plant by a pest known to be in the vicinity.

Definitive evidence for the existence of such defence mechanisms in plants was first provided by Engelberth et al. (2004). These authors observed that the effects of green leaf volatiles on the defence responses of neighbouring plants are weaker, more transient and involve the induction of a limited selection of defence genes in the receiving plant compared to direct exposure to methyl jasmonate or herbivory. However, they demonstrate that *Z. mays* seedlings treated with green leaf volatiles respond both more rapidly and more strongly to a subsequent challenge by mechanical damage or induction with *S. exigua*, regurgitant. This enhanced response involves production of increased amounts of jasmonic acid and volatile sesquiterpenes. By studying the effects on gene expression, volatile emission and insect behaviour Ton et al. (2007) have also shown that green leaf volatiles released from the Mediterranean brocade caterpillar, *Spodoptera littoralis*, infested plants prime neighbouring plants for defence against the same insects. The volatile emissions induced faster and greater enhancement of expression of several defence genes on caterpillar attack, but did not induce the gene expression directly. The volatile profile was enhanced, caterpillar growth negatively effected and there were positive effects on the third trophic level. Priming has also been demonstrated with other plant species, for example volatile compounds released from detached sagebrush, *Artemisia tridentate*, leaves increase the defence response of *N. tobacum* plants to *M. sexta* attack (Kessler et al., 2006). The within plants effects described above in poplar are also described as priming, and here the advantage to the plant of the speed of the transmission of the wound signal are self evident (Frost et al., 2007; Heil and Ton, 2008).

The underlying mechanisms responsible for priming are currently speculative but are the focus of much scientific investigation. It has been suggested that signalling metabolites or transcription factors may be increased upon exposure to the priming agent. Potentially this could represent the increased accumulation of an inactive precursor (glycoside for example) that could be converted to an active defence compound more rapidly than the production of the defence compound by *de novo* biosynthesis. Epigenetic modifications could also occur, such that after priming a genetic change is instigated that enables the transcriptional events involved in defence to proceed more rapidly upon induction (Bruce et al., 2007).

Conclusions

The interactions between plants and insects are highly complex and are subject to continuing co-evolutionary changes and developments (Wheat et al., 2007). Plants have successfully exploited the wide range of natural secondary metabolites that they produce to defend themselves against insects. Although these strategies include the accumulation or release of compounds that are directly toxic or repellent to herbivores, there are also many examples of the ability of plants to further enhance their defensive capacity by the attraction of natural enemies of herbivorous insects. In turn, however, pest populations have managed to circumvent, or even benefit from, many plant defence products by detoxification or sequestration, enabling some specialist feeders to develop their own independent ecological niche.

The important challenge for the future is to establish to what extent these interactions can be extended or manipulated to develop environmentally compatible pest control. This could involve either traditional breeding or genetic engineering to introduce defence enhancing traits into plants. Such traits could include spatially advantageous enhanced constitutive or induced expression of defence compounds or more rapid, sensitive priming responses. At this stage field studies to test the feasibility of such approaches are at preliminary stages but there is clear evidence that plant released compounds can influence pest behaviour in natural situations (De Moraes et al., 1998; Rasmann et al., 2005; Steppuhn et al., 2004; Thaler, 1999). Fast through-put molecular techniques, genome mapping and both plant and insect whole genome sequence information are now becoming available. It can be anticipated that with the application of these technologies the regulatory mechanisms underlying plant defence responses will be unraveled enabling progress towards this goal to be achieved. However, it remains to be established whether the pressures of co-evolution leading to pest resistance can be overcome.

References

Adjei-Afriyie F, Kim CS, Takemura M, Ishikawa M, Horiike M (2000), Isolation and identification of the probing stimulants in the rice plant for the white-back planthopper, *Sogatella furcifera* (Homoptera: Delohacidae). Biosci Biotechnol Biochem 64:443–446

Agelopoulos NG, Keller MA (1994) Plant-natural enemy association in the tritrophic system, *Cotesia rubecula–Pieris rapae–Brassiceae (Cruciferae)* III: collection and identification of plant and frass volatiles. J Chem Ecol 20:1955–1967

Agrawal AA, Kurashige NS (2003) A role for isothiocyanates in plant resistance against the specialist herbivore *Pieris rapae*. J Chem Ecol 29:1403–1415

Aharoni A, Giri AP, Deuerlein S, Griepink F, de Kogel W-J, Verstappen FWA, Verhoeven HA, Jongsma MA, Schwab W, Bouwmeester HJ (2003) Terpenoid metabolism in wild-type and transgenic Arabidopsis plants. Plant Cell 15:2866–2884

Akhkha A, Kusel J, Kennedy M, Curtis RHC (2002) Effects of phytohormones on the surfaces of plant-parasitic nematodes. Parasitology 125:165–175

Alborn HT, Turlings TCJ, Jones TH, Stenhagen G, Loughrin JH, Tumlinson JH (1997) An elicitor of plant volatiles from beet armyworm oral secretion. Science 276:945–949

Aliabadi A, Renwick JAA, Whitman DW (2002) Sequestration of glucosinolates by harlequin bug *Murgantia histrionica*. J Chem Ecol 28:1749–1762

Aplin RT, d'Arcy Ward R, Rothschild M (1975) Examination of the large white and small white butterflies (Pieris spp.) for the presence of mustard oil and mustard oil glycosides. J Entomol 30:73–78

Argandona VH, Luza LC, Niemeyer HM, Corcuera LJ (1980) Role of hydroxamic acids in the resistance of cereals to aphids. Phytochemistry 19:1665–1668

Arimura G, Ozawa R, Nishioka T, Boland W, Koch T, Kuhnemann F, Takabayashi J (2002) Herbivore-induced volatiles induce the emission of ethylene in neighboring lima bean plants. Plant J 29:87–98

Baldwin IT, Schultz JC (1983) Rapid changes in tree chemistry induced by damage: evidence for communication between plants. Science 221:277–279

Baldwin IT, Halitschke R, Paschold A, von Dahl CC, Preston CA (2006) Volatile signaling in plant-plant interactions: "Talking Trees" in the Genomics Era. Science 10:311812–815

Bate NJ, Rothstein SJ (1998) C6-volatiles derived from the lipoxygenase pathway induce a subset of defense-related genes. Plant J 16:561–569

Baumeler A, Hesse M, Werner C (2000) Benzoxazinoids–cyclic hydroxamic acids, lactams and their corresponding glucosides in the genus *Aphelandra* (Acanthaceae). Phytochemistry 53:213–222

Beale MH, Birkett MA, Bruce TJA, Chamberlain K, Field LM, Huttly AK, Martin JL, Parker R, Phillips AL, Pickett JA, Prosser IM, Shewry PR, Smart LE, Wadham LJ, Woodcock CM, Zhang Y (2006) Aphid alarm pheromone produced by transgenic plants affects aphid and parasitoid behaviour. Proc Natl Acad Sci USA 6:10509–10513

Beckers GJM, Spoel, SH (2006) Fine-tuning plant defence signalling: salicylate versus jasmonate. Plant Biol 8: 1–10

Bengtsson M, Jaastad G, Knudsen G, Kobro S, Bäckman A-C, Eva Pettersson, Witzgall P (2005) Plant volatiles mediate attraction to host and non-host plant in apple fruit moth, *Argyresthia conjugella*. Ent. Exp. Et App. 118:77–85

Bennet RN, Wallsgrove RM (1994) Secondary metabolites in plant defence mechanisms. New Phytol 127:617–633

Bernays E, Edgar JA, Rothchild M (1977) Pyrrolizidine alkaloids sequestered and stored by the aposematic grasshopper *Zonocerus variegates*. J Zool 182:85–87

Birkett MA, Campbell CAM, Chamberlain K, Guerrieri E, Hick AJ, Martin JL, Matthes M, Napier JA, Pettersson J, Pickett JA, Poppy GM, Pow EM, Pye PJ, Smart LE, Wadhams GH, Wadhams LJ, Woodcock CM (2000) New roles for *cis*-jasmone as an insect semiochemical and in plant defence. Proc Natl Acad Sci USA 97:9329–9334

Birkett MA, Bruce TJA, Martin JL, Smart LE, Oakley J, Wadhams LJ (2004) Responses of female orange wheat blossom midge, Sitodiplosis mosellana, to wheat panicle volatiles. J Chem Ecol 30:1319–1328

Bisio A, Corallo A, Gastaldo P, Romussi G, Ciarallo G, Fontana N, de Tommasi N, Profumo P (1999) Glandular hairs and secreted material in *Salvia blepharophylla* Brandegee ex Epling grown in Italy. Ann Bot 83:441–452

Blight MM, Pickett JA, Smith MC, Wadhams LJ (1984) An aggregation pheromone of *Sitona lineatus*. Naturwissenschaften 71:S 480

Blight MM, Pickett JA, Wadhams LJ, Woodcock CM (1989) Antennal response of *Ceutorhynchus assimilis and Psylliodes chrysocephala* to volatiles from oil seed rape. Aspect Appl Biol 23:329–334

Bones AM, Rossiter JT (1996) The myrosinase-glucosinolate system, its organization and biochemistry. Physiol Plant 97:194–208

15

Bostock RM (2005) Signal crosstalk and induced resistance: straddling the line between cost and benefit. Annu Rev Phytopathol 43:545–580

Broadway RM, Duffey SS (1988) The effect of plant protein quality on insect digestive physiology and the toxicity of plant proteinase inhibitors. J Insect Physiol 34:1111–1117

Brower LP (1969) Ecological chemistry. Sci Am 220:22–29

Brown KS Jr (1984) Adult-obtained pyrrolizidine alkaloids defend ithomiine butterflies against a spider predator. Nature 309:707–709

Bruce TJA, Wadhams LJ, Woodcock CM (2005) Insect host location: a volatile situation. Trends Plant Sci 10:269–274

Bruce TJA, Matthes MC, Napier JA, Pickett JA (2007) Stressful "memories" of plants: Evidence and possible mechanisms. Plant Science 173:603–608

Chamberlain K, Guerrieri E, Pennacchio F, Pettersson J, Pickett JA, Poppy GM, Powell W, Wadhams LJ, Woodcock CM (2002) Can aphid-induced plant signals be transmitted aerially and through the rhizosphere? Biochem Syst Ecol 29:1063–1074

Chapman RF (2003) Contact chemoreception in feeding by phytophagous insects. Annu Rev Entomol 48:455–484

Combrinck S, Du Plooy GW, McCrindle RI, Botha BM (2007) Morphology and Histochemistry of the Glandular Trichomes of *Lippia scaberrima (Verbenaceae)*. Ann Bot 99:1111–1119

Conn EE (1981) Cyanogenic glycosides. In: Conn EE (ed) The Biochemistry of Plants. A Comprehensive Treatise, Vol 7. Secondary Plant Products. Academic, New York, pp 479–499

Conrath U, Beckers GJM, Flors V, Garcia-Agustin P, Jakab G, Mauch F, Newman MA, Pieterse CMJ, Poinssot P, Pozo MJ, Pugin A, Schaffrath J, Ton J, Wendehenne D, Zimmerli L, Mauch-Mani B (2006) Priming: getting ready for battle. Mol Plant-Microbe Interact 19:1062–1071

Creelman RA, Mullet JE (1997) Biosynthesis and action of jasmonates in plants. Annu Rev Plant Physiol Plant Mol Biol 48:355–381

Curtis RHC (2007) Do phytohormones influence nematode invasion and feeding site establishment? Nematology 9:155–160

Dawson GW, Griffiths DC, Pickett JA, Wadhams LJ, Woodcock CM (1987) Plant-derived synergists of alarm pheromone from turnip aphids, *Lipaphis (Hyadaphis) erysimi (Homoptera, Aphididae)*. J Chem Ecol 13:1663–1671

De Boer G, Hanson FE (1987) Feeding responses to Solanaceous allelochemicals by larvae of the tobacco hornworm, Manduca sexta. Entomol Exp Appl 45:123–131

Delphia CM, Mescher MC, De Moraes CM (2007) Induction of plant volatiles by herbivores with different feeding habits and the effects of induced defenses on host-plant selection by thrips. J Chem Ecol 33: 997–1012

De Luca V, St Pierre B (2000) The cell and developmental biology of alkaloid biosynthesis. Trends Plant Sci 5:168–173

De Moraes CM, Lewis WJ, Paré PW, Alborn HT, Tumlinson JH (1998) Herbivore-infested plants selectively attract parasitoids. Nature 393:570–573

de Vos M, Kim JH, Jander G (2007) Biochemistry and molecular biology of Arabidopsis-aphid interactions. Bioessays 29:879–883

Devine KJ, Jones PW (2000) Purification and partial, characterisation of hatching factors for the potato cyst nematode *Globodera rostochiensis* from potato root leachate. Nematology 2:231–236

Devine KJ, Jones PW (2003) Investigations into the chemoattraction of the potato cyst nematodes Globodera rostochiensis and G. pallida towards fractionated potato root leachate. Nematology 5:65–75

Dicke M, Bruin J (2001) Chemical information transfer between plants: back to the future. Biochem Syst Ecol 29:981–994

Dicke M, Dijkman H (2001) Within-plant circulation of systemic elicitor of induced defence and release from roots of elicitor that affects neighbouring plants. Biochem Syst Ecol 29:1075–1087

Dicke M, Sabelis MW, Takabayashi J, Bruin J, Posthumus MA (1990) Plant strategies of manipulating predator-prey interactions through allelochemicals: prospects for application in pest control. J Chem Ecol 16:3091–3118

Dicke M, Takabyashi J, Posthumus MA, Schütte C, Krips OE (1998) Plant-phytoseiid interactions mediated by prey-induced plant volatiles: variation in production of cues and variation in responses of predatory mites. Exp Appl Acarol 22: 311–333

Dicke M, Van Poecke RMP, De Boer JG (2003) Inducible indirect defence of plants: from mechanisms to ecological functions. Basic Appl Ecol 4: 27–42

Ding H, Lamb RJ, Ames N (2000) Inducible production of phenolic acids in wheat and antibiotic resistance to *Sitodiplosis mosellana*. J Chem Ecol 26:969–985

Doughty KJ, Kiddle GA, Pye BJ, Wallsgrove RM, Pickett JA (1995) Selective induction of glu-

cosinolates in oilseed rape leaves by methyl jasmonate. Phytochemistry 38:347–350
Downs CT, McDonald PM, Brown K, Ward D (2003) Effects of *Acacia* condensed tannins on urinary parameters, body mass, and diet choice of an *Acacia* specialist rodent, *Thallomys nigricauda*. J Chem Ecol 29:845–857
Du YJ, Poppy GM, Powell W (1996) Relative importance of semiochemicals from first and second trophic levels in host foraging behavior of *Aphidius ervi*. J Chem Ecol 22:1591–1605
Du YJ, Poppy GM, Powell W, Pickett J A, Wadhams LJ, Woodcock CM (1998) Identification of semiochemicals released during aphid feeding that attract parasitoid *Aphidius ervi*. J Chem Ecol 24:1355–1368
Edgar JA, Boppré M, Kaufmann E (2007) Insect-synthesised retronecine ester alkaloids: precursors of the common Arctiine (Lepidoptera) pheromone hydroxydanaidal. J Chem Ecol 33:2266–2280
Elzen GW, Williams HJ, Bell AA, Stipanovic RD, Vinson SB (1985) Quantification of volatile terpenes of glanded and glandless *Gossypium hirsutum L.* cultivars and lines by gas chromatography. J Agric Food Chem 33:1079–1082
Endo N, Abe M, Sekine T, Matsuda K (2004) Feeding stimulants of solanaceae-feeding lady beetle, Epilachna vigintioctomaculata (*Coleoptera:Coccinellidae*) from potato leaves. Appl Entomol Zool 39:411–416
Engelberth J, Alborn HT, Schmelz EA, Tumlinson JH (2004) Airborne signals prime plants against insect herbivore attack. Proc Natl Acad Sci USA 101:1781–1785
Esen A (1992) Purification and partial characterization of maize (Zea mays L.) beta glucosidase Plant Physiol 98:174–182
Faccoli M, Blaženec M, Schlyter F (2005) Feeding response to host and non-host compounds by males and females of the spruce bark beetle Ips typographus in a tunnelling microassay. J Chem Ecol 31:745–759
Faccoli M, Schlyter F (2007) Conifer phenolic resistance markers are bark beetle antifeedant semiochemicals. Agric Forest Entomol 9:237–245
Fahey JW, Zalcmann AT, Talalay P (2001) The chemical diversity and distribution of glucosinolates and isothiocyanates among plants. Phytochemistry 56:5–51
Farag MA. Pare PW (2002) C6-Green leaf volatiles trigger local and systemic VOC emissions in tomato. Phytochemistry 61:545–554
Farmer EE (2001) "Surface-to-air signals". Nature 411: 854–856
Farmer EE, Almeras E, Krishnamurthy V (2003) Jasmonates and related oxylipins in plant responses to pathogenesis and herbivory. Curr Opin Plant Biol 6:372–378
Feeny P (1976) Plant apparency and chemical defense. In: Wallace JW, Mansell RL (eds) Biochemical Interaction Between Plants and Insects. Recent Advances in Phytochemistry, Vol 10. Plenum Press New York 1–40
Fernandez P, Hilker M (2007) Host plant location by Chrysomelidae. Basic and Applied Ecology 8:97–116
Fraser AM, Mechaber WL, Hildebrand JG (2003) Electroantennographic and behavioral responses of the sphinx moth *Manduca sexta* to host plant headspace volatiles. J Chem Ecol 29: 1813–1833
Frey M, Chomet P, Glawischnig E, Stettner C, Grün S, Winklmair A, Eisenreich W, Bacher A, Meeley RB, Briggs SP, Simcox K, Gierl A (1997). Analysis of a chemical defense mechanism in grasses. Science 277:696–699
Friebe A, Rith U, Kuck P, Schnabl H, Schulz M (1997) Effects of 2,4-dihydroxy-1,4-benzoxazin-3-ones on the activity of the plasma membrane H^+-ATPase Phytochemistry 44: 979–983
Frost CJ, Appel HM, Carlson JE, De Moraes CM, Mescher MC, Schultz JC (2007) Within-plant signalling via volatiles overcomes vascular constraints on systemic signalling and primes responses against herbivores. Ecology Letters 10:490–498
Gershenzon J, Turlings TCG, Rasmann S, Köllner TG, Degenhardt J, Hiltpold I, Toepfer S, Kuhlmann U, Gershenzon J, Turlings T (2005) Recruitment of entomopathogenic nematodes by insect-damaged maize roots ct-damaged maize roots. Nature 434:732–737
Gianoli E, Niemeyer HM (1998) Allocation on herbivory induced hydroxamic acids in the wild wheat *Triticum uniaristatum*. Chemecology 8:19–23
Gibson RW, Pickett JA (1983) Wild potato repels aphids by release of aphid alarm pheromone. Nature 302:608–609
Gierl A, Frey M (2001) Evolution of benzoxazinone biosynthesis and indole production in maize. Planta 213:493–498
Ginzel MD, Hanks LM (2005) Role of host plant volatiles in mate location for three species of longhorned beetles. J Chem Ecol 31:213–217
Glendinning JI, Nelson NM, Bernays EA (2000) How do inositol and glucose modulate feeding in *Manduca sexta* caterpillars? J Exp Biol 203:1299–315

15

Glinwood R, Pettersson J, Ahmed E, Ninkovic V, Birkett MA, Pickett JA (2003) Change in Acceptability of Barley Plants to Aphids After Exposure to Allelochemicals from Couch-Grass (*Elytrigia repens*). J Chem Ecol 29:261–274

Grubb DG, Abel S (2006) Glucosinylate metabolism and its control. Trends in Plant Sci 11:89–100

Guerrero A, Feixas J, Pajares J, Wadhams LJ, Pickett JA, Woodcock CM (1997) Semiochemically induced inhibition of behaviour of *Tomicus destruens* (Woll) (Coleoptera: Scolytidae). Naturwissenschaften 84:155–157

Guerrieri E, Poppy GM, Powell W, Tremblay E, Pennacchio F (1999) Induction and systemic release of herbivore-induced plant volatiles mediating in-flight orientation of *Aphidius ervi*. J Chem Ecol 25:1247–1261

Guerrieri E, Poppy GM, Powell W, Rao R, Pennacchio F (2002) Plant-to-plant communication mediating in-flight orientation of Aphidius ervi. J Chem Ecol 28:1703–1715

Halitschke R, Schittko U, Boland W, Baldwin IT (2001) Molecular interactions between the specialist herbivore *Manduca sexta* (*Lepidoptera, Sphingidae*) and its natural host *Nicotiana attenuata*. III. Fatty acid–amino acid conjugates in herbivore oral secretions are necessary and sufficient for herbivore-specific plant responses. Plant Physiol 125:711–717

Halkier BA, Gershenzon J (2006) Biology and biochemistry of glucosinolates. Annu Rev Plant Biol 57:303–333

Hallahan DL (2000) Monoterterpenoid biosynthesis in glandular trichomes of Labiate plants. Adv Bot Res 31:77–120

Hardie J, Isaacs R, Pickett JA, Wadhams LJ, Woodcock CM (1994) Methyl salicylate and (-)-(1r,5s)-myrtenal are plant-derived repellents for black bean aphid, *Aphis-fabae* scop (Homoptera, Aphididae). J Chem Ecol 20:2847–2855

Harmel N, Létocart E, Cherqui A, Giordanengo P, Mazzucchelli G, Guillonneau F,. De Pauw E, Haubruge E, Francis F Identification of aphid salivary proteins: a proteomic investigation of *Myzus persicae*. Insect mol Biol 17:165–174

Hatanaka A (1993) The biogeneration of green odour by green leaves. Phytochemistry 34:1201–1218

Heil M (2008) Indirect defence via tritrophic interactions. New Phytologist 178:41–61

Heil M, Ton J (2008) Long-distance signalling in plant defence. Trends in Plant Science 13:264–272

Hori M (1996) Settling inhibition and insecticidal activity of garlic and onion oils against *Myzus persicae* (Sulzer) (Homoptera, Aphidae). Applied Entomology and Zoology (Japan) 31:605–612

Hori M, Kamatsu H (1997) Repellancy of rosemary oil and its components against onion aphid *NeotoxopteraH formosana* (Takahashi) (Homoptera, Aphidae). Applied Entomology and Zoology (Japan) 32:303–310

Itier V, Bertrand D (2001). Neuronal nicotinic receptors: from protein structure to function. FEBS letters 504:118–25

Jassbi AR, Gase K, Hettenhausen C, Schmidt A, Baldwin IT (2008) Silencing Geranylgeranyl Diphosphate Synthase in *Nicotiana attenuata* Dramatically Impairs Resistance to Tobacco Hornworm. Plant Physiol 146: 974–986

Jassbi AR, Zamanizadehnajari S, Kessler D, Baldwin IT (2006) A new acyclic diterpene glycoside from Nicotiana attenuata with a mild deterrent effect on feeding *Manduca sexta* larvae. Z Naturforsch 61b:1138–1142

Jordan AT, Jones TH, Conner WE (2005) If you've got it, flaunt it: ingested alkaloids affect corematal display behavior in the salt marsh moth, Estigmene acrea. Journal of Insect Science (Tucson) 5:Art1

Kalberer NM, Turlings TCJ, Rahier M (2001) Attraction of a leaf beetle (*Oreina cacaliae*) to damaged host plants. J Chem Ecol 27:647–6

Karban R, Baldwin IT (1997) Induced Responses to Herbivory. University of Chicago Press, Chicago, IL

Karban R, Maron J, Felton GW, Ervin G, Eichenseer H (2003) Herbivore damage to sagebrush induces resistance in wild tobacco: evidence for eavesdropping between plants. Oikos 100:325–332

Keinänen M, Oldham NJ, Baldwin IT (2001) Rapid HPLC Screening of Jasmonate-Induced Increases in Tobacco Alkaloids, Phenolics, and Diterpene Glycosides in *Nicotiana attenuate*. J Agric Food Chem 49:3553–3558

Kendrick AP, Raffa KF (2006) Sources of insect and plant volatiles attractive to cottonwood leaf beetles feeding on hybrid poplar. J Chem Ecol 32:2585–2594

Kessler A, Halitschke R, Diezel C, Baldwin IT (2006) Priming of plant defense responses in nature by airborne signaling between Artemisia tridentata and Nicotiana attenuata. Oecologia 148:280–292

Khan HMM, Khan ZR, Mueke JM, Hassanali A, Kairu E, Pickett JA (2007) Behaviour and biology of *Chilo partellus* (Swinhoe) on *Striga hermonthica* (Del.) Benth. infested and uninfested maize plants. Crop Prot 26:998–1005

Kim JH, Jander G (2007) *Myzus persicae* (green peach aphid) feeding on Arabidopsis induces

the formation of a deterrent indole glucosinolate. The Plant Journal 49:1008–1019
Kim M, Koh HS, Fukami H (1985) Isolation of C-glucosylflavones as probing stimulants of planthoppers in the rice plant. J Chem Ecol 11:441–452
Kluge M, Schneider B, Sicker D (1997) Diastereoselective synthesis of the benzoxazinone acetal glucoside ent-GDIMBOA: The first enantiomer of a natural acetal glucoside Carbohydrate Research 298:147–152
Klun JA, Robinson JF (1969) Concentration of two 1,4-benzoxazinones in dent corn at various stages of development of the plant and its relation to resistance of the host plant to the European corn borer. J Econ Entomol 62:214
Koroleva OA, Davies A, Deeken R, Thorpe MR, Tomos AD, Hedrich R (2000) Identification of a New Glucosinolate-Rich Cell Type in Arabidopsis Flower Stalk. Plant Physiol. 124: 599–608
Lambrix V, Reichelt M, Mitchell-Olds T, Kliebenstein DJ, Gershenzon J (2001) The Arabidopsis Epithiospecifier Protein Promotes the Hydrolysis of Glucosinolates to Nitriles and Influences *Trichoplusia ni* herbivory. Plant Cell 13:2793–2807
Landholt PJ, Phillips TW (1997) Host plant influences on sex pheromone behaviour of phytophagous insects. Annual review of Entomology 42:371–391
Leon J, Rojo E, Sanchez-Serrano JJ (2001) Wound signalling in plants. J Exp Bot 52:1–9
Li Q, Eigenbrode SD, Stringam GR, Thiagarajah MR (2000) Feeding and Growth of Plutella xylostella and Spodoptera eridania on Brassica juncea with Varying Glucosinolate Concentrations and Myrosinase Activities. J Chem Ecol 26:2291–2305
Loughrin JH, Manukian A, Heath RR, Turlings TCJ, Tumlinson JH (1994) Diurnal cycle of emission of induced volatile terpenoids herbivore-injured cotton plants. Proc Natl Acad Sci USA 91:11836–11840
Lunau K (1992) Evolutionary aspects of perfume collection in male euglossine bees (Hymenoptera) and of nest deception in bee-pollinated flowers. Chemeoecology 3:65–73
Macias FA, Galindo JLG, Galindo JCG (2007) Evolution and current status of ecological phytochemistry. Phytochemistry 68:2917–2936
Mahroof RM, Phillips TW (2008) Responses of stored-product Anobiidae to pheromone lures and plant-derived volatiles. Journal of applied Entomology 132:161–167
Matsuda K (1978) Feeding stimulation of flavonoids for various leaf beetles (Chrysomeliae:coleoptera) Appl Ent Zool 13:228–230
Mattiacci L, Dicke M, Posthumus MA (1994) Induction of parasitoid attracting synomone in Brussels sprouts plants by feeding of *Pieris brassicae* larvae: role of mechanical damage and herbivore elicitor. J Chem Ecol 20:2229–2247
McKey D (1979) The distribution of secondary compounds within plants In: Rosenthal GA, Janzen DH, Orlando (eds) Herbivores: Their Interaction with Secondary Metabolites. Academic, Orlando, FL, pp. 56–134
Miller DR (2006) Ethanol and (-)-alpha-pinene: attractant kairomones for some large wood-boring beetles in southeastern USA. J Chem Ecol 32:779–794
Montgomery ME, Arn H (1972) Feeding response of *Aphis pomi, Myzus persicae*, and *Amphorophora agathonica* to phlorizin. J Insect Physiol 20:413–421
Moraes MCB, Birkett MA, Gordon-Weeks R, Smart LE, Martin JL, Pye BJ, Bromilow R, Pickett JA (2008) cis-Jasmone induces accumulation of defence compounds in wheat, *Triticum aestivum*. Phytochemistry 69:9–17
Moran PJ, Cheng Y, Cassell JL, Thompson GL (2007) Gene expression profiling of Arabidopsis thaliana in compatible plant-aphid interactions. Arch Insect Biochem Physiol 51:182–203
Morewood WD, Simmonds KE, Wilson IM, Borden JH, Mcintosh RL (2002) α-Pinene and ethanol: key host volatiles for *Xylotrechus longitarsis* (Coleoptera: Cerambycidae). J Entomol Soc B C 99:117–122
Morimoto S, Suemori K, Moriwaki J, Taura F, Tanaka H, Aso M, Tanaka M, Hiroshi Suemune H, Shimohigashi Y, Shoyama Y (2001) Morphine metabolism in the Opium poppy and its possible physiological function. J Biol Chem 2001 276:38179–38184
Müller C, Agerbirk N, Olsen CE, Boevé J-L, Schaffner U, Brakefield PB (2001) Sequestration of host plant glucosinolates in the defensive hemolymph of the sawfly *Athalia rosae*. J Chem Ecol 27: 2505–2516
Neal JJ, Tingey WM, Steffens JC (1990) Sucrose esters of carboxylic-acids in glandular trichomes of *Solanum berthaultii* deter settling and probing by green peach aphid. J Chem Ecol 16:487–497
Netherlands Organization For Scientific Research (2001) Environmentally-friendly pesticide to combat potato cyst nematodes. Science Daily
Nickolson RL, Hammerschmidt R (1992) Phenolic compounds and their role in disease resistance. Annu Rev Phytopathol 30: 369–389

15

Niemeyer, H.M (1988) Hydroxamic acids (4-hydroxy-1,4-benzoxazin-3-ones) defence chemicals in the Gramineae. Phytochemistry 27:3349–3358

Niemeyer HM, Pesel E, Copaja SV, Bravo HR, Franke S, Franke W (1989) Changes in hydroxamic acid levels of wheat plants induced by aphid feeding. Phytochemistry 28:447–449

Nishida R (2002) Sequestration of defensive substances from plants by Lepidoptera, Annu Rev Entomol 47:57–92

Nishida R, Schulz S, Kim CS, Fukami H, Kuwahara Y, Honda K, Hayashi N (1996) Male sex pheromone of a giant danaine butterfly, *Idea leuconoe*. J Chem Ecol 22: 949–72

Nottingham SF, Hardie J, Dawson GW, Hick AJ, Pickett JA, Wadhams LJ, Woodcock CM (1991) Behavioral and electrophysiological responses of aphids to host and nonhost plant volatiles. J Chem Ecol 17:1231–1242

O'Donnell PJ, Calvert C, Atzorn R, Wasternack C, Leyser HMO, Bowles DJ (1996) Ethylene as a signal mediating the wound response of tomato plants. Science 274:1914–1917

Ohta N, Mori N, Kuwahara Y, Nishida R (2006) A hemiterpene glucoside as a probing deterrent of the bean aphid, *Megoura crassicauda*, from a non-host vetch, *Vicia hirsute*. Phytochemistry 67:584–588

Omar S, Marcotte M, Fields P, Sanchez PE, Poveda L, Mata R, Jimenez A, Durst T, Zhang J, MacKinnon S, Leaman D, Arnason JT, Philogène BJR (2007) Antifeedant activities of terpenoids isolated from tropical Rutales. J Stored Prod Res 43:92–96

Onyilagha JC, Lazorko J, Gruber MY, Soroka JG, Erlandson MA (2004) Effect of flavonoids on feeding preference and development of the crucifer pest *Mamestra configurata* Walker. J Chem Ecol 30:109–124

Osbourn A (1996) Saponins and plant defence-a soapy story. Trends Plant Sci 1:4–9

Paschold A, Halitschke R, Baldwin IT (2006) Using 'mute' plants to translate volatile signals. Plant J 45:275–291

Peacock L, Herrick S, Brain P (1999) Spatio-temporal dynamics of willow beetle (*Phratora vulgatissima*) in short-rotation coppice willows grown as monocultures or a genetically diverse mixture. Agric Forest Entomol 1:287–296

Peacock L, Lewis M, Powers S (2001) Volatile compounds from Salix spp. Varieties differing in susceptibility to three willow beetle species J Chem Ecol 27:1943–1951

Peña-Cortés H, Sánchez-Serrano JJ, Mertens R, Willmitzer L (1989) Abscisic acid is involved in the wound-induced expression of the proteinase inhibitor II gene in potato and tomato. Proc Natl Acad Sci USA 86:9851–9855

Perry RN (2005) An evaluation of types of attractants enabling plant-parasitic nematodes to locate plant roots. Russ J Nematol 13:83–88

Pettersson J, Pickett JA, Pye BJ, Quiroz A, Smart LE, Wadhams LJ, Woodcock CM (1994) Winter host component reduces colonization by bird-cherry oat aphid, *Rhopalosiphum padi* (L) (Homoptera, Aphididae), and other aphids in cereal fields. J Chem Ecol 20:2565–2574

Pettersson J, Ninkovic V, Ahmed A (1999) Volatiles from different barley cultivars affect aphid acceptance of neighbouring plants. Acta Agric Scand B Soil Plant Sci 49:152–157

Pickett A (1990) Gas chromatography-mass spectrometry in insect pheromone identification: three extreme case histories. In: McCaffrey ID, Wilson AR (eds) Chromatography and Isolation of Insect Hormones and Pheromones. Plenum, New York, pp 281–288

Pickett JA, Glinwood RT (2007) Chemical ecology. In: van Emden H, Harrington R (eds) Aphids as Crop Pests. CAB International, Wallingford, pp 235–260

Pickett JA, Wadhams LJ, Woodcock CM (1998) Insect supersense: mate and host location by insects as model systems for exploiting olfactory interactions. The Biochemist 20:8–13

Pickett JA, Smiley DWM, Woodcock CM (1999) Secondary metabolites in plant-insect interactions: dynamic systems of induced and adaptive responses. Adv Bot Res 30:91–115

Pickett JA, Rasmussen HB, Woodcock CM, Matthes W, Napier JA (2003) Plant stress signalling: understanding and exploiting plant-plant interactions. Biochem Soc Trans 31:123–127

Pope TW, Kissen R, Grant M, Pickett JA, Rossiter JA, Powell G (2008) Comparative innate responses of the aphid parasitoid Diaeretiella rapae to alken glucosinolate derived isothiocyanates, nitriles, and epithionitriles. J Chem Ecol 34:1302–1310

Pureswaran DS, Gries R, Borden JH (2004) Quantitative variation in monoterpenes in four species of conifers. Biochem Syst Ecol 32:1109–1136

Raghu S, Clarke AR (2003) Sexual selection in a tropical fruit fly: role of a plant derived chemical in mate choice. Entomol Exp Appl 108: 53–58

Rasmann S, Köllner TG, Degenhardt J, Hiltpold I, Toepfer S, Kuhlmann U, Gershenzon J, Turlings TC (2005) Recruitment of entomopathogenic nematodes by insect-damaged maize roots. Nature 434:732–737

Ratzka A, Vogel H, Kliebenstein DJ, Mitchell-Olds T, Kroymann J (2002) Disarming the mustard oil bomb. Proc Natl Acad Sci USA 99:11223–11228

Rees CJC (1969) Chemoreceptor specificity associated with choice of feeding site by the beetle Chrysolina brunsvicensis on its foodplant, *Hypericum hirsutum*. Entomol Exp Appl 12:565–83

Rhoades DF, Cates RG (1976) Toward a general theory of plant antiherbivore chemistry. In: Wallace JW, Mansell RL (eds) Biochemical Interaction Between Plants and Insects, Recent Advances in Phytochemistry, Vol 10. Plenum, New York, pp 168–213

Robinson AF (2002) Soil and plant interaction impact on plant-parasitic nematode host finding and recognition In: Lewis EE, Cambell LF, Sukhdeo MVK (eds) The Biochemical Ecology of Parasites. CAB International, Wallingford, pp 89–110

Roeske CN, Seiber JN, Brower LP, Moffitt CM (1975). Milkweed cardenolides and their comparative processing by monarch butterflies (*Danaus plexippus*). Recent Adv Phytochem 10:93–16

Rolf RN, Barrett J, Perry RN (2000) Analysis of chemosensory responses of second stage juveniles of *Globodera rostochiensis* using electrophysiological techniques. Nematology 2:523–533

Rothschild M (1972) Secondary plant substances and warning colouration in insects. Symp R Entomol Soc London 6:59–83

Rühm R, Dietsche E, Harloff H-J, Lieb M, Franke S, Aumann J (2003) Characterisation and partial purification of a white mustard kairomone that attracts the beet cyst nematode, *Heterodera schachtii*. Nematology 5:17–22

Ruther J, Kleier S (2005) Plant–plant signaling: ethylene synergizes volatile emission in Zea mays induced by exposure to (Z)-3-hexen-1-ol. J Chem Ecol 31:2217–2222

Sadek MM, Anderson P (2007) Modulation of reproductive behaviour of Spodoptera littoralis by host and non-host plant leaves. Basic Appl Ecol 8:444–452

Schie CCN, van Haring MA, Schuurink, RC (2007) Tomato linalool synthase is induced in trichomes by jasmonic acid. Plant Mol Biol 64:251–263

Schindek R, Hilker M (1996) Influence of larvae of *Gastrophysa viridula* on the distribution of conspecific adults in the field. Ecol Entomol 21:370–376

Schoonhoven LM, van Loon JJA (2002) An inventory of taste in caterpillars: each species its own key. Acta Zool Acad Sci Hung 48:215–63

Schuler MA (1996) The role of cytochrome P450 monooxygenase in plant defence interactions. Plant Physiol 112:1411–1419

Schullehner K, Dick R, Vitzthum F, Schwab W, Brandt W, Frey M, Gierl A (2008) Benzoxazinoid biosynthesis in dicot plants. Phytochemistry 69:2668–2677

Seigler DS (1991) Cyanide and cyanogenic glycosides. In: Rosenthal GA, Berenbaum MR (eds) Herbivores: Their Interaction with Secondary metabolites. Academic, Orlando, FL, pp 35–77

Seigler DS (1998) Cyanogenic glycosides and cyanolipids. In: Siegler DS (ed) Plant Secondary Metabolism. Kluwer, Boston, MA/Dordrecht/London, pp 273–299

Shiojiri K, Takabayashi J, Yano S, Takafuji A (2000) Herbivore-species-specific interactions between crucifer plants and parasitic wasps (Hymenoptera: Braconidae) that are mediated by infochemicals present in areas damaged by herbivores. Appl Entomol Zool 35:519–524

Shiojiri K, Takabayashi J, Yano S, Takafuji A (2001) Infochemically mediated tritrophic interaction webs on cabbage plants. Popul Ecol 43:23–29

Sicker D, Frey M, Schulz M, Gierl A (2000) Role of benzoxazinones in the survival strategy of plants. Int Rev Cytol 198:319–347

Simpson CL, Simpson SJ, Abisgold JD (1990) The role of various amino acids in the protein compensatory response of *Locusta migratoria*. Symp Biol Hung 39:39–52

Snook ME, Johnson AW, Severson RF, Teng Q, White RA Jr, Sisson VA, Jackson DM (1997) Hydroxygeranyllinalool glycosides from tobacco exhibit antibiosis activity in the tobacco budworm, *Heliothis virescens* (F.). J Agric Food Chem 45: 2299–2308

Spencer KC (1988) Chemical mediation of coevolution in the *Passiflora–Heliconius* interaction. In: Spencer KC (ed) Chemical Mediation of Coevolution. Academic, New York, pp 167–240

Steppuhn A, Gase K, Krock B, Halitschke R, Ian T. Baldwin IT (2004) Nicotine's defensive function in nature. PLoS Biol 2: E217

Stamp N (2003) Out of the quagmire of plant defence hypotheses. Quart Rev Biol 78: 23–55

Sue M, Yamazaki K, Yajima S, Nomura T, Matsukawa T, Iwamura H, Miyamoto T (2006)

Molecular and structural characterization of hexameric β-D-glucosidases in wheat and rye. Plant Physiol 141:1237–1247

Takabayashi J, Dicke M (1996) Plant-carnivore mutualism through herbivore-induced carnivore attractants. Trends Plant Sci 1:109–113

Takabayashi J, Dicke M, Posthumus MA (1994) Volatile herbivore-induced terpenoids in plant–mite interactions: variation caused by biotic and abiotic factors. J Chem Ecol 20:1329–1354

Takemura M, Nishida R, Mori N, Kuwahara Y (2002) Acylated flavonol glycosides as probing stimulants of a bean aphid, *Megoura crassicauda*, from *Vicia angustifolia* . Phytochemistry 61:135–140

Textor S, Bartram S, Kroymann J, Falk KL, Hick A, Pickett JA, Jonathan Gershenzon J (2004) Biosynthesis of methionine-derived glucosinolates in *Arabidopsis thaliana*: recombinant expression and characterization of methylthioalkylmalate synthase, the condensing enzyme of the chain-elongation cycle. Planta 218:1026–1035

Thaler J S (1999) Jasmonate-inducible plant defences cause increased parasitism of herbivores. Nature 399:686–688

Thackaray DJ, Wratten SD, Edwards PJ, Niemeyer HM (1990) Resistance to the aphids *Sitobion avenae* and *Rhopalosiphum padi* in Gramineae in relation to hydroxamic acid levels. Ann Appl Biol 116:573–582

Thompson GA, Goggin FL (2006) Transcriptomics and functional genomics of plant defence induction by phloem feeding insects. J Exp Bot 57:755–766

Ton J, D'Alessandro M, Jourdie V, Jakab G, Karlen D, Held M, Mauch Mani B, Turlings TCJ (2007) Priming by airborne signals boosts direct and indirect resistance in maize. Plant J 49:16–26

Tscharntke T, Thiessen S, Dolch R, Boland W (2001) Herbivory, induced resistance, and interplant signal transfer in *Alnus glutinosa*. Biochem Syst Ecol 29:1025–1047

Turlings TCJ, Wäckers FL (2004) Recruitment of predators and parasitoids by herbivore-damaged plants. Cambridge University Press Cambridge

Turlings TC, Loughrin JH, McCall PJR, Lewis WJ, Tumlinson JH (1995) How caterpillar-damaged plants protect themselves by attracting parasitic wasps. Proc Natl Acad Sci USA 92:4169–4174

van Drongelen W (1979) Contact chemoreception of host plant specific chemicals in larvae of various Yponomeuta species (Lepidoptera). J Comp Physiol 134:265–279

Velozo JA, Alvarez RI, Wächter GA, Timmermann BN Corcuera LJ (1999) Increase in gramine content in barley infested by the aphid *Schizaphis graminum* R. Phytochemistry 52:1059–1061

Visser JH (1988) Host-plant finding by insects – orientation, sensory input and search patterns. J Insect Physiol 34:259–268

Walling LL (2000) The myriad plant responses to herbivores. J Plant Growth Regul 19:195–216

Wasternack C (2007) Jasmonates: an update on biosynthesis, signal transduction and action in plant stress response, growth and development Ann Bot 100:681–697

Wheat CW, Vogel H, Wittstock U, Braby MF, Underwood D, Mitchell-Olds T (2007) The genetic basis of a plant–insect coevolutionary key innovation. Proc Natl Acad Sci USA:104:20427–20431

Wieczorek H, Koppl R (1978) Effect of sugars on the labellar water cell receptor of the fly. J Comp Physiol 126:131–36

Williams III L, Rodriguez-Saona C, Pare PW, Crafts-Brandner SJ (2005) The piercing–sucking herbviores *Lygus hesperus* and *Nezara viridula* induce volatile emissions in plants. Arch Insect Biochem Physiol 58:84–96

Wink M (2000) Interference of alkaloids with neuroreceptors and ion channels. In: Atta-Ur-Rahman X (ed) Bioactive Natural Products. Elsevier, Amsterdam, pp 3–129

Wink M (2003) Evolution of secondary metabolites from an ecological and molecular phylogenetic perspective. Phytochemistry 64:3–19

Winz RA, Baldwin IT (2001) Molecular interactions between the specialist herbivore *Manduca sexta* (Lepidoptera, Sphingidae) and its natural host *Nicotiana attenuata*. IV. Insect-induced ethylene reduces jasmonate-induced nicotine accumulation by regulating putrescine N-methyltransferase transcripts. Plant Physiol 125:2189–2202

Wittstock U, Gershenzon J (2002) Constitutive plant toxins and their role in defence against herbivores and pathogens. Curr Opin Plant Biol 5:1–8

Wittstock U, Lichtnow KH, Teuschere M (1997) Effects of cicutoxin and related polyacetylenes from *Cituta virosa* on neuronal action potentials: a comparative study on the convulsive action. Planta Med 63:120–124

Witzgall P, Stelinski L, Gut L, Thomson L (2008) Codling moth management and chemical ecology. Annu Rev Entomol 53: 503–522

Wu H, Haig T, Pratley J, Lemerle D, An M (2001) Allelochemicals in wheat (Triticum aestivum L.): production and exudation of 2,4-dihydroxy-7-methoxy-1,4-benzoxazin-3-one. J Chem Ecol 27:1691–1700

Wuyts N, Swennen R, De Waele D (2006) Effects of plant phenylpropanoid pathway products and selected terpenoids and alkaloids on the behaviour of the plant-parasitic nematodes *Radopholus similis, Pratylenchus penetrans* and *Meloidogyne incognita*. Nematology 8:89–101

Yan ZG, Wang CZ (2006) Wound-induced green leaf volatiles cause the release of acetylated derivatives and a terpenoid in maize. Phytochemistry 67: 34–42

Zarate SI, Kempema LA, Walling LL (2007) Silverleaf whitefly induces salicylic acid responses and represses effectual jasmonic responses in Arabidopsis. Plant Physiol 143: 866–875

Zenk MH, Juenger M (2007) Evolution and current status of the phytochemistry of nitrogenous compounds. Phytochemistry 68:2757–2772

Oligosaccharide Signalling Molecules 16

Robert A. Field

Abstract The signalling potential of plant carbohydrates had largely been overlooked until the last 2 decades. Whether trehalose- and sucrose-derived compounds, or oligosaccharides emanating from cell wall fragmentation, carbohydrate signals are thought to play key roles in plant growth and development at one level, and in activating plant defences against microbial infection at another. However, the literature is less than definitive, in many cases, about how *in vitro* observations correlate, or not, with true *in vivo* function. Nonetheless, a better understanding of the developmental regulation, allelopathic and defence-inducing properties of 'oligosaccharin' signalling carbohydrates offers scope not only for better understanding plant biology *per se*, but also presents novel opportunities for commercial exploitation. This chapter highlights oligosaccharide structures that have been reported to possess the properties of signalling molecules; although in many cases physiological relevance has not been demonstrated there is potential for exploitation.

This chapter outlines the main classes of plant-derived oligosaccharide molecules that have been reported to possess signalling capabilities. Comment is made on reports of the physiological function and/or the action of such compounds when fed exogenously. In many cases, whilst biological activity has been demonstrated its relevance to physiological function remains to be established.

R.A. Field (✉)
Department of Biological Chemistry, John Innes Centre, Colney Lane, Norwich NR4 7UH, UK
e-mail: rob.field@bbsrc.ac.uk

Carbohydrate Sensing and Signalling

Primary Metabolism

The sensing and signalling systems that are involved in the regulation of carbon metabolites in plants form a complex network of metabolites, transporters, enzymes, transcription factors and hormones [1]. Carbohydrate metabolism, secondary metabolism, nitrogen assimilation and photosynthesis are all under the control of these systems, along with fundamental developmental processes such as embryogenesis, germination, sprouting, pollen development and senescence. Understanding the cross-talk between carbon metabolite and other major signalling systems in plant cells, such as those associated with phytohormones [2], is key to manipulating this fundamentally important aspect of metabolic regulation for crop improvement (*viz.* carbon allocation to storage processes) [3, 4]. It has yet to be established exactly which metabolites are sensed, but likely candidates include sucrose [5] and trehalose-6-phosphate [6].

A.E. Osbourn and V. Lanzotti (eds.), *Plant-derived Natural Products*,
DOI 10.1007/978-0-387-85498-4_16,

16

Exogenous Carbohydrate

Beyond primary metabolism *per se*, carbohydrate signalling is found in a variety of different contexts. Trehalose, a non-reducing disaccharide that functions as a storage carbohydrate and osmoprotectant in yeast, fungi and certain insects, is also present in trace amounts in flowering plants, where metabolites derived from trehalose are necessary for embryo development [7]. In contrast to its role in endogenous metabolism, exogenous trehalose can be toxic to some higher plants, as evidenced by a severe inhibition of root and cotyledon growth [8]. Feeding exogenous trehalose to liquid cultures of *Arabidopsis* seedlings rapidly alters disaccharide levels and induces detoxification and stress response proteins. These findings support the idea that exogenous trehalose can function as an elicitor of genes involved in biotic and abiotic stress. In a similar vein, flavonoid and anthocyanin biosynthetic pathways in *Arabidopsis* are strongly up-regulated in response to exogenous sucrose in a highly selective manner [7, 8]. This suggests that carbon metabolites, including glycans, may play key signalling roles in plant biology in both an intracellular and extracellular manner.

Lectins: Mammals and Plants

It has been noted that, in mammalian systems, "proteins are often unduly portrayed as (the) decisive hardware. In contrast, and actually prominent among the biochemical systems to store information, carbohydrate epitopes of cellular glycoconjugates favourably combine high-density coding with strategic positioning, rendering them readily accessible for interactions with adaptor molecules. The interaction with lectins is the ignition key to start glycoconjugate-mediated signalling" [9]. Plant lectins comprise a widespread group of carbohydrate-binding proteins that show a marked heterogeneity with respect to their structure, sugar-binding specificity and temporal and spatial regulation. Until recently, the role of most lectins was thought to be associated with their binding to foreign glycans in either recognition and/or defence-related phenomena. Over the past few years, evidence has accumulated to support the idea that when plants are stimulated by specific biotic or abiotic stimuli they respond through the expression of cytoplasmic and/or nuclear lectins [10]. The location and the regulation of the expression of these lectins indicate that they are involved in specific endogenous protein–carbohydrate interactions. These and other findings [11] led to the proposition that lectins might be involved in cellular regulation and signalling in plants, in a manner similar to that described in mammals (for review material on the background to this area see [12–16]).

Oligosaccharins: Background

Cell signalling in plants can be mediated by so-called 'oligosaccharin' glycans, some of which are produced by the partial digestion of cell walls [17, 18]. This may occur as part of normal developmental processes, such as fruit ripening, but it also occurs in response to infection by microorganisms. The literature provides more clear-cut evidence for a role for oligosaccharins in infection than in development. The infection process, whether by symbiont or pathogen, can result in the microorganisms themselves releasing glycans (typically chitooligosaccharides, peptidoglycan or β-glucans) that serve as signals to the plant. This latter topic is beyond the scope of this book, but readers are referred to the following review material [19–28]. Following pioneering work in the area [17, 18], plant-derived oligosaccharins have been much reviewed over the years [19, 29–35]. However, our understanding of the specifics of signal perception and resulting signal transduction remain rather limited. The structural requirements for oligosaccharin biological activity vary in stringency depending on the plant species being studied, the glycan class under investigation and the biological response being assessed. The mechanisms by which structurally diverse glycans are perceived by plants and cause response are also varied, ranging from specific receptor-mediated processes to general membrane polarization effects, for instance. The following sections outline the structure and action of plant cell wall-derived and other plant glycans in signalling processes.

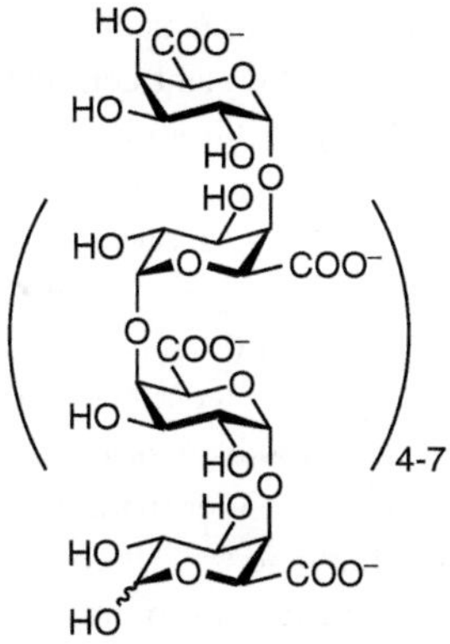

Fig. 1 A typical pectin-derived oligogalacturonide oligosaccharin [35]

Oligosaccharide Signals: Cell Wall-Derived Oligosaccharides

The majority of plant carbohydrate signals described to date derive from cell wall polysaccharides, although this is not always the case (*vide infra*)(for relevant general review material on plant cell walls see [36–39].

Oligogalacturonides

Pectin constitutes a large family of polysaccharides present in all plant primary cell walls [35]. The complex structure of pectic polysaccharides and the ubiquitous nature of this large class of glycans, suggests that these polysaccharides may have multiple functions in plant growth and development [34]. Digestion of cell wall pectin, either by plant or microbial enzymes, gives rise to the oligogalacturonides, which were amongst the first plant-derived oligosaccharins to be reported [40, 41]. Oligomers α-1,4-linked galacturonic acid (Fig. 1), with d.p. ~10–16, are well documented for their ability to activate plant defence responses, leading to phytoalexin and lignin production (reviewed in [30, 33, 35]). Regulatory roles in non-diseased plants have also been proposed. Endopolygalacturonase, which occurs in fruit, could in principle generate the oligogalacturonides that have been shown to promote fruit ripening [34]. For a comprehensive review of pectin structure, biosynthesis, and oligogalacturonide-related signalling see [35], which follows on from earlier critiques of the latter [30, 33].

In terms of oligogalacturonide perception, recent studies show that the extra-cellular domain of the *Arabidopsis thaliana* cell wall-associated kinase, WAK1, forms complexes with pectins in a calcium-dependent manner [42]. Interaction studies with WAK1 mutants have identified at least five specific amino acids involved in the interaction with homogalacturonan dimers and multimers, which may help shed light on receptor-mediated oligogalacturonide signalling [43]. The down-stream effects of oligogalacturonide signalling can result in repression of gibberellin-dependent processes, suggesting cross-talk between the defence, hormonal and metabolic signalling pathways [44]. In addition, proteomic analysis of Arabidopsis seedlings responding to oligogalacturonides identified changes in the nuclear proteome: proteins responding to the oligogalacturonide treatment were mainly involved in the protein translation and regulation, suggesting a general reprogramming of the plant cell metabolism in response to oligogalacturonides [45]. Of potential commercial interest, grapevine (*Vitis vinifera* L.) is vulnerable to a variety of pathogenic fungi, including the grey mould producing *Botrytis cinerea*. This fungus, which if not kept at bay with fungicides, would be responsible for substantial yield losses worldwide. There is scope to begin to consider 'natural' alternatives to current fungicides: oligogalacturonides induce defence responses in grapevine, including stimulation of anti-fungal chitinase and β-1,3-glucanase activities, and protect against pathogenic fungi by about 60% [46].

Rhamnogalacturonan I-Derived Disaccharide (Lepidimoide)

The sodium salt of 2-*O*-rhamnopyranosyl-4-deoxy-*threo*-hex-4-enopyranosiduronate, was first isolated from mucilage of germinated cress seeds [47]. This compound, known as lepidimoide (Fig. 2), is widespread in plants, occurring in substantial quantities in the exudates of germinated seeds of sunflower and buckwheat. It has also been detected in those of rice, lettuce, slender amaranth, leek, Persian speedwell and *Arabidopsis* [48–50]. *In vitro*, lepidimoide promotes plant shoot growth at low micromolar concentration, but inhibits root growth at concentrations >100 μM. The growth-promoting activity in hypocotyls is ~25 times higher than that of gibberellic acid. Structure–activity relationship of lepidimoide and its analogues have been established [51].

Whilst lepidimoide is plant-produced, it can also be generated from plant cell walls by the action of endophytic fungi of the genus Colletotrichum, which produce the requisite galactosidase, rhamnogalacturonan lyase and esterase activities [52, 53]. Gene clusters responsible for related rhamnogalacturonan depolymerisation have also been identified in saprophytic *Bacillus subtilis* strains [54].

It is not clear if lepidimoide generation in nature has allelopathic impact. The concentrations of lepidimoide required for activity in *in vitro* studies (10s–100s μM) may be somewhat higher than is achieved *in vivo*.

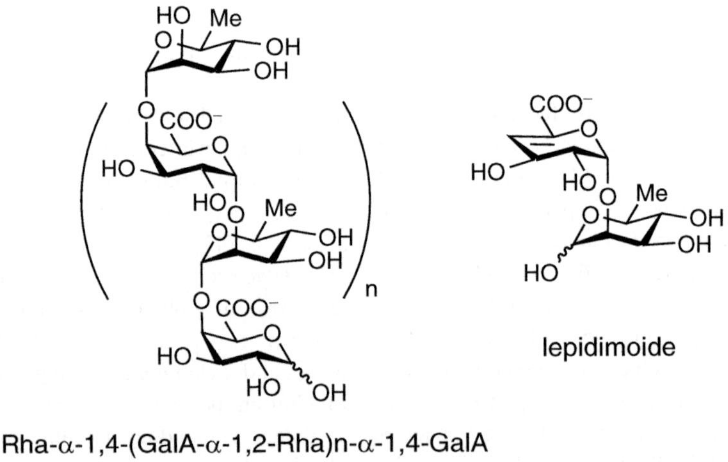

Fig. 2 Rhamnogalacturonan I and cleavage fragment lepidimoide [47, 51]

Hemicellulose Fragments

The xyloglucan nonasaccharide oligosaccharin XXFG (Fig. 3) [31], which antagonises the growth-promoting effect of auxin, arises *in vivo* by breakdown of pre-formed xyloglucan polysaccharide [55] (for xyloglucan nomenclature see [56]). When the reduced form of this oligosaccharin was applied to excised pea shoots, it inhibited growth at a site remote from its administration, suggesting that biologically active xyloglucan oligosaccharides can move throughout the stem [57]. A synthetic pentasaccharide related to XXFG, L-Fuc-(α-1,2)-D-Gal-(β-1,2)-D-Xyl-(α-1,6)-D-Glc-(β-1,4)-D-Glc (Fig. 3), has been found to suppress elongation of isolated pumpkin cotyledons at nanomolar concentration. Further, addition of the pentasaccharide into the culture medium of wheat embryos growing *in vitro* induced rhizogenesis and stimulated the formation of callus and meristematic zones [58]. In terms of potential physiological control of xyloglucan signalling, a developmentally regulated α-fucosidase that inactivates xyloglucan oligosaccharin has also been found in pea stems [59]. Despite the activities reported with exogenous compounds, evidence to support a physiological signalling role for xyloglucans has yet to be reported.

Frost resistance in wheat seedlings is proposed to involve, among other cellular responses, activate degradation of cell wall polysaccharides. This probably liberates hemicellulose fragments that may play a role in the cold adaptation process [60, 61]. However, a link between structures and physiological action is currently lacking.

Galactoglucomannans

Poplar wood galactoglucomannan oligosaccharides with d.p. 4–8 (Fig. 4) have been reported to impact on elongation of pea and spruce stem segments [62–64]. These biological effects are noted to be dependent on the concentration of oligosaccharide (10^{-5}–10^{-10} M range), as well as plant species. The potency of the effects reported may be consistent with biological function, although this still needs to be confirmed.

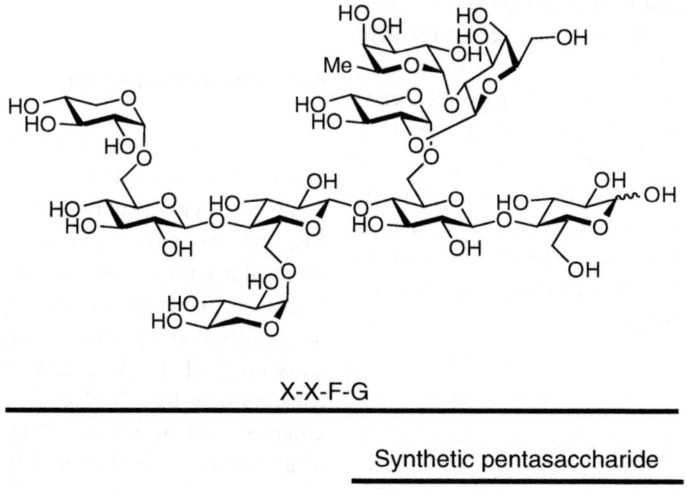

Fig. 3 Natural and synthetic xyloglucans [31, 58]

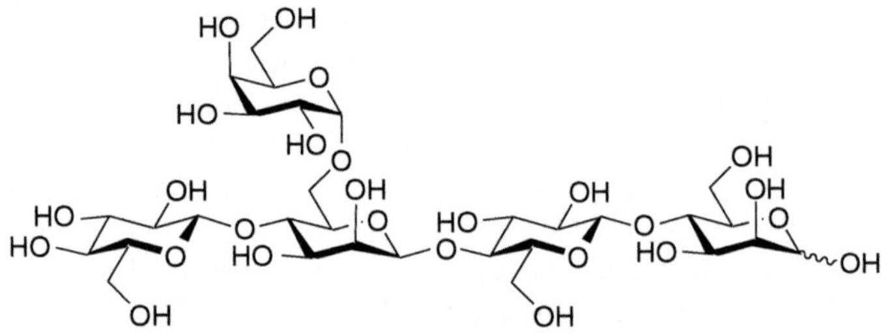

Glc-β-1,4-(Gal-α-1,6-)Man-β-1,4-Glc-β-1,4-Glc

Fig. 4 Representative galactoglucomannan oligosaccharide [64]

Glyco-inositols

In mammals, soluble inositolphosphoglycans [65] related to the conserved glycosyl-phosphatidyl-inositol (GPI) core structure [66] mediate insulin action. Such structures have similarity to the major uronic acid-containing non-reducing trisaccharide oligosaccharin produced by cultures of rose and sycamore, namely α-D-mannopyranosyl-(1,4)-α-D-glucuronopyranosyl-(1,2)-*myo*-inositol (Fig. 5). This glyco-inositol signal is capable of inhibiting leucine incorporation into protein and promotes isodityrosine formation when fed to rose cell cultures [67, 68].

Although the *in vivo* effect of such compounds remains to be established, the structural similarities between the plant and mammalian glycol-inositol systems, and further similarity to phytoglycolipid structures, suggests potentially related biosynthetic routes to these classes of signalling molecule (discussed in [69]).

Gentiobiose

Two naturally occurring neutral disaccharides, gentiobiose (β-Glc-(1,6)-Glc) and nigerose (α-Glc-(1,3)-Glc) (Fig. 6), have been identified in tomato pericarp and locule. Whilst gentiobiose stimulates the onset of ripening when vacuum-infiltrated into green tomato fruit, nigerose alone does not. In contrast, the combination of nigerose plus gentiobiose was more effective than either sugar fed on its own.

When gentiobiose was vacuum-infiltrated into unripe or ripe fruit, the disaccharide remained intact for at least 1 h but was largely degraded within 24 h, consistent with the notion that gentiobiose is an oligosaccharin that rapidly turns over [69].

Protein-Free *N*-linked Glycans

Protein-bound asparagine-linked (*N*-linked) glycans are subjected to extensive modification as glycoproteins mature and pass through the ER, via the Golgi complex, to their final destinations inside and outside the cell. The *N*-glycans play a pivotal role in protein folding, oligomerisation, quality control, sorting, and transport [70]. In yeast and mammalian systems, protein-free *N*-glycans can arise from glucosidase-I deficiency [71] or as a result of glucosidase-I inhibition [72]. In plant cells, *N*-glycans [73] free from protein are present at micromolar concentrations during differentiation, growth and maturation, and might play a

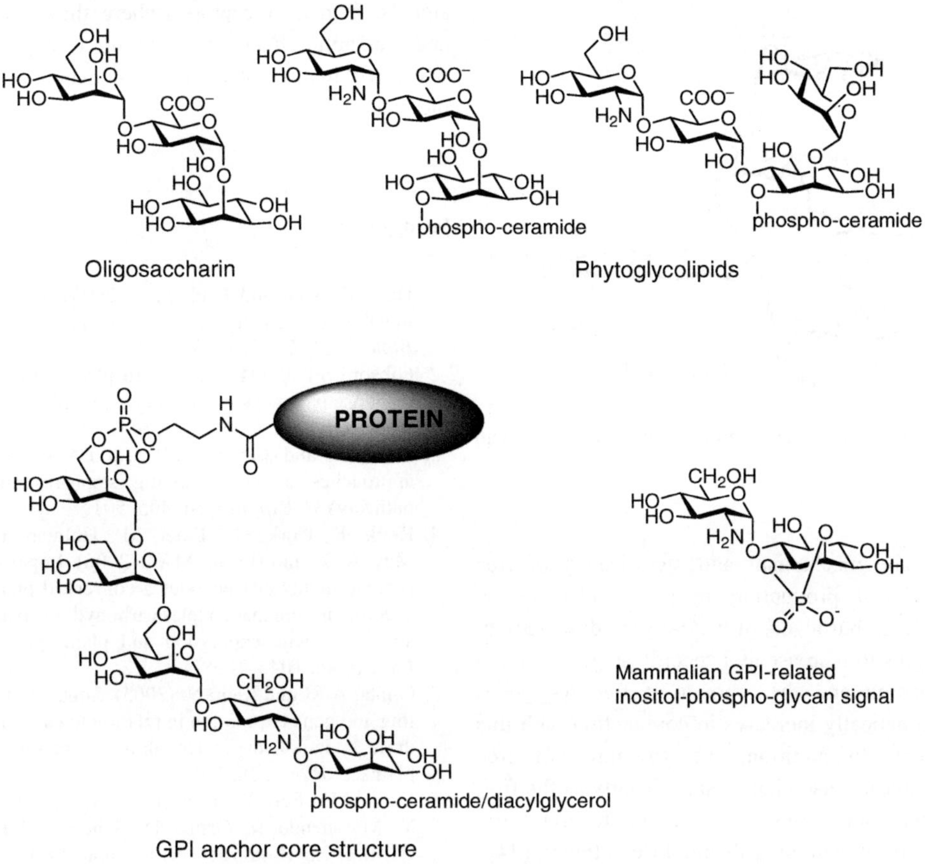

Fig. 5 Glyco-inositol oligosaccharin, related phytoglyolipids, conserved GPI anchor core structiure and mammalian inositol-phospho-glycan involved in insulin action [34]

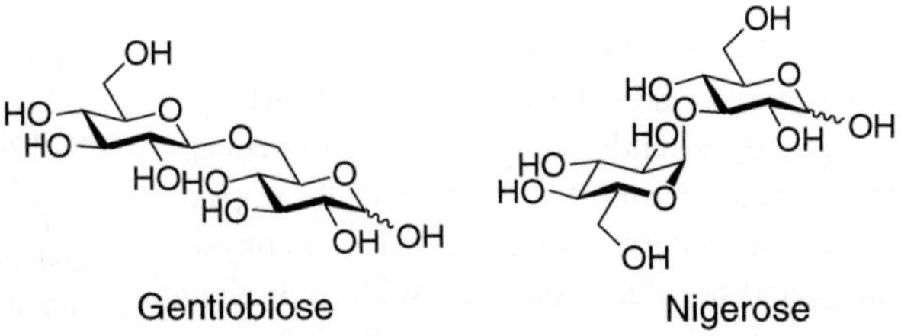

Fig. 6 Oligosaccharin gentiobiose and the isomeric nigerose [69]

role in processes such as seed germination and fruit ripening [74]. For instance, $Man_5GlcNAc$ (Fig. 7), one of ten unconjugated *N*-glycans purified from tomato pericarp tissue, may have a key role in triggering and modulating tomato fruit ripening [75, 76].

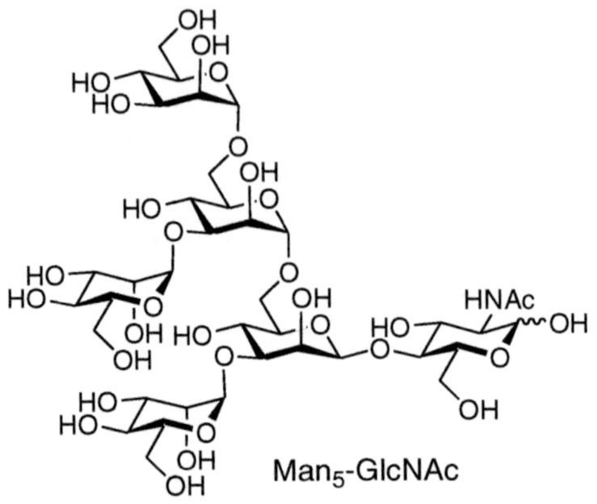

Fig. 7 Representative protein-free *N*-linked glycan [74]

The mechanism and significance of free *N*-glycan production in plant cells remain unclear, but levels of *endo*-glycosidase activity begins to increase at a specific stage of tomato ripening and the amount of free *N*-glycans dramatically increases in conjunction with this event. In addition, the structures of free *N*-glycans also change significantly as the fruit ripens, suggesting a possible role for free sugar chains in plant growth and differentiation [74].

Concluding Remarks

Whilst the signalling properties of plant oligosaccharides have been known for nearly 30 years, only recently have we begun to scratch the surface in terms of understanding the impact of such plant-derived oligosaccharins on physiology and defence responses. With improving analytical and structural characterisation tools, this field is set to move forward at pace. Key issues to be addressed revolve around the physiological function, or otherwise, of oligosaccharide structure that show *in vitro* signalling properties. No doubt the repertoire of glycan signals, cognate receptors (where they exist) and metabolic effects will continue to expand, which will in turn potentially open up new commercial opportunities.

References

1. Halford, N.G. and Paul, M.J. (2003). Carbon metabolite sensing and signalling. *Plant Biotechnol. J.* 1, 381–398.
2. Gibson, S.I. (2004). Sugar and phytohormone response pathways: navigating a signalling network. *J. Exp. Bot.* 55, 253–264.
3. Rook, F. and Bevan, M.W. (2003). Genetic approaches to understanding sugar-response pathways. *J. Exp. Bot.* 54, 495–501.
4. Rook, F., Corke, F., Baier, M., Holman, R., May, A.G., and Bevan, M.W. (2006). Impaired sucrose induction1 encodes a conserved plant-specific protein that couples carbohydrate availability to gene expression and plant growth. *Plant J.* 46, 1045–1058.
5. Gupta, A.K. and Kaur, N. (2005). Sugar signalling and gene expression in relation to carbohydrate metabolism under abiotic stresses in plants. *J. Biosci.* 30, 761–776.
6. Lunn, J.E., Feil, R., Hendriks, J.H.M., Gibon, Y., Morcuende, R., Osuna, D., Scheible, W.R., Carillo, P., Hajirezaei, M.R., and Stitt, M. (2006). Sugar-induced increases in trehalose 6-phosphate are correlated with redox activation of ADPglucose pyrophosphorylase and higher rates of starch synthesis in *Arabidopsis thaliana*. *Biochem. J.* 397, 139–148.
7. Bae, H., Herman, E., and Sicher, R. (2005). Endogenous trehalose promotes non-structural carbohydrate accumulation and induces chemical detoxification and stress response proteins in *Arabidopsis thaliana* grown in liquid culture. *Plant Sci.* 168, 1293–1301.
8. Solfanelli, C., Poggi, A., Loreti, E., Alpi, A., and Perata, P. (2006). Sucrose-specific induction of the anthocyanin biosynthetic pathway in *Arabidopsis*. *Plant Physiol.* 140, 637–646.
9. Villalobo, A., Nogales-Gonzalez, A., and Gabius, H.J. (2006). A guide to signaling pathways connecting protein-glycan interaction with the emerging versatile effector functionality of mammalian lectins. *Trends Glycosci. Glycotechnol.* 18, 1–37.

10. van Damme, E.J.M., Barre, A., Rouge, P., and Peumans, W.J. (2004). Cytoplasmic/nuclear plant lectins: a new story. *Trends Plant Sci.* 9, 484–489.
11. Barre, A., Herve, C., Lescure, B., and Rouge, P. (2002). Lectin receptor kinases in plants. *Crit. Rev. Biochem.* 21, 379–399.
12. Dwek, R. A. (1996). Glycobiology: toward understanding the function of sugars. *Chem. Rev.* 96, 683–720.
13. Varki, A., Cummings, R., Esko, J., Freeze, H., Hart, G., and Marth, J., Editors. (2009). *Essentials of Glycobiology*, 2nd edition, Cold Spring Harbor Laboratory Press, New York.
14. Gabius, H.J., Siebert, H. C., Andre, S., Jimenez-Barbero, J., and Rudiger, H. (2004). Chemical biology of the sugar code. *ChemBioChem* 5, 741–764.
15. Sharon, N. and Lis, H. (2004). History of lectins: from hemagglutinins to biological recognition molecules. *Glycobiology* 14, 53R–62R.
16. Ambrosi, M., Cameron, N.R., and Davis, B.G. (2005). Lectins: tools for the molecular understanding of the glycocode. *Org. Biomol. Chem.* 3, 1593–1608.
17. Albersheim, P. and Darvill, A.G. (1985). Oligosaccharins: novel molecules that can regulate growth, development, reproduction, and defense against disease in plants. *Sci. Am.* 253, 58–64.
18. Albersheim, P., Augur, C., Cheong, J.-J., Eberhard, S., Hahn, M.G., Marfa, V., Mohnen, D., O'Neill, M.A., Spiro, M.D., York, W.S., and Darvill, A.G. (1992). Oligosaccharins -oligosaccharide regulatory molecules. *Acc. Chem. Res.* 25, 77–83.
19. Ozeretskovskaya, O.L. and Romenskaya, I.G. (1996). Oligosaccharins as regulatory molecules of plants. *Russ. J. Plant Physiol.* 43, 648–655.
20. D'Haeze, W. and Holsters, M. (2002). Nod factor structures, responses, and perception during initiation of nodule development. *Glycobiology* 12, 79R–105R.
21. Rose, J.K., Ham, K.S., Darvill, A.G., and Albersheim P. (2002). Molecular cloning and characterization of glucanase inhibitor proteins: coevolution of a counterdefense mechanism by plant pathogens. *Plant Cell* 14, 1329–1345.
22. Laus, M.C. and Kijne, W.C. (2004). A fixer's dress code: Surface polysaccharides and host-plant-specificity in the root nodule symbiosis. *Trends Glycosci. Glycotechnol.* 16, 281–290.
23. Spaink, H.P. (2004). Specific recognition of bacteria by plant LysM domain receptor kinases. *Trends Microbiol.* 12, 201–204.
24. Becker, A., Fraysse, N., and Sharypova, L. (2005). Recent advances in studies on structure and symbiosis-related function of rhizobial K-antigens and lipopolysaccharides. *Mol Plant-Microbe Interact* 18, 899–905.
25. Geurts, R., Fedorova, E., and Bisseling, T. (2005). Nod factor signalling genes and their function in the early stages of Rhizobium infection. *Curr. Opin. Plant Biol.* 8, 346–352.
26. Mulder, L., Lefebvre, B., Cullimore, J., and Imberty, A. (2006). LysM domains of *Medicago truncatula* NFP protein involved in Nod factor perception. Glycosylation state, molecular modeling and docking of chitooligosaccharides and Nod factors. *Glycobiology* 16, 801–809.
27. Kaku, H., Nishizawa, Y., Ishii-Minami, N., Akimoto-Tomiyama, C., Dohmae, N., Takio, K., Minami, E., and Shibuya, N. (2006). Plant cells recognize chitin fragments for defense signaling through a plasma membrane receptor. *Proc. Natl. Acad. Sci USA* 103, 11086–11091.
28. Damasceno, C.M.B., Bishop, J.G., Ripoll, D.R., Win, J., Kamoun, S., Rose, J.K.C. (2008). Structure of the glucanase inhibitor protein (GIP) family from Phytophthora species suggests coevolution with plant endo-β-1,3-glucanases. *Mol. Plant-Microbe Interact.* 21, 820–830.
29. Aldington, S., McDougall, G.J., and Fry, S.C. (1991). Structure-activity relationships of biologically active oligosaccharides. *Plant Cell Environ.* 14, 625–636.
30. Cote, F. and Hahn, M.G. (1994). Oligosacharins: structures and signal transduction. *Plant Mol. Biol.* 26, 1379–1411.
31. Fry, S.C. (1996). Oligosaccharin mutants. *Trends Plant Sci.* 1, 326–328.
32. John, M., Rohrig, H., Schmidt, J., Walden, R., and Schell, J. (1997). Cell signalling by oligosaccharides. *Trends Plant Sci.* 2, 111–115.
33. Cote, F., Ham, K.-S., Hahn, M.G., and Bergman, C.W. (1998). Oligosaccharide elicitors in host-pathogen interactions: generation, perception and signal transduction. In: Biswas, B.B. and Das, H. (Eds), *Subcellular Biochemistry*, Vol. 29, Plant-microbe interactions. Plenum Press, New York, pp. 385–431.
34. Dumville, J.C. and Fry, S.C. (2000). Uronic acid-containing oligosaccharins: their biosynthesis, degradation and signalling roles in non-diseased plant tissues. *Plant Physiol. Biochem.* 38, 125–140.

35. Ridley, B.L., O'Neill, M.A., and Mohnen, D.A. (2001). Pectins: structure, biosynthesis, and oligogalacturonide-related signalling. *Phytochemistry* 57, 929–967.
36. Brett, C. and Waldron, K. (1996). *Physiology and Biochemistry of Plant Cell Walls*, 2nd Edition, Chapman & Hall, London.
37 Carpita, N. and McCann, M. (2000). in: Buchanan, B.B., Gruissem, W., and Jones, R.L. (Eds) *Biochemistry and Molecular Biology of Plants*, American Society of Plant Biologists, Rockville, MD, pp. 52–109.
38. Somerville, C., Bauer, S., Brininstool, G., Facette, M., Hamann, T., Milne, J., Osborne, E., Paredez, A., Persson, S., Raab, T., Vorwerk, S., and Youngs, H. (2004). Toward a systems approach to understanding plant-cell walls. *Science* 306, 2206–2211.
39. Fry, S.C. (2004). Primary cell wall metabolism: tracking the careers of wall polymers in living plant cells. *New Phytologist* 161, 641–675.
40. Bishop, P.D., Makus, D.J., Pearce, G., and Ryan, C.A. (1981). Proteinase inhibitor-inducing factor activity in tomato leaves resides in oligosaccharides enzymatically released from cell walls. *Proc. Natl. Acad. Sci. USA* 78, 3536–3540.
41. Hahn, M.G., Darvill, A.G., and Albersheim, P. (1981). Host-pathogen interactions XIX. The endogenous elicitor, a fragment of a plant cell wall polysaccharide that elicits phytoalexin accumulation in soybeans. *Plant Physiol.* 68, 1161–1169.
42. Decreux, A. and Messiaen, J. (2005). Wall-associated kinase WAK1 interacts with cell wall pectins in a calcium-induced conformation. *Plant Cell Physiol.* 46, 268–278.
43. Decreux, A., Thomas, A., Spies, B., Brasseur, R., van Cutsem, P., and Messiaen, J. (2006). *In vitro* characterization of the homogalacturonan-binding domain of the wall-associated kinase WAK1 using site-directed mutagenesis. *Phytochemistry* 67, 1068–1079.
44. Loreti, E., Bellincampi, D., Millet, C., Alpi, A., and Perata, P. (2002). Elicitors of defence responses repress a gibberellin signalling pathway in barley embryos. *J. Plant Physiol.* 159, 1383–1386.
45. Casasoli, M., Meliciani, I., Cervone, F., De Lorenzo, G., and Mattei, B. (2007). Oligogalacturonide-induced changes in the nuclear proteome of *Arabidopsis thaliana*. *Int. J. Mass Spectrom.* 268, 277–282.
46. Aziz, A., Heyraud, A., and Lambert, B. (2004). Oligogalacturonide signal transduction, induction of defense-related responses and protection of grapevine against *Botrytis cinerea*. *Planta* 218, 767–774.
47. Hasegawa, K., Mitzutani, J., Kosemura, S., and Yamamura, S. (1992). Isolation and identification of lepidimoide, a new allelopathic substance from mucilage of germinated cress seeds. *Plant Physiol.* 100, 1059–1061.
48. Yamada, K., Anai, T., and Hasegawa, K. (1995). Lepidimoide, an allelopathic substance in the exudates from germinated seeds. *Phytochemistry* 39, 1031–1032.
49. Yamada, K., Kosemura, S., Yamamura, S., and Hasegawa, K. (1997). Exudation of an allelopathic substance lepidimoide from seeds during germination. *Plant Growth Regul.* 22, 189–192.
50. Yokotani-Tomita, K., Goto, N., Kosemura, S., Yamamura, S., and Hasegawa, K. (1998). Growth-promoting allelopathic substance exuded from germinating Arabidopsis thaliana seeds. *Phytochemistry* 47, 1–2.
51. Yamada, K., Anai, T., Kosemura, S., Yamamura, S., and Hasegawa, K. (1996). Structure activity relationship of lepidimoide and its analogues. *Phytochemistry* 41, 671–673.
52. Tanaka, M., Yoshimura, M., Suto, M., Yokota, A., Asano, K., Sukara, E., and Tomita, F. (2002). Production of lepidimoide by an endophytic fungus from polysaccharide extracted from Abelmoschus sp.: identification of the product and the organism producing it. *J. Biosci. Bioeng.* 93, 531–536.
53. Saranpuetti, C., Tanaka, M., Sone, T., Asano, K., and Tomita, F. (2006). Determination of enzymes from Colletotrichum sp AHU9748 essential for lepidimoide production from okra polysaccharide. *J. Biosci. Bioeng.* 102, 452–456.
54. Ochiai, A., Itoh, T., Kawamata, A., Hashimoto, W., and Murata, K. (2007). Plant cell wall degradation by saprophytic *Bacillus subtilis* strains: Gene clusters responsible for rhamnogalacturonan depolymerization. *Appl. Environ. Microbiol.* 73, 3803–3813.
55. McDougall, G.J. and Fry, S.C. (1991). Xyloglucan nonasaccharide, a naturally occurring oligosaccharin, arises *in vivo* by polysaccharide breakdown. *J. Plant. Physiol.* 137, 332–336.
56. Fry, S.C., York, W.S., Albersheim, P., Darvill, A., Hayashi, T., Joseleau, J.P., Kato, Y.,

Lorences, E.P., Maclachlan, G.A., McNeil, M., Mort, A.J., Reid, J.S.G., Seitz, H.U., Selvendran, R.R., Voragen, A.G.J., and White, A.R. (1993). An unambiguous nomenclature for xyloglucan-derived oligosaccharides. *Physiol. Plant.* 89, 1–3.

57. Warneck, H.M., Fulton, D.C., Seitz, H.U., and Fry, S.C. (1998). Transport, degradation and cell wall-integration of XXFGol, a growth-regulating nonasaccharide of xyloglucan, in pea stems. *Planta* 204, 78–85.
58. Pavlova, Z.N., Ash, O.A., Vnuchkova, V.A., Babakov, A.V., Torgov, V.I., Nechaev, O.A., Usov, A.I., and Shibaev, V.N. (1992). Biological activity of a synthetic pentasaccharide fragment of xyloglucan. *Plant Sci.* 85, 131–134.
59. Augur, C., Benhamou, N., Darvill, A., and Albersheim, P. (1993). Purification, characterization and cell wall localization of an α-fucosidase that inactivates a xyloglucan oligosaccharin. *Plant J.* 3, 415–426.
60. Zabotin, A.I., Barisheva, T.S., Larskaya, I.A., Toroshina, T.E., Trofimova, O.V., Hahn, M.G., and Zabotina, O.A. (2005). Oligosaccharin – a new systemic factor in the acquisition of freeze tolerance in winter plants. *Plant Biosyst.* 139, 36–41.
61. Zabotina, O.A., Ayupova, D.A., Toroshchina, T.E., and Zabotin, A.I. (2003). Involvement of oligosaccharides in adaptation of winter wheat seedlings to sub-zero temperatures. *Biol. Bull.* 30, 464–467.
62. Auxtova, O., Liskova, D., Kakoniova, D., Kubackova, M., Karacsonyi, S., and Bilisics, L. (1995). Effect of galactoglucomannan-derived oligosaccharides on elongation growth of pea and spruce stem segments stimulated by auxin. *Planta* 196, 420–424.
63. Kollarova, K., Liskova, D., and Capek, P. (2006). Further biological characteristics of galactoglucomannan oligosaccharides. *Biol. Plant.* 50, 232–238.
64. Kollarova, K., Liskova, D., and Lux, A. (2007). Influence of galactoglucomannan oligosaccharides on root culture of *Karwinskia humboldtiana*. *Plant Cell Tissue Organ Cult.* 91, 9–19.
65. VarelaNieto, I., Leon, Y., and Caro, H.N. (1996). Cell signalling by inositol phosphoglycans from different species. *Comp. Biochem. Physiol. B.* 115, 223–241.
66. Ferguson, M.A.J. and Williams, A.F. (1988). Cell-surface anchoring of proteins via glycosyl-phosphatidylinositol structures. *Annu. Rev. Biochem.* 57, 285–320.
67. Smith, C.K., Hewage, C.M., Fry, S.C., and Sadler, I.H. (1999). α-D-Mannopyranosyl-(1,4)-α-D-glucuronopyranosyl-(1,2)-*myo*-inositol, a new and unusual oligosaccharide from cultured rose cells. *Phytochemistry* 52, 387–396.
68. Smith, C.K. and Fry, S.C. (1999). Biosynthetic origin and longevity *in vivo* of α-D-mannopyranosyl-(1,4)-α-D-glucuronopyranosyl-(1,2)-myo-inositol, an unusual extracellular oligosaccharide produced by cultured rose cells. *Planta* 210, 150–156.
69. Dumville, J.C. and Fry, S.C. (2003a, b). Gentiobiose: a novel oligosaccharin in ripening tomato fruit. *Planta* 216, 484–495; Planta 217, 346–348.
70. Helenius, A. and Aebi, M. (2001). Intracellular functions of *N*-linked glycans. *Science* 291, 2364–2369.
71. Durrant, C. and Moore, S.E.H. (2002). Perturbation of free oligosaccharide trafficking in endoplasmic reticulum glucosidase I-deficient and castanospermine-treated cells. *Biochem. J.* 365, 239–247.
72. Alonzi, D.S., Neville, D.C.A., Lachmann, R.H., Dwek, R.A., and Butters, T.D. (2008). Glucosylated free oligosaccharides are biomarkers of endoplasmic reticulum α-glucosidase inhibition. *Biochem. J.* 409, 571–580.
73. Wilson, I.B.H. (2002). Glycosylation of proteins in plants and invertebrates. *Curr. Opin. Struct. Biol.* 12, 569–577.
74. Megumi, M. and Yoshinobu, K. (2005). *N*-Glycan metabolism and plant cell differentiation and growth. *Trends Glycosci. Glycotechnol.* 17, 205–214.
75. Yunovitz, H. and Gross, K. C. (1994). Effect of tunicamycin on metabolism of unconjugated *N*-glycans in relation to regulation of tomato fruit ripening. *Phytochemistry* 37, 663–668.
76. Yunovitz, H., Livsey, J.N., and Gross, K.C. (1996). Unconjugated $Man_5GlcNAc$ occurs in vegetative tissues of tomato. *Phytochemistry* 42, 607–610.

Biological Activity of Allelochemicals

17

Franck E. Dayan and Stephen O. Duke

Abstract All plants produce compounds that are phytotoxic to another plant species at some concentration. In some cases, these compounds function, at least in part, in plant/plant interactions, where a phytotoxin donor plant adversely affects a target plant, resulting in an advantage for the donor plant. This review discusses how such an allelochemical role of a phytotoxin can be proven and provides examples of some of the more studied phytochemicals that have been implicated in allelopathy. These include artemisinin, cineoles, β-triketones, catechin, sorgoleone, juglone and related quinones, rice allelochemicals, benzoxazinoids, common phenolic acids, L-DOPA, and *m*-tyrosine. Mechanisms of avoiding autotoxicity in the donor species are also discussed.

Introduction

Most scientists consider allelopathy to be the phenomenon of chemical warfare between plants, although some have defined it much more broadly (e.g. [1]), in some cases encompassing all of chemical ecology. We prefer the more accepted definition and, for the purposes of this review, we define the word allelochemical as a phytotoxin produced by a plant that can negatively influence vegetation in the vicinity of the producing plant. Some toxins can be stimulatory at low doses, while being very toxic at higher doses. This phenomenon of hormesis (stimulatory effect of a low dose of a toxin) is common with allelochemicals (e.g. [2, 3]). But, compared to the more drastic negative effects, the sometimes stimulatory (not necessarily beneficial) effects are minor. Many phytotoxins from plants are referred to as allelochemicals with little evidence that this is a role of the compound in nature.

In this chapter we will discuss the biological activity of some of the more well-studied allelochemicals. Many other allelochemicals and putative allelochemicals have not been included because of space limitations. Where possible, we will discuss the molecular target sites affected by these compounds.

Proof of Allelopathy

Proof of allelopathy should require proof of production of a phytotoxin or phytotoxins by a donor species, movement of that compound or compounds to a target plant at concentrations proven to cause harm to the target species and

F.E. Dayan and S.O. Duke (✉)
Natural Products Utilization Research Unit,
Agricultural Research Service,
United States Department of Agriculture,
P. O. Box 8048, University, MS 38677, USA
e-mails: fdayan@olemiss.edu; sduke@olemiss.edu

A.E. Osbourn and V. Lanzotti (eds.), *Plant-derived Natural Products*,
DOI 10.1007/978-0-387-85498-4_17, © Springer Science+Business Media, LLC 2009

measurement of a harmful effect to the target plant caused by the compound(s) in the environment containing both the donor and target plants. Furthermore, allelopathy can be due to release of phytotoxins from living plants or to release of phytotoxins from decaying plant material (e.g. [4]). In the latter case, there may be no evolutionary advantage to the producing species if it is an annual. An additional complication is that the compound produced by the plant might be converted to a more phytotoxic compound by soil microbes. This aspect has seldom been examined. The interactions of compounds in soil (e.g. antagonism or synergism or other effects by direct interactions with soil components or metabolic conversion) are potentially complex (e.g. [5]). To prove allelopathy in a natural setting is daunting.

The history of allelopathy research is fraught with controversy over the validity of allelopathy claims. For example, the high profile paper in which Müller et al. [6] claimed vegetation patterning around aromatic shrubs was due to release of phytotoxic volatile compounds was later strongly disputed by Bartholomew [7], who explained the phenomenon by animal activity around the shrubs. Such controversies persist (see the catechin section below), but solid evidence for an allelochemical role of some phytochemicals is growing. Perhaps the most thoroughly studied cases of allelopathy, in terms of biosynthesis, release to the environment, soil interactions, and uptake by target plants is that of the benzoxanoids [8], yet there is still much to learn, even for this system.

The question of how one proves allelopathy has been somewhat controversial because so few published studies even come close to considering all of the parameters mentioned above. The literature of allelopathy has many papers that define phytotoxic phytochemicals as allelochemicals. Most of these papers do not consider the activity of the compound(s) in soil or the actual concentration or availability of the compound that one might expect in soil in a field situation. The situation is further confounded by the lack of information on the half-life of most allelochemicals in soil and the kinetics of the putative allelochemical pool in soil.

Blum et al. [9] reviewed and discussed the intricacies of soil and soil biota interaction with natural phytotoxins. For example, the soil solution concentration of a compound with low water solubility is not useful in estimating its potential activity. Even with relatively water soluble phenolic compounds, Blum et al. [10] found the proportion of the compound in soil solution to be a small fraction of that bound to soil particles. Many soil-applied, synthetic herbicides (e.g. dinitroanilines such as trifluralin) are very lipophilic in part to minimize leaching out of the root zone of weeds [11]. When applied to soil, the aqueous concentration of these synthetic herbicides is very low, but the herbicidal activity is high. This is because of rapid flux of the compound in soil water as the compound moves from adsorption to soil particles, to soil water, to lipophilic domains of plant cells. One might expect the best allelochemicals to also be lipophilic for the same reason. The continuous movement of new allelochemicals from the producing plant species increases the complexity of the dynamics of an allelochemical in soil. Unlike the soil-applied herbicide, for which the concentration is greatest immediately after application, the allelochemical pool is continuously replenished, albeit at a changing rate as the physiological status, age, and number of producing plants change. Thus, the dose received by the target species is sustained over a relatively long time period, which at low doses at any point in time may result in chronic symptoms, rather than acute toxicity as farmers expect with herbicides. But, a slow accumulation of such symptoms in competing vegetation may be quite favorable to the allelopathic donor species. Cheng [12] discusses many of the problems mentioned above in a systems approach analysis of allelopathic processes.

Allelopathy may be invoked with confidence when soil concentrations of an allelochemical are at or above levels applied to soil that cause phytotoxicity. When concentrations are below such levels, allelopathy may still be taking place due to the continuous input of the compounds by the producing plant. This would be very difficult to imitate by exogenous application of the compounds. Furthermore, some have invoked synergism between natural phytotoxins to support the view that weakly phytotoxic phytochemicals are allelochemicals (e.g. [13]), but others have disputed such claims (e.g. [14, 15]). This topic is discussed in more detail in the Section on Common Phenolic Acids.

So, unless a putative allelochemical is clearly at phytotoxic levels in the soil, its role in allelopathy is questionable. Despite this, the literature is full of papers using the term allelochemical for compounds for which there is very little support for such a role.

A more definitive approach to this question is to link the symptoms of the target plant to those caused by the putative allelochemical, just as environmental toxicologists do to prove that phytotoxicity symptoms are due to exposure to a particular herbicide or herbicide class. For example, if one suspects phytotoxicity might be due to exposure to the widely used herbicide glyphosate, elevated levels of shikimic acid in the plant would confirm the hypothesis, as no other herbicide is known to have such a physiological effect [16]. Unfortunately, the molecular target site of most allelochemicals is unknown or unclear, and some of them apparently have more than one target site (e.g. sorgoleone and juglone, Sections on Sorgoleones and Juglone and Related Quinones).

Nevertheless, the molecular target sites of natural phytotoxins, including allelochemicals, are being discovered at an increasing rate (e.g. [17, 18]), so the approach of linking an allelochemical to a particular physiological, transcriptional, or translational response is a reasonable one; however, we are unaware of such an approach being used in a plant/plant interaction system.

Allelochemical Examples

Artemisinin

Plant sesquiterpenes comprise a large class of natural products with a number of biological activities, including some with interesting herbicidal and allelopathic potentials [19]. Artemisinin (Fig. 1) and its structural analog arteether (Fig. 1) are sesquiterpene endoperoxide lactones with strong phytotoxic activity isolated from the shoots of *Artemisia annua* [20, 21].

Artemisinin affects root growth and chlorophyll content, and is accompanied with a reduction of mitosis [22]. Abnormal multipolar configurations indicate that this sesquiterpene disrupts the formation of microtubule-organizing centers. However, it is not likely that inhibition of mitosis is the sole mechanism of action of sesquiterpene endoperoxide lactones.

While sesquiterpene lactones may cause strong herbicidal activity in the laboratory, these compounds perform poorly in the field. The herbicidal potential of artemisinin has been tested in field experiments. However, weed control was not satisfactory even when applied at rates greater than 1 kg/ha (Dr. Hans-Peter Fischer, personal communication, 1998).

Other studies incorporating dried leaves of *Artemisia* spp. in soil provided good weed control via inhibition of seed germination [23–25]. However, in the case of *A. annua*, the level of herbicidal activity was independent of the concentration of artemisinin in soil, suggesting that other factors may play important roles [23]. This may be due to the fact that *Artemisia* species produce a large number of phytotoxic terpenes that may act together to cause a herbicidal response [26].

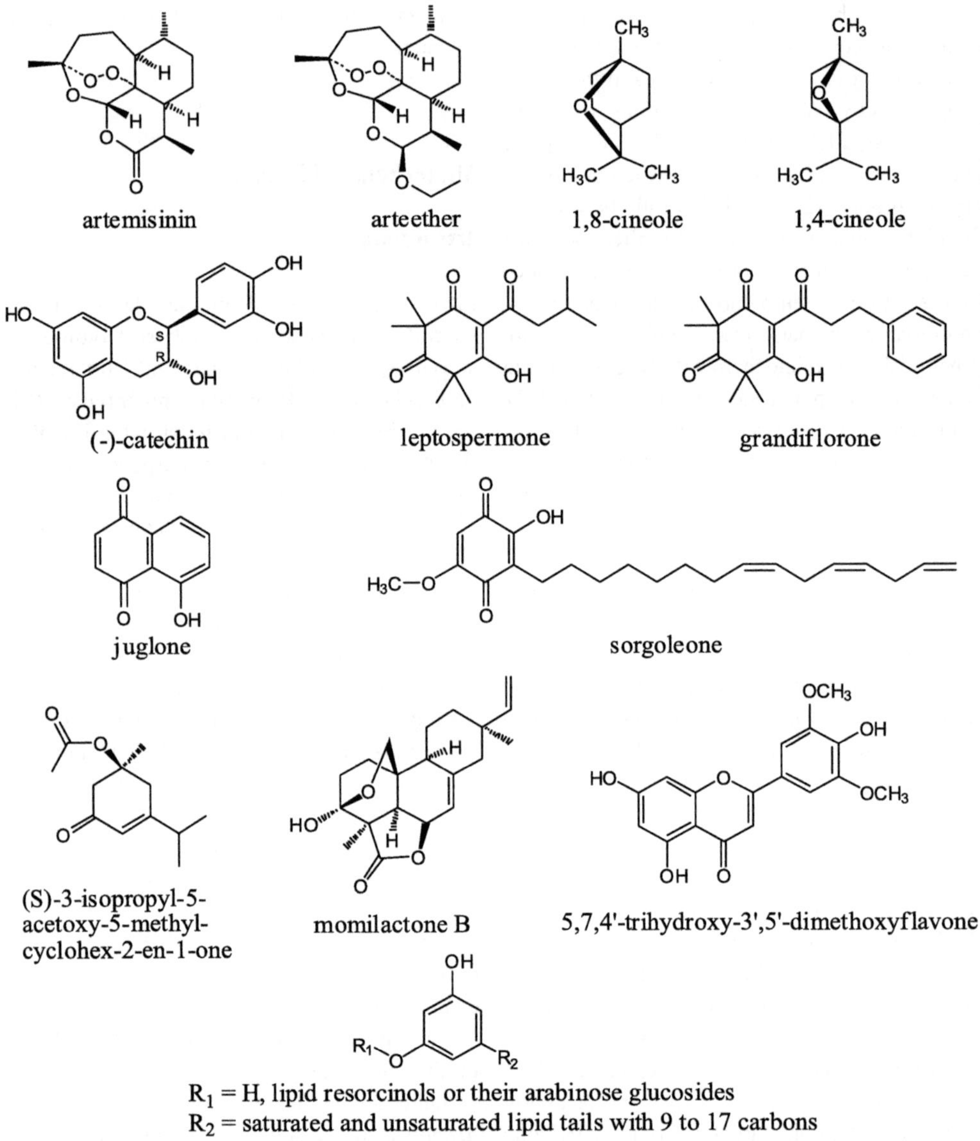

Fig. 1 Chemical structures of the natural phytotoxins mentioned in the text from Sections Artemisinin to Rice Allelochemicals

Several studies have aimed at identifying the molecular target site of this compound as well as the structural requirements for herbicidal activity [22, 27–29]. However, the target site remains to be identified.

Artemisinin is an excellent drug against ampicomplexan parasites, including *Plasmodium* spp.. These protozoan microbes contain a plastid, the apicoplast, that encodes many of the same genes of the plant plastid (reviewed by [30]), which could increase the chances that its mode of action as a pharmaceutical and that as a phytotoxin are related. Although the search for a target site for this compound as a pharmaceutical has been extensive, its mode of action remains unclear. Papers have been published associating its activity with reactive oxygen species and radical production after interaction with heme or iron [31–33] and disruption of calcium homeostasis by inhibition of sarcoplasmic-endoplasmic reticulum calcium ATPase [34, 35]. We are aware of no studies to link this information to its possible mode of action in plants.

On the other hand, studies on the structural requirements for herbicidal activity have shown that the presence of the endoperoxide bridge is important for activity [22, 27–29].

Cineoles

Cineoles are monoterpenes present in the essential oils of many aromatic plants (e.g. *Laurus nobilis* L., *Salvia* spp., *Eucalyptus* spp., *Xanthoxylum rhetsa* D. C., and *Artemisia* spp). Many classes of volatile monoterpenes inhibit plant growth (e.g. [36]). Growth inhibition of grasses by volatile monoterpenes released by *Salvia leucophylla* is most effective during seedling development and establishment [6]. 1,8-Cineole (Fig. 1) is a major constituent of these plant essential oils and is one of the most potent allelochemicals released by *Artemisia* spp. [37]. However, the potential allelopathic role of this monoterpene is not clearly established. Other factors such as competitiveness, reduced seed predation, and/or drought tolerance may be more important in the survival of the species [37].

While 1,8-cineole is the most abundant phytotoxic cineole in these species, 1,4-cineole (Fig. 1) is often present in much lower concentrations. This monoterpene analog is also phytotoxic [38] and may either contribute to or actually cause many of the reported symptoms. Several studies have suggested that cineoles, are potent inhibitors of mitosis [38]. Analysis of mitotic index shows that both 1,4-cineole and 1,8-cineole inhibited prophase of mitosis [39]. However, 1,4-cineole did not inhibit any other mitotic stage when compared to controls. Conversely, 1,8-cineole inhibited all stages of mitosis. Cells treated with 1,8-cineole were primarily in interphase with only a few undergoing any stages of mitosis. While both compounds cause decreased root growth, 1,4-cineole also inhibited shoot growth, and caused much greater photosynthetic stress on the plant than does 1,8-cineole. Conversely, 1,8-cineole decreased germination and severely inhibited mitosis. Because these compounds often occur together in plant species, it is conceivable that they may both contribute to any allelopathic properties associated with the producing plants. The phytotoxicity of cineoles and several other monoterpenes appears to be associated with the presence of their epoxide ring [40]. However, in spite of the strong structural similarity between 1,8- and 1,4-cineole, differences in the position of the oxygen bridge affect the overall conformation of the structures to give them distinct modes of action.

The herbicide cinmethylin has been reported to be derived from the monoterpene backbone of the cineole class of natural products. Indeed, cinmethylin is identical to 1,4-cineole, with the addition of a benzyl ether moiety that was added to lower the volatility of the natural product. Cinmethylin was in fact developed by Shell Chemical during a biorational synthetic program to discover new insecticides that targeted the glycerol-3-phosphate shuttle. A phytotoxic dioxalane intermediate was identified. Optimization of the monocyclic structures led to rigid bicyclic structures with improved herbicidal activity. Some of the bicyclic structures possessed the basic backbone of the monoterpene cineoles found in nature.

17

β-Triketones

The essential oils of several plant species (e.g. *Leptospermum*, *Eucalyptus*, *and Callistemon* spp.) contain relatively large amounts of natural *β*-triketones (e.g. leptospermone, flavesone, agglomerone, tasmonone and grandiflorone, see Fig. 1) [41, 42]. Some of these β-triketones are very phytotoxic and may be involved in allelopathy, as can be observed by the lack of vegetation surrounding stands of the bottlebrush plant (*Callistemon* spp.). Seedlings growing near this species are stunted and their foliage lacks chlorophylls and carotenoids. This observation has led the agrochemical companies to develop and commercialize structural analogues of leptospermone as herbicides [43].

Synthetic β-triketones inhibit *p*-hydroxyphenylpyruvate dioxygenase (HPPD), a key enzyme in the biosynthesis of prenylquinones and tocopherols. Therefore, the phytotoxicity and mode of action of the β-triketone-rich essential oil of manuka (*Leptospermum scoparium*), and its components leptospermone, grandiflorone and flavesone, were recently tested [44]. Plants exposed to manuka oil or its purified β-triketone components were bleached and had the same phenotype as those exposed to sulcotrione, a synthetic HPPD inhibitor. A triketone-rich fraction and purified leptospermone were approximately ten times more active than that of the crude manuka oil suggesting that the phytotoxic activity of the oil is due to the presence of the triketones. The effect of these samples caused similar reduction in chlorophyll and carotenoid levels.

Dose-response curves against the enzyme activity of HPPD demonstrated that grandiflorone was the most active compound in the essential oil (Fig. 2). While leptospermone was less active than grandiflorone, it is by far the most abundant β-triketone in the oil. Therefore, most of the activity against HPPD can be attributed to leptospermone. Flavesone was not active. Analysis of this simple structure-activity relationship suggests that the size and lipophilicity of the side chain affected the potency of the compounds. Computational analysis of the catalytic domain of HPPD determined that the catalytic site of HPPD includes a lipophilic domain proximate from the Fe^{2+} that increases the binding of lipophilic ligands.

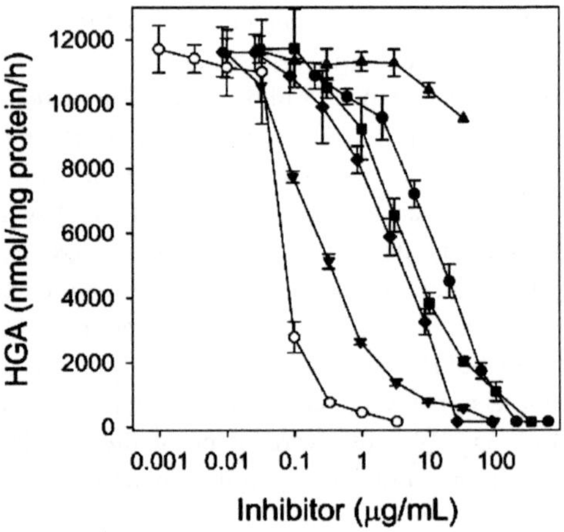

Fig. 2 Inhibition of *p*-hydroxyphenylpyruvate dioxygenase by manuka oil (●), the *β*-triketone-rich fraction (■), and its individual *β*-triketone components: leptospermone (♦), flavesone (▲), and grandiflorone (▼). The synthetic herbicide sulcotrione (○) was added as positive control. Each data point represents the mean of two independent experiment ± 1 SD (N = 6) (Figure from [44]. With permission)

Therefore, the high activity of grandiflorone, relative to leptospermone, is due to the greater lipophilicity of its side chain. The affinity of leptospermone is not as great because its side chain is less lipophilic, and flavesone does not inhibit at all because it does not have a lipophilic side chain. This is in agreement with other QSAR studies on synthetic HPPD inhibitors which show that two distinct binding interactions, bidentate coordination and π–stacking, are involved in high affinity for HPPD [45].

Catechins

Allelopathic processes via (–)-catechin (2S,3R-enantiomer) (Fig. 1) have been invoked to partially explain the success of *Centaurea maculosa* (spotted knapweed) in North America [46–48]. Through bioassays, root exuded (–)-catechin was identified as the putative allelochemical of spotted knapweed, and multiple studies have examined various facets of the role of this compound as an allelochemical [46–51]. The 2R,3S-(+)-catechin enantiomer is also produced by the plant, but was reported essentially to be non-phytotoxic. This work supported the 'novel weapons hypothesis,' which states that the success of some exotic invasive plant species may be due to the production of allelochemicals that native species have never encountered and, thus, to which they have not evolved defenses [52].

Blair et al. [53, 54] questioned the findings of (–)-catechin as the allelochemical responsible for the successful invasion of spotted knapweed. They found that the methods described in these previous papers did not extract catechin from liquid media or soil. With appropriate methods, they quantified levels of production up to two orders of magnitude less than previously reported. A previously reported catechin-sensitive grass species *Festuca idahoensis* (Idaho fescue) was only slightly affected by catechin concentrations ten-fold higher than that previously reported to cause 100% mortality. Lastly, no measurable catechin of any type could be detected in soil samples from two spotted knapweed sites in Montana, USA. The last issue was countered with the argument that catechins can only be found at certain times in these soils, and that Blair et al. [53] had sampled at the wrong times of the year [55]. While Blair et al. [53] did not detect catechins in field-collected soils, it was noted that the soil samples were collected in October, 2004, when plants are less likely to be actively producing the root exudate. Later, Perry et al. [56] found little or no catechins in 402 soil samples from eleven *C. maculosa* sites. These authors were less firm in their hypothesis that (–)-catechin plays a role in allelopathy. In fact, molecules such as catechins, catechol, and L-DOPA are quite unstable in soil, especially at high pH [57]. In a more recent paper [58] those who previously strongly supported an allelopathic role of (–)-catechin indicate that because of the instability of catechin in soil and its ability to form complexes in soil, it is extremely difficult to conclusively attribute allelopathic significance to catechins without improved methods. Tharayil et al. [5] found the presence of phenolic acids in soil to increase the half life of (±)-catechin (50:50 racemic mixture). But, they also found (±)-catechin to polymerize in soil to form a procyanidin dimer that has a short soil half life. They found (±)-catechin to be essentially non-phytotoxic in their bioassays with both monocots and dicots, even though they observed the same root browning that Bais et al. [46] attributed to necrosis. Tharayil et al. [5] attributed the root browning to enzyme-mediated polymerization of catechin to brown polymers, rather than to necrosis.

Errata were published to the papers by Bais et al. [47] and Veluri et al. [49] in which the claims of differential phytotoxicity of (–) and (+) catechin and the degree of phytotoxicity of (–)-catechin were substantially reduced. The same laboratory [59] then reported that low doses of catechins that might at times be present in soil can stimulate growth of *Arabidopsis thaliana*. They attributed this effect to induction of low levels of reactive oxygen species (ROS). They also reported that these doses induced pathogen resistance, an effect observed with low doses of other chemical inducers of ROS [60]. Most plants are not very sensitive to (–)-catechin, and some may be even more resistant by quenching of ROS with oxalate according to Weir et al. [61].

17

In one of the original papers of the Vivanco laboratory [46], the mechanism of action of (–)-catechin was linked to induction of ROS. This is debatable since catechins are well known antioxidants (e.g. [62]). In fact, more than 1500 papers mention catechin as an antioxidant (SciFinder® search, as of February 2008). Finally, Qin et al. [63] found no evidence of root exudate-mediated allelopathy with *Centaurea solstitialis*, but the authors apparently did not determine if this species produces (±)-catechin. If it were to produce similar amounts of (±)-catechin as *C. maculosa*, it would be strong evidence of the lack of allelopathic influence of these compounds.

Although many compounds can be shown to be allelopathic in the laboratory or greenhouse, it is ultimately necessary to demonstrate the presence of the compound under natural field conditions in concentrations high enough to have an impact on neighboring plants [12, 64]. In our studies, we find essentially no phytotoxicity of either enantiomeric form of catechin in soil (unpublished data), however, in solution catechins can be phytotoxic to plants, including algae [65]. In summary, catechin is not likely to play a significant role in allelopathy of any terrestrial species, nor do we think that catechin-mediated phytotoxicity can be invoked to support the novel weapons hypothesis. We include a discussion of this compound because of the very high profile that this work has been given and to point out that care must be taken in invoking allelopathic properties of phytochemicals.

Sorgoleones

The allelopathic property of sorghum (*Sorghum bicolor*) has been observed for many years by its negative effect on the growth of other crops grown in rotation (e.g. [66]). This weed suppressing potential can be exploited by planting sorghum as a green manure or as a cover crop [67]. Certain species of *Sorghum* (e.g. sudan grass, *S. sudanese*) can easily be grown as weed-free monocultures without additional herbicide application. Other *Sorghum* species, such as the noxious weed johnsongrass (*S. halepense*) also have allelopathic properties that may be associated with the release of such specialized phenolic lipids [68–69].

Sorghum releases a substantial amount of phenolic lipids including sorgoleone, a 3-pentadecatriene benzoquinone that is the primary source of the allelopathy properties of sorghums (Fig. 1) [70].

The fate of sorgoleone in soil is not well characterized, but the dynamics of its degradation in soil has been studied [71]. Sorgoleone inhibits growth of many weeds (e.g. [70, 72–74]) and acts as a strong inhibitor of PSII *in vitro* [73, 75, 76]. Sorgoleone also inhibits mitochondrial functions [77], inhibits *p*-hydroxyphenylpyruvate dioxygenase (HPPD) [78], and interferes with root H + -ATPase and water uptake [79]. While the primary mechanism of action of sorgoleone in allelopathy remains to be determined, having multiple potential modes of action is desirable from the standpoint of slowing evolution of resistance in target species.

Sorghum accumulates sorgoleone and its analogues in mature root hairs [80–82] and the production (approximately 18 mg/g of root dry weight) is optimum at temperatures ranging from 25°C to 35°C. The biosynthesis of these molecules appears to be stimulated by the presence of root exudates of other plants, suggesting that the allelopathic potential of sorghum may be enhanced in the presence of competitors [81].

Sorgoleone is secreted only from root hairs in droplets in which about 90% of the material is sorgoleone and its 1,4-hydroquinone form. The droplets also contain several minor congeners varying in the substitutions in the aromatic ring, and/or in number of carbon and the level of unsaturation in the side chain [73, 83, 84]. All of these intermediates are derived from the same biosynthetic pathway and contribute to the overall allelopathic potential of sorghum [84].

Labeling studies showed that biosynthesis of sorgoleone involves the convergence of the fatty acid and polyketide pathways [83, 85]. The hydrophobic side chain is derived from a 16:3 fatty acid intermediate synthesized by the combined action of fatty acid synthase and desaturases. The ring is derived from action of a polyketide synthase that produces 5-pentadecatriene resorcinol as an intermediate identified in small amounts in root extracts. This lipid resorcinol intermediate is then methylated by a S-adenosylmethionine-dependent O-methyltransferase and dihydroxylated by a P450 monooxygenase to yield the reduced form of sorgoleone [85]. Every step of the pathway has been tested in isolated root hair preparation [86]. The genes for this pathway have been tentatively identified and the substrate specificity of the enzymes that they encode has been verified for the O-methyltransferase and the desaturases [87–90].

Juglone and Related Quinones

Juglone (Fig. 1) is an allelochemical present in walnut (*Juglans* spp). It represses the growth of many weed species, including wild mustard (*Sinapis arvensis* L.), creeping thistle (*Cirsium arvense* L.), field poppy (*Papaver rhoeas* L.), and henbit (*Lamium amplexicaule* L.) [91]. The phytotoxicity of juglone is more pronounced on the elongation of the shoot than on the development of roots [91, 92]. When assayed by bioautographic thin layer chromatography, juglone and its structurally related quinone plumbagin completely inhibited the growth of lettuce (*Lactuca sativa*) and *Arabidopsis thaliana* seeds at micromolar concentrations [93]. Juglone is also a very potent phytotoxin on plants grown hydroponically. Overall, soybean (*Glycine max* L. Merr.) is more sensitive to juglone than maize (*Zea mays* L.), but both crops are negatively affected, suggesting that intercropping these species within alleys in a walnut (*J. nigra* L.) grove may not be feasible [94, 95].

Growth and physiological processes (leaf photosynthesis, transpiration, stomatal conductance, and leaf and root respiration) of maize and soybean seedlings are strongly inhibited by micromolar concentrations of juglone [96]. It inhibits several physiological and metabolic processes, which suggests that it may have more than one molecular target site in plants. Several quinones, including juglone, inhibit the enzyme *p*-hydroxyphenylpyruvate dioxygenase, the target site of triketones herbicides. However, unlike compounds with a triketone backbone, these natural quinones did not bind tightly to this enzyme [78].

Juglone also affects uptake of molecular oxygen by excised roots of maize and inhibits state III oxidation rate in mitochondrial respiration [97]. The inhibition of state III O_2 uptake (ADP-dependent respiration) by purified mitochondria preparations is commensurate with the inhibition observed *in vivo*. Exposing soybean seedlings to juglone also causes changes in the activities of root peroxidases and the overall lignification of the roots [98].

Studies on the effect of juglone on the growth and photosynthetic capacity of *Lemna minor* L. (duckweed) suggest that this naphthoquinone disrupts chloroplast and mitochondrial functions [96], which is fairly common for compounds from the quinone family (such as sorgoleone) [73, 84]. More recently, it was reported that juglone also shared sorgoleone's ability to disrupt water uptake and acid efflux by interfering with root plasma membrane function. In particular, juglone inhibited H^+-ATPase activity, which would consequently cause many of the physiological effects observed to date [99].

Biochemical and molecular information on the juglone biosynthetic pathway is lacking. While many labeled precursors, such as glucose, acetate and shikimic acid, were successfully incorporated in juglone, *o*-succinylbenzoic acid appears to be the immediate precursor of the naphthoquinone [100], which is a key step in the formation of more oxidized naphthoquinones [101].

Juglone accumulates to phytotoxic levels in the soil of walnut groves. Though this compound can be degraded in soil via bacterial and abiotic means, the rate of degradation is not sufficient to completely detoxify the soil [102, 103]. While Jose and Gillespie [94,95] suggested that the soil within alleys in a walnut grove may contain too much juglone to grow soybean and corn as intercrops, another analysis of the release, accumulation, and degradation of juglone in soils from a black walnut orchard showed that levels of the allelochemical in the soil water reach toxic levels for sensitive species, but may not be sufficiently high to affect the growth of maize and soybeans [103].

However, one must be careful not to equate soil degradation with detoxification. For instance, the soil degradation of arbutin, a hydroquinone glycoside produced by *Polygonella myriophylla* (Small's jointweed), yields the free hydroquinone aglycon which is further metabolized into the benzoquinone. The benzoquinone is ultimately degraded via nonmicrobial oxidative processes. The phytotoxicity and soil persistence of each of the degradation steps must be considered when evaluating the overall allelopathic potential of a compound released by the donor plant. Therefore, metabolic and chemical transformations of allelochemicals occurring in non-sterile soils may play important roles in mediating the allelopathic effects of these compounds [104] (also discussed in Section Benzoxazinoids).

Rice Allelochemicals

Many cultivars and ancestral lines of rice have significant allelopathic potential in the field, and several scientific programs to develop commercial allelopathic rice varieties are on-going [105–107]. Up to 4% of rice varieties can suppress important paddy field weeds such as *Echinochloa crus-galli* (barnyardgrass) (e.g. [106, 108]) and *Cyperus difformis* (variable flatsedge) [109]. While the level of weed management is not as good as that obtained with conventional herbicide treatments, allelopathic rice varieties can be used to reduce the herbicide application rates in paddy fields (Fig. 3) [110].

Fig. 3 Field study at the Dale Bumpers National Rice Research Center in Stuttgart, Arkansas, showing the effect of allelopathic rice (*right*) on the growth of barnyardgrass compared to a non-allelopathic rice (*left*). Photographs by F. E. Dayan

A multinational study conducted with more than 100 rice cultivars showed that the allelopathic rice cultivars behaved similarly in different geographical locations, indicating that the traits were more influenced by genetics than environmental conditions. This allelopathic trait appears to be inherited [111]. A study on broad-sense heritability of the ability of rice to reduce the root growth of barnyardgrass produced a value of 0.85, which is above the threshold required by breeders to screen for a trait. Quantitative Trait Loci (QTL) analysis of allelopathic cultivars indicates that a third of the total variation for the allelopathic phenotype in a large population of inbred rice lines was accounted for by four QTLs located on three separate chromosomes [108].

Suppression of weed growth by some rice varieties is often associated with their tillering characteristics and increase root biomass, which complicates the differentiation of allelopathic effects from competition for resources or superior vigor. In the case of rice, some of the most allelopathic cultivars indeed have larger root masses than non-allelopathic cultivars, but none of the QTLs associated with allelopathy were linked to QTLs associated with root morphology.

The relatively high occurrence of allelopathy in rice (about 4%) is not common in commercial crops. This may be due to the fact that the genetic background of most post-green revolution rice cultivars includes traits of the highly allelopathic cultivar Taichung Native 1 [112]. Since the allelopathic trait is inheritable, many of these cultivars may possess some of the allelopathic character of Taichung Native 1, or that of other allelopathic lines in their background.

A number of phytotoxic compounds have been identified in rice and there is still some ambiguity as to the primary compounds responsible for the allelopathic potential of rice. However, the following compounds are examples of the phytotoxin classes found in root exudates of allelopathic rice varieties: momilactone B, lipid resorcinols, flavones, benzoxazinoids, and their respective glycosides, and a cyclohexenone [107, 109, 113, 114] (Fig. 1). More than one type of phytotoxin is likely to play a role in combating weeds in the most allelopathic varieties of rice.

Most of these compounds have additional biological activities involved in plant defense responses. For example, momilactone and the structurally related oryzalexins and phytocassanes have well characterized antimicrobial activities, and their synthesis is stimulated upon infection with pathogenic fungi [115]. Similarly, cyclohexenone and flavone both inhibit the germination of common rice fungal pathogens (e.g. *Pyricularia oryzae* and *Rhizoctonia solani*). The 5-alkyl resorcinols have antibacterial and fungistatic properties and have been associated with plant defense mechanisms against fungal infections [116]. Furthermore, lipid resorcinols have molluscicidal activities, with their potency being inversely proportional to the level of unsaturation of the alkyl tail [116].

Additionally, the allelopathic potential of rice plants can be stimulated in the presence of other plants by increasing the production of allelochemicals that may help them repress the growth of competitors. The synthesis of some rice allelochemicals, such as flavone and cyclohexenone, can be stimulated in the presence of barnyardgrass nearby [117]. This has not been demonstrated for alkyl resorcinols; however, our interest in other lipid resorcinol-derived allelochemicals, such as those produced in sorghum [85], has led our research group to identify, clone and characterize the substrate specificity of the key enzymes involved in the synthesis of the rice lipid resorcinols (unpublished data).

Benzoxazinoids

Many grass species (e.g. wheat, rye, and maize) and a few non-Poaceae species produce a variety of benzoxazinoids known to be herbicidal,

17

fungicidal, antimicrobial, and insecticidal (reviewed by [8] and [118]). Numerous papers on their properties and biological activities, including their possible roles in allelopathy are available. As stated earlier, the roles of these compounds in allelopathy, including synthesis, and behavior in soil is perhaps better understood than any other group of allelochemicals. Extensive research on the benzoxazinoids by several European laboratories has recently been highly funded by the European Union [119].

Benzoxazinone glucosides are exuded from roots of the producing species into soil where the sugars are hydrolyzed, releasing the generally more active benzoxazolinone aglycones [120, 121] (Fig. 4). Many papers studied the activity of these compounds in the absence of soil (e.g. [87, 122]). Some of these (e.g. BOA and DIBOA, Fig. 4) have soil activity at high application rates [123], and there is a high correlation between root exudation of DIBOA and DIMBOA and allelopathic activity of different wheat varieties [124, 125].

Belz [8] provides an excellent summary of the source, soil conversions and half-lives, and phytotoxicity of most of the benzoxazinoid compounds (Fig. 4). Perhaps the most important recent finding is the discovery that APO, a degradation product of BOA, is a very phytotoxic and has a relatively long soil half-life, an indication that it may account for most of the phytotoxicity of benzoxazinoids (Fig. 4) [126].

The herbicidal mode of action of benzoxazinoids is unknown. Sánchez-Moreiras et al. [127] reviewed what was known of the mode of action of BOA up until about 2003. BOA can inhibit mitochondrial function by interfering with both electron transport and mitochondrial ATPase activity [128]. A correlation between inhibition of plasma membrane H^+-ATPase activity and inhibition of growth by BOA and DIBOA [129], suggests that their effects on nutrient uptake and electrolyte leakage could be caused by this primary effect [130]. BOA reduces the number of dividing meristematic cells [127], which could be an indirect effect of ATPase inhibition. DIMBOA stimulates its own degradation by horseradish peroxidase, whereas DIBOA and MBOA did not have such an effect [131]. MBOA (Fig. 4) could inhibit or slow seed germination by repressing the induction of α-amylase [132]; however, almost any germination inhibitor would have such an effect. We agree with the conclusions of Macías et al. [120] that there have been no definitive mode of action studies with this group of compounds, nor has there been evidence that the different phytotoxic members of this chemical family have a common mode of action.

Baerson et al. [133] examined whole genome transcriptome changes induced by sub-lethal concentrations of BOA on Arabidopsis. The transcriptional effects were too complex to provide a clear clue to the mode of action. However, the up-regulation of genes encoding detoxification enzymes was profound. Target plants can detoxify benzoxazinoids by both glucosylation and hydroxylation [118, 134]. Enzymatic products predicted by the transcriptome responses in Arabidopsis were identified (Table 1) [133]. Different detoxification rates might account for differential sensitivity between species.

Common Phenolic Acids

There are probably more papers claiming that common phenolic acids derived from the shikimate pathway (e.g. cinnamic acid and some of its derivatives such as ferulic and *p*-coumaric acids, see Fig. 5) than for any other class of compound. These compounds are ubiquitous to higher plants and are generally very weakly phytotoxic, making it difficult to view them as allelochemicals. One reason for the abundance of papers on the 'allelopathic' properties of simple phenolic acids is that these compounds are easily extracted and identified. In case after case, papers trying to correlate the activity of crude extracts with phenolic acid content were

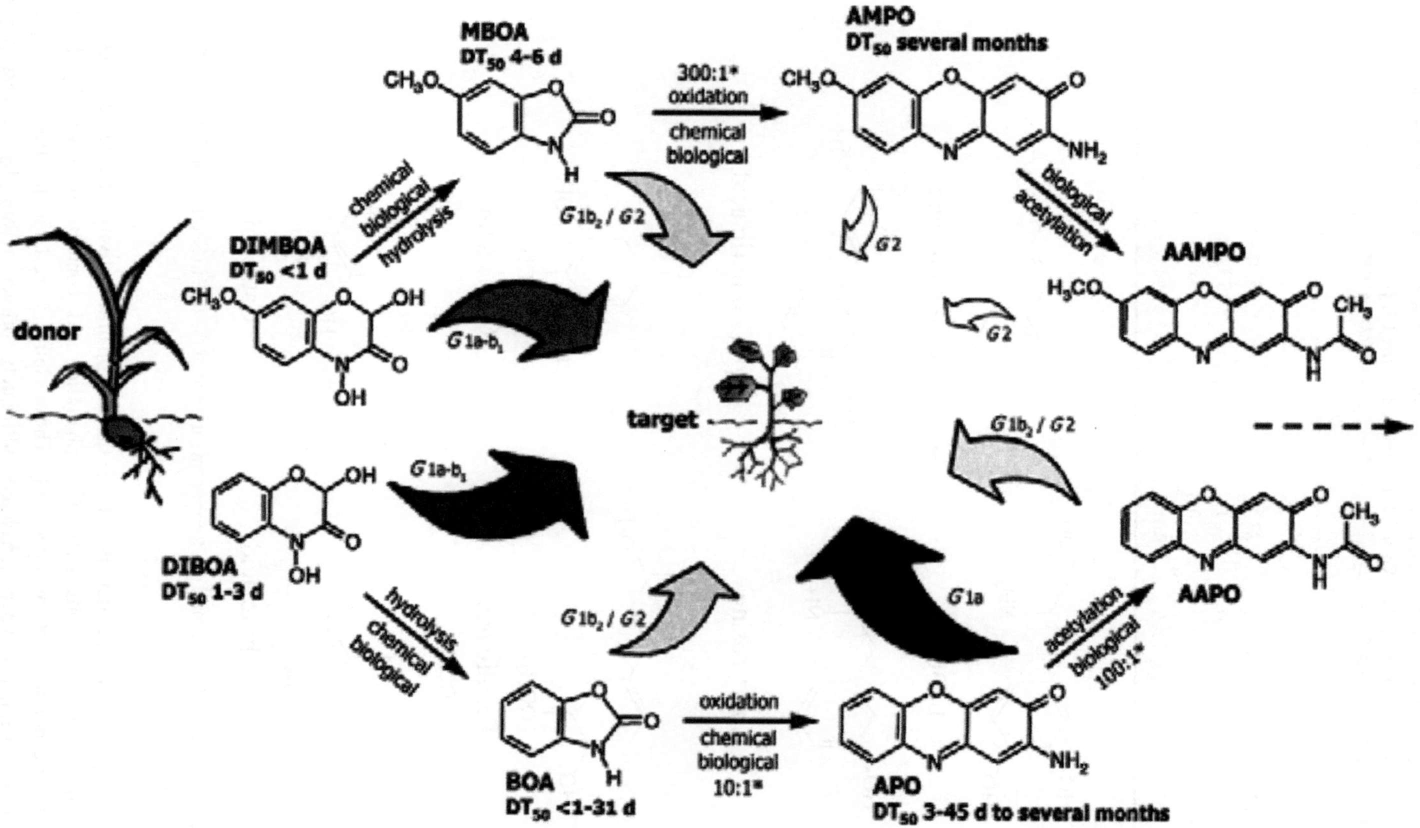

Fig. 4 Chemical structures of benzoxinoids mentioned in the text with their time for loss of 50% of the compounds (DT_{50} values), degradation and transformation pathways, and relative activity (the darker the arrow, the more phytotoxic). The G codes are activity classes according to [126, 120]. Figure from [8]. (Copyright by the Society of Chemical Industry. Reproduced with permission from Wiley on behalf of the SCI)

Table 1 Metabolites of BOA found in 3-week-old Arabidopsis plants exposed to different concentrations of BOA for 24 h (Adapted from [133])

	Metabolite (nmol/g fresh weight)		
[BOA]	BOA-6-OH	BOA-6-O-glucoside	Glucoside carbamate
10 μM	20.3 ± 4.0	23.8 ± 5.3	n.d.
100 μM	66.00 ± 10.02	121.5 ± 41.85	33.5 ± 32.8
250 μM	174.8 ± 58.0	249.3 ± 90	38.7 ± 20.0
500 μM	212.3 ± 35.6	381.3 ± 147.7	54.8 ± 10.0

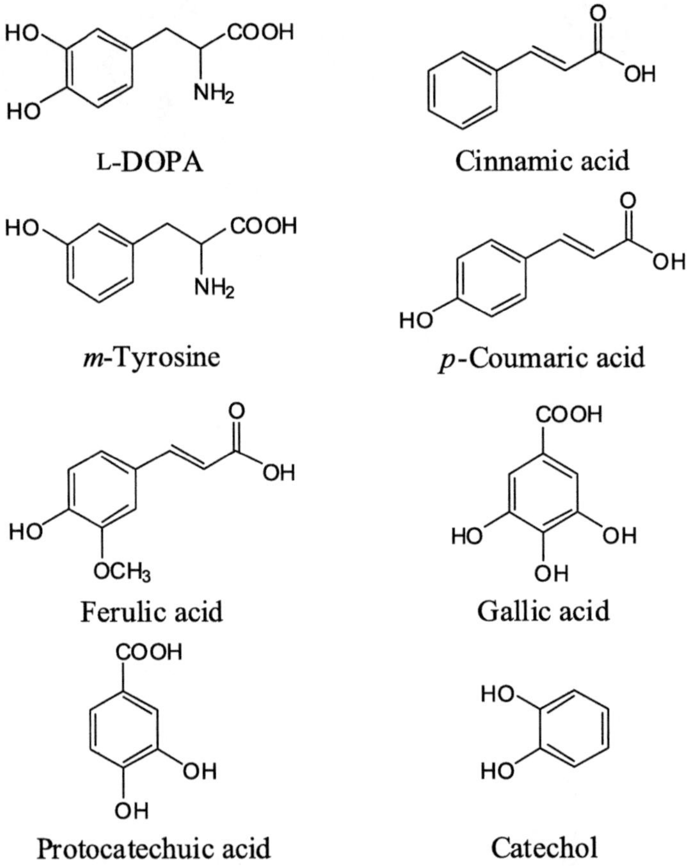

Fig. 5 Chemical structures of the natural phytotoxins mentioned in the text in Section on Common Phenolic Acids to *m*-Tyrosine

later debunked by papers showing that the activity was due almost entirely to more phytotoxic compounds that are not so easily isolated and identified. For example, many papers correlated the allelopathic activity of rice varieties with phenolic acid content (reviewed by [135]). These claims were later questioned [136] and ultimately proven wrong by several papers identifying much more likely allelochemicals (discussed in detail in Section Rice Allelochemicals). Unfortunately, bioassay-guided isolation, followed by structural identification of active phytotoxic compounds has been under-utilized in allelopathy studies.

Blum [137] concluded that the levels of individual phenolic acids in soils of apparently allelopathic species were insufficient for plant growth inhibition, but that mixtures of phenolic acids or of phenolic acids with other compounds could be inhibitory. As mentioned above, some have invoked synergism to support the view that these weak phytotoxins are allelochemicals (e.g. [13, 138–140]). However, when experiments are done properly, the phytotoxic activities of phenolic acids are most often antagonistic or, at best, additive [14, 15, 141–148]. Even though synergism might be observed in rare cases, depending on the growth parameter measured [142, 147], data from papers with good mathematical determinations of phytotoxin interactions indicate that one cannot use synergism as leverage to explain allelopathy in terms of simple phenolic acids. Furthermore, the uptake of common phenolic acids by plants is antagonized by combining the compounds [149], and increases in phenolic acid-utilizing soil bacteria by phenolic acid mixtures can further decrease their already low phytotoxic effect [150]. Lastly, in combination, these compounds compete for soil binding sites quite differently, each changing the availability to plants of the others [151].

Blum et al. [142] observed that effects of these simple phenolic compounds on cucumber leaf expansion were reversible, and that normal growth resumed upon removal of these compounds. Preconditioning plants with water stress or exposure to phenolic acids renders the plants less sensitive to the compounds [152]. The effects of these compounds are similar to those of water stress [14, 142, 153]. They can also inhibit ion uptake by plant roots [153]. There is some evidence that at least some of the effects take place at the root surface rather than within living cells [154]. This hypothesis is supported by the finding of Blum and Gerig [155] that the effects of exogenous phenolic acids are reduced as the compounds are taken up, reducing the concentration at the root surface. In summary, we doubt that common cinnamate-derived phenolic acids from the shikimate pathway alone or in combination play a significant role in allelopathy.

Benzoic acids such as gallic or protocatechuic acids (Fig. 5) are also not very phytotoxic, even though they have been invoked as allelochemicals. Gallic acid is highly unstable in soil, but there is evidence that gallic acid might be transformed by soil microbes into a more potent phytotoxin [104].

L-DOPA

The laboratory of Fujii [156] found L-3,4-dihydroxyphenylalanine (L-DOPA) (Fig. 5) to be the phytotoxin produced by velvetbean (*Mancuna pruriens*) that appeared to give green manure crops of this plant their ability to suppress other plant species. L-DOPA is not a particularly potent phytotoxin, but velvetbean produces it in high quantity. Comparing effects of exogenous L-DOPA and that produced by velvetleaf roots, Nishihara et al. [157] concluded that L-DOPA accounted for the allelopathic influence of velvetleaf on lettuce growth. Some species such as those in the Gramineae are apparently less sensitive to L-DOPA due to enzymatic detoxification by enzymes exuded by their roots [158]. Hachinohe et al. [159] reached similar conclusions, comparing resistant barnyardgrass with sensitive lettuce.

Fujii [156] hypothesized that the mechanism of action of L-DOPA might be inhibition of lipoxygenase. L-DOPA is structurally related to catechol (Fig. 5), which is a weak inhibitor of lipoxygenase. L-DOPA is an even weaker lipoxygenase inhibitor [156], so this hypothesis is

17

difficult to accept. However, more recent work by Hachinohe and Matsumoto [160, 161] found that reactive oxygen species generated during conversion of L-DOPA to melanin in the plant cell was responsible for L-DOPA phytotoxicity. Lipid peroxides and melanin increased in parallel in affected plant tissues. Phytotoxicity could be stopped by addition of the antioxidants ascorbate and α-tocopherol. They speculated that differences in polyphenol oxidase, an enzyme that can initiate conversion of L-DOPA to melanin, may be responsible for species differences in susceptibility to L-DOPA as a phytotoxin [161].

Soares et al. [162] found that L-DOPA increases lignification of soybean roots and that this effect correlated with root growth inhibition. It also increased the extractable levels of phenolic compounds and of phenylalanine ammonia lyase and peroxidase. No primary target site was indicated.

L-DOPA is quite unstable in soil, especially those with a high pH [57, 163]. The effect is more rapid in some soil types than others. Therefore, the role of L-DOPA as an allelochemical is questionable in some soil types without near root to root contact between donor and target species.

m-Tyrosine

The non-protein amino acid *m*-tyrosine (Fig. 5) is responsible for most, if not all, of the allelopathic activity of a *Festuca rubra* (red fescue) subspecies [164, 165]. The compound is exuded into the soil from roots of the fescue plant. While some non-protein amino acids have functions in plant primary metabolism (e.g. δ-aminolevulinic acid), others are thought to be involved in protection of plants from a variety of biotic threats, particularly herbivores. The modes of action of these compounds range from direct neurotoxicity, such as produced by β-*N*-methylamino-L-alanine [166], to incorporation into proteins to produce aberrant molecules, leading to multiple physiological problems (e.g. [167]).

Non-protein amino acids have previously been implicated in allelopathy. For example, mimosine has been associated with allelopathy of the legume tree *Leucaena leucocephala* [168]. L-DOPA, a compound structurally related to *m*-tyrosine (Fig. 5), has been implicated in allelopathy of *Mucuna pruriens* ([158] and Section Rice Allelochemicals). Roots of pea (*Pisum sativa*) exude *β*-(3-isoxazolin-5-on-2yl)-alanine which inhibits root growth on non-legume plant species [169], although this non-protein amino acid is much less phytotoxic than *m*-tyrosine.

Circumstantial evidence suggests that *m*-tyrosine substitutes in proteins for at least one protein amino acid (apparently phenylalanine) during translation, resulting in dysfunctional proteins [165]. Demonstration of significant loss of specific activity of phenylalanine-containing enzymes would support this hypothesis. An alternative hypothesis is that *m*-tyrosine is converted to L-DOPA, a known phytotoxin (see Section L-DOPA). Bertin et al. [165] state that this is unlikely because L-DOPA is significantly less phytotoxic than *m*-tyrosine. however, *m*-tyrosine might be taken up more readily by plant cells than L-DOPA, leaving conversion to L-DOPA as a potentially more limiting step. Furthermore, *m*-tyrosine is apparently more stable in soil than L -DOPA [57, 165].

It is not known if *m*-tyrosine can be converted to L-DOPA by a cell-free extract of a species susceptible to *m*-tyrosine. If so, would the process be highly efficient *in vivo*? Synthetic proherbicides, such as diclofop-methyl [11], that are inactive at the molecular target site are much more effective when applied to intact plants than the active molecule to which they are converted *in vivo*. This is due to superior cuticular and cellular uptake of the proherbicide. Some potent natural phytotoxins from microbial origin, such as hydantocidin and 2,5-anhydro-D-glucitol, are protoxins [170–171].

We do not know how the plant synthesizes *m*-tyrosine. L-phenylalanine is a precursor of *m*-tyrosine synthesis in at least some animal systems [172]. If only one enzyme is involved, overexpression of the gene for this enzyme could be a simpler approach to transgenically producing weed-fighting plants than genetically engineering whole biosynthetic pathways (e.g. [173]).

Autotoxicity and Its Avoidance

How do plants producing phytotoxins avoid poisoning themselves? This topic has been previously reviewed to some extent [174]. However, there are many cases in which plant species appear to poison themselves [175], although differentiating between autotoxicity and buildup of host-specific plant pathogens in plant monocultures or high densities of a plant species is not trivial. In many of these cases, the apparent autotoxicity develops over a period of years.

There are multiple potential mechanisms for avoidance of autotoxicity, including resistance at the molecular target site, metabolic detoxification, sequestration, and exudation. The last three mechanisms all keep the active compound from reaching the molecular target site.

In general, plants are highly resistant to the allelochemical that they produce. However, we are aware of no case in which the mechanism of resistance is due to a resistant molecular target site. The lack of examples may be due to our ignorance of the molecular target sites of most known allelochemicals. Nonetheless, there is evidence for the other mechanisms.

So, what mechanism does the producing plant use to avoid the effects seen on other species? Does the plant avoid accumulation of the compound by secreting it almost as quickly as it produced, in a manner similar to that of *Sorghum* species that produce the allelochemical sorgoleone only in root hairs that secrete it rapidly. Apparently this is not always the mechanism, as Bertin et al. [165] indicate that although the dry weight of the fescue root exudate consists of up to 43% *m*-tyrosine, it is also a relatively abundant metabolite of the root. Is the compound sequestered into intra- or intercellular locations where it can do little or no harm? This is the case for artemisinin, which is highly toxic to *A. annua* [21], but the plant sequesters it entirely within the subcuticular spaces of glandular trichomes, where it is not in contact with living cellular contents [176–178]. Bertin et al. [164] found intracellular bodies in roots that might be associated with *m*-tyrosine sequestration, but this remains to be determined.

Sequestration is facilitated by attaching sugars, amino groups, and acyl groups to the phytotoxin, facilitating their import and storage into vacuoles. For example, benzoxazinones are stored in tissues as glucosides [118], as is the phytotoxin podophyllotoxin [85, 179].

Parting Thoughts

We have sought to provide the reader with a feel for the current state of allelopathy research. Although this subcategory of chemical ecology is clearly a factor in nature and agriculture, it has had a credibility problem due to many weak papers (and a few high profile papers) that over-reach the presented data and make unsubstantiated or poorly supported claims. This review critically covers some of this less definitive information in the context of more substantive literature in order to give the reader a view of current strategies of allelopathy research, including the many questions that are still unanswered. Weak papers in this area of research continue to be published, but the number of rigorous papers employing robust chemistry, molecular biology, and ecology methods is increasing. Methods such as selective silencing of genes in the biochemical pathway of a putative allelochemical will be very helpful in determining the presence and

17

significance of allelopathy. Such methods are already providing answers to similar questions in other facets of chemical ecology (e.g. [180]).

References

1. Rice, E.L. (1984) Allelopathy, 2nd Edition. Academic Press, Orlando, FL
2. Duke, S.O. et al. (2006) Hormesis: Is it an important factor in herbicide use and allelopathy? *Outlooks Pest. Manag.* 17, 29–33
3. Belz, R.G. et al. (2007) Dose/response relationships in allelopathy research. In *Allelopathy: New concepts and methodology* (Fuji, Y. and Hiradate, S. eds). Science Publishers, Enfield, NH, pp. 3–29
4. Bonanomi, G. et al. (2006) Phytotoxicity dynamics of decaying plant materials. *New Phytol.* 169, 571–578
5. Tharayil, N. et al. (2008) Bioavailability of allelochemicals as affected by companion compounds in soil matrices. *J. Agric. Food Chem.* 56, 3706–3713
6. Müller, C.H. et al. (1964) Volatile growth inhibitors produced by aromatic shrubs. *Science* 143, 471–473
7. Bartholomew, B. (1970) Bare zone between California shrub and grassland communities: The role of animals. *Science* 170, 1210–1212
8. Belz, R.G. (2007) Allelopathy in crop/weed interactions – an update. *Pest Manag. Sci.* 63, 308–326
9. Blum, U. et al. (1999) Evidence for inhibitory allelopathic interactions involving phenolic acids in field soils: concepts vs. an experimental model. *Crit. Rev. Plant Sci.* 18, 673–693
10. Blum, U. et al. (1991) Phenolic acid content of soils from wheat-no till, wheat-conventional till, and fallow-conventional till soybean cropping systems. *J. Chem. Ecol.* 17, 1045–1068
11. Devine, M.D. et al. (1993) Physiology of herbicide action. Prentice Hall, Englewood Cliffs, NJ
12. Cheng, H.H. (1995) Characterization of the mechanisms of allelopathy: Modeling and experimental approaches. *Am. Chem. Soc. Symp. Ser.* 582, 132–141
13. Einhellig, F.A. et al. (1983) Synergistic effects of four cinnamic acid compounds on grain sorghum. *J. Plant Growth Regul.* 1, 251–258
14. Duke, S.O. et al. (1983) Interaction of moisture stress and three phenolic compounds on lettuce seed germination. *Ann. Bot.* 52, 923–926
15. Inderjit et al. (2002) Joint action of phenolic acid mixtures and its significance in allelopathy research. *Physiol. Plant.* 114, 422–428
16. Duke, S.O. et al. (2003) Herbicides: Glyphosate. In *Encyclopedia of Agrochemicals* (Plimmer, J.R., Gammon, D.W., Ragsdale, N.N. eds). Wiley, New York http://www.mrw.interscience.wiley.com/eoa/articles/agr119/frame.html
17. Duke, S.O. and Oliva, A. (2004) Mode of action of phytotoxic terpenoids. In *Allelopathy: Chemistry and mode of action of allelochemicals* (Macías, F.A. et al. eds). CRC Press, Boca Raton, FL, pp. 201–216
18. Duke, S.O. and Dayan, F.E. (2006) Modes of action of phytotoxins from plants. In *Allelopathy: A physiological process with ecological implications* (Reigosa, M. et al., eds), Springer, Amsterdam, The Netherlands, pp. 511–536
19. Tellez, M.R. et al. (2002) Terpenoid-based defense in plants and other organisms. In *Lipid Biotechnology* (Kuo, T.M. and Gardner, H.W. eds). Marcel Dekker, New York, pp. 319–355
20. Bagchi, G.D. et al. (1997) Arteether: a potent plant growth inhibitor from *Artemisia annua*, *Phytochemistry* 45, 1131–1133
21. Duke, S.O. et al. (1987) Artemisinin, a constituent of annual wormwood (*Artemisia annua*), is a selective phytotoxin. *Weed Sci.* 35, 499–505
22. Dayan, F.E. et al. (1999) Comparative phytotoxicity of artemisinin and several sesquiterpene analogues. *Phytochemistry* 50, 607–614
23. Lydon, J. et al. (1997) Allelopathic activity of annual wormwood (*Artemisia annua*) and the role of artemisinin. *Weed Sci.* 45, 807–811
24. Escudero, A. et al. (2000) Inhibitory effects of *Artemisia herba-alba* on the germination of the gypsophyte *Helianthemum squamatum*. *Plant Ecol.* 148, 71–80
25. Karban, R. (2007) Experimental clipping of sagebrush inhibits seed germination of neighbours. *Ecol. Lett.* 10, 791–797
26. Barney, J.N. et al. (2005) Isolation and characterization of allelopathic volatiles from mugwort (*Artemisia vulgaris*). *J. Chem. Ecol.* 31, 247–265
27. Duke, S.O. et al. (1988) Terpenoids from the genus *Artemisia* as potential pesticides. *Amer. Chem. Soc. Symp. Ser.* 380, 318–334
28. Chen, P.K. and Leather, G.R. (1990) Plant growth regulatory activities of artemisinin and its related compounds. *J. Chem. Ecol.* 16, 1867–1876

29. DiTomaso, J.M. and Duke, S.O. (1991) Is polyamine biosynthesis a possible site of action of cinmethylin and artemisinin? *Pestic. Biochem. Physiol.* 39, 158–167
30. Bajsa, J. et al. (2007) A survey of synthetic and natural phytotoxic compounds and phytoalexins as potential antimicrobial compounds. *Biol. Pharm. Bull.* 30, 1740–1744
31. Drew, M.G.B. et al. (2006) Reactions of artemisinin and arteether with acid: Implications for stability and mode of action of antimalarial action. *J. Med. Chem.* 49, 6065–6073
32. Posner, G.H. and O'Neill, P.M. (2004) Knowledge of the proposed mechanism of action and cytochrome P450 metabolism of antimalarial trioxanes like artemisinin allows rational design of new antimalarial peroxides. *Accounts Chem. Res.* 37, 397–404
33. Stocks, P.A. et al. (2007) Evidence for a common non-heme chelatable-iron-dependent activation mechanism for semisynthetic and synthetic endoperoxide antimalarial drugs. *Angew. Chem.* 46, 6278–6283
34. Mishina, Y.V. et al. (2007) Artemisinins inhibit *Trypanosoma cruzi* and *Trypanosoma brucei rhodesiense in vitro* growth. *Antimicrob. Agents Chemother.* 51, 1852–1854
35. Nagamune, K. et al. (2007) Artemisinin induces calcium-dependent protein secretion in the protozoan parasite *Toxoplasma gondii*. *Eukar. Cell* 6, 2147–2156
36. Müller, W.H. and Müller, C.H. (1964) Volatile growth inhibitors produced by Salvia species. *Bull. Torrey Bot. Club* 91, 327–330
37. Halligan, J.P. (1975) Toxic terpenes from *Artemisia californica*. *Ecology* 56, 999–1003
38. Vaughn, S.F. and Spencer, G.F. (1993) Volatile monoterpenes as potential parent structures for new herbicides. *Weed Sci.* 41, 114–119
39. Romagni, J.G. et al. (2000) Allelopathic effects of volatile cineoles on two weedy plant species. *J. Chem. Ecol.* 26, 303–313
40. Vaughn, S.F. and Spencer, G.F. (1996) Synthesis and herbicidal activity of modified monoterpenes structurally similar to cinmethylin. *Weed Sci.* 44, 7–11
41. Douglas, M.H. et al. (2004) Essential oils from New Zealand manuka: triketone and other chemotypes of *Leptospermum scoparium*. *Phytochemistry* 65, 1255–1264
42. Hellyer, R.O. (1968) The occurrence of β-triketones in the steam-volatile oils of some myrtaceous Australian plants. *Austral. J. Chem.* 21, 2825–2828
43. Lee, D.L. et al. (1997) The discovery and structural requirements of inhibitors of *p*-hydroxyphenylpyruvate dioxygenase. *Weed Sci.* 45, 601–609
44. Dayan, F.E. et al. (2007) *p*-Hydroxyphenylpyruvate dioxygenase is a herbicidal target site for β-triketones from *Leptospermum scoparium*. *Phytochemistry* 68, 2004–2014
45. Neidig, M.L. et al. (2005). Spectroscopic and computational studies of NTBC bound to the non-heme iron enzyme (4-hydroxyphenyl) pyruvate dioxygenase: active site contributions to drug inhibition. *Biochem. Biophys. Res. Commun.* 338, 206–214
46. Bais, H.P. et al. (2003) Allelopathy and exotic plant invasion: from molecules and genes to species interaction. *Science* 301, 1377–1380
47. Bais, H.P. et al. (2005) Enantiomeric-dependent phytotoxic and anti-microbial activity of (±)-catechin. A rhizosecreted racemic mixture from spotted knapweed. *Plant Physiol.* 128, 1173–1179
48. Weir, T.L. et al. (2003) Intraspecific and interspecific interactions mediated by a phytotoxin (–)-catechin, secreted by the roots of *Centauria maculosa* (spotted knapweed). *J. Chem. Ecol.* 29, 2397–2412
49. Veluri, R. et al. (2004) Phytotoxic and antimicrobial activities of catechin derivatives. *J. Agric. Food Chem.* 52, 1077–1082
50. Perry, L.G. and Vivanco, J.M. (2005) Dual role for an allelochemical: (±)-catechin from *Centauria maculosa* root exudates regulates conspecific seedling establishment. *J. Ecol.* 93, 1126–1135
51. Thelen, C.C. et al. (2005) Insect herbivory stimulates allelopathic exudation by an invasive plant and the suppression of natives. *Ecol. Lett.* 8, 209–217
52. Callaway, R.M. and Ridenour, W.M. (2004) Novel weapons: invasive success and the evolution of increased competitive ability. *Front. Ecol. Environ.* 2, 436–443
53. Blair, A.C. et al. (2005) New techniques and findings in the study of a candidate allelochemical implicated in invasion success. *Ecol. Lett.* 8, 1039–1047
54. Blair, A.C. et al. (2006) A lack of evidence for an ecological role of the putative allelochemical (±)-catechin in *Centauria maculosa* invasion process. *J. Chem. Ecol.* 32, 2327–2331
55. Inderjit et al. (2006) Can plant biochemistry contribute to understanding of invasion ecology. *Trends Plant Sci.* 11, 574–580
56. Perry, L.G. et al. (2007) Concentrations of the allelochemical (±)-catechin in *Centaurea maculosa* soils. *J. Chem. Ecol.* 33, 2337–2344

57. Furubayashi, A. et al. (2007) Role of catechol structure in the adsorption and transformation reactions of L-DOPA in soils. *J. Chem. Ecol.* 33, 239–250
58. Broeckling, C.D. and Vivanco, J.M. (2008) A selective, sensitive, and rapid in-field assay for soil catechin, an allelochemical of *Centauria maculosa. Soil Biol. Biochem.* 40, 1189–1196
59. Prithiviraj, B. et al. (2007) Chemical facilitation and induced pathogen resistance mediated by root-secreted phytotoxin. *New Phytol.* 173, 852–860
60. Duke, S.O. et al. (2007) Interactions of synthetic herbicides with plant disease and microbial herbicides. In *Novel biotechnologies for biocontrol agent enhancement and management* (Vurro, M. and Gressel, J. eds)., Springer, Dordrecht, The Netherlands, pp. 277–296
61. Weir, T.L. et al. (2006) Oxalate contributes to the resistance of *Gaillardia grandiflora* and *Lupinus sericeus* to a phytotoxin produced by *Centaurea maculosa. Planta* 223, 785–795
62. Almajano, M.P. et al. (2007) Albumin causes a synergistic increase in the antioxidant activity of green tea catechins in oil-in-water emulsions. *Food. Chem.* 102, 1375–1382
63. Qin, B. et al. (2007) No evidence for root-mediated allelopathy in *Centaurea solstitialis*, a species in a commonly allelopathic genus. *Biolog. Invasions* 9, 897–907
64. Romeo, J.T. (2000) Raising the beam: moving beyond phytotoxicity. *J. Chem. Ecol.* 26, 2011–2014
65. D'Abrosca, B. et al. (2006) Chemical constituents of the aquatic plant *Schoenoplectus lacustris*: Evaluation of phytotoxic effects on the green alga *Selanastrum capricornutum. J. Chem. Ecol.* 32, 81–96
66. Breazeale, J.F. (1924) The injurious after-effects of sorghum. *J. Am. Soc. Agron.* 16, 689–700
67. Weston, L.A. (1996) Utilization of allelopathy for weed management in agroecosystems. *Agron. J.* 88, 860–866
68. Czarnota, M.A. et al. (2003) Anatomy of sorgoleone-secreting root hairs of *Sorghum* species. *Internat. J. Plant. Sci.* 164, 861–866
69. Vasilakoglou, I. et al. (2005) Allelopathic potential of bermudagrass and johnsongrass and their interference with cotton and corn. *Agron. J.* 97, 303–313
70. Netzly, D.H. and Butler, L.G. (1986) Roots of sorghum exude hydrophobic droplets containing biologically active components. *Crop Sci.* 26, 775–778
71. Gimsing, A.L. et al. (2009) Mineralization of the allelochemical sorgoleone in soil, *Chemosphere*, in press
72. Einhellig, F.A. and Souza, I.F. (1992) Phytotoxicity of sorgoleone found in grain sorghum root exudates. *J. Chem. Ecol.* 18, 1–11
73. Rimando, A.M. et al. (1998) A new photosystem II electron transfer inhibitor from *Sorghum bicolor. J. Nat. Prod.* 61, 927–930
74. de Almeida Barbosa, L.C. et al. (2001) Preparation and phytotoxicity of sorgoleone analogues. *Quim. Nova* 24, 751–755
75. Einhellig, F.A. et al. (1993) Effects of root exudate sorgoleone on photosynthesis. *J. Chem. Ecol.* 19, 369–375
76. Gonzalez, V.M. et al. (1997) Inhibition of a photosystem II electron transfer reaction by the natural product sorgoleone. *J. Agric. Food Chem.* 45, 1415–1421
77. Rasmussen, J.A. et al. (1992) Sorgoleone from root exudate inhibits mitochondrial functions. *J. Chem. Ecol.* 18, 197–207
78. Meazza, G. et al. (2002) The inhibitory activity of natural products on plant *p*-hydroxyphenylpyruvate dioxygenase. *Phytochemistry* 59, 281–288
79. Hejl, A.M. and Koster, K.L. (2004) The allelochemical sorgoleone inhibits root H++--ATPase and water uptake. *J. Chem. Ecol.* 30, 2181–2191
80. Czarnota, M.A. et al. (2003) Evaluation of seven sorghum (*Sorghum* sp.) accessions. *J. Chem. Ecol.* 29, 2073–2083
81. Dayan, F.E. (2006) Factors modulating the levels of the allelochemical sorgoleone in *Sorghum bicolor. Planta* 224, 339–346.
82. Yang, X. et al. (2004) Manipulation of root hair development and sorgoleone production in sorghum seedlings. *J. Chem. Ecol.* 30, 199–213
83. Fate, G.D. and Lynn, D.G. (1996) Xenognosin methylation is critical in defining the chemical potential gradient that regulates the spatial distribution in *Striga* pathogenesis. *J. Am. Chem. Soc.* 118, 11369–11376
84. Kagan, I.A. et al. (2003) Chromatographic separation and *in vitro* activity of sorgoleone congeners from the roots of *Sorghum bicolor. J. Agric. Food Chem.* 51, 7589–7595
85. Dayan, F.E. et al. (2003) *Podophyllum peltatum* possesses a β-glucosidase with high substrate specificity for the aryltetralin lignan podophyllotoxin. *Biochem. Biophys. Act.* 1646, 157–163

86. Dayan, F.E. et al. (2007) Biosynthesis of lipid resorcinols and benzoquinones in isolated secretory plant root hairs. *J. Exp. Bot.* 58, 3263–3272
87. Baerson, S.R. et al. (2006) A functional genomics approach for the identification of genes involved in the biosynthesis of the allelochemical sorgoleone. *Amer. Chem. Symp. Ser.* 927, 265–276
88. Baerson, S.B. et al. (2008) A functional genomics investigation of allelochemical bioysynthesis in *Sorghum bicolor* root hairs. *J. Biol. Chem.* 83, 3231–3247
89. Cook, D. et al. (2007) Molecular and biochemical characterization of a novel polyketide synthase likely to be involved in the biosynthesis of sorgoleone. *Am. Chem. Soc. Symp. Ser.* 955, 141–151
90. Pan, Z. et al. (2007) Functional characterization of desaturases involved in the formation of the terminal double bond of an unusual $16{:}3\Delta^{9,12,15}$ fatty acid isolated from *Sorghum bicolor* root hairs. *J. Biol. Chem.* 282, 4326–4335
91. Topal, S. et al. (2007) Herbicidal effects of juglone as an allelochemical. *Phyton* 46, 259–269
92. Reynolds, T. (1987) Comparative effects of alicyclic compounds and quinones on inhibition of lettuce fruit germination. *Ann. Bot.* 60, 215–223
93. Meyer, J.J.M. et al. (2007) Identification of plumbagin epoxide as a germination inhibitory compound through a rapid bioassay on TLC. *South Afric. J. Bot.* 73, 654–656
94. Jose, S. and Gillespie, A.R. (1998) Allelopathy in black walnut (*Juglans nigra* L.) alley cropping. I. Spatio-temporal variation in soil juglone in a black walnut-corn (*Zea mays* L.) alley cropping system in the midwestern USA. *Plant Soil* 203, 191–197
95. Jose, S. and Gillespie, A.R. (1998) Allelopathy in black walnut (*Juglans nigra* L.) alley cropping. II. Effects of juglone on hydroponically grown corn (*Zea mays* L.) and soybean (*Glycine max* L. Merr.) growth and physiology. *Plant Soil* 203, 199–205
96. Hejl, A.M. et al. (1993) Effects of juglone on growth, photosynthesis, and respiration. *J. Chem. Ecol.* 19, 559–568
97. Koeppe, D.E. (1972) Reactions of isolated corn mitochondria influenced by juglone. *Physiol. Plant.* 27, 89–94
98. Bohm, P.A.F. et al. (2006) Peroxidase activity and lignification in soybean root growth-inhibition by juglone. *Biol. Plant* 50, 315–317
99. Hejl, A.M. and Koster, K.L. (2004) Juglone disrupts root plasma membrane H+-ATPase activity and impairs water uptake, root respiration, and growth in soybean (*Glycine max*) and corn (*Zea mays*). *J. Chem. Ecol.* 30, 453–471
100. Dansette, P. and Azerad, R. (1970) New intermediate in naphthoquinone and menaquinone biosynthesis. *Biochem. Biophysic. Res. Comm.* 40, 1090–1095
101. Müller, W.U. and Leistner, E. (1976) Naphthoquinone, an intermediate in juglone (5-hydroxy-1,4-naphthoquinone) biosynthesis. *Phytochemistry* 15, 407–410
102. Jose, S. (2002) Black walnut allelopathy: current state of the science. In *Chemical Ecology of Plants: Allelopathy in aquatic and terrestrial ecosystems* (Inderjit and Mallik, A.U. eds). Birkhauser- Verlag AG, Basel, Swizerland, pp. 149–172
103. von Kiparski, G.R. et al. (2007) Occurrence and fate of the phytotoxin juglone in alley soils under black walnut trees. *J. Environ. Qual.* 36, 709–717
104. Weidenhamer, J.D. and Romeo, J.T. (2004) Allelochemicals of *Polygonella myriophylla*: Chemistry and soil degradation. *J. Chem. Ecol.* 30, 1067–1082
105. Olofsdotter, M. et al. (2002) Improving crop competitive ability using allelopathy – an example from rice. *Plant Breed.* 121, 1–9
106. Dilday, R.H. et al. (2001) Allelopathic potential in rice germplasm against ducksalad, redstem and barnyardgrass. *J. Crop Prod.* 4, 287–301
107. Kong, C. et al. (2002) Using specific secondary metabolites as markers to evaluate allelopathic potentials of rice varieties and individual plants. *Chin. Sci. Bull.* 47, 839–843
108. Jensen, L.B. et al. (2001) Locating genes controlling allelopathic effects against barnyardgrass in upland rice. *Agron. J.* 93, 21–26
109. Kong, C. et al. (2004) Two compounds from allelopathic rice accession and their inhibitory activity on weeds and fungal pathogens. *Phytochemistry* 65, 1123–1128
110. Gealy, D.R. et al. (2003) Rice cultivar differences in suppression of barnyardgrass (*Echinochloa crus-galli*) and economics of reduced propanil rates. *Weed Sci.* 51, 601–609
111. Courtois, B. and Olofsdotter, M. (1998) Incorporating the allelopathy trait in upland

17

rice breeding programs. In *Proceedings of the Workshop on Allelopathy in Rice* (Olofsdotter, M., ed). Manila, Philippines, pp. 57–68
112. Olofsdotter, M. (2001) Getting closer to breeding for competitive ability and the role of allelopathy – An example from rice (*Oriza sativa*). *Weed Technol.* 15, 798–806
113. Kato-Noguchi, H. and Ino, I. (2004) Release level of momilactone B from rice plants. *Plant Product. Sci.* 7, 189–90
114. Bouillant, M.L. et al. (1994) Identification of 5-(12-heptadecenyl)-resorcinol in rice root exudates. *Phytochemistry* 35, 769–771
115. VanEtten, H.D. et al. (1994) Two classes of plant antibiotics: phytoalexins versus 'phytoanticipins'. *Plant Cell* 6, 1191–1192
116. Kozubek, A. and Tyman, J.H.P. (1999) Resorcinolic lipids, the natural non-isoprenoid phenolic amphiphiles and their biological activity. *Chem. Rev.* 99, 1–25
117. Kong, C. et al. (2004) Release and activity of allelochemicals from allelopathic rice seedlings. *J. Agric. Food Chem.* 52, 2861–2865
118. Sicker, D. et al. (2004) Benzoxazolin-2(3H)-ones - Generation, effects and detoxification in the competition among plants. In *Allelopathy: Chemistry and mode of action of allelochemicals* (Macías, F.A. et al., eds). CRC Press, Boca Raton, FL, pp. 77–102
119. Fomsgaard, I.S. (2006) Chemical ecology in wheat plant-pest interactions. How the use of modern techniques and a multidisciplinary approach can throw new light on a well-known phenomenon: Allelopathy. *J. Agric. Food Chem.* 54, 987–990
120. Macías, F.A. et al. (2007) Allellopathy – a natural alternative for weed control. *Pest Manag. Sci.* 63, 327–348
121. Schulz, M. et al. (1994) Allelopathic effects of living quackgrass (*Agropyron repens* L.). Identification of inhibitory allelochemicals exuded fro rhizome borne roots. *Angew. Bot.* 68, 195–200
122. Chiapusio, G. et al. (1997) Do germination indices adequately reflect allelochemical effects on the germination process? *J. Chem. Ecol.* 23, 2445–2453
123. Barnes, J.P. and Putnam, A.R. (1986) Allelopathic activity of rye (*Secale cereale L.*). In *The science of allelopathy* (Putnam, A.R. and Tang, C.-S. eds). Wiley-Interscience, New York, pp. 271–286
124. Belz, R.G. and Hurle, K. (2005) Differential exudation of two benzoxazinoids: some of the determining factors for seedling allelopathy of Triticeae species. *J. Agric. Food Chem.* 53, 250–261
125. Huang, Z. et al. (2003) Correlation between phytotoxicity on annual grasses (*Lolium rigidum*) and production dynamics of allelochemicals within root exudates of an allelopathic wheat. *J. Chem. Ecol.* 29, 2263–2279
126. Macías, F.A. et al. (2006) Structure-activity relationship (SAR) studies of benzoxazinones, their degradation products and analogues. Phytotoxicity on target species (STS). *J. Agric. Food Chem.* 53, 538–548
127. Sánchez-Moreiras, A.M. et al. (2004) Mode of action of the hydoxamic acid BOA and other related compounds. In *Allelopathy: Chemistry and mode of action of allelochemicals* (Macías, F.A. et al. eds). CRC Press, Boca Raton, FL, pp. 239–252
128. Niemeyer, H.M. et al. (1987) Inhibition of energy metabolism by benzoxazolin-2-one. *Comp. Biochem. Physiol.* 87B, 35–39
129. Freibe, A. et al. (1997) Effects of 2,4-dihydroxy-1,4-benzoxazin-3-ones on the activity of plasma membrane H+-ATPase. *Phytochemistry* 44, 979–983
130. Reigosa, M.J. et al. (2001) Comparison of physiological effects of allelochemicals and commercial herbicides. *Allelopathy J.* 8, 211–220
131. Rojas, M.C. et al. (1997) Stimulatory effect of DIMBOA on NADH oxidation catalyzed by horseradish peroxidase. *Phytochemistry* 46, 11–15
132. Kato-Noguchi, H. and Macías, F.A. (2005) Effect of 6-methoxy-2-benzoxazolinone on the germination and α-amylase activity in lettuce seeds. *J. Plant Physiol.* 162, 1304–1307
133. Baerson, S.R. et al. (2005) Detoxification and transcriptome response in *Arabidopsis* seedlings exposed to the allelochemical benzoxazolin-2(3H)-one (BOA). *J. Biol. Chem.* 280, 21867–21881
134. Wieland, I. et al. (1999) Detoxification of benzoxazolin-2(*3H*)-one in higher plants. In *Recent advances in allelopathy* Vol. 1 (Macías, F.A. et al., eds). Servicio e Publicaciones-Univ. Cádiz, Spain, pp. 47–56
135. Rimando, A.M. and Duke, S.O. (2003) Rice allelopathy. In *Rice production: Origin, history, and technology* (Smith, C.S. and Dilday, R.H. eds). Wiley, New York, pp. 221–244
136. Olofsdotter, M. et al. (2002) Why phenolic acids are unlikely primary allelochemicals in rice. *J. Chem. Ecol.* 28, 229–242

137. Blum, U. (1996) Allelopathic interactions involving phenolic acids. *J. Nematol.* 28, 259–267
138. Einhellig, F.A. and Rasmussen, J.A. (1978) Synergistic inhibitory effects of vanillic and *p*-hydroxybenzoic acids on radish and grain sorghum. *J. Chem. Ecol.* 4, 425–436
139. Rasmussen, J.A. and Einhellig, F.A. (1977) Synergistic inhibitory effects of *p*-coumaric and ferulic acids on germination and growth of grain sorghum. *J. Chem. Ecol.* 3, 197–205
140. Rasmussen, J.A. and Einhellig, F.A. (1979) Inhibitory effects of combinations of three phenolic acids on grain sorghum germination. *Plant Sci. Lett.* 14, 69–74
141. Blum, U. et al. (1984) Effects of ferulic acid and some of its microbial metabolic products on radicle growth of cucumber. *J. Chem. Ecol.* 10, 1169–1191
142. Blum, U. et al. (1985) Effects of various mixtures of ferulic acid and some of its microbial metabolic products on cucumber leaf expansion and dry matter in nutrient culture. *J. Chem. Ecol.* 11, 619–641
143. Blum, U. et al. (1985) Effects of ferulic and *p*-coumaric acids in nutrient culture on cucumber leaf expansion as influenced by pH. *J. Chem. Ecol.* 11, 1567–1582
144. Blum, U. et al. (1989) Effects of mixtures of phenolic acids on leaf area expansion of cucumber seedlings grown in different pH Portsmouth A1 soil materials. *J. Chem. Ecol.* 15, 2413–2423
145. Gerig, T.M. and Blum, U. (1991) Effects of mixtures of four phenolic acids on leaf area expansion of cucumber grown in Portsmouth B1 soil materials. *J. Chem. Ecol.* 17, 29–40
146. Gerig, T.M. et al. (1989) Statistical analysis of the joint inhibitory action of similar compounds. *J. Chem. Ecol.* 15, 2403–2412
147. Jia, C. et al. (2006) Joint action of benzoxazinone derivatives and phenolic acids. *J. Agric. Food Chem.* 54, 1049–1057
148. Lyu, S.W. et al. (1990) Effects of mixtures of phenolic acids on phosphorus uptake by cucumber seedlings. *J. Chem. Ecol.* 16, 2559–2567
149. Shann, J.R. and Blum, U. (1987) The uptake of ferulic and *p*-hydroxybenzoic acids by *Cucumis sativus*. *Phytochemistry* 26, 2959–2964
150. Blum, U. et al. (2000) Induction and/or selection of phenolic acid-utilizing bulk-soil and rhizosphere bacteria and their influence on phenolic acid phytotoxicity. *J. Chem. Ecol.* 26, 2059–2078
151. Tharayil, N. et al. (2006) Preferential sorption of phenolic phytotoxins to soil: implications for altering the availability of allelochemicals. *J. Agric. Food Chem.* 54, 3033–3040
152. Lehman, M.E. and Blum, U. (1999) Influence of pretreatment stresses on inhibitory effects of ferulic acid, an allelopathic phenolic acid. *J. Chem. Ecol.* 25, 1517–1529
153. Booker, F.L. et al. (1993) Short-term effects of ferulic acid on ion uptake and water relations in cucumber seedlings. *J. Exp. Bot.* 43, 649–55
154. Lehman, M.E. and Blum, U. (1999) Evaluation of ferulic acid uptake as a measurement of allelochemical dose: effective concentration. *J. Chem. Ecol.* 25, 2585–2600
155. Blum, U. and Gerig, T.M. (2005) Relationships between phenolic acid concentrations, transpiration, water utilization, leaf area expansion, and uptake of phenolic acids: Nutrient culture studies. *J. Chem. Ecol.* 31, 1907–1932
156. Fujii, Y. (1999) Allelopathy of velvetbean: Determination and identification of L-DOPA as a candidate of allelopathic substances. In *Biologically Active Natural Products: Agrochemicals* (Cutler, H.G. and Cutler, S.J. eds). CRC Press, Boca Raton, FL, pp. 33–47
157. Nishihara, E. et al. (2005) l-3-(3,4-Dihydroxyphenyl)alanine (L-DOPA), an allelochemical exuded from velvetbean (*Mucuna pruriens*) roots. *Plant Growth Regul.* 45, 113–120
158. Nishihara, E. et al. (2004) Germination growth response of different plant species to the allelochemical L-3,4-dihydroxyphenylalanine (L-DOPA). *Plant Growth Regul.* 42, 181–189
159. Hachinohe, M. et al. (2004) absorption, translocation and metabolism of L-DOPA in barnyardgrass and lettuce: Their involvement in species-selective phytotoxic action. *Plant Growth Regul.* 43, 237–243
160. Hachinohe, M. and Matsumoto, H. (2005) Involvement of reactive oxygen species generated from melanin synthesis pathway in phytotoxicity of L-DOPA. *J. Chem. Ecol.* 31, 237–246
161. Hachinohe, M. and Matsumoto, H. (2007) Mechanism of selective phytotoxicity of L-3,4-dihydroxyphenylalanine (L-DOPA) in barnyardgrass and lettuce. *J. Chem. Ecol.* 33, 1919–1926

162. Soares, A.R. et al. (2007) L-DOPA increases lignification associated with *Glycine max* root growth-inhibition. *J. Chem. Ecol.* 33, 265–275
163. Hiradate, S. et al. (2005) Changes in chemical structure and biological activity of L-DOPA as influence by andosol and its components. *Soil Sci. Plant Nut.* (Japan) 51, 477–484
164. Bertin, C. et al. (2003) Laboratory assessment of the allelopathic effects of fine leaf fescues. *J. Chem. Ecol.* 29, 1919–1937
165. Bertin, C. et al. (2007) Grass roots chemistry: meta-tyrosine, an herbicidal nonprotein amino acid. *Proc. Natl. Acad. Sci. USA* 104, 16964–16969
166. Schneider, D. et al. (2002) Cycads: their evolution, toxins, herbivores and insect pollinators. *Naturwissenschaften* 89, 281–294
167. Lambein, F. et al. (2001) Non-protein amino acids and food safety. Special Publication – *Royal Soc. Chem.* 269, 580–583
168. Xuan, T.D. et al. (2006) Mimosine in *Leucaena* as a potent bio-herbicide. *Agron. Sustain. Devel.* 26, 89–97
169. Schenk, S.U. and Werner, D. (1991) β-(3-Isoxazolin-5-on-2-yl)-alanine from *Pisum*: allelopathic properties and antimycotic bioassay. *Phytochemistry* 30, 467–470
170. Fonné-Pfister, R. et al. (1996) The mode of action and the structure of a herbicide in complex with its target: binding of activated hydantocidin to the feedback regulation site of adenylosuccinate synthetase. *Proc. Natl. Acad. Sci. USA* 93, 9431–9436
171. Dayan, F.E. et al. (2002) Bioactivation of the fungal phytotoxin 2,5-anhydro-D-glucitol by glycolytic enzymes is an essential component of its mechanism of action. *Z. Naturforsch.* 57c, 645–653
172. Ishimitsu, S. et al. (1980) Formation of *m*-tyrosine and *o*-tyrosine from L-phenylalanine by rat brain homogenate. *Chem. Pharm. Bull.* (Tokyo) 28, 1653–1655
173. Duke, S.O. (2003) Weeding with transgenes. *Trends Biotechnol.* 21, 182–195
174. Duke, S.O. et al. (1999) Tissue localization and potential uses of phytochemicals with biological activity. In *Recent advances in allelopathy* Vol. 1 (Macías, F.A. et al., eds). Servicio e Publicaciones-Univ. Cádiz, Spain, pp. 211–218
175. Singh, H.P. et al. (1999) Autotoxicity: concept, organisms, and ecological significance. *Crit. Rev. Plant Sci.* 18, 757–772
176. Duke, S.O. and Paul, R.N. (1993) Development and fine structure of the glandular trichomes of *Artemisia annua* L. *Int. J. Plant Sci.* 154, 107–118
177. Duke, M.V. et al. (1994) Localization of artemisinin and artemisitene in foliar tissues of glanded and glandless biotypes of *Artemisia annua*. *Int. J. Plant Sci.* 155, 365–373
178. Tellez, M.R. et al. (1999) Differential accumulation of isoprenoids in glanded and glandless *Artemisia annua* L. *Phytochemistry* 52, 1035–1040
179. Oliva, A. et al. (2002) Aryltertralin lignans inhibit plant growth by affecting formation of mitotic microtubular organizing centers. *Pestic. Biochem. Physiol.* 72, 45–54
180. Zavala, J.A. et al. (2004) Constitutive and inducible trypsin proteinase inhibitor production incurs large fitness costs in *Nicotiana attenuata*. *Proc. Natl. Acad. Sci. USA* 101, 1607–1612

Health Benefits of Dietary Plant Natural Products 18

Maria Traka and Richard F. Mithen

Abstract During the last few decades it has become widely reported that diets rich in fruit and vegetables reduce the risk of chronic disease such as cancer and cardiovascular disease, and that these beneficial effects are at least partially mediated by secondary metabolites that occur in these foods. Recent prospective epidemiological studies have provided further support for the protective effects of diets rich in fruits and vegetables towards cardiovascular disease, but, in general, less support for protective effects towards cancer, with some notable exceptions such as diets that are rich in cruciferous vegetables. Here, we review the epidemiological and experimental evidence for health benefits of diets rich in fruits and vegetables and certain classes of secondary metabolites, and then focus on the role of flavonoids, which are wide spread in fruits and vegetables, in providing protection against cardiovascular disease, and glucosinolates and their derivatives, which, within food plants, are largely restricted to the Brassicaceae, in reducing the risk of cancer.

M. Traka and R.F. Mithen (✉)
Institute of Food Research, Colney Lane, Norwich, NR4 7UA, UK
e-mail: maria.traka@bbsrc.ac.uk, richard.mithen@bbsrc.ac.uk

Dietary Phytochemicals and Health Benefits: The Epidemiological Evidence

Many epidemiological studies have sought to quantify the association between diets that are rich in fruits and vegetables with reduced risk of chronic illness such as cardiovascular disease, cancer and cognitive decline. The majority of studies undertaken in the 1980s–2000 were retrospective case control studies, and many of these reported inverse associations between diets rich in fruits and vegetables in general, or particular classes of secondary metabolites that occur in these foods and risk of chronic disease. Case control studies suffer from two major problems. Firstly, they do not give any indication of absolute risk, and thus can give a misleading impression of health benefits. Secondly, selection of the control group is complex and challenging, particularly as diets rich in fruits and vegetables are often highly correlated with other lifestyle attributes and socioeconomic status. Furthermore, due to the large number of studies there is likely to be bias in the reporting of studies that have a positive outcome, as opposed to those that find no association between diet and health. More recently, several long term prospective studies involving many thousands of volunteers have provided further evidence for the associations between diet and health. In general, and with acknowledging that there are several

A.E. Osbourn and V. Lanzotti (eds.), *Plant-derived Natural Products*,
DOI 10.1007/978-0-387-85498-4_18,

exceptions, longer term prospective studies and pooled analyses of smaller cohort studies have supported the association between diets rich in fruits and vegetables and a reduction in risk of cardiovascular disease [1, 2] and, although there are fewer studies, a reduction in risk of cognitive decline and neurological disorders [3] and chronic obstructive pulmonary disease [4]. Concurrent with the reporting of results from prospective studies has been the establishment of databases of the phytochemical content of fruits and vegetables and other food stuffs, particularly with regard flavonoids [5–7]. Integrating these two datasets has provided evidence for the association of flavonoids, particularly anthocyanidins and proanthocyanidins, with reduction in vascular disease [2]. However, these larger studies have provided less supporting evidence for an association between diets rich in fruit and vegetables and a reduction in cancer risk [8–13], with some notable exceptions. For example, diets rich in cruciferous vegetables, and therefore rich in glucosinolates, have been associated with a reduction in risk of the development of aggressive prostate cancer [14], diets rich in soy, and therefore rich in isoflavones, have been associated with reduction in risk of localised prostate cancer [15], and diets rich in vegetables in general have been associated with reduction in risk of gastric cancer [16].

Dietary Phytochemicals and Health Benefits: Evidence from Cell and Animal Studies

Cell and animal models have been used to provide further experimental evidence for the association between phytochemicals and reduction in risk of chronic disease. These models have largely been developed for toxicological research and caution needs to be exercised in extrapolation of results from these models to humans consuming a normal diet. For example, many cell lines that are routinely used have been in culture for several years and were originally derived from cancerous tissues. Both of these factors result in differences in expression of genes involved in, for example, cell proliferation compared to 'normal' cells. This is of particular importance if the main interest is in disease prevention rather than cure. Secondly, there is often insufficient appreciation of the bio-transformations and metabolism that dietary phytochemicals undergo within the gastrointestinal tract prior to absorption, either due to mammalian metabolism or colonic fermentation, and, following absorption, within epithelial cells and hepatic tissue. Thus, frequently, cells are exposed to plant metabolites that would not be found within the GI tract or in the systemic circulation. Thirdly, the concentration of dietary phytochemicals to which cells and animal models are exposed is frequently significantly greater (sometimes by a few orders of magnitude) than that which would result from normal dietary exposure. Fourthly, cells and rodent models may have different functional alleles and lack the equivalent allelic polymorphisms that mediate the interaction between phytochemicals and, for example, gene expression. Lastly, even a single item of food consists of a complex mixture of phytochemicals, which may interact, possibly synergistically, with each other.

Despite these reservations, cell and animal models are effective in identifying particular bioactive phytochemicals that are good candidates to underpin the health benefits of fruit and vegetables, even if the nature of the mechanistic basis is questionable.

Dietary Phytochemicals and Health Benefits: Evidence from Human Intervention Studies

Obtaining experimental evidence for the health benefit of plant natural products from human intervention studies is challenging. Classic placebo controlled doubled blinded designs, as used for clinical trials of pharmaceuticals, are not possible except for the use of dietary supplements, in which specific phytochemicals are provided at relatively high doses. Likewise, most of the evidence for the protective effect of plant natural products is towards forms of chronic diseases that are associated with aging. These may take many years to develop and preclude the use of clinical endpoints in intervention trials. Thus, researchers are forced to evaluate 'biomarkers' of risk. While these are well defined for vascular disease, there are few biomarkers appropriate for evaluating cognitive decline and cancer risk.

The positive results of associations obtained from case control studies supported by mechanistic studies with cell models led to the establishment of intervention trials with dietary supplements. However, these have frequently either not resulted in supporting the association between phytochemicals and health, at least with the dose that were provided which tended to be higher than dietary intake, or suggested that the supplements may increase risk. For example, while certain epidemiological studies with carotenoid rich foods have suggested that β-carotene and vitamins A and E may have protective effects against lung cancer, intervention trials have found either no protective effect or an increase in cancer risk [17, 18]. In a similar manner, in a prospective epidemiological study (i.e. not an intervention study) it was found that men who were heavy users of multivitamin supplements had enhanced risk of advanced and fatal prostate cancer [19].

In contrast, many relatively short intervention trials with either supplements or foods have provided supporting evidence for the benefits of flavonoids for vascular health. The majority of these studies are of an acute nature in which relatively high doses are provided and various biomarkers, including, for example, platelet aggregation [20], flow mediated arterial dilation and nitric oxide metabolism [21], and plasma LDL cholesterol [22] are quantified over the following hours or days. Some studies have also demonstrated increased cerebral blood flow and cognitive function [23]. The major challenge is to design intervention studies in which health benefits of normal dietary intake can be assessed over several months or possibly a few years, and to go beyond the assessment of 'biomarkers' to probe the underlying mechanisms *in vivo*.

Dietary Phenolics, Polyphenolics, Tannins: Structure and Human Metabolism

The chemistry and biochemistry of flavonoids and related compounds has recently been extensively reviewed [24]. Over 8,000 compounds with phenolic structures have been reported in plants, ranging from low molecular weight simple phenolic acids to high molecular weight tannins. They can be conveniently divided into two major groups, the flavonoids, which are compounds comprising two aromatic rings connected by a three carbon bridge, and the non flavonoids, including simple phenolic acids, hydroxycinammates and stilbenes. Flavonoids can further be divided into six major subclasses: the flavanols, flavan-3-ols (comprising both monomers and the polymeric proanthocyanidins or condensed tannins), isoflavones, flavones, flavanones and anthocyanidins (Figs. 1 and 2). The principle phenolic acid is gallic acid which is the base unit of gallotannins, whereas gallic acid and hexahydroxydiphenoyl are subunits of ellagitannins. Both forms of tannins are collectively known as hydrolysable tannins. The most common hydroxycinnamates are caffeic, *p*-coumaric and ferulic acids. Quinic acid conjugates of caffeic acid such as 5-O-caffeoylquinic acid (chlorogenic acid) are common components of fruit and vegetables, with coffee being a major source in the diet. Resveratrol, an important polyphenolic found in red wine, is the most common stillbene, and has frequently been associated with health benefits (Fig. 3).

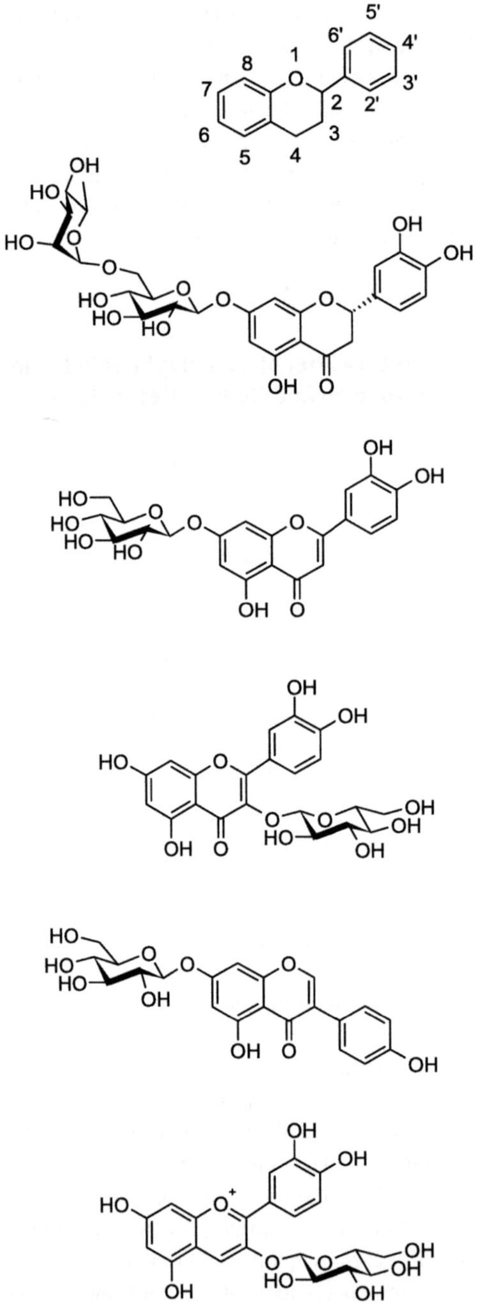

C6-C3-C6 flavanoid skeleton

hesperitn-7-rutinoside- a flavanone

luteolin-7-glucoside- a flavone

quercetin-3-glucoside- a flavonol

genistein-7-glucoside- an isoflavone

cyanidin-3-glucoside- an anthocyanin

Fig. 1 The basic C6-C3-C6 flavonoid skeleton and examples of common glycosylated flavonoids that are found in foods

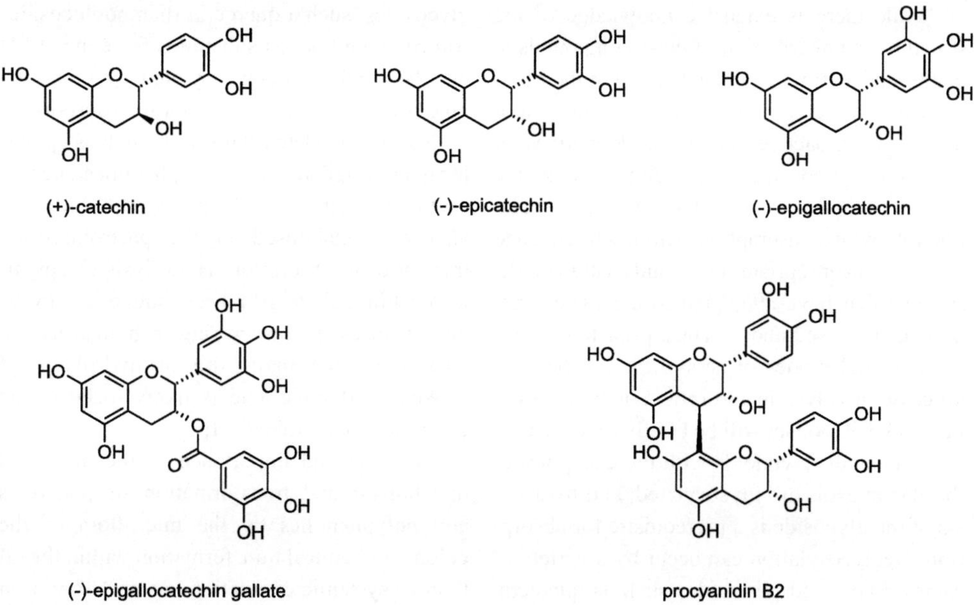

Fig. 2 Examples of flavan-3-ols

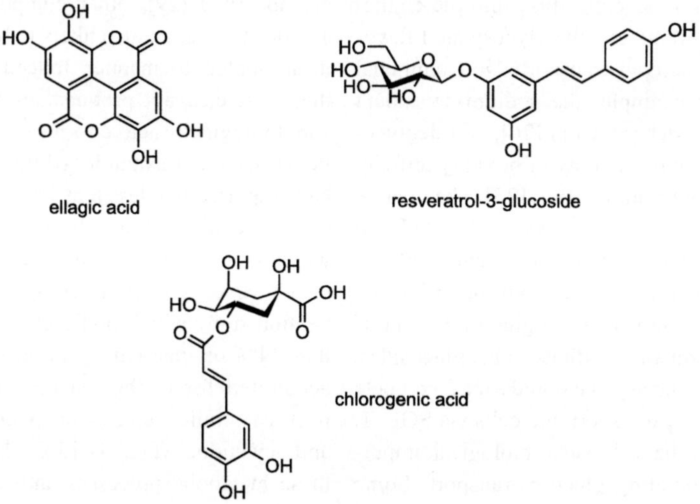

Fig. 3 Examples of common non flavonoid phenolics that are found in foods and are associated with health benefits

While there is extensive knowledge of the structure and distribution of these compounds in plants, and a growing understanding of their biosynthesis, there is a relatively poor understanding of the bioactive functional derivatives of these compounds that occur either within the gastrointestinal tract or in the systemic circulation following consumption. This has often led to the use of inappropriate compounds within model systems that have attempted to elucidate function, as discussed above. Thus, prior to discussing potential modes of action, the biochemical processes involved in absorption, biotransformation and metabolism will be briefly reviewed.

Almost all flavonoids with the exception of the flavan-3-ols are glycosylated, and hydrolysis of the glycoside is a prerequisite for absorption. Deglycosylation can occur by a variety of routes. Flavanoid glycosides such as quercetin glucoside are hydrolysed by the endogenous β-glucosidases lactase phloridzin hydrolyase (LPH) present on the brush border of small intestine epithelial cells [25]. The resulting aglycone can passively diffuse into the epithelial cells. Alternatively, the glycosylated flavonoids may be actively transported into epithelial cells by, for example, the sodium-dependent glucose transorter (SGLT1) [26], and deglycosylated through the activity of broad specificity cytosolic β-glucosides (CBG) [27]. The relative importance of these two routes is likely to depend upon the position and extent of glycosylation, which determines both the efficiency of transport into the cells by glucose transporters and subsequent specificity of the glucosides. Glycosylated dietary flavonoids may compete with glucose for transport into cells via SGLT1, and this may have broader biological importance in modulating glucose transport. Some studies have suggested that polyphenol rich diets may reduce the post prandial surge in plasma glucose, and thus effectively reduce the glycaemix index of foods consumed as part of a polyphenol rich diet [28]. This may in itself have potential health benefits. Some flavanoid glycosides, such a quercetin rhamnoglucosides (rutin) cannot act as substrates for either LPH or CBG and are deglycosylated by microbial glucosidases in the colon, prior to absorption.

Non glycosylated flavanoids such as epigallocatechin gallate and other phenolics such as 5-O-caffeoylquinic acid (chlorogenic acid) may also be metabolised in the gastrointestinal tract prior to absorption. Hydrolysis of epigallocatechin gallate has been suggested to be due to an esterase occurring in human saliva, whereupon the main site of hydrolysis of 5-*O*-caffeoylquinic acid is likely to be in the colon due to microbial activity.

The role and importance of the microbial metabolism and transformation of phenolics and polyphenolics by the microflora of the colon, or chemical transformation within the GI tract or systemic circulation has probably been underestimated. While it is increasingly likely that oligomeric and polymeric proanthocyanidins are extensively metabolised in the colon to produce a range of small phenolic acids that are absorbed [29], other compounds such as anthocyanidins are also likely to be metabolised in an analogous manner. Indeed it is possible that these cleavage products are the most relevant biologically active forms of dietary phenolics *in vivo*. For example, while several studies have reported that less that 1% of anthocyanins that are consumed are absorbed and excreted, which is difficult to reconcile with the reported health benefits of these compounds, an intervention study with blood orange juice reported that 44% of ingested cyanidin glycoside was accounted for in the plasma in the form of protocatechuic acid, a cleavage product of anthocyanidins (Fig. 4) [30, 31]. Elucidating these metabolic processes and identifying the precise metabolites that are absorbed are an important prerequisite for understanding biological function and the mechanistic basis of the health benefits of phenolics and polyphenolics.

Once absorbed, flavanoids and phenolic acid derivatives undergo extensive metabolism.

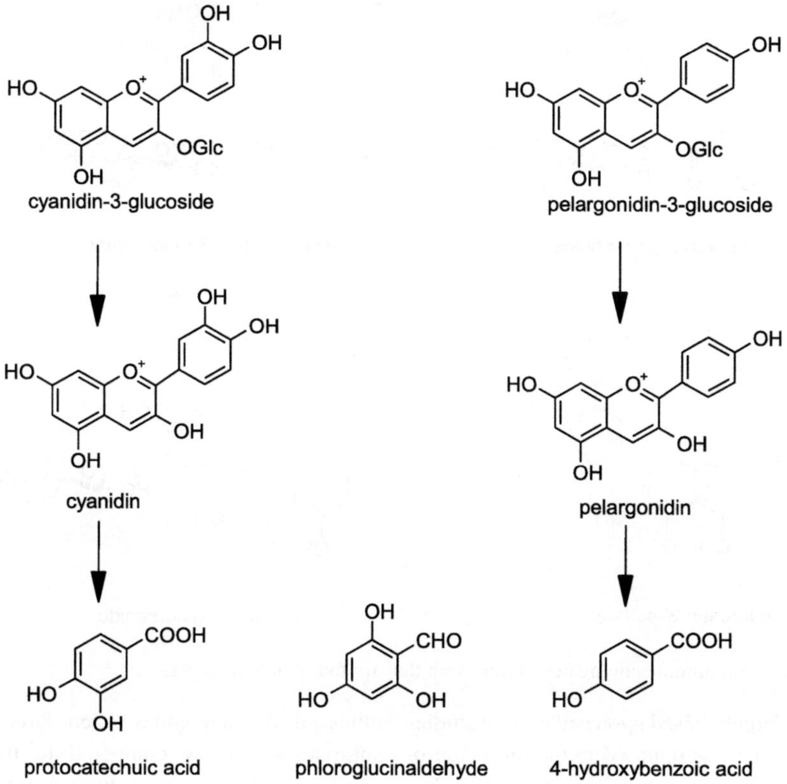

Fig. 4 Cleavage products from anthocyanins (Redrawn from Ref. [31])

Initially extensive glucuronidation and some methylation of the aglycone occur in the cells of the small intestine, with further glucuronidation, methylation and sulphation occurring in the liver. Thus, following ingestion and deglycosylation a single polyphenolic may give rise to several conjugated metabolites. For example, following ingestion of quercetin glycoside, at least 12 different glucuronide and sulphate conjugates of quercetin or methylquercetin are found in plasma (Fig. 5) [32, 33]. Likewise, several glucuronide and sulphate conjugates are observed following ingestion of other polyphenols, such as the isoflavones [34], and it is likely that similar compounds are also derived from other flavonoids. Non flavanoid phenolics follow similar routes of absorption and metabolism. For example, hydroxycinnamic acids such as caffeic, ferulic and coumaric acids, are rapidly absorbed in the small intestine and glucuronidated and sulphated in a similar manner to the flavanoids [35].

Mechanistic Basis of Health Promoting Activity of Dietary Phenolics and Polyphenolics

Polyphenols as Antioxidants

The antioxidant activity of phenolics and polyphenolics is often regarded to be the basis of their health promoting activity. The evidence

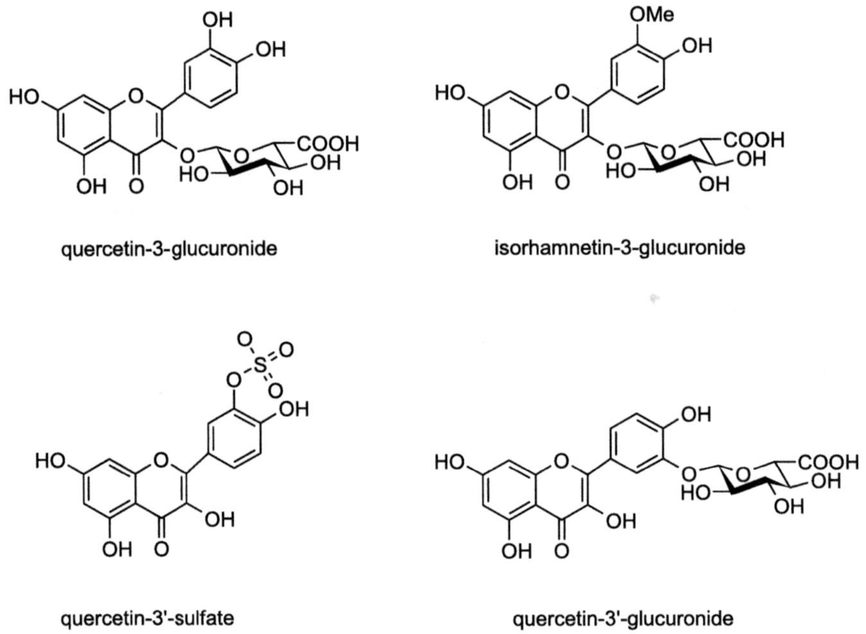

Fig. 5 The most common conjugates of quercetin that are found in human plasma

for this is largely based upon cell culture studies with aglycones, as opposed to the more appropriate conjugated metabolites that occur *in vivo*. Several studies have also attempted to demonstrated that a polyphenol rich diet results in an increase in plasma antioxidant capacity, but few, if any, have demonstrated a significant increase [36]. This is due to two main factors: Firstly, the endogenous phenolic and ascorbate concentrations in the plasma is between 159 and 380 μM. The additional concentrations that can be obtained from dietary sources is relatively low, probably less that 1% from average diets and up to 5% for diets that are particularly rich in polyhenols [36]. Moreover, these additional marginal increases are transient and it is difficult to envisage how these changes can have a significant impact upon health. However, it is conceivable that in certain elderly populations in which the plasma ascorbate levels can become depleted heavy consumption of tea and coffee may have a significant effect on plasma antioxidant activity. Secondly, the conjugated metabolites of the aglycones often have reduced antioxidant activity compared to the parent agylcone, with the precise activity dependant upon the nature and position of conjugation [37, 38]. For example, sulphation has been shown to reduce the antioxidant activity of isoflavones [39]. In addition, the interaction between polyphenols and plasma proteins can reduce their antioxidant activity [40].

Counter to these arguments, there are two factors that may lead to the antioxidant activity of human metabolites of dietary polyphenols being underestimated. Firstly, as described above, there is some uncertainty to the precise metabolic derivatives of polyphenols and thus the plasma concentration of the active metabolites may be underestimated. Secondly, it is possible that there may be local deglucuronidation at, for example, sites of inflammation through β-glucuronidase activity to release the biological active aglycone [41]. However, despite the possible factors in mitigation, while the epidemiological evidence for health promoting activity of polyphenolic rich diets has increased, it is unlikely that this is mediated by enhancement of antioxidant activity of plasma.

Polyphenols and Vascular Disease

Partially due to the epidemiological evidence that has associated diets rich in polyhenols with reduction in risk of cardiovascular disease, several studies have specifically investigated the effects of polyphenolics on risk factors for cardiovascular health. Inflammation plays a key event in the initiation of atherosclerosis and the development of atherothrombotic events, which are leading causes for CVD. Adhesion of circulating monocytes to the endothelium and subsequent migration into the vascular wall are critical events in these processes. The binding of monocytes to the vascular endothelium is mediated by cross linkage of cell adhesion molecules, such as intercellular adhesion molecule 1 (ICAM-1) and vascular cell adhesion molecule-1 (VCAM-1), and the expression of these molecules is greatly enhanced at sites of atherosclerosis. The expression of these adhesion molecules is enhanced through several risk factors for CVD, such as increased plasma cholesterol and pro-inflammatory cytokines, such as TNF-α and IL-1, mediated through activation of the NF-κB and AP-1 transcription factors. Through the use of HUVEC primary cells, it has been shown that quercetin conjugates at physiologically appropriate concentrations inhibit cell surface expression of VCAM-1, and that quercetin 3-glucuronide, but not other conjugates, also inhibit ICAM-1 expression. Interestingly, these effects were not observed at higher concentrations [42]. The precise mechanism involved in inhibition of cell adhesion molecules is not understood, but may involve inhibition of NF-κB-mediated signalling. Several other studies with different cell lines and polyphenols, including flavanones, isoflavones, anthocyanidins and catechins, are consistent with these results, although many of these use only the aglycones, as opposed the physiologically appropriate metabolites [43, 44]. Moreover, it has been shown that red wine consumption in humans reduces TNF-α induced adhesion of monocytes to endothelial cells *ex vivo* [45].

In addition to changes in expression of adhesion molecules, polyphenols have been shown to affect other risk factors for CVD. For example, several polyphenols, including quercetin, epigallocatechin gallate and resveratrol have been shown to inhibit vascular smooth muscle cell proliferation [46–49] and can modulate the response of platelets to thrombin and other agents [50–52].

Polyphenols and Plasma Protein Interactions

The oral sensation of dryness known as astringency is probably the most familiar example of a polyphenol–protein interaction. Salivary proteins contain multiple binding sites for polyphenols, and when sufficiently high concentrations of polyphenols, and in particular tannins, are present the polyphenol–protein complexes precipitate leading to the taste sensation [53]. Several polyphenols are also known to inactivate digestive enzymes in the gut [54]. Polyphenols can bind with plasma proteins by hydrophobic interactions, hydrogen bonds, and covalent bonds. The nature of the interaction depends upon the structure of the polyphenol, so that there may be a combination of non specific binding, and binding to specific proteins. It has been shown that quercetin will bind with plasma albumins [55], epigallocatechin-gallate binds with plasma fibronectin and fibrinogen [56], and wine catechins with Apo-A1 and transferrin [57]. It is likely that these interactions are far more extensive, and while they have mainly been considered in the context of polyphenol transport in plasma, they may themselves mediate the biological activity of polyphenol metabolites, either through effects such as reducing or enhancing antioxidant activity [40, 57], or interaction with, for example, pro-inflammatory cytokines.

Interactions with ligand receptor proteins have largely been restricted to the isoflavones which are structural mimics of estrogens [58]. Estrogen hormones influence the growth and functioning of many tissues of the male and female reproductive systems. Isoflavones, and in particular genistein, are structural mimics of estrogens and can tightly bind to the estrogen receptors α and β, and act as estrogen agonists [59, 60]. This phenomenon has been offered for an explanation for the possible protective effects of diets rich in soy towards breast and prostate cancer, but it is likely other mechanisms are also of importance.

Glucosinolates and Isothiocyanates: Structure and Human Metabolism

Epidemiological studies have suggested that diets rich in cruciferous vegetables, such as broccoli, may reduce the risk of cancer and myocardial infarction. It is widely thought that isothiocyanates, derived from glucosinolates that accumulate in cruciferous vegetables, are the active component although indole degradation products from tryptophan-derived glucosinolates and other polyphenolic compounds, as discussed above, that are found in these vegetables may also play a role. The glucosinolate molecule consists of a β-thioglucose moiety, a sulfonated oxime moiety and a variable side chain, derived from an amino acid. Glucosinolates with more than 120 side chain structures have been described [61], although only about 16 of these are commonly found within crop plants. Seven of these 120 side chain structures correspond directly to a protein amino acid (alanine, valine, leucine, isoleucine, phenylalanine, tyrosine and tryptophan). The remaining glucosinolates have side chain structures which arise in three ways. Firstly, many glucosinolates are derived from chain-elongated forms of protein amino acids, notably from methionine, but also from phenylalanine and branch chain amino acids. Secondly, the structure of the side chain may be modified after amino acid elongation and glucosinolate biosynthesis by, for example, the oxidation of the methionine sulfur to sulfinyl and sulfonyl, and by the subsequent loss of the ω-methylsulfinyl group to produce a terminal double bond. Subsequent modifications may also involve hydroxylation and methoxylation of the side chain. Chain elongation and modification interact to result in several homologous series of glucosinolates, such as those with methylthioalkyl side chains ranging from $CH_3S\ (CH_2)_3$- to $CH_3(CH_2)_8$-, and methylsulfinylalkyl side chains ranging from $CH_3SO(CH_2)_3$- to $CH_3SO(CH_2)_{11}$-. Thirdly, some glucosinolates occur which contain relatively complex side chains such as o-(α-L-rhamnopylransoyloxy)-benzyl glucosinolate in *Reseda odorata* and glucosinolates containing a sinapoyl moiety in *Raphanus sativus*. Several comprehensive reviews of glucosinolate structure and biosynthesis have recently been published [62–66].

Despite the potential large number of glucosinolates, the major cruciferous crops have a restricted range of glucosinolates. All of these have a mixture of indolylmethyl and N-methoxyindolylmethyl glucosinolates, derived from tryptophan, and either a small number of methionine-derived or phenylalanine-derived glucosinolates. The greatest diversity within a species is found in *B. oleracea*, which includes such crops as broccoli, cabbages, Brussels sprouts and kales. These contain indolyl glucosinolates combined with a small number of methionine-derived glucosinolates. For example, broccoli (*B. oleracea* var. *italica*) accumulates 3-methylsuphinylpropyl and 4-methylsulfinylbutyl glucosinolates, while other botanical forms of *B. oleracea* have mixtures of 2-propenyl, 3-butenyl and 2-hydroxy-3-butenyl. Some cultivars of cabbage and Brussels sprouts also contain significant amounts of methylthiopropyl and methylthiobutyl glucosinolates. *B. rapa* (Chinese cabbage, Bok Choi, turnips etc.) and *B. napus* (Swedes) contains 3-butenyl and sometimes 4-pentenyl glucosinolates and often their hydroxylated homologues. In addition to methionine-derived glucosinolates, phenylethyl glucosinolate usually occurs in low levels in many vegetables. Several surveys of glucosinolate variation between cultivars of *Brassica* species have been reported, for example *B. rapa* [67, 68] and *B. oleracea* [69, 70].

The distinctive taste of many minor horticultural cruciferous crops is due to their glucosinolate content. For example, watercress accumulates large amounts of phenylethyl glucosinolate, combined with low levels of 7-methylsulfinylheptyl and 8-methylsulfinyloctyl glucosinolates, rockets (*Eruca* and *Diplotaxis* species) possess 4-methylthiobutyl glucosinolate, and cress (*Lepidium* spp) contains benzyl glucosinolate.

In the intact plant, glucosinolates are probably located in the vacuole of many cells but may also be concentrated within specialized cells. Following tissue disruption, glucosinolates are hydrolysed by thioglucosidases, known as myrosinases (Fig. 6). Myrosinase activity results in the cleavage of the thio-glucose bond to give rise to unstable thiohydroximate-*O*-sulfonate. This aglycone spontaneously rearranges to produce several products [71]. Most frequently, it undergoes a Lossen rearrangement to produce an isothiocyanate (ITC). Aglycones from glucosinolates which contain β-hydroxylated side chains, such as 2-hydroxy-3-butenyl ('progoitrin') found in the seeds of oilseed rape and some horticultural brassicas, such as Brussels sprouts and Chinese cabbage, spontaneously cyclise to form the corresponding oxazolidine-2-thiones. If the isothiocyanate contains a double bond, and in the presence of an epithiospecifier protein (ESP), the isothiocyanate may rearrange to produce an epithionitrile [72, 73]. ESP is also likely to be involved in the production of nitriles from glucosinolates such as methylsulfinylalkyls [74]. Cooking can denature both ESP and myrosinase [75], in which case intact glucosinolates can be metabolized by microbial thioglucosidases in the colon to generate isothiocyanates [76].

Following ingestion of cruciferous vegetables with intact myrosinase, ITCs will be formed in the mouth and rapidly absorbed in the upper GI and subsequently metabolized [77, 78]. When myrosinases in the plant tissue are deactivated, e.g. by excess cooking, then glucosinolates are hydrolysed in the distal gut by microbial activity, and the resulting ITCs are absorbed from the lower GI tract [78]. Conjugation with glutathione occurs spontaneously but may be further promoted by gluthatione transferases (GST) within the epithelial cells of the GI tract. The glutathione conjugate is then exported to the systemic circulation via the multidrug resistance associated protein-1 (MRP1), MRP2 and P-glycoprotein-1 (Pgp-1) [79, 80].

The ITC-glutathione conjugate is metabolised via the mercapturic acid pathway in which the glutathione conjugate undergoes further enzymatic modifications including cleavage of glutamine, which yields cysteine-glycine- conjugates, cleavage of glycine, yielding cysteine-conjugates and finally acetylation to produce N-acetylcysteine (NAC)-conjugates that are excreted in urine (Fig. 6) [81]. However, it has been shown that 45% of ingested sulforaphane (SF, the major isothiocyanate derived from 4-methylsulphinylbutyl glucosinolate that accumulates in broccoli) in the plasma occurs as the free ITC, as opposed to thiol conjugates, and it has been speculated that the ITC-glutathione conjugate may be cleaved in the plasma to release the free, and biologically active ITC, possibly through GSTM1 activity [77]. The peak concentration of SF and its thiol conjugates following consumption of a standard portion of

18

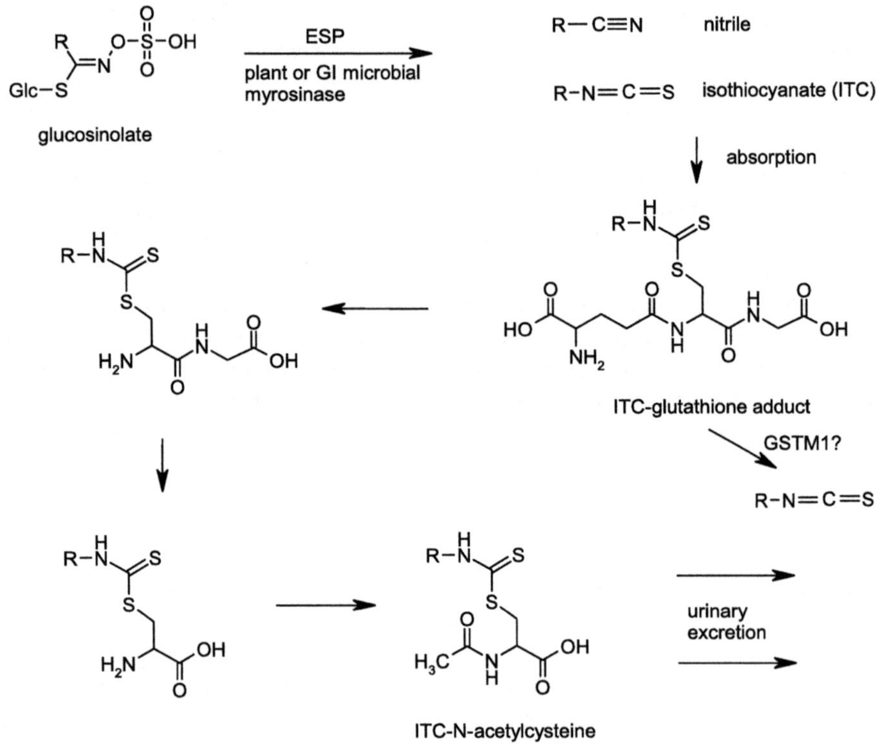

Fig. 6 Glucosinolate structure and metabolism, and side chains (R) occurring in common cruciferous vegetables

broccoli is less than 2 μM, falling to low (nM) levels within a few hours [77].

Diet-Gene Interactions, and the Role of GSTM1 Genotype

Epidemiological evidence both from prospective cohort studies and retrospective case-control studies suggest that there is an inverse association between consumption of cruciferous vegetables and the risk of lung, stomach, colorectal, breast and prostate cancer [14, 82–91]. Several of these studies suggest that the protective effects of crucifer consumption are modulated by GST polymorphisms, and in particular GSTM1 genotype. Fifty percent of the population have a homozygous deletion of the GSTM1 allele. Studies of US populations have suggested that individuals with GSTM1 positive genotype benefit more from consumption of brassica vegetables compared to those that have a homozygous GSTM1 deletion (i.e. GSTM1 null) [82, 85, 91, 92]. However, similar studies conducted on Asian populations have found the converse; with GSTM1 nulls gaining greatest benefit [87, 88]. It has been speculated that this may be due to the contrasting types of vegetables being consumed; broccoli is the major crucifer consumed in the US, whereas in Asia the major cruciferous vegetable consumed is Chinese cabbage. These two vegetables have contrasting types of glucosinolates which may interact with GSTM1 in different ways, and as previously described for structural similar ITCs [93]. Polymorphism also occurs at the GSTT1 locus, with about 20% of Caucasians and up to 60% of people of Asiatic descent having a homozygous deletion. GSTT1 genotype has been associated with modulating the reduction in cancer risk through cruciferous vegetable consumption in some studies, but not others [85, 94, 95]. Although GSTT1 has not been extensively studied there is some evidence that a combination of GSTM1 and GSTT1 genotype might influence cancer risk following brassica consumption [87]. Recently, consumption of cruciferous vegetables was also associated with a lower risk of myocardial infarction among those individuals with a functional GSTT1 allele [96].

Complementary to the epidemiological data, a few experimental studies have shown that GSTM1 null genotypes excrete a greater proportion of ingested SF via mercaturic acid metabolism than those with at least one functional allele [77, 97]. As it is the latter who gain more protection this suggests that there may be other metabolic fates for ITC. Further studies are required.

Mechanistic Basis of Health Promoting Activity of Dietary Glucosinolates and their Derivatives

The anticarcinogenic activity of cruciferous vegetables has largely been attributed to the biological activity of isothiocyanates, although degradation products from indole glucosinolates may also play a role. The activity of sulforaphane, the ITC derived from broccoli has recently been comprehensively reviewed [98]. Other ITCs have similar activity. Sulforaphane has been shown to be protective against carcinogen-induced tumorigenesis at a variety of sites in rodents, including breast, colon, skin, lung, stomach and prostate. It is effective in reducing and even preventing the formation of preneoplastic lesions in tissues resulting from carcinogen administration [99–102] and can also suppress the growth of tumours in spontaneous or xenograft mouse cancer models [103, 104]. The chemopreventive effect of SF is likely to involve multiple mechanisms, which are likely to interact together to reduce risk of carcinogenesis [98]. These include: inhibition of phase 1 enzymes, induction of phase 2 metabolism enzymes, antioxidant functions through increased tissue GSH levels, apoptosis-inducing properties, induction of cell cycle arrest, anti-inflammatory properties and inhibition of angiogenesis. As discussed above, certain caution is required in interpreting these results as these mechanistic studies are usually undertaken with far higher concentrations of ITCs than that which would occur following normal dietary consumption of cruciferous vegetables.

During phase 1 metabolism molecules, including dietary and environmental carcinogens, are converted into highly reactive intermediates that can potentially be harmful by binding to critical macromolecules such as DNA, RNA and protein. SF potently decreased enzyme activities of several cytochrome P450 enzymes (CYPs), which catalyse phase 1 biotransformation, in intact human and rat hepatocytes [105, 106]. Activated carcinogens generated from phase 1 metabolism are subsequently converted into inactive metabolites during the phase 2 metabolism and can readily be excreted from the body. SF has received much attention over the past decade as it was found to be the most potent naturally-occurring inducer of phase 2 enzymes such as quinone reductase (NQO1), GST and genes related to glutathione biosynthesis in both animals and humans [107–110]. Induction of phase 2 enzymes is mediated by the nrf2/Keap1 pathway and exposure of cells to SF leads to dissociation of the Nrf2/Keap1 complex and subsequent nuclear translocation of Nrf2 where it activates cancer protective genes [111].

There are a few studies in humans that have sought to provide evidence for the induction of phase II enzymes *in vivo* following consumption of glucosinolate or isothiocyanate rich diets. A three week diet rich in cruciferous vegetables has been shown to elevate plasma levels of GSTs and reduced the level of 8-oxo-7, 8-dihydro-2'-deoxyguanosine in urine, a marker of oxidative damage [112, 113]. However, an intervention study in Qidong region in the People's Republic of China, where the residents are at high risk of developing hepatocellular carcinoma, partly due to consumption of aflatoxin-contaminated foods, and partly due to exposure to air pollutants such as phenanthrene, did not produce any evidence of differences in excretion of phase 2 metabolic products of aflatoxin and phenanthrene between a diet rich in glucosinolates (via broccoli sprouts) and the placebo control [114]. The large intra-individual variation in excretion of ITC metabolites, which may be partially caused by GST polymorphisms, was noted in this study. A further study quantified global gene expression in gastric mucosa samples after broccoli consumption. While there was evidence for the induction of phase 2 gene expression following consumption of a broccoli that had elevated levels of glucosinolates [115], there was no evidence for induction following consumption of standard broccoli [116]. However, there is evidence that topical application of ITCs to skin can induce phase 2 enzymes [117].

Using cancer cell models ITCs have also been shown to induce apoptosis and cell cycle arrest through a variety of pathways depending on the origin of cells used. The mechanism for such an effect involves induction of several members of the caspase family, responsible for the execution of apoptosis in higher eukaryotes, as well as the induction of the pro-apoptotic Bcl-2 family members. Alternatively, SF can also induce apoptosis through induction of any of the three parallel MAPK cascades identified in mammalian cells, extracellular signal-regulated kinase (ERK), c-Jun N-terminal kinase (JNK) and p38, depending on tissue and dose applied. In line with cancer suppressing properties SF has been shown to arrest the cell cycle through regulation of cyclin levels and induction of p21, a potent inhibitor of cell cycle progression. Additionally, SF down-regulates important mediators of the pro-inflammatory response such as iNOS, Cox-2 and NF-κB, and also inhibits angiogenesis, both processes thought to be mechanistically linked with carcinogenesis.

This brief review has focused on the two classes of plant natural products – flavonoids and glucosinolates – for which the epidemiological evidence for health benefits is strongest. This is, of course, largely due to the frequent occurrences of these compounds in crops and foods. Other natural products that have a limited distribution in certain foods may indeed have far greater health benefits, but these would not be evident from epidemiological studies. The greatest challenge in this field of research is to design human intervention studies of sufficient length coupled with analyses of target tissues to provide a mechanistic understanding of the activity of plant natural products *in vivo*. The use of genetically modified plants with specific alterations in natural product profiles will greatly facilitate these studies.

Glossary

Case control studies

Prospective studies

Flavonoids

Glucosinolates

Antioxidant

Atherosclerosis

Isothiocyanates

Phase 2 metabolism

References

1. Arts, I.C. and Hollman, P.C. (2005) Polyphenols and disease risk in epidemiologic studies. *Am J Clin Nutr* 81 (1 Suppl), 317S–325S
2. Mink, P.J. et al. (2007) Flavonoid intake and cardiovascular disease mortality: a prospective study in postmenopausal women. *Am J Clin Nutr* 85 (3), 895–909
3. Gao, X. et al. (2007) Prospective study of dietary pattern and risk of Parkinson disease. *Am J Clin Nutr* 86 (5), 1486–1494
4. Varraso, R. et al. (2007) Prospective study of dietary patterns and chronic obstructive pulmonary disease among US women. *Am J Clin Nutr* 86 (2), 488–495
5. Anon. (2003) USDA database for the flavanoid content of selected foods. *http://www.nal.usda.gov/fnic/foodcomp/Data/Flav/flav.html*
6. Anon. (2004) USDA database for the proanthocyanidin content of selected foods. *http://www.nalusda.gov.fnic/foodcomp/Data/PA/PA.html*
7. Anon. (1999) USDA-Iowa State University database on the isoflavone content of foods. *http://www.nal.usda.gov/fnic/foodcomp/Data/isoflav/isoflav.html*
8. Koushik, A. et al. (2007) Fruits, vegetables, and colon cancer risk in a pooled analysis of 14 cohort studies. *J Nat Cancer Inst* 99 (19), 1471–1483
9. Oba, S. et al. (2007) Soy product consumption and the risk of colon cancer: a prospective study in Takayama, Japan. *Nutr Cancer* 57 (2), 151–157
10. Rohrmann, S. et al. (2007) Fruit and vegetable consumption and lymphoma risk in the European Prospective Investigation into Cancer and Nutrition (EPIC). *Cancer Causes Control* 18 (5), 537–549
11. Holick, C.N. et al. (2007) Prospective study of intake of fruit, vegetables, and carotenoids and the risk of adult glioma. *Am J Clin Nutr* 85 (3), 877–886
12. McCullough, M.L. et al. (2007) A prospective study of fruits, vegetables, and risk of endometrial cancer. *Am J Epidemiol* 166 (8), 902–911
13. Nishio, K. et al. (2007) Consumption of soy foods and the risk of breast cancer: findings from the Japan Collaborative Cohort (JACC) Study. *Cancer Causes Control* 18 (8), 801–808
14. Kirsh, V.A. et al. (2007) Prospective study of fruit and vegetable intake and risk of prostate cancer. *J Natl Cancer Inst* 99 (15), 1200–1209
15. Kurahashi, N. et al. (2007) Soy isoflavone consumption is not associated with increased risk of advanced prostate cancer. *Cancer Epidemiol Biomarkers Prev* 16 (10), 2169
16. Larsson, S.C. et al. (2006) Fruit and vegetable consumption and incidence of gastric cancer: a prospective study. *Cancer Epidemiol Biomarkers Prev* 15 (10), 1998–2001
17. Vivekananthan, D.P. et al. (2003) Use of antioxidant vitamins for the prevention of cardiovascular disease: meta-analysis of randomised trials. *Lancet* 361 (9374), 2017–2023
18. Omenn, G.S. (2007) Chemoprevention of lung cancers: lessons from CARET, the beta-carotene and retinol efficacy trial, and prospects for the future. *Eur J Cancer Prev* 16 (3), 184–191
19. Lawson, K.A. et al. (2007) Multivitamin use and risk of prostate cancer in the National Institutes of Health-AARP Diet and Health Study. *J Nat Cancer Inst* 99 (10), 754–764
20. O'Kennedy, N. et al. (2006) Effects of tomato extract on platelet function: a double-blinded crossover study in healthy humans. *Am J Clin Nutr* 84 (3), 561–569
21. Sies, H. et al. (2005) Cocoa polyphenols and inflammatory mediators. *Am J Clin Nutr* 81 (1 Suppl), 304S–312S
22. Baba, S. et al. (2007) Plasma LDL and HDL cholesterol and oxidized LDL concentrations are altered in normo- and hypercholesterolemic humans after intake of different levels of cocoa powder. *J Nutr* 137 (6), 1436–1441
23. Francis, S.T. et al. (2006) The effect of flavanol-rich cocoa on the fMRI response to a cognitive task in healthy young people. *J Cardiovasc Pharmacol* 47 (Suppl 2), S215–220
24. OM, A. and KR, M. (2006) *Flavonoids: Chemistry, biochemistry and applications*, Taylor & Francis, Boca Raton, FL
25. Day, A.J. et al. (2000) Dietary flavonoid and isoflavone glycosides are hydrolysed by the lactase site of lactase phlorizin hydrolase. *FEBS Lett* 468 (2–3), 166–170
26. Gee, J.M. et al. (2000) Intestinal transport of quercetin glycosides in rats involves both deglycosylation and interaction with the hexose transport pathway. *J Nutr* 130 (11), 2765–2771
27. Day, A.J. et al. (1998) Deglycosylation of flavonoid and isoflavonoid glycosides by human small intestine and liver beta-glucosidase activity. *FEBS Lett* 436 (1), 71–75
28. Thompson, L.U. et al. (1984) Relationship between polyphenol intake and blood glucose response of normal and diabetic individuals. *Am J Clin Nutr* 39 (5), 745–751

29. Gonthier, M.P. et al. (2003) Metabolism of dietary procyanidins in rats. *Free Radic Biol Med* 35 (8), 837–844
30. Vitaglione, P. et al. (2007) Protocatechuic acid is the major human metabolite of cyanidin-glucosides. *J Nutr* 137 (9), 2043–2048
31. Sadilova, E. et al. (2007) Thermal degradation of anthocyanins and its impact on color and in vitro antioxidant capacity. *Mol Nutr Food Res* 51 (12), 1461–1471
32. Mullen, W. et al. (2004) Flavonoid metabolites in human plasma and urine after the consumption of red onions: analysis by liquid chromatography with photodiode array and full scan tandem mass spectrometric detection. *J Chromatogr A* 1058 (1–2), 163–168
33. Mullen, W. et al. (2006) Absorption, excretion and metabolite profiling of methyl-, glucuronyl-, glucosyl- and sulpho-conjugates of quercetin in human plasma and urine after ingestion of onions. *Br J Nutr* 96 (1), 107–116
34. Clarke, D.B. et al. (2002) Measurement of intact sulfate and glucuronide phytoestrogen conjugates in human urine using isotope dilution liquid chromatography-tandem mass spectrometry with [13C(3)]isoflavone internal standards. *Anal Biochem* 309 (1), 158–172
35. Kahle, K. et al. (2007) Polyphenols are intensively metabolized in the human gastrointestinal tract after apple juice consumption. *J Agric Food Chem* 55 (26), 10605–10614
36. Clifford, M.N. (2004) Diet-derived phenols in plasma and tissues and their implications for health. *Planta Med* 70 (12), 1103–1114
37. Janisch, K.M. et al. (2004) Properties of quercetin conjugates: modulation of LDL oxidation and binding to human serum albumin. *Free Radic Res* 38 (8), 877–884
38. Shirai, M. et al. (2006) Effect of a conjugated quercetin metabolite, quercetin 3-glucuronide, on lipid hydroperoxide-dependent formation of reactive oxygen species in differentiated PC-12 cells. *Free Radic Res* 40 (10), 1047–1053
39. Rimbach, G. et al. (2004) Sulfation of genistein alters its antioxidant properties and its effect on platelet aggregation and monocyte and endothelial function. *Biochim Biophys Acta* 1670 (3), 229–237
40. Arts, M.J. et al. (2001) Masking of antioxidant capacity by the interaction of flavonoids with protein. *Food Chem Toxicol* 39 (8), 787–791
41. Shimoi, K. et al. (2001) Deglucuronidation of a flavonoid, luteolin monoglucuronide, during inflammation. *Drug Metab Dispos* 29 (12), 1521–1524
42. Tribolo, S. et al. (2007) Comparative effects of quercetin and its predominant human metabolites on adhesion molecule expression in activated human vascular endothelial cells. *Atherosclerosis* [Epub ahead of print]
43. Zhang, F.L. et al. (2006) Selective inhibition by grape seed proanthocyanidin extracts of cell adhesion molecule expression induced by advanced glycation end products in endothelial cells. *J Cardiovasc Pharmacol* 48 (2), 47–53
44. Lo, H.M. et al. (2007) Tea polyphenols inhibit rat vascular smooth muscle cell adhesion and migration on collagen and laminin via interference with cell-ECM interaction. *J Biomed Sci* 14 (5), 637–645
45. Sacanella, E. et al. (2007) Down-regulation of adhesion molecules and other inflammatory biomarkers after moderate wine consumption in healthy women: a randomized trial. *Am J Clin Nutr* 86 (5), 1463–1469
46. Nie, L. et al. (2006) Mechanism by which avenanthramide-c, a polyphenol of oats, blocks cell cycle progression in vascular smooth muscle cells. *Free Radic Biol Med* 41 (5), 702–708
47. Wang, Z. et al. (2006) Regulation of proliferation and gene expression in cultured human aortic smooth muscle cells by resveratrol and standardized grape extracts. *Biochem Biophys Res Commun* 346 (1), 367–376
48. Hofmann, C.S. and Sonenshein, G.E. (2003) Green tea polyphenol epigallocatechin-3 gallate induces apoptosis of proliferating vascular smooth muscle cells via activation of p53. *FASEB J* 17 (6), 702–704
49. Araim, O. et al. (2002) Inhibition of vascular smooth muscle cell proliferation with red wine and red wine polyphenols. *J Vasc Surg* 35 (6), 1226–1232
50. Rechner, A.R. and Kroner, C. (2005) Anthocyanins and colonic metabolites of dietary polyphenols inhibit platelet function. *Thromb Res* 116 (4), 327–334
51. Kaneider, N.C. et al. (2004) Inhibition of thrombin-induced signaling by resveratrol and quercetin: effects on adenosine nucleotide metabolism in endothelial cells and platelet-neutrophil interactions. *Thromb Res* 114 (3), 185–194
52. Olas, B. et al. (2002) Effect of resveratrol, a natural polyphenolic compound, on platelet activation induced by endotoxin or thrombin. *Thromb Res* 107 (3–4), 141–145
53. Lesschaeve, I. and Noble, A.C. (2005) Polyphenols: factors influencing their sensory properties and their effects on food and beverage

preferences. *Am J Clin Nutr* 81 (1 Suppl), 330S–335S
54. McDougall, G.J. and Stewart, D. (2005) The inhibitory effects of berry polyphenols on digestive enzymes. *Biofactors* 23 (4), 189–195
55. Manach, C. et al. (1995) Quercetin metabolites in plasma of rats fed diets containing rutin or quercetin. *J Nutr* 125 (7), 1911–1922
56. Sazuka, M. et al. (1996) Evidence for the interaction between (-)-epigallocatechin gallate and human plasma proteins fibronectin, fibrinogen, and histidine-rich glycoprotein. *Biosci Biotechnol Biochem* 60 (8), 1317–1319
57. Brunet, M.J. et al. (2002) Human apo A-I and rat transferrin are the principal plasma proteins that bind wine catechins. *J Agric Food Chem* 50 (9), 2708–2712
58. Miksicek, R.J. (1995) Estrogenic flavonoids: structural requirements for biological activity. *Proc Soc Exp Biol Med* 208 (1), 44–50
59. Kuiper, G.G. et al. (1998) Interaction of estrogenic chemicals and phytoestrogens with estrogen receptor beta. *Endocrinology* 139 (10), 4252–4263
60. Mueller, S.O. et al. (2004) Phytoestrogens and their human metabolites show distinct agonistic and antagonistic properties on estrogen receptor alpha (ERalpha) and ERbeta in human cells. *Toxicol Sci* 80 (1), 14–25
61. Fahey, J.W. et al. (2001) The chemical diversity and distribution of glucosinolates and isothiocyanates among plants. *Phytochemistry* 56 (1), 5–51
62. Mithen, R. (2001) Glucosinolates and their degradation products. *Adv Botanical Res* 35, 214–262
63. Mithen, R.F. (2001) Glucosinolates and their degradation products. *Adv Botanical Res* 35, 213–262
64. Mithen, R.F. et al. (2000) The nutritional significance, biosynthesis and bioavailability of glucosinolates in human foods. *J Sci Food Agric* 80 (7), 967–984
65. Halkier, B.A. and Gershenzon, J. (2006) Biology and biochemistry of glucosinolates. *Annu Rev Plant Biol* 57, 303–333
66. Yan, X. and Chen, S. (2007) Regulation of plant glucosinolate metabolism. *Planta* 226 (6), 1343–1352
67. Carlson, D.G. et al. (1987) Glucosinolates in turnip tops and roots – cultivars grown for greens and or roots. *J Am Soc Horticultural Sci* 112 (1), 179–183
68. Hill, C.B. et al. (1987) Variation in glucosinolates in oriental Brassica vegetables. *J Am Soc Horticultural Sci* 112 (2), 309–313
69. Carlson, D.G. et al. (1987) Glucosinolates in crucifer vegetables - Broccoli, brussels- sprouts, cauliflower, collards, kale, mustard greens, and kohlrabi. *J Am Soc Horticultural Sci* 112 (1), 173–178
70. Kushad, M.M. et al. (1999) Variation of glucosinolates in vegetable crops of Brassica oleracea. *J Agric Food Chem* 47 (4), 1541–1548
71. Bones, A.M. and Rossiter, J.T. (1996) The myrosinase-glucosinolate system, its organisation and biochemistry. *Physiologia Plantarum* 97 (1), 194–208
72. Foo, H.L. et al. (2000) Purification and characterisation of epithiospecifier protein from Brassica napus: enzymic intramolecular sulphur addition within alkenyl thiohydroximates derived from alkenyl glucosinolate hydrolysis. *FEBS Lett* 468 (2–3), 243–246
73. Lambrix, V. et al. (2001) The Arabidopsis epithiospecifier protein promotes the hydrolysis of glucosinolates to nitriles and influences Trichoplusia ni herbivory. *Plant Cell* 13 (12), 2793–2807
74. Matusheski, N.V. et al. (2006) Epithiospecifier protein from broccoli (Brassica oleracea L. ssp. italica) inhibits formation of the anticancer agent sulforaphane. *J Agric Food Chem* 54 (6), 2069–2076
75. Sarikamis, G. et al. (2006) High glucosinolate broccoli: a delivery system for sulforaphane. *Mol Breed* 18 (3), 219–228
76. Rouzaud, G. et al. (2003) Influence of plant and bacterial myrosinase activity on the metabolic fate of glucosinolates in gnotobiotic rats. *Brit J Nutri* 90, 395–404
77. Gasper, A.V. et al. (2005) Glutathione S-transferase M1 polymorphism and metabolism of sulforaphane from standard and high-glucosinolate broccoli. *Am J Clin Nutr* 82 (6), 1283–1291
78. Conaway, C.C. et al. (2000) Disposition of glucosinolates and sulforaphane in humans after ingestion of steamed and fresh broccoli. *Nutr Cancer* 38 (2), 168–178
79. Payen, L. et al. (2001) Reactive oxygen species-related induction of multidrug resistance-associated protein 2 expression in primary hepatocytes exposed to sulforaphane. *Biochem Biophys Res Commun* 282 (1), 257–263
80. Zhang, Y. and Callaway, E.C. (2002) High cellular accumulation of sulphoraphane, a dietary anticarcinogen, is followed by rapid transporter-mediated export as a glutathione conjugate. *Biochem J* 364 (Pt 1), 301–307
81. Brusewitz, G. et al. (1977) The metabolism of benzyl isothiocyanate and its cysteine conjugate. *Biochem J* 162 (1), 99–107

18

82. Lin, H.J. et al. (1998) Glutathione transferase null genotype, broccoli, and lower prevalence of colorectal adenomas. *Cancer Epidemiol Biomarkers Prev* 7 (8), 647–652
83. Cohen, J.H. et al. (2000) Fruit and vegetable intakes and prostate cancer risk. *J Nat Cancer Inst* 92 (1), 61–68
84. London, S.J. et al. (2000) Isothiocyanates, glutathione S-transferase M1 and T1 polymorphisms, and lung-cancer risk: a prospective study of men in Shanghai, China. *Lancet* 356 (9231), 724–729
85. Spitz, M.R. et al. (2000) Dietary intake of isothiocyanates: Evidence of a joint effect with glutathione S-transferase polymorphisms in lung cancer risk. *Cancer Epidemiol Biomarkers Prev* 9 (10), 1017–1020
86. Zhao, B. et al. (2001) Dietary isothiocyanates, glutathione S-transferase-M1,-T1 polymorphisms and lung cancer risk among Chinese women in Singapore. *Cancer Epidemiol Biomarkers Prev* 10 (10), 1063–1067
87. Seow, A. et al. (2002) Dietary isothiocyanates, glutathione S-transferase polymorphisms and colorectal cancer risk in the Singapore Chinese Health Study. *Carcinogenesis* 23 (12), 2055–2061
88. Fowke, J.H. et al. (2003) Urinary isothiocyanate levels, brassica, and human breast cancer. *Cancer Res* 63 (14), 3980–3986
89. Giovannucci, E. et al. (2003) A prospective study of cruciferous vegetables and prostate cancer. *Cancer Epidemiol Biomarkers Prev* 12 (12), 1403–1409
90. Ambrosone, C.B. et al. (2004) Breast cancer risk in premenopausal women is inversely associated with consumption of broccoli, a source of isothiocyanates, but is not modified by GST genotype. *J Nutr* 134 (5), 1134–1138
91. Wang, L.I. et al. (2004) Dietary intake of Cruciferous vegetables, Glutathione S-transferase (GST) polymorphisms and lung cancer risk in a Caucasian population. *Cancer Causes Control* 15 (10), 977–985
92. Joseph, M.A. et al. (2004) Cruciferous vegetables, genetic polymorphisms in glutathione s-transferases m1 and t1, and prostate cancer risk. *Nutr Cancer* 50 (2), 206–213
93. Zhang, Y. et al. (1995) Reversible conjugation of isothiocyanates with glutathione catalyzed by human glutathione transferases. *Biochem Biophys Res Commun* 206 (2), 748–755
94. Lin, H.J. et al. (2002) Glutathione transferase GSTT1, broccoli, and prevalence of colorectal adenomas. *Pharmacogenetics* 12 (2), 175–179
95. Moore, L. et al. (2007) Glutathione S-transferase polymorphisms, cruciferous vegetable intake, and cancer risk in the Central and Eastern European Kidney Cancer Study. *Carcinogenesis* 28:1960–1964
96. Cornelis, M.C. et al. (2007) GSTT1 genotype modifies the association between cruciferous vegetable intake and the risk of myocardial infarction. *Am J Clin Nutr* 86 (3), 752–758
97. Steck, S.E. et al. (2007) GSTM1, GSTT1, GSTP1, and GSTA1 polymorphisms and urinary isothiocyanate metabolites following broccoli consumption in humans. *J Nutr* 137 (4), 904–909
98. Juge, N. et al. (2007) Molecular basis for chemoprevention by sulforaphane: a comprehensive review. *Cell Mol Life Sci* 64: 1105–1127
99. Gerhauser, C. et al. (1997) Cancer chemopreventive potential of sulforamate, a novel analogue of sulforaphane that induces phase 2 drug-metabolizing enzymes. *Cancer Res* 57 (2), 272–278
100. Chung, F.L. et al. (2000) Chemoprevention of colonic aberrant crypt foci in Fischer rats by sulforaphane and phenethyl isothiocyanate. *Carcinogenesis* 21 (12), 2287–2291
101. Fahey, J.W. et al. (2002) Sulforaphane inhibits extracellular, intracellular, and antibiotic-resistant strains of Helicobacter pylori and prevents benzo a pyrene-induced stomach tumors. *Proc Nat Acad Sci USA* 99 (11), 7610–7615
102. Conaway, C.C. et al. (2005) Phenethyl isothiocyanate and sulforaphane and their N-acetylcysteine conjugates inhibit malignant progression of lung adenomas induced by tobacco carcinogens in A/J mice. *Cancer Res* 65 (18), 8548–8557
103. Singh, A.V. et al. (2004) Sulforaphane induces caspase-mediated apoptosis in cultured PC-3 human prostate cancer cells and retards growth of PC-3 xenografts in vivo. *Carcinogenesis* 25 (1), 83–90
104. Khor, T.O. et al. (2006) Pharmacogenomics of cancer chemopreventive isothiocyanate compound sulforaphane in the intestinal polyps of ApcMin/+ mice. *Biopharm Drug Dispos* 27 (9), 407–420
105. Barcelo, S. et al. (1996) CYP2E1-mediated mechanism of anti-genotoxicity of the broccoli constituent sulforaphane. *Carcinogenesis* 17 (2), 277–282
106. Maheo, K. et al. (1997) Inhibition of cytochromes P-450 and induction of glutathione S-transferases by sulforaphane in primary human and rat hepatocytes. *Cancer Res* 57 (17), 3649–3652

107. Prochaska, H.J. et al. (1992) Rapid detection of inducers of enzymes that protect against carcinogens. *Proc Natl Acad Sci USA* 89 (6), 2394–2398
108. Zhang, Y. et al. (1992) A major inducer of anticarcinogenic protective enzymes from broccoli: isolation and elucidation of structure. *Proc Natl Acad Sci USA* 89 (6), 2399–2403
109. Talalay, P. (2000) Chemoprotection against cancer by induction of phase 2 enzymes. *Biofactors* 12 (1–4), 5–11
110. Traka, M. et al. (2005) Transcriptome analysis of human colon caco-2 cells exposed to sulforaphane. *J Nutr* 135 (8), 1865–1872
111. McMahon, M. et al. (2003) Keap1-dependent proteasomal degradation of transcription factor Nrf2 contributes to the negative regulation of antioxidant response element-driven gene expression. *J Biol Chem* 278 (24), 21592–21600
112. Bogaards, J.J. et al. (1994) Consumption of Brussels sprouts results in elevated alpha-class glutathione S-transferase levels in human blood plasma. *Carcinogenesis* 15 (5), 1073–1075
113. Verhagen, H. et al. (1995) Reduction of oxidative DNA-damage in humans by brussels sprouts. *Carcinogenesis* 16 (4), 969–970
114. Cornblatt, B.S. et al. (2007) Preclinical and clinical evaluation of sulforaphane for chemoprevention in the breast. *Carcinogenesis* 28: 1485–1490
115. Mithen, R. et al. (2003) Development of isothiocyanate-enriched broccoli, and its enhanced ability to induce phase 2 detoxification enzymes in mammalian cells. *Theoret Appl Genet* 106 (4), 727–734
116. Gasper, A.V. et al. (2007) Consuming broccoli does not induce genes associated with xenobiotic metabolism and cell cycle control in human gastric mucosa. *J Nutr* 137 (7), 1718–1724
117. Dinkova-Kostova, A.T. et al. (2007) Induction of the phase 2 response in mouse and human skin by sulforaphane-containing broccoli sprout extracts. *Cancer Epidemiol Biomarkers Prev* 16 (4), 847–851

Floral Scents and Fruit Aromas Inspired by Nature

19

Florence Negre-Zakharov, Michael C. Long, and Natalia Dudareva

Abstract Plants use floral and fruit volatiles as chemical cues to interact with their environment by attracting pollinators and seed dispersers, thus ensuring plant reproductive success. These volatiles also have a significant economic value as they contribute directly to the quality, and indirectly to the yield, of crops. The scent of flowers and the aroma of fruits are composed of complex mixtures of tens or sometimes hundreds of volatile compounds, many of which are found in both flowers and fruits. Arising from diverse biochemical pathways, floral and fruit volatiles can be divided into four major classes according to their metabolic origin: terpenoids, phenylpropanoids/benzenoids, fatty acid derivatives and amino acid derivatives. Recent discoveries of genes and enzymes responsible for the formation of volatile compounds have facilitated the investigation of the regulation of the biosynthesis of flower and fruit volatiles. Our growing understanding of the plant volatile network, together with pioneering attempts for fragrance modification, provide a platform for future metabolic engineering of floral scent and fruit aroma for plant improvement and human enjoyment.

F. Negre-Zakharov
Department of Plant Sciences, University of California, Davis, CA 95616
e-mail: fnegre@plantsciences.ucdavis.edu

M.C. Long and N. Dudareva (✉)
Department of Horticulture and Landscape Architecture, Purdue University, West Lafayette, IN 47907, USA
e-mails: longmc@purdue.edu; dudareva@purdue.edu

Abbreviations

AADC	aromatic amino acid decarboxylase
AAT	alcohol acyltransferase
ABC	adenosine triphosphate binding cassette
ADH	alcohol dehydrogenase
AHCT	anthocyanin O-hydroxycinnamoyltransferase
AOS	allene oxide synthase
BA	benzoic acid
BAMT	benzoic acid carboxyl methyl transferase
BEAT	acetyl-CoA:benzyl alcohol acetyltransferase
BPBT	benzyl alcohol/phenylethanol benzoyl transferase
BSMT	benzoic acid/salicylic acid carboxyl methyltransferase
CA	cinnamic acid
CCD	carotenoid cleavage dioxygenase
CFAT	coniferyl alcohol acyltransferase
CoA	coenzyme-A
DAHP	3-deoxy-D-arabino-heptulosonate 7-phosphate
DAT	deacetylvindoline 4-O-acetyltransferase
DMAPP	dimethylallyl diphosphate

A.E. Osbourn and V. Lanzotti (eds.), *Plant-derived Natural Products*,
DOI 10.1007/978-0-387-85498-4_19, © Springer Science+Business Media, LLC 2009

EGS	eugenol synthase
Ery4P	erythrose 4-phosphate
F6P	fructose 6-phosphate
FPP	farnesyl diphosphate
FPPS	farnesyl pyrophosphate synthase
GA-3P	glyceraldehyde-3-phosphate
G6P	glucose 6-phosphate
GC-EAD	gas chromatography coupled with electroantennogram detection
GES	geraniol synthase
GGPP	geranylgeranyl pyrophosphate
GGPPS	GGPP synthase
GPP	geranyl diphosphate
GPPS	GPP synthase
HCBT	anthranilate N-hydroxycinnamoyl/ benzoyltransferase
HPL	hydroperoxyde lyase
IDI	isopentenyl diphosphate isomerase
IGS	isoeugenol synthase
IGL	indole-3-glycerol phosphate lyase
Indole-3GP	indole 3-glycerol phosphate
IPP	isopentenyl diphosphate
JMT	jasmonic acid carboxyl methyl transferase
LIS	(*S*)-linalool synthase
LOX	lipoxygenase
LTP	lipid transfer proteins
MEP	methyl-erythritol-phosphate
MVA	mevalonic acid
PAAS	phenylacetaldehyde synthase
PAL	L-phenylalanine ammonia-lyase
Pap1	production of anthocyanin pigment 1
PEP	phosphoenolpyruvate
Phe	L-phenylalanine
RhAAT	rose alcohol acyltransferase
SAAT	strawberry alcohol acyltransferase
TPS	terpene synthase

Introduction

Within the wide array of natural products found in plants, volatile metabolites can be considered as the plant's interface with the surrounding environment. These volatile compounds can be released from flowers, fruits, leaves, and roots into the atmosphere or soil, allowing the plant to interact with other organisms while remaining anchored to the ground. Volatiles are low molecular weight metabolites (less than 300 Da) with diverse chemical structures (hydrocarbons, alcohols, aldehydes, ketones, ethers, and esters), and originate from several biosynthetic pathways, including the terpenoid, phenylpropanoid/benzenoid, and lipoxygenase pathways. They typically exist as lipophilic liquids with high vapor pressures and can cross cell membranes, walls and cuticles to be released into the atmosphere [1, 2].

Volatiles have been shown to play a role in many aspects of a plant's life cycle, and their functions are extensively reviewed in Chapters 17 and 18 of this book. Leaf- or root-emitted volatiles can be involved in the plant's defense by directly repelling herbivores or pathogens, or recruiting enemies of their aggressors to limit or eliminate further damage [3–6]. Volatile compounds emitted from flowers provide chemical cues to pollinators, thereby ensuring plant reproductive and evolutionary success [7]. Although the exact olfactory information used by pollinators to recognize and locate flowers is still unclear, it was recently shown that honeybees are capable of using all emitted volatiles to discriminate subtle, quantitative differences in the scent [8]. In addition to providing guidance to pollinators [9], floral volatiles could serve as part of a defense mechanism against florivores or protect the flowers from abiotic stress, as was reported for many volatiles released from vegetative tissues. Emission of volatiles with antimicrobial activity can also protect the carbohydrate-rich nectar by inhibiting microbial growth. However, some floral volatiles could attract herbivores, resulting in a challenging tradeoff between pollinator and herbivore attraction [10, 11].

Volatiles emitted from fruits determine their overall aroma properties and flavor which are perceived through olfactory detection of multiple distinct sensory inputs processed to generate the overall sensation [12]. Thus, fruit volatiles could play a role in the attraction of animal seed dispersers. For example, fruit bats can detect and locate fruits, as well as assess their ripeness, based on odor alone [13, 14]. Olfactory cues emanating from the food source are especially important in food location for nocturnal foragers which have limited access to visual cues, as was suggested for night monkeys and lemurs [15–17].

For humans, aromas and scents constitute an important part of social and cultural interactions. Although humans do not strictly depend on olfactory cues for survival, our sense of smell lets us enjoy fragrances, foods and beverages, and can also warn us of dangers like poisons or smoke. The important impact of olfaction in human society may explain the existence of entire industries dedicated to the extraction, production and sale of natural or artificial aromas and fragrances, constituting a $16 billion market worldwide [18]. In this sense, floral and fruit volatiles are rich biological resources of significant importance from an economical and commercial standpoint. Volatiles emitted from flowers determine their characteristic scent, a trait that is becoming an increasingly important factor for consumers in assessing the overall quality of ornamentals in the cut flower industry [19]. The aroma volatiles produced during fruit and vegetable ripening contribute significantly to their sensory attributes and greatly determine consumer preference to them [20].

In this chapter, we review the biosynthetic pathways responsible for the formation of floral and fruit volatiles, the regulation and localization of scent and aroma biosynthesis, and the recent advances in biotechnological manipulation of floral fragrances and fruit aromas.

Diversity, Biosynthesis and Regulation of Floral and Fruit Volatiles

To date, 1,700 compounds were identified in the scent of flowers belonging to 90 plant families [9] in addition to 700 flavor volatiles known to be present in aromas of fruits and vegetables [21] with many compounds found in both flowers and fruits. In general, the scent of flowers and the aroma of fruits are composed of complex mixtures of tens or sometimes hundreds of volatile compounds arising from diverse biochemical pathways (Figs. 1 and 2). Although different flowers as well as fruits share many aroma constituents, each flower/fruit has a distinctive aroma that is a result of the relative abundances and interactions between emitted volatiles in addition to the presence or absence of unique components. In contrast to scent bouquets which usually contain between 20 to 60 different compounds in most plant species [22], aromas of fruits are generally more complex and often consist of more than 150 volatiles [23]. While 400 different volatiles have been found in tomato fruits [24, 25], 360 in ripening strawberry [26], 350 in banana [27], 225 in grapes [28] and over 200 in red raspberries [29], a relatively small subset of these compounds, based on their concentrations and odor thresholds for human organoleptic perception, determines the "aroma fingerprint" of each particular fruit. Compounds with very low odor thresholds can be perceived by humans at extremely low concentrations. A good example is β-ionone (Fig. 2) which is present only in minute quantities (4 nL L^{-1}) in tomato fruits, but represents the second most important aroma compound in this fruit due to its extremely low threshold (0.007 nL L^{-1}) [30]. Similarly, the characteristic strawberry fruit aroma depends on the presence of small quantities of volatiles with low threshold values including the pleasant-smelling esters γ– and δ-lactones and methyl anthranilate as well as the fruit's most important aroma compounds,

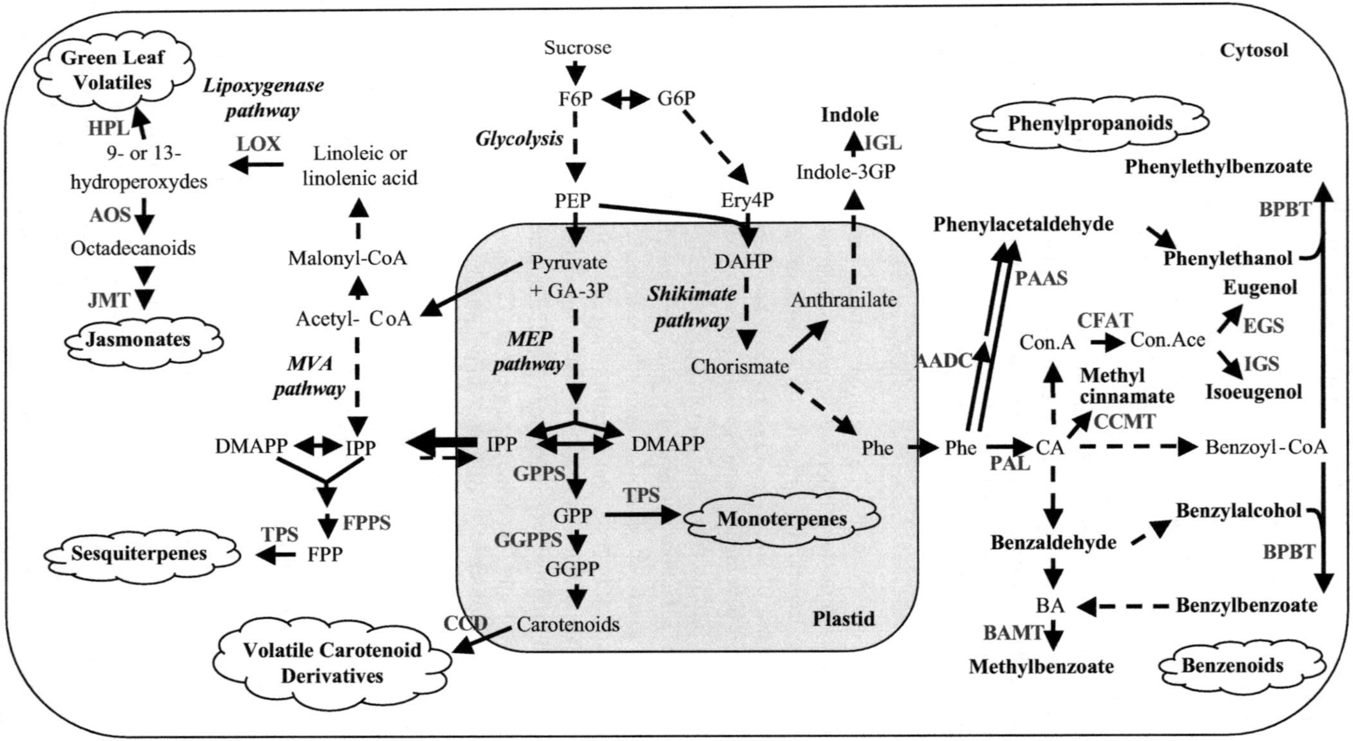

Fig. 1 Metabolic routes involved in the formation of the major classes of scent and aroma compounds in flowers and fruits. Volatiles are shown in bold, enzymes are shadowed, and pathway names are italicized. Abbreviations: AADC, amino acid decarboxylase; AOS, allene oxide synthase; BA, benzoic acid; BAMT, benzoic acid carboxyl methyl transferase; BPBT, benzyl alcohol/phenylethanol benzoyl transferase; CA, cinnamic acid; CCD, carotenoid cleavage dioxygenase; CCMT, cinnamate/*p*-coumarate carboxyl methyltransferase; CFAT, coniferyl alcohol acyltransferase; CoA, coenzyme-A; Con.A, coniferyl alcohol; Con. Ace, coniferyl acetate; DAHP, 3-deoxy-D-arabino-heptulosonate 7-phosphate; DMAPP, dimethylallyl diphosphate; EGS, eugenol synthase; Ery4P, erythrose 4-phosphate; F6P, fructose 6-phosphate; FPP, farnesyl diphosphate; FPPS, FPP synthase; GA-3P, glyceraldehyde-3-phosphate; G6P, glucose 6-phosphate; GGPP, geranylgeranyl diphosphate; GGPPS, GGPP synthase; GPP, geranyl diphosphate; GPPS, GPP synthase; HPL, fatty acid hydroperoxide lyase; IGL, indole-3-glycerol phosphate lyase; IGS, isoeugenol synthase; Indole-3GP, indole 3-glycerol phosphate; IPP, isopentenyl diphosphate; JMT, jasmonic acid carboxyl methyl transferase; LOX, lipoxygenase; MEP, 2-C-methyl-D-erythritol 4-phosphate; MVA, mevalonate; PAAS, phenylacetaldehyde synthase; PAL,

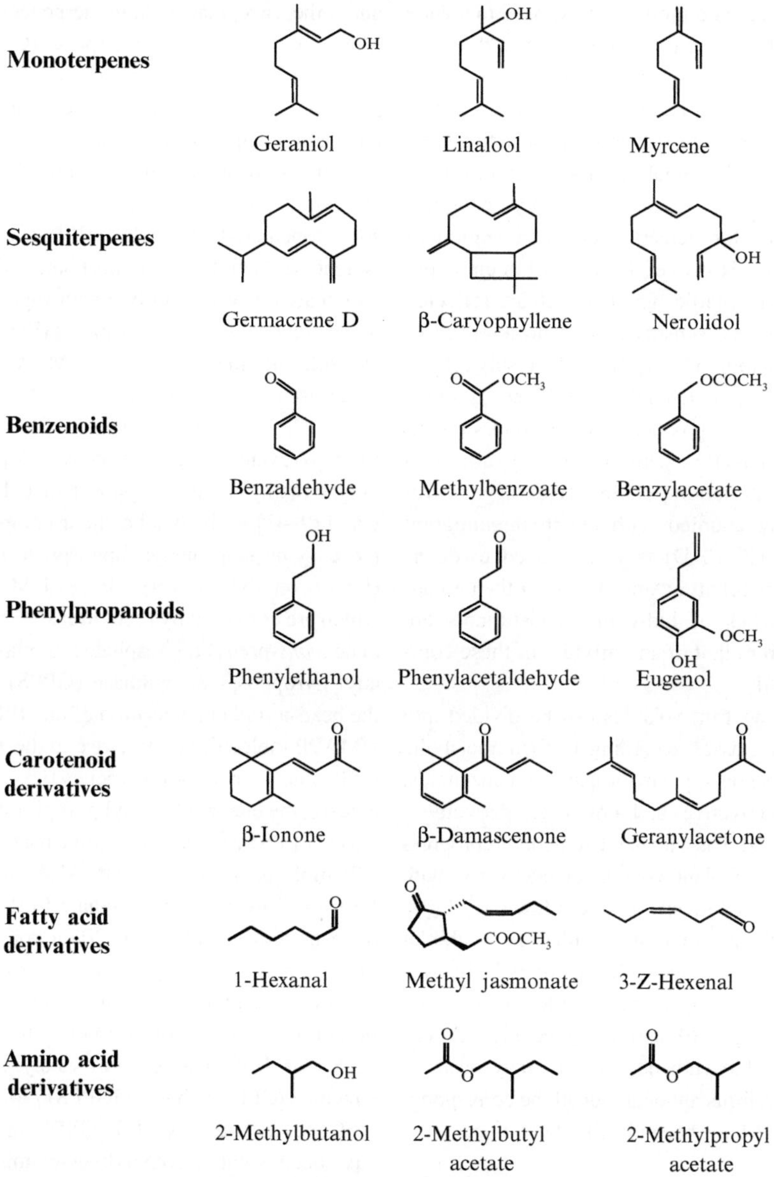

Fig. 2 Representative volatile compounds found in floral scents and fruit aromas

the caramel-like furaneol and its methoxy derivatives (methoxyfuraneol and mesifurane).

Although fewer volatiles contribute to the entire scent profile of flowers, very little is known about their perception by insects and humans, whose odor detection capability is much poorer than that of insects and is often subjective [31, 32]. Unlike humans, insects rely on accurate monitoring of their chemical environment for survival. During foraging, insects including bees and moths, are exposed to a wide range of odor concentrations even from the same flower, ranging from small amounts of volatiles at a distance to an enormous abundance during the actual feeding [33]. Such exposure to a large array of volatile signals exerts pressure on the insect olfactory system to correctly recognize the stimulus for detection and localization of food sources. However, flowers often emit species-specific volatile signals which are reflected in the olfactory sensitivities or preferences of their pollinators [34, 35]. Indeed, closely related plant species, which rely on different types of insect pollinators, produce different odors and at different intensities, with higher levels for moth versus bee attraction [35, 36]. To date, gas chromatography coupled with electroantennogram detection (GC-EAD) is widely used to determine key insect attractants, however the perceptual properties of individual constituents are different from that of any mixture of these compounds [36].

Floral and fruit volatiles can be divided into four major classes according to their metabolic origin: terpenoids, phenylpropanoids/benzenoids, fatty acid derivatives and amino acid derivatives. The last decade has witnessed tremendous progress in the area of plant volatile biochemistry, with studies centered mostly around a few model systems for floral and fruit volatiles. A wealth of information is also available regarding the biosynthetic routes to vegetative volatiles reviewed in (see also Chapter 6), and it is generally believed that similar biosynthetic reactions may occur in flowers and fruits although not all the corresponding enzymes have been identified yet.

Biosynthetic Pathways

Terpenoids

The most abundant class of volatiles is derived from the terpenoid pathways which give rise to mono-, sesqui- and diterpenes, apocarotenoids and other irregular volatile terpenes (Fig. 2). Many terpenoids, e.g. hemiterpenes (C5), monoterpenes (C10), sesquiterpenes (C15), homoterpenes (C11 and C16), and some diterpenes (C20) have a high vapor pressure allowing their release into the atmosphere. The building blocks of all terpenoids are the universal five carbon precursors, isopentenyl diphosphate (IPP) and its allelic isomer dimethylallyl diphosphate (DMAPP), which are derived from two pathways operating in different cellular compartments (Fig. 1). In the cytosol, the mevalonic acid (MVA) pathway gives rise to IPP from three molecules of acetyl-CoA [37, 38], while in plastids, IPP is derived from pyruvate and glyceraldehyde-3-phosphate via the methyl-erythritol-phosphate (MEP) pathway [39–41]. In both subcellular compartments, the enzyme isopentenyl diphosphate isomerase (IDI) reversibly converts IPP to DMAPP [42], which are both used by prenyltransferases to produce *trans*-prenyl diphosphates. In plastids, geranyl pyrophosphate synthase (GPPS) catalyzes the head-to-tail condensation of one IPP and one DMAPP molecules to give rise to the 10-carbon GPP, precursor of monoterpenes [43, 44]. In the cytosol, the enzyme farnesyl pyrophosphate synthase (FPPS) catalyzes the condensation of two IPP molecules with one DMAPP molecule to form the 15-carbon intermediate FPP, precursor of sesquiterpenes [45]. The 20-carbon precursor of diterpenes, geranylgeranyl pyrophosphate (GGPP), is synthesized in plastids by condensation of one molecule of DMAPP with three molecules of IPP in a reaction catalyzed by the enzyme GGPP synthase (GGPPS) [44, 46].

Genes encoding GPPS, FPPS, and GGPPS have been isolated from a diverse range of plant species [47–50]. While both FPPS and GGPPS are functional homodimers, the situation with GPPS is more complex. The GPPSs of Arabidopsis [51] and *Abies grandis* [48] are also homodimers, whereas those reported from peppermint leaves [48] and the flowers of snapdragon and *Clarkia breweri* [47] are unusual heterodimeric enzymes in which both subunits are absolutely required for prenyltransferase activity. Although the subcellular compartmentalization of the MVA and MEP pathways allows them to operate independently, metabolic "cross-talk" between these two pathways mediated by specific metabolite transporters [52] was recently discovered [53], particularly in the direction from plastids to cytosol [54, 55]. In snapdragon flowers, it was shown that the MEP pathway alone provides IPP precursors for both plastidial monoterpene and cytosolic sesquiterpene biosynthesis [56].

The tremendous diversity of volatile monoterpenes, sesquiterpenes and diterpenes arises from enzymatic modification of the non-volatile prenyl diphosphate intermediates GPP, FPP and GGPP through the action of terpene synthases (TPS) [57, 58], many of which have the distinctive ability to catalyze the formation of multiple products from a single prenyl diphosphate substrate [59–61]. A recently isolated monoterpene synthase from *Nicotiana suaveolens* flowers was found to produce a blend of 5 cyclic and 2 acyclic monoterpenes, all of which are components of *N. suaveolens* floral scent [62]. In Arabidopsis, two terpene synthases account for the biosynthesis of nearly all sesquiterpenes found in the floral volatile blend [63]. In addition to sesquiterpenes, Arabidopsis flowers also emit monoterpenes dominated by β-myrcene and (*S*)-linalool. While β-myrcene could likely be synthesized by multi-product monoterpene synthases [64, 65], one single-product monoterpene synthase may be solely responsible for the emission of (*S*)-linalool [65]. In snapdragon, two specialized single-product monoterpene synthases are responsible for the biosynthesis of the two floral monoterpenes, myrcene and (*E*)-β-ocimene [66]. The reaction mechanism of TPSs involves the formation of carbocationic intermediates which can then be differentially metabolized to form multiple products [57, 58].

To date, the TPS gene family consists of more than 100 members (with about a third isolated from flowers and fruits) which have been isolated and characterized from many plant species. This gene family has been divided into seven subfamilies (designated TPS-a through TPS-g) based on sequence relatedness, functional assessment, and gene architecture [59, 60, 67, 68]. However, as TPSs from related plant species tend to cluster together more than enzymes of similar function, substrate/product predictions based on sequence similarities remain problematic [59].

In addition to a wide range of volatile terpenoids formed directly by terpene synthases, terpenoid diversity is further increased by other enzymes which are capable of modifying the TPS products via hydroxylation, dehydrogenation, acylation, or other reactions thus increasing their volatility and altering their olfactory properties [69]. In caraway fruits, the monoterpene limonene is hydroxylated by a cytochrome-P450 enzyme to *trans*-carveol, which is further oxidized by a nonspecific dehydrogenase to form the key aromatic compound carvone (Fig. 3a) [70, 71]. The monoterpene myrtenol, which contributes to the typical aroma of wild strawberry, is formed by the C10 hydroxylation of α-pinene by another P450 monooxygenase (Fig. 3b) [72]. The monoterpene geraniol serves as a precursor for a number of other volatiles: in rose, an acetyltransferase esterifies geraniol to geranyl acetate [73], while in grape, geraniol is reduced to dihydrogeraniol [(*S*)-citronellol] the precursor of the potent odorant rose oxide [74].

Another important class of volatile compounds originating from terpenoids are the cleavage products of carotenoids which have carbon skeletons ranging from C8 to C18 (Fig. 2) [9, 75]. Many of these volatile carotenoid derivatives are key aroma constituents of fruits and flowers due to their extremely low aroma thresholds. For example, β-damascenone, found in numerous fruits, has a floral/fruity odor with an aroma threshold of a few parts per trillion, making it one of the most potent odorants known

19

Fig. 3 Biosynthesis of monoterpene derivatives in caraway fruits (**a**) and wild strawberries (**b**)

[76]. The biosynthesis of carotenoid-derived volatiles is thought to proceed via three steps: an initial dioxygenase cleavage yielding apocarotenoids, followed by enzymatic transformations of these apocarotenoids to the polar aroma precursors, which finally undergo acid-catalyzed conversions to volatile compounds [77]. In some cases including tomato, petunia and melon, the dioxygenase cleavage step itself can yield a volatile product, such as α- and β-ionone, geranylacetone, and pseudoionone, from an array of carotenoid pigments (Fig. 2) [78–80].

Phenylpropanoids and Benzenoids

Phenylpropanoids and benzenoids constitute the second most ubiquitous class of plant volatile compounds [9] and are derived from the amino acid L-phenylalanine (Phe) (Fig. 2). Despite their abundance, diversity, and importance in the scent and aroma of countless plant species, the precise biochemical pathways leading to the formation of volatile phenylpropanoids and their derivatives are still mostly unknown. The first committed step through the phenylpropanoid/benzenoid pathway is the conversion of L-Phe to *trans*-cinnamic acid by the enzyme L-phenylalanine ammonia-lyase (PAL) (Fig. 1). *Trans*-cinnamic acid can undergo methylation to form methylcinnamate [81], a known component of floral scent, or can serve as the precursor for a plethora of further intermediates such as hydroxycinnamic acids, aldehydes and alcohols, which are formed by a series of hydroxylation and methylation reactions common with the lignin/lignan biosynthetic pathway [82]. One of these intermediates, coniferyl alcohol, has been recently shown to serve as the precursor for the synthesis of isoeugenol in petunia and *Clarkia breweri* flowers (Fig. 4) [83, 84]. This monolignol is first esterified to coniferyl acetate by the enzyme coniferyl alcohol acyltransferase [85], a member of the BAHD family, named for the first four biochemically characterized enzymes of this family, BEAT (acetyl-CoA:benzyl alcohol acetyltransferase), AHCT (anthocyanin *O*-hydroxycinnamoyltransferase), HCBT (anthranilate *N*-hydroxycinnamoyl/ benzoyltransferase) and DAT (deacetylvindoline 4-*O*-acetyltransferase) [86]. Isoeugenol is then produced from coniferyl acetate in a reaction catalyzed by isoeugenol synthase [83]. Although eugenol and isoeugenol differ exclusively by the position of the double bond in the propene side chain, in petunia their formation is mediated by two different and highly divergent NADPH-dependent reductases [84]. In contrast, *C. breweri* contains three distinct NADPH-dependent reductases, two of which are responsible for eugenol formation and the third possessing isoeugenol synthase activity. While one eugenol synthase (EGS) isoform is closely related to isoeugenol

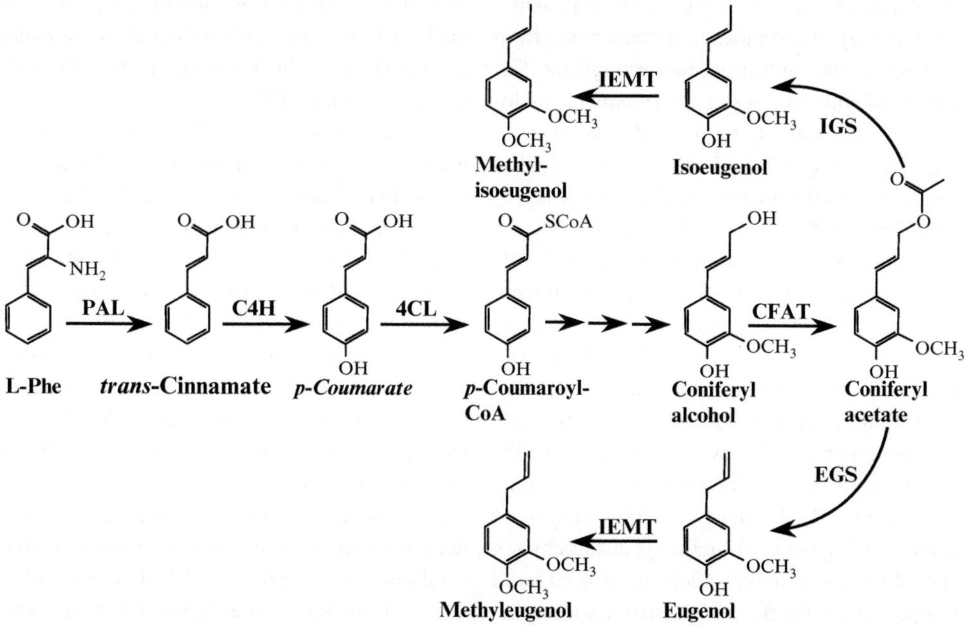

Fig. 4 Branch of the phenylpropanoid pathway leading to isoeugenol/eugenol and their derivatives. CFAT, coniferyl alcohol acyltransferase; C4H, cinnamate 4-hydroxylase; 4CL, 4-hydroxycinnamoyl CoA ligase; EGS, eugenol synthase; IEMT, (iso)eugenol *O*-methyltransferase; IGS, isoeugenol synthase; PAL, phenylalanine ammonia lyase

synthase (IGS), the other is highly diverged [84]. Interestingly, the synthesis of chavicol, another volatile phenylpropene biochemically related to eugenol, has been hypothesized to proceed via a similar mechanism where the monolignol coumaryl alcohol would undergo esterification followed by reduction to chavicol [87]. Eugenol, isoeugenol and chavicol can be further methylated by *O*-methyltransferases to form the volatiles methyleugenol, isomethyleugenol (Fig. 4) and methylchavicol, respectively. Such enzymes have been isolated and characterized from sweet basil [88] and *Clarkia* [89].

Another branch of the phenylpropanoid pathway gives rise to benzenoid compounds, the formation of which requires the shortening of the side chain of *trans*-cinnamic acid by two carbons. Although the precise biochemical steps are still under investigation, this process could occur via a CoA-dependent β-oxidative pathway, a CoA-independent non-β-oxidative pathway, or a combination of both routes. In the β-oxidative pathway, the synthesis of benzenoids from *trans*-cinnamic acid proceeds through the formation of four CoA-ester intermediates analogous to that underlying β-oxidation of fatty acids. In the non-β-oxidative pathway, hydration of the *trans*-cinnamic acid to 3-hydroxy-3-phenylpropionic acid is followed by a reverse aldol reaction which shortens the propyl side chain and gives rise to benzaldehyde. Recent investigations of the phenylpropanoid/benzenoid pathway using *in vivo* stable isotope labeling and computer-assisted metabolic flux analysis revealed that both the β-oxidative and non-β-oxidative pathways are involved in the formation of benzenoid compounds in petunia flowers [90, 91]. Similarly, the biosynthesis of vanillin (4-hydroxy-3-methoxybenzaldehyde), undoubtedly the most widely used benzenoid compound in the flavor industry, requires the shortening of the side chain of coumaric acid by two carbons and could proceed via a β-oxidative or non-β-oxidative pathway. The operation of a non-β-oxidative pathway was supported by isolation of a 4-hydroxybenzaldehyde synthase, which catalyzes the formation of 4-hydroxybenzaldehyde from coumaric acid [92]. In this pathway, 4-hydroxybenzaldehyde is then hydroxylated at position 3 on the ring to yield 3,4-dihydroxybenzaldehyde, the direct precursor of vanillin, but the enzyme responsible for this reaction has not yet been identified. The following methylation leading to the synthesis of vanillin may be catalyzed by a multifunctional *O*-methyltransferase isolated from vanilla beans, albeit 3,4-dihydroxybenzaldehyde was not its preferred substrate *in vitro* [93].

Methyl anthranilate is a benzenoid compound formed independently from Phe and is responsible for the characteristic aroma of Concord grapes (*Vitis labrusca*). Despite its striking structural similarity to other methyl esters like methylbenzoate and methylsalicylate, which are formed by carboxyl methyltransferase [94], it is produced from anthraniloyl-CoA and methanol by an anthraniloyl-CoA:methanol acyltransferase which belongs to BAHD acyltransferase family [95].

The biosynthesis of other volatile phenylpropanoid-related compounds such as phenylacetaldehyde and 2-phenylethanol, does not occur via *trans*-cinnamic acid and competes with PAL for Phe utilization [90, 96, 97]. Phenylacetaldehyde biosynthesis from Phe requires the removal of both the carboxyl and amino groups. A classical sequential two-step removal is believed to occur in tomato where Phe was shown to be first converted to phenylethylamine by aromatic amino acid decarboxylase (AADC) and further required the action of a hypothesized amine oxidase, dehydrogenase, or transaminase for phenylacetaldehyde formation [97]. On the other hand, in petunia, one bifunctional enzyme, phenylacetaldehyde synthase (PAAS) catalyzes the unprecedented efficient coupling of Phe decarboxylation to oxidation resulting in phenylacetaldehyde formation [96]. While in tomato phenylacetaldehyde is then converted to 2-phenylethanol by the action of 2-phenylacetaldehyde reductase [97], in petunia it is not the only precursor for 2-phenylethanol based on feeding experiments with deuterium-labeled Phe [90]. Instead, the largest contribution to 2-phenylethanol formation appeared to come from a different biosynthetic route, possibly through phenylpyruvate and phenyllactic acid as has been recently reported in rose flowers [90, 91, 98]. While the exact biochemical steps leading to the formation of phenylpropanoids and benzenoids are still mostly unknown, much progress has been achieved in the discovery of enzymes responsible for the final steps in the biosynthesis of volatiles in this network [69].

Fatty Acid and Amino Acid Derivatives

Volatile fatty acid derivatives such as *cis*-3-hexenol, 1-hexanal, nonanal, and methyl jasmonate constitute another important group of plant volatiles which are present in the scent and aroma of numerous flowers and fruits. Volatile aliphatic C6 compounds provide fruits and vegetables with characteristic "fresh green" aroma. These compounds originate from C18 unsaturated fatty acids (linoleic or linolenic acids), which enter the "lipoxygenase pathway" (Fig. 1). The first step of this pathway is the dioxygenation of unsaturated fatty acids, catalyzed by lipoxygenase enzymes (LOX) (Fig. 5) [99]. LOX enzymes belong to a large family of nonheme iron containing fatty acid dioxygenases. They catalyze the oxygenation of polyenoic fatty acids at C9

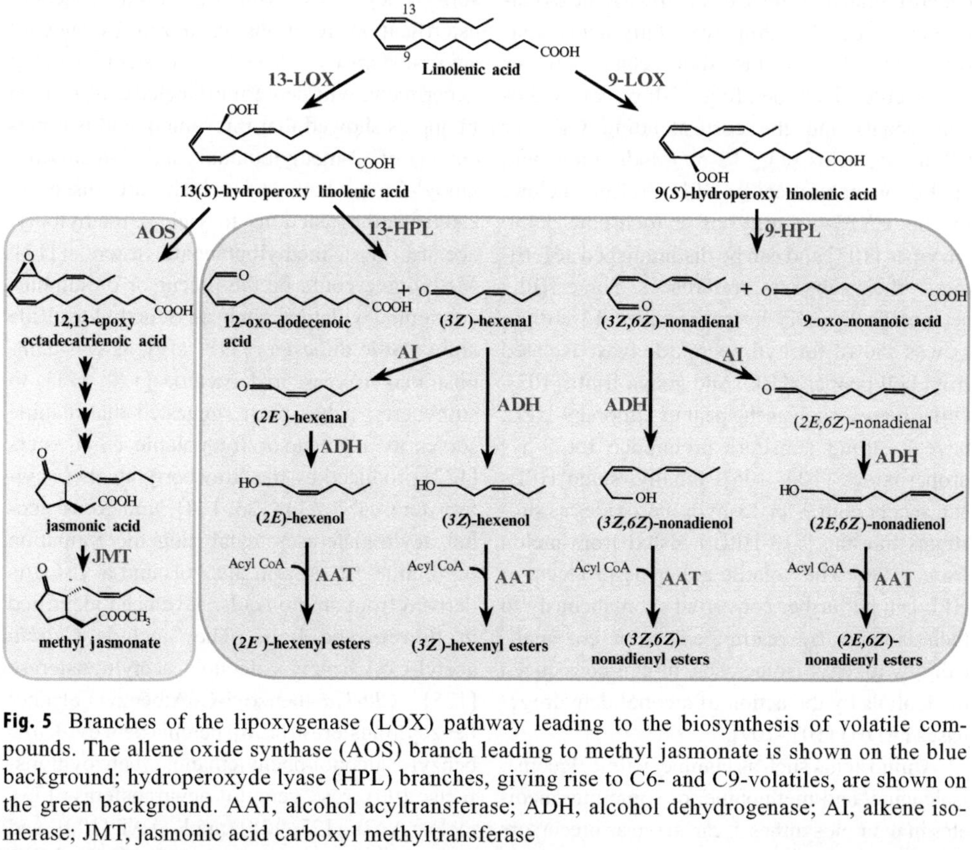

Fig. 5 Branches of the lipoxygenase (LOX) pathway leading to the biosynthesis of volatile compounds. The allene oxide synthase (AOS) branch leading to methyl jasmonate is shown on the blue background; hydroperoxyde lyase (HPL) branches, giving rise to C6- and C9-volatiles, are shown on the green background. AAT, alcohol acyltransferase; ADH, alcohol dehydrogenase; AI, alkene isomerase; JMT, jasmonic acid carboxyl methyltransferase

or C13 positions (the enzyme is then referred to as 9-LOX or 13-LOX, respectively), yielding two groups of compounds, the 9-hydroperoxy and the 13-hydroperoxy derivatives of polyenoic fatty acids. These derivatives can be further metabolized through different pathway branches, according to their ultimate metabolic fate. These two pathway branches, with the first committed step catalyzed by allene oxide synthase (AOS) or hydroperoxide lyase (HPL), give rise to volatile compounds (Figs. 1 and 5). In the AOS branch, 13-hydroxyperoxy linolenic acid is converted to 12,13-epoxy octadecatrienoic acid by AOS [99]. A series of subsequent enzymatic reactions lead to the formation of jasmonic acid, which can in turn be esterified to the volatile methyl jasmonate by jasmonic acid carboxyl methyltransferase [100, 101]. In the HPL branch of the LOX pathway, the oxidative cleavage of hydroperoxy fatty acids catalyzed by HPL yields short chain C6- or C9-volatile aldehydes (e.g., 3-hexenal or 3,6-nonadienal) and the corresponding C12- or C9-ω fatty acids (e.g., 12-oxo-dodecenoic acid or 9-oxononanoic acid). HPL enzymes belong to the CYP74 family of cytochrome P450 enzymes [102] and can be distinguished according to their substrate preferences. Some HPLs act specifically on 13-hydroperoxides (13-HPLs), as was shown for hydroperoxide lyase isolated from bell peppers [102] and guava fruits [103]. Other lyases, such as the pear or almond 9-HPL, have a strong substrate preference for 9-hydroperoxides [104, 105]. Finally, some HPLs can accept both 9- or 13-hydroperoxides as substrates like the 9/13-HPL isolated from melon fruits [106]. The volatile aldehyde products of HPL can be further converted spontaneously to their isomers by rearrangement or enzymatically by alkenyl isomerases, or can be reduced to alcohols by the action of alcohol dehydrogenases (ADH) [107–109].

Amino acids such as alanine, valine, leucine, isoleucine, and methionine, or some intermediates in their biosynthesis, can serve as precursors for many floral and fruit volatiles including aldehydes, alcohols, esters, acids, and nitrogen- and sulfur-containing volatiles. The biosynthesis of amino acid derived volatiles is thought to proceed in a similar way as that in bacteria or yeast, where these pathways have been studied more extensively [110–112]. Amino acids can undergo an initial deamination or transamination leading to the formation of the corresponding α-keto acid. Subsequent decarboxylation followed by reductions, oxidations and/or esterifications give rise to aldehydes, acids, alcohols and esters [113]. The catabolism of the branched chain amino acids leucine, isoleucine and valine gives rise to branched chain volatile alcohols, aldehydes and esters which are important constituents of the aroma of fruits such as banana, apple, strawberry and tomato (Table 1) [12, 114–116]. Amino acids can also be the precursors of acyl-CoAs, which are used in alcohol esterification reactions catalyzed by alcohol acyltransferases (AATs). Precursor feeding experiments with deuterium labeled L-isoleucine in apples showed that this amino acid is a precursor of 2-methylbutanol and 2-methylbutanoyl-CoA, both of which are used in esterification reactions to yield 2-methylbutyl acetate or 2-methylbutanoate esters [117]. Methionine could be the precursor of sulphur-containing volatiles such as dimethyldisulfide and volatile thioesters [118, 119], as was demonstrated in yeast and bacteria [120, 121]. In strawberry, it has been suggested that alanine serves as a precursor for volatile ethyl esters [122] produced by the strawberry alcohol acyltransferase (SAAT) [123, 124]. Numerous alcohol acyltransferases, catalyzing the formation of volatile esters from alcohols and acyl-CoAs derived from amino acids, have been identified in flowers and fruits. They include *Clarkia* acetyl-CoA:benzyl alcohol acetyltransferase [125], *Clarkia* benzoyl-CoA:benzyl alcohol benzoyltransferase [126], petunia benzoyl-CoA: benzyl alcohol/phenylethanol benzoyltransferase [90], rose geraniol acyltransferase [73], melon AATs [127, 128], apple AAT [129], and wild strawberry and banana AATs [124]. Amino

Table 1 Amino acid precursors of volatile aroma compounds in fruits

Amino acid precursor	Amino acid-derived volatiles	Fruit	References
Leucine	3-methylbutanol 3-methylbutanoate 3-mehtlybutyl esters 2-ketoisocaproate	Banana	[202–204]
Valine	2-methylpropanol 2-methylpropionate 2-methylpropyl esters 2-ketoisovalerate	Banana	[202, 204]
Isoleucine	2-methylbutanol 2-methylbutyl esters 2-methylbutanoate esters 2-methylbut-2-enyl esters	Banana, Apple, Strawberry Apple	[202, 114–116]
Alanine	Ethyl esters	Strawberry	[122]
Methionine	Thioesters	Melon	[205, 118–119]
Phenylalanine	1-nitro-2-phenylethane, phenylacetaldehyde, phenylethanol	Tomato	[97]

acid precursors can also serve as intermediates in the biosynthesis of volatile compounds. For example, the cyclic volatile compound indole, present in the scent of many flowers, originates in maize from the enzymatic cleavage of indole-3-glycerol phosphate, the direct precursor of the amino acid tryptophan [130].

Cellular and Subcellular Localization of Fruit and Flower Volatile Formation

Volatile compounds have been found to be synthesized *de novo* in the cells of the plant tissues from which they are emitted [131, 132]. In flowers, their biosynthesis occurs in the epidermal cell layers, allowing easy escape of volatiles into the atmosphere [133–135]. While in some species, including *Clarkia breweri* and some rose varieties (e.g. *Rosa* x *hybrida* 'Baronne Edmond de Rothshild', 'Anna', 'The Mac Cartney Rose'), the biosynthesis of volatile constituents of scent is localized on either side of the petals [136, 137], in others (snapdragon, *Stephanotis floribunda*, *Nicotiana suaveolens* and *Rosa* x *hybrida* 'Lady Hillingdon'), their biosynthesis occurs preferentially on the side of the petal epidermis which faces or comes in contact with pollinators [133]. In contrast to floral volatiles, even less is known about the cellular localization of fruit aroma formation. In fruits like citrus, volatiles accumulate in specialized structures adapted to contain high levels of volatile oils [138]. In others, aroma compounds are formed in the mesocarp and exocarp, as was shown in apples [139], or in parenchyma tissues as was found in red ripe strawberry [131].

Although much progress has been made in the understanding of plant volatile biosynthesis (see above), the mechanisms of their transport from the site of biosynthesis and their emission into the atmosphere remain largely unknown. Volatile emission from any plant tissue is likely to involve four general steps: (i) trafficking within the cell; (ii) export from the plasma membrane and subsequent transport across the cell wall; (iii) transfer through the cuticle; and (iv) evaporation at the surface of the cuticle [2]. In general, the rate of release of any volatile compound depends on its specific volatility as well as on the physical properties of the cellular and intracellular membranes through which it has to diffuse and which likely have different permeability to different volatiles. The emitted compounds may also be associated with an entirely different secretory compartment to stored volatiles which accumulate in cells. In any case, emission of volatiles may proceed via a cytologically organized excretory process involving the Golgi apparatus, vesicular transport or protein-mediated movement [140], or be a consequence of the endogenous concentrations of compounds [141]. The export of volatiles across the plasma membrane and their transport through the cell wall may involve exocytosis or could be facilitated by specialized proteins such as the adenosine triphosphate (ATP) binding cassette (ABC) transporters or lipid transfer proteins (LTPs) [2]; however, experimental evidence for their involvement is still lacking.

Although it is possible to predict the subcellular localization of a protein based on its sequence, to date only a few studies have investigated and confirmed the exact subcellular localization of the enzymes involved in the biosynthesis of scent and aroma compounds. Immunogold labeling studies revealed that the biosynthesis of methylbenzoate, a benzenoid ester, takes place in the cytoplasm of the conical cells of the inner epidermal layer of snapdragon flowers [133]. The biosynthesis of monoterpene compounds within the snapdragon scent bouquet was shown to occur in the leucoplast where the small subunit of GPP synthase responsible for GPP formation was localized using immunogold labeling [47]. In contrast to snapdragon myrcene and ocimene synthases which possess plastid-targeting peptides [66], in cultivated strawberry fruits a dual nerolidol/linalool synthase lacks N-terminal targeting sequences and was localized in the cytosol using green fluorescent protein localization studies [72]. This suggests that ripe strawberries contain sufficient levels of GPP and FPP in the cytosol to produce roughly equal amounts of linalool and nerolidol, respectively. Also, using protein import experiments it was shown in tomato fruits that carotenoid cleavage dioxygenases (CCDs) are localized in the cytosol but attached to the outer chloroplastic membrane, thus breaking membrane-associated carotenoids and producing β-ionone and geranylacetone [78].

Similar to terpenoid biosynthesis, the lipoxygenase pathway may encompass several cellular compartments as several LOX isoforms may localize to distinct parts of the cell, presumably according to their specific metabolic function [142]. A plethora of LOX locations include cytosol, mitochondria, plastid, and vacuole [143–146] or the enzyme may be associated with microsomal/ plasma membranes and lipid-protein bodies [147–150]. In tomato fruits, LOX enzymes involved in fruit aroma formation have been found in chloroplasts [151] while in receptacle parenchyma tissue of strawberry they were mostly associated with cellular membranes (probably ER, from which they disaggregate in the cytoplasm at the ripe stage) and the surface of lipid-protein bodylike structures [152]. The targeting of the C9 volatile aldehyde forming enzyme HPL to the endomembrane system and lipid bodies was found in almond seeds [105]. While scattered information is available to date about the subcellular localization of the biosynthesis of volatiles in flowers and fruits, virtually nothing is presently known about metabolite trafficking between various subcellular compartments as well as its contribution to the regulation of volatile emission.

Regulation of Floral and Fruit Volatile Formation

Emission of volatile compounds from flowers and fruits depends on environmental factors such as light, temperature and moisture status, and is spatially and temporally regulated [153, 154]. Of all plant organs, flowers in scented species and fruits generally produce the most diverse blends of volatile compounds, and in highest quantities. Scent and aroma production generally increases during the early stages of organ development peaking when flowers are ready for pollination and fruits are fully ripe [70, 136, 155]. The discovery of genes involved in plant volatile biosynthesis has provided a starting point for the investigation of the regulation of volatile formation and emission. During flower or fruit development, the expression of genes from one or more metabolic pathways is mostly transcriptionally regulated in a coordinated manner to create a particular fragrance or aroma bouquet [128, 132, 156], suggesting the existence of master regulators which orchestrate volatile formation upstream of metabolic pathways. The first transcription factor ODORANT1 involved in the regulation of the production of volatile benzenoids in petunia was recently isolated [157]; however, transcription factors that regulate multiple, distinct metabolic networks involved in volatile biosynthesis have yet to be discovered.

In flowers, volatile emission often exhibits distinct diurnal or nocturnal patterns [158] which generally coincide with pollinator activity [36] (see also Chapter 15). These oscillating patterns of scent emission have been shown to be controlled by the endogenous circadian clock in snapdragon, *Nicotiana suaveolens*, *Stephanotis floribunda* and *Rosa* x *hybrida* 'Honesty' [62, 159, 160] or regulated by light intensity in *Odontoglossum constrictum*, *Trifolium repens* L. and *Rosa* x *hybrida* 'Fragrant Cloud' [140, 153, 161]. Even within the same fragrance bouquet some compounds could be under the control of a circadian clock while others are light regulated [79, 161–163]. In many flowers, intricate levels of regulation, including transcriptional control of scent biosynthetic genes and/or substrate availability (discussed below), have been shown to contribute to the rhythmic emission of floral volatiles [62, 159, 161]. Moreover, when several different biosynthetic routes are involved in the biosynthesis of the same volatile end product, their relative contribution within the biochemical network may vary according to the photoperiod, as was shown using computer-assisted metabolic flux analysis in petunia flowers [91]. Additionally, the rhythmic emission of terpenoid compounds in snapdragon scent bouquet is the result of the circadian-controlled rhythmicity of the flux through the MEP, but not the MVA, pathway [56].

The scent of many flowers is markedly reduced soon after pollination. Such quantitative and/or qualitative post-pollination changes in floral bouquets, shown mostly in orchids [164–167], lower the attractiveness of pollinated flowers as well as increase the overall reproductive success of the plant by directing pollinators to unpollinated ones [168]. The decrease in scent emission begins after successful fertilization as was shown in snapdragon and petunia flowers [169]. The plant hormone ethylene was found to play a major role in regulating volatile emission after pollination through the downregulation of expression of scent biosynthetic genes in petunia [163, 169]. Ethylene can also induce a decrease in floral volatile emission in cut flowers such as sweet peas and carnations [170, 171], indicating a general role for this plant hormone in the regulation of the floral scent output. In contrast to floral emission, in fruits (melon and apple) ethylene has positive regulatory effects on the expression of genes involved in the final step of aroma biosynthesis [127, 172]. While the role of ethylene in aroma regulation has been demonstrated in climacteric fruits [128, 129, 173], little is known about the regulation of aroma formation in non-climacteric fruits such as strawberries, which typically produce little to no ethylene. Several genes implicated in aroma volatile biosynthesis in strawberry were shown to be developmentally regulated during fruit ripening [72, 123]; but the molecular mechanisms of their regulation remain unclear. However, recently, a gene encoding a UDP-glucose:cinnamate glucosyltransferase, which catalyzes the formation of the biogenetic precursors of cinnamate esters in strawberry, was shown to be negatively regulated by auxin or induced by oxidative stress and exogenous application of hydroxycinnamic acids [174].

In addition to the transcriptional regulation of genes involved in scent and aroma formation, the availability of substrates for volatile biosynthesis can control the level of emitted volatiles. For example, in snapdragon flowers the internal pool of benzoic acid, an immediate precursor of methylbenzoate, controls the level of emission of this volatile ester during flower development and a daily light/dark cycle [159]. The availability of substrate can also determine the nature of the volatile blend produced, especially when the reaction is catalyzed by an enzyme with broad substrate specificity (e.g. some carboxyl methyltransferases and acyltransferases) [94, 123]. Indeed, in petunia benzoic acid/salicylic acid carboxyl methyltransferase (BSMT), while having a high catalytic efficiency towards salicylic acid, uses benzoic acid as a substrate and produces methylbenzoate due to the lack of salicylic acid in the petal tissue [169]. Similarly, aroma formation can be controlled by substrate availability and may also be limited when substrates and their corresponding enzymes are not localized in the same compartment or subcellular organelle [70, 72]. On the other hand, by directing nearly identical bifunctional enzymes to more than one cellular compartment plants can extend the range of available substrates for enzyme utilization, thus increasing the diversity of produced volatiles [175]. The role of substrate availability was further demonstrated in recent metabolic engineering studies [124, 176], which highlighted the importance of enzyme biochemical characterization along with metabolite profiling for the successful manipulation of plant volatile emission [177]. Scent and aroma compounds may be present in bound, non-volatile forms, such as β-glucosides in rose, white clover [178–181] and strawberry [182] and their emission could be regulated by the activity of β-glucosidase which can hydrolyze glucosides and release volatile compounds from plant tissue.

Metabolic Engineering

To date many cultivars have lost their characteristic fragrance/aroma properties possibly due to breeding programs which often focused on improving plant yield, resistance to pests and appearance. Recently several attempts have been made to restore and/or modify the scent of flowers and aroma of fruits via metabolic engineering [183–186]. In general, the bioengineering of the volatile spectrum can be achieved either through the modification of existing pathways (e.g., up- or down-regulation of one or more steps and/or redirection of flux to a desirable compound via blockage of competing pathways) or by the introduction of new gene(s) or branchways normally not found in the host plant.

The ectopic expression of the *Clarkia breweri* (*S*)-linalool synthase (*LIS*), which catalyzes the conversion of GPP to (*S*)-linalool [132] in petunia and carnations, both of which lack this monoterpene, represents the first attempt to modify the floral scent bouquet via the introduction of a new gene [187]. Although linalool production was achieved in these transgenic plants, its emission was undetectable by humans in both cases due to different reasons. While in petunia most of the linalool was sequestered as a non-volatile linalool glycoside by the action of endogenous glucosyltransferase [187], in carnations much of the linalool was further oxidized by an endogenous enzyme to *cis*- and *trans*-linalool oxide which together with linalool constituted almost 10% of the total volatiles emitted from the transgenic carnation flowers (Fig. 6). However, this amount was either below the threshold for human perception or masked by other volatiles [188]. These studies revealed unexpected problems that can be encountered in the genetic engineering of flower

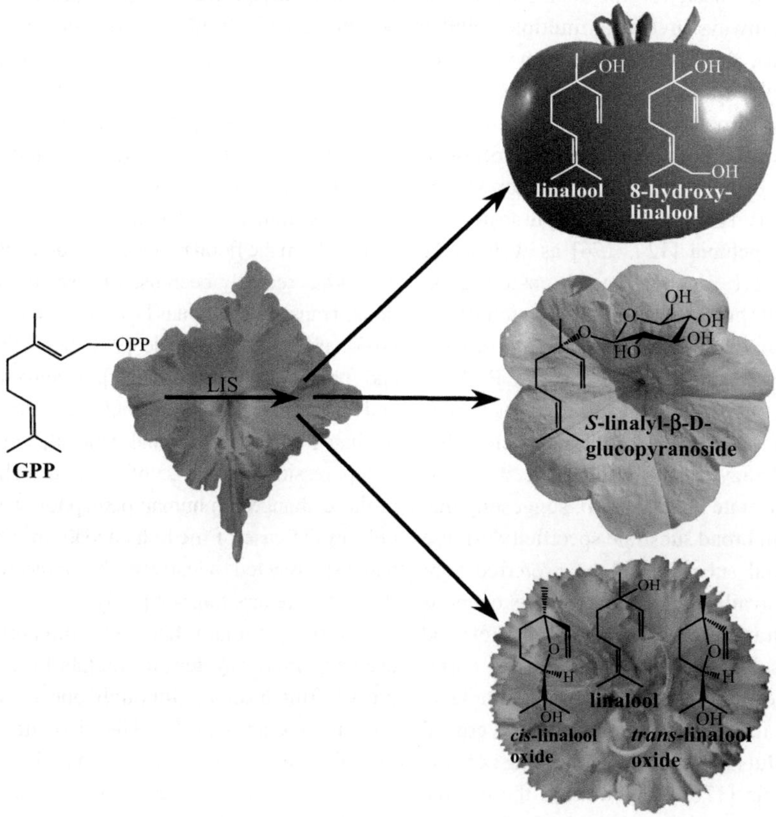

Fig. 6 Different species-specific fates of linalool produced via ectopic expression of *Clarkia breweri* linalool synthase (LIS) [132] in tomato [200], petunia [187] and carnation [188]. GPP, geranyl diphosphate

fragrance: modification of the scent compound into a non-volatile form, masking by other volatiles, or the emitted amount being insufficient for olfactory detection by humans.

Generation of transgenic tobacco plants ectopically expressing three lemon monoterpene synthases (γ-terpinene cyclase, (+)-limonene cyclase, and (-)-β- pinene cyclase) is an example of introducing multiple genes for modification of the volatile profile [189]. Although the total level of monoterpenes increased greatly (10- to 25-fold) in these transgenic plants which led to drastic changes in both leaf and flower volatiles, humans detected changes in leaves only [190]. The introduction of the mint limonene-3-hydroxylase, which catalyzes the hydroxylation of (+)-limonene with the formation of (+)-*trans*-isopiperitenol, led to further modification of the scent profile of transgenic tobacco plants possessing three integrated transgenes [191]. Since monoterpene synthases and limonene-3-hydroxylase are localized in different subcellular compartments (plastids and ER, respectively), the obtained results revealed that multiple gene

19

transfer can be used for the manipulation of metabolic pathways involving multiple cellular compartments, thereby facilitating the production of desirable molecules in transgenic plants.

Attempts at metabolic engineering of floral volatiles have not been restricted to terpenoids. The ectopic expression of both the strawberry (SAAT) and rose alcohol acyltransferases (RhAAT) in petunia [124, 176] as well as *C. breweri* benzyl alcohol acetyltransferase in lisianthus [192] resulted in no production of the expected compounds due to the lack of available substrates. However, in the case of rose RhAAT, the enzyme, while showing the highest preference to geraniol and citronellol, used phenylethyl alcohol and benzyl alcohol and produced the corresponding acetate esters *in vivo*, suggesting that enzymes with broad substrate specificity can use other potential substrates if the preferred substrate is not available. Moreover, the exogenous feeding of transgenic flowers with a preferred substrate resulted in the formation of the corresponding acetate esters suggesting that the level of precursor for the introduced enzyme is crucial for successful metabolic engineering of the volatile profile [176, 192]. Indeed, the ectopic expression of tomato phenylacetaldehyde reductases, which catalyze the conversion of phenylacetaldehyde to 2-phenylethanol, in petunia flowers with high levels of phenylacetaldehyde led to the production of significantly higher levels of 2-phenylethanol with a subsequent decrease in phenylacetaldehyde levels [193].

The redirection of flux to the target pathway represents an alternative molecular genetic approach for enhancement of flower fragrance as was recently demonstrated in transgenic carnations with the antisense suppression of the flavanone 3-hydroxylase, a key enzyme in the anthocyanin pathway [194]. While losing their original orange/reddish color these transgenic flowers produced increasing amounts of methylbenzoate due to the diversion of the flux from the competing anthocyanin pathway to benzoic acid, the precursor of methylbenzoate. The flux to volatile phenylpropanoid compounds could also be enhanced via the upregulation of the transcription factor ODORANT1 [157] or Pap1 (production of anthocyanin pigment 1), which activates the phenylpropanoid pathway leading to increases in both anthocyanin accumulation and volatile phenylpropanoid emission [195, 196].

The elimination of individual volatile compounds from the floral bouquet is another approach which has recently been used for scent modifications. Transgenic petunias lacking methylbenzoate [163], phenylacetaldehyde [96], benzylbenzoate and phenylethylbenzoate [83], isoeugenol [85] and β-ionone [78] were obtained via RNAi-mediated posttranscriptional gene silencing or by cosuppression in the case of β-ionone. The effect of these changes on human perception was tested only in the case of methylbenzoate in which the panelists reacted negatively by indicating that flowers were less fragrant [163].

To date, tomato has predominantly been used as a model system for metabolic engineering of fruit aromas, although one example is known in grapes [197]. The first attempts to modify fruit flavor were accomplished via the overexpression of a yeast Δ-desaturase or nonspecific tomato alcohol dehydrogenase (ADH) and resulted in changes in the levels and/or ratios of aroma-determining short-chain aldehydes and alcohols [108, 198, 199]. Similar results were obtained in grapes overexpressing grapevine alcohol dehydrogenase [197]. Although neither of these manipulations introduced new aroma compounds, fruits with elevated ADH activity and higher levels of alcohols were identified as having a more intense "ripe fruit" flavor in taste trials [198].

Modification of existing biochemical pathways leading to tomato fruit aroma was also achieved by down-regulation of carotenoid cleavage dioxygenases 1 (LeCCD1A and LeCCD1B) [78] and lipoxygenase (TomloxC) [151] or via the up-regulation of aromatic amino acid decarboxylases (LeAADC1 and LeAADC2) [97]. Although in all cases significant changes in the aroma profile were obtained, the impact of these alterations on human perception has not yet been studied.

As with floral scent, the first attempt to introduce new compounds in fruit aroma was the overexpression of *Clarkia breweri* LIS gene in tomato under the control of the tomato late-ripening-specific *E8* promoter [200]. Transgenic tomato fruits accumulated > 50-fold more linalool than wild-type plants, without a concomitant decrease in the levels of other non-volatile terpenoids (Fig. 6). Although some linalool was further oxidized to the volatile 8-hydroxylinalool, the levels of both in ripe fruits of transgenic plants were sufficient for olfactory detection by humans [200]. Another modification of the tomato fruit flavor profile was recently achieved by introducing the basil geraniol synthase gene (GES) under the control of the fruit ripening-specific tomato polygalacturonase promoter [201]. These transgenic plants produced a large amount of geraniol and its derivatives, causing profound changes in fruit flavor, which were recognized and favored over that of nontransgenic fruits by a majority of members of three untrained panels. The redirection of the metabolic flux toward monoterpene formation in these transgenic plants led to the reduction of compounds (including a 50% drop in the level of lycopene and 70–90% decrease in phytoene levels) formed from common precursors. Taken together these results clearly show that aroma and flavor of fruits as well as scent of flowers, could be improved via genetic engineering thus restoring lost fragrant properties of plants.

Conclusion

Research on floral scent and fruit aroma has witnessed tremendous progress in the past decade. During this period, numerous genes have been identified however the information about the biosynthesis of a vast majority of known volatile compounds is still missing. Nevertheless, the discovered genes have expanded our understanding of biosynthetic pathways, their regulation and localization, leading to some successful biotechnological manipulations of floral fragrances and fruit aromas. These pioneering metabolic engineering attempts have also revealed some unexpected problems such as the formation of non-volatile byproducts or the inability to olfactorily detect target compounds due to their low levels, thus demonstrating the complexity of plant volatile networks and the lack of a comprehensive understanding of their regulation and the human perception in a blend. A more successful approach will use the power of modern genomic, proteomic and metabolomic tools in combination with metabolic flux analysis and computer modeling to generate a library of targets for metabolic engineering, identify competing pathways and their possible compartmentation, and decipher the regulatory properties of the pathways of interest. The integration of these tools will also bring precision and predictability in attaining our biotechnological goals.

The understanding of the intracellular metabolite trafficking from the site of biosynthesis to the site of emission as well as the mechanism of the volatile release will significantly improve our success in metabolic engineering of plant volatiles. Discoveries of transcription factors that coordinately regulate the formation of volatiles derived from an array of pathways will advance our capabilities for fragrance and aroma enhancement. The knowledge obtained could ultimately be translated into commercial applications to increase product appeal for customers by enhancing or creating novel scents or aromas, or for the industrial production of natural flavors and fragrances.

Acknowledgements Work in ND's lab is supported by the U.S. National Science Foundation (grant numbers MCB-0615700 and MCB-0331333), the U.S. Department of Agriculture (grant numbers 2003–35318–13619 and 2005–35318–16207) and

19

the Fred Gloeckner Foundation, Inc. Work in FNZ's lab is supported by the College of Agricultural and Environmental Sciences at UC Davis and the California Melon Research Board.

References

1. Pichersky, E., et al. (2006) Biosynthesis of plant volatiles: Nature's diversity and ingenuity. *Science* 311, 808–811
2. Jetter, R. (2006) Examination of the processes involved in the emission of scent volatiles from flowers. In *Biology of Floral Scent* (Dudareva, N., and Pichersky, E., eds), 125–143, CRC/Taylor & Francis, Boca Raton, FL
3. Kessler, A., and Baldwin, I.T. (2001) Defensive function of herbivore-induced plant volatile emissions in nature. *Science* 291, 2141–2144
4. Kessler, A., and Halitschke, R. (2007) Specificity and complexity: the impact of herbivore-induced plant responses on arthropod community structure. *Curr. Opin. Plant Biol.* 10, 409–414
5. Birkett, M.A., et al. (2003) Volatiles from whitefly-infested plants elicit a host-locating response in the parasitoid, *Encarsia formosa*. *J. Chem. Ecol.* 29, 1589–1600
6. Rasmann, S., et al. (2005) Recruitment of entomopathogenic nematodes by insect-damaged maize roots. *Nature* 434, 732–737
7. Reinhard, J., et al. (2004) Floral scents induce recall of navigational and visual memories in honeybees. *J. Exp. Biol.* 207, 4371–4381
8. Wright, G.A., et al. (2005) Intensity and the ratios of compounds in the scent of snapdragon flowers affect scent discrimination by honeybees (*Apis mellifera*). *J. Comp. Physiol. A -Neuroethol. Sens. Neural Behav. Physiol* 191, 105–114
9. Knudsen, J.T., et al. (2006) Diversity and distribution of floral scent. *Bot. Rev.* 72, 1–120
10. Andrews, E.S., et al. (2007) Pollinator and herbivore attraction to *Cucurbita* floral volatiles. *J. Chem. Ecol.* 33, 1682–1691
11. Theis, N. (2006) Fragrance of canada thistle (*Cirsium arvense*) attracts both floral herbivores and pollinators. *J. Chem. Ecol.* 32, 917–927
12. Goff, S.A., and Klee, H.J. (2006) Plant volatile compounds: Sensory cues for health and nutritional value? *Science* 311, 815–819
13. Luft, S., et al. (2003) The use of olfaction in the foraging behaviour of the golden-mantled flying fox, *Pteropus pumilus*, and the greater musky fruit bat, *Ptenochirus jagori* (Megachiroptera: *Pteropodidae*). *Naturwissenschaften* 90, 84–87
14. Hodgkison, R., et al. (2007) Chemical ecology of fruit bat foraging behavior in relation to the fruit odors of two species of paleotropical bat-dispersed figs (*Ficus hispida* and *Ficus scortechinii*). *J. Chem. Ecol.* 33, 2097–2110
15. Bolen, R.H., and Green, S.M. (1997) Use of olfactory cues in foraging by owl monkeys (*Aotus nancymai*) and capuchin monkeys (*Cebus apella*). *J. Comp. Psychol.* 111, 152–158
16. Bicca-Marques, J.C., and Garber, P.A. (2004) Use of spatial, visual, and olfactory information during foraging in wild nocturnal and diurnal anthropoids: A field experiment comparing *Aotus*, *Callicebus*, and *Saguinus*. *Am. J. Primatol.* 62, 171–187
17. Siemers, B.M., et al. (2007) Sensory basis of food detection in wild *Microcebus murinus*. *Int. J. Primatol.* 28, 291–304
18. Frey, C. (2005) Natural flavors and fragrances: chemistry, analysis, and production. In *Natural flavors and fragrances: Chemistry, Analysis, and Production* (Frey, C., and Rouseff, R.L., eds), 3–19, American Chemical Society, Washington DC
19. Chandler, S., and Tanaka, Y. (2007) Genetic modification in floriculture. *Crit. Rev. Plant Sci.* 26, 169–197
20. Waldron, K.W., et al. (2003) Plant cell walls and food quality. *Compr. Rev. Food Sci. Food Saf.* 2, 101–119
21. Acree, T., Arn, H. (2004) Flavornet and human odor space. http://www.flavornet.org/index.html Accessed 10 Jan. 2008
22. Knudsen, J.T., and Gershenzon, J. (2006) The chemical diversity of floral scent. In *Biology of Floral Scent* (Dudareva, N., and Pichersky, E., eds), 27–52, CRC/Taylor & Francis, Boca Raton, FL
23. Nursten, H.E., and Williams, A.A. (1967) Fruit aromas – a survey of components identified. *Chem. Indust.* 12, 486–497
24. Buttery, R.G., et al. (1971) Characterization of additional volatile components of tomato. *J. Agric. Food Chem.* 19, 524–529
25. Petró-Turza, M. (1986) Flavor of tomato and tomato products. *Food Rev. Int.* 2, 309–352
26. Leahy, M.M., and Roderick, R.G. (1999) Fruit flavor biogenesis. In *Flavor Chemistry: Thirty*

Years of Progress (Teranishi, R., et-al, eds), 275–285, Kluwer/Plenum, New York
27. Tressl, R., et al. (1969) Gas chromatography of aroma constituents of bananas. *Z. Nat B.* 24, 781–783
28. Schreier, P., et al. (1976) Identification of volatile constituents from grapes. *J. Agric. Food Chem.* 24, 331–336
29. Klesk, K., et al. (2004) Aroma extract dilution analysis of cv. Meeker (*Rubus idaeus* L.) red raspberries from Oregon and Washington. *J. Agric. Food Chem.* 52, 5155–5161
30. Buttery, R.G., et al. (1987) Fresh tomato aroma volatiles – a quantitative study. *J. Agric. Food Chem.* 35, 540–544
31. Stockhorst, U., and Pietrowsky, R. (2004) Olfactory perception, communication, and the nose-to-brain pathway. *Physiol. Behav.* 83, 3–11
32. Vosshall, L.B. (2000) Olfaction in *Drosophila*. *Curr. Opin. Neurobiol.* 10, 498–503
33. Carlsson, M.A., and Hansson, B.S. (2006) Detection and coding of flower volatiles in nectar-foraging insects. In *Biology of Floral Scent* (Dudareva, N., and Pichersky, E., eds), 243–262, CRC/Taylor & Francis, Boca Raton, FL
34. Henderson, A. (1986) A review of pollination studies in the *Palmae*. *Bot. Rev.* 52, 221–259
35. Raguso, R.A., and Pichersky, E. (1995) Floral volatiles from *Clarkia breweri* and *C. concinna* (Onagraceae) – recent evolution of floral scent and moth pollination. *Plant Syst. Evol.* 194, 55–67
36. Hoballah, M.E., et al. (2005) The composition and timing of flower odour emission by wild *Petunia axillaris* coincide with the antennal perception and nocturnal activity of the pollinator *Manduca sexta*. *Planta* 222, 141–150
37. McCaskill, D., and Croteau, R. (1995) Monoterpene and sesquiterpene biosynthesis in glandular trichomes of peppermint (*Mentha* X *Piperita*) rely exclusively on plastid-derived isopentenyl diphosphate. *Planta* 197, 49–56
38. Newman, J.D., and Chappell, J. (1999) Isoprenoid biosynthesis in plants: Carbon partitioning within the cytoplasmic pathway. *Crit. Rev. Biochem. Mol. Biol.* 34, 95–106
39. Lichtenthaler, H.K. (1999) The 1-deoxy-D-xylulose-5-phosphate pathway of isoprenoid biosynthesis in plants. *Annu. Rev. Plant Physiol Plant Mol. Biol.* 50, 47–65
40. Rohmer, M. (1999) The discovery of a mevalonate-independent pathway for isoprenoid biosynthesis in bacteria, algae and higher plants. *Nat. Prod. Rep.* 16, 565–574
41. Eisenreich, W., et al. (1998) The deoxyxylulose phosphate pathway of terpenoid biosynthesis in plants and microorganisms. *Chem. Biol.* 5, R221-R233
42. Phillips, M.A., et al. (2008) The *Arabidopsis thaliana* type I isopentenyl diphosphate isomerases are targeted to multiple subcellular compartments and have overlapping functions in isoprenoid biosynthesis. *Plant Cell* 20, 677–696
43. Poulter, C.D., and Rilling, H.C. (1981) Prenyl transferases and isomerase. In *Biosynthesis of Isoprenoid Compounds* (Porter, J.W., and Spurgeon, S.L., eds), 161–224, Wiley, New York
44. Ogura, K., and Koyama, T. (1998) Enzymatic aspects of isoprenoid chain elongation. *Chem. Rev.* 98, 1263–1276
45. McGarvey, D.J., and Croteau, R. (1995) Terpenoid metabolism. *Plant Cell* 7, 1015–1026
46. Koyama, T., and Ogura, K. (1999) Isopentenyl diphosphate isomerase and prenyltransferases. In *Comprehensive Natural Product Chemistry. Isoprenoids Including Carotenoids and Steroids* (Cane, D.E., ed), 69–96, Pergamon, Oxford
47. Tholl, D., et al. (2004) Formation of monoterpenes in *Antirrhinum majus* and *Clarkia breweri* flowers involves heterodimeric geranyl diphosphate synthases. *Plant Cell* 16, 977–992
48. Burke, C.C., et al. (1999) Geranyl diphosphate synthase: Cloning, expression, and characterization of this prenyltransferase as a heterodimer. *Proc. Natl. Acad. Sci. USA* 96, 13062–13067
49. Gershenzon, J., and Kreis, W. (1999) Biochemistry of terpenoids: Monoterpenes, sesquiterpenes, diterpenes, sterols, cardiac glycosides and steroid saponins. In *Biochemistry of Plant Secondary Metabolism* (Wink, M., ed), 222–299, CRC Press, Boca Raton, FL
50. Wang, K., and Ohnuma, S. (1999) Chain-length determination mechanism of isoprenyl diphosphate synthases and implications for molecular evolution. *Trends Biochem. Sci.* 24, 445–451
51. Bouvier, F., et al. (2000) Molecular cloning of geranyl diphosphate synthase and compartmentation of monoterpene synthesis in plant cells. *Plant J.* 24, 241–252
52. Bick, J.A., and Lange, B.M. (2003) Metabolic cross talk between cytosolic and plastidial pathways of isoprenoid biosynthesis: unidirectional transport of intermediates across the chloroplast envelope membrane. *Arch. Biochem. Biophys.* 415, 146–154
53. Schuhr, C.A., et al. (2003) Quantitative assessment of crosstalk between the two isoprenoid

biosynthesis pathways in plants by NMR spectroscopy. *Phytochemistry Reviews* 2, 3–16
54. Laule, O., et al. (2003) Crosstalk between cytosolic and plastidial pathways of isoprenoid biosynthesis in *Arabidopsis thaliana. Proc. Natl. Acad. Sci. USA* 100, 6866–6871
55. Hemmerlin, A., et al. (2003) Cross-talk between the cytosolic mevalonate and the plastidial methylerythritol phosphate pathways in Tobacco Bright Yellow-2 cells. *J. Biol. Chem.* 278, 26666–26676
56. Dudareva, N., et al. (2005) The nonmevalonate pathway supports both monoterpene and sesquiterpene formation in snapdragon flowers. *Proc. Natl. Acad. Sci. USA* 102, 933–938
57. Wise, M.L., and Croteau, R. (1999) Monoterpene biosynthesis. In *Comprehensive Natural Products Chemistry: Isoprenoids Including Carotenoids and Steroids* (Cane, D.E., ed), 97–153, Pregamon, Oxford
58. Cane, D.E. (1999) Sesquiterpene biosynthesis: Cyclization mechanisms. In *Comprehensive Natural Products Chemistry. Isoprenoids Including Carotenoids and Steroids* (Cane, D.E., ed), 155–200, Pergamon, Oxford
59. Bohlmann, J., et al. (1998) Plant terpenoid synthases: molecular biology and phylogenetic analysis. *Proc. Natl. Acad. Sci. USA* 95, 4126–4133
60. Martin, D.M., et al. (2004) Functional characterization of nine Norway spruce TPS genes and evolution of gymnosperm terpene synthases of the TPS-d subfamily. *Plant Physiol.* 135, 1908–1927
61. Steele, C.L., et al. (1998) Sesquiterpene synthases from grand fir (*Abies grandis*) – Comparison of constitutive and wound-induced activities, and cDNA isolation, characterization and bacterial expression of δ-selinene synthase and γ-humulene synthase. *J. Biol. Chem.* 273, 2078–2089
62. Roeder, S., et al. (2007) Regulation of simultaneous synthesis of floral scent terpenoids by the 1,8-cineole synthase of *Nicotiana suaveolens*. *Plant Mol. Biol.* 65, 107–124
63. Tholl, D., et al. (2005) Two sesquiterpene synthases are responsible for the complex mixture of sesquiterpenes emitted from *Arabidopsis* flowers. *Plant J.* 42, 757–771
64. Bohlmann, J., et al. (2000) Terpenoid secondary metabolism in *Arabidopsis thaliana*: cDNA cloning, characterization, and functional expression of a myrcene/(E)-β-ocimene synthase. *Arch. Biochem. Biophys.* 375, 261–269
65. Chen, F., et al. (2003) Biosynthesis and emission of terpenoid volatiles from *Arabidopsis* flowers. *Plant Cell* 15, 481–494
66. Dudareva, N., et al. (2003) (E)-β-ocimene and myrcene synthase genes of floral scent biosynthesis in snapdragon: Function and expression of three terpene synthase genes of a new terpene synthase subfamily. *Plant Cell* 15, 1227–1241
67. Trapp, S.C., and Croteau, R.B. (2001) Genomic organization of plant terpene synthases and molecular evolutionary implications. *Genetics* 158, 811–832
68. Aubourg, S., et al. (2002) Genomic analysis of the terpenoid synthase (AtTPS) gene family of *Arabidopsis thaliana. Mol. Genet. Genomics* 267, 730–745
69. Dudareva, N., et al. (2004) Biochemistry of plant volatiles. *Plant Physiol.* 135, 1893–1902
70. Bouwmeester, H.J., et al. (1998) Biosynthesis of the monoterpenes limonene and carvone in the fruit of caraway – I. Demonstration of enzyme activities and their changes with development. *Plant Physiol.* 117, 901–912
71. Bouwmeester, H.J., et al. (1999) Cytochrome P-450 dependent (+)-limonene-6-hydroxylation in fruits of caraway (*Carum carvi*). *Phytochemistry* 50, 243–248
72. Aharoni, A., et al. (2004) Gain and loss of fruit flavor compounds produced by wild and cultivated strawberry species. *Plant Cell* 16, 3110–3131
73. Shalit, M., et al. (2003) Volatile ester formation in roses. Identification of an acetyl-coenzyme A: Geraniol/citronellol acetyltransferase in developing rose petals. *Plant Physiol.* 131, 1868–1876
74. Luan, F., et al. (2005) Metabolism of geraniol in grape berry mesocarp of *Vitis vinifera* L. cv. Scheurebe: demonstration of stereoselective reduction, E/Z-isomerization, oxidation and glycosylation. *Phytochemistry* 66, 295–303
75. Auldridge, M.E., et al. (2006) Characterization of three members of the *Arabidopsis* carotenoid cleavage dioxygenase family demonstrates the divergent roles of this multifunctional enzyme family. *Plant J.* 45, 982–993
76. Buttery, R.G., et al. (1990) Quantitative and sensory studies on tomato paste volatiles. *J. Agric. Food Chem.* 38, 336–340
77. Winterhalter, P., and Rouseff, R.L. (2002) Carotenoid-derived aroma compounds: an introduction. In *Carotenoid-derived aroma compounds* (Winterhalter, P., and Rouseff, R.L.,

eds), 1–17, American Chemical Society, Washington DC
78. Simkin, A.J., et al. (2004) The tomato carotenoid cleavage dioxygenase 1 genes contribute to the formation of the flavor volatiles β-ionone, pseudoionone, and geranylacetone. *Plant J.* 40, 882–892
79. Simkin, A.J., et al. (2004) Circadian regulation of the PhCCD1 carotenoid cleavage dioxygenase controls emission of β-Ionone, a fragrance volatile of petunia flowers. *Plant Physiol.* 136, 3504–3514
80. Ibdah, M., et al. (2006) Functional characterization of CmCCD1, a carotenoid cleavage dioxygenase from melon. *Phytochemistry* 67, 1579–1589
81. Kapteyn, J., et al. (2007) Evolution of cinnamate/*p*-coumarate carboxyl methyltransferases and their role in the biosynthesis of methylcinnamate. *Plant Cell* 19, 3212–3229
82. Humphreys, J.M., and Chapple, C. (2002) Rewriting the lignin roadmap. *Curr. Opin. Plant Biol.* 5, 224–229
83. Koeduka, T., et al. (2006) Eugenol and isoeugenol, characteristic aromatic constituents of spices, are biosynthesized via reduction of a coniferyl alcohol ester. *Proc. Natl. Acad. Sci. USA* 103, 10128–10133
84. Koeduka, T., et al. (2008) The multiple phenylpropene synthases in both *Clarkia breweri* and *Petunia hybrida* represent two distinct protein lineages. *Plant J.* 54, 362–374
85. Dexter, R., et al. (2007) Characterization of a petunia acetyltransferase involved in the biosynthesis of the floral volatile isoeugenol. *Plant J.* 49, 265–275
86. D'Auria, J.C. (2006) Acyltransferases in plants: a good time to be BAHD. *Curr. Opin. Plant Biol.* 9, 331–340
87. Vassao, D.G., et al. (2006) Chavicol formation in sweet basil (*Ocimum basilicum*): cleavage of an esterified C9 hydroxyl group with NAD(P)H-dependent reduction. *Org. Biomol. Chem.* 4, 2733–2744
88. Gang, D.R., et al. (2002) Characterization of phenylpropene *O*-methyltransferases from sweet basil: Facile change of substrate specificity and convergent evolution within a plant *O*-methyltransferase family. *Plant Cell* 14, 505–519
89. Wang, J.H., et al. (1997) Floral scent production in *Clarkia breweri* (Onagraceae). 2. Localization and developmental modulation of the enzyme *S*-adenosyl-L-methionine-(iso)eugenol *O*-methyltransferase and phenylpropanoid emission. *Plant Physiol.* 114, 213–221
90. Boatright, J., et al. (2004) Understanding in vivo benzenoid metabolism in petunia petal tissue. *Plant Physiol.* 135, 1993–2011
91. Orlova, I., et al. (2006) Reduction of benzenoid synthesis in petunia flowers reveals multiple pathways to benzoic acid and enhancement in auxin transport. *Plant Cell* 18, 3458–3475
92. Podstolski, A., et al. (2002) Unusual 4-hydroxybenzaldehyde synthase activity from tissue cultures of the vanilla orchid *Vanilla planifolia*. *Phytochemistry* 61, 611–620
93. Pak, F.E., et al. (2004) Characterization of a multifunctional methyltransferase from the orchid *Vanilla planifolia*. *Plant Cell Rep.* 22, 959–966
94. Effmert, U., et al. (2005) Floral benzenoid carboxyl methyltransferases: From in vitro to in planta function. *Phytochemistry* 66, 1211–1230
95. Wang, J., and De Luca, V. (2005) The biosynthesis and regulation of biosynthesis of Concord grape fruit esters, including 'foxy' methylanthranilate. *Plant J.* 44, 606–619
96. Kaminaga, Y., et al. (2006) Plant phenylacetaldehyde synthase is a bifunctional homotetrameric enzyme that catalyzes phenylalanine decarboxylation and oxidation. *J. Biol. Chem.* 281, 23357–23366
97. Tieman, D., et al. (2006) Tomato aromatic amino acid decarboxylases participate in synthesis of the flavor volatiles 2-phenylethanol and 2-phenylacetaldehyde. *Proc. Natl. Acad. Sci. USA* 103, 8287–8292
98. Watanabe, S., et al. (2002) Biogenesis of 2-phenylethanol in rose flowers: Incorporation of [$^{2}H_{8}$]L-phenylalanine into 2-phenylethanol and its β-D-glucopyranoside during the flower opening of *Rosa* 'Hoh-Jun' and *Rosa damascena* Mill. *Biosci. Biotechnol. Biochem.* 66, 943–947
99. Feussner, I., and Wasternack, C. (2002) The lipoxygenase pathway. *Annu. Rev. Plant Biol.* 53, 275–297
100. Seo, H.S., et al. (2001) Jasmonic acid carboxyl methyltransferase: A key enzyme for jasmonate-regulated plant responses. *Proc. Natl. Acad. Sci. USA* 98, 4788–4793
101. Song, M.S., et al. (2005) Isolation and characterization of a jasmonic acid carboxyl methyltransferase gene from hot pepper (*Capsicum annuum* L.). *J. Plant Biol.* 48, 292–297
102. Matsui, K., et al. (1996) Bell pepper fruit fatty acid hydroperoxide lyase is a cytochrome P450 (CYP74B). *FEBS Lett.* 394, 21–24

103. Tijet, N., et al. (2000) Purification, molecular cloning, and expression of the gene encoding fatty acid 13-hydroperoxide lyase from guava fruit (*Psidium guajava*). *Lipids* 35, 709–720

104. Kim, I.S., and Grosch, W. (1981) Partial purification and properties of a hydroperoxide lyase from fruits of pear. *J. Agric. Food Chem.* 29, 1220–1225

105 Mita, G., et al. (2005) Molecular cloning and characterization of an almond 9-hydroperoxide lyase, a new CYP74 targeted to lipid bodies. *J. Exp. Bot.* 56, 2321–2333

106. Tijet, N., et al. (2001) Biogenesis of volatile aldehydes from fatty acid hydroperoxides: Molecular cloning of a hydroperoxide lyase (CYP74C) with specificity for both the 9-and 13-hydroperoxides of linoleic and linolenic acids. *Arch. Biochem. Biophys.* 386, 281–289

107. Grechkin, A. (1998) Recent developments in biochemistry of the plant lipoxygenase pathway. *Prog. Lipid Res.* 37, 317–352

108. Prestage, S., et al. (1999) Volatile production in tomato fruit with modified alcohol dehydrogenase activity. *J. Sci. Food Agr.* 79, 131–136

109. Akacha, N.B., et al. (2005) Production of hexenol in a two-enzyme system: kinetic study and modelling. *Biotechnol. Lett.* 27, 1875–1878

110. Dickinson, J.R., et al. (2000) An investigation of the metabolism of isoleucine to active amyl alcohol in *Saccharomyces cerevisiae*. *J. Biol. Chem.* 275, 10937–10942

111. Beck, H.C., et al. (2002) Metabolite production and kinetics of branched-chain aldehyde oxidation in *Staphylococcus xylosus*. *Enzyme Microb. Technol.* 31, 94–101

112. Tavaria, F.K., et al. (2002) Amino acid catabolism and generation of volatiles by lactic acid bacteria. *J. Dairy Sci.* 85, 2462–2470

113. Reineccius, G. (2006) *Flavor chemistry and technology, Second edition*. CRC Press, Boca Raton, FL

114. Hansen, K., and Poll, L. (1993) Conversion of L-isoleucine into 2-methylbut-2-enyl esters in apples. *Lebensmittel-Wissenschaft Technologie* 26, 178–180

115. Rowan, D.D., et al. (1996) Biosynthesis of 2-methylbutyl, 2-methyl-2-butenyl, and 2-methylbutanoate esters in Red Delicious and Granny Smith apples using deuterium-labeled substrates. *J. Agric. Food Chem.* 44, 3276–3285

116. Pérez, A.G., et al. (2002) Biosynthesis of strawberry aroma compounds through amino acid metabolism. *J. Agric. Food Chem.* 50, 4037–4042

117. Matich, A., and Rowan, D. (2007) Pathway analysis of branched-chain ester biosynthesis in apple using deuterium labeling and enantioselective gas chromatography-mass spectrometry. *J. Agric. Food Chem.* 55, 2727–2735

118. Wyllie, S.G., et al. (1995) Key aroma compounds in melons - their development and cultivar dependence. *Fruit Flav.* 596, 248–257

119. Wyllie, S.G., and Leach, D.N. (1992) The role of sulfur volatiles in melon aroma. *J. Agric. Food Chem.* 40, 253–256

120. Amarita, F., et al. (2004) Identification and functional analysis of the gene encoding methionine-γ-lyase in *Brevibacterium linens*. *App. Environ. Microbiol.* 70, 7348–7354

121. Bondar, D.C., et al. (2005) Involvement of a branched-chain aminotransferase in production of volatile sulfur compounds in *Yarrowia lipolytica*. *Appl. Environ. Microbiol.* 71, 4585–4591

122. Pérez, A.G., et al. (1992) Aroma components and free amino-acids in strawberry variety Chandler during ripening. *J. Agric. Food Chem.* 40, 2232–2235

123. Aharoni, A., et al. (2000) Identification of the SAAT gene involved in strawberry flavor biogenesis by use of DNA microarrays. *Plant Cell* 12, 647–661

124. Beekwilder, J., et al. (2004) Functional characterization of enzymes forming volatile esters from strawberry and banana. *Plant Physiol.* 135, 1865–1878

125. Dudareva, N., et al. (1998) Acetyl-CoA: benzylalcohol acetyltransferase – an enzyme involved in floral scent production in *Clarkia breweri*. *Plant J.* 14, 297–304

126. D'Auria, J.C., et al. (2002) Characterization of an acyltransferase capable of synthesizing benzylbenzoate and other volatile esters in flowers and damaged leaves of *Clarkia breweri*. *Plant Physiol.* 130, 466–476

127. Yahyaoui, F.E.L., et al. (2002) Molecular and biochemical characteristics of a gene encoding an alcohol acyl-transferase involved in the generation of aroma volatile esters during melon ripening. *Eur. J. Biochem.* 269, 2359–2366

128. El-Sharkawy, I., et al. (2005) Functional characterization of a melon alcohol acyl-transferase gene family involved in the biosynthesis of ester volatiles. Identification of the crucial role of a threonine residue for enzyme activity. *Plant Mol. Biol.* 59, 345–362

129. Defilippi, B.G., et al. (2005) Relationship of ethylene biosynthesis to volatile production, related enzymes, and precursor availability in apple peel and flesh tissues. *J. Agric. Food Chem.* 53, 3133–3141
130. Frey, M., et al. (2000) An herbivore elicitor activates the gene for indole emission in maize. *Proc. Natl. Acad. Sci. USA* 97, 14801–14806
131. Raab, T., et al. (2006) FaQR, required for the biosynthesis of the strawberry flavor compound 4-hydroxy-2,5-dimethyl-3(2H)-furanone, encodes an enone oxidoreductase. *Plant Cell* 18, 1023–1037
132. Dudareva, N., et al. (1996) Evolution of floral scent in *Clarkia*: novel patterns of *S*-linalool synthase gene expression in the *C. breweri* flower. *Plant Cell* 8, 1137–1148
133. Kolosova, N., et al. (2001) Cellular and subcellular localization of *S*-adenosyl-L-methionine:benzoic acid carboxyl methyltransferase, the enzyme responsible for biosynthesis of the volatile ester methylbenzoate in snapdragon flowers. *Plant Physiol.* 126, 956–964
134. Scalliet, G., et al. (2006) Role of petal-specific orcinol *O*-methyltransferases in the evolution of rose scent. *Plant Physiol.* 140, 18–29
135. Effmert, U., et al. (2006) Localization of the synthesis and emission of scent compounds within the flower. In *Biology of Floral Scent* (Dudareva, N., and Pichersky, E., eds), 105–124, CRC/Taylor & Francis, Boca Raton, FL
136. Dudareva, N., and Pichersky, E. (2000) Biochemical and molecular genetic aspects of floral scents. *Plant Physiol.* 122, 627–633
137. Bergougnoux, V., et al. (2007) Both the adaxial and abaxial epidermal layers of the rose petal emit volatile scent compounds. *Planta* 226, 853–866
138. Turner, G.W., et al. (1998) Schizogenous secretory cavities of *Citrus limon* (L.) Burm. f. and a reevaluation of the lysigenous gland concept. *Int. J. Plant Sci.* 159, 75–88
139. Li, D.P., et al. (2006) Molecular cloning and expression of a gene encoding alcohol acyltransferase (MdAAT2) from apple.(cv. Golden Delicious). *Phytochemistry* 67, 658–667
140. Altenburger, R., and Matile, P. (1990) Further observations on rhythmic emission of fragrance in flowers. *Planta* 180, 194–197
141. Oyama-Okubo, N., et al. (2005) Emission mechanism of floral scent in *Petunia axillaris*. *Biosci. Biotechnol. Biochem.* 69, 773–777
142. Perez, A.G., et al. (1999) Lipoxygenase and hydroperoxide lyase activities in ripening strawberry fruits. *J. Agric. Food Chem.* 47, 249–253
143. Braidot, E., et al. (2004) Biochemical and immunochemical evidence for the presence of lipoxygenase in plant mitochondria. *J. Exp. Bot.* 55, 1655–1662
144. Matsui, K., et al. (1992) Developmental changes in lipoxygenase activity in cotyledons of cucumber seedlings. *Plant Sci.* 85, 23–32
145. Tranbarger, T.J., et al. (1991) The soybean 94-kilodalton vegetative storage protein is a lipoxygenase that is localized in paraveinal mesophyll cell vacuoles. *Plant Cell* 3, 973–987
146. Vernooy-Gerritsen, M., et al. (1984) Intracellular localization of lipoxygenases-1 and -2 in germinating soybean seeds by indirect labeling with protein A-colloidal gold complexes. *Plant Physiol.* 76, 1070–1078
147. Feussner, I., and Kindl, H. (1992) A lipoxygenase is the main lipid body protein in cucumber and soybean cotyledons during the stage of triglyceride mobilization. *FEBS Lett.* 298, 223–225
148. Feussner, I., and Kindl, H. (1994) Particulate and soluble lipoxygenase isoenzymes – comparison of molecular and enzymatic properties. *Planta* 194, 22–28
149. Georgalaki, M.D., et al. (1998) Characterization of an 13-lipoxygenase from virgin olive oil and oil bodies of olive endosperms. *Lipid-Fett.* 100, 554–560
150. Nellen, A., et al. (1995) Lipoxygenase forms located at the plant plasma membrane. *Z. Nat. C* 50, 29–36
151. Chen, G.P., et al. (2004) Identification of a specific isoform of tomato lipoxygenase (TomloxC) involved in the generation of fatty acid-derived flavor compounds. *Plant Physiol.* 136, 2641–2651
152. Leone, A., et al. (2006) Lipoxygenase involvement in ripening strawberry. *J. Agric. Food Chem.* 54, 6835–6844
153. Jakobsen, H.B., and Olsen, C.E. (1994) Influence of climatic factors on emission of flower volatiles in-situ. *Planta* 192, 365–371
154. MacTavish, H.S., et al. (2000) Emission of volatiles from brown boronia flowers: Some comparative observations. *Ann. Bot.* 86, 347–354
155. Sharon-Asa, L., et al. (2003) Citrus fruit flavor and aroma biosynthesis: isolation, functional characterization, and developmental regulation of Cstps1, a key gene in the production of the

sesquiterpene aroma compound valencene. *Plant J.* 36, 664–674
156. Dudareva, N., et al. (2000) Developmental regulation of methyl benzoate biosynthesis and emission in snapdragon flowers. *Plant Cell* 12, 949–961
157. Verdonk, J.C., et al. (2005) ODORANT1 regulates fragrance biosynthesis in petunia flowers. *Plant Cell* 17, 1612–1624
158. Matile, P., and Altenburger, R. (1988) Rhythms of fragrance emission in flowers. *Planta* 174, 242–247
159. Kolosova, N., et al. (2001) Regulation of circadian methyl benzoate emission in diurnally and nocturnally emitting plants. *Plant Cell* 13, 2333–2347
160. Helsper, J., et al. (1998) Circadian rhythmicity in emission of volatile compounds by flowers of *Rosa hybrida* L. cv. Honesty. *Planta* 207, 88–95
161 Hendel-Rahmanim, K., et al. (2007) Diurnal regulation of scent emission in rose flowers. *Planta* 226, 1491–1499
162. Loughrin, J.H., et al. (1991) Circadian rhythm of volatile emission from flowers of *Nicotiana sylvestris* and *N. suaveolens*. *Physiol. Plant.* 83, 492–496
163. Underwood, B.A., et al. (2005) Ethylene-regulated floral volatile synthesis in petunia corollas. *Plant Physiol.* 138, 255–266
164. Arditti, J. (1979) Aspects of the physiology of orchids. In *Advances in Botanical Research* (Woolhouse, H.W., ed), 422–697, Academic Press, Washington DC
165. Schiestl, F.P., et al. (1997) Variation of floral scent emission and postpollination changes in individual flowers of *Ophrys sphegodes* subsp. *sphegodes*. *J. Chem. Ecol.* 23, 2881–2895
166. Tollsten, L. (1993) A multivariate approach to postpollination changes in the floral scent of *Platanthera bifolia* (Orchidaceae). *Nord. J. Bot.* 13, 495–499
167. Tollsten, L., and Bergstrom, J. (1989) Variation and post-pollination changes in floral odors released by *Platanthera bifolia* (Orchidaceae). *Nord. J. Bot.* 9, 359–362
168. Schiestl, F.P., and Ayasse, M. (2001) Post-pollination emission of a repellent compound in a sexually deceptive orchid: a new mechanism for maximising reproductive success? *Oecologia* 126, 531–534
169. Negre, F., et al. (2003) Regulation of methylbenzoate emission after pollination in snapdragon and petunia flowers. *Plant Cell* 15, 2992–3006
170. Schade, F., et al. (2001) Fragrance volatiles of developing and senescing carnation flowers. *Phytochemistry* 56, 703–710
171. Sexton, R., et al. (2005) Aroma production from cut sweet pea flowers (*Lathyrus odoratus*): the role of ethylene. *Physiol. Plant.* 124, 381–389
172. Schaffer, R.J., et al. (2007) A Genomics approach reveals that aroma production in apple is controlled by ethylene predominantly at the final step in each biosynthetic pathway([w]). *Plant Physiol.* 144, 1899–1912
173. Manriquez, D., et al. (2006) Two highly divergent alcohol dehydrogenases of melon exhibit fruit ripening-specific expression and distinct biochemical characteristics. *Plant Mol. Biol.* 61, 675–685
174. Lunkenbein, S., et al. (2006) Cinnamate metabolism in ripening fruit. Characterization of a UDP-glucose: Cinnamate glucosyltransferase from strawberry. *Plant Physiol.* 140, 1047–1058
175. Nagegowda, D.A., et al. (2008) Two nearly identical terpene synthases catalyze the formation of nerolidol and linalool in snapdragon flowers. *Plant J.* 55, 224–239
176. Guterman, I., et al. (2006) Generation of phenylpropanoid pathway-derived volatiles in transgenic plants: Rose alcohol acetyltransferase produces phenylethyl acetate and benzyl acetate in petunia flowers. *Plant Mol. Biol.* 60, 555–563
177. Dudareva, N., and Negre, F. (2005) Practical applications of research into the regulation of plant volatile emission. *Curr. Opin. Plant Biol.* 8, 113–118
178. Ackermann, I.E., et al. (1989) β-glucosides of aroma components from petals of rosa species – assay, occurrence, and biosynthetic implications. *J. Plant Physiol.* 134, 567–572
179. Jakobsen, H.B., and Christensen, L.P. (2002) Diurnal changes in the concentrations of 2-phenylethyl β-D-glucopyranoside and the corresponding volatile aglycone in the tissue and headspace of *Trifolium repens* L. florets. *Plant Cell Environ.* 25, 773–781
180. Francis, M.J.O., and Allcock, C. (1969) Geraniol β-D-glucoside; Occurrence and synthesis in rose flowers. *Phytochemistry* 8, 1339–1347
181. Sakai, M., et al. (2008) Purification and characterization of β-glucosidase involved in the emission of 2-phenylethanol from rose flowers. *Biosci. Biotechnol. Biochem.* 72, 219–221

182. Lunkenbein, S., et al. (2006) Up- and down-regulation of *Fragaria x ananassa* *O*-methyltransferase: impacts on furanone and phenylpropanoid metabolism. *J. Exp. Bot.* 57, 2445–2453
183. Aharoni, A., et al. (2005) Volatile science? Metabolic engineering of terpenoids in plants. *Trends Plant Sci.* 10, 594–602
184. Degenhardt, J., et al. (2003) Attracting friends to feast on foes: engineering terpene emission to make crop plants more attractive to herbivore enemies. *Curr. Opin. Biotechnol.* 14, 169–176
185. Dudareva, N., and Pichersky, E. (2006) Floral scent metabolic pathways: their regulation and evolution. In *Biology of Floral Scent* (Dudareva, N., and Pichersky, E., eds), 55–78, CRC/Taylor & Francis, Boca Raton, FL
186. Lücker, J., et al. (2006) Molecular engineering of floral scent. In *Biology of Floral Scent* (Dudareva, N., and Pichersky, E., eds), 321–337, CRC/Taylor & Francis, Boca Raton, FL
187. Lücker, J., et al. (2001) Expression of *Clarkia* *S*-linalool synthase in transgenic petunia plants results in the accumulation of *S*-linalyl-β-D-glucopyranoside. *Plant J.* 27, 315–324
188. Lavy, M., et al. (2002) Linalool and linalool oxide production in transgenic carnation flowers expressing the *Clarkia breweri* linalool synthase gene. *Mol. Breed.* 9, 103–111
189. Lücker, J., et al. (2004) Increased and altered fragrance of tobacco plants after metabolic engineering using three monoterpene synthases from lemon. *Plant Physiol.* 134, 510–519
190. El Tamer, M.K., et al. (2003) The influence of monoterpene synthase transformation on the odour of tobacco. *J. Biotechnol.* 106, 15–21
191. Lücker, J., et al. (2004) Metabolic engineering of monoterpene biosynthesis: two-step production of (+)-trans-isopiperitenol by tobacco. *Plant J.* 39, 135–145
192. Aranovich, D., et al. (2007) Post-harvest enhancement of aroma in transgenic lisianthus (*Eustoma grandiflorum*) using the *Clarkia breweri* benzyl alcohol acetyltransferase (BEAT) gene. *Postharv. Biol. Technol.* 43, 255–260
193. Tieman, D.M., et al. (2007) Tomato phenylacetaldehyde reductases catalyze the last step in the synthesis of the aroma volatile 2-phenylethanol. *Phytochemistry* 68, 2660–2669
194. Zuker, A., et al. (2002) Modification of flower color and fragrance by antisense suppression of the flavanone 3-hydroxylase gene. *Mol. Breed.* 9, 33–41
195. Ben Zvi, M.M., et al. (2008) Interlinking showy traits: co-engineering of scent and colour biosynthesis in flowers. *Plant Biotechnol. J.* 6, 403–415
196. Zvi, M.M.B., et al. (2006) Navigating the network of floral scent production. *Acta Hort. (ISHS)* 714, 143–154
197. Torregrosa, L., et al. (2008) Manipulation of VvAdh to investigate its function in grape berry development. *Plant Sci.* 174, 149–155
198. Speirs, J., et al. (1998) Genetic manipulation of alcohol dehydrogenase levels in ripening tomato fruit affects the balance of some flavor aldehydes and alcohols. *Plant Physiol.* 117, 1047–1058
199. Wang, C.L., et al. (1996) Changes of fatty acids and fatty acid-derived flavor compounds by expressing the yeast δ-9 desaturase gene in tomato. *J. Agric. Food Chem.* 44, 3399–3402
200. Lewinsohn, E., et al. (2001) Enhanced levels of the aroma and flavor compound *S*-linalool by metabolic engineering of the terpenoid pathway in tomato fruits. *Plant Physiol.* 127, 1256–1265
201. Davidovich-Rikanati, R., et al. (2007) Enrichment of tomato flavor by diversion of the early plastidial terpenoid pathway. *Nat. Biotechnol.* 25, 899–901
202. Wyllie S.G., Fellman, J.K. (2000) Formation of volatile branched chain esters in bananas (*Musa sapientum* L.). *J. Agric. Food Chem.* 48, 3493–3496
203. Myers, M.J., et al. (1970) L-Leucine as a precursor of isoamyl alcohol and isoamyl acetate, volatile aroma constituents of banana fruit discs. *Phytochemistry* 9, 1693–1700
204. Tressl, R., Drawert, F. (1973) Biogenesis of banana volatiles. *J. Agric. Food Chem.* 21, 560–565
205. Homatidou, V.I. et al. (1992) Determination of total volatile components of *Cucumis melo* L. variety Cantaloupensis. *J. Agric. Food Chem.* 40, 1385–1388

Part IV

New Trends

Bioengineering 20

Maxim Itkin and Asaph Aharoni

Abstract Plants produce a myriad of secondary metabolites (SMs), which constantly contribute to plants' interaction with the surroundings. Since ancient times and up to this day mankind has been using SMs as sources for medicines, spices, fragrances, pesticides, poisons, hallucinogens, stimulants, dyes, perfumery and countless more purposes. The shared value for both humans and plants makes SMs important targets for bioengineering. The formation of certain SM compounds may be restricted to single plant species, specific plant organs, cells or even particular cell compartments. Bioengineering can modulate the levels, time and site of production of natural products in plants. In this chapter we review the state of the art in the bioengineering of natural products at the whole plant level. In the first part of this review, we summarize the current and emerging bioengineering strategies and methods, including the use of the riboswitches, immunomodulation, synthetic microRNAs and Zinc-finger nucleases. The second and major part of this chapter provides examples from different fields of bioengineering in plants including: (a) the production of nutraceuticals, (b) modifying volatiles and pigments (in fruit and flowers), (c) production of medicinal agents and (d) aiding plants in the fight against biotic stresses. The experiments described here were conducted either in target plants, usually crop species, or as a form of a "proof of concept" in model plant species (e.g. Arabidopsis). Future challenges for SM bioengineering include reducing unwanted effects on plant fitness, transfer of knowledge from models to crops, the reduction of genomic position effects and the capacity to predict the outcome of bioengineering. These aspects are also discussed. The large and rapid advances made during the last decade in our understanding of the molecular genetic control of SM production and biological function provide an excellent foundation for successful bioengineering of these small molecules in plants.

Introduction

The production of tens of thousands of natural products or secondary metabolites (SMs) by plants is a consequence of metabolic diversification coupled with selection. Numerous roles of these substances in the plant life cycle have now been described. These include diverse functions associated with both the interaction of the plant with the environment (both biotic and abiotic aspects) and the assurance of successful reproduction. The importance of these molecules to the plant often receives relatively little attention in comparison with the focus on the

M. Itkin and A. Aharoni (✉)
Department of Plant Sciences,
Weizmann Institute of Science, Israel
e-mail: maxitkin@gmail.com,
asaph.aharoni@weizmann.ac.il

A.E. Osbourn and V. Lanzotti (eds.), *Plant-derived Natural Products*,
DOI 10.1007/978-0-387-85498-4_20,

significance of these compounds for humankind. SMs are not only essential ingredients of the human diet but also influence both pre- and post- harvest traits of many agricultural crops. The inventory of plant SMs can be divided into three major classes including: (I) phenolic and polyphenolic compounds, (II) terpenoids (or isoprenoids) (III) nitrogen- (i.e. alkaloids) and sulfur-containing compounds.

Not every plant can produce every SM, and the formation of certain compounds is restricted to single species or to related groups, where many SMs are found only in certain specific plant organs, in just one type of cell and even in a certain cell compartment. They are often generated only during a specific developmental period in the plant life or under particular stress conditions. Thus, it is frequently difficult to extract such substances to high purity and obtain the required concentrations either in the plant itself or in an isolated form.

Bioengineering, in most cases through the modification of metabolic pathways (i.e. metabolic engineering), could make SM-related applications more efficient, relevant and extensive. Three main approaches are usually exploited for bioengineering of plant SMs. These include the use of microorganisms, production in plant cell cultures (e.g. biotransformation techniques) and the application at the whole plant level. In this report we will focus on SM bioengineering at the whole plant level although examples of the other two approaches will also be referred to. The first part of this chapter will describe a "toolbox for the bioengineer" including recently introduced bioengineering strategies in whole plants. This section will also point to the limitations and key issues for consideration when carrying out bioengineering of SMs in plants. The second and main part of this chapter will provide examples from different fields of bioengineering including: (a) engineering nutraceuticals in plants, (b) engineering of volatiles and pigments (in fruit and flowers), (c) production of medicinal agents in plants and (d) combating biotic stress with plant-derived natural products.

While some bioengineering activities have been carried out in the target plant species (normally a crop plant), others have involved "proof of concept", in model plant species (e.g. Arabidopsis). Examples of both kinds of approach will be described and the difficulties in the transfer of successful bioengineering strategies from model plants to crops will be discussed. The diversity of SM structures can be divided into three major groups including flavonoids and related phenolic and polyphenolic compounds, terpenoids (or isoprenoids), and nitrogen-containing alkaloids and sulfur compounds. Bioengineering of these different classes of SMs will be described here, but we will exclude mention of compounds that are intermediates between primary and secondary metabolism, such as plant hormones (e.g. gibberellic acid, auxin, ethylene and abscisic acid) and compounds involved in cell wall synthesis (e.g. cinnamic acid and its polymeric derivative, lignin). The topic of bioengineering is very wide and it is therefore impossible to cover all the research in this area in a single review. Thus, we apologize for any omissions and hope that this chapter will demonstrate the state of the art in bioengineering of SMs in plants and its future projections.

A Toolbox for the Bioengineer

Since the development of basic plant molecular biology techniques including plant regeneration and gene transfer in the mid-1980s, significant progress has been made in the molecular dissection of metabolic pathways in plants and the use of newly cloned genes to engineer plant metabolism. Although there are numerous success stories in which the expected results have been obtained, there have been many examples of unanticipated outcomes. Most attempts to Bioengineer plant SMs have focused on modifying the expression of a single gene in order to alter committed steps in the biosynthesis of a particular metabolite.

Common Bioengineering Strategies

Tissue- and Organ-specific Manipulation of Gene Expression using Specific and Inducible Promoters

Metabolic engineering may often have a dramatic effect on plant fitness, growth and morphology. This is particularly true when strong promoters are used to drive gene expression with consequent pleiotropic effects. One way to overcome this problem is through utilization of tissue- or organ-specific promoters. The use of chemically-inducible promoters offers another convenient way to modulate the expression of genes where stable overexpression or silencing may result in harmful effects or lethality. In the absence of an inducer, a chemically-inducible promoter is inactive, while application of the inducer allows controlled activation of gene expression at particular developmental stages and for specific periods [1, 2]. Moreover, by expressing the chemical-responsive gene under the control of a tissue/organ specific promoter one can even limit the target gene expression to specific cell types, tissues or plant organs [3].

Substituting the Subcellular Location of Enzymes

The plant cell is a complex biochemical factory in which biosynthetic processes take place in different compartments. It is possible that the substrate for one enzyme will be produced in more than one cell compartment. In the case of promiscuous enzymes, two different potential substrates could be located in different cellular compartments. This can provide opportunities for bioengineering to form a new product or to increase the levels of current products. One can simply replace, add or remove an existing localization signal in order to modify the target compartment of a particular engineered protein. Recent work by Kappers et al. (2005) [4] showed that production of sesquiterpenes could be effectively increased through redirection of the terpene synthase FaNES1 to the mitochondria rather than to the cytosol (where sesquiterpenes are normally produced). Interestingly, the sesquiterpene formed in the mitochondria (i.e. nerolidol) was further metabolized to the signaling molecule 4,8-dimethyl-1,3(*E*),7-nonatriene ((*E*)-DMNT). Since FaNES1 is a dual monoterpene/sesquiterpene synthase that could generate both the monoterpene linalool and the sesquiterpene nerolidol, its targeting to a third cell location, the chloroplast, resulted in high level production of linalool [5]. Wu et al. [6] presented another example in which they substituted the subcellular location of enzymes in tobacco, overexpressing an avian farnesyl diphosphate synthase and a sesquiterpene synthase, fused with the Arabidopsis transit peptide signal sequence of the RUBISCO small unit protein, therefore directing the recombinant proteins to the plastid. This strategy resulted in an increase of more than 1,000-fold of the sesquiterpenes patchoulol and amorpha-4,11-diene.

Manipulation of Several Steps in Metabolic Pathways by Simultaneous Introduction of Multiple Genes and the use of Transcription Factors

Introduction or silencing of a single gene will often result in minor or unaltered levels of the desired metabolite. This might be due to the complexity and redundancy in the metabolic network. Thus, introduction of several genes from the target biosynthetic pathway may be desirable, as carried out by Lorenc-Kukula et al. [7] in order to increase the antioxidative properties of flax (*Linum usitatissimum* L.). The authors transformed flax with three genes, encoding key enzymes of the flavonoid biosynthetic pathway from petunia (i.e. chalcone synthase (*CHS*), chalcone isomerase (*CHI*), and dihydroflavonol reductase (*DFR*)).

Transgenic plants that simultaneously and constitutively expressed all three genes exhibited an increase in the levels of phenolic acids up to 26% compared to the control plants, increasing the antioxidant potential up to 6-fold in seeds and up to 91-fold in leaves.

Altering multiple steps in one metabolic pathway can be also carried out by overexpression or silencing of a transcription factor that regulates several genes belonging to same pathway [8]. The maize *Lc* (leaf color) MYC-type transcription factor is one of the most studied regulatory genes that control the flavonoid and anthocyanin pathway [9]. Expression of *Lc* in alfalfa induced the accumulation of *CHS*, flavanone 3β-hydroxylase (*F3H*) and leucocyanidin reductase (*LCR*) transcripts, under low temperature or high light stress conditions [10]. The elevation in the expression of these genes was accompanied by production of anthocyanins and proanthocyanidins. Bioengineering of the flavonoid biosynthetic pathway by expressing the maize *Lc* gene was also described in apple (*Malus domestica* Borkh.) [11]. The transgenic plants showed increased mRNA levels of multiple genes including phenylalanine ammonia-lyase (*PAL*), *CHS*, *F3H*, *DFR*, *LCR*, anthocyanidin synthase (*ANS*) and anthocyanidin reductase (*ANR*). Leaves of transgenic lines accumulated higher levels of the anthocyanin idaein, the flavan 3-ol epicatechin, the isomeric catechin, and some distinct dimeric proanthocyanidins. Hence, the introduction of a single regulator as demonstrated in apple could serve as an ideal alternative to the activation of multiple genes in a given pathway.

The Use of a Chimeric Repressor Protein

Genetic redundancy frequently interferes with attempts of gene silencing. Even when similar regions in a few redundant genes are used for co-silencing it is difficult to predict the silencing effect on a single gene. Overexpression of a chimeric repressor transcription factor might circumvent problems of redundancy. In this approach, a fusion protein is generated, bearing a repression domain (the EAR-motif), derived from native plant transcription factors (e.g. SUPERMAN; Hiratsu et al. [12]) and typically six amino acid residues in length, attached to the full-length transcriptional activator of interest [13]. Overexpression of the chimeric protein results in suppression of the transcription factor target gene/s expression and furthermore prevents the binding and transcriptional activation by redundant activators [14]. Chimeric repressor proteins (e.g. ETHYLENE-INSENSITIVE3 (EIN3), CUP-SHAPED COTYLEDON1 (CUC1), PRODUCTION of ANTHOCYANIN PIGMENT1 (PAP1), and AtMYB23) were demonstrated to posses dominant loss-of-function phenotypes in transgenic plants [15] that effect metabolic [16], physiological [17] and developmental [18] processes.

Expression of Feedback Insensitive Enzymes

A major hurdle in engineering of metabolic pathways in some cases are feedback mechanisms in which, for example, inhibition of enzymes by end-products of their biosynthetic pathways occurs. The use of a heterologous feedback-insensitive enzyme can solve such a problem, as demonstrated by Zhu and Galili (2004) [19]. To elucidate the relative significance of lysine synthesis and catabolism in determining lysine level in plant seeds, they expressed a bacterial feedback-insensitive dihydrodipicolinate synthase (DHPS) in a seed-specific manner in wild type Arabidopsis. Transgenic plants, expressing the bacterial DHPS, contained approximately 12-fold higher levels of seed-free lysine compared to untransformed plants.

Virus induced Gene Silencing and Overexpression

Virus induced gene silencing (VIGS) is a commonly used technique allowing systemic silencing of genes in various organs of the plant [20]. A recent example of the applicability of *Tobacco Rattle Virus* (*TRV*)-based VIGS for modifying floral scent metabolites was demonstrated in petunia [21]. This experiment involved several genes required for floral scent production including the MYB transcription factor ODORANT1 (ODO1) [22]. VIGS-silenced *ODO1* flowers exhibited a strong reduction in levels of volatile compounds typically emitted by non-infected petunia plants.

Lindbo (2007) reported on a *Tobacco Mosaic Virus* (*TMV*)-based transient expression vector which can express very high levels of foreign proteins in plants [23]. In order to increase the efficiency of agroinfection the vector contained a *TMV* replicon under the control of the *Cauliflower Mosaic Virus* (*CaMV*) 35S promoter and lacked the *TMV* coat protein gene sequence. The advantage of such a technique is in the possibility to transiently express foreign proteins at relatively high levels (3–5 mg per gram fresh weight of plant tissue).

Recent Bioengineering Strategies

Recent discoveries regarding the role of small and non-coding RNAs not only impacted our understanding of regulatory mechanisms but also promoted the development of novel methods for the bioengineering of plants (Table 1). As described below, RNA molecules can be exploited for manipulating gene and protein activity in plants, either by introducing modified native molecular systems or by introducing complete synthetic structures that mimic the action of natural regulatory mechanisms. A major concern in plant bioengineering is that due to the low efficiency of homologous recombination, gene targeting is restricted, and our capability to control for activity of the introduced transgene is limited [24]. Thus, the future potential for precise bioengineering largely depends on advancements in the understanding of homologous recombination in plants and on how to elevate its efficiency, and/or the development of new tools for gene targeting in plants.

Engineering Riboswitches for Gene Control and Metabolite Sensing

Riboswitches are natural RNA sensors capable of controlling gene expression by using their ability to bind specific small molecule ligands such as vitamins, nucleotides, amino acids and various enzyme co-factors. These elements have been described mainly in bacterial systems [25–29] and, very recently, in plants [30, 31]. Typically riboswitches are composed of a sensing domain, which directly binds a metabolite and thereby undergoes structural modification, and an adjacent "expression platform", which converts the metabolite binding information to gene control. These RNA elements are present in untranslated regions of genes encoding enzymes and/or transporters, related to a particular biosynthetic pathway. Once the level of a pathway intermediate or end-product rises, ligand binding induces a conformational change in the riboswitch, and alters the flanking "expression platform". This affects, in a sort of "negative feedback", the transcription or translation of the corresponding mRNA, and thereby diminishes the metabolite level. For example, in Gram-negative bacteria, the thiamine pyrophosphate ligand binding to the RNA molecule renders conformational changes that mask the Shine-Dalgarno sequence of the a thiamine biosynthetic polycistron and as a result, ribosomes fail to initiate translation [26]. While only a dozen different riboswitch ligands have been identified up to date in both prokaryotic and eukaryotic organisms, techniques to develop synthetic RNA elements that could bind a particular small molecule at high specificity (called aptamers) have been available for a long time [32, 33]. Recent studies have demonstrated that

Table 1 Recent bioengineering strategies

Bioengineering strategy	Mode of action	Plant species engineered (aim of the experiment)	References
Riboswitches for gene control and metabolite sensing	Riboswitches are RNA sensors capable of controlling gene expression by binding specific small molecule ligands. Riboswitches could be engineered in plants to control expression of a gene of interest or, when coupled to a reporter gene, to serve as metabolite sensors	Arabidopsis (detection of Thiamine levels in plants)	[30] [31]
Immunomodulation	Explicit sequestration of a metabolite, protein or RNA by the ectopic expression of genes encoding antibodies or antibody fragments	Tobacco (reducing ABA levels)	[35] [36] [37]
		Arabidopsis (reducing GA levels)	[38]
		Tobacco (reducing Jasmonate levels)	[39]
		Tobacco (herbicide resistance – picloram)	[41]
		Arabidopsis (herbicide resistance – chlorpropham)	[40]
		Arabidopsis (herbicide resistance – carbamate)	[46]
		Tobacco (heat shock protein loss-of-function mutant)	[43]
		Tobacco (inhibition of polyamine biosynthesis)	[44]

Overexpression of cleavage-resistant gene targets, native and synthetic microRNAs (miRs)	Synthetic miRs are overexpressed as part of a native miRs precursor and could target either conserved or non-conserved regions of genes. This system provides an elegant way to simultaneously down regulate the expression levels of multiple members of a gene family	Arabidopsis (resistance to virus)	[61]
		Tobacco (resistance to virus)	[62]
		Arabidopsis, Tomato, Tobacco (interference with developmental processes)	[59]
		Arabidopsis (interference with developmental processes)	[63]
		Arabidopsis (interference with glucosinolate biosynthesis)	[268]
Zinc-finger Nucleases	Generation of targeted double-strand DNA breaks by the use of specific Zinc-finger endonuclease induces homologous recombination that facilitates gene targeting. In the future it will be possible to conduct *de novo* design of Zinc-finger endonuclease that will target genes to specific location in the genome	Tobacco protoplasts (gene targeting)	[65]
		Maize (gene targeting)	[66]

riboswitches can be engineered in whole plants, thus opening the way for using them to control expression of a gene of interest and sensing metabolite levels *in vivo* [34].

Bioengineering Through Immunomodulation

Immunomodulation is a technique that allows explicit sequestration of a metabolite, protein or RNA by the ectopic expression of genes encoding antibodies or antibody fragments. Several studies reported the use of immunomodulation as a tool in the research of phytohormones (e.g. ABA [35–37], GA [38] and jasmonate [39]), herbicide resistance [40–42], heat shock proteins [43], and of primary metabolism and cell cycle regulation [44]. The advantage of the technique is in the ability to allow the inactivation of the end-product of a specific biosynthetic pathway without disturbance of precursors, as often occurs when specific pathway inhibitors or RNAi techniques are employed. Several options for antibody design exist, including (a) antibody attachment to a specific enzyme, and as a result disruption or alteration in the enzyme function, (b) antibody attachment to a specific metabolite thereby reducing its level, (c) antibody attachment to a specific RNA molecule, thus altering its stability or tertiary structure. By utilizing specific promoters or leader peptides it will be possible to deliver such an antibody to a specific tissues, cells or even organelles.

Overexpression of Cleavage-resistant Gene Targets, Native and Synthetic microRNAs

In recent years, it has become evident that microRNAs (miRs) play a significant role in the regulation of plant development [45–51], biotic and abiotic stress response and, in a few cases, central (primary) metabolism [52–58]. Expression of miR target genes could be reduced by overexpressing the corresponding miRs, or elevated by generating a miR-resistant version of the target gene through modification of the miR binding site. The use of such techniques is limited to known miRs and target genes. Recently, synthetic miRs that target gene regions that do not correspond to native miRs binding sites have been described [59, 60]. In this method, the designed miRs are overexpressed as part of a native miR precursor and can target either conserved or non-conserved regions of genes, providing an elegant way to simultaneously downregulate the expression levels of multiple members of a gene family. Recently, several successful experiments utilizing synthetic miRs have been shown to give increased resistance to plant viruses [61, 62] and have demonstrated the potential for interfere with plant developmental processes [59, 63]. Constructing artificial specific microRNA against regulators or enzymes involved in SM production will provide an efficient and precise tool for the bioengineer. An example for the use of a synthetic miR for down regulation of the carotenoid pathway gene, phytoene desaturase (*PDS*), is shown in Fig. 1.

Zinc-Finger Nucleases

The inability to perform efficient gene targeting in plants is a major obstacle for optimizing bioengineering experiments since the various transgenes generated through a single gene construct posses a range of transgene penetration levels. Although some progress in gene targeting was reported in the last years [24, 64], these techniques have not been sufficiently robust or efficient to be effective. Recently, a proof of concept of a relatively efficient targeted genome optimization in plants was demonstrated. The researchers generated targeted double-strand DNA breakage by the use of a rare and specific endonuclease and could induce 10% homologous recombination in transformed tobacco protoplasts [65]. A different group used

Fig. 1 The use of synthetic microRNAs (miRs) as a tool for plant bioengineering. (**a**) Transgenic Arabidopsis plant expressing a synthetic microRNA (miR) for silencing the carotenoid biosynthesis gene phytoene desaturase (*PDS*) under the control of the *APETALA1* promoter. Arrow indicates the bleaching of the inflorescence. (**b**) Putative folding of the synthetic precursor-miRNA-*PDS* constructed on the backbone of miRNA164a. Nucleic acids shadowed black are the 21-mer synthetic miR sequence. The images were kindly provided by John Paul Alvarez (Weizmann Institute of Science, Israel)

a similar technique to demonstrate a large number of precisely engineered homologous recombination events in maize in which DNA integration occurred very accurately [66]. Recent advances in the design of the architecture of specific zinc-finger endonucleases [67–69] and the availability of novel user-friendly computational tools [70] are expected to allow efficient and simple modification of plant genomes.

Engineering of Nutraceuticals in Plants

Increasing the Pool of Vitamins

The level of vitamins in plant-derived food products is an issue of great awareness due to the importance of vitamins for our diet. The three most extensively studied plant vitamin biosynthetic pathways are those for provitamin

A, vitamin C and vitamin E. The metabolic pathways leading to the formation of these compounds have been largely elucidated, and up to date several attempts to metabolically engineer these pathways in plants have been carried out with promising results.

Engineering of *β*-Carotene (Provitamin A)

The engineering of *β*-carotene (provitamin A) content in rice endosperm [71] was one of the landmarks of bioengineering of SMs in plants. Vitamin A deficiency is a serious health issue in many parts of the developing world. Rice is a major food source in developing countries, but it is a poor resource of many essential vitamins and minerals, including *β*-carotene. The production of *β*-carotene was achieved in rice endosperm by the simultaneous expression of three carotenoid biosynthetic genes (two derived from plants and one from a bacterium). The first step required was the introduction of phytoene synthase (*PSY*) from daffodil (*Narcissus pseudonarcissus*). This gene was expressed under the endosperm-specific glutelin (Gt1) promoter and contained a plastid targeting signal. For the conversion of phytoene into lycopene, a bacterial (*Erwinia uredovora*) bi-functional carotene desaturase gene (*CrtI*), that encodes a protein capable of performing two reactions that are normally done in plants by two enzymes (phytoene desaturase and ζ-carotene desaturase), was used. Overexpression of *CrtI* that also encoded a plastid targeted protein was carried out using the constitutive *CaMV* 35S promoter. In the third engineering step, a plastid targeted lycopene-cyclase encoded by the daffodil *LCY* gene was introduced to rice plants. Ye et al. (2000) [71] used *Agrobacterium*-mediated transformation to introduce the triple-step carotenoid pathway into rice endosperm. The first generation transgenic rice produced yellow endosperm (so-called "Golden Rice") containing *β*-carotene at levels that would provide 10% of the recommended daily human consumption with an average daily rice intake. Subsequent manipulation and creation of "Golden Rice 2" [72], allowed the vitamin A recommended daily dose to be approached with a daily rice intake and potentially provide a relief from vitamin A deficiency for millions worldwide. The "Golden Rice" trait is now being introduced into agronomically superior local rice varieties [73–75], which can be cultivated in South Eastern Asia, where it is needed the most.

Engineering of Vitamin C Content

Vitamin C (L-ascorbic acid) is a powerful antioxidant and it also serves as cofactor in several biological reactions in plants and animals. Humans and several animals cannot synthesize vitamin C, which must therefore be consumed as part of their diet. The biosynthesis of vitamin C in plants has only recently been properly understood [76]. The L-galactose pathway, one of a few reported routes to vitamin C biosynthesis in which GDP-D-mannose is converted into L-ascorbate by six enzymatic reactions [77–80], is now fully elucidated [81]. Overexpression of several enzymes of the L-galactose pathway in plants appears to be an efficient way to create plants with higher vitamin C content. For example, overexpression of the last enzyme in the pathway, the L-galactono-1,4-lactone dehydrogenase under the control of the *CaMV* 35S promoter in tobacco BY-2 cell culture resulted in over production of vitamin C, in addition to a slower cell senescence rate and higher resistance to oxidative stress [79]. Moreover, endogenous expression of the recently discovered enzyme, L-galactose guanyltransferase in Arabidopsis plants resulted in a threefold increase in leaf ascorbic acid content [80]. Vitamin C content in Arabidopsis plants was also increased through an alternative pathway for vitamin C biosynthesis by overexpression of a D-galacturonic acid reductase [82]. In a different approach for engineering vitamin C content, heterologues expression of the rice dehydroascorbate reductase (*DHAR*) that reduces oxidized ascorbic acid back into its native state enhanced resistance to salt stress in Arabidopsis [83]. Thus, healthier and stress resistant plants that are rich with vitamin C can be created by means of plant bioengineering.

Engineering of Tocochromanols, the Vitamin E Family of Compounds

Four tocotrienols and four tocopherols (together termed tocochromanols) are lipid-soluble antioxidants that constitute the vitamin E family of compounds in plants, and are an essential part of the mammalian diet. The α-tocopherol form is the most biologically active among the vitamin E family. Tocochromanols are synthesized in many plant organs, particularly in seeds, by fusion of intermediates from the shikimate and isoprenoid pathways. The polar head of the molecule is derived from the cytosolic homogentisic acid (HGA), while the non-polar tail is derived from phytyl-PP (PDP) or geranylgeranyl-PP (GGPP), both synthesized by the plastidic isoprenoid biosynthetic pathway [84].

Many of the attempts to engineer plant vitamin E content have focused on seeds, which is where most of the vitamin is concentrated. Almost a decade ago, the *VTE4* gene (γ-tocopherol methyltransferase) was expressed in a seed-specific manner in Arabidopsis under the control of the *DC3* gene promoter from carrot [85]. While the total level of vitamin E stayed unaltered in transgenic plants, an almost complete shift from γ-tocopherol to α-tocopherol and from δ-tocopherol to β-tocopherol was evident. This suggested that the same enzyme is responsible for those two conversions, and demonstrated that it is possible to produce plants with a more active form of vitamin E. In another example, the *VTE3* (2-methyl-6-phytylbenzoquinol (MPBQ) methyltransferase) from Arabidopsis was heterologously co-expressed with *VTE4* (γ-tocopherol methyltransferase) in soybean. The transgenic seeds accumulated more than 95% of α-tocopherol and exhibited five times more vitamin E activity (biopotency for animals and humans) [86]. A successful complementation to the latter experiment was achieved by identification of the rate-limiting reactions in the tocochromanol biosynthetic pathway, and application of this knowledge for engineering tocochromanol biosynthesis in soybean by heterologous seed-specific co-expression of pathway-limiting genes [87]. The resulting line, which showed a 15-fold increase in total tocochromanols (94% of which were tocotrienols), was crossed with the transgenic high α-tocopherol soybean line generated by Van Enennaam et al. (2003) [86]. The seed vitamin E activity in the best performing F_2 line was found to be approximately 11-fold higher than in wild type soybean seed [87]. The recent advances in understanding the role of vitamin E in the human diet should accelerate the efforts towards increasing total tocochromanol content in various edible plant organs.

Carotenoids

The isoprenoid class of carotenoids is important for plant membrane stability, photosynthesis, protection against oxidative stress, and in development and biotic stress responses [88–91]. Carotenoids are also significant for the diet of human and livestock, acting as colorants, antioxidants and possibly in the prevention of heart disease and cancer [92–96]. The engineering of β-carotene (provitamin A) content in the rice endosperm (described above with relation to vitamin engineering) was one of many attempts carried out along the years to engineer compounds belonging to this class. The plant carotenoid (C_{40}) biosynthetic pathway is sited in plastids, where two molecules of GGPP are fused by the enzyme phytoene synthase (PSY) in the first committed step. The resulting phytoene is subjected to several oxygenation, ketonisation, desaturation and glycosylation reactions, leading to new carotenoid species [97].

Engineering Lycopene Levels in Plants

The red colored carotenoid lycopene has emerged as a focus of scientific interest in recent years. Several reports present evidence for or against the role of lycopene as a natural product that can be used in prevention of various diseases, including cancer [94, 98–100]. So far, several high lycopene content tomato lines and mutants have been characterized [101, 102], but none of these accumulate lycopene to levels that satisfy the pharmaceutical needs for supplemental drug production. It appears that light signaling-regulated genes influence lycopene levels in tomato plants and fruit. For example, the UV-damage DNA-binding protein 1 (*DDB1*) gene was found responsible for the *high pigment 1* (*hp1*) lycopene rich phenotype [103] and a defect in the *DEETIOLATED1* (*DET1*) gene results in the lycopene-rich *high pigment 2* (*hp2*) mutant phenotype [104]. Davuluri et al. (2005) [105] attempted to increase the nutritional value of tomato fruit by suppressing the *DET1* gene, using fruit-specific promoters combined with RNA interference (RNAi) technology. Interestingly, both carotenoid (*β*-carotene and lycopene) and flavonoid contents were increased significantly whereas other parameters of fruit quality were largely unchanged.

Engineering of Ketocarotenoids

Only a minority of plant species possesses the ability to form ketocarotenoids, but bioengineering could provide opportunities for production of these compounds *in planta*. The ketocarotenoid astaxanthin is formed from *β*-carotene by the introduction of keto- and hydroxyl moieties at the 4,4' and 3,3' positions of the *β*-ionone rings, respectively. In addition to its use as an antioxidant in human nutrition [106, 107], astaxanthin is also being used as a natural food additive for farmed salmon, coloring its flesh in pink to orange-red color [108]. The microscopic green alga *Haematococcus pluvialis* is used to produce astaxanthin at levels of ~4–5% dry weight, but the use of this alga as a source of astaxanthin is limited due to slow growth and the requirement for high light intensity [109, 110]. Successful bioengineering of astaxanthin and other keto-carotenoids in transgenic plants has been demonstrated several times [111–114]. Ralley et al. (2004) [114], reconstructed an astaxanthin pathway by transforming the marine bacteria 3,3'-β-hydroxylase (*CrtW*) and 4,4'-β-oxygenase (*CrtZ*) genes as a polycistronic construct into tobacco and tomato. In the nectary tissues of transgenic tobacco flowers, a quantitative increase in carotenoids (tenfold) and the presence of astaxanthin, canthaxanthin and 4-ketozeaxanthin were evident.

Engineering Flavonoids and the Related Isoflavonoids and Stilbenes

Flavonoids are synthesized through the phenylpropanoid pathway in which 4-coumaroyl-CoA (produced from phenylalanine) is combined with malonyl-CoA to yield chalcones, i.e. flavonoid precursors, with two phenyl rings. Conjugate ring-closure of chalcones results in a three-ring structure, the typical form of flavonoids. The metabolic pathway continues through a series of enzymatic modifications to yield several flavonoid classes including the flavonols, dihydroflavonols and anthocyanins. Many other products can be formed along this pathway, including the flavan-3-ols, proanthocyanidins (tannins) and additional polyphenolics. Flavonoids have been suggested to protect against oxidative stress, heart diseases and certain cancers and they are considered to bear potential benefits for human and livestock health [115–121]. The pathways leading to flavonoid biosynthesis are well characterized [122], hence, there are multiple strategies for their modification through bioengineering. Apart from the use of structural genes for engineering, the relative extensive knowledge regarding transcription factors, controlling various branches of the pathways, made regulatory proteins an excellent tool for manipulation of flavonoids.

Modification of Flavonols

Attempts to increase the levels of flavonols have been carried out in several plant species, but predominantly in tomato fruit. Flavonoid pathway gene expression and biochemical analyses in tomato fruit peel suggested that the low levels of chalcone isomerase (*CHI*) might be the limiting point for increasing the levels of flavonols [123]. Heterologous expression of the petunia *CHI* under the control of the *CaMV* 35S promoter resulted in a 78-fold increase in peel flavonols (mainly quercetin rutinoside and glucoside) in the best engineered line. In a complementary approach, heterologous expression of the MYC type transcription factor *LEAF COLOR* (*Lc*) gene and the MYB type factor *COLORLESS* (*C1*) in tomato fruit resulted in an up to 60-fold increase in levels of the flavonol kaempferol in the flesh tissue [124]. These experiments pointed out that success in the engineering of plant secondary metabolism is largely dependent on the endogenous gene expression patterns and activity of pathway enzymes in the different tissue types, as the same strategy was not appropriate for increasing flavonol levels in the fruit flesh tissue. It also highlights the need for the rigorous understanding of metabolic pathways in order to design successful bioengineering strategies.

Modification of Proanthocyanidins

Proanthocyanidins (PAs) or condensed tannins are beneficial for the health of humans [125–128] and livestock [129, 130]. In plants, they are important in defense against diseases and in seed dormancy [131], and are present in many plant products, particularly wine, fruit juices, and teas, contributing to their health benefits in addition to taste. Most of our current knowledge on structural and regulatory genes involved in the metabolism of PAs comes from the study of the Arabidopsis seed coat mutants [132, 133]. The biosynthetic route towards the formation of the polymeric PAs branches out from the pathway leading to the biosynthesis of anthocyanins when the first committed steps forming PA precursors are involve the conversion of anthocyanidins to the flavan-3-ols catechin or epicatechin [134]. Xie et al. (2006) [135] co-expressed the *PAP1* transcription factor (typically inducing anthocyanin accumulation; Borevitz et al. 2000 [136]) in tobacco and also the *Medicago truncatula* anthocyanidin reductase (*ANR*) gene. ANR generates epicatechin from cyanidin. In leaves of transgenic plants the high anthocyanin levels typical of the *PAP1* transformants were significantly reduced while production of PA monomers and oligomers, in which epicatechin units predominated, was increased. These results demonstrate that the combination of a transcription factor to increase the level of the anthocyanidin substrate coupled with a PA biosynthesis gene to divert that substrate into the PA pathway is an appropriate approach for bioengineering of PAs in plants.

Modification of Isoflavonoids

In contrast to the wide distribution of flavonoids, the production of isoflavonoids is mostly restricted to legumes. In these species, the formation of isoflavonoids branches out from the first step of the flavonoid pathway. The flavanone generated by CHI is converted by 2-hydroxyisoflavanone synthase (2HIS) to hydroxyisoflavanone, which is subsequently transformed by hydroxyisoflavanone dehydratase (2HID) to the typical legume isoflavonoids, genistein and daidzein [137]. In humans, isoflavonoids may have various health-promoting activities that are associated with their antioxidant and estrogenic properties [138–140]. Attempts to bioengineer isoflavonoids have been carried out in three different ways: ectopic expression of *2HIS*; co-expression of *2HIS* and *CHI*; and co-expression of *2HIS*, early pathway genes or transcription factors along with suppression of competitive pathway genes (reviewed in Ref. [137]). All three approaches generated isoflavone conjugates (but not aglycones) in different tissues of both legume and non-legume plant species.

Engineering of Stilbenes

Stilbenes are produced by a number of plant families including the *Fagaceae*, *Liliaceae*, *Mirtaceae* and the *Vitaceae* [141]. While plants produce stilbenoids as a response to pathogen attack, this group of phytoalexins also displays multiple beneficial effects on human health [142–145]. Particularly, the compound 3,5,4'-trihydoxystilbene, known as resveratrol (synthesized for example by grapes and peanuts), has been investigated intensively over the past 2 decades for its health-related properties [146]. Bioengineering of resveratrol in plants is of high interest both due to its potential as a nutraceutical food supplement or cosmetic but also for increasing resistance of plants against diseases.

Stilbene synthase (STS) generates resveratrol from precursors that are also used by chalcone synthase (CHS), the key enzyme of flavonoid biosynthesis in plants. Therefore, engineering of resveratrol in plants is considered a relatively straightforward approach since the introduction of a single gene is required for resveratrol formation and a pool of precursors is in most cases available. Following the hallmark study of Hain and co-workers (1993) [147], in which the production of resveratrol was demonstrated in tobacco as a result of the *STS* gene transfer, a similar approach has been adopted for a wide range of plant species including kiwi [148], apple [149, 150], papaya [151], pea [152], tomato [153], hop [154] and alfalfa [155]. In some cases, the transgenic plants were found to synthesize not only resveratrol, but also a resveratrol-3-β-D-glucopyranoside (piceid) [156], while in others, only the glucosylated form accumulated [155]. Although engineering of resveratrol was successful, and even led to a success in terms of increasing plant disease resistance [157] (see below the section "combating biotic stress with plant-derived natural products"), the commercial value of high-resveratrol plants as a health promoting product remains to be seen. This is due to the low cost of its extraction and the ability to synthesize it chemically.

Engineering of Plant Volatiles and Pigmentation (Fruit and Flowers)

Being sessile organisms, plants make massive use of volatile compounds as mediators for their interaction with the surroundings. Volatiles are released from every plant organ including aerial green tissues, reproductive tissues (various flower parts and fruit) and even roots, which emit volatiles into the rhizosphere. Apart from roots, attempts to bioengineer volatiles in different aerial organs have been carried out multiple times [158]. As in the case of engineering of synthesis of non-volatile compounds, the outcome was often unexpected, as activity of endogenous plant enzymes resulted in the formation of different derivatives of the target molecule [159]. The numerous reports regarding the importance of volatiles, emitted from vegetative tissues for plant defense, promoted attempts to enhance plant protection by modulating the volatile spectrum. The results of such experiments will be described in section "combating biotic stress with plant-derived natural products" whereas in this section we will focus on bioengineering of the volatiles produced by flowers and fruit. In the second part of this section attempts to bioengineer pigmentation, mainly in flowers will be reviewed.

Engineering of Fruit and Flower Volatile Compounds

Flavor and aroma of fruit and scent of flowers is represented by a mixture of dozens of volatile components [160, 161]. Most of these volatiles are derived from three chemical groups, namely isoprenoids, phenylpropanoids and aliphatics. Several volatiles in the mixture will have a dominant effect on the overall fruit aroma or flower scent that is independent from their relative levels. Namely, trace levels of a certain component might have a substantial effect on the typical aroma and scent associated with a particular fruit or flower. Due to promiscuity of enzymes associated with volatile compounds metabolism, a relatively large number of volatile components could be altered by modifying a single gene. For instance, overexpression of the (*E*)-β-caryophyllene synthase from rice in Arabidopsis results in emission of a bouquet of terpenoid volatiles, including (*E*)-β-caryophyllene, β-elemene and α-humulene [162]. In addition, bioengineering experiments carried out to date have demonstrated the accumulation of non-volatile compounds formed by modification of volatiles (e.g. glucosylation) [163]. These issues make the precise engineering of fruit and flower volatiles exceptionally complex. Accordingly, attempts to bioengineer volatiles have been successful in altering the chemical profiles of both fruit and flowers. However, in most cases the effects on scent and flavor could not be predicted in advance.

Modification of Fruit Volatiles

Nowadays, when crop shelf life is of considerable importance to retailers and consumers, several hybrids of fruit-ripening mutants with commercial lines are grown and supplied to the stores. Since fruit flavor develops during ripening, such hybrid lines often suffer from poor taste and aroma. While the knowledge regarding the composition of fruit flavor volatiles is extensive and spans multiple species [164], only in recent years has there been great progress in understanding the molecular genetics governing fruit flavor. Genes associated with pathways generating flavor volatiles have been cloned and functionally characterized from plants such as strawberry [165, 166]. However, attempts to modify the flavor of fruit by means of bioengineering have been carried out mainly in tomato.

The first successful alterations to fruit flavor involved modification of the composition of aliphatic volatiles in tomato by introduction of an alcohol dehydrogenase [167]. The subsequent introduction of linalool synthase from *Clarkia breweri* under the control of the tomato late-ripening E8 promoter resulted in accumulation of (*S*)-linalool and 8-hydroxylinalool in ripening tomato fruits [168]. Recently, redirection of the early plastidic terpenoid pathway by overexpression of the geraniol synthase (GES) from lemon basil driven by the tomato polygalacturonase promoter resulted in a substantial increase in the levels of geraniol and its derivatives, as well as elevation in the levels of various monoterpenes [169]. Surprisingly the content of carotenoid-derived volatiles such as β-ionone and other norisoprenes did not change substantially, even though these volatiles have isoprenoid precursors in common with the described monoterpenes. Tests with an untrained human taste panel showed that 68% of panel members described transgenic fruits as 'more aromatic' and over 60% of the panel preferred the flavor of transgenic fruit to control fruit.

20

Modification of Floral Scent

Although humans agree that a pleasant aroma enhances the perception of beauty in flowers, flower breeding programs have given priority breeding to color, shape and extended shelf life. The advances in our knowledge of the biochemistry and molecular biology of fruit flavors have also been matched with major progress in our understanding of the molecular basis of floral scent [170]. Biosynthetic and regulatory genes involved in the production of scent components and their rate-limiting steps have been studied. For example, Lucker et al. (2001) [163] transformed petunia with (*S*)-linalool synthase from *Clarkia breweri*. Surprisingly, almost no free (*S*)-linalool was identified due to its efficient conversion to the non-volatile (*S*)-linalyl-β-D-glucopyranoside by endogenous glucosyltransferase activity. In another experiment, Lucker et al. (2004) [171] engineered tobacco plants using three different monoterpene synthases (γ-terpinene synthase, limonene synthase and β-pinene synthase) from lemon (*Citrus limon*). The subsequent combination of all three genes into one plant through crossing resulted in transgenic lines emitting β-pinene, limonene, and γ-terpinene along with several side products of the introduced monoterpene synthases, from its leaves and flowers. Redirecting metabolic flux is an additional approach that can be used to modify floral scent. Zuker, et al. (2002) [172] suppressed the flavanone 3-hydroxylase (*F3H*), a key gene in the anthocyanin pathway, in carnations, and this resulted in elevated emission of methyl benzoate (Fig. 2). The increase in the levels of emitted methyl benzoate in transgenic plants occurred as a consequence of altering the flux in the shikimate pathway, from which both benzoic acid and anthocyanins are derived. While the transgenic flowers lost their anthocyanin pigments, the scent of methyl benzoate was strong enough to be detected by a human sensory panel. Verdonk et al. (2005) [22] reported on a MYB family transcription factor ODORANT1 (ODO1) that regulates volatile benzenoids production in petunia flowers. Silencing of *ODO1* using RNAi greatly reduced volatile benzenoid levels in transgenic petunia. Thus, regulatory genes such as *ODO1* might be used in future attempts to modify the composition of benzenoids and the related phenylpropanoids that often contribute to scents of flowers.

Engineering Fruit and Flower Pigmentation

The major plant pigments include carotenoids, anthocyanins and other flavonoids, betalains and chlorophylls. Most of these pigments are also highly nutritious, and the bioengineering of astaxanthin and lycopene for human health reasons is described in section "engineering of nutraceuticals in plants". In the present section, attempts to bioengineer plant or particularly fruit and flower pigmentation in order to alter appearance are described. Since almost the entire metabolic pathways for flavonoid and carotenoid biosynthesis and their regulation (in the case of flavonoids) have been elucidated, modification of these types of pigment molecules has become a relatively easy target for bioengineering. Up to date, most experiments have targeted the alteration of flavonoid accumulation (particularly anthocyanins) within flowers, although it has also been demonstrated that flower color can be determined and modified by changing the shape of petal cells [173, 174] or vacuolar pH [175, 176].

Modification of Anthocyanins and Other Flavonoids Pigments

Starting nearly 2 decades ago, attempts to engineer flower coloration were carried out by overexpressing or down regulating expression of the genes in the flavonoid biosynthetic pathway [177–179]. While these early attempts used basic strategies for modifying

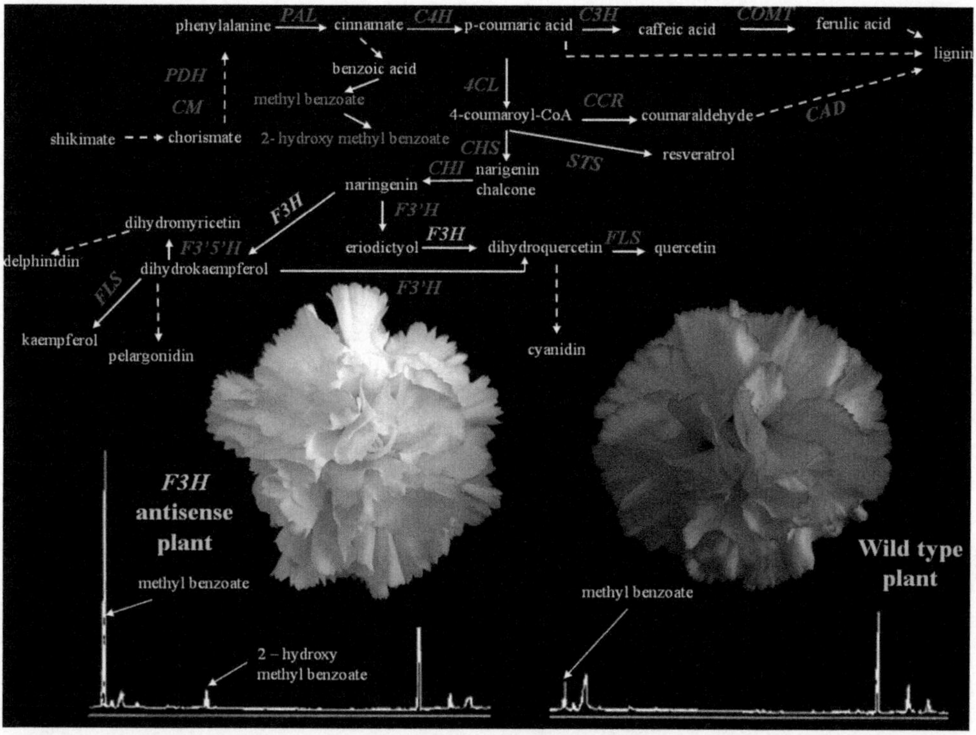

Fig. 2 Bioengineering of floral scent through modification of the flux to anthocyanin production in carnation flowers. The main enzymes of the pathway are marked by red. Dashed arrows represent multiple biosynthetic steps. The left image is a flower of transgenic carnation plant in which expression of flavanone 3-hydroxylase (*F3H*; marked in yellow) was silenced by an antisense approach [172]. The image on the right shows a wild type carnation flower. The volatile profiles of transgenic and wild type flowers are presented below the corresponding flowers, (volatiles that were altered in the mutant are marked with green in the pathway map). 4CL, 4-coumarate:CoA ligase; C3H, p-coumarate 3-hydroxylase; C4H, cinnamate 4-hydroxylase; CAD, cinnamyl-alcohol dehydrogenase; CCR, cinnamoyl-CoA reductase; CHI, chalcone isomerase; CHS, chalcone synthase; CM, chorismate mutase; COMT, caffeic acid O-methyl transferase; F3'5'H, flavonoid 3',5'-hydroxylase; F3H, flavanone 3-hydroxylase; F3'H, flavonoid 3'-hydroxylase; FLS, flavonol synthase; PAL, phenylalanine ammonia-lyase; PDH, prephenate dehydrogenase; STS, stilbene synthase. The flowers and the profiles of volatiles were kindly provided by Alexander Vainstein (Hebrew University of Jerusalem, Israel)

flavonoid-type pigments (i.e. modifying a single gene of the core pathway), recent studies have focused on more complex approaches for engineering pigment modification. Today's challenges are in modifying more extraordinary pigments and plant species, shifting between classes of pigments, enhancing the precursor pool by blocking side branches and by altering stability or localization of pigments.

20

Overexpression of the *Vinca major* flavonoid 3',5'-hydroxylase gene in petunia flowers shifted the accumulation from the main red cyanidin pigment to the blue pigment delphinidin [180]. In *Torenia hybrida*, one of the most commercially important bedding plants, silencing of the *CHS* gene eliminated accumulation of malvidin, a compound that is responsible for the blue color of the petals [181]. Co-expression of 4'-O-glycosyltransferase and aureusidin synthase in the same plant resulted in the accumulation of aureusidin 6-O-glucoside [182]. This was the first demonstration of manipulating aurone-type flavonoids in plants, compounds that confer a bright yellow color on flowers such as those of snapdragon (*Antirrhinum majus*) and dahlia (*Dahlia variabilis*). Modification of three genes was used in order to create pelargonidin- accumulating flowers of tobacco instead of the typical cyanidin pigments. This was carried out by suppression of the flavonol synthase and flavonoid 3'-hydroxylase genes expression while overexpressing the dehydroflavonol 4-reductase from gerbera [183]. Modification of flower color in the flavonoid pathway can be readily altered through modification of transcription factors [8, 184–186]. Very recently, this strategy has been employed in apple fruit in order to up-regulate anthocyanin accumulation in skin and flesh tissues [187–189].

Modification of Carotenoid Pigments

In many plant species flower and fruit coloration is determined by the carotenoid content of the tissue. Most of the carotenoids in nature are colorful molecules ranging from pale yellow (e.g. ζ-carotene) to purple (e.g. astaxanthin). Several studies dealing with altering plant carotenoid pigments have been published in the last decade. For example, constitutive overexpression of the *CrtW* gene (encoding β-carotene ketolase) from *Agrobacterium aurantiacum* altered the color of *Lotus japonicus* flowers through modification of the carotenoid biosynthetic pathway. Most of the transgenic plants exhibited a change in the color of flower petals from light yellow to deep yellow or orange due to the accumulation of novel ketocarotenoids [190]. Cleavage of carotenoids by carotenoid cleavage dioxygenase (CCD) genes could be used to prevent the accumulation of carotenoids and as a mean of creating white flowers [191]. Blocking the carotenoid biosynthesis pathways by silencing transcription factors such as the fruit *RIPENING INHIBITOR* (*RIN*) gene [192] is also an applicable option, albeit with a reduction in the fruit nutritional value. The deposition and accumulation of pigments in storage compartments are important factors in plant organ coloration. For instance, the *Or* gene from cauliflower triggers the differentiation of etioplasts into chromoplasts - the storage place for carotenoids, including β-carotene. Transgenic cauliflower and potato tubers overexpressing *Or* exhibit increased levels of the orange carotenoid pigment β-carotene [193].

Production of Medicinal Agents in Plants

For several thousands of years plants have been used as a source for medicine and to date, nearly a quarter of the modern medications are of plant origin (World Health Organization, 2003 - http://www.who.int/mediacen-tre/factsheets/fs134/en/). The modern industry requires plants with higher level of pharmaceutically active compounds since these compounds are usually only found in plants at low levels. In the following section we present several studies in which plants bearing the most prominent medicinal agents have been modified through bioengineering approaches.

Isoprenoid Type Medicinal Agents

Production of Artemisinin

Malaria is one of the most severe infectious diseases in the world, and more than 500 million people are infected by it annually (World Health Organization, 2007 - http://www.who.int/mediacentre/factsheets/fs094/en/). The plant *Artemisia annua* has been used in chinese medicine for centuries as a natural antimalarial drug because it produces artemisinin, which is a potent antimalarial agent [194]. Since chemical synthesis of artemisinin is very complex, plants remain the principal supply of this compound. The artemisinin content in *A. annua* is relatively low and it is therefore important to create a more efficient system for its production. Artemisinin is a sesquiterpene lactone containing an endoperoxide bridge that is synthesized in the cytosol from isoprenoid, which is derived from the mevalonate pathway [195]. Attempts to increase the levels of precursors for artemisinin production through bioengineering were carried out by overexpressing either the native or the *Gossypium* farnesyl diphosphate synthase genes in *A. annua* [196]. In both cases an increase in the levels of artemisinin was detected (0.8–1% of dry leaves), reaching up to three times higher levels of artemisinin in transgenes compared with the untransformed plant. In an attempt to reconstruct part of the artemisinin pathway in a model plant, the amorpha-4,11-diene synthase (*AMS*) from *A. annua* was expressed in tobacco, resulting in an accumulation of relatively low levels of amorpha-4,11-diene [197]. On the other hand, redirection of isoprenoid precursors and overexpression of *AMS* in the chloroplast resulted in a more than 40,000-fold increase in amorpha-4,11-diene content in tobacco (up to 2.5% of dry leaves weight) [6]. In other experiments, a surprising 30–70% increase in artemisinin content was detected in *A. annua* plants overexpressing the isopentenyl transferase gene, which encodes an enzyme that is required for generation of precursors for cytokinin synthesis [198]. Interestingly, the rise in artemisinin content was positively correlated with increased chlorophyll and cytokinin levels in the transgenic plants.

Production of Taxol

The diterpene paclitaxel (mostly known by its commercial name taxol) is produced in the bark and the needles of yew trees (*Taxus* sp.) and is one of the most powerful chemotherapeutic agents known to date. Taxol and taxol-derived drugs have been used in recent decades against a wide range of cancers and in cardiovascular therapies [199]. Obtaining taxol from yew trees is problematic since it is produced in relatively small quantities, the growth rate of the trees is slow and harvesting process is destructive to the tree. *Taxus* cell cultures can be induced to yield high levels of the taxol in large scale fermentations in bioreactors with methyl jasmonate treatment [200, 201], but these are costly and inefficient procedures. Since the demands for taxol have greatly exceeded its availability from natural sources, the drug is now produced by semi-synthetic processes that involve many reactions using more abundant taxol biosynthesis precursors such as 10-deacetylbaccatin III and baccatin III [202]. The highly complex taxol biosynthetic pathway is still not fully revealed, although major progress in its elucidation has been made in recent years [203]. Eight of the 15 enzymatic steps leading to the biosynthesis of baccatin III have been recreated in the yeast *Saccharomyces cerevisiae* by functionally expressing the genes in episomal vectors [204]. Successful reconstruction of the entire taxol biosynthetic pathway in transgenic plants would represent a considerable advance in our capability to meet the demands of the drug industry. The first committed step in taxol biosynthesis is catalyzed by the enzyme taxadiene synthase (TXS) which converts the diterpene geranylgeranyl pyrophosphate (GGPP) to taxa-4-(5),11(12)-diene (taxadiene). Induced expression of *TXS* in Arabidopsis using the two component glucocorticoid system [205] resulted in a 30 times increase in levels of taxadiene as compared to constitutive (i.e. using *CaMV* 35S) expression of the same gene [206]. A cross between Arabidopsis plants expressing *TXS* under the *CaMV* 35S promoter with plants expressing 1-hydroxy-2-methyl-2-(*E*)-butenyl 4-diphosphate reductase (*HDR*), encoding an enzyme of the core isoprenoid pathway, resulted in up to 13-fold increase in taxadiene levels in several F_2 lines [207]. Recently, a yellow-fruited tomato line that lacks the ability to utilize geranylgeranyl diphosphate (GGPP) for carotenoid synthesis was transformed with taxadiene synthase. The re-routing of GGPP resulted in the production of 470 μg/g dry weight of taxadiene in ripe fruit [208]. The latter work proved the concept of reproducing functional steps of taxol biosynthesis in plants other then *Taxus* species. In a different experiment, Rontein et al. (2008) attempted to produce early components of the taxol biosynthetic pathway in *Nicotiana sylvestris*, expressing yew cDNAs encoding TXS and TXS 5α-hydroxylase (a cytochrome p450 oxygenase) in trichome cells [209]. The unexpected production of 5(12)-oxa-3(11)-cyclotaxane (OCT) has emphasized our limited understanding of the taxol biosynthetic pathway.

Alkaloid Type Medicinal Agents

Production of Tropane Alkaloids

The tropane alkaloids (TA), synthesized mainly in roots of several *Solanaceous* plants, are derived from ornithine, arginine and phenylanine [122, 210, 211]. A number of TAs, including atropine, hyoscyamine, scopolamine and the more active form, 6,7-β-epoxide scopolamine, are used in medicine, primarily as anticholinergic drugs. The enzyme hyoscyamine 6-β-hydroxylase (H6H) catalyzes the last step of scopolamine biosynthesis from hyoscyamine. Heterologous expression of *H6H* from *Hyoscyamus niger* [212] driven by the *CaMV* 35S promoter in *Atropa belladonna*, resulted in the transgenic plants with elevated levels of alkaloid content, mainly composed of scopolamine [213, 214]. Expression of *H6H* in *H. muticus* hairy roots led to an about one hundred fold elevation in scopolamine levels in addition to accumulation of relatively low levels of hyoscyamine [215]. In a different experiment, hairy root cultures of tobacco were transformed with *H6H* and shown to convert externally applied hyoscyamine to scopolamine while releasing 85% of the generated scopolamine to the medium [216]. Overexpression of both *H6H* and *PMT* (putrescine N-methyltransferase, the enzyme responsible for the first committed step in the biosynthesis of tropane alkaloids), resulted in massive production of scopolamine in *H. niger* root cultures [217].

A pair of tropinone reductases, TR-I and TR-II, constitutes a branching point in the biosynthesis of TAs [218, 219]. While TR-I diverts the pathway towards the production of scopolamine, TR-II diverts it towards the biosynthesis of calystegine. Overexpression of *TR-I* or *TR-II* under constitutive promoters altered the ratio of tropine-derived alkaloids versus pseudotropine-derived alkaloids in root culture lines of *Atropa belladonna* [220].

Production of Morphinan Alkaloids

Opium poppy (*Papaver somniferum*), one of the oldest cultivated plants has been used in medicine since the beginning of civilization [221]. To date, it is the only commercial source of the narcotic drug morphine, the cough suppressant codeine and the antitumor agent noscapine. In addition to the substances mentioned above, poppy plants contain about eighty different alkaloids [222] including thebaine, which by itself encompasses a mild opioid effect, and is currently being used as starting substance for the synthesis of opioid drugs (e.g. naloxon, naltrexone and buprenorphine). In the last few decades the demand for poppy-derived alkaloids has rise continuously - in particular the need for thebaine, which has increased by ~67% in the last 6 years (The International Narcotics Control Board, 2006 – http://www.incb.org/incb/en/narcotic_drugs_reports.html). In poppy, benzylisoquinoline alkaloids, and morphinan alkaloids among them, are localized to laticifers - a vessel-like series of long cells, which accompany vascular tissues throughout the plant and contain a milky sap called latex. The biosynthetic gene transcripts and enzymes are almost exclusively restricted to sieve elements of the phloem and their companion cells [223–225].

Recently, an industrial elite line of poppy was genetically engineered to modify the alkaloid content of the plant by overexpression of the (*S*)-*N*-methylcoclaurine 3′-hydroxylase [226]. The resulting transgenic plants contained up to 450% more alkaloids in their latex compared to untransformed plants. Overexpression of codeinone reductase in poppy led to a 30% increase of morphinan alkaloid content on a dry weight basis in transgenic plants [227]. In a different experiment, the berberine bridge enzyme (bbe), which is the first enzyme of the sanguinarine biosynthesis branch, was silenced in poppy [228]. This resulted in an increase in the concentration of several pathway intermediates from all biosynthetic branches of benzylisoquinoline alkaloids in the latex of transgenic plants. Surprisingly, the transgenic plants also exhibited an increase in concentration of (*S*)-scoulerine. Silencing of *BBE* in California poppy (*Eschscholzia californica*) resulted in increased levels of (*S*)-reticuline [229]. Since industry uses only pure extracts of poppy-derived alkaloids that are devoid of genetic material it is expected that a transgenic source of these substances will be accepted by consumers.

Production of Terpenoid Indole Alkaloids

Terpenoid indole alkaloids (TIAs) are used for a vast variety of medicinal purposes, including as anti-cancer agents. Most of the progress in characterization of the biosynthesis of these compounds has been made in *Catharanthus roseus* (Madagascar periwinkle), a member of the *Apocynaceae* family that produces more than 100 types of TIAs. In the first committed step of TIA biosynthesis, tryptamine (monoamine alkaloid), derived from the shikimate pathway, and secologanin (an isoprenoid) are coupled by the enzyme strictosidine synthase (STR) [230, 231] to form strictosidine. All TIAs in periwinkle are produced by various reactions from the latter substance [232]. Periwinkle leaves and stems produce vinblastine and vincristine, used as anticancer drugs and for chemotherapies. The roots, on the other hand, accumulate the antihypertensive alkaloids ajmalicine and serpentine. An elevation in TIA levels was demonstrated by overexpression of *STR* in the cell cultures of periwinkle under the control of *CaMV* 35S promoter. Alkaloid accumulation by highly productive transgenic cell lines showed considerable instability and was strongly influenced by culture conditions, such as the hormonal composition of the medium and the availability of precursors [233]. Biosynthesis of TIAs and their precursor primary metabolites in periwinkle is regulated by ORCA3, a jasmonate inducible member of the APETALA2 (AP2)/Ethylene Response Factor (ERF) transcription factor family [234]. Genes belonging to the isoprenoid pathway (1-deoxy-D-xylulose synthase, cytochrome P450-reductase), tryptophan biosynthesis and catabolism (anthranilate synthase and tryptophan decarboxylase) as well as genes from the TIAs biosynthesis pathway (strictosidine-β-D-glycosidase and desacetoxy-vindoline 4-hydroxylase) were upregulated by ORCA3 [234]. ORCA3 was demonstrated to bind specifically to and activate gene expression via a jasmonate- and elicitor-responsive element (JERE) located in the promoter of strictosidine synthase [235].

20

Moreover, ORCA3 regulated also the biosynthesis of other alkaloids including scopolamine, morphine and sanguinarine [236]. Thus, upregulating the entire spectrum of TIAs by the ORCA3 regulator is a promising way to increase yields of these important medicinal substances.

Combating Biotic Stress with Plant-Derived Natural Products

Being immobile organisms, plants evolved in various ways to fight biotic stress imposed by numerous pathogens such as bacteria, fungi, viruses and herbivores. A large body of evidence clearly shows that many plant substances, derived from different classes of SMs including alkaloids, cyanogenic glycosides, glucosinolates, terpenes, saponins and tannins are toxic or deterrent to animals (insects, vertebrates), and several of these groups of compound display antibiotic activities. Organs such as flowers, fruit and seeds that are important for survival and multiplication are nearly always a rich source of defense chemicals. However, production of chemicals as defenses against herbivores and pathogens is not necessarily constitutive and they are often produced or activated upon wounding and infection [237–239]. Plants also use indirect defense against herbivores by emitting particular volatiles in order to attract the natural enemies of the herbivore [240–242]. Bioengineering of both direct and indirect defense mechanisms in plants are promising tools to provide plants that are better capable to fight the numerous pathogens attacking them. Yet, as often a certain substance could deter or be toxic to one type of pathogen but serve as an attractant to another, it is not unexpected that the use of such approaches for improving plant defense will be problematic in terms of their specificity.

Engineering Volatile Terpenoids for Fighting Herbivory

The sesquiterpene patchoulol is a constituent of the essential oil of *Pogostemon cablin* (patchouli) [243], acting as an insect repellent [244]. By co-expressing the patchoulol synthase (from *P. cablin*) with the avian farnesyl diphosphate synthase gene of the core isoprenoid pathway, and targeting both proteins to the plastids, Wu et al. (2006) [6] created tobacco plants that accumulated patchoulol up to 30 μg/g fresh weight. When examined in choice preference tests, leaves of the transgenes exhibited reduced damage by tobacco hornworms. In a different experiment, Arabidopsis plants were transformed with a dual linalool/nerolidol synthase (*FaNES1*) gene [5]. Apart from the production of the monoterpene linalool and its derivatives, the leaves of transgenes also emitted the sesquiterpene nerolidol, though at a low level. Exploring the capacity of the newly produced volatiles to alter insect behavior through dual-choice assays demonstrated that the transgenic lines significantly repelled *Myzus persicae* aphids. Introducing the *FaNES1* gene to potato plants under the control of the strong *Chrysanthemum* RUBISCO small subunit promoter resulted in a dramatic effect on plant height, which was accompanied by an extreme bleaching effect (Fig. 3; Aharoni et al., 2006 [245]). The latter observations pointed once again to the importance of taking into consideration the possibility of influencing metabolic pathways that share the same precursors when engineering SMs.

It appears that indirect defense mechanisms can also be engineered through production of terpenoid-type volatile signals. One such signaling terpenoid is the C11 homoterpene (3*E*)-4,8-dimethyl-1,3,7-nonatriene ((*E*)-DMNT) shown to be synthesized from the sesquiterpene (*E*)-nerolidol [240, 246]. Targeting the strawberry linalool/nerolidol synthase (FaNES1) to mitochondria

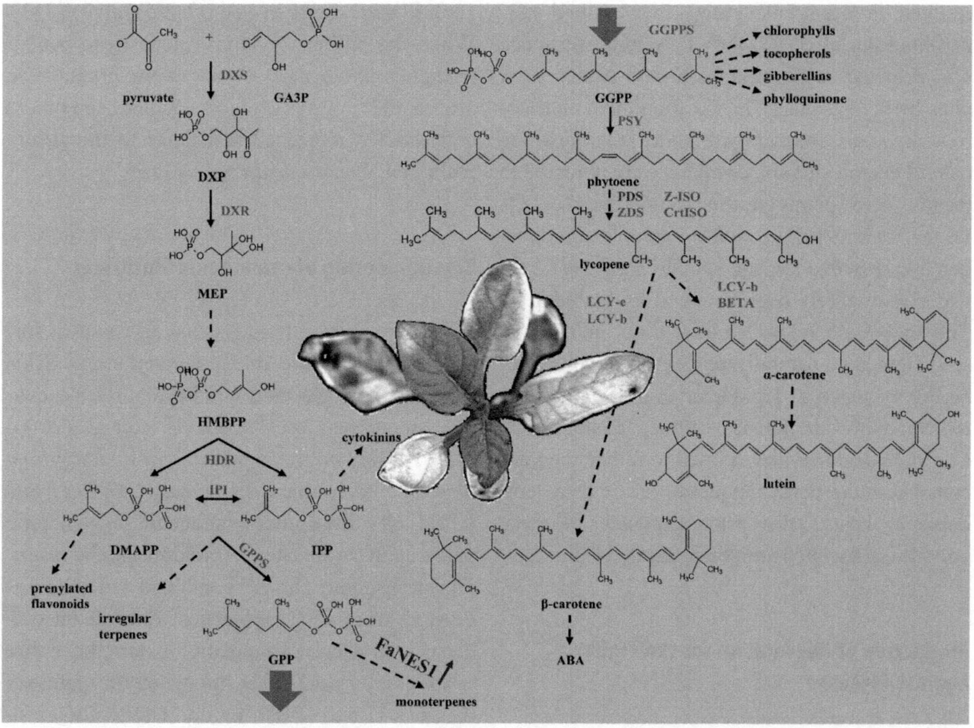

Fig. 3 Retardation of growth and bleaching due to the high level of monoterpene (an isoprenoid) production in potato plants. Monoterpenes are produced through the plastidic isoprenoid pathway that is depicted in the figure. The main enzymes of the pathway are marked in green, dashed arrows represent multiple biosynthetic steps, the red thick arrows represent the continuation of the pathway. In the center, image of a transgenic potato plant expressing the strawberry *FaNES1* gene (indicated with blue in the pathway) under the control of the super-strong *Chrysanthemum* RUBISCO small subunit promoter [245]. These plants produced high levels of the monoterpene linalool. Upon transfer from tissue culture to the greenhouse leaves of plants with high *FaNES1* expression levels turned white and were retarded in growth. The dramatic phenotypes were possibly a result of depleting the pool of geranyl diphosphate (GPP) that is a precursor for multiple branches of the isoprenoid pathway including carotenoids, chlorophylls and gibberellins metabolism. ABA, abscisic acid; BETA, chromoplast-specific lycopene *β*-cyclase; CrtISO, carotene isomerase; DMAPP, dimethylallyl diphosphate; DXP, 1-deoxy-D-xylulose 5-phosphate; DXR, 1-deoxy-D-xylulose 5-phosphate reductoisomerase; DXS, 1-deoxy-D-xylulose 5-phosphate synthase; FaNES1, *Fragaria X ananasa* (strawberry) linalool/nerolidol synthase; GA3P, glyceraldehyde 3-phosphate; GGPP, geranylgeranyl diphosphate; GGPPS, geranylgeranyl diphosphate synthase; GPP, geranyl diphosphate; GPPS, geranyl diphosphate synthase; HDR, 1-hydroxy-2-methyl-2-(*E*)-butenyl 4-diphosphate reductase; HMBPP, 1-hydroxy-2-methyl-2-(*E*)-butenyl 4-phosphate; IPI, isopentenyl pyrophosphate isomerase; IPP, isopentenyl pyrophosphate; LCY-b, lycopene *β*-cyclase; LCY-e, lycopene *ε*-cyclase; MEP, 2-C-methyl-D-erythritol 4-phosphate; PDS, phytoene desaturase; PSY, phytoene synthase; ZDS, *ζ*-carotene desaturase; Z-ISO, 15-cis *ζ*-carotene isomerase

resulted in transgenic plants that emitted (*E*)-nerolidol but also (*E*)-DMNT. Neither substance was detected in the untransformed plants, even after herbivore attack. It was found that carnivorous predatory mites *Phytoseiulus persimilis*, the natural enemies of spider mites, were attracted to the transgenic plants emitting (*E*)-nerolidol or (*E*)-DMNT or both [4]. In another example of indirect defense, a maize terpene synthase gene (*TPS10*) was constitutively overexpressed in Arabidopsis. The transgenic plants emitted high amounts of sesquiterpenes, naturally released by maize after herbivory attack by lepidopterian larvae. In olfactometer assays the females of the parasitoid wasp *Cotesia marginiventris* learned to utilize the generated sesquiterpene bouquet to locate their lepidopteran hosts after prior exposure to these volatiles, associated with their hosts [247].

Production of Resveratrol for the Fight Against Diseases

As in the case of a large number of SMs, the phenylpropanoid resveratrol is not only beneficial for human health (see above section "engineering of nutraceuticals in plants"), but is important for defense in plants [157, 248, 249]. Several papers describing the increased resistance to fungal pathogens of transgenic plants, producing elevated levels of resveratrol or its derivatives, have been published in the past years. For example, resistance to *Botrytis cinerea* was reported in tobacco [147] and barley [250] lines that expressed stilbene synthases from *Vitis vinifera*, while expression of these enzymes in tomato and wheat resulted in enhanced resistance to *Phytophthora infestans, Botrytis cinerea, Alternaria solani* [251], and to *Septoria nodorum* and *Puccinia recondita* f. sp. *tritici* [252], respectively. Expression of the *Vitis vinifera* stilbene synthase driven by its own pathogen-induced promoter in papaya (*Carica papaya*) resulted in increased resistance to *Phytophthora palmivora* [151], and to *Pyricularia oryzae* in rice (*Oryza sativa*) [253]. When the stilbene synthase gene from peanut (*Arachis hypogaea*) was overexpressed in alfalfa, the transgenic plants accumulated piceid and showed increased resistance to the alfalfa pathogen, *Phoma medicaginis* [155].

Bioengineering of Cyanogenic Glucosides

Several important food crops (e.g. cassava, sorghum, almonds) are able to convert amino acids to cyanogenic glucosides [254] that release cyanide upon their breakdown. Bak et al. (2000) [255], described the expression of two *Sorghum bicolor* cytochrome P450 genes (*CYP79A1* and *CYP71E1*) from the cyanogenic biosynthetic pathway in tobacco and Arabidopsis. The resulting transgenic lines contained metabolites derived from intermediates in dhurrin biosynthesis and were cyanogenic. Later, the entire cyanogenic glucosides biosynthetic pathway (CYP79A1 and CYP71E1 and UDP-Glucose:p-hydroxymandelonitrile-O-glucosyltransferase (sbHVNGT)) was transferred from *Sorghum bicolor* into Arabidopsis [256]. The transgenic plants showed increased resistance to the flea beetle *Phyllotreta nemorum*, a natural pest of many crucifers. Recently, the entire cyanogenic glucoside biosynthetic pathway was transformed into *Vitis vinifera* hairy root culture [257]. Suprizingly, *in vitro* co-culture of a transgenic line that was proved to be cyanogenic, and the grapevine pest *Daktulosphiara vivifoliae* did not show any evidence for reduced infestation of the plant tissue. Another cyanogenic glucoside, epiheterodendrin, is produced by barley in the leaf epidermis. Barley is not cyanogenic due to the lack of the cyanide-releasing enzyme *β*-D-glucosidase. Expression of the cyanide-releasing dhurrinase from *Sorghum bicolor* in barley resulted in up to 60% reduction in colonization rate by an obligate parasite *Blumeria graminis* f. sp. *hordei*, the causal agent of barley powdery mildew [258]. The bioengineering of commercial lines containing cyanogenic glucosides should be performed with caution, as cyanide is toxic to humans and animals.

Bioengineering of Glucosinolates

Glucosinolates are sulfur-containing compounds produced by several plant families including the *Brassicaceae* and contribute to their resistance to pests. Plants producing glucosinolates also contain myrosinases (thioglucosidases, TGG) – enzymes, that cleave the sugar moiety from glucosinolates upon tissue damage thereby forming various aglycons, which are further decomposed to various toxic products such as nitriles, isothiocyanates and cyanates [259]. Levy et al. (2005) [260], screened a T-DNA activation tagged Arabidopsis lines collection [261] for plants displaying altered leaf glucosinolate content. The *IQD1* gene, encoding a calmodulin-binding nuclear protein, was identified in the screen. Plants overexpressing *IQD1* showed a dramatic reduction in insect herbivory, as revealed in dual-choice assays with the generalist phloem-feeding peach aphid *Myzus persicae*, and in weight-gain assays with the cabbage looper *Trichoplusia ni*, a generalist-chewing lepidopteran [260]. A study of Arabidopsis plants mutated in two myrosinase encoding genes (*TGG1* and *TGG2*), demonstrated that maturing leaves of the double mutant line had significantly increased glucosinolate levels [262]. Feeding assays with insect herbivores on double mutant plants showed that the weight gain of the generalist *Trichoplusia ni* and the facultative *Solanaceae*-specialist *Manduca sexta*, was significantly increased in comparison to control plants. Thus, future manipulation of these genes might be valuable in the generation of plants that are resistant to herbivory.

The recent identification of several MYB-type transcription factors, that regulate tryptophan and methionine derived glucosinolate biosynthesis in Arabidopsis [263–268] opens up new opportunities for engineering the biosynthesis of these potent insect deterrents in crops. A double knockout Arabidopsis mutant of the MYB28 and MYB29 transcription factors (depleted of any methionine-derived glucosinolates) was subjected to herbivory tests [269]. Larvae of the lepidopteran insect *Mamestra brassicae* gained up to 2.6-fold weight while feeding on the double mutant as compared to larvae feeding on untransformed plants. In addition, the damage to the plants correlated inversely to the measured levels of methionine-derived glucosinolates. Malitsky et al. (2008) [268], subjected transgenic lines overexpressing either one of the methionine-derived glucosinolates regulators *MYB29* and *MYB76* or the tryptophan-derived glucosinolates regulators *MYB34* (known as *ATR1*; Smolen and Bender 2002 [270]) and *MYB51* factors to choice assays with the whitefly *Bemisia tabaci*. The *B. tabaci* females laid significantly lower number of eggs on the transgenic plants overexpressing either one of the transcription factors that accumulate increased levels of the corresponding glucosinolates, as compared to untransformed plants. In a no-choice experiment, the number of eggs oviposited per female on untransformed plants was significantly higher than the number of eggs, oviposited on the *MYB29* and *MYB76* overexpressing plants that accumulate methionine-derived glucosinolates. In the same assay, no difference in oviposition was observed between *MYB34* and *MYB51* expressing plants (accumulating tryptophan-derived glucosinolates) and untransformed plants. The latter experiment pointed out the importance of selecting target metabolites in order to bioengineer increased resistance to a particular herbivore.

Engineered Caffeine for Plant Defense

The biosynthetic pathway of the alkaloid caffeine (1,3,7-trimethylxanthine) has been fully elucidated [122, 271]. In addition to its use by humans as a component in various drugs and food products, caffeine has also been reported to exhibit pesticide and insecticide activities [272–276]. To date, only one report has demonstrated the generation of pest repellent transgenic plants by engineering caffeine production. The simultaneous expression of N-methyltransferases that are responsible for converting xanthosine into caffeine in tobacco plants resulted in the production of over 5µg per gram fresh weight caffeine. A choice assay showed that tobacco cutworms *Spodoptera litura* fed on untransformed leaf discs but almost did not feed on leaf discs from caffeine-producing transgenes [277].

Concluding Remarks

One of the most characteristic features of plant natural products is that in most cases these substances are not only produced in specific *taxa* but also in specialized plant organs and cells or under particular environmental conditions. Thus, there are plenty of opportunities to breed traits governed by SMs through bioengineering of crop plants. A major problem associated with engineering natural products is the indirect effects on metabolic networks. Plants of the future will not have wide acceptance if they have reduced fitness. On the contrary, being grown at the extreme weather conditions caused by the era of global warming, they will be required to survive in more harsh environments. Hence, future bioengineering strategies should have minimal interference with branching metabolic pathways, possibly through induced expression of genes and targeting of proteins to specific tissues and subcellular locations. Despite long years of metabolic pathway discovery, including metabolites, enzymes and genes that take part in the production of SMs, we are still far from obtaining complete maps of metabolic networks. This lack of information strongly hampers our ability to predict the outcome of bioengineering experiments. The promiscuous activity of enzymes, particularly those involved in secondary metabolism, makes this problem even more severe. At this point, metabolomics (see Box 1) and trancriptomics technologies that allow broad coverage of metabolites and gene expression in a single analysis should prove very useful in pathway discovery. They will also permit the understanding of unexpected results obtained by metabolic engineering and improving engineering strategies. Although bioengineering approaches are in essence targeted towards the modification of known metabolites that are present in nature they can also result in the formation of novel metabolites that have not previously been detected in any plant species. The promiscuous activity of enzymes, both in terms of acceptance of different substrates and stereo-specificity, and the chemical diversity available in nature allow the bioengineering of non-natural metabolites in plants through the so-called combinatorial biochemistry approach.

As the number of successful examples increases exponentially, products of bioengineering will no doubt play a dominant role in the future of natural products biotechnology. While the current policy in some countries is to avoid products of bioengineering, the wide

Box 1 Metabolomics and plant bioengineering

The term **"Metabolome"** was first defined by Oliver et al. (1998) [278] as the quantitative complement of all of the low molecular weight molecules present in cells in a particular physiological or developmental state. In recent years there has been an increasing effort to analyze the Metabolome in large-scale and to integrate the information as a component of

(continued)

Box 1 (continued)

functional genomics, systems biology and metabolic engineering approaches [279–282]. As such, the determination of metabolic profiles through **Metabolomics** complements other profiling methods such as transcriptional profiling and proteomics. Ideally, Metabolomics will quantify all of the metabolites in a cellular system. It is currently estimated that a typical eukaryotic organism contains 4,000 to 20,000 metabolites [283] while prokaryotes are estimated to contain much less, in the order of several hundreds [284]. Various complementary analytical tools have been developed in recent years, which could be employed for Metabolomics. The most widely used method for Metabolomics includes Gas Chromatography – Mass Spectrometry (GC-MS) analysis that by itself is a method limited to the analysis of volatile and heat stable compounds. Many biological compounds are not sufficiently volatile as such to be separated by GC-MS and therefore must undergo prior chemical derivatization. The main components identified and reliably detected by the GC-MS derivatization method are primary metabolites including sugars, amino acids, organic acids, or sugar alcohols [285], selected secondary metabolites such as shikimate, ascorbate and tocopherol [286] and fatty acids, fatty alcohols, sterols and aliphatics within the polar phase [287]. Another method, which could extend the amount of metabolites detected (in particularly secondary metabolites), is the one using Liquid-Chromatography – MS (LC-MS) [288–290]. LC-MS could detect non-volatile compounds, which do not derivatize easily or do not contain good chromophores for conventional HPLC. Using an additional fragmentation step in MS/MS could provide structural information, important for the identification of unknowns. In the same manner, and even with improved potential for mass resolution ($>10^6$) and accuracy, another type of mass analyzer, namely, Fourier Transform Ion Cyclotron Mass Spectrometry (FTICR-MS) can be utilized [280, 291]. The main alternative to mass spectrometry based approaches is the use of Nuclear Magnetic Resonance (NMR) based techniques [292]. Despite limitations in sensitivity, the power in NMR lies in the unambiguous identification of metabolites and the fact NMR approaches have been used for a long time and were subjected for exhaustive validation. As in the case of transcriptome or proteome profiling, data analysis is a major issue both in terms of storage and the use of algorithms to convert the complex data into biological knowledge. Metabolite identification is a key issue in Metabolomics that is largely limited by the availability of reference compounds. Thus, relatively low percentage of the total metabolites detected in Metabolomics experiments can be unambiguously identified.

Metabolomics emerges as an essential tool for both deciding on the strategy of metabolic engineering as well as for analyzing the outcome of such experiments and the subsequent improvement of engineering strategies. Our current knowledge on metabolic networks in plants, particularly in the case of secondary metabolism, is far from being complete. This gap in knowledge of metabolic pathways is a major hurdle in the design of engineering strategies and solving unexpected problems in engineering experiments. As more metabolites and branches of metabolic pathways are being discovered and our analytical capability to detect and identify a large number of metabolites progresses, the likelihood of planning a successful metabolic engineering strategy or improving the results of an existing one is greatly increased.

20

array of useful and bioactive substances from plants ought to increase the world-wide acceptance of such products.

Acknowledgements A.A. is an incumbent of the Adolfo and Evelyn Blum Career Development Chair. The research in A.A. laboratory is supported by the William Z. and Eda Bess Novick Young Scientist Fund, the Y. Leon Benoziyo Institute for Molecular Medicine, the EU project 'METAPHOR', contract number FOODCT-2006–036220, the AAFC/BARD research project C-9105–06 and the Minerva foundation. We would like to thank Alexander Vainstein and John Paul Alvarez for the contribution of images.

GLOSSARY

Bioengineering: an interdisciplinary field dealing with the application of analytical and experimental engineering methods, design and techniques to tissues, cells and molecules of living system. Rooted in mathematics, physical, chemical and the life sciences, the bioengineering is used to solve problems in medicine, daily life and biology by understanding the function of the live organism and the application of engineering technologies to design new tools and techniques.

Immunomodulation: a technique that allows explicit sequestration of a metabolite, protein or RNA by the ectopic expression of genes encoding antibodies or antibody fragments.

Metabolic engineering: targeted improvement of cellular properties or metabolite production via manipulation of specific metabolic or signal transduction pathways.

Nutraceuticals: extracts of foods claimed to have a medicinal effect on human health.

Riboswitch: a natural RNA sensor capable of controlling gene expression by using its ability to bind specific small molecule ligands such as vitamins, nucleotides, amino acids and various enzyme co-factors.

Virus-induced gene silencing (VIGS): a technology that exploits an RNA-mediated antiviral defense mechanism. This technique is used in plants for the analysis of gene function and has been adapted for high-throughput functional genomics.

References

1. Baudry, A. et al. (2004) TT2, TT8, and TTG1 synergistically specify the expression of *BANYULS* and proanthocyanidin biosynthesis in *Arabidopsis thaliana*. *Plant J.* 39, 366–380
2. Liu, C. et al. (2008) Direct interaction of AGL24 and SOC1 integrates flowering signals in Arabidopsis. *Development* 135, 1481–1491
3. Brand, L. et al. (2006) A versatile and reliable two-component system for tissue-specific gene induction in Arabidopsis. *Plant Physiol.* 141, 1194–1204
4. Kappers, I.F. et al. (2005) Genetic engineering of terpenoid metabolism attracts bodyguards to Arabidopsis. *Science* 309, 2070–2072
5. Aharoni, A. et al. (2003) Terpenoid metabolism in wild-type and transgenic Arabidopsis plants. *Plant Cell* 15, 2866–2884
6. Wu, S. et al. (2006) Redirection of cytosolic or plastidic isoprenoid precursors elevates terpene production in plants. *Nat. Biotech.* 24, 1441–1447
7. Lorenc-Kukula, K. et al. (2005) Pleiotropic effect of phenolic compounds content increases in transgenic flax plant. *J. Agri. Food Chem.* 53, 272–281
8. Mathews, H. et al. (2003) Activation tagging in tomato identifies a transcriptional regulator of anthocyanin biosynthesis, modification, and transport. *Plant Cell* 15, 1689–1703
9. Ludwig, S.R. et al. (1989) Lc, a member of the maize R gene family responsible for tissue-specific anthocyanin production, encodes a protein similar to transcriptional activators and contains the *myc*-homology region. *Proc. Natl. Acad. Sci. U.S.A.* 86, 7092–7096
10. Ray, H. et al. (2003) Expression of anthocyanins and proanthocyanidins after transformation of alfalfa with maize Lc. *Plant Physiol.* 132, 1448–1463

11. Li, H. et al. (2007) Maize Lc transcription factor enhances biosynthesis of anthocyanins, distinct proanthocyanidins and phenylpropanoids in apple (*Malus domestica* Borkh.). *Planta* 226, 1243–1254
12. Hiratsu, K. et al. (2002) The SUPERMAN protein is an active repressor whose carboxy-terminal repression domain is required for the development of normal flowers. *FEBS Lett.* 514, 351–354
13. Hiratsu, K. et al. (2004) Identification of the minimal repression domain of SUPERMAN shows that the DLELRL hexapeptide is both necessary and sufficient for repression of transcription in Arabidopsis. *Biochem. Biophys. Res. Commun.* 321, 172–178
14. Ohta, M. et al. (2001) Repression domains of class II ERF transcriptional repressors share an essential motif for active repression. *Plant Cell* 13, 1959–1968
15. Hiratsu, K. et al. (2003) Dominant repression of target genes by chimeric repressors that include the EAR motif, a repression domain, in Arabidopsis. *Plant J.* 34, 733–739
16. Matsui, K. et al. (2004) Suppression of the biosynthesis of proanthocyanidin in Arabidopsis by a chimeric *PAP1* repressor. *Plant Biotech. J.* 2, 487–493
17. Ogo, Y. et al. (2008) A novel NAC transcription factor, IDEF2, that recognizes the iron deficiency-responsive element 2 regulates the genes involved in iron homeostasis in plants. *J. Biol. Chem.* 283, 13407–13417
18. Xu, Y. et al. (2006) Floral organ identity genes in the orchid *Dendrobium crumenatum*. *Plant J.* 46, 54–68
19. Zhu, X. and Galili, G. (2004) Lysine metabolism is concurrently regulated by synthesis and catabolism in both reproductive and vegetative tissues. *Plant Physiol.* 135, 129–136
20. Dinesh-Kumar, S.P. et al. (2003) Virus-induced gene silencing. *Methods Mol. Biol.* 236, 287–294
21. Spitzer, B. et al. (2007) Reverse genetics of floral scent: application of *tobacco rattle virus*-based gene silencing in petunia. *Plant Physiol.* 145, 1241–1250
22. Verdonk, J.C. et al. (2005) ODORANT1 regulates fragrance biosynthesis in petunia flowers. *Plant Cell* 17, 1612–1624
23. Lindbo, J.A. (2007) TRBO: a high-efficiency tobacco mosaic virus RNA-based overexpression vector. *Plant Physiol.* 145, 1232–1240
24. Shaked, H. et al. (2005) High-frequency gene targeting in Arabidopsis plants expressing the yeast *RAD54* gene. *Proc. Natl. Acad. Sci. U.S.A.* 102, 12265–12269
25. Grundy, F.J. and Henkin, T.M. (1998) The S box regulon: a new global transcription termination control system for methionine and cysteine biosynthesis genes in gram-positive bacteria. *Mol. Microbiol.* 30, 737–749
26. Winkler, W. et al. (2002) Thiamine derivatives bind messenger RNAs directly to regulate bacterial gene expression. *Nature* 419, 952–956
27. Epshtein, V. et al. (2003) The riboswitch-mediated control of sulfur metabolism in bacteria. *Proc. Natl. Acad. Sci. U.S.A.* 100, 5052–5056
28. Barrick, J.E. et al. (2004) New RNA motifs suggest an expanded scope for riboswitches in bacterial genetic control. *Proc. Natl. Acad. Sci. U.S.A.* 101, 6421–6426
29. Winkler, W.C. and Breaker, R.R. (2005) Regulation of bacterial gene expression by riboswitches. *Annu. Rev. Microbiol.* 59, 487–517
30. Bocobza, S. et al. (2007) Riboswitch-dependent gene regulation and its evolution in the plant kingdom. *Genes Dev.* 21, 2874–2879
31. Wachter, A. et al. (2007) Riboswitch control of gene expression in plants by splicing and alternative 3' end processing of mRNAs. *Plant Cell* 19, 3437–3450
32. Burke, D.H. and Gold, L. (1997) RNA aptamers to the adenosine moiety of S-adenosyl methionine: structural inferences from variations on a theme and the reproducibility of SELEX. *Nucleic Acids Res.* 25, 2020–2024
33. Famulok, M. and Mayer, G. (1999) Aptamers as tools in molecular biology and immunology. *Curr. Top. Microbiol. Immunol.* 243, 123–136
34. Bocobza, S. and Aharoni, A. (2008) Switching the light on plant riboswitches. *Trends Plant Sci.* 13, 526–533
35. Artsaenko, O. et al. (1995) Expression of a single-chain Fv antibody against abscisic acid creates a wilty phenotype in transgenic tobacco. *Plant J.* 8, 745–750
36. Phillips, J. et al. (1997) Seed-specific immunomodulation of abscisic acid activity induces a developmental switch. *EMBO J.* 16, 4489–4496
37. Wigger, J. et al. (2002) Prevention of stomatal closure by immunomodulation of endogenous abscisic acid and its reversion by abscisic acid treatment: physiological behaviour and morphological features of tobacco stomata. *Planta* 215, 413–423
38. Suzuki, Y. et al. (2008) Immunomodulation of bioactive gibberellin confers gibberellin-deficient phenotypes in plants. *Plant Biotechnol. J.* 6, 355–367

20

39. ten-Hoopen, P. et al. (2007) Immunomodulation of jasmonate to manipulate the wound response. *J. Exp. Bot.* 58, 2525–2535
40. Eto, J. et al. (2003) Anti-herbicide single-chain antibody expression confers herbicide tolerance in transgenic plants. *FEBS Lett.* 550, 179–184
41. Almquist, K.C. et al. (2004) Immunomodulation confers herbicide resistance in plants. *Plant Biotechnol. J.* 2, 189–197
42. Weiss, Y. et al. (2006) Herbicide-resistance conferred by expression of a catalytic antibody in *Arabidopsis thaliana*. *Nat. Biotechnol.* 24, 713–717
43. Miroshnichenko, S. et al. (2005) Immunomodulation of function of small heat shock proteins prevents their assembly into heat stress granules and results in cell death at sublethal temperatures. *Plant J.* 41, 269–281
44. Nolke, G. et al. (2005) Immunomodulation of polyamine biosynthesis in tobacco plants has a significant impact on polyamine levels and generates a dwarf phenotype. *Plant Biotechnol. J.* 3, 237–247
45. Aukerman, M.J. and Sakai, H. (2003) Regulation of flowering time and floral organ identity by a microRNA and its *APETALA2*-like target genes. *Plant Cell* 15, 2730–2741
46. Chen, X. (2004) A microRNA as a translational repressor of *APETALA2* in Arabidopsis flower development. *Science* 303, 2022–2025
47. Palatnik, J.F. et al. (2003) Control of leaf morphogenesis by microRNAs. *Nature* 425, 257–263
48. Juarez, M.T. et al. (2004) MicroRNA-mediated repression of *rolled leaf1* specifies maize leaf polarity. *Nature* 428, 84–88
49. Ko, J.H. et al. (2006) Developmental and seasonal expression of *PtaHB1*, a Populus gene encoding a class III HD-Zip protein, is closely associated with secondary growth and inversely correlated with the level of microRNA (miR166). *New Phytol.* 169, 469–478
50. Kidner, C. and Timmermans, M. (2006) In situ hybridization as a tool to study the role of microRNAs in plant development. *Methods Mol. Biol.* 342, 159–179
51. Kutter, C. et al. (2007) MicroRNA-mediated regulation of stomatal development in Arabidopsis. *Plant Cell* 19, 2417–2429
52. Lu, S. et al. (2005) Novel and mechanical stress-responsive microRNAs in Populus trichocarpa that are absent from Arabidopsis. *Plant Cell* 17, 2186–2203
53. Navarro, L. et al. (2006) A plant miRNA contributes to antibacterial resistance by repressing auxin signaling. *Science* 312, 436–439
54. Ru, P. et al. (2006) Plant fertility defects induced by the enhanced expression of microRNA167. *Cell Res.* 16, 457–465
55. Reyes, J.L. and Chua, N.H. (2007) ABA induction of miR159 controls transcript levels of two MYB factors during Arabidopsis seed germination. *Plant J.* 49, 592–606
56. Liu, P.P. et al. (2007) Repression of *AUXIN RESPONSE FACTOR10* by microRNA160 is critical for seed germination and post-germination stages. *Plant J.* 52, 133–146
57. Zhang, J.F. et al. (2008) The disturbance of small RNA pathways enhanced abscisic acid response and multiple stress responses in Arabidopsis. *Plant Cell Environ.* 31, 562–74
58. Allen, E. et al. (2005) microRNA-directed phasing during trans-acting siRNA biogenesis in plants. *Cell* 121, 207–221
59. Alvarez, J.P. et al. (2006) Endogenous and synthetic microRNAs stimulate simultaneous, efficient, and localized regulation of multiple targets in diverse species. *Plant Cell* 18, 1134–1151
60. Warthmann, N. et al. (2008) Highly specific gene silencing by artificial miRNAs in rice. *PLoS ONE* 3, e1829
61. Niu, Q.W. et al. (2006) Expression of artificial microRNAs in transgenic *Arabidopsis thaliana* confers virus resistance. *Nat. Biotechnol.* 24, 1420–1428
62. Qu, J. et al. (2007) Artificial microRNA-mediated virus resistance in plants. *J. Virol.* 81, 6690–6699
63. Choi, K. et al. (2007) Arabidopsis homologs of components of the SWR1 complex regulate flowering and plant development. *Development* 134, 1931–1941
64. Terada, R. et al. (2007) Gene targeting by homologous recombination as a biotechnological tool for rice functional genomics. *Plant Physiol.* 144, 846–856
65. Wright, D.A. et al. (2005) High-frequency homologous recombination in plants mediated by zinc-finger nucleases. *Plant J.* 44, 693–705
66. D'Halluin, K. et al. (2008) Homologous recombination: a basis for targeted genome optimization in crop species such as maize. *Plant Biotechnol. J.* 6, 93–102
67. Dhanasekaran, M. et al. (2005) Designer Zinc finger proteins: tools for creating artifitial DNA-

binding functional proteins. *Acc. Chem. Res.* 39, 45–51
68. Mani, M. et al. (2005) Design, engineering, and characterization of zinc finger nucleases. *Biochem. Biophys. Res. Commun.* 335, 447–457
69. Miller, J.C. et al. (2007) An improved zinc-finger nuclease architecture for highly specific genome editing. *Nat. Biotechnol.* 25, 778–785
70. Mandell, J.G. and Barbas, C.F., 3rd (2006) Zinc finger tools: custom DNA-binding domains for transcription factors and nucleases. *Nucleic Acids Res.* 34, W516–523
71. Ye, X. et al. (2000) Engineering the provitamin A (beta-carotene) biosynthetic pathway into (carotenoid-free) rice endosperm. *Science* 287, 303–305
72. Paine, J.A. et al. (2005) Improving the nutritional value of Golden Rice through increased pro-vitamin A content. *Nat. Biotechnol.* 23, 482–487
73. Hoa, T.T. et al. (2003) Golden Indica and Japonica rice lines amenable to deregulation. *Plant Physiol.* 133, 161–169
74. Datta, K. et al. (2003) Bioengineered 'golden' indica rice cultivars with beta-carotene metabolism in the endosperm with hygromycin and mannose selection systems. *Plant Biotechnol. J.* 1, 81–90
75. Baisakh, N. et al. (2006) Marker-free transgenic (MFT) near-isogenic introgression lines (NIILs) of 'golden' indica rice (cv. IR64) with accumulation of provitamin A in the endosperm tissue. *Plant Biotechnol. J.* 4, 467–475
76. Ishikawa, T. and Shigeoka, S. (2008) Recent advances in ascorbate biosynthesis and the physiological significance of ascorbate peroxidase in photosynthesizing organisms. *Biosci. Biotechnol. Biochem.* 72, 1143–1154
77. Conklin, P.L. et al. (2006) *Arabidopsis thaliana VTC4* encodes L-galactose-1-P phosphatase, a plant ascorbic acid biosynthetic enzyme. *J. Biol. Chem.* 281, 15662–15670
78. Wheeler, G.L. et al. (1998) The biosynthetic pathway of vitamin C in higher plants. *Nature* 393, 365–369
79. Tokunaga, T. et al. (2005) Generation and properties of ascorbic acid-overproducing transgenic tobacco cells expressing sense RNA for l-galactono-1,4-lactone dehydrogenase. *Planta* 220, 854–863
80. Laing, W.A. et al. (2007) The missing step of the L-galactose pathway of ascorbate biosynthesis in plants, an L-galactose guanyltransferase, increases leaf ascorbate content. *Proc. Natl. Acad. Sci. U.S.A.* 104, 9534–9539
81. Giovannoni, J. (2007) Completing a pathway to plant vitamin C synthesis. *Proc. Natl. Acad. Sci. U.S.A.* 104, 9109–9110
82. Agius, F. et al. (2003) Engineering increased vitamin C levels in plants by overexpression of a D-galacturonic acid reductase. *Nat. Biotechnol.* 21, 177–181
83. Ushimaru, T. et al. (2006) Transgenic Arabidopsis plants expressing the rice dehydroascorbate reductase gene are resistant to salt stress. *J. Plant Physiol.* 163, 1179–1184
84. Dellapenna, D. and Pogson, B. (2006) Vitamin synthesis in plants: tocopherols and carotenoids. *Annu. Rev. Plant Biol.* 57, 711–738
85. Shintani, D. and DellaPenna, D. (1998) Elevating the vitamin E content of plants through metabolic engineering. *Science* 282, 2098–2100
86. Van Eenennaam, A.L. et al. (2003) Engineering vitamin E content: from Arabidopsis mutant to soy oil. *Plant Cell* 15, 3007–3019
87. Karunanandaa, B. et al. (2005) Metabolically engineered oilseed crops with enhanced seed tocopherol. *Metab. Eng.* 7, 384–400
88. Holt, N.E. et al. (2005) Carotenoid cation formation and the regulation of photosynthetic light harvesting. *Science* 307, 433–436
89. Havaux, M. and Niyogi, K.K. (1999) The violaxanthin cycle protects plants from photooxidative damage by more than one mechanism. *Proc. Natl. Acad. Sci. U.S.A.* 96, 8762–8767
90. Davison, P.A. et al. (2002) Overexpression of b-carotene hydroxylase enhances stress tolerance in Arabidopsis. *Nature* 418, 203–206
91. Schwartz, S.H. et al. (2003) Elucidation of the indirect pathway of abscisic acid biosynthesis by mutants, genes, and enzymes. *Plant Physiol.* 131, 1591–1601
92. Eriksson, C.E. and Na, A. (1995) Antioxidant agents in raw materials and processed foods. *Biochem. Soc. Symp.* 61, 221–234
93. Giovannucci, E. (2005) Tomato products, lycopene, and prostate cancer: a review of the epidemiological literature. *J. Nutr.* 135, 2030S-2031S
94. Rao, A.V. and Rao, L.G. (2007) Carotenoids and human health. *Pharmacol. Res.* 55, 207–216
95. Sies, H. and Stahl, W. (2003) Non-nutritive bioactive constituents of plants: lycopene, lutein and zeaxanthin. *Int. J. Vitam. Nutr. Res.* 73, 95–100
96. Stahl, W. and Sies, H. (2003) Antioxidant activity of carotenoids. *Mol. Aspects Med.* 24, 345–351

97. Wang, F. et al. (2007) Progress on molecular breeding and metabolic engineering of biosynthesis pathways of C(30), C(35), C(40), C(45), C(50) carotenoids. *Biotechnol. Adv.* 25, 211–222
98. Kavanaugh, C.J. et al. (2007) The U.S. Food and Drug Administration's evidence-based review for qualified health claims: tomatoes, lycopene, and cancer. *J. Natl. Cancer Inst.* 99, 1074–1085
99. Voutilainen, S. et al. (2006) Carotenoids and cardiovascular health. *Am. J. Clin. Nutr.* 83, 1265–1271
100. Rao, A.V. et al. (2006) Lycopene. *Adv. Food Nutr. Res.* 51, 99–164
101. Long, M. et al. (2006) Metabolite profiling of carotenoid and phenolic pathways in mutant and transgenic lines of tomato: identification of a high antioxidant fruit line. *Phytochemistry* 67, 1750–1757
102. Frusciante, L. et al. (2007) Antioxidant nutritional quality of tomato. *Mol. Nutr. Food Res.* 51, 609–617
103. Liu, Y. et al. (2004) Manipulation of light signal transduction as a means of modifying fruit nutritional quality in tomato. *Proc. Natl. Acad. Sci. U.S.A.* 101, 9897–9902
104. Levin, I. et al. (2003) The tomato dark green mutation is a novel allele of the tomato homolog of the *DEETIOLATED1* gene. *Theor. Appl. Genet.* 106, 454–460
105. Davuluri, G.R. et al. (2005) Fruit-specific RNAi-mediated suppression of *DET1* enhances carotenoid and flavonoid content in tomatoes. *Nat. Biotechnol.* 23, 890–895
106. Guerin, M. et al. (2003) *Haematococcus* astaxanthin: applications for human health and nutrition. *Trends Biotechnol.* 21, 210–216
107. Hussein, G. et al. (2006) Astaxanthin, a carotenoid with potential in human health and nutrition. *J. Nat. Prod.* 69, 443–449
108. Torrissen, O.J. et al. (1989) Pigmentation of salmonids carotenoid deposition and metabolism. *Aquatic Sci.* 2, 209–225
109. Yuan, J.P. and Chen, F. (2000) Purification of trans-astaxanthin from a high-yielding astaxanthin ester-producing strain of the microalga *Haematococcuss pluvialis*. *Food Chem.* 68, 443–448
110. Sarada, R. et al. (2006) An efficient method for extraction of astaxanthin from green alga *Haematococcus pluvialis*. *J. Agri. Food Chem.* 54, 7585–7588
111. Gerjets, T. et al. (2007) Metabolic engineering of ketocarotenoid biosynthesis in leaves and flowers of tobacco species. *Biotechnol. J.* 2, 1263–1269
112. Mann, V. et al. (2000) Metabolic engineering of astaxanthin production in tobacco flowers. *Nat. Biotechnol.* 18, 888–892
113. Stalberg, K. et al. (2003) Synthesis of ketocarotenoids in the seed of *Arabidopsis thaliana*. *Plant J.* 36, 771–779
114. Ralley, L. et al. (2004) Metabolic engineering of ketocarotenoid formation in higher plants. *Plant J.* 39, 477–486
115. Clair, R.S. and Anthony, M. (2005) Soy, isoflavones and atherosclerosis. *Handb. Exp. Pharmacol.* 170, 301–323
116. Gonzalez-Gallego, J. et al. (2007) Anti-inflammatory properties of dietary flavonoids. *Nutr. Hosp.* 22, 287–293
117. Patel, D. et al. (2007) Apigenin and cancer chemoprevention: progress, potential and promise (review). *Int. J. Oncol.* 30, 233–245
118. Rice-Evans, C. (2001) Flavonoid antioxidants. *Curr. Med. Chem.* 8, 797–807
119. Tipoe, G.L. et al. (2007) Green tea polyphenols as an anti-oxidant and anti-inflammatory agent for cardiovascular protection. *Cardiovasc. Hematol. Disord. Drug Targets* 7, 135–144
120. Wang, H.K. (2000) The therapeutic potential of flavonoids. *Expert Opin. Investig. Drugs* 9, 2103–2119
121. Korkina, L.G. (2007) Phenylpropanoids as naturally occurring antioxidants: from plant defense to human health. *Cell Mol. Biol.* 53, 15–25
122. Seigler, D.S. (2002) Plant secondary metabolism. Kluwer, Boston
123. Muir, S.R. et al. (2001) Overexpression of petunia chalcone isomerase in tomato results in fruit containing increased levels of flavonols. *Nat. Biotechnol.* 19, 470–474
124. Bovy, A. et al. (2002) High-flavonol tomatoes resulting from the heterologous expression of the maize transcription factor genes *LC* and *C1*. *Plant Cell* 14, 2509–2526
125. Scalbert, A. et al. (2000) Proanthocyanidins and human health: systemic effects and local effects in the gut. *Biofactors* 13, 115–120
126. Ishida, K. et al. (2006) Influence of tannins from *Stryphnodendron adstringens* on growth and virulence factors of *Candida albicans*. *J. Antimicrob. Chemother.* 58, 942–949
127. Sadava, D. et al. (2007) The green tea polyphenol, epigallocatechin-3-gallate inhibits telomerase and induces apoptosis in drug-resistant lung cancer cells. *Biochem. Biophys. Res. Commun.* 360, 233–237

128. Ramljak, D. et al. (2005) Pentameric procyanidin from *Theobroma cacao* selectively inhibits growth of human breast cancer cells. *Mol. Cancer Ther*. 4, 537–546
129. Brunet, S. and Hoste, H. (2006) Monomers of condensed tannins affect the larval exsheathment of parasitic nematodes of ruminants. *J. Agri. Food Chem*. 54, 7481–7487
130. Iqbal, Z. et al. (2007) Direct and indirect anthelmintic effects of condensed tannins in sheep. *Vet.Parasitol*. 144, 125–131
131. Debeaujon, I. et al. (2000) Influence of the testa on seed dormancy, germination, and longevity in Arabidopsis. *Plant Physiol*. 122, 403–414
132. Abrahams, S. et al. (2002) Identification and biochemical characterization of mutants in the proanthocyanidin pathway in Arabidopsis. *Plant Physiol*. 130, 561–576
133. Pourcel, L. et al. (2005) *TRANSPARENT TESTA10* encodes a laccase-like enzyme involved in oxidative polymerization of flavonoids in Arabidopsis seed coat. *Plant Cell* 17, 2966–2980
134. Bogs, J. et al. (2007) The grapevine transcription factor VvMYBPA1 regulates proanthocyanidin synthesis during fruit development. *Plant Physiol*. 143, 1347–1361
135. Xie, D.Y. et al. (2006) Metabolic engineering of proanthocyanidins through co-expression of anthocyanidin reductase and the PAP1 MYB transcription factor. *Plant J*. 45, 895–907
136. Borevitz, J.O. et al. (2000) Activation tagging identifies a conserved MYB regulator of phenylpropanoid biosynthesis. *Plant Cell* 12, 2383–2394
137. Tian, L. et al. (2007) Biosynthesis and genetic engineering of proanthocyanidins and (iso)flavonoids. *Phytochem. Rev*. 7, 445–465
138. Dixon, R.A. and Sumner, L.W. (2003) Legume natural products: understanding and manipulating complex pathways for human and animal health. *Plant Physiol*. 131, 878–885
139. Wuttke, W. et al. (2007) Isoflavones - safe food additives or dangerous drugs? *Ageing Res. Rev*. 6, 150–188
140. Branca, F. and Lorenzetti, S. (2005) Health effects of phytoestrogens. *Forum Nutr*. 57, 100–111
141. Giorcelli, A. et al. (2004) Expression of the stilbene synthase (StSy) gene from grapevine in transgenic white poplar results in high accumulation of the antioxidant resveratrol glucosides. *Transgenic Res*. 13, 203–214
142. Aggarwal, B.B. et al. (2004) Role of resveratrol in prevention and therapy of cancer: preclinical and clinical studies. *Anticancer Res*. 24, 2783–2840
143. Ulrich, S. et al. (2005) Molecular mechanisms of the chemopreventive effects of resveratrol and its analogs in carcinogenesis. *Mol. Nutr. Food Res*. 49, 452–461
144. Shankar, S. et al. (2007) Chemoprevention by resveratrol: molecular mechanisms and therapeutic potential. *Front. Biosci*. 12, 4839–4854
145. de Lange, D.W. et al. (2007) Polyphenolic grape extract inhibits platelet activation through PECAM-1: an explanation for the French paradox. *Alcohol Clin. Exp. Res*. 31, 1308–1314
146. Halls, C. and Yu, O. (2008) Potential for metabolic engineering of resveratrol biosynthesis. *Trends Biotechnol*. 26, 77–81
147. Hain, R. et al. (1993) Disease resistance results from foreign phytoalexin expression in a novel plant. *Nature* 361, 153–156
148. Kobayashi, S. et al. (2000) Kiwifruits (*Actinidia deliciosa*) transformed with a *Vitis* stilbene synthase gene produce piceid (resveratrol-glucoside). *Plant Cell Reports* 19, 904–910
149. Ruhmann, S. et al. (2006) Piceid (resveratrol glucoside) synthesis in stilbene synthase transgenic apple fruit. *J. Agri. Food Chem*. 54, 4633–4640
150. Szankowski, I. et al. (2003) Transformation of apple (*Malus domestica* Borkh.) with the stilbene synthase gene from grapevine (*Vitis vinifera* L.) and a PGIP gene from kiwi (*Actinidia deliciosa*). *Plant Cell Rep*. 22, 141–149
151. Zhu, Y.J. et al. (2004) Expression of the grapevine stilbene synthase gene *VST1* in papaya provides increased resistance against diseases caused by *Phytophthora palmivora*. *Planta* 220, 241–250
152. Richter, A. et al. (2006) Transgenic peas (*Pisum sativum*) expressing polygalacturonase inhibiting protein from raspberry (*Rubus idaeus*) and stilbene synthase from grape (*Vitis vinifera*). *Plant Cell Rep*. 25, 1166–1173
153. Giovinazzo, G. et al. (2005) Antioxidant metabolite profiles in tomato fruit constitutively expressing the grapevine stilbene synthase gene. *Plant Biotechnol. J*. 3, 57–69
154. Schwekendiek, A. et al. (2007) Constitutive expression of a grapevine stilbene synthase gene in transgenic hop (*Humulus lupulus* L.) yields resveratrol and its derivatives in substantial quantities. *J. Agri. Food Chem*. 55, 7002–7009

20

155. Hipskind, J.D. and Paiva, N.L. (2000) Constitutive accumulation of a resveratrol-glucoside in transgenic alfalfa increases resistance to *Phoma medicaginis*. *Mol. Plant Microbe Interact.* 13, 551–562
156. Regev-Shoshani, G. et al. (2003) Glycosylation of resveratrol protects it from enzymic oxidation. *Biochem. J.* 374, 157–163
157. Jeandet, P. et al. (2002) Phytoalexins from the *Vitaceae*: biosynthesis, phytoalexin gene expression in transgenic plants, antifungal activity, and metabolism. *J. Agri. Food Chem.* 50, 2731–2741
158. Dudareva, N. and Pichersky, E. (2008) Metabolic engineering of plant volatiles. *Curr. Opin. Biotechnol.* 19, 181–189
159. Aharoni, A. et al. (2005) Volatile science? Metabolic engineering of terpenoids in plants. *Trends Plant Sci.* 10, 594–602
160. Aubert, C. and Bourger, N. (2004) Investigation of volatiles in Charentais cantaloupe melons (*Cucumis melo* var. *cantalupensis*). Characterization of aroma constituents in some cultivars. *J. Agri. Food Chem.* 52, 4522–4528
161. Picone, J.M. et al. (2004) Rhythmic emission of floral volatiles from *Rosa damascena semperflorens* cv. 'Quatre Saisons'. *Planta* 219, 468–478
162. Cheng, A.X. et al. (2007) The rice (*E*)-β-caryophyllene synthase (OsTPS3) accounts for the major inducible volatile sesquiterpenes. *Phytochemistry* 68, 1632–1641
163. Lucker, J. et al. (2001) Expression of Clarkia *S*-linalool synthase in transgenic petunia plants results in the accumulation of *S*-linalyl-β-D-glucopyranoside. *Plant J.* 27, 315–324
164. Rouseff, R.L. and Leahy, M.M. (1995) Fruit flavors. Biogenesis, characterization, and authentication. American Chemical Society, Washington
165. Lunkenbein, S. et al. (2006) Cinnamate metabolism in ripening fruit. Characterization of a UDP-glucose:cinnamate glucosyltransferase from strawberry. *Plant Physiol.* 140, 1047–1058
166. Aharoni, A. et al. (2004) Gain and loss of fruit flavor compounds produced by wild and cultivated strawberry species. *Plant Cell* 16, 3110–3131
167. Speirs, J. et al. (1998) Genetic manipulation of alcohol dehydrogenase levels in ripening tomato fruit affects the balance of some flavor aldehydes and alcohols. *Plant Physiol.* 117, 1047–1058
168. Lewinsohn, E. et al. (2001) Enhanced levels of the aroma and flavor compound *S*-linalool by metabolic engineering of the terpenoid pathway in tomato fruits. *Plant Physiol.* 127, 1256–1265
169. Davidovich-Rikanati, R. et al. (2007) Enrichment of tomato flavor by diversion of the early plastidial terpenoid pathway. *Nat. Biotechnol.* 25, 899–901
170. Gang, D.R. (2005) Evolution of flavors and scents. *Annu. Rev. Plant Biol.* 56, 301–325
171. Lucker, J. et al. (2004) Increased and altered fragrance of tobacco plants after metabolic engineering using three monoterpene synthases from lemon. *Plant Physiol.* 134, 510–519
172. Zuker, A. et al. (2002) Modification of flower color and fragrance by antisense suppression of the. flavanone 3-hydroxylase gene. *Mol. Breeding* 9, 33–41
173. Baumann, K. et al. (2007) Control of cell and petal morphogenesis by R2R3 MYB transcription factors. *Development* 134, 1691–1701
174. Noda, K. et al. (1994) Flower colour intensity depends on specialized cell shape controlled by a MYB-related transcription factor. *Nature* 369, 661–664
175. Fukada-Tanaka, S. et al. (2000) Colour-enhancing protein in blue petals. *Nature* 407, 581
176. Iida, S. et al. (2004) Genes encoding proteins regulating the pH of vacuoles. In: *Patent US, United States*
177. Meyer, P. et al. (1987) A new petunia flower colour generated by transformation of a mutant with a maize gene. *Nature* 330, 677–678
178. Tanaka, Y. et al. (1998) Metabolic engineering to modify flower color. *Plant Cell Physiol.* 39, 1119–1126
179. Forkmann, G. and Martens, S. (2001) Metabolic engineering and applications of flavonoids. *Curr. Opin. Biotechnol.* 12, 155–160
180. Mori, S. et al. (2004) Heterologous expression of the flavonoid 3',5'-hydroxylase gene of *Vinca major* alters flower color in transgenic *Petunia hybrida*. *Plant Cell Rep.* 22, 415–421
181. Fukusaki, E. et al. (2004) Flower color modulations of *Torenia hybrida* by downregulation of chalcone synthase genes with RNA interference. *J. Biotechnol* 111, 229–240
182. Ono, E. et al. (2006) Yellow flowers generated by expression of the aurone biosynthetic pathway. *Proc. Natl. Acad. Sci. U.S.A.* 103, 11075–11080
183. Nakatsuka, T. et al. (2007) Production of red-flowered plants by genetic engineering of

multiple flavonoid biosynthetic genes. *Plant Cell Rep*. 26, 1951–1959
184. Selinger, D.A. and Chandler, V.L. (1999) A mutation in the *pale aleurone color1* gene identifies a novel regulator of the maize anthocyanin pathway. *Plant Cell* 11, 5–14
185. Hoballah, M.E. et al. (2007) Single gene-mediated shift in pollinator attraction in petunia. *Plant Cell* 19, 779–790
186. Quattrocchio, F. et al. (1999) Molecular analysis of the *ANTHOCYANIN2* gene of petunia and its role in the evolution of flower color. *Plant Cell* 11, 1433–1444
187. Ban, Y. et al. (2007) Isolation and functional analysis of a MYB transcription factor gene that is a key regulator for the development of red coloration in apple skin. *Plant Cell Physiol*. 48, 958–970
188. Espley, R.V. et al. (2007) Red colouration in apple fruit is due to the activity of the MYB transcription factor, MdMYB10. *Plant J*. 49, 414–427
189. Takos, A.M. et al. (2006) Light-induced expression of a MYB gene regulates anthocyanin biosynthesis in red apples. *Plant Physiol*. 142, 1216–1232
190. Suzuki, S. et al. (2007) Flower color alteration in *Lotus japonicus* by modification of the carotenoid biosynthetic pathway. *Plant Cell Rep*. 26, 951–959
191. Ohmiya, A. et al. (2006) Carotenoid cleavage dioxygenase (CmCCD4a) contributes to white color formation in chrysanthemum petals. *Plant Physiol*. 142, 1193–1201
192. Vrebalov, J. et al. (2002) A MADS-box gene necessary for fruit ripening at the tomato ripening-inhibitor (*rin*) locus. *Science* 296, 343–346
193. Lu, S. et al. (2006) The cauliflower *Or* gene encodes a DnaJ cysteine-rich domain-containing protein that mediates high levels of beta-carotene accumulation. *Plant Cell* 18, 3594–3605
194. Balint, G.A. (2001) Artemisinin and its derivatives: an important new class of antimalarial agents. *Pharmacol. Ther*. 90, 261–265
195. Liu, C. et al. (2006) Artemisinin: current state and perspectives for biotechnological production of an antimalarial drug. *Appl. Microbiol. Biotechnol*. 72, 11–20
196. Matsushita, Y. et al. (1996) Cloning and analysis of a cDNA encoding farnesyl diphosphate synthase from *Artemisia annua*. *Gene* 172, 207–209
197. Wallaart, T.E. et al. (2001) Amorpha-4,11-diene synthase: cloning and functional expression of a key enzyme in the biosynthetic pathway of the novel antimalarial drug artemisinin. *Planta* 212, 460–465
198. Sa, G. et al. (2001) Effects of ipt gene expression on the physiological and chemical characteristics of *Artemisia annua* L. *Plant Sci*. 160, 691–698
199. Kingston, D.G. and Newman, D.J. (2007) Taxoids: cancer-fighting compounds from nature. *Curr. Opin. Drug Discov. Devel*. 10, 130–144
200. Ketchum, R.E. et al. (2003) Taxus metabolomics: methyl jasmonate preferentially induces production of taxoids oxygenated at C-13 in *Taxus* x *media* cell cultures. *Phytochemistry* 62, 901–909
201. Navia-Osorio, A. et al. (2002) Production of paclitaxel and baccatin III in a 20-L airlift bioreactor by a cell suspension of *Taxus wallichiana*. *Planta Med*. 68, 336–340
202. Doi, T. et al. (2006) A formal total synthesis of taxol aided by an automated synthesizer. *Chem. Asian J*. 1, 370–383
203. Guo, B.H. et al. (2005) Taxol synthesis. *African J. Biotechnol*. 5, 15–20
204. Dejong, J.M. et al. (2006) Genetic engineering of taxol biosynthetic genes in *Saccharomyces cerevisiae*. *Biotechnol. Bioeng*. 93, 212–224
205. Aoyama, T. and Chua, N.H. (1997) A glucocorticoid-mediated transcriptional induction system in transgenic plants. *Plant J*. 11, 605–612
206. Besumbes, O. et al. (2004) Metabolic engineering of isoprenoid biosynthesis in Arabidopsis for the production of taxadiene, the first committed precursor of taxol. *Biotechnol. Bioeng*. 88, 168–175
207. Botella-Pavia, P. et al. (2004) Regulation of carotenoid biosynthesis in plants: evidence for a key role of hydroxymethylbutenyl diphosphate reductase in controlling the supply of plastidial isoprenoid precursors. *Plant J*. 40, 188–199
208. Kovacs, K. et al. (2007) Redirection of carotenoid metabolism for the efficient production of taxadiene [taxa-4(5),11(12)-diene] in transgenic tomato fruit. *Transgenic Res*. 16, 121–126
209. Rontein, D. et al. (2008) CYP725A4 from yew catalyzes complex structural rearrangement of taxa-4(5),11(12)-diene into the cyclic ether 5(12)-oxa-3(11)-cyclotaxane. *J. Biol. Chem*. 283, 6067–6075
210. Suzuki, K. et al. (1999) An *Atropa belladonna* hyoscyamine 6beta-hydroxylase gene is

differentially expressed in the root pericycle and anthers. *Plant Mol. Biol.* 40, 141–152
211. Palazon, J. et al. (2003) Alkaloid production in *Duboisia* hybrid hairy roots and plants overexpressing the h6h gene. *Plant Sci.* 165, 1289–1295
212. Matsuda, J. et al. (1991) Molecular cloning of hyoscyamine 6 beta-hydroxylase, a 2-oxoglutarate-dependent dioxygenase, from cultured roots of *Hyoscyamus niger*. *J. Biol. Chem.* 266, 9460–9464
213. Yun, D.J. et al. (1992) Metabolic engineering of medicinal plants: transgenic *Atropa belladonna* with an improved alkaloid composition. *Proc. Natl. Acad. Sci. U.S.A.* 89, 11799–11803
214. Hashimoto, T. et al. (1993) Production of tropane alkaloids in genetically engineered root cultures. *Phytochemistry* 32, 713–718
215. Jouhikainen, K. et al. (1999) Enhancement of scopolamine production in *Hyoscyamus muticus* L. hairy root cultures by genetic engineering. *Planta* 208, 545–551
216. Hakkinen, S.T. et al. (2005) Enhanced secretion of tropane alkaloids in *Nicotiana tabacum* hairy roots expressing heterologous hyoscyamine-6beta-hydroxylase. *J. Exp. Bot.* 56, 2611–2618
217. Zhang, L. et al. (2004) Engineering tropane biosynthetic pathway in *Hyoscyamus niger* hairy root cultures. *Proc. Natl. Acad. Sci. U.S.A.* 101, 6786–6791
218. Nakajima, K. et al. (1999) Structures and expression patterns of two tropinone reductase genes from *Hyoscyamus niger*. *Biosci. Biotechnol. Biochem.* 63, 1756–1764
219. Drager, B. (2006) Tropinone reductases, enzymes at the branch point of tropane alkaloid metabolism. *Phytochemistry* 67, 327–337
220. Richter, U. et al. (2005) Overexpression of tropinone reductases alters alkaloid composition in *Atropa belladonna* root cultures. *J. Exp. Bot.* 56, 645–652
221. Brownstein, M.J. (1993) A brief history of opiates, opioid peptides, and opioid receptors. *Proc. Natl. Acad. Sci. U.S.A.* 90, 5391–5393
222. Ye, K. et al. (1998) Opium alkaloid noscapine is an antitumor agent that arrests metaphase and induces apoptosis in dividing cells. *Proc. Natl. Acad. Sci. U.S.A.* 95, 1601–1606
223. Bird, D.A. et al. (2003) A tale of three cell types: alkaloid biosynthesis is localized to sieve elements in opium poppy. *Plant Cell* 15, 2626–2635
224. Facchini, P.J. and De Luca, V. (1995) Phloem-specific expression of tyrosine/dopa decarboxylase genes and the biosynthesis of isoquinoline alkaloids in opium poppy. *Plant Cell* 7, 1811–1821
225. Samanani, N. et al. (2006) The role of phloem sieve elements and laticifers in the biosynthesis and accumulation of alkaloids in opium poppy. *Plant J.* 47, 547–563
226. Frick, S. et al. (2007) Metabolic engineering with a morphine biosynthetic P450 in opium poppy surpasses breeding. *Metab. Eng.* 9, 169–176
227. Larkin, P.J. et al. (2007) Increasing morphinan alkaloid production by over-expressing codeinone reductase in transgenic *Papaver somniferum*. *Plant Biotechnol. J.* 5, 26–37
228. Frick, S. et al. (2004) Transformation of opium poppy (*Papaver somniferum* L.) with antisense berberine bridge enzyme gene (anti-bbe) via somatic embryogenesis results in an altered ratio of alkaloids in latex but not in roots. *Transgenic Res.* 13, 607–613
229. Fujii, N. et al. (2007) Knockdown of berberine bridge enzyme by RNAi accumulates (*S*)-reticuline and activates a silent pathway in cultured California poppy cells. *Transgenic Res.* 16, 363–375
230. Barleben, L. et al. (2007) Molecular architecture of strictosidine glucosidase: the gateway to the biosynthesis of the monoterpenoid indole alkaloid family. *Plant Cell* 19, 2886–2897
231. McCoy, E. et al. (2006) Substrate specificity of strictosidine synthase. *Bioorg. Med. Chem. Lett.* 16, 2475–2478
232. Liu, D.H. et al. (2007) Terpenoid Indole Alkaloids biosynthesis and metabolic engineering in *Catharanthus roseus*. *J.Integrat. Plant Biol.* 49, 961–974
233. Canel, C. et al. (1998) Effects of over-expression of strictosidine synthase and tryptophan decarboxylase on alkaloid production by cell cultures of *Catharanthus roseus*. *Planta* 205, 414–419
234. van der Fits, L. and Memelink, J. (2000) ORCA3, a jasmonate-responsive transcriptional regulator of plant primary and secondary metabolism. *Science* 289, 295–297
235. van der Fits, L. and Memelink, J. (2001) The jasmonate-inducible AP2/ERF-domain transcription factor ORCA3 activates gene expression via interaction with a jasmonate-responsive promoter element. *Plant J.* 25, 43–53
236. Memelink, J. et al. (2001) ORCAnization of jasmonate-responsive gene expression in alkaloid metabolism. *Trends Plant Sci.* 6, 212–219

237. De Moraes, C.M. et al. (2001) Caterpillar-induced nocturnal plant volatiles repel conspecific females. *Nature* 410, 577–580
238. Kessler, A. and Baldwin, I.T. (2001) Defensive function of herbivore-induced plant volatile emissions in nature. *Science* 291, 2141–2144
239. Delphia, C.M. et al. (2007) Induction of plant volatiles by herbivores with different feeding habits and the effects of induced defenses on host-plant selection by thrips. *J. Chem. Ecol.* 33, 997–1012
240. Bouwmeester, H.J. et al. (1999) Spider mite-induced (3*S*)-(*E*)-nerolidol synthase activity in cucumber and lima bean. The first dedicated step in acyclic C11-homoterpene biosynthesis. *Plant Physiol.* 121, 173–180
241. Turlings, T.C. and Fritzsche, M.E. (1999) Attraction of parasitic wasps by caterpillar-damaged plants. *Novartis Found. Symp.* 223, 21–32; discussion 32–38
242. Van Poecke, R.M. et al. (2001) Herbivore-induced volatile production by *Arabidopsis thaliana* leads to attraction of the parasitoid Cotesia rubecula: chemical, behavioral, and gene-expression analysis. *J. Chem. Ecol.* 27, 1911–1928
243. Akila, A. and Tewari, R. (1984) Chemistry of patchouli oil: a review. *Curr. Res. Med. Arom. Plants* 6, 38–54
244. Zhu, B.C. et al. (2003) Toxicity and repellency of patchouli oil and patchouli alcohol against Formosan subterranean termites *Coptotermes formosanus* Shiraki (*Isoptera*: *Rhinotermitidae*). *J. Agri. Food Chem.* 51, 4585–4588
245. Aharoni, A. et al. (2006) Metabolic engineering of terpenoid biosynthesis in plants. *Phytochem. Rev.* 5, 49–58
246. Degenhardt, J. and Gershenzon, J. (2000) Demonstration and characterization of (*E*)-nerolidol synthase from maize: a herbivore-inducible terpene synthase participating in (3*E*)-4,8-dimethyl-1,3,7-nonatriene biosynthesis. *Planta* 210, 815–822
247. Schnee, C. et al. (2006) The products of a single maize sesquiterpene synthase form a volatile defense signal that attracts natural enemies of maize herbivores. *Proc. Natl. Acad. Sci. U.S.A.* 103, 1129–1134
248. Schulze, K. et al. (2005) Inhibiting effects of resveratrol and its glucoside piceid against *Venturia inaequalis*, the causal agent of apple scab. *J. Agri. Food Chem.* 53, 356–362
249. Torres, P. et al. (2003) Antioxidant and insect growth regulatory activities of stilbenes and extracts from *Yucca periculosa*. *Phytochemistry* 64, 463–473
250. Leckband, G. and Lorz, H. (1998) Transformation and expressoin of a stilbene synthase gene of *Vitis vinifera* L. in barley and wheat for increased fungal resistance. *Theor. Appl. Genet.* 96, 1004–1012
251. Thomzik, J.E. et al. (1997) Synthesis of a grapevine phytoalexin in transgenic tomatoes (*Lycopersicon esculentum* Mill.) conditions resistance against *Phytophthora infestans*. *Physiol. Mol. Plant Pathol.* 51, 265–278
252. Serazetdinova, L. et al. (2005) Expression of transgenic stilbene synthases in wheat causes the accumulation of unknown stilbene derivatives with antifungal activity. *J. Plant Physiol.* 162, 985–1002
253. Stark-Lorenzen, P. et al. (1997) Transfer of a grapevine stilbene synthase gene to rice (*Oryza sativa* L.) *Plant Cell Rep.* 16, 668–673
254. Vetter, J. (2000) Plant cyanogenic glycosides. *Toxicon* 38, 11–36
255. Bak, S. et al. (2000) Transgenic tobacco and Arabidopsis plants expressing the two multifunctional sorghum cytochrome P450 enzymes, CYP79A1 and CYP71E1, are cyanogenic and accumulate metabolites derived from intermediates in dhurrin biosynthesis. *Plant Physiol.* 123, 1437–1448
256. Tattersall, D.B. et al. (2001) Resistance to an herbivore through engineered cyanogenic glucoside synthesis. *Science* 293, 1826–1828
257. Franks, T.K. et al. (2006) Consequences of transferring three sorghum genes for secondary metabolite (cyanogenic glucoside) biosynthesis to grapevine hairy roots. *Transgenic Res.* 15, 181–195
258. Nielsen, K.A. et al. (2006) Reconstitution of cyanogenesis in barley (*Hordeum vulgare* L.) and its implications for resistance against the barley powdery mildew fungus. *Planta* 223, 1010–1023
259. Mithen, R. (2001) Glucosinolates - biochemistry, genetics and biological activity. *Plant Growth Regulat.* 34, 91–103
260. Levy, M. et al. (2005) Arabidopsis IQD1, a novel calmodulin-binding nuclear protein, stimulates glucosinolate accumulation and plant defense. *Plant J.* 43, 79–96
261. Weigel, D. et al. (2000) Activation tagging in Arabidopsis. *Plant Physiol.* 122, 1003–1013
262. Barth, C. and Jander, G. (2006) Arabidopsis myrosinases TGG1 and TGG2 have redundant

function in glucosinolate breakdown and insect defense. *Plant J.* 46, 549–562
263. Hirai, M.Y. et al. (2007) Omics-based identification of Arabidopsis MYB transcription factors regulating aliphatic glucosinolate biosynthesis. *Proc. Natl. Acad. Sci. U.S.A.* 104, 6478–6483
264. Gigolashvili, T. et al. (2007) The transcription factor HIG1/MYB51 regulates indolic glucosinolate biosynthesis in *Arabidopsis thaliana*. *Plant J.* 50, 886–901
265. Gigolashvili, T. et al. (2008) HAG2/MYB76 and HAG3/MYB29 exert a specific and coordinated control on the regulation of aliphatic glucosinolate biosynthesis in *Arabidopsis thaliana*. *New Phytol.* 177, 627–642
266. Gigolashvili, T. et al. (2007) The R2R3-MYB transcription factor HAG1/MYB28 is a regulator of methionine-derived glucosinolate biosynthesis in *Arabidopsis thaliana*. *Plant J.* 51, 247–261
267. Sonderby, I.E. et al. (2007) A systems biology approach identifies a R2R3 MYB gene subfamily with distinct and overlapping functions in regulation of aliphatic glucosinolates. *PLoS ONE* 2, e1322
268. Malitsky, S. et al. (2008) The "Inner" and "Outer" circles of the transcriptome and metabolome affected by the two clades of Arabidopsis glucosinolate biosynthesis regulators. *Plant Physiol.*, 148, 2021–2049
269. Beekwilder, J. et al. (2008) The impact of the absence of aliphatic glucosinolates on insect herbivory in Arabidopsis. *PLoS ONE* 3, e2068
270. Smolen, G. and Bender, J. (2002) Arabidopsis cytochrome P450 cyp83B1 mutations activate the tryptophan biosynthetic pathway. *Genetics* 160, 323–332
271. Ashihara, H. et al. (1996) Biosynthesis of caffeine in leaves of coffee. *Plant Physiol.* 111, 747–753
272. Russell, D.W. et al. (1991) Caffeine, a naturally occurring acaricide. *J. Allergy Clin. Immunol.* 87, 107–110
273. Nathanson, J.A. (1984) Caffeine and related methylxanthines: possible naturally occurring pesticides. *Science* 226, 184–187
274. Hollingsworth, R.G. et al. (2002) Caffeine as a repellent for slugs and snails. *Nature* 417, 915–916
275. Guerreiro Filho, O. and Mazzafera, P. (2003) Caffeine and resistance of coffee to the berry borer *Hypothenemus hampei* (*Coleoptera, Scolytidae*). *J. Agri. Food Chem.*51, 6987–6991
276. Araque, P. et al. (2007) Insecticidal activity of caffeine aqueous solutions and caffeine oleate emulsions against *Drosophila melanogaster* and *Hypothenemus hampei*. *J. Agri. Food Chem.* 55, 6918–6922
277. Uefuji, H. et al. (2005) Caffeine production in tobacco plants by simultaneous expression of three coffee N-methyltrasferases and its potential as a pest repellant. *Plant Mol. Biol.* 59, 221–227
278. Oliver, S. et al. (1998) Systematic functional analysis of the yeast genome. *Trends Biotechnol.* 16, 373–378
279. Fiehn, O. et al. (2002) Metabolomics - the link between genotypes and phenotypes. *Plant Mol. Biol.* 48, 155–171
280. Aharoni, A. et al. (2002) Non-targeted metabolome analysis by use of Fourier transform ion cyclotron mass spectrometry. *Omics* 6, 217–234
281. Kell, D. B. (2004) Metabolomics and systems biology: making sense of the soup. *Curr. Opin. Microbiol.* 7, 296–307
282. Sweetlove, L.J. et al. (2008) Getting to grips with the plant metabolic network. *Biochem. J.* 409, 27–41
283. Saito, K. et al. (2008) Decoding genes with coexpression networks and metabolomics - 'majority report by precogs' *Trends Plant Sci.* 13, 36–43
284. Nobeli, I. et al. (2003) A structure-based anatomy of the *E.coli* metabolome. *J Mol Biol.* 334, 697–719
285. Fiehn et al. (2000) Metabolite profiling for plant functional genomics. *Nat. Biotechnol.* 18, 1157–1161
286. Roessner, U. et al. (2001) High-resolution metabolic phenotyping of genetically and environmentally diverse potato tuber systems. Identification of phenocopies. *Plant Physiol.* 127, 749–764
287. Bao, X. et al. (2000) Understanding *in vivo* carbon precursor supply for fatty acid synthesis in leaf tissue. *Plant J.* 22, 39–50
288. von Roepenack-Lahaye, E. et al. (2004) Profiling of Arabidopsis secondary metabolites by capillary liquid chromatography coupled to electrospray ionization quadrupole time-of-flight mass spectrometry. *Plant Physiol.* 134, 548–559
289. Moco, S. et al. (2007) Tissue specialization at the metabolite level is perceived during the development of tomato fruit. *J. Exp. Bot.* 58, 4131–4146

290. Mintz-Oron, S. et al. (2008) Gene expression and metabolism in tomato fruit surface tissues. *Plant Physiol.*147, 823–851
291. Suzuki, H. et al. (2008) Metabolic profiling of flavonoids in *Lotus japonicus* using liquid chromatography Fourier transform ion cyclotron resonance mass spectrometry. *Phytochemistry* 69, 99–111
292. Choi, H. K. et al. (2004) Metabolic fingerprinting of wild type and transgenic tobacco plants by 1H NMR and multivariate analysis technique. *Phytochemistry* 65, 857–864

Genome Wide Approaches in Natural Product Research

21

Jürgen Ehlting, Björn Hamberger, Jean-François Ginglinger, and Danièle Werck-Reichhart

Summary Within the last decade, the strategy for pathway discovery in plant secondary metabolism has reversed from a metabolite to protein to gene approach to a gene to pathway strategy. Different genome-wide "omics" strategies apply to gene discovery in model plants with sequenced genomes and plants of pharmaceutical or industrial interest. In this chapter, we provide a brief description as well as a few examples of the main approaches that have so far been applied to plant metabolism. Combination of such global approaches leads to the new field of 'integrative biology', which highlights metabolic networks connecting the different branches of primary and "secondary" metabolism.

J. Ehlting, J.-F. Ginglinger, and D. Werck-Reichhart(✉)
Department of Plant Metabolic Responses,
Institute for Plant Molecular Biology (IBMP),
C.N.R.S., 28 rue Goethe,
67000 Strasbourg, France
e-mails: jean-francois.ginglinger@ibmp-ulp.u-strasbg.fr; daniele.werck@ibmp-ulp.u-strasbg.fr

J. Ehlting
Centre for Forest Biology, University of Victoria,
P.O. Box 3020 STN CSC,
Victoria BC V8W 3N5, Canada
e-mail: je@uvic.ca

B. Hamberger
Michael Smith Laboratories,
University of British Columbia, 2185, East Mall,
Vancouver BC V6T 1Z4, Canada
e-mail: bjoernh@interchange.ubc.ca

Introduction

The elucidation of biochemical pathways leading to plant natural products traditionally started by feeding radiolabelled precursors of compounds of interest. Proposition of a sequence of intermediates was followed by a characterization of the enzymes catalyzing their conversion. Enzyme purification from protein extracts was most often time and material consuming. Many purification procedures had to be evaluated and optimized until a "pure" enzyme fraction was available for determination of partial sequence or generation of antibodies allowing gene discovery and molecular studies (as indicated with blue arrows in Fig. 1). Despite this, a good insight into major pathways, e.g. the phenylpropanoid pathway leading to lignin and flavonoids, or those leading to carotenoids or cyanogenic glucosides has been achieved. The golden age of this approach was the sixties of the last century. The biochemical work was then complemented and extended by the use of classical, or forward genetic screens. Examples include the Arabidopsis *transparent testa* (tt) and *reduced epidermal fluorescence* (*ref*) mutants, which are impaired in flavonoid/(pro-)anthocyanin and phenylpropanoid biosynthesis respectively [1]. Also in non-model plants genetic screens highlighted important metabolic pathways such as the triterpenoid pathway leading to avenacin in oat [2]. The strategy for pathway discovery was

A.E. Osbourn and V. Lanzotti (eds.), *Plant-derived Natural Products*,
DOI 10.1007/978-0-387-85498-4_21, © Springer Science+Business Media, LLC 2009

then abruptly reversed with the ascent of plant molecular biology and genomics. Within the last decade, large-scale genome and cDNA sequencing projects have led to an explosion of sequence information available in digital databases. This not only simplified identification of homologous sequences, but it also revealed an amazing number of paralogues, the existence of very large families of enzymes and regulators. Those pointed to an unexpected complexity of plant metabolism.

The first investigations carried out on a genomic basis led to the revision of canonical pathways and to the accelerated analysis of new routes (e.g. biosynthesis of brassinosteroids [3–5]). They have also revealed new and so far overlooked pathways and an extensive crosstalk in plant metabolism (e.g. [6–8]).

A major goal of the post-genomic era is now to identify biochemical and physiological functions of the products of orphan genes, using a reversed approach from gene to function (Fig. 1), or to directly connect groups of genes, proteins and metabolites. Another goal is to get a better appraisal of the regulation and crosstalk of the different branches of metabolism, taking into account post-transcriptional, post-translational and biochemical regulations. To this end, many tools for reverse genetics and biochemistry are being developed on the whole genome, transcriptome, proteome, and metabolome level. Such tools include web interfaces to retrieve and search genomic information, platforms for simultaneous identification and quantification of thousands of transcripts and proteins, collections of knock-out and knock-down mutants, full length cDNA collections for recombinant protein expression, and technologies for medium or large scale screening of enzyme activity. In parallel, a big step was achieved with the development of new tools for the analysis and identification of metabolites, sometimes from very

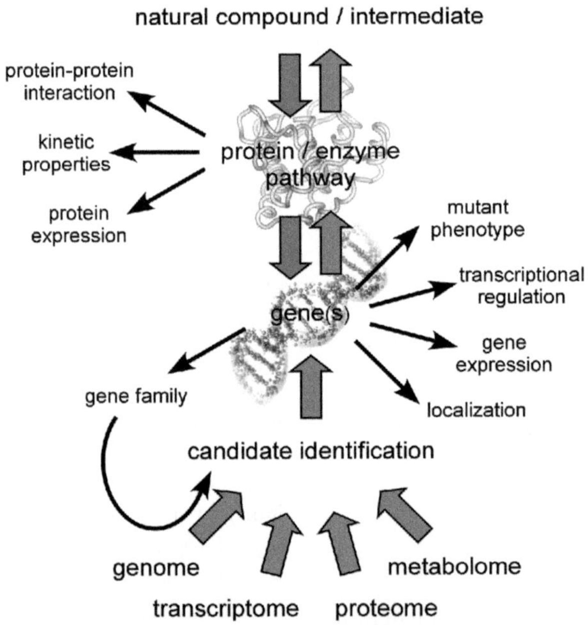

Fig. 1 Schematic workflow comparing classic biochemical approaches (blue arrows) with functional genomic approaches (red arrows). Black arrows denote experimental approaches leading to the given information

small samples. Combination of these tools has opened the era of "systems biology", in which interactions between genes, RNA, proteins and metabolites are investigated. Extraction of relevant data from these investigations largely relies on bioinformatics.

In this chapter, we will summarize the current status of plant genome and transcriptome projects, describe strategies and examples that exploit the resulting information in the context of plant natural product research, and give an overview of how proteome and metabolome analyses are being used to promote the field.

Genome and Transcript Sequencing Projects in Plants

The first plant genome project, targeting the model plant *Arabidopsis thaliana*, was initiated in 1996, and the 125-megabase genome was completed 4 years later [9]. In parallel, almost 700,000 *Arabidopsis* cDNA clones were at least partially sequenced giving rise to expressed sequence tags (ESTs) and full length cDNA sequences (Table 1). To date (April 2008), six additional higher plant genomes have been elucidated completely. These include two monocot crops, rice (*Oryza sativa*) and sorghum (*Sorghum bicolor*) [10, 11], and four dicots, namely the tree poplar (*Populus trichocarpa*) [12], the grape vine (*Vitis vinifera*) [13], the castor bean (*Ricinus communis*) [14] and the papaya (*Carica papaya*) [15]. Genome sequencing projects are ongoing for a number of other plant species (Table 2). More crop plants genomic data were generated in industrial restricted context. With the development of new very high throughput sequencing technologies [16] it can be expected that many more will follow soon. In addition, large data sets of partial EST or full length cDNA sequences have been generated for hundreds of plants. Compilations of overlapping sequences, or transcript assemblies, can be found e.g. from the 'TIGR' and 'PlantGDB' websites (Table 2).

Upon assembly of a (partial) genome sequence, the uppermost important task is to identify and annotate the genes encoded within this sequence. In a first step, *ab initio* gene finding algorithms are used to identify putative gene coding regions or gene models, which are refined by mapping thousands of EST or full length cDNA sequences to the genome sequence. These predicted structures are then visualized using a searchable web based graphical interface (Table 2). These genome browsers may also highlight other features of the genome, such as homologous sequences from other plants. In addition, some genome portals combine genome and cDNA sequence information from diverse species and provide tools for comparative genomics, such as alignment and phylogenies of gene families (Table 2). Although this provides a crucial framework, it must be kept in mind that accurate predictions are not always possible, and that predictions may vary depending on the gene finding algorithm that has been used. Therefore some projects (i.e. those from the JGI, Table 2) provide gene predictions from several algorithms, which might appear confusing at first sight, but allow the user to select the most appropriate model for a given gene of interest. With increasing numbers of full length cDNAs, and based on manual curation by the community, gene models change over time and become more accurate. The functional annotation of gene models is primarily based on a comparison of the DNA and deduced protein sequences with other sequences previously described. Ultimately, all functional annotations are based on experimentally characterized genes (or on similarity to these). Keeping in mind that very small sequence differences may result in divergent biochemical activities, it is important to know the type of evidence serving as a basis for the annotation. It is also important to keep in mind that, even for the best described model *Arabidopsis*, 'only' 4,790 genes (17%) are annotated based on immediate experimental

21

Table 1 Vascular plants with completed genome sequencing projects

Species	Estimated size (Mb)	Sequenced (Mb)	Gene models	cDNA sequences ESTs	cDNA sequences Full length	Genes with cDNA
Arabidopsis thaliana	135[a]	119[a]	28,152[b,c]	624,151[c]	73,527[c]	22,032[c]
Oryza sativa	389[d]	372[e]	53,461[e]	962,258[e]	34,946[e]	31,439[e]
Populus trichocarpa	485[f]	473[g]	45,555[g]	89,198[h]	4,664[f]	–
Vitis vinifera	475[i]	487[i]	30,434[i]	48,239[i]	1,672[i]	5,038[i]
Sorghum bicolor	770[j]	–	49,826[j]	203,575[k]	125[k]	–
Ricinus communis	400[l]	350[m]	31,300[m]	~50,000[m]	–	5,300 [m]
Carica papaya	372[n]	271[n]	28,629[n]	1,283[o]	297[o]	9,760[n]
Medicago truncatula	525[p]	211[q]	42,358[r]	217,148[k]	690[k]	13,841[r]

[a]http://www.arabidopsis.org/portals/genannotation/gene_structural_annotation/agicomplete.jsp
[b]Excluding pseudogenes
[c]http://www.arabidopsis.org/portals/genannotation/gene_structural_annotation/ annotation_data.jsp
[d][10]
[e]http://rapdb.dna.affrc.go.jp/rap2_statistics.html
[f][12]
[g]http://genome. jgi-psf.org/ poptr1_1/poptr1_1.info.html
hhttp://plantta.tigr.org/cgi-bin/plantta_release.pl
[i][13]
[j]http://www.phytozome.net/sorghum
[k]http://plantta.tigr.org/cgi-bin/plantta_release.pl
[l]http://castor bean.tigr.org
[m]Pablo Rabinowicz (University of Maryland), 2007
[n][15]
[o]http://plantgdb.org, but ming et al. [15] reported 16,362 unique est assemblies
[p]http://mips. gsf.de/proj/plant/jsf/medi/index.jsp
[q]Focus on gene-rich euchromatin, http://www.medicago.org/ genome/genome_stats.php
[r]Klaus Mayer (mips/ibi inst. for bioinformatics), 2007

evidence, i.e. direct assays, expression patterns, mutant phenotypes, physical interaction, or genetic interaction (http://www.arabidopsis.org/portals/genAnnotation/functional_annotation/go.jsp).

In order to describe the (putative) function(s) of genes and group them in broader categories, systems of hierarchical, but cross-connected descriptive terms have been established. Examples are the GeneOntology (GO; http://www.geneontology.org) and the KEGG BRITE (http://www.genome.ad.jp/kegg/brite.html) systems. For example, within the GO system, phenylalanine ammonia lyase (PAL) catalyzes a 'phenylpropanoid biosynthetic process', which is a 'secondary metabolic process', which is a 'metabolic process', which is a 'biological process'. In the KEGG BRITE 'Enzymes' database, PAL encodes 'phenylalanine ammonia-lyase', which is an 'ammonia-lyase', which is a 'carbon-nitrogen lyase', which is a 'lyase'. Such systems are crucial to place gene information into a functional context. Additional information on putative genes comes from the presence of conserved protein domains. The InterPro database (http://www.ebi.ac.uk/interpro) combines protein domain information from many different protein structural tools like PROSITE,

Table 2 Selected plant genome information portals

Portal	Species	Selected features	URL (http://…)
TAIR: The Arabidopsis Information Resource	*Arabidopsis*	Genome browser; gene models; cDNAs/EST mapping; insertion mutants; polymorphisms; seed and DNA stocks including phenotype information; functional annotations including evidence; literature references; metabolic pathway mapping; community annotated gene families; genetic and physical maps; search for sequence similarity (BLAST), gene name or function, protein domains, or genetic markers	arabidopsis.org
RAP-DB: Rice Annotation Project Database	*Oryza*	Genome browser, gene models, cDNAs/EST mapping; functional annotations; insertion mutants; homologs from other monocots	rapdb.dna.affrc.go.jp
MIPSPlantsDB: Munich Information Center for Protein Sequences Plants Databases	*Arabidopsis*, *Zea*, *Oryza*, *Medicago*, *Lotus*	Genome browser; gene models; cDNAs; functional annotations; literature references, protein domains; BLAST; syntenic regions in Medicago, lotus, rice, and maize; similar genes in other organisms (SIMAP), links to external images of mutants, reporter gene lines	mips.gsf.de/proj/plant/jsf/index.jsp
TIGR: J. Craig Venter Institute (The Institute for Genomic Research) Plant Genomics	*Arabidopsis*, *Oryza*, *Zea*, *Ricinus*, *Medicago*, *Solanum*, *Triticum*, others	Genome browser, gene models with functional annotations, EST/cDNA assemblies and mapping, insertion mutants (rice/Arabidopsis); protein features, EST frequency data, BLAST; community annotated gene families, synteny information; plant transcript assemblies	tigr.org/plantProjects.shtml
JGI: Department of Energy Joint Genome Institute	*Populus*, *Physcomitrella*, *Chlamydomonas*, *Ostreococcus*, *Volvox*	Genome Browser with diverse gene models, functional annotation, EST/cDNA mapping (also from other species), BLAST, search by functional category or protein domains	genome.jgi-psf.org/ euk_home.html
Grape Genome Browser	*Vitis*	Genome browser with gene models; EST/cDNA mapping (also from other species); automated functional predictions; BLAST	www.genoscope.cns.fr/externe/English/Projets/Projet_ML
The Hawaii Papaya Genome Project	*Carica*	Genome browser, not yet released for public access (Sept. 2008)	aspgp.mhpcc.hawaii.edu/papaya

(continued)

Table 2 (continued)

Portal	Species	Selected features	URL (http://…)
Gramene: A Resource for Comparative Grass genomics	*Oryza*, *Zea*, other *Poaceae*	Genome browser, BLAST search, rice-maize synteny, cDNAs/EST mapping (also from other Poaceae); functional annotations; genetic markers, quantitative trait loci, biochemical pathways	gramene.org
Phytozome	*Sorghum*, *Oryza*, *Arabidopsis*, *Populus*, *Physcomitrella*	Genome Browser, BLAST search, diverse functional annotations, clusters of homologous genes with alignments.	phytozome.net
PlantGDB: Resource for Plant Comparative Genomics	*Arabidopsis*, *Oryza*, *Zea*, *Hordeum*, *Medicago*, *Lotus*, *Populus*, *Solanum*, *Glycine*, *Brassica*, *Triticum*, *Sorghum*, others	EST-contigs from more than 100 plant species, genome browser for 12 species; gene models; transcript mapping; BLAST.	plantgdb.org

PRINTS, SMART, Pfam, and ProDom [17]. For example, the InterPro domain IPR005922 ('Phenylalanine ammonia-lyase') was found in 177 protein sequences from higher plants (http://www.ebi.ac.uk-/interpro/IEntry?ac = IPR005922).

These genome sequencing and annotation projects have revealed that many gene families encoding enzymes that potentially act in plant natural product biosynthesis are frequently drastically enlarged compared to mosses and green algae, but also in comparison to other eukaryotes such as fungi and mammals (Table 3). These gene families encode enzymes acting in diverse metabolic pathways. The largest code for cytochrome P450 monooxygenases and 2-oxoglutarate/iron (II) dioxygenases (Table 3), which catalyze a wide range of oxygenation reactions in plants [18]. Also very large is the family of the UDP-glycosyl transferases (UGT), which act on a wide array of compounds, to produce glycosylated natural products to modulate biodisponibility, storage and transport [19]. Some databases dedicated to the functional and structural annotation of gene families related to plant natural products have been established (Table 4). The large size of these gene families, and the fact that only very few of these have been characterized biochemically, highlights the vast diversity of plant natural product biosynthesis and shows that our understanding of the underlying pathways is just at its infancy. The first challenge of the postgenomic era is to assign biochemical, physiological and ecological roles to these "orphan" genes and to draw a more comprehensive map of plant metabolism.

Selection and Functional Characterization of Candidate Genes

The selection of candidate genes for functional characterization may either be guided by a pathway, by a biological function (e.g. deduced from a pattern of expression), or by a gene family of interest. In each case the strategy for a functional characterization is similar, and includes the identification of a candidate gene followed by molecular and biochemical characterization of the corresponding protein and mutant plants (Fig. 1). Candidate selection may either be based on sequence similarity (functional analysis of closely related paralogues in gene families) or employs approaches that are based on transcript levels detected *in planta* via transcriptome analysis or EST sequencing. Functional characterizations then usually is based on reverse genetics, which consists of eliminating, decreasing or enhancing gene expression and observing the resulting phenotypes at various level, and reverse biochemistry, which employs heterologous expression of protein (families) followed by testing of biological or enzymatic activity.

Similarity Based Approaches

In similarity based approaches, genome (annotation) and transcript assembly databases are searched using sequence information of characterized proteins of interest. Based on structural features and phylogenetic reconstructions, groups of genes likely encoding the same or highly similar activities can be found from diverse plant species. Such analyses of orthologous genes broaden our knowledge about the encoded enzymes in terms of their biochemical properties, which may explain the broad range of unique chemical makeups in different evolutionary plant lineages and even within individual species. This is well exemplified by the cyanogenic glucoside pathway, in which functional analysis of homologous genes in different taxa revealed evolution of specific metabolic branches derived from different amino acids [20]. It is also well illustrated by the different classes of phylogenetically related P450 enzymes involved in oxylipin metabolism to

Table 3 Size of gene families encoding potential functions in plant natural product biosynthesis from selected organisms with complete genome sequences

Gene Family	InterPro domain[a]	Number of genes with domain in species						
		Arath[b]	Orysa[c]	Poptr[d]	Phypa[e]	Chlre[f]	Sacce[g]	Homsa[h]
Acyl transferases	IPR003480	64	127	111	14	0	1	0
UDP glycosyltransferases	IPR002213	129	254	138	12	3	1	17
O-methyltransferase, family 2	IPR001077	20	34	40	3	2	0	2
Carboxyl methyltransferase	IPR005299	27	20	27	3	0	0	0
NAD-dependent dehydratase	IPR001509	86	90	30	33	9	9	21
Glutathione *S*-transferase	IPR010987/IPR004045	81	93	93	36	18	5	23
Glycoside hydrolase, family 1	IPR001360	62	36	56	13	4	2	9
Alcohol dehydrogenase	IPR002085	56	50	67	35	19	0	0
Aldehyde dehydrogenase	IPR002086	22	20	69	64	46	12	46
AMP-dependent ligase	IPR000873	58	44	77	34	23	11	32
Terpene synthase	IPR001906	34	30	52	1	0	0	0
Cytochrome P450 monooxygenase	IPR001128	273	314	362	74	79	4	77
2-oxo-/Fe(II) dioxygenase	IPR005123	147	103	154	63	23	1	19
Haem peroxidase	IPR002016	96	143	155	88	29	7	15
Ascorbate peroxidase	IPR002207	18	11	12	7	6	1	0

[a]http://www.ebi.ac.uk/interpro; the presence of the given interpro domains were searched against genome annotations[b-h], and the number of gene models found to contain the domain is given

[b]http://www.arabidopsis.org/servlets/search?action = new_search&type = protein (*arabidopsis thaliana* [tair7])

[c]http://rapdb.dna.affrc.go.jp/tools/search (*oryza sativa*, [rap release 2])

[d]http://genome. jgi-psf.org:8080/annotator/servlet/jgi.annotation.annotation?pdb = poptr1_1 (*populus trichocarpa* [jgi v1.1])

[e]http://genome.jgi-psf.org:8080/annotator/servlet/jgi.annotation.annotation?pdb = phypa1_1 (*physcomitrella patens* [jgi v1.1]

[f]http://genome.jgi-psf.org:8080/annotator/servlet/jgi. annotation.annotation?pdb = chlre3 (*chlamydomonas reinhartii* [jgi v3.0])

[g]http://www.ensembl. org/biomart/martview (*saccharomyces cerevisiae* [ensemble 47, sgd1.01])

[h]http://www.ensembl. org/biomart/martview (*homo sapiens* [ensemble 47, ncbi36])

Table 4 Gene family information portals related to natural product metabolism

Resource title	Description	URL (http://...)
CAZy: Carbohydrate-Active enZymes	Protein sequence and structure links for glycoside hydrolases, glycosyl transferase, and other carbohydrate-active enzymes from all species	www.cazy.org
Arabidopsis Glycosyltransferase Family 1	Alignments, phylogenies, and genome maps of UGPs from *Arabidopsis thaliana*	www.p450.kvl.dk/UGT.shtml
Cytochrome P450 Homepage	Collection of P450 sequences from all species	drnelson.utmem.edu/CytochromeP450.html
Arabidopsis Cytochromes P450	Gene and protein sequences, alignments, phylogenetic trees, and intron maps of P450s from *Arabidopsis thaliana*	www.p450.kvl.dk/p450.shtml
P450s in Plants	Functions of P450s from plants, reference links	members.shaw.ca/P450sinPlants
CYPedia: Cytochrome P450 Expression Database using *Arabidopsis*	Functions of Arabidopsis P450, large scale expression profiles, and pathway predictions based on co-expression analyses	ibmp.u-strasbg.fr/~CYPedia
PeroxyBase: The peroxidase database	Curated collection of haem and non haem peroxidases from all species; text and sequence search; cellular localization, tissue types, inducers, and repressor	peroxibase.isb-sib.ch
The *Arabidopsis* Lipid Gene Database	Annotation and structures of genes involved in acyl lipid metabolism including oxylipin and wax metabolism, evidence for activity, pathway schemes, tissue expression profiling	lipids.plantbiology.msu.edu/
AMPL: Arabidopsis Membrane Protein Library	Structural and functional annotation of membrane proteins including transporters, expression data	wardlab.cbs.umn.edu/arabidopsis/
Cell Wall Navigator	Structural annotation of gene families involved in cell wall generation (lignin biosynthesis excluded)	bioweb.ucr.edu/Cellwall/

generate signaling compounds or precursors for biopolymers such as cutin or suberin [21, 22]. A striking example is the systematic analysis of the CYP90 and CYP85 gene families that was a major step in the elucidation of the brassinosteroid pathway [23, 24]. Diversification of a gene family within a species does not necessarily imply different biochemical activities, but isoforms may channel products into divergent subpathways. Divergent *Arabidopsis* isoforms encoding 4-coumarate:CoA ligase appear to diverge precursors into the flavonoid and monolignol pathways, respectively [25, 26]. The analysis of related gene family members may also identify novel roles. For example, members of the cytochrome P450 subfamily CYP79 were known to be involved in cyanogenic glucoside biosynthesis. It was hypothesized that members of the same subfamily may function in glucosinolate biosynthesis [20]. This was validated by a yeast complementation screen [27]. Expression of recombinant enzymes and analysis of knock-out and over-expression lines then showed that several *Arabidopsis* CYP79 family members encode divergent functions in glucosinolate biosynthesis [27–30].

Gene Expression Based Approaches

In Plants with a Sequenced Genome

The primary structure of genes allows classification of the encoded proteins in broad functional categories (e.g. P450 monooxygenases, dioxygenases, glycosyltransferases, *O*-methyltransferases, ABC transporters, etc.), but, in most cases, this does not lead to hypotheses regarding the substrate(s) they are acting on. With the publication of more and more extensive expression data, it is now possible to generate hypotheses based on co-expression analysis [31]. The underlying idea is that genes acting in the same pathway share a similar expression profile. This has been shown for gene sets involved in secondary cell wall biosynthesis as well as for the shikimate and phenylpropanoid pathways [32–35]. In particular for *Arabidopsis*, thousands of microarray expression data sets are available, which have been used to generate web based expression analysis tools. These include 'digital Northern' and 'electronic Fluorescence Protein' browsers, which can be used to visualize the expression profile of individual genes (Table 5). In extension, tools have been developed to compare expression patterns of two genes of interest or to identify other genes co-expressed with a bait gene of interest [32, 36, 37]. Based on the functional annotation of co-expressed genes (and the expression pattern *per se*), a hypothesis regarding the biochemical pathway and the physiological context the bait acts in can be generated. Employing this approach, a CoA ligase-like gene was identified that showed a highly similar expression profile with OPR3 and many other genes encoding enzymes of jasmonate biosynthesis. Biochemical characterization of the recombinant protein and the reduced ability of knock-out plants to produce jasmonate showed that it codes for a CoA-activating enzyme in the octadecanoid pathway [38]. Co-expression analysis was also employed successfully with larger gene families. Gachon et al. [34] compared the expression patterns of glycosyltransferase genes with characterized genes from *Arabidopsis* involved in the shikimate, phenylpropanoid, flavonoid, and indole pathways, and identified strong candidates for glycosyltransferases involved in glucosinolate and anthocyanin metabolism. The latter, UGT75C1, was at the same time found up-regulated in plants over-expressing the *PAP1* MYB transcription factor, a known transcriptional regulator of flavonoid biosynthesis, and metabolic profiles of *ugt75c1* knock-out plants indicated a lack of 5-glucosylation activity of anthocyanins [39]. Using a set of structural and regulatory genes involved in flavonoid biosynthesis as baits, seven additional UGTs were identified (including two previously characterized UGTs) as involved in flavonoid metabolism [40]. One of these, UGT89C1, was shown to have rhamnosyltransferase activity with flavonol-glycosides when expressed in *E. coli*, and *ugt89c1* knock-out lines lack the major kaempferol-rhamnosides [40]. An extensive analysis of the expression of Arabidopsis P450 genes compared to other genes of plant metabolism was recently made available as a website [41]. This analysis predicts new, and sometimes unexpected, functions for P450 genes, and provides lists of coexpressed genes and predicted pathways. Some of these predictions were just validated by Field and Osbourn [42] that used another strategy based on physical gene clustering to identify P450 genes in triterpenoid metabolism. The co-expression approach was also successfully extended to include transcriptional regulators of metabolic genes. *MYB28* and *MYB29* are co-expressed with genes involved in aliphatic glucosinolate biosynthesis [43] and *myb28* knock-out mutants displayed a reduced content of aliphatic glucosinolates and repressed expression of the biosynthetic genes. Over-expression of *MYB28* had the reverse effect, showing that this transcription factor is indeed a positive regulator of aliphatic glucosinolate biosynthesis.

These examples show that this type of predictive approach, which has the potential to highlight whole pathways including regulatory

Table 5 Web based gene expression analysis tools

Resource title	Description	URL
BAR: The Bio-Array Resource for *Arabidopsis* Functional Genomics	Digital Northern based on AtGenExpress datasets; graphical visualization of organs, tissues, and treatments with overlaid expression data (eFP browser); co-expression analysis tool comparing all genes on array with a given gene (Expression Angler), other tools	bar.utoronto.ca
Genevestigator	Digital Northern tools with summaries of expression in organ and tissues, during development, in response to stimuli, and in mutants; based on Affymetrix arrays for *Arabidopsis thaliana* (>3,000 arrays) and *Hordeum vulgare* (>700 arrays); additional tools with subscription	www.genevestigator.ethz.ch
ATTED II: Arabidopsis thaliana trans-factor and cis-element prediction database	Visualization of gene expression based on the Arabidopsis AtGenExpress data; co-expression analysis with visualization as network; promoter analysis of co-expressed genes	www.atted.bio.titech.ac.jp/
AVT: AtGenExpress Visualization Tool	Digital Northern database based on the AtGenExpress data set for Arabidopsis	jsp.weigelworld.org/expviz/expviz.jsp
NASC Array	Data repository for Arabidopsis Affymetrix arrays; digital Northern, other array data-mining tools	affymetrix.arabidopsis.info/narrays/experimentbrowse.pl
GEO: Gene Expression Omnibus	Data repository for microarrays from all organisms; queries by gene or datasets; cluster analysis, expression visualization in each data set	www.ncbi.nlm.nih.gov/geo/
ArrayExpress	Data repository from all organisms, visualization of expression in individual data sets	www.ebi.ac.uk/microarray-as/aer

genes, is one of the most promising for filling open gaps in plant metabolism.

When the Genome is not Sequenced...

When the genome of the plant harboring a candidate pathway is not sequenced and annotated, forward genetics remains a useful approach for plants with well documented collections of mutants and genetic variants such as crop plants. In other cases, a suitable shortcut to the elucidation and engineering of the pathway is usually the creation and analysis of ESTs from plant tissues treated so as to selectively induce the relevant metabolism, or from the plant tissues accumulating the compound of interest. When relevant, EST collections can be enriched in the genes of interest via subtraction of house keeping or redundant genes. The development of new sequencing technologies (e.g. pyrosequencing) will encourage new EST projects in diverse plants for gene function discovery. EST approaches have been successful, for example, in characterizing genes and enzymes involved in the biosynthesis of essential oil components in glandular trichomes of aromatic herbs such as mint or sweet basil [44, 45] or of floral fragrances, for example from rose petals [46]. Similar strategies were used to explore biosynthesis of bioactive drugs in medicinal plants, for example anticancer taxoids in jasmonate treated

Taxus cuspidata cells [47], ginsenosides in similarly induced *Panax ginseng* hairy roots cultures [48], or antimalarial artemisinin in *Artemisia annua* trichomes [49–51]. In the latter case, candidate gene selection was also supported by conservation of ESTs from different *Asteraceae*.

Reverse Genetics

Reverse genetics aims at identifying the functions of genes *in planta*. Transgenic plants are most often generated by *Agrobacterium*-mediated T-DNA insertion. Transgenic lines can then be characterized phenotypically at the morphological, molecular or biochemical level. Increasing the level of expression is achieved by the construction of over-expression lines, where the target gene is fused to a promoter that drives strong ectopic, tissue specific or conditional expression. Decreasing or abolishing the expression of a target gene can be achieved via gene silencing (also referred to as RNA interference) or via T-DNA/transposon insertion. RNA interference is obtained by plant transformation with constructs triggering *in planta* the production of specific short double-stranded RNAs. The latter suppresses expression of a target protein by stimulating the specific degradation of the target mRNA [52]. An optimal design strategy for RNA interference constructs can be found for example at the 'WMD2 – Web MicroRNA Designer' webpage (http://wmd2.weigelworld.org/cgi-bin/mirnatools.pl?page = 1). The advantage of insertion compared to RNA interference is more specific inactivation of a single gene rather than of a group of closely related genes, and, when insertion occurs in the coding sequence or near upstream, complete gene inactivation. Conversely, interference might be preferred for simultaneous inactivation of redundant genes. In addition, while insertion may lead to extreme phenotypes in the case of essential genes, silencing can trigger a range of more exploitable alterations. Large knock-out collections with characterized insertion locations are now available for a few plants, such as *Arabidopsis* or rice [53–56], which can be analyzed for developmental, physiological or chemical phenotypes. Chemical profiling of appropriate tissues of the mutant plants might then point to the absence of an end-pathway product and/or the accumulation of precursor compounds (e.g. [42]). Phenotypes and metabolic profiles must, however, be interpreted with care, since disruption in one branch pathway can lead to far reaching perturbations in metabolic fluxes, which may blur the view on modification of the target pathway [8, 57].

The aforementioned methods, however, do not apply to all types of plants. Acute need for genetic tools in crop or pharmaceutical plants has thus led to the development of more generic approaches for large scale gene inactivation. Targeting Induced Local Lesions in Genomes (TILLING) combines chemical or radiation mutagenesis with mutation screens of pooled PCR products [58]. TILLING mutants carry point missense, nonsense or small deletion alleles, which can be selected on the basis of a target gene, or of a desired phenotype. TILLING has the (socio-economic) advantage not to involve transgenic manipulation, to be applicable to any plant and to generate allelic series of mutations including hypomorphic alleles. High density of chemical mutations, however, requires extensive backcrosses to ensure single mutation [59]. Several TILLING-based projects with focus on secondary metabolism have been initiated.

Reverse Biochemistry

Biochemical activities of candidate genes may also be tested directly using heterologous expression of the cDNA. Recombinant proteins can then be used for testing enzyme activity. Using collections of recombinant proteins, this strategy can be set up for medium to high throughput assays with collections of potential substrates. The large number of gene family members, and the even larger group of potential substrates, however limit in most cases a comprehensive screening. This approach thus needs to be complemented by the approaches described above, to guide the selection of candidate genes and of the class of substrates.

In general, full length cDNA for individual target genes or gene family members may be obtained via RT-PCR from mRNA or from large scale projects [60], some of which are focused on the isolation of open reading frames (ORFs). Such ORFeome collections [60, 61], http://urgv.evry.inra.fr/orfeome, are available for transfer into divergent destination vectors designed for plant transformation, or for production of recombinant proteins from appropriate expression systems, such as bacterial, yeast, or insect cells. The latter may then be used to generate libraries of recombinant proteins of interest. The choice of the appropriate expression system depends on the class of protein analyzed. Soluble proteins are usually expressed in *E. coli*, membrane proteins in yeast or insect cells. Functional screening of transporters often requires expression in *Xenopus* eggs. Screening procedures can be adjusted to each protein activity. But it is important to keep in mind that the efficiency of generating and maintaining active enzymes or transporters may vary drastically within closely related families, and frequently relies on manual optimization of the expression systems. Comprehensive approaches are thus less straightforward than theoretically predicted. Nonetheless, systematic expression followed by individual biochemical characterization has been very successful for several gene families. For example, Katsuyama et al. [62] screened the rice genome for type II polyketide synthases (PKS) and tested more than 30 recombinant enzymes for activity with 4-coumaroyl-CoA and malonyl-CoA using thin layer chromatography. This screening led to the characterization of a novel PKS that is capable of synthesizing curcuminoids, which are characteristic polyphenols from turmeric (*Curcuma longa*), but which have not yet been found in rice. Other examples take advantage of optimized expression and enzyme activity testing systems to systematically assign functions to complete gene families. Although *Arabidopsis* was originally considered to be basically devoid of terpenoids, the genome sequence revealed the existence 32 putative mono-/ sesqui-/ diterpene synthase (*TPS*) genes and eleven genes encoding divergent triterpene synthases (TTPS). To date, nine of the TPS were shown to synthesize more than 15 mono- and 18 sequiterpenoids, and most of these were also found *in planta* [63–68]. Likewise, 10 of the 11 *TTPS* genes have been characterized to produce almost 40 divergent triterpene structures, most of which, though, have not yet been found in *Arabidopsis* via metabolic profiling experiments. This has mainly been achieved by exploiting an optimized yeast expression system, which accumulates the substrate of these enzymes, oxidosqualene [69–75]. These successes are also due to the fact that a limited number of canonical precursors (present in optimized yeast strains) are used as substrates by these enzymes. Functional screening is more challenging when enzyme families use diverse sets of plant-specific substrates. The major challenge in this approach is that substrates are chemically diverse, most often not predictable, nor commercially available. To overcome the problem resulting from this chemical diversity, simple activity detection methods are required, which are independent of the substrate used. For example, a collection of more than 100 recombinant glycosyltransferases (UGTs) has been expressed in *E. coli* to identify diverse glucose acceptors ranging from benzoic acid derivatives to the phytohormones cytokinin and abscisic acid [76–78]. An *in vivo* system has been developed that uses modified *E. coli* cells accumulating the precursor UDP-glucose. These cells were transformed with the UGT expression constructs and fed with potential substrates. The produced glycosides were then cleaved enzymatically and the released glucose was detected colorimetrically [79].

Although involving a more complex experimental set-up, plants (whole plants, plant cells, leaf infiltration) can also be considered as platform for recombinant expression and screening. This might be particularly useful since they provide the appropriate precursors and functional context for the genes and enzymes to be characterized. An example of such an approach is provided by the work of Kruse et al. [80], who tested Arabidopsis expressed cytochromes P450 to identify a 8-methoxypsoralen metabolizing enzyme. This also led to the demonstration that the hydroxylation product was further glycosylated by the plant.

Another example of substrate independent medium or large scale enzyme assay is the ATP depletion test that can be employed to detect activity of adenylate forming enzymes or of any ATP dependent enzyme, such as coenzymeA (CoA) ligases or synthetases. An ATP dependent reaction is carried out using a number of potential substrates. A second reaction then uses the ATP not consumed in the first reaction to form a luminescent product. If ATP was consumed in the first reaction this leads to reduced luminescence generated by the second assay. Employing this approach, Schneider et al. [81] identified medium chain fatty acids, but also the jasmonate precursor 12-oxo-phytodienoic acid as an *in vitro* substrate for a 4-coumarate:CoA-ligase-like gene.

Cytochrome P450 enzymes usually require co-expression with plant P450 reductase in the membranes of yeast or insect cells. The catalytic activity of P450s or of any oxygen consuming enzyme (e.g. dioxygenases) can be theoretically monitored by measuring the depletion of oxygen during the reaction. Based on this assumption, a microtiter plate assay has been recently designed, using an oxygen-quenched fluorochrome, detecting oxygen consumption during the reaction instead of conversion of the specific substrates of each enzyme [82]. It led to the demonstration that an orphan Arabidopsis P450, CYP72A8, catalyzes the hydroxylation of quinidine, a major component of the most common antimalarial drugs. This approach is however effective only for well coupled reactions and can be hampered by the presence of natural substrates (e.g. lipids) in the recombinant cell membranes. Such examples of substrate independent assay show the potential of reverse biochemistry.

The Next Level: Proteomics and Metabolomics

Proteins and their modifications, localizations and interactions represent another level of complexity connecting transcription and translation of metabolic pathway genes with the respective natural compounds.

Quantitative changes in transcript often show only minor correlation to protein levels due to numerous post-transcriptional regulation mechanisms, which can be as low as 20–40% in yeast [83]. Systematic comparative studies in plants remain rare, yet it has been demonstrated that genome or transcriptome information cannot be extrapolated to the proteome. In *Medicago* (Medicago truncatula) a correlation of EST to protein abundance was estimated to be 50% [84]. In *Arabidopsis,* Jones et al. [85] reported widespread changes to the proteome during the defense response before significant transcriptional reprogramming occurred. Compared to the proteome, the plant metabolome confronts the researcher with even more challenges. Within the metabolome there is an enormous range of chemical and physical properties, concentrations, absolute number of the compounds, and unlike the proteome, the metabolome cannot be mapped onto the genome.

Proteomics: the Connecting Link

Proteomics, in its core (idealistic) definition is the quantitative analysis of the full set of proteins of a given organism. Complexity of the sample and diversity of the proteins (in concentration and physico-chemical properties) prevent comprehensive analysis of the full proteome in a standard experiment. More pragmatically, the term proteomics applies to the broad spectrum of large scale protein biochemistry, including analysis of protein structure, post-translational modifications, protein localization, protein-protein interaction and complex formation. Plant proteomics has received considerable buoyancy from rapidly expanding resources in genomics, with increasing numbers of genome sequences and transcript assemblies (Table 2) facilitating identification of proteins in model plants. On the other hand, in non-model plants, proteomics exploiting differences between samples (e.g. comparisons between treatment and control, developmental stages or in genotypic variation) have been successfully applied.

Proteomic analyses commonly consist of four steps: (i) Sample preparation, fractionation and protein extraction; (ii) protein separation, detection and quantification; (iii) protein characterization and identification; and (iv) integration of results into the respective context. A simplified general workflow outlining key steps and technologies is shown in Fig. 2. For a comprehensive, more in-depth coverage of the numerous plant proteomic approaches we refer

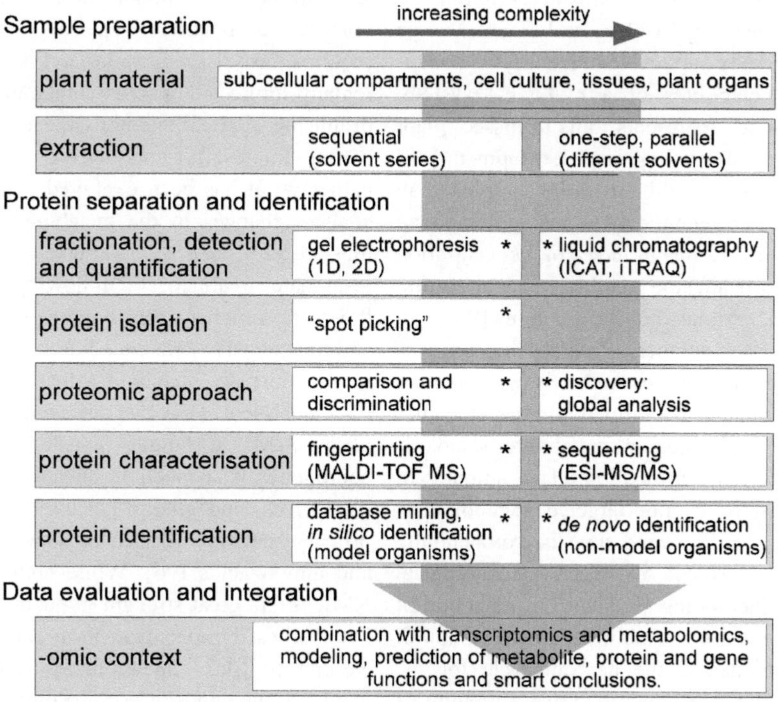

Fig. 2 Schematic workflow in plant proteomic studies. Methods on each level are interchangeable and depend on the experimental approach. Asterisks denote selected examples further described in the text or glossary. Descriptions of key technologies can be found in the Table 6

to recent reviews providing an overview of technologies, plant protein preparation, general advances, and practical details [86–88]. In short, the sheer complexity and dynamic range of the mixture of proteins obtained from plant samples threatens to overpower proteomics separation and detection methods. Therefore, fractionation steps are crucial to deconvolve the samples, either before or throughout the protein extraction (by protein solubility, chromatography, organelle separation or subcellular compartment enrichment), or downstream, in the protein separation stage (Fig. 2)

Metabolomics: The Highest Level of Complexity

As for proteomics, a realistic definition of metabolomics cannot claim to be nonselective and universally applicable to identify and quantify all metabolites of a biological system. Metabolomics represents tools to dissect plant metabolism, physiology and development [89] in an analytical field far more complex than proteomics or genomics. The metabolome may cover up to 12 magnitudes of concentration (fmol-mmol) and the number of plant natural products is estimated in the six digits [90]. Not only the sheer number of metabolites remains challenging, but the researcher is confronted at the same time with a limited number of available isotopically-labeled metabolite standards for absolute quantification. The amount of data generated by non-targeted metabolomics approaches at the same time is exponentially increasing, driven by recent technological advances facilitating simultaneous detection of a large number of metabolites. Identification of the metabolites is the next obstacle for the metabolomic researcher, who may find himself in a similar position as early genomics researchers. Despite its challenges, the metabolome has received a lot of attention as it represents the most basal level in a biological system, reflecting developmental and adaptive aspects in a direct manner. Hence, metabolomics has been suggested to be the greatest "omics" of all [91]. The general metabolomic methodology of sample preparation, extraction, separation and detection as well as an introduction into the major technologies are comprehensively reviewed by Hall [92] and Hagel and Facchini [93]. Again, fractionation steps together with a combination of experimental approaches were the key to untangle complex mixtures and to balance sensitivity, throughput and identification power. A promising method for obtaining global snapshots of a broad spectrum of metabolites is high resolution nuclear magnetic resonance (NMR) spectroscopy, which has the advantage of providing information on the relative amounts of the metabolites and their chemical structures, without limitation due to volatility or polarity [94]. This method still requires fairly large samples and its sensitivity remains limited compared to other methods, but promising applications are expected for comparing samples, flux analyses and elucidation of pathways. It has been exploited to effectively analyze changes in the metabolome of field mustard (*Brassica rapa*) after elicitation with jasmonate treatment or following herbivory [95–97]. A rather large number of replicates and an extensive reference library were necessary to identify metabolites that accumulated after treatment. However, the unbiased nature of the NMR technique revealed changes in compounds from phenolic metabolism, indole derivatives, and glucosinolates in addition to primary metabolites such as sucrose, glucose and amino acids [96]. While high resolution NMR offers great strength in quantification of metabolites, Fourier Transform Ion Cyclotron Resonance Mass Spectrometry (FTICR-MS) resides at the opposite extreme of metabolomic applications. FTICR-MS has the prospective to become a powerful qualitative technique for high-throughput metabolomic analysis, as its mass resolution, detection sensitivity and the ultra high mass accuracy is exceptional. Aharoni et al. demonstrated the power of the method by analyzing crude strawberry fruit and tobacco flower extracts, in which metabolites were classified using the elemental composition based on the accurate mass [98].

Combination of FTICR-MS with other advanced technologies has helped to address some limitations. Iijama et al. studied tomato fruit metabolites with LC-FTICR-MS to separate isomers of the same elemental composition [99]. Using mass spectral features, together with predicted molecular formulae and putative structures, 869 metabolites were tentatively identified, of which 70 and 93 were assigned to the flavonoid and glycoalkaloid groups, respectively.

Despite substantial gains in popularity of FTICR-MS for plant metabolic profiling this analytical method is still far from being established as routine technique and public resources, such as the KNApSAcK metabolite database with accurate metabolite masses will require substantial input from the metabolomics community in the future (22,573 metabolite entries, updated 2008/06/20; http://kanaya.naist.jp/KNApSAcK/).

Examples of Integrative Approaches to Study Specific Metabolite Classes in Model Plant Systems

Prominent examples for model systems of natural product classes that were analyzed on proteome and metabolome levels are from the *Fabaceae* (legume family) and the *Brassicaceae* (mustard family). These are outlined below, together with examples for studies in trees.

The model legume *Medicago* has attracted considerable attention for two classes of secondary metabolites: isoflavonoids and triterpene saponins. To cover a large portion of metabolomic changes following elicitation (biotic and abiotic) of *Medicago* cell culture, Broeckling et al. [100] used a combination of techniques (gas chromatography-mass spectrometry [GC-MS], capillary electrophoresis, and liquid chromatography-mass spectrometry [LC-MS]). Beside an increase in saponin and isoflavonoid levels, an impact of the elicitation on primary metabolism, both distant and proximal to secondary metabolic branch points was reported. As one of the earliest and most extensive plant proteome projects, a survey of alfalfa (*Medicago sativa*) organ-/tissue-specific proteomes was performed applying LC-MS/MS to proteins extracted from leaves, stems, roots, flowers, seed pods, and cell suspension cultures and separated by 2-dimensional SDS-polyacrylamide gel electrophoresis (2D-SDS-PAGE) [84]. A range of enzymes involved in secondary metabolism were identified in stems, flowers and in roots, where they constituted up to 8% of the identified proteins. Examples include cinnamoyl-CoA reductase (phenylpropanoids), and proteins of the flavonoid/isoflavonoid biosynthetic pathway.

Glucosinolates represent a class of plant defense thioglucoside compounds which are mainly produced by members of the mustard family, including the model plant *Arabidopsis*. Cruciferous vegetables are especially rich in these natural compounds, of great interest because of a variety of beneficial effects in the human diet, such as chemoprotection and anticancer activities [101]. Hirai et al. [102, 103] studied gene-to-metabolite networks in primary and secondary metabolism by combining transcript profiling with metabolite detection in sulfur deprived *Arabidopsis*. Using LC-FTICR-MS (Table 6) for non-targeted metabolic profiling and HPLC for targeted profiling 2,123 metabolites were detected. Based on transcript and metabolite accumulation, the authors suggested coordinated regulation of glucosinolates metabolism in response to sulfur deficiency. Furthermore, the approach allowed identification

Table 6 Glossary of separation and detection technologies

Name	Description
SDS-PAGE	Sodium Dodecyl Sulfate Polyacrylamide Gel Electrophoresis is a technique to separate proteins according to their physical properties, such as their molecular weight or their charge at a given pH. If used consecutively, a two-dimensional (2D-PAGE) separation of the protein extract can be achieved.
MALDI, MALDI-TOF and MALDI-LDI	Matrix-Assisted Laser Desorption/Ionization is a soft ionisation technique that deals well with thermolabile, non-volatile organic compounds and is used successfully in biochemical areas for the analysis of proteins, peptides, glycoproteins, oligosaccharides, and oligonucleotides, as well as for smaller metabolites. The sample is pre-mixed with a highly absorbing matrix, which transforms the laser energy into excitation energy for the sample, leading to sputtering of analyte and matrix ions from the surface of the mixture. MALDI-MS therefore represents a technology applicable for metabolite and protein analyses. In proteomics, MALDI Time-Of-Flight mass spectrometry allows measuring the masses of peptides resulting from an excised and digested protein spot from a 2D-PAGE gel. The resulting peptide mass fingerprint is used to assign the identity by querying against patterns predicted from cDNA, protein and genomic DNA datasets. MALDI MS and laser desorption/ionisation (LDI) imaging is a further development of this technique where the sample is moved on the x/y axis and a mass spectrum is generated at each position, creating a two-dimensional MS image of the metabolic profile of a sample.
ESI-MS/MS or LC-MS/MS	Electro-Spray Ionization/Liquid Chromatography tandem Mass Spectrometry technology finds applications in both proteomic and metabolomic fields. The electrospray ionization technique creates ions derived from the analyte without fragmentation. After the initial ionisation, precursor ions (peptides/metabolites) are separated in the first MS and selected target ions undergo then fragmentation before the separation in the second MS. Specifically in proteomics, ESI-MS/MS is used for direct protein sequencing by the two coupled MS separation steps. The secondary ions are analyzed after separation by MS2 yielding a characteristic pattern for the peptide, from which the sequence can be deduced.
ICAT	The Isotope Coded Affinity Tags method is used to compare and quantify relative protein levels between different samples. Proteins of one sample are labeled with a heavy, the other with a light isotope of the reagent. The samples are combined, digested and used in a first round of MS to quantify the relative abundances of each peptide pair (heavy and light). Subsequent fragmentation determines the peptide amino-acid sequences. Dunkley et al. [122] successfully applied ICAT for organelle proteome studies in Arabidopsis.

iTRAQ	Isobaric Tags for Relative and Absolute Quantitation represents recent technology advancement, where four specific reagents label peptides in up to four samples. While the labeled peptides have the same mass, each label fragments into a different sized, characteristic tag that can be identified, and quantified in the tandem mass spectra. In *Arabidopsis* iTRAQ has been used to map the organellar proteomes [123]. Another study applied iTRAQ to describe early changes to the phosphoproteome of *Arabidopsis* during defense response [124].
NMR	Classically, nuclear magnetic resonance spectroscopy was applied together with MS and other spectroscopic methods for the structure determination of low molecular weight metabolites. Lately NMR has become a promising tool method for the identification, structure elucidation and quantification of metabolites in mixtures. In metabolomics, structures, and relative abundance of metabolites are directly analysed by NMR without preceding derivatisation. The inherent disadvantages of NMR of relatively low resolution and overcrowding of signals in important regions of the spectra are at least partially conquered through two- and higher dimensional experiments and correlation techniques. Liang et al. [96, 125] used two-dimensional NMR spectroscopy to analyse metabolomic changes in *Brassica rapa* leaves after methyl jasmonate treatment. A combination of HPLC with a number of NMR experiments allowed Liang et al. to structurally identify and quantify five phenylpropanoids that accumulate after induction. Event though NMR has been applied for the analysis of the macromolecular structure of proteins, the technique is as of its intrinsic characteristics unsuitable for the analysis of mixtures of proteins.
FTICR-MS	Fourier transformed ion cyclotron resonance mass spectrometry is a newly emerging ultra-high resolution and accuracy technique. Ions are trapped in a magnetic field, accelerated in the cyclotron and the mass-to-charge ratio (m/z) of the ions is determined based on the cyclotron frequency. The path length of the cyclotron allows for separation of a large number of natural compounds or protein fragments per analysis. The first reported application in plants targeted strawberry fruits at different ripening stages [98].

21

of specific metabolite-to-gene networks and suggestions for gene-functions.

The complex regulation of xylogenesis (wood formation) in trees was also addressed on the proteome level in the gymnosperm maritime pine (*Pinus pinaster*) [104] and the angiosperm eucalyptus (*Eucalyptus gunii*) [105]. Both studies used 2D-LC-MS/MS to identify a range of proteins with the help of public data repositories. Many targets identified are involved in phenylpropanoid metabolism, especially in lignin biosynthesis, and associated proteins such as *S*-adenosylmethionine synthetase, which are playing an important role in the methylation of monolignol precursors. While metabolite profiling of Douglas-fir (*Pseudotsuga menziesii*) grown in the field showed that intermediates in the synthesis of aromatic amino acids, flavonoids, and a range of other secondary metabolites (e.g. shikimic acid, coniferin and quinic acid) accumulate in the developing xylem [106].

Systems Biology Approaches in Plants Without Large-Scale Genomic Resources

Pharmacologically active natural compounds frequently accumulate to only low levels in plants, which, combined with laborious extraction methods and limited source plant availability, results in exorbitant market prices. Diverse biotechnological approaches (such as conquering rate limiting steps, suppressing the shunt of intermediates in competitive pathways, reducing catabolism, and over-expression of regulatory genes) have great potential to lead to cost-effective production of natural compounds through large scale plant cell cultures and/or metabolic engineering. Yet, this requires knowledge of the biosynthetic pathways involved. Functional genomics approaches for most plant species producing the most active pharmacological agents are still in their infancy. Proteomics and metabolomics are thus key technologies in non-model plants for linking genes involved in the respective pathways with biosynthesis and accumulation of the natural compound. Huge efforts are presently invested in the elucidation of the biosynthetic pathways of anticancer (e.g. vinblastin in *Catharanthus roseus* or taxoids in *Taxus baccata*), antimalarial (artemisinin in *Artemisia annua*) drugs, or painkillers (e.g. morphine in the opium poppy) on different continents, supported by national and international programs.

Differential genomics has been successfully combined with metabolomics for gene discovery in plant secondary metabolism in a number of cases. Targeted metabolite analyses of tobacco cells (*Nicotiana tabacum*) treated with jasmonate were used to identify several defense compounds such as nicotine alkaloids and polyamines [107]. These analyses were correlated with gene transcription using cDNA-amplified fragment length polymorphisms (cDNA-AFLP) to identify candidate genes. This method provides quantitative expression profiles, does not require prior sequence information to be set up, and, most importantly, allows the identification of novel genes.

It was therefore applied to Madagascar periwinkle (*Catharanthus roseus*), a source of the tubulin-binding anti-mitotic drugs vinblastine and vincristine, which are used to treat various forms of cancer [108]. cDNA-AFLP was combined with metabolic profiling of elicited *C. roseus* cell cultures to identify a collection of not previously described transcript tags and metabolites associated with terpenoid indole alkaloids. Gene-to-gene and gene-to-metabolite networks were drawn based on correlations between expression profiles of 417 gene tags and the accumulation profiles of 178 metabolite peaks.

Opium poppy (*Papaver somniferum*), producing a large number of alkaloids including the narcotics morphine and codeine, may be the prime example for a non-model plant on the verge of becoming a model. Even though the vast majority of knowledge stems from classical molecular biological approaches, EST resources and a cDNA microarray platform has recently been developed. Together with proteomics, this will allow the development of systems biology approaches [109]. Decker et al. [110] and Ounaroon et al. [111] used proteomic analyses of *P. somniferum* latex (2D gel electrophoresis, microsequencing) to create proteomic maps. Beside enzymes from the primary metabolism, a codeinone reductase was identified based on homology to known isoforms [110]. In addition, *Papaver somniferum* sequences coding for reticuline 7-*O*-methyltransferase and norcoclaurine 6-*O*-methyltransferase were isolated based on peptide sequences and the respective methyl transfer enzymes of alkaloid biosynthesis were characterized [111].

The peltate glandular trichomes are highly specialized for production of a variety of metabolites, including phenylpropanoids and terpenoids and these cells can be isolated from the plant, allowing biochemical and molecular investigations of a single, fully differentiated cell type. Gang et al. established these gland cells of sweet basil (*Ocimum basilicum* L.) as model system to study plant metabolic processes and their regulation [112]. An EST database build from four basil lines with distinct product profiles provided the sequence foundation required for comparative proteomic studies [113]. Analysis of the proteomic data for the four lines showed a surprisingly low number of matches and limited abundance correlations between the EST and proteome data sets. As suggested by the authors, more research will be required to explain the biological significance of the findings and to reveal the complex levels of regulation of metabolic pathways.

The metabolome and metabolic changes are intrinsically linked with quality traits and economic value in crops. Rising concerns over loss of genetic diversity through selection and breeding, together with genomics-assisted breeding for crop improvement were recently driving initial integrated approaches in crop species. Schauer and Semel et al. [114, 115] determined by GC/MS the relative contents of 74 primary metabolites with known chemical structure in tomato introgression lines containing chromosome segments of a wild species in the genetic background of a cultivated variety. It was shown that 50% of the metabolites measured were associated with at least one yield-associated trait, however without a sequenced genome, a large number of complex QTLs were used for in-depth association studies.

In second generation biofuel crops (e.g. willow, poplar, *Miscanthus* and switchgrass) various approaches are pursued to increase efficiency in the production of cellulosic ethanol through reduction of the pre-processing costs [116]. Current research aims to reduce lignin formation, modify cellulose, increase biomass, and to produce degradation enzymes *in planta* [117]. Since for most species employed genomic resource are scarce, this young multidisciplinary area is in urgent need of metabolomic and proteomic platforms to survey the large and diverse array of varieties generated. The genomes of the recently sequenced poplar [12] and switchgrass (JGI, sequencing pending; http://www.jgi.doe.gov/sequencing/allinoneseqplans.php) will greatly assist with this task.

Challenges for the Future

Despite rapid methodological advances and benefits provided by technology transfer from the medical and animal field, practical limitations still persist, due for the most part to the complexity of plant metabolism and resulting problems in sample preparation, fractionation and protein/metabolite identification. Due to the enormous variety of plants, tissues and cell types, each with specific requirements and restrictions such as interfering metabolites (e.g. polysaccharides and polyphenols), proteomic approaches have still to be established and optimized on an individual plant, or even tissue basis.

The membrane-bound cytochrome P450 enzymes represent key players in secondary metabolite pathways as they catalyze vital, often rate-limiting steps. However, basic, hydrophobic and membrane-spanning proteins remain an intrinsic limitation for 2D-PAGE [118]. Adoption of gel free proteomic approaches such as iTRAQ (Table 6) may help to overcome this bottleneck.

The analysis from small cell- or tissue-specific samples provides other challenges. While transcript profiling has been successfully established from minuscule plant samples, proteome analyses from such samples remain rare. Schad et al. [119] provided a first example of protein identification based on LC-MS/MS using extracts isolated from *Arabidopsis* vascular bundles captured using laser micro dissection. Four hundred vascular bundles were shown to be sufficient for the identification of 33 specific proteins. The same group demonstrated feasibility of metabolite profiling of only 100 vascular bundles equivalent to not more than 5,000 cells, by standard GC-TOF-MS measurements. A total of 68 metabolites were identified [120]. MALDI imaging is a second, rapidly developing field that emerged from medical research and that can be used to address the spatial distribution of compounds in heterogenous tissues at cellular to multicellular resolution without the need of pre-analytical extraction and separation steps. Examples in plants are scarce, but Cha et al. [121] demonstrate feasibility of MALDI-MS imaging using colloidal graphite as assisting material to overcome some of the obstacles for plant metabolite imaging. It was possible to analyze accumulation of flavonoids in leaves, flowers and stems of Arabidopsis through MS ion selective imaging.

Global or focused analytical approaches, in particular NMR and FTICR-MS, can be coupled with plant feeding with isotope-labeled metabolic precursors for reconstruction of metabolic sequences, providing information on metabolic networks and dynamic monitoring of fluxes reorganization in response to environmental perturbations. Such dynamic approaches, usually referred to as "fluxomics", are a very promising field of investigation. They however add an additional level of complexity to the analysis of the data. Development of new bioinformatic tools is essential to efficiently deal with the increasing amount of information, statistic analyses and visualization of the correlations between genomic, proteomic and metabolic networks.

Building a complete inventory of plant proteins and metabolites remains a major challenge. It is now obvious that even in a model plant such as *Arabidopsis* this inventory is very far from comprehensive, and that only the most abundant/extractable constituents have been described. Due to an additional level of complexity, the complete description of the main plant polymers will take several decades. The last step for closing the loop will be to better understand how these small metabolites impact gene and protein expression/activity, and plant homeostasis.

References

1. Lepiniec, L., Debeaujon, I., Routaboul, J.M., Baudry, A., Pourcel, L., Nesi, N., and Caboche, M. 2006. Genetics and biochemistry of seed flavonoids. Annu Rev Plant Biol 57: 405–430.
2. Trojanowska, M.R., Osbourn, A.E., Daniels, M.J., and Threlfall, D.R. 2001. Investigation of avenacin-deficient mutants of *Avena strigosa*. Phytochem 56: 121–129.
3. Humphreys, J.M. and Chapple, C. 2002. Rewriting the lignin roadmap. Curr Opin Plant Biol 5: 224–229.
4. Nomura, T. and Bishop, G.J. 2006. Cytochrome P450s in plant steroid hormone synthesis and metabolism. Phytochem Rev 5: 421–432.
5. Takei, K., Yamaya, T., and Sakakibara, H. 2004. *Arabidopsis CYP735A1* and *CYP735A2* encode

cytokinin hydroxylases that catalyze the biosynthesis of trans-zeatin. J Biol Chem 279: 41866–41872.
6. Abdulrazzak, N., Pollet, B., Ehlting, J., Larsen, K., Asnaghi, C., Ronseau, S., Proux, C., Erhardt, M., Seltzer, V., Renou, J.P., Ullmann, P., Pauly, M., Lapierre, C., and Werck-Reichhart, D. 2006. A coumaroyl-ester-3-hydroxylase insertion mutant reveals the existence of nonredundant meta-hydroxylation pathways and essential roles for phenolic precursors in cell expansion and plant growth. Plant Physiol 140: 30–48.
7. Aubourg, S., Lecharny, A., and Bohlmann, J. 2002. Genomic analysis of the terpenoid synthase (*AtTPS*) gene family of *Arabidopsis thaliana*. Mol Genet Genomics 267: 730–745.
8. Hemm, M.R., Ruegger, M.O., and Chapple, C. 2003. The Arabidopsis ref2 mutant is defective in the gene encoding CYP83A1 and shows both phenylpropanoid and glucosinolate phenotypes. Plant Cell 15: 179–194.
9. The Arabidopsis Genome Initiative 2000. Analysis of the genome sequence of the flowering plant *Arabidopsis thaliana*. Nature 408:716–815
10. International Rice Genome Sequencing Project 2005. The map-based sequence of the rice genome. Nature 436: 793–800.
11. Paterson, A., Bowers, J., Kresovich, S., Hash, C., Messing, J., Peterson, D., Schmutz, J., and Rokhsar, D. Phytozome *Sorghum bicolor*. http://www.phytozome.net/sorghum. 2007.
12. Tuskan, G.A., DiFazio, S., Jansson, S., Bohlmann, J., Grigoriev, I., Hellsten, U., Putnam, N., Ralph, S., Rombauts, S., Salamov, A., Schein, J., Sterck, L., Aerts, A., Bhalerao, R.R., Bhalerao, R.P., Blaudez, D., Boerjan, W., Brun, A., Brunner, A., Busov, V., Campbell, M., Carlson, J., Chalot, M., Chapman, J., Chen, G.L., Cooper, D., Coutinho, P.M., Couturier, J., Covert, S., Cronk, Q., Cunningham, R., Davis, J., Degroeve, S., Dejardin, A., dePamphilis, C., Detter, J., Dirks, B., Dubchak, I., Duplessis, S., Ehlting, J., Ellis, B., Gendler, K., Goodstein, D., Gribskov, M., Grimwood, J., Groover, A., Gunter, L., Hamberger, B., Heinze, B., Helariutta, Y., Henrissat, B., Holligan, D., Holt, R., Huang, W., Islam-Faridi, N., Jones, S., Jones-Rhoades, M., Jorgensen, R., Joshi, C., Kangasjarvi, J., Karlsson, J., Kelleher, C., Kirkpatrick, R., Kirst, M., Kohler, A., Kalluri, U., Larimer, F., Leebens-Mack, J., Leple, J.C., Locascio, P., Lou, Y., Lucas, S., Martin, F., Montanini, B., Napoli, C., Nelson, D.R., Nelson, C., Nieminen, K., Nilsson, O., Pereda, V., Peter, G., Philippe, R., Pilate, G., Poliakov, A., Razumovskaya, J., Richardson, P., Rinaldi, C., Ritland, K., Rouze, P., Ryaboy, D., Schmutz, J., Schrader, J., Segerman, B., Shin, H., Siddiqui, A., Sterky, F., Terry, A., Tsai, C.J., Uberbacher, E., Unneberg, P., Vahala, J., Wall, K., Wessler, S., Yang, G., Yin, T., Douglas, C., Marra, M., Sandberg, G., Van de Peer, Y., and Rokhsar, D. 2006. The genome of black cottonwood, *Populus trichocarpa* (Torr. & Gray). Science 313: 1596–1604.
13. The French-Italian Public Consortium for Grapevine Genome Characterization 2007. The grapevine genome sequence suggests ancestral hexaploidization in major angiosperm phyla. Nature 449: 463–467.
14. Rabinowicz, P., Ravel, J., Chan, A., Melake, A., Wortman, J., Zhao, Q., Orvis, J., and Puiu, D. TIGR: Castor Bean Genome Database. http://castorbean.tigr.org/. 2007.
15. Ming, R., Hou, S., Feng, Y., Yu Qonne-Laporte, A., Saw, J.H., Senin, P., Wang, W., Ly, B.V., Lewis, K.L., Salzberg, S.L., Feng, L., Jones, M.R., Skelton, R.L., Murray, J.E., Chen, C., Qian, W., Shen, J., Du, P., Eustice, M., Tong, E., Tang, H., Lyons, E., Paull, R.E., Michael, T.P., Wall, K., Rice, D.W., Albert, H., Wang, M.L., Zhu, Y.J., Schatz, M., Nagarajan, N., Acob, R.A., Guan, P., Blas, A., Wai, C.M., Ackerman, C.M., Ren, Y., Liu, C., Wang, J., Wang, J., Na, J.K., Shakirov, E.V., Haas, B., Thimmapuram, J., Nelson, D., Wang, X., Bowers, J.E., Gschwend, A.R., Delcher, A.L., Singh, R., Suzuki, J.Y., Tripathi, S., Neupane, K., Wei, H., Irikura, B., Paidi, M., Jiang, N., Zhang, W., Presting, G., Windsor, A., Navajas-Perez, R., Torres, M.J., Feltus, F.A., Porter, B., Li, Y., Burroughs, A.M., Luo, M.C., Liu, L., Christopher, D.A., Mount, S.M., Moore, P.H., Sugimura, T., Jiang, J., Schuler, M.A., Friedman, V., Mitchell-Olds, T., Shippen, D.E., dePamphilis, C.W., Palmer, J.D., Freeling, M., Paterson, A.H., Gonsalves, D., Wang, L., and Alam, M. 2008. The draft genome of the transgenic tropical fruit tree papaya (*Carica papaya* Linnaeus). Nature 452: 991–996.
16. Bentley, D.R. 2006. Whole-genome re-sequencing. Curr Opin Genet Dev 16: 545–552.
17. Mulder, N.J., Apweiler, R., Attwood, T.K., Bairoch, A., Bateman, A., Binns, D., Bork, P., Buillard, V., Cerutti, L., Copley, R., Courcelle, E., Das, U., Daugherty, L., Dibley, M., Finn, R., Fleischmann, W., Gough, J., Haft, D., Hulo, N., Hunter, S., Kahn, D., Kanapin, A., Kejariwal, A., Labarga, A., Langendijk-Genevaux, P.S.,

Lonsdale, D., Lopez, R., Letunic, I., Madera, M., Maslen, J., McAnulla, C., McDowall, J., Mistry, J., Mitchell, A., Nikolskaya, A.N., Orchard, S., Orengo, C., Petryszak, R., Selengut, J.D., Sigrist, C.J.A., Thomas, P.D., Valentin, F., Wilson, D., Wu, C.H., and Yeats, C. 2007. New developments in the InterPro database. Nucleic Acids Res 35: D224–D228.

18. Nelson, D.R., Schuler, M.A., Paquette, S.M., Werck-Reichhart, D., and Bak, S. 2004. Comparative genomics of rice and *Arabidopsis*. Analysis of 727 cytochrome P450 genes and pseudogenes from a monocot and a dicot. Plant Physiol 135: 756–772.
19. Vogt, T. and Jones, P. 2000. Glycosyltransferases in plant natural product synthesis: characterization of a supergene family. Trends Plant Sci 5: 380–386.
20. Bak, S., Nielsen, H., and Halkier, B. 1998. The presence of CYP79 homologues in glucosinolate-producing plants shows evolutionary conservation of the enzymes in the conversion of amino acid to aldoxime in the biosynthesis of cyanogenic glucosides and glucosinolates. Plant Mol Biol 38: 725–734.
21. Kandel, S., Sauveplane, V., Olry, A., Diss, L., Benveniste, I., and Pinot, F. 2006. Cytochrome P450-dependent fatty acid hydroxylases in plants. Phytochem Rev 5: 359–372.
22. Stumpe, M. and Feussner, I. 2006. Formation of oxylipins by CYP74 enzymes. Phytochem Rev 5: 347–357.
23. Bishop, G., Nomura, T., Yokota, T., Montoya, T., Castle, J., Harrison, K., Kushiro, T., Kamiya, Y., Yamaguchi, S., Bancos, S., Szatmari, A.M., and Szekeres, M. 2006. Dwarfism and cytochrome P450-mediated C-6 oxidation of plant steroid hormones. Biochem Soc Trans 34: 1199–1201.
24. Ohnishi, T., Szatmari, A.M., Watanabe, B., Fujita, S., Bancos, S., Koncz, C., Lafos, M., Shibata, K., Yokota, T., Sakata, K., Szekeres, M., and Mizutani, M. 2006. C-23 hydroxylation by *Arabidopsis* CYP90C1 and CYP90D1 reveals a novel shortcut in brassinosteroid biosynthesis. Plant Cell 18: 3275–3288.
25. Ehlting, J., Buttner, D., Wang, Q., Douglas, C.J., Somssich, I.E., and Kombrink, E. 1999. Three 4-coumarate:coenzyme A ligases in *Arabidopsis thaliana* represent two evolutionarily divergent classes in angiosperms. Plant J 19: 9–20.
26. Hamberger, B. and Hahlbrock, K. 2004. The 4-coumarate:CoA ligase gene family in *Arabidopsis thaliana* comprises one rare, sinapate-activating and three commonly occurring isoenzymes. Proc Natl Acad Sci USA 101: 2209–2214.
27. Hull, A.K., Vij, R., and Celenza, J.L. 2000. Arabidopsis cytochrome P450s that catalyze the first step of tryptophan-dependent indole-3-acetic acid biosynthesis. Proc Natl Acad Sci USA 97: 2379–2384.
28. Chen, S., Glawischnig, E., Jorgensen, K., Naur, P., Jorgensen, B., Olsen, C.E., Hansen, C.H., Rasmussen, H., Pickett, J.A., and Halkier, B.A. 2003. CYP79F1 and CYP79F2 have distinct functions in the biosynthesis of aliphatic glucosinolates in *Arabidopsis*. Plant J 33: 923–937.
29. Mikkelsen, M.D., Hansen, C.H., Wittstock, U., and Halkier, B.A. 2000. Cytochrome P450 CYP79B2 from *Arabidopsis* catalyzes the conversion of tryptophan to indole-3-acetaldoxime, a precursor of indole glucosinolates and indole-3-acetic acid. J Biol Chem 275: 33712–33717.
30. Wittstock, U. and Halkier, B.A. 2000. Cytochrome P450 CYP79A2 from *Arabidopsis thaliana* L. catalyzes the conversion of L-phenylalanine to phenylacetaldoxime in the biosynthesis of benzylglucosinolate. J Biol Chem 275: 14659–14666.
31. Saito, K., Hirai, M.Y., and Yonekura-Sakakibara, K. 2008. Decoding genes with coexpression networks and metabolomics - 'majority report by precogs'. Trends Plant Sci 13: 36–43.
32. Aoki, K., Ogata, Y., and Shibata, D. 2007. Approaches for extracting practical information from gene co-expression networks in plant biology. Plant Cell Physiol 48: 381–390.
33. Ehlting, J., Mattheus, N., Aeschliman, D.S., Li, E., Hamberger, B., Cullis, I.F., Zhuang, J., Kaneda, M., Mansfield, S.D., Samuels, L., Ritland, K., Ellis, B.E., Bohlmann, J., and Douglas, C.J. 2005. Global transcript profiling of primary stems from *Arabidopsis thaliana* identifies candidate genes for missing links in lignin biosynthesis and transcriptional regulators of fiber differentiation. Plant J 42: 618–640.
34. Gachon, C., Langlois-Meurinne, M., Henry, Y., and Saindrenan, P. 2005. Transcriptional co-regulation of secondary metabolism enzymes in *Arabidopsis*: functional and evolutionary implications. Plant Mol Biol 58: 229–245.
35. Persson, S., Wei, H., Milne, J., Page, G.P., and Somerville, C.R. 2005. Identification of genes required for cellulose synthesis by regression analysis of public microarray data sets. Proc Natl Acad Sci USA 102: 8633–8638.
36. Toufighi, K., Brady, S.M., Austin, R., Ly, E., and Provart, N.J. 2005. The Botany Array

Resource: e-northerns, expression angling, and promoter analyses. Plant J 43: 153–163.
37. Zimmermann, P., Hirsch-Hoffmann, M., Hennig, L., and Gruissem, W. 2004. GENEVESTIGATOR. Arabidopsis microarray database and analysis toolbox. Plant Physiol 136: 2621–2632.
38. Koo, A.J.K., Chung, H.S., Kobayashi, Y., and Howe, G.A. 2006. Identification of a peroxisomal acyl-activating enzyme involved in the biosynthesis of jasmonic acid in *Arabidopsis*. J Biol Chem 281: 33511–33520.
39. Tohge, T., Nishiyama, Y., Hirai, M.Y., Yano, M., Nakajima, J.I., Awazuhara, M., Inoue, E., Takahashi, H., Goodenowe, D.B., Kitayama, M., Noji, M., Yamazaki, M., and Saito, K. 2005. Functional genomics by integrated analysis of metabolome and transcriptome of *Arabidopsis* plants over-expressing an MYB transcription factor. Plant J 42: 218–235.
40. Yonekura-Sakakibara, K., Tohge, T., Niida, R., and Saito, K. 2007. Identification of a flavonol 7-O-rhamnosyltransferase gene determining flavonoid pattern in *Arabidopsis* by transcriptome coexpression analysis and reverse genetics. J Biol Chem 282: 14932–14941.
41. Ehlting, J., Sauveplane, V., Olry, A., Ginglinger, J.F., Provart, N.J., and Werck-Reichhart, D. 2008. An extensive (co-)expression analysis tool for the cytochrome P450 superfamily in *Arabidopsis thaliana*. BMC Plant Biol 8: 47.
42. Field, B. and Osbourn, A.E. 2008. Metabolic diversification - independent assembly of operon-like gene clusters in different plants. Science 320: 543–547.
43. Hirai, M.Y., Sugiyama, K., Sawada, Y., Tohge, T., Obayashi, T., Suzuki, A., Araki, R., Sakurai, N., Suzuki, H., Aoki, K., Goda, H., Nishizawa, O.I., Shibata, D., and Saito, K. 2007. Omics-based identification of *Arabidopsis* Myb transcription factors regulating aliphatic glucosinolate biosynthesis. Proc Natl Acad Sci USA 104: 6478–6483.
44. Gang, D.R., Wang, J., Dudareva, N., Nam, K.H., Simon, J.E., Lewinsohn, E., and Pichersky, E. 2001. An investigation of the storage and biosynthesis of phenylpropenes in Sweet Basil. Plant Physiol 125: 539–555.
45. Lange, B.M., Wildung, M.R., Stauber, E.J., Sanchez, C., Pouchnik, D., and Croteau, R. 2000. Probing essential oil biosynthesis and secretion by functional evaluation of expressed sequence tags from mint glandular trichomes. Proc Natl Acad Sci USA 97: 2934–2939.
46. Guterman, I., Shalit, M., Menda, N., Piestun, D., fny-Yelin, M., Shalev, G., Bar, E., Davydov, O., Ovadis, M., Emanuel, M., Wang, J., Adam, Z., Pichersky, E., Lewinsohn, E., Zamir, D., Vainstein, A., and Weiss, D. 2002. Rose scent: genomics approach to discovering novel floral fragrance-related genes. Plant Cell 14: 2325–2338.
47. Jennewein, S., Wildung, M.R., Chau, M., Walker, K., and Croteau, R. 2004. Random sequencing of an induced Taxus cell cDNA library for identification of clones involved in Taxol biosynthesis. Proc Natl Acad Sci USA 101: 9149–9154.
48. Choi, D.-W., Jung, J., Ha, Y.I., Park, H.-W., In, D.S., Chung, H.-J., and Liu, J.R. 2005. Analysis of transcripts in methyl jasmonate-treated ginseng hairy roots to identify genes involved in the biosynthesis of ginsenosides and other secondary metabolites. Plant Cell Reports 23: 557–566.
49. Bertea, C.M., Voster, A., verstappen, F.W., Maffei, M., Beekwilder, J., and Bouwmeester, H.J. 2006. Isoprenoid biosynthesis in *Artemisia annua*: cloning and heterologous expression of a germacrene A synthase from a glandular trichome cDNA library. Arch Biochem Biophys 448: 3–12.
50. Ro, D.K., Paradise, E.M., Ouellet, M., Fisher, K.J., Newman, K.L., Ndungu, J.M., Ho, K.A., Eachus, R.A., Ham, T.S., Kirby, J., Chang, M.C., Withers, S.T., Shiba, Y., Sarpong, R., and Keasling, J.D. 2006. Production of the antimalarial drug precursor artemisinic acid in engineered yeast. Nature 440: 940–943.
51. Teoh, K.H., Polichuk, D.R., Reed, D.W., Nowak, G., and Covello, P.S. 2006. *Artemisia annua* L. (Asteraceae) trichome-specific cDNAs reveal CYP71AV1, a cytochrome P450 with a key role in the biosynthesis of the antimalarial sesquiterpene lactone artemisinin. FEBS Lett 580: 1411–1416.
52. Small, I. 2007. RNAi for revealing and engineering plant gene functions. Curr Opin Biotechnol 18: 148–153.
53. Alonso, J.M., Stepanova, A.N., Leisse, T.J., Kim, C.J., Chen, H., Shinn, P., Stevenson, D.K., Zimmerman, J., Barajas, P., Cheuk, R., Gadrinab, C., Heller, C., Jeske, A., Koesema, E., Meyers, C.C., Parker, H., Prednis, L., Ansari, Y., Choy, N., Deen, H., Geralt, M., Hazari, N., Hom, E., Karnes, M., Mulholland, C., Ndubaku, R., Schmidt, I., Guzman, P., guilar-Henonin, L., Schmid, M., Weigel, D., Carter, D.E., Marchand, T., Risseeuw, E., Brogden, D., Zeko, A., Crosby, W.L., Berry, C.C., and Ecker, J.R. 2003. Genome-wide insertional mutagenesis of *Arabidopsis thaliana*. Science 301: 653–657.

54. Li, Y., Rosso, M.G., Viehoever, P., and Weisshaar, B. 2007. GABI-Kat SsimpleSearch: an *Arabidopsis thaliana* T-DNA mutant database with detailed information for confirmed insertions. Nucleic Acids Res 35: D874–D878.
55. Miyao, A., Tanaka, K., Murata, K., Sawaki, H., Takeda, S., Abe, K., Shinozuka, Y., Onosato, K., and Hirochika, H. 2003. Target site specificity of the *Tos17* retrotransposon shows a preference for insertion within genes and against insertion in retrotransposon-rich regions of the genome. Plant Cell 15: 1771–1780.
56. Sessions, A., Burke, E., Presting, G., Aux, G., McElver, J., Patton, D., Dietrich, B., Ho, P., Bacwaden, J., Ko, C., Clarke, J.D., Cotton, D., Bullis, D., Snell, J., Miguel, T., Hutchison, D., Kimmerly, B., Mitzel, T., Katagiri, F., Glazebrook, J., Law, M., and Goff, S.A. 2002. A high-throughput Arabidopsis reverse genetics system. Plant Cell 14: 2985–2994.
57. Bak, S., Tax, F.E., Feldmann, K.A., Galbraith, D.W., and Feyereisen, R. 2001. CYP83B1, a cytochrome P450 at the metabolic branch point in auxin and indole glucosinolate biosynthesis in Arabidopsis. Plant Cell 13: 101–111.
58. Comai, L. and Henikoff, S. 2006. TILLING: practical single-nucleotide mutation discovery. Plant J 45: 684–694.
59. Slade, A.J. and Knauf, V.C. 2005. TILLING moves beyond functional genomics into crop improvement. Transgenic Res 14: 109–115.
60. Yamada, K., Lim, J., Dale, J.M., Chen, H., Shinn, P., Palm, C.J., Southwick, A.M., Wu, H.C., Kim, C., Nguyen, M., Pham, P., Cheuk, R., Karlin-Newmann, G., Liu, S.X., Lam, B., Sakano, H., Wu, T., Yu, G., Miranda, M., Quach, H.L., Tripp, M., Chang, C.H., Lee, J.M., Toriumi, M., Chan, M.M.H., Tang, C.C., Onodera, C., Deng, J.M., Akiyama, K., Ansari, Y., Arakawa, T., Banh, J., Banno, F., Bowser, L., Brooks, S., Carninci, P., Chao, Q., Choy, N., Enju, A., Goldsmith, A.D., Gurjal, M., Hansen, N.F., Hayashizaki, Y., Johnson-Hopson, C., Hsuan, V.W., Iida, K., Karnes, M., Khan, S., Koesema, E., Ishida, J., Jiang, P.X., Jones, T., Kawai, J., Kamiya, A., Meyers, C., Nakajima, M., Narusaka, M., Seki, M., Sakurai, T., Satou, M., Tamse, R., Vaysberg, M., Wallender, E.K., Wong, C., Yamamura, Y., Yuan, S., Shinozaki, K., Davis, R.W., Theologis, A., and Ecker, J.R. 2003. Empirical Analysis of Transcriptional Activity in the *Arabidopsis* genome. Science 302: 842–846.
61. Gong, W., Shen, Y.P., Ma, L.G., Pan, Y., Du, Y.L., Wang, D.H., Yang, J.Y., Hu, L.D., Liu, X.F., Dong, C.X., Ma, L., Chen, Y.H., Yang, X.Y., Gao, Y., Zhu, D., Tan, X., Mu, J.Y., Zhang, D.B., Liu, Y., Lnesh-Kumar, S.P., Li, Y., Wang, X.P., Gu, H.Y., Qu, L.J., Bai, S.N., Lu, Y.T., Li, J.Y., Zhao, J.D., Zuo, J., Huang, H., Deng, X.W., and Zhu, Y.X. 2004. Genome-wide ORFeome cloning and analysis of Arabidopsis transcription factor genes. Plant Physiol 135: 773–782.
62. Katsuyama, Y., Matsuzawa, M., Funa, N., and Horinouchi, S. 2007. *In vitro* synthesis of curcuminoids by type III polyketide synthase from *Oryza sativa*. J Biol Chem 282: 3770–3779.
63. Bohlmann, J., Martin, D., Oldham, N.J., and Gershenzon, J. 2000. Terpenoid secondary metabolism in *Arabidopsis thaliana*: cDNA cloning, characterization, and functional expression of a myrcene/(E)-beta-ocimene synthase. Arch Biochem Biophys 375: 261–269.
64. Chen, F., Tholl, D., D'Auria, J.C., Farooq, A., Pichersky, E., and Gershenzon, J. 2003. Biosynthesis and emission of terpenoid volatiles from *Arabidopsis* flowers. Plant Cell 15: 481–494.
65. Chen, F., Ro, D.K., Petri, J., Gershenzon, J., Bohlmann, J., Pichersky, E., and Tholl, D. 2004. Characterization of a root-specific *Arabidopsis* terpene synthase responsible for the formation of the volatile monoterpene 1, 8-cineole. Plant Physiol 135: 1956–1966.
66. Fäldt, J., Arimura, G.-I., Gershenzon, J., Takabayashi, J., and Bohlmann, J. 2003. Functional identification of AtTPS03 as (E)-ß-ocimene synthase: a monoterpene synthase catalyzing jasmonate- and wound-induced volatile formation in *Arabidopsis thaliana*. Planta 216: 745–751.
67. Ro, D.K., Ehlting, J., Keeling, C.I., Lin, R., Mattheus, N., and Bohlmann, J. 2006. Microarray expression profiling and functional characterization of AtTPS genes: duplicated *Arabidopsis thaliana* sesquiterpene synthase genes At4g13280 and At4g13300 encode root-specific and wound-inducible (Z)-gamma-bisabolene synthases. Arch Biochem Biophys 448: 104–116.
68. Tholl, D., Chen, F., Petri, J., Gershenzon, J., and Pichersky, E. 2005. Two sesquiterpene synthases are responsible for the complex mixture of sesquiterpenes emitted from *Arabidopsis* flowers. Plant J 42: 757–771.

69. Ebizuka, Y., Katsube, Y., Tsutsumi, T., Kushiro, T., and Shibuya, M. 2003. Functional genomics approach to the study of triterpene biosynthesis. Pure Appl Chem 75: 369–374.
70. Fazio, G.C., Xu, R., and Matsuda, S.P.T. 2004. Genome mining to identify new plant triterpenoids. J Am Chem Soc 126: 5678–5679.
71. Husselstein-Muller, T., Schaller, H., and Benveniste, P. 2001. Molecular cloning and expression in yeast of 2, 3-oxidosqualene-triterpenoid cyclases from *Arabidopsis thaliana*. Plant Mol Biol 45: 75–92.
72. Kolesnikova, M.D., Wilson, W.K., Lynch, D.A., Obermeyer, A.C., and Matsuda, S.P.T. 2007. *Arabidopsis* camelliol C synthase evolved from enzymes that make pentacycles. Org Lett 9: 5223–5226.
73. Lodeiro, S., Xiong, Q., Wilson, W.K., Kolesnikova, M.D., Onak, C.S., and Matsuda, S.P.T. 2007. An oxidosqualene cyclase makes numerous products by diverse mechanisms: A challenge to prevailing concepts of triterpene biosynthesis. J Am Chem Soc 129: 11213–11222.
74. Xiang, T., Shibuya, M., Katsube, Y., Tsutsumi, T., Otsuka, M., Zhang, H., Masuda, K., and Ebizuka, Y. 2006. A new triterpene synthase from *Arabidopsis thaliana* produces a tricyclic triterpene with two hydroxyl groups. Org Lett 8: 2835–2838.
75. Xiong, Q., Wilson, W.K., and Matsuda, S.P. 2006. An *Arabidopsis* oxidosqualene cyclase catalyzes iridal skeleton formation by Grob fragmentation. Angew Chem Int Ed Engl 45: 1285–1288.
76. Hou, B., Lim, E.K., Higgins, G.S., and Bowles, D.J. 2004. N-glucosylation of cytokinins by glycosyltransferases of *Arabidopsis thaliana*. J Biol Chem 279: 47822–47832.
77. Lim, E.K., Doucet, C.J., Li, Y., Elias, L., Worrall, D., Spencer, S.P., Ross, J., and Bowles, D.J. 2002. The activity of *Arabidopsis* glycosyltransferases toward salicylic acid, 4-hydroxybenzoic acid, and other benzoates. J Biol Chem 277: 586–592.
78. Lim, E.K., Doucet, C.J., Hou, B., Jackson, R.G., Abrams, S.R., and Bowles, D.J. 2005. Resolution of (+)-abscisic acid using an *Arabidopsis* glycosyltransferase. Tetrahedron: Asymmetry 16: 143–147.
79. Weis, M., Lim, E.K., Bruce, N., and Bowles, D. 2006. Regioselective glucosylation of aromatic compounds: Screening of a recombinant glycosyltransferase library to identify biocatalysts. Angew Chemie Int Ed 45: 3534–3538.
80. Kruse, T., Ho, K., Yoo, H.D., Johnson, T., Hippely, M., Park, J.H., Flavell, R., and Bobzin, S. 2008. *In planta* biocatalysis screen of P450s identifies 8-methoxypsoralen as a substrate for the CYP82C subfamily, yielding original chemical structures. Chem Biol 15: 149–156.
81. Schneider, K., Kienow, L., Schmelzer, E., Colby, T., Bartsch, M., Miersch, O., Wasternack, C., Kombrink, E., and Stuible, H.P. 2005. A new type of peroxisomal acyl-coenzyme A synthetase from *Arabidopsis thaliana* has the catalytic capacity to activate biosynthetic precursors of jasmonic acid. J Biol Chem 280: 13962–13972.
82. Olry, A., Schneider-Belhaddad, F., Heintz, D., and Werck-Reichhart, D. 2007. A medium-throughput screening assay to determine catalytic activities of oxygen-consuming enzymes: a new tool for functional characterization of cytochrome P450 and other oxygenases. Plant J 51: 331–340.
83. Brockmann, R., Beyer, A., Heinisch, J., and Wilhelm, T. 2007. Posttranscriptional expression regulation: What determines translation rates? PLoS Comp Biol 3: e57.
84. Watson, B.S., Asirvatham, V.S., Wang, L., and Sumner, L.W. 2003. Mapping the proteome of barrel medic (*Medicago truncatula*). Plant Physiol 131: 1104–1123.
85. Jones, A.M., Thomas, V., Bennett, M.H., Mansfield, J., and Grant, M. 2006. Modifications to the *Arabidopsis* defense proteome occur prior to significant transcriptional change in response to inoculation with *Pseudomonas syringae*. Plant Physiol 142: 1603–1620.
86. Chen, S. and Harmon, A.C. 2006. Advances in plant proteomics. Proteomics 6: 5504–5516.
87. Rose, J.K., Bashir, S., Giovannoni, J.J., Jahn, M.M., and Saravanan, R.S. 2004. Tackling the plant proteome: practical approaches, hurdles and experimental tools. Plant J 39: 715–733.
88. Thelen, J.J. and Peck, S.C. 2007. Quantitative proteomics in plants: choices in abundance. Plant Cell 19: 3339–3346.
89. King, J. 2002. Recollections and reflections of a plant physiologist. Trends Plant Sci 7: 278–280.
90. Fiehn, O. 2002. Metabolomics - the link between genotypes and phenotypes. Plant Mol Biol 48: 155–171.
91. Ryan, D. and Robards, K. 2006. Metabolomics: The greatest omics of them all? Anal Chem 78: 7954–7958.
92. Hall, R.D. 2006. Plant metabolomics: from holistic hope, to hype, to hot topic. New Phytol 169: 453–468.

93. Hagel, J.M. and Facchini, P. 2008. Plant metabolomics: analytical platforms and integration with functional genomics. Phytochem Rev 7: 479–497.
94. Eisenreich, W. and Bacher, A. 2007. Advances of high-resolution NMR techniques in the structural and metabolic analysis of plant biochemistry. Phytochem 68: 2799–2815.
95. Abdel-Farid, I.B., Kim, H.K., Choi, Y.H., and Verpoorte, R. 2007. Metabolic characterization of Brassica rapa leaves by NMR spectroscopy. J Agric Food Chem 55: 7936–7943.
96. Liang, Y.S., Choi, Y.H., Kim, H.K., Linthorst, H.J., and Verpoorte, R. 2006. Metabolomic analysis of methyl jasmonate treated Brassica rapa leaves by 2-dimensional NMR spectroscopy. Phytochem 67: 2503–2511.
97. Widarto, H.T., van der Meijden, E., Lefeber, A.W., Erkelens, C., Kim, H.K., Choi, Y.H., and Verpoorte, R. 2006. Metabolomic differentiation of Brassica rapa following herbivory by different insect instars using two-dimensional nuclear magnetic resonance spectroscopy. J Chem Ecol 32: 2417–2428.
98. Aharoni, A., Ricd., V, Verhoeven, H.A., Maliepaard, C.A., Kruppa, G., Bino, R., and Goodenowe, D.B. 2002. Nontargeted metabolome analysis by use of Fourier Transform Ion Cyclotron Mass Spectrometry. Omics 6: 217–234.
99. Iijima, Y., Nakamura, Y., Ogata, Y., Tanaka, K., Sakurai, N., Suda, K., Suzuki, T., Suzuki, H., Okazaki, K., Kitayama, M., Kanaya, S., Aoki, K., and Shibata, D. 2008. Metabolite annotations based on the integration of mass spectral information. Plant J 54: 949–962.
100. Broeckling, C.D., Huhman, D.V., Farag, M.A., Smith, J.T., May, G.D., Mendes, P., Dixon, R.A., and Sumner, L.W. 2005. Metabolic profiling of *Medicago truncatula* cell cultures reveals the effects of biotic and abiotic elicitors on metabolism. J Exp Bot 56: 323–336.
101. Higdon, J.V., Delage, B., Williams, D.E., and Dashwood, R.H. 2007. Cruciferous vegetables and human cancer risk: epidemiologic evidence and mechanistic basis. Pharmacol Res 55: 224–236.
102. Hirai, M.Y., Yano, M., Goodenowe, D.B., Kanaya, S., Kimura, T., Awazuhara, M., Arita, M., Fujiwara, T., and Saito, K. 2004. Integration of transcriptomics and metabolomics for understanding of global responses to nutritional stresses in *Arabidopsis thaliana*. Proc Natl Acad Sci U S A 101: 10205–10210.
103. Hirai, M.Y., Klein, M., Fujikawa, Y., Yano, M., Goodenowe, D.B., Yamazaki, Y., Kanaya, S., Nakamura, Y., Kitayama, M., Suzuki, H., Sakurai, N., Shibata, D., Tokuhisa, J., Reichelt, M., Gershenzon, J., Papenbrock, J., and Saito, K. 2005. Elucidation of gene-to-gene and metabolite-to-gene networks in arabidopsis by integration of metabolomics and transcriptomics. J Biol Chem 280: 25590–25595.
104. Gion, J.M., Lalanne, C., Le, P.G., Ferry-Dumazet, H., Paiva, J., Chaumeil, P., Frigerio, J.M., Brach, J., Barre, A., de, D.A., Claverol, S., Bonneu, M., Sommerer, N., Negroni, L., and Plomion, C. 2005. The proteome of maritime pine wood forming tissue. Proteomics 5: 3731–3751.
105. Fiorani Celedon, P.A., de Andrade A., Xavier Meireles, K.G., Gallo de Carvalho, M.C., Gomes Caldas, D.G., Moon, D.H., Carneiro, R.T., Franceschini, L.M., Oda, S., and Labate, C.A. 2007. Proteomic analysis of the cambial region in juvenile Eucalyptus grandis at three ages. Proteomics 7: 2258–2274.
106. Robinson, A.R., Ukrainetz, N.K., Kang, K.Y., and Mansfield, S.D. 2007. Metabolite profiling of Douglas-fir (*Pseudotsuga menziesii*) field trials reveals strong environmental and weak genetic variation. New Phytol 174: 762–773.
107. Goossens, A., Hakkinen, S.T., Laakso, I., Seppanen-Laakso, T., Biondi, S., De, S., Lammertyn, F., Nuutila, A.M., Soderlund, H., Zabeau, M., Inze, D., and Oksman-Caldentey, K.M. 2003. A functional genomics approach toward the understanding of secondary metabolism in plant cells. Proc Natl Acad Sci USA 100: 8595–8600.
108. Rischer, H., Oresic, M., Seppanen-Laakso, T., Katajamaa, M., Lammertyn, F., rdiles-Diaz, W., Van Montagu, M.C., Inze, D., Oksman-Caldentey, K.M., and Goossens, A. 2006. Gene-to-metabolite networks for terpenoid indole alkaloid biosynthesis in *Catharanthus roseus* cells. Proc Natl Acad Sci USA 103: 5614–5619.
109. Facchini, P.J., Hagel, J.M., Liscombe, D.K., Loukanina, N., MacLeod, B.P., Samanani, N., and Zulak, K.G. 2007. Opium poppy: blueprint for an alkaloid factory. Phytochem Rev 6: 97–124.
110. Decker, G., Wanner, G., Zenk, M.H., and Lottspeich, F. 2000. Characterization of proteins in latex of the opium poppy (*Papaver somniferum*) using two-dimensional gel electrophoresis

and microsequencing. Electrophoresis 21: 3500–3516.
111. Ounaroon, A., Decker, G., Schmidt, J., Lottspeich, F., and Kutchan, T.M. 2003. (R, S)-Reticuline 7-O-methyltransferase and (R, S)-norcoclaurine 6-O-methyltransferase of *Papaver somniferum* - cDNA cloning and characterization of methyl transfer enzymes of alkaloid biosynthesis in opium poppy. Plant J 36: 808–819.
112. Gang, D.R., Wang, J., Dudareva, N., Nam, K.H., Simon, J.E., Lewinsohn, E., and Pichersky, E. 2001. An investigation of the storage and biosynthesis of phenylpropenes in sweet basil. Plant Physiol 125: 539–555.
113. Xie, Z., Kapteyn, J., and Gang, D.R. 2008. A systems biology investigation of the MEP/terpenoid and shikimate/phenylpropanoid pathways points to multiple levels of metabolic control in sweet basil glandular trichomes. Plant J 54: 349–361.
114. Schauer, N., Semel, Y., Roessner, U., Gur, A., Balbo, I., Carrari, F., Pleban, T., Perez-Melis, A., Bruedigam, C., Kopka, J., Willmitzer, L., Zamir, D., and Fernie, A.R. 2006. Comprehensive metabolic profiling and phenotyping of interspecific introgression lines for tomato improvement. Nat Biotechnol 24: 447–454.
115. Schauer, N., Semel, Y., Balbo, I., Steinfath, M., Repsilber, D., Selbig, J., Pleban, T., Zamir, D., and Fernie, A.R. 2008. Mode of inheritance of primary metabolic traits in tomato. Plant Cell 20: 509–523.
116. Sticklen, M.B. 2008. Plant genetic engineering for biofuel production: towards affordable cellulosic ethanol. Nat Rev Genet 9: 433–443.
117. Rubin, E.M. 2008. Genomics of cellulosic biofuels. Nature 454: 841–845.
118. Santoni, V., Molloy, M., and Rabilloud, T. 2000. Membrane proteins and proteomics: un amour impossible? Electrophoresis 21: 1054–1070.
119. Schad, M., Lipton, M.S., Giavalisco, P., Smith, R.D., and Kehr, J. 2005. Evaluation of two-dimensional electrophoresis and liquid chromatography - tandem mass spectrometry for tissue-specific protein profiling of laser-microdissected plant samples. Electrophoresis 26: 2729–2738.
120. Schad, M., Mungur, R., Fiehn, O., and Kehr, J. 2005. Metabolic profiling of laser microdissected vascular bundles of *Arabidopsis thaliana*. Plant Methods 1: 2.
121. Cha, S., Zhang, H., Ilarslan, H.I., Wurtele, E.S., Brachova, L., Nikolau, B.J., and Yeung, E.S. 2008. Direct profiling and imaging of plant metabolites in intact tissues by using colloidal graphite-assisted laser desorption ionization mass spectrometry. Plant J 55: 348–360.
122. Dunkley, T.P., Dupree, P., Watson, R.B., and Lilley, K.S. 2004. The use of isotope-coded affinity tags (ICAT) to study organelle proteomes in *Arabidopsis thaliana*. Biochem Soc Trans 32: 520–523.
123. Dunkley, T.P., Hester, S., Shadforth, I.P., Runions, J., Weimar, T., Hanton, S.L., Griffin, J.L., Bessant, C., Brandizzi, F., Hawes, C., Watson, R.B., Dupree, P., and Lilley, K.S. 2006. Mapping the *Arabidopsis* organelle proteome. Proc Natl Acad Sci USA 103: 6518–6523.
124. Jones, A.M., Bennett, M.H., Mansfield, J.W., and Grant, M. 2006. Analysis of the defence phosphoproteome of Arabidopsis thaliana using differential mass tagging. Proteomics 6: 4155–4165.
125. Liang, Y.S., Kim, H.K., Lefeber, A.W., Erkelens, C., Choi, Y.H., and Verpoorte, R. 2006. Identification of phenylpropanoids in methyl jasmonate treated Brassica rapa leaves using two-dimensional nuclear magnetic resonance spectroscopy. J Chromatogr A 1112: 148–155.

Metabolomics and the Detection of Unintended Effects in Genetically Modified Crops

22

Laetitia Shintu, Gwénaëlle Le Gall, and Ian J. Colquhoun

Abstract The chapter describes current procedures for the safety assessment of genetically modified crops and foods. The concepts of substantial equivalence, the conventional comparator, and intended and unintended effects are introduced. Most published examples of substantial equivalence testing deal with crops that have been modified for insect resistance or herbicide tolerance. A standard procedure has developed based on broadly similar field trial designs, sampling schemes and targeted analyses of a consensus set of compounds for each crop. The main characteristics of the procedure are summarised with reference to published analyses of this type of crop and different statistical approaches to judging 'equivalence' are discussed.

There is a current trend towards development of crops with enhanced nutritional properties or health-related benefits through genetic modification of metabolic pathways. These more complex modifications have greater potential for introducing unpredictable unintended effects, and it may be advisable to supplement current targeted analysis procedures with metabolomics methods. The second part of the chapter discusses the application of metabolomics to substantial equivalence testing. As yet there is no standard procedure for this approach so individual studies, which differ greatly in size and scope, are discussed. The major analytical techniques (GC/MS, LC/MS and NMR) are briefly described and examples of their use are given: a few studies have shown how the massive amounts of data produced by non-targeted profiling methods may be treated to judge equivalence. Some limitations need to be overcome before metabolomics can be adopted as part of the official safety assessment procedure.

Abbreviations

2D	Two-dimensional
ANOVA	Analysis of Variance
AOAC	Association of Analytical Communities
COSY	Correlation Spectroscopy
DIMS	Direct Injection Mass Spectrometry
DP	Degree of Polymerisation
EFSA	European Food Safety Authority
ESI	Electrospray Ionisation
FAO/WHO	Food and Agriculture Organisation/ World Health Organisation
FIE-MS	Flow Injection Electrospray Mass Spectrometry

L. Shintu, G. Le. Gall, and I.J. Colquhoun (✉)
Institute of Food Research,
Norwich Research Park,
Norwich NR4 7UA, UK
e-mail: laetitia.shintu@bbsrc.ac.uk;
gwenaelle.legall@bbsrc.ac.uk;
ian.colquhoun@bbsrc.ac.uk

A.E. Osbourn and V. Lanzotti (eds.), *Plant-derived Natural Products*,
DOI 10.1007/978-0-387-85498-4_22, © Springer Science+Business Media, LLC 2009

FT-ICR-MS	Fourier Transform Ion Cyclotron Resonance Mass Spectrometry
FTIR	Fourier Transform Infrared
GC/FID	Gas Chromatography/ Flame Ionisation Detector
GC/MS	Gas Chromatography/ Mass Spectrometry
GC-TOF-MS	Gas Chromatography-Time of Flight-Mass Spectrometry
GM	Genetically Modified
HMBC	Heteronuclear Multiple Bond Correlation
HPLC	High Performance Liquid Chromatography
HSQC	Heteronuclear Single Quantum Coherence
ILSI	International Life Sciences Institute
LC/MS	Liquid Chromatography/ Mass Spectrometry
LDA	Linear Discriminant Analysis
MAS	Magic Angle Spinning
NMR	Nuclear Magnetic Resonance
OECD	Organisation for Economic Cooperation and Development
PC	Principal Component
PCA	Principal Component Analysis
PLS	Partial Least Squares
PLS-DA	Partial Least Squares-Discriminant Analysis
RT	Retention Time
SD	Standard Deviation
SPE	Solid Phase Extraction
TOCSY	Total Correlation Spectroscopy
UV	Ultraviolet

Substantial Equivalence and its Role in the Safety Assessment of Genetically Modified Plants

If the composition and properties of a GM plant, measured or evaluated for a selected group of components and features, are not significantly different from those of a conventional counterpart (or comparator – see Box 1) then the two are judged to be substantially equivalent. When the test of substantial equivalence is met it gives reassurance that the GM plant is at least as safe and nutritious as the comparator, a plant with a history of safe use as a source of human food or animal feed.

Before the advent of GM technology the safety testing of new crop varieties which were products of classical breeding or mutagenesis programmes had never been an official requirement for their introduction to the market. When the introduction of the first commercial GM food crops was about to take place the establishment of rigorous safety assessment procedures was seen as a necessity in order to allay concerns that the use of biotechnology might lead to new and unforeseen hazards. Established testing procedures were available for food additives, colours etc. which involved adding measured doses of the individual, well characterised, pure substances to rodent diets in order to detect the level at which toxic effects were observed in the rodents and hence to determine safe intake levels for humans (Renwick, 2004). For practical reasons such direct procedures are not applicable to the safety testing of whole foods or crops (although adapted animal testing procedures have subsequently been developed for the purpose).

Application of the concept of substantial equivalence, as one element of a safety assessment procedure that could be applied to whole foods, was first suggested in an OECD report (OECD, 1993): the role and effectiveness of substantial equivalence testing have since been questioned (Millstone et al., 1999), re-evaluated (FAO/WHO, 2000) and reviewed (Cellini et al., 2004). Compositional analysis is just one part, albeit an important one, of the full safety assessment process. Its place in an integrated assessment scheme is illustrated in Fig. 1 (König et al., 2004) and such schemes now form the basis of official procedures (EFSA, 2006).

The term substantial equivalence is used for two reasons. First, there will always be some

Box 1 Key concepts

Hazard: the intrinsic *potential* of a material to cause adverse health effects.

Risk: the likelihood that, under particular conditions of exposure, an intrinsic hazard will represent a threat to human health.

Risk assessment is a scientific procedure consisting of the following steps: (i) hazard identification; (ii) hazard characterisation; (iii) exposure assessment; (iv) risk characterisation.

Genetically modified (GM) or **Genetically Engineered (GE)** plants and foods are derived by application of

(i) *In vitro* nucleic acid techniques including recombinant DNA and direct injection of nucleic acid into cells or organelles or
(ii) Fusion of cells beyond the taxonomic family

thus overcoming natural physiological reproductive or recombinant barriers.

The term **novel food** has been used in the same sense although it is also applied to foods outside the above definition.

Conventional counterpart or comparator (of the GM food): a related organism/ variety, its components and/ or products for which there is experience of establishing safety based on common use as food. It is required for substantial equivalence testing.

Substantial equivalence testing forms one part of the safety assessment and is concerned with hazard identification. The safety assessment includes a comparison between the GM food/ crop and its conventional counterpart focusing on determination of similarities and differences.

The comparison of the GM food and the conventional counterpart should

(i) *Take account of intended and unintended effects of the modification*
(ii) *Seek to identify new or altered hazards*
(iii) *Identify changes, relevant to human health, in key nutrients*

Intended Effects of genetic engineering are those that are targeted to occur from the introduction of the gene(s) in question and which fulfil the original objectives of the genetic modification process.

Unintended Effects represent a statistically significant difference in the phenotype, response or composition of the GM plant compared with the parent from which it is derived, but taking the expected effects of the target gene into account. Such comparisons should be made when GM and non-GM plants are grown under the same regimes and environments.

Definitions of intended and unintended effects are from Cellini et al. (2004).

difference between the genetically modified plant and its parent since the motivation is to introduce a specific beneficial change via the modification. Secondly, the compositional analysis of the GM plant and its comparator is carried out on a selected group of compounds. It would never be practically possible to analyse every compound in the plant to show that the GM plant and its parent were completely equivalent (apart from the intended change) because plants contain thousands of different substances, many of them only present in minute quantities. There is no complete list of all the metabolites present in any plant (the metabolome) and there are

22

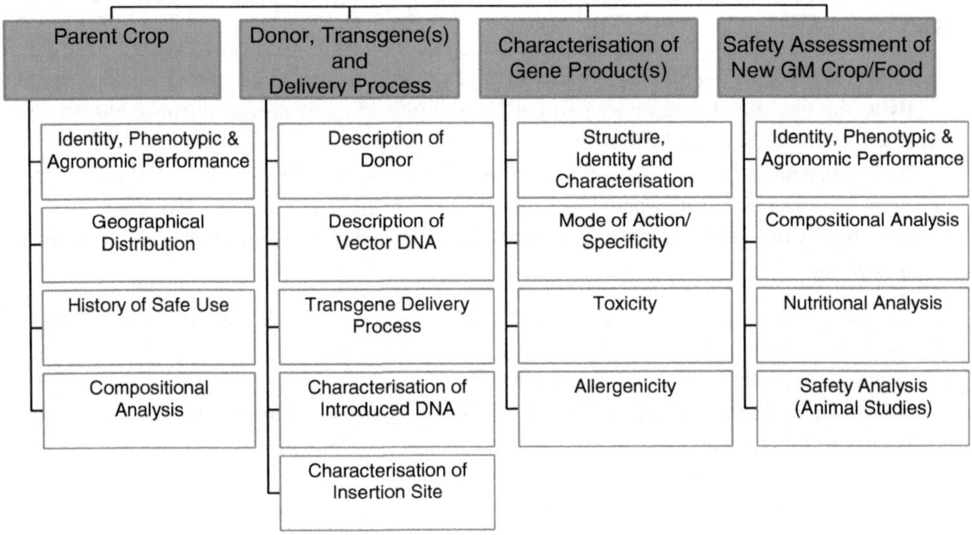

Fig. 1 An integrated and iterative approach to the hazard assessment and characterisation of new GM varieties. Compositional Analysis (the subject of this chapter) is an essential component but represents only one part of the overall scheme. (Reprinted fromKönig et al., 2004. © 2004. With permission from Elsevier)

undoubtedly substances present that are not yet fully characterised, even in the major food crops. Therefore indications for the occurrence of unintended effects (see Box 1) are sought from a comparison of the agronomic characteristics of the GM plant and its conventional counterpart based on the usual plant breeding criteria (appearance, yield etc.) plus a statistical comparison of the compositional data for each of the selected compounds. The next section discusses how this compositional analysis has been done in practice.

Substantial Equivalence Testing in Practice

Introduction

It has always been agreed that safety assessment of GM plants should be carried out on a case by case basis. For that reason the various official guidelines and guidance documents have emphasised principles rather than laid down hard and fast rules to cover all cases. Early discussions of the topic were rather abstract because of a lack of actual case studies, certainly in the peer reviewed literature. Since 2000 that situation has changed. Numerous peer reviewed articles have now been published on the substantial equivalence testing of GM crops and it is possible to see clearly how these studies are conducted and presented as scientific reports and as integral parts of applications for regulatory approval. Some examples from the literature are listed in Table 1. The following sections discuss factors to be considered in carrying out such studies using these examples from Table 1 as evidence of current approaches.

Choice of Comparators

The availability of a suitable comparator is essential to the substantial equivalence approach. The preferred comparator, if appropriate, is the near isogenic parental line. For plants that reproduce sexually the ideal comparison is between a line that is homozygous in the transgene and its matched azygous line. Commercial GM lines

Table 1 Examples of peer reviewed substantial equivalence studies (2000–2007) based on targeted analyses of samples from field trials

			Multiple		
Plant	Genetic modification	Parameters[a]	Sites	Years	Reference
Corn	Herbicide tolerance (glyphosate, *mesps*)	Pr, Fi, AA, FA, M	✓	✓	Sidhu et al. (2000)
Potato	Insect and viral resistance *cry3A* and *orf1/orf2* or *PVYcp*	Pr, total solids, soluble protein, Glc, Suc, AA, M, vitamins C, B_6, niacin, glycoalkaloids	✓	✓	Rogan et al. (2000)
Corn	Herbicide tolerance (glyphosate, *cp4 esps*)	Pr, Fi, AA, FA, M, vitamin E, phytic ac., TI, ferulic ac., *p*-coumaric ac., raffinose	✓	✓	Ridley et al. (2002)
Wheat	Herbicide tolerance (glyphosate, *cp4 esps*)	Pr, AA, FA, M, starch, sugars, niacin, vitamins B_1, B_2, B_6, E, folic ac., phytic ac., ferulic ac., *p*-coumaric ac., oxalic ac.	✓	✓	Obert et al. (2004)
Corn	Insect resistance (*cry1F*)	Pr, AA, FA, M, vitamins A, B_1, E, total tocopherols, folic ac., inositol, raffinose, *p*-coumaric ac., ferulic ac., phytic ac., TI	✓	x	Herman et al. (2004)
Corn	Insect resistance (modified *cry3Bb1*)	Pr, Fi, AA, FA, M, folic ac., vitamins B_1, B_2, E, ferulic ac., *p*-coumaric ac., inositol, raffinose, TI	✓	✓	George et al. (2004)
Rice	Herbicide tolerance (glufosinate, *bar*)	Pr, Fi, AA, FA, M, niacin, pantothenic ac., vitamins B_1, B_2, E, phytic ac., oryzanol, tocopherols, tocotrienols, lectin, TI, prolamins	✓	✓	Oberdoerfer et al. (2005)
Soy	Herbicide tolerance (glyphosate, *cp4 esps*)	Pr, lectin, TI, isoflavones	✓	✓	McCann et al. (2005)
Potato	Modified metabolism/ development (*fk*, *w2gbss*, *Mal1*, *SamDC*)	Dry matter, tuber number, sugars, glycoalkaloids, total nitrogen, vitamin C, TI, FA	x	x	Shepherd et al. (2006)
Soy	Herbicide tolerance (glyphosate, *cp4 esps*)	Pr, Fi, AA, FA, isoflavones, raffinose, stachyose, TI, lectin, phytic ac.	✓	x	Harrigan et al. (2007)
Corn	Insect resistance (*cry34Ab1* and *cry35Ab1*) and herbicide tolerance (glufosinate, *pat*)	Pr, Fi, M,AA, FA, β-carotene, vitamins B_1, B_2, folic ac. tocopherols, ferulic ac., *p*-coumaric ac., inositol, raffinose, phytic ac., TI	✓	✓	Herman et al. (2007)
Rice	Insect resistance (*sck* and *cry1Ac*)	Pr, AA, FA, M, vitamins B_1, B_2, E, phytic ac., TI, lectin	x	x	Li et al. (2007)

[a]pr: proximates (protein, total fat, carbohydrates, moisture, ash); fi: fibre; aa: (total) amino acids; fa: fatty acids; m: minerals; ti: trypsin inhibitor

are often produced by crossing the original GM plant with a different non-transgenic line. In that case the chosen comparator is a cross between the same two lines, both non-transgenic (e.g. Ridley et al., 2002). Empty vector and tissue culture lines, in addition to wild type parental varieties, have been used as controls for vegetatively propagated plants such as potato (Shepherd et al., 2006). The GM test plant and its near isogenic comparator are grown side by side in field trials (see next section).

It is usual practice to grow in the same field trials a number of different commercial lines of the same crop for analysis. This gives a range of compositional variation for the targeted compounds for plants grown in the same environment and at the same time as the test plant and its principal comparator. Literature sources and, more recently, on-line databases on the composition of conventional crops provide additional information on natural variation. These sources cover a wider range of locations and seasons than are provided by individual field trials and may justifiably be used providing the analytical results are known to be of sufficient quality. For example the ILSI crop composition database (Ridley et al., 2004; http://www.cropcomposition.org/) for corn, soy and cotton is a cumulative compilation of data from studies such as those listed in Table 1.

Design of Field Trials

Field trial designs differ somewhat from each other in details but follow the same general principles. The GM crop and its control are planted in randomised blocks within a plot, often with three or four replicate blocks each. Recent studies of herbicide tolerant GM plants have included three treatment groups: the control and transgenic plants are treated with conventional herbicides and in a separate block the transgenic plant is sprayed with the herbicide to which it is tolerant (Herman et al., 2004; Oberdoerfer et al., 2005). One pooled sample is taken for analysis from each block. As mentioned, additional blocks may be planted with several commercial varieties for an extended comparison. Other points to note from Table 1 are that the whole plot is usually replicated at several locations which are representative of areas where the crop will be grown (sometimes with a very wide geographical spread e.g. N. America, S. America, Europe); the number of locations is not fixed but studies with 6 or more locations are quite common; and field trials are usually conducted over two successive growing seasons, sometimes more than this.

Choice of Compounds for Targeted Analyses

The choice of compounds for analysis is a key decision in any substantial equivalence study. It has to take into account nutritional and toxicological issues including the intended use (food or animal feed), the importance of the crop to the diets of particular groups of consumers and the presence of known natural toxins and anti-nutrient factors.

- Analysis is most often carried out on the raw plant material (e.g. grain, forage) although processed or partly processed samples may also be important (rough rice, brown rice, milled rice etc.). A range of analytical methods is used, each one specific for one component or one class of compounds. Established and validated analytical methods are preferred (with AOAC official methods used where available).
- An early criticism of the substantial equivalence procedure was that there was a lack of consistency in the selection of compounds for analysis. The OECD Task Force for the Safety of Novel Foods and Feeds has issued a series of Consensus Documents for major crops that provide surveys on existing knowledge of the crop composition and provide recommendations of key components

to be analysed for the safety assessment of new varieties. The first document (on canola) was published in 2001 and subsequent documents have appeared on soybean, sugar beet, potato, maize, wheat, rice, cotton, barley, alfalfa and other legumes, cultivated mushroom and sunflower. All these may be downloaded from the OECD website (http://www.oecd.org/document/9/0,3343,en_2649_201185_1812041_1_1_1_1,00.html).

- The analytes for each crop (listed as 'Parameters' in Table 1) include macro- and micronutrients. Proximates, (total) amino acids, fatty acids and a range of minerals are routinely analysed for every crop while analysis of vitamins and other minor components (toxins, anti-nutrients) is based on their known presence and importance in the particular crop. The OECD recommendations should be regarded as providing a basic set of components that may need to be augmented depending upon the details of the modification. In practice all those publications listed in Table 1, where work was also carried out as part of a regulatory approval application, adhered closely to the OECD lists of recommended components.

Considerations for Statistical Analyses

The results of field trials to test substantial equivalence are typically presented (e.g. Ridley et al., 2002) as summary statistical tables that give for each component

- The mean amount and range (minimum-maximum) for that component in the modified plant and its non-transgenic control. The means are taken across the different sites and blocks within a site (where appropriate) but results from each growing season are presented separately. Components for which a significant difference is found (e.g. $p < 0.05$) are indicated. Confidence intervals for the differences in means are also calculated but only occasionally discussed as the basis for the equivalence test (Oberdoerfer et al., 2005)
- A tolerance interval and range for each component calculated for the conventional commercial varieties that are grown in the same field trials. The tolerance interval is calculated to include a chosen proportion (e.g. 99%) of all samples from the population with a specified degree of confidence $100(1-\alpha)\%$, where, e.g. $\alpha = 0.05$
- Minimum-maximum ranges for the component for conventional varieties taken from similar earlier field trials or from the literature. A graphical presentation has also been adopted (Herman et al., 2004) in which, for each component, the mean values for each treatment are plotted in separate columns against a background of the literature range. The mean values from each site are plotted with a different symbol. It is possible to see immediately the spread in values across sites, whether any sites are unusual, and if transgenic and control treatments behave similarly when values fall outside the literature limits.

Significance testing is done by analysis of variance with the treatment as a fixed effect and the sites, blocks and (site x treatment) interaction as random effects (Ridley et al., 2002) or with all factors as fixed effects: if a (site x treatment) interaction is evident an examination of the data on a site by site basis is of interest (Oberdoerfer et al., 2005). Significant differences were found for a few components in all of the transgenics in Table 1. A number of significant differences are inevitable simply because evaluation of each crop involves testing of over 50 components. However the differences found were relatively small, usually not repeated across growing seasons and the mean value for the transgenic invariably fell within the tolerance interval for existing commercial varieties. Therefore the

22

differences that were found have all been judged 'not biologically relevant' and the transgenics in Table 1 were judged to be at least as safe and nutritious as their conventional counterparts.

The use of t-tests and related methods in the context of substantial equivalence has been questioned (Hothorn and Oberdoerfer, 2006). It has been argued that the statistical 'Equivalence test' (Limentani et al., 2005) would be more appropriate. Since detailed recommendations for statistical analysis are not included in the official documents (e.g. EFSA, 2006) we include a discussion of some of the issues in Box 2.

Box 2 Proof of hazard or proof of safety? The t-test and the equivalence test

The standard approach to significance testing for substantial equivalence based on the two-sample *t*-test and applied in the majority of studies in Table 1, has been questioned (Hothorn and Oberdoerfer, 2006). In these studies the hypothesis test is framed as follows:

Null hypothesis H_0: $\mu_{\text{modified}} = \mu_{\text{control}}$ modified crop is 'safe'
Alternative hypothesis H_1: $\mu_{\text{modified}} \neq \mu_{\text{control}}$ modified crop is 'hazardous'

where μ is the expected value (mean) for a particular component. If we take $p < 0.05$ as the level at which to reject H_0 we accept a false positive (Type I error) rate of 5%, i.e. that is the proportion of occasions on which it will be concluded that the treatment has an effect (modified crop is hazardous) when in fact it does not. Hothorn and Oberdoerfer argue that in the context of substantial equivalence it is the false negative (Type II error) rate that is more important and that the hypothesis test above should be recast to provide 'Proof of Safety' rather than 'Proof of Hazard'. A false negative conclusion is that the treatment has no effect (modified crop is safe) when in reality there is a difference between GM and control: to the consumer an erroneous conclusion of safety is of more concern than an erroneous conclusion of hazard. They show that the false negative rate can vary greatly among components as it depends on the variance of the component as well as on the number of replicates: it will almost certainly be undesirably high for at least some 'non-significant' components with the typical low number of replicates in a field trial.

To provide 'Proof of Safety' they propose a hypothesis test which reverses the null and alternative hypotheses in the customary test:

Null hypothesis H_0: $|\mu_{\text{modified}} - \mu_{\text{control}}| \geq \delta$ modified crop is 'hazardous'.
Alternative hypothesis H_1: $|\mu_{\text{modified}} - \mu_{\text{control}}| < \delta$ modified crop is 'safe'

where δ (> 0) is the 'minimally relevant safety difference'. This approach (equivalence test) has the advantage that the statistical test can be applied at a chosen level, α^{safety} which controls the rate of erroneous conclusions that the modified crop is safe when it is not. However, compared with the 'Proof of Hazard', it presents the new task of finding a value for δ, the smallest change that is not biologically relevant, based on the data from comparators and their accepted history of safe use. The authors show how to conduct the statistical tests and calculate confidence intervals for both the difference and the ratio between means ($\mu_{\text{modified}}/\mu_{\text{control}}$). Figure 2 illustrates how different conclusions can be reached using the *t*-test and equivalence

(continued)

Box 1 (continued)

test with confidence intervals created by the two methods (Limentani et al., 2005). With the *t*-test the two means are declared equal if the confidence interval includes zero, as in cases (a) – (c). With the equivalence test the means are only declared equivalent if the confidence interval is completely included within the [-δ, δ] interval, as in cases (a) and (d).

Hothorn and Oberdoerfer (2006) discuss the limitations of using a fixed safety margin for all components and show how component specific values may be derived for δ, from either the non-transgenic control data or from tolerance limits for commercial lines. Statistical analysis of data from multiple sites; multiple treatments (herbicide sprays); and the integration of tests on multiple components into a global equivalence statement are also considered. Practical examples of all these calculations, based on field trial data for genetically modified oilseed rape seeds, are provided in the paper. Other papers have made partial use of the Proof of Safety approach (Oberdoerfer et al., 2005; Li et al., 2007).

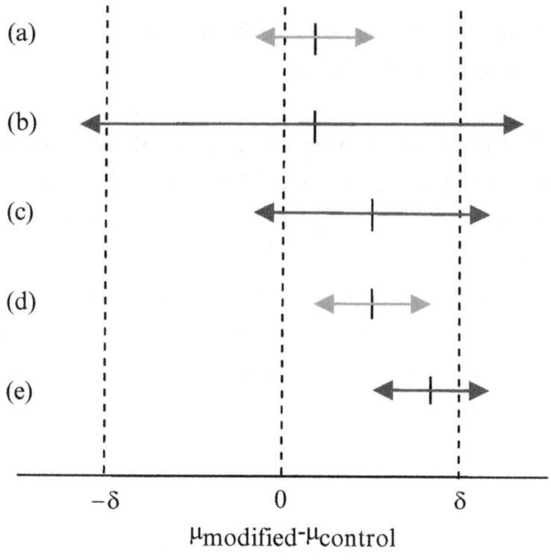

Fig. 2 The two sample *t*-test and the equivalence test can lead to different conclusions for the same difference between means ($\mu_{modifed}$ – $\mu_{control}$) and confidence interval (shown by the double-headed arrow). By the *t*-test the means are concluded to be equal in **(a)** – **(c)** as the confidence intervals include zero, but unequal in **(d)** – **(e)**. By the equivalence test the means of each of the two data sets are concluded to be equivalent in (a) and (d), but not (b) or (c), because the confidence intervals fall entirely within the acceptance range [-δ, δ] for (a) and (d). (Adapted from Limentani et al (2005))

Profiling Methods. Metabolomics

Introduction

Over the last decade the use of non-targeted approaches, involving different profiling methods, has made constant progress in the analysis of complex systems such as plants, human tissues and fluids. These methods employ the so-called post-genomic or '-omics' technologies (Fig. 3). On their different biological levels the '-omics' technologies all share the same strategy and purpose: identification of the differences between two or more sets of samples without a prior selection of the genes, proteins or compound families to analyse. In their review, Cellini et al. (2004) detailed precisely the different principles, possibilities and limits of each '-omics' method in regard to the detection of unintended effects in genetically modified crops.

In this chapter, we will focus on the potential of metabolomics to detect these unintended effects. This method enables the analysis of the overall metabolite composition employing some of the traditional analytical techniques used for the targeted approach. These include GC and

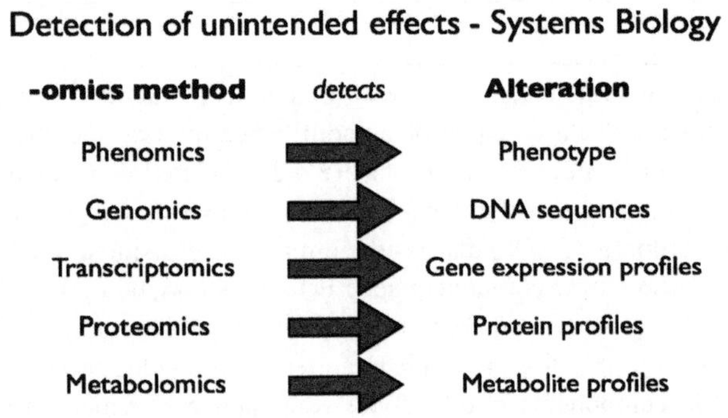

Fig. 3 'Omics' methods and their areas of application. Alterations can be detected at several biological levels in the organism. Transcriptomics, proteomics and metabolomics have all been employed to detect unintended effects but most published examples relate to metabolomics

HPLC, both generally coupled with mass spectrometry for metabolomics, plus additional techniques including NMR and FT-ICR-MS. The spectroscopic instrumentation provides the ability to *measure and identify unknown* as well as to measure known compounds.

Of all the post-genomics technologies, metabolomics offers the best compromise between practical performance and cost per sample. It gives increased scope but reduced accuracy compared with targeted analysis (Fiehn, 2002). Unfortunately, it is impossible to analyse all the different metabolites with a single method or to employ the optimum conditions for every detected analyte. Thus, it is necessary to make some compromises in deciding on the extraction step(s) and technique(s) or the most appropriate combinations of these prior to the analysis. Factors to consider include the polarity and concentration of the analytes. Polarity affects both the choice of extraction solvent and the separation (chromatography) method, and concentration may dictate the detection method since mass spectrometry, for example, has much greater sensitivity than NMR.

Experimental Design and Reporting of Metabolomics Data

The purpose of substantial equivalence analysis is the detection and identification of metabolic differences between a genetically modified organism and its "safe" counterpart or control grown under the same conditions. Indeed, the crop composition is very sensitive to the environment and differences in the growth conditions can lead to variations in metabolic content independent of those arising from genetic modification. Even when the growth conditions, the harvesting of plants and the sample preparation are meticulously controlled, the biological variability remains considerable. Thus, in order to separate the effects of biological variability from the metabolic changes arising from the genetic transformation, the number of replicates needs to be established. The field trial designs described earlier that provided samples for targeted analysis should be equally applicable in metabolomics, although few of the studies mentioned below were carried out on this scale.

The chemical analysis working group of the Metabolomics Standards Initiative proposed a set of minimum reporting standards related to the chemical analysis aspects of metabolomics together with some (non-prescriptive) recommendations on replicate sampling (Sumner et al., 2007). They advised analysts to prefer biological replicates (repetitive analyses of samples obtained from different individuals or pooled individuals from a population) over analytical replicates (repetitive analyses of samples obtained from the same individual or pooled individuals) as biological variance almost always exceeds analytical variance. This report is part of a general community effort to assure the reliability and trustworthiness of metabolomics data and to enable data sharing between different laboratories. One of the most important aspects is the systematic documenting of 'metadata', information about experimental design, sample growth and harvesting, extraction and measurement, quality control measures and data processing (Fiehn et al., 2008).

Sampling and Extraction Procedures

Metabolic profiling can be done on freshly harvested, freeze-dried or frozen stored material. This decision depends on the type of tissue to be analysed (whole plants, roots, leaves, fruits, seeds) and on the analytical techniques used. An efficient extraction method should be reproducible and lead to the extraction of a wide range of constituents across the different chemical classes, taking care to suppress enzyme activity that would change the metabolite composition. A common solvent is the methanol/water mixture, which has been used in different proportions (80/20, 70/30, 50/50 or 20/80) in the case of polar extraction. For instance, De Vos et al. (2007) described a protocol for LC/MS of plant materials and proposed that 75% methanol (v/v) and 0.1% formic acid is the most efficient solvent for extraction of a wide range of compounds including secondary metabolites. Ternary solvent mixtures, for example chloroform/methanol/water (Gullberg et al., 2004; Lisec et al., 2006; Shepherd et al., 2007) or isopropanol/acetonitrile/water (Fiehn et al., 2008), have been recommended for extraction of plant tissues for GC/MS analysis whether the interest is only in polar compounds or in both polar and non-polar compounds. Frenzel et al. (Frenzel et al., 2002) developed a metabolic profiling methodology for substantial equivalence work using rice as model crop. Their approach was based on consecutive extraction of lipids and polar compounds and subsequent fractionations of both extracts. Each fraction was analysed by GC/FID and GC/MS. This method was presented as an unbiased profiling method that allowed isolation of a broad range of compounds from the sample matrix. Even though the method is time-consuming it shows good reproducibility and efficiency in the detection of major and minor compounds, which can increase the chance to detect unintended effects.

Analytical Techniques

A short overview is presented here of the analytical techniques commonly used in metabolomics for the detection of unintended effects in genetically modified crops. More details about these techniques are presented in the Chapter 4 by Hill and Wang.

GC/MS

GC/MS based metabolite profiling is nowadays an integral part of plant functional genomics (Fernie et al., 2004; Fiehn et al., 2000; Roessner et al., 2002). It is a non-biased method which has long been used for metabolic profiling. This technique allows the identification and quantification of low-molecular-weight metabolites (up to ~1,000 Da) and offers the advantage of the availability of stable protocols for the machine set up, sample preparation and chromatography evaluation and interpretation (Lisec et al., 2006). The identification of the compounds is aided by the availability of shared mass spectral and retention time index libraries and databases (Kopka et al., 2005; Schauer et al., 2005; Fiehn et al., 2008). Recently GC-TOF-MS instruments have come to be preferred for metabolic profiling over the more traditional quadrupole mass spectrometers (Santos and Galceran, 2003). The TOF (time of flight) MS presents the advantage of fast scan times which allows the improvement of the deconvolution procedure, the reduction of run times and in some instruments a higher mass accuracy.

HPLC/MS

The second major separation technique is High Performance Liquid Chromatography (HPLC). It is more appropriate than GC for thermolabile or large molecules including bis- and tris-phosphates, CoA adjuncts and lipids (Fiehn, 2002) as well as a large range of secondary metabolites (including alkaloids, flavonoids, glucosinolates, isoprenoids, oxylipins, phenylpropanoids, pigments, saponins). Ultra Performance Liquid Chromatography (UPLC) is used increasingly in metabolomics because it has greater chromatographic resolution and allows shorter run times than HPLC (Grata et al., 2008).

HPLC-ESI-MS (Electrospray Ionisation mass spectrometry) uses a soft ionisation method, which yields the quasi molecular ion, in positive or negative ionisation mode or cation adducts, but a very restricted amount of fragmentation. The identification of unknown compounds is a major challenge in plant metabolomics. Unlike GC/MS, LC/MS libraries have been almost non-existent, especially for secondary metabolites. HPLC-TOF mass spectrometers enable accurate mass determination to several decimal places. This allows candidate empirical formulae to be assigned to peaks with masses of a few hundred or so; the number of candidates is significantly reduced if isotope abundance ratios are available also (Kind and Fiehn, 2006). FTICR-MS (Fourier Transform Ion Cyclotron Resonance mass spectrometry) instruments provide even greater mass accuracy than TOFs. A scheme has been demonstrated, with tomato as an example, for annotation of metabolites using LC/FTICR-MS (Iijima et al., 2008). Each entry in the database of almost 900 metabolites (of which nearly 500 are novel) is annotated with the accurate mass and derived formula plus additional information including the MS/MS spectrum. Further development and public availability of such databases will facilitate characterisation of novel compounds and pathways.

DIMS (Direct Injection Mass Spectrometry) or FIE-MS (Flow Injection Electrospray-MS)

Direct injection of crude extracts can be done into the ESI source of the mass spectrometer without chromatographic separation. A fingerprint spectrum is obtained with a single dominant peak for each metabolite and peaks are separated according to the molecular mass allowing the fingerprint to be used as a classification tool. The technique is very rapid but is quantitatively less reliable than methods that incorporate a separation step. It is at its most powerful on spectrometers with very high mass accuracy such as FT-ICR instruments (Aharoni et al., 2002). A metabolic profiling scheme that uses direct injection FT-ICR/MS has been described (Oikawa et al., 2006). With suitable precautions reliable accurate masses are obtained that can be used to search the

KNApSAcK database for metabolite identification; the signal intensities are sufficiently repeatable for input to statistical classification programmes.

NMR

In the last decade, many reviews have been written on the application of NMR spectroscopy to the metabolic profiling of plants and crops (Colquhoun, 2007; Defernez and Colquhoun, 2004; Holmes et al., 2006; Krishnan et al., 2005).

NMR spectroscopy enables the direct analysis of crude extracts in solution or, using the 'magic angle spinning' (MAS) technique, of intact tissues. Potentially NMR spectroscopy can detect any molecule containing one or more atoms with a non-zero nuclear magnetic moment such as ^{1}H, ^{13}C, ^{14}N, ^{15}N or ^{31}P. The technique is non-destructive and non-invasive and allows the analysis of *in vivo* samples. In routine profiling only one-dimensional (1D) solution state ^{1}H NMR spectra of extracts are acquired. In addition some two-dimensional (2D) NMR spectra are acquired (^{1}H-^{1}H COSY, ^{1}H-^{1}H TOCSY, ^{1}H-^{13}C HSQC, ^{1}H-^{13}C HMBC are the most common experiments) from a few representative samples in order to determine ^{1}H and ^{13}C chemical shifts of constituents in the extracts. These can then be compared with data from chemical shift libraries of pure compounds for identification.

The sensitivity of NMR is several orders of magnitude lower than mass spectrometry. It is impossible to detect compounds in very low concentration. Signal overlap is reduced and sensitivity improved by working at high magnetic field. Most profiling experiments are carried out on 500 or 600 MHz instruments but the cost of going to higher fields is substantial. Nevertheless, the sensitivity can also be improved by using cryogenic probe heads which deliver a three- to fourfold increase in signal to noise ratio by cooling the detection system (but not the sample) to 20 K.

Data Analysis

The aim of the data analysis is to determine whether the genetically modified samples can be differentiated from their controls, and if so, to identify the factors underlying the discrimination. Compared with the conventional (targeted) substantial equivalence studies discussed above (Section 1.2) multivariate statistical methods are of much greater importance in non-targeted analysis, especially for the 'fingerprinting' techniques, DIMS and NMR. The first step of the data analysis is to transform the analytical results into computable numerical data as input for statistical analyses. Several software packages are available to convert the spectroscopic or chromatographic data:

- LC/MS and GC/MS data are three-dimensional in nature with retention time (RT), *m/z* and intensity axes. Recent developments in processing software for unbiased mass peak extraction and retention time alignment across multiple samples include publicly available software packages such as MetAlign, XCMS and MZmine plus commercial packages from all the major instrument manufacturers (see Katajamaa and Orešič, 2007 for a more complete list). These offer the possibility of an untargeted approach which aims to gather information on as many metabolites as possible in each extract analysed. The outputs are typically tables arranged with one sample per row and a combined mass_RT variable per column. The cells of the table contain the integrated value of each peak.
- For NMR spectra, the data can be analysed either at the full acquired resolution or more often after a primary data reduction step (called 'binning' or 'bucketing') whereby

each spectrum is divided into integrated regions over fixed or variable spectral intervals and normalised.

Because of the complexity of the data, it is difficult to make any meaningful comparison of individual spectra or chromatograms simply by inspection. Nevertheless, a useful first step in NMR is to compare average spectra of GM and conventional samples and attempt to identify the chemical origins of the most obvious differences.

- Univariate methods, such as t-test or ANOVA, may be applied to identify significant differences in metabolite levels between GM and controls (Le Gall et al., 2003). If the compounds are explicitly quantified as with GC/MS or LC/MS a direct compositional comparison is possible enabling an extended substantial equivalence study, using data that has been collected in a non-targeted way. A proportion of the signals quantified in GC/MS or LC/MS experiments will be associated with unknown compounds. The indexing by *m/z* value and retention time gives assurance that the same entity is being measured across the different samples and thus statistical comparison can be made. It is, of course, desirable to identify unequivocally any unknown compound for which a significant difference is found. The disadvantage of the univariate approach is that it is difficult to recognise any underlying relationships in the changes for the many compounds measured.
- Multivariate statistical analyses give a much better overall picture of how a given sample relates to other samples. The purpose is to treat all the compounds or peaks simultaneously despite significant challenges in processing such data, mainly related to the fact that the number of variables (data points, peaks or compounds) usually exceeds the number of samples. The mathematical difficulties associated with treating the whole data set can be avoided by using methods that reduce the dimensionality of the data (i.e. reduce the number of variables).
- The main technique used for the compression of data is Principal Component Analysis (PCA) which is an unsupervised statistical method (Kemsley et al., 2007). Its purpose is to transform the original data into a set of "scores" (coordinates) for each sample on the principal component (PC) axes. The principal components are orthogonal to each other and explain progressively less variance in the data set. The relationship between the initial variables and the PC axes is expressed through the "loadings" which are the weights each of the initial variables have in the linear combination that defines the new axes. Scatter plots of the scores and the loadings on the first few PC axes provide a good means of visualizing the data. The scores plots will show any clustering of similar samples, separation of different classes (GM and controls) or the presence of outliers. The loadings plots show which compounds are responsible for the group discrimination, as compounds with high loadings values are most significant. Other statistical methods can also be used such as Hierarchical Cluster Analysis, Partial Least Squares Regression (PLS) (Wold, 1984) or linear discriminant analysis (LDA).

Applications of Metabolomics

Introduction

The applications of targeted metabolite analyses to first generation commercial GM crops, with modified input requirements, were described above. Here we will focus on "state-of-art" metabolite profiling studies for the determination of possible unintended effects, earlier applications in this area having already been reported (Cellini et al., 2004). The emphasis has moved to plants with modified output traits that will bring nutritional or other benefits to consumers. The remainder of the chapter deals mainly with plants where output traits have been modified by direct intervention in the plant metabolism by introduction of novel (foreign) enzymes or by changing the activity of endogenous enzymes. Engineering of plant metabolic pathways to produce high value natural products has been reviewed. Whilst there have been notable successes, introduction of a multi-step pathway into a heterologous target plant may have unexpected outcomes (Dixon, 2005). Metabolomic analysis of transgenic plants can demonstrate the absence of unexpected results or, if they do occur, can help to understand their cause.

An experiment in which a multi-step pathway from *Sorghum bicolor* was introduced into *Arabidopsis thaliana* illustrates these points. It was demonstrated that introduction of the full pathway produced the desired product without accompanying side effects, whereas partial introduction of the pathway might or might not have an adverse impact on the plant (Kristensen et al., 2005). These observations could largely be rationalised by metabolomic and transcriptomic analyses of the 1*x*, 2*x*, 3*x* transgenics (see Fig. 4 for definitions) and comparison with wild-type *A. thaliana*. The pathway by which dhurrin, a cyanogenic glycoside (metabolite S3), is produced from tyrosine (A1) and its relationship to the endogenous *A. thaliana* glucosinolate pathway are shown in Fig. 4. Introduction of the entire pathway led to accumulation of dhurrin in line 3*x* with little additional discernible effect on the phenotype, metabolome or transcriptome. It was argued that the three enzymes E1-E3 form a complex (metabolon) that directs flux through the intermediates S1 and S2 towards dhurrin formation.

Line 1*x* also had a normal phenotype and typical phenylpropanoid profile but accumulated additional glucosinolates derived from tyrosine, evidently through transformation of intermediate S1 by endogenous *A. thaliana* enzymes of the glucosinolate pathway (aided by possible co-localisation of E1 and E5). In contrast the 2*x* line exhibited a stunted phenotype and unusual metabolic profile with reduced

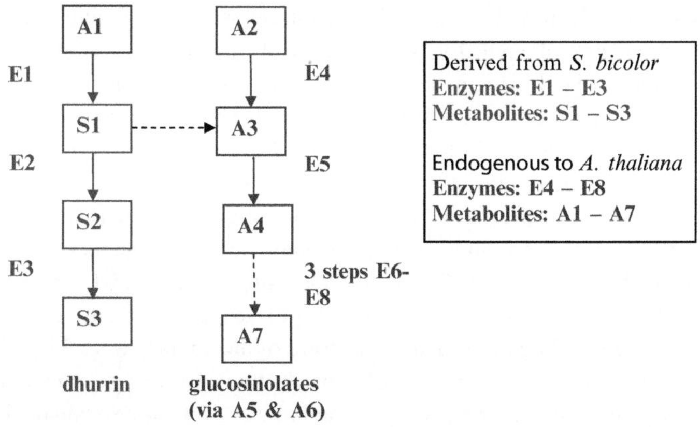

Fig. 4 Schematic representation of the synthesis of dhurrin (S3) from tyrosine (A1) in transgenic *A. thaliana* via the pathway introduced from *S. bicolor*. The other pathway shows the endogenous *A. thaliana* pathway for general synthesis of glucosinolates (A7) from amino acids (A2). Transgenic lines of three *A. thaliana* genotypes 1*x*, 2*x*, 3*x* resulting from introduction respectively of the genes for E1 only, (E1 + E2), and (E1 + E2 + E3) were studied. The 1*x* and 3*x* plants exhibited the same phenotype as wild-type *A. thaliana* but the 2*x* plants were stunted (Kristensen et al., 2005)

levels of the UV-protective compound sinapoyl malate and its precursor sinapoylglucose, and of kaempferol glucosides. Additional compounds that appeared in 2*x* (glucosides of *p*-hydroxybenzoate) were proposed to be S2 detoxification products. This side reaction also led to loss of the important sinapate esters, from a pathway that was not obviously related. It was argued that the pleiotropic effects in line 2*x* resulted from the association of enzymes E1 and E2 which directs flux through S1 towards S2 rather than into the glucosinolate pathway, the route for disposal of S1 in line 1*x*. This example illustrates the delicate balance that determines whether predicted or unexpected results are obtained when introduced and endogenous pathways interact and when factors such as metabolic channelling (metabolons) have to be considered.

Potato

Flow Injection Electrospray-MS (FIE-MS) and GC-TOF-MS have been used to compare the metabolite profiles of tubers of two series of transgenic Desirée potato lines (3 lines for each series) with two parental lines (Desirée wild-type and tissue culture controls) plus four other, non-Desirée, conventional cultivars, making 12 genotypes in all (Catchpole et al., 2005; Enot et al., 2007). In the first series the *1-SST* transgene for the enzyme sucrose:sucrose 1-fructosyltranferase was expressed inducing the production of the trisaccharide 1-kestose and oligofructans (Fig. 5); in the second series both *1-SST* and the fructan:fructan 1-fructosyl-transferase gene (*1-FFT*) were expressed leading to formation of inulin polymers from 1-kestose and oligofructans. The transgenes were originally from globe artichoke and the presence of fructose-containing carbohydrates in the diet has been proposed to have possible beneficial effects for the prevention of colon cancer (Van Loo et al., 1999). The crops were grown in a designed field trial (one location, 2 years) and the analytical samples were taken from individual tubers.

In the first part of the study (Catchpole et al., 2005) 600 potato tuber samples from the 12 genotypes were analysed by FIE-MS. PCA, LDA and decision tree multivariate analysis methods were applied successively to the MS data. Each statistical method enabled discrimination of three groups: the conventional cultivars (including both types of Desirée control), the *SST* and the *SST/FFT* samples. A further differentiation between the individual conventional cultivars could be achieved by the two supervised methods. Display of the results in the form of a 'confusion matrix' showed how many independent test samples were assigned to the correct genotype and what the predicted genotype was in cases when the prediction was incorrect. The authors found that fructose containing compounds (fructans with DP3–7, i.e. the intended products of the modification) were the source of the ions most responsible for the discrimination between GM and non-GM samples. The data analysis was then repeated with these top-ranking ions omitted from the initial data set. Although there was an increase of confusion between SST lines and Desiree controls, suggesting equivalence of GM and control genotypes, a good discrimination between SST/FFT lines and Desirée controls was still observable, implying non-equivalence and the presence of additional discriminating metabolites.

In order to identify these metabolites, Catchpole et al. decided to use GC-TOF-MS, analysing 2,182 tubers (~180 replicates per genotype) for 252 metabolite peaks. Roughly a third of metabolites were fully identified, a third partly known (class of compound) and a third unknown. Univariate statistical analyses were carried out systematically for all 252 metabolites, in addition to multivariate analyses using the methods already described. Two metabolites found in GM samples were not present in any of the conventional controls. The large number of

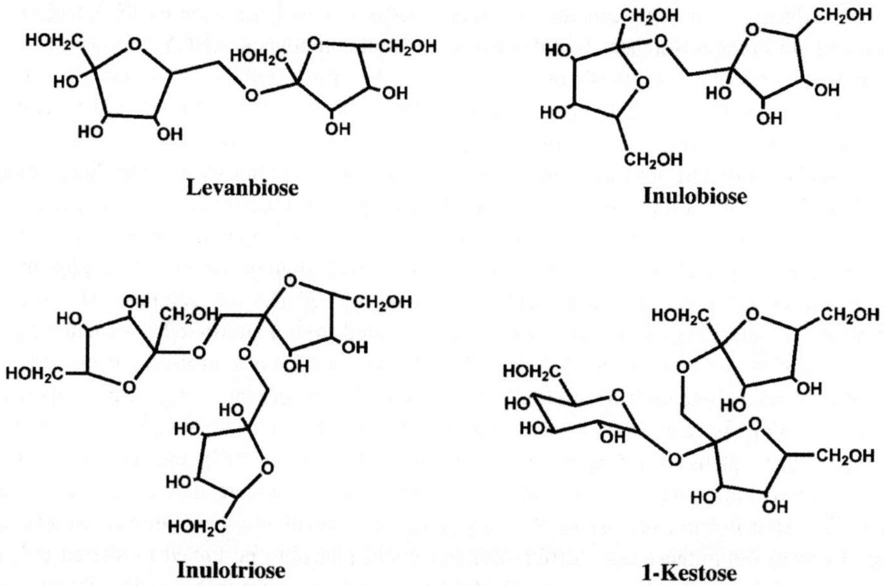

Fig. 5 Chemical structures of simplest fructose-containing oligosaccharides present in GM potatoes in which the *1-SST* gene or both *1-SST* and *1-FFT* genes were expressed

replicates led to rather precise estimates for the mean and standard deviation (SD) of the metabolite concentrations for all the other metabolites in the non-GM cultivars (cv). A "safety range" was defined for each individual metabolite running from a minimum of (mean - SD)$_{cvA}$ to a maximum of (mean + SD)$_{cvB}$ where the minimum and maximum values were derived from the data on all the non-GM cultivars (cf. discussion of parameter δ in Box 2). "Out of range" metabolites, of which the concentrations in the transgenic lines were outside the "safe" range defined with the controls, were sought.

Only six "out of range" markers corresponding to fructans (DP2: levanbiose, inulobiose 1 and 2, and DP3: inulotriose 1 and 2 and 1-kestose (see Fig. 5)), were identified, of which the DP3 saccharides were already known from FIE-MS analysis. Multivariate analysis on the data set, from which these oligosaccharides were removed, failed to separate the GM lines from the Desirée controls although the conventional cultivars were still separated from each other and from the Desirée GM/non-GM group. Thus the metabolite composition of field-grown fructan- and insulin-producing potatoes was within the "safe limits" defined by conventional cultivars, with the exception of the compounds produced by the introduced genes. A second study (Enot et al., 2007) based on the same samples plus further samples from a second harvest demonstrated that consistent results could be obtained by FIE-MS fingerprinting even using two different instruments for analysis of the 2 years' crops.

The metabolite profiles of 40 GM potato lines and controls belonging to Desirée and Record varieties modified in primary carbon metabolism, starch synthesis, glycoprotein processing or polyamine/ethylene metabolism were analysed (Defernez et al., 2004). The ^{1}H NMR and HPLC-UV profiles of the tuber extracts of these samples were recorded and PCA was performed on both datasets: the most obvious difference was between the varieties. Significant differences were also found between two of the four GM Desiree lines with modified polyamine metabolism and their controls. These two lines presented a very abnormal phenotype and the authors pointed out that these samples would not have passed the primary substantial equivalence test based on agronomic performance and phenotype. However, even for these extreme lines the changes in the metabolite concentrations remained modest and the compounds found to be responsible for the GM/control discrimination, such as proline, trigonelline or choline, have elevated concentrations in GM lines, most likely reflecting a general response to osmotic stress. In the other GM series, few significant compositional changes were observed and in these cases differences in mean values did not exceed a two- to threefold change which was small relative to overall variability.

As part of a UK Food Standards Agency project (G02001), metabolomics studies of these GM lines continued with crops grown over three successive years near Dundee, Scotland (the first year under containment, then 2 years' field trials). The transgenic lines studied were the same as the Desirée lines in Defernez et al (2004) with the addition of series called sampat containing the potato S-adenosylmethionine decarboxylase (*SamDC*) gene in sense and antisense orientation under control of a tuber specific promoter (Shepherd et al, 2006). The sense and antisense transformants had respectively increased and decreased levels of polyamines but only the sense lines showed an altered phenotype (higher proportion of small tubers but no change in total yield). Numerous controls were grown at the same time including Desirée wild type, empty vector and tissue culture lines. In the last year's field trial over 20 additional varieties (current and former commercial cultivars, landraces) were grown to assess natural variability. Samples were analysed by NMR (all years) and GC/MS and LC/MS (final year). Statistical analyses were by PCA and, for individual compounds, ANOVA.

An example of the PCA analysis is shown in Fig. 6 (NMR data, 2002 field trial). Overall, there were no gross differences between GM lines and controls: some antisense mal1 lines (modified glycoprotein processing, show stunted growth in field trials) and the antisense sampat line were separated slightly on the PCA plot from the majority of Desirée controls. The separation resulted from a multiplicity of small differences between GM and controls, rarely exceeding twofold (differences involving other compounds but of the same small magnitude were detected by GC/MS and LC/MS) and attributable to both intended and unintended effects. For example changes involving some amino acids (Met, Asn) could plausibly be linked to altered polyamine metabolism; other changes (Phe, Tyr) were more difficult to explain. Much more extensive variation in metabolite concentrations was found among the different commercial and breeding varieties studied than between the Desirée GM lines and controls.

Metabolic profiling by LC/MS led to the discovery of a class of phenolic polyamine conjugates (kukoamines) not previously identified in potato, or in other solanaceous species or common foods (Parr et al., 2005). One of these compounds, N,N'-*bis* (Dihydrocaffeoyl) spermine (Fig. 7), was present at a higher level in the sampat antisense line than in the Desirée controls, or in any of the non-Desirée cultivars. Since *SamDC* is involved in polyamine biosynthesis, changes in phenolic amine conjugates might be said to represent primary effects of the genetic transformation. The kukoamines (which have attracted interest for possible biological activity) are obviously safe to consume in potato as they were found in all the commercial cultivars examined. Nevertheless they are an example of a class of compounds which would never have been targeted for analysis in a conventional substantial equivalence study.

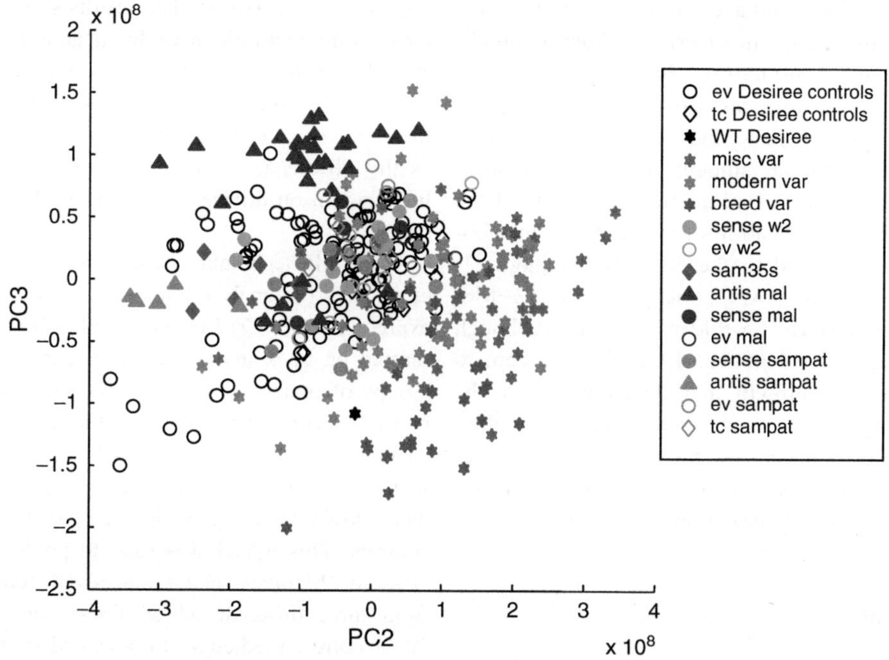

Fig. 6 PCA scores plot of tuber extracts from a field trial in 2002 (NMR data). The Desirée based GM lines (mal1, sampat, sam35s and w2gbss) are shown as filled symbols, open symbols of the same colour are empty vector (ev) or tissue culture (tc) controls generated at the same time as the GM lines and share the same history. Additional Desirée controls are in black. Varieties other than Desirée are shown as stars and classified as modern, breeding or miscellaneous varieties (listed in Parr et al, 2005). There are four replicates per line (each replicate made up of pooled material from one of four plots in the same location). Some constructs (e.g. mal1) are represented by several lines, others (e.g. antisense sampat) by only one. The plot shows that most antisense mal1 lines and antisense sampat are peripheral against the background of the remaining GM lines, controls and other varieties. A division occurs between the Desirée potatoes and many other varieties and there is a further subdivision between modern and breeding cultivars

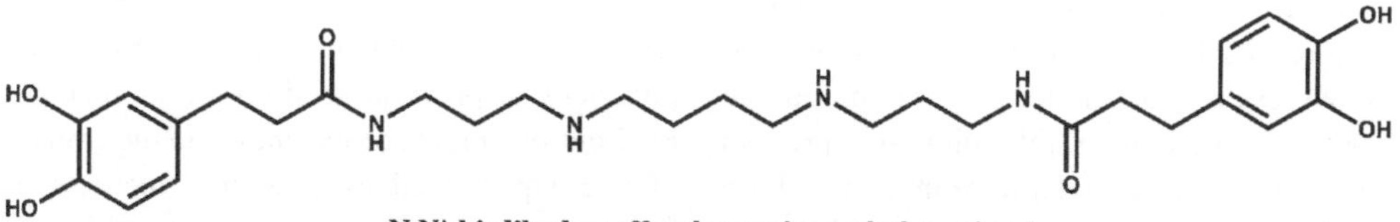

N,N'-bisdihydrocaffeoylspermine or kukoamine A

Fig. 7 Chemical structure of N,N′-bisdihydrocaffeoylspermine (kukoamine A), one member of a family of compounds revealed to be present in potato for the first time by LC/MS metabolite profiling

The identification of metabolite changes in transgenic potatoes containing an altered rice anthranilate synthase α-subunit gene (*OASA1D*) has been investigated (Matsuda et al., 2005). In order to determine the effects of this gene on levels of free tryptophan and anthranilate synthase activity, HPLC-UV profiles of leaf and tuber extracts from the modified and non-modified lines were recorded. The chromatograms showed that the expression of this gene induced a 2- to 20-fold increase in the amount of free tryptophan. They did not reveal unintended effects of the genetic modification in the levels of metabolites compared to the controls but indole-3-acetic acid (IAA), analysed by LC/MS, increased up to 2.5-fold in leaf tissue. Although a higher level of free IAA did not seem to affect the growth and development of the transgenic plants, its accumulation in other species induced by the mutation of genes related to IAA homeostasis resulted in phenotypic abnormalities (Ljung et al., 2002).

Cereals

The substantial equivalence of three genetically modified wheats expressing additional high molecular-weight subunit genes and two corresponding parental lines was assessed (Baker et al., 2006). ^{1}H NMR metabolic profiling was carried out on polar extracts of ground grains (flour) from samples grown in replicate field trials on two UK sites (Rothamsted and Long Ashton) over 3 years. The comparison of the ^{1}H NMR spectra of the modified and non-modified samples by PCA showed a stronger influence of site and year than of genotype. It also revealed increased levels of maltose and/or sucrose and differences in free amino acid content in the line with the highest expression of the transgene. In order to examine more precisely the variations in the amino acid composition, the authors used a GC/MS technique on selected materials (grown in 2000). The most noticeable difference was that the samples grown at Rothamsted contained higher amounts of glutamine, glutamic acid, asparagine and aspartic acid and lower amounts of GABA and proline compared to the samples grown at Long Ashton. The results show that the environment influences the metabolic variation between samples more than the genetic modification and that differences between GM lines and controls remain within the same range as the differences observed between control lines grown on different sites and in different years.

The NMR technique was also used (Manetti et al., 2004) to analyse transgenic maize seeds expressing the *Cry1Ab* gene. PLS-DA performed on bucketed ^{1}H NMR spectra of two groups of maize seeds (one GM line and its non-GM counterparts) enabled the authors to characterise the transgenic seeds as samples with lower concentrations in choline, asparagine, histidine and trigonelline compared to the controls. This model was used to predict the class for 12 other seed samples (all controls from three different inbred lines) and gave 100% correct prediction. In a second study of *Cry1Ab* transgenic maize, the same authors introduced a complex protocol to investigate the metabolite pathways affected by the genetic modification (Manetti et al., 2006). However it is difficult to draw any general conclusions about equivalence or non-equivalence from small scale studies such as this.

Transgenic rice seeds expressing the *OASA1D* gene inducing the accumulation of tryptophan in calli and leaves were analysed using HPLC-UV (Wakasa et al., 2006). The results obtained in this investigation are similar to these obtained on the transgenic potatoes (see previous section), i.e. an increase of about two orders of magnitude in the amount of free Trp as well as a twofold increase of indole-3-acetic acid compared to wild type samples. The HPLC profiles did not highlight substantial changes for the other phenolics. It was demonstrated that *OASA1D* genetic modification could be used to increase the tryptophan content in the seeds of rice as a means of improving its nutritional value for human consumption or animal feed.

A very interesting study (Jacobs et al., 2007) has made a comparison of the same genetic modification (expression of a sodium pumping ATPase, *PpENA1*) in both rice and barley. The transgenic barley and rice lines showed large differences in the effects on their GC/MS metabolic profiles (relative to appropriate null-segregants) although the same modification was involved. On the one hand the genetic modification of the barley samples led to a significant increase of free amino acids (e.g. Ala (8-fold), GABA (9-fold), Gln (22-fold), Pro (85-fold), Gly, Homoserine and Phe); fatty acids (mainly 9,12 (Z,Z) octadecanoic acid, tetradecanoic acid and ethanolamine); salicylic acid and 4-hydroxycinnamic acid. They also observed the decrease of sugars (glucose-6-phosphate, fructose-6-phosphate); citric and isocitric acids, glycolytic and TCA intermediates. On the other hand, the transgenic rice showed a decrease of the free amino acid content (mainly Leu, β-Ala, Ileu, Pro, Thr, Tyr, Val) and an increase of quinic, salicylic (also seen in barley), and glyceric acids. They also noticed the decrease of citric and isocitric acid, as seen in the barley samples, but no effect was observed for the sugars and fatty acids. PCA has been applied to the metabolite profiles and a clear separation between species could be observed on PC1 indicating that the greatest metabolite changes were between species rather than due to transgene expression. The other PCs showed that overall differences are larger between transgenic and control rice leaves than in barley. The observed differences between the two species would be difficult to predict based solely on the known function of a Na^+ pumping ATPase which highlights the efficiency of metabolic profiling in assessing the unexpected effects of transgene introduction. This study shows that it is just as important to investigate the metabolic consequences of a genetic modification when the gene is involved in a general biochemical process such as ion homeostasis as it is when the gene encodes a biosynthetic enzyme.

Tomato

The metabolic profiling of tomato using ^{1}H NMR spectroscopy has been applied in order to detect the unintended effects of the introduction of two maize transcription factors (*LC* and *C1*) inducing the accumulation of flavonols (Le Gall et al., 2003) and in another study, to identify the effects on tomato fruits engineered to accumulate spermine and spermidine (Mattoo et al., 2006).

Alcoholic extracts of freeze-dried *LC/C1* tomatoes and their null-segregants were analysed by NMR (Le Gall et al., 2003) for both types of samples at different stages of maturity. PLS and ANOVA methods, applied to the ^{1}H NMR profiles allowed the identification of the effects of the genetic modification for the mature tomatoes. In addition to the large expected increase of naringenin and kaempferol glycosides (increased at least tenfold), the transgenic lines were characterised by increased amounts of Gln, Asn and trigonelline and a decrease of GABA, Phe, Val, some nucleotides, malic and citric acids and sucrose. However, these changes in mean values were relatively minor (two- or threefold) and remained in the range of the natural variability.

In the study of tomatoes engineered to produce spermine and spermidine (Mattoo et al., 2006) cluster analysis, linear discriminant analysis and PCA were applied to ^{1}H NMR spectra of transgenic and control (wild type and azygous) samples from fruit at three different stages of ripeness (red, pink and green + breaker). Only pink and red fruits showed clear distinction between GM and non-GM samples and significant changes were found in levels of choline, Gln, Asn, Asp, Val, citrate, malate and fumarate and an unidentified compound (δ = 2.78 ppm). However, no novel compounds were detected and the metabolic profiling suggested that spermine and spermidine are perceived as nitrogenous metabolites by fruit cells, which results in the stimulation of carbon sequestration.

Other Vegetables

Once again, NMR spectroscopy has been the technique of choice for the metabolic profiling of transgenic pea leaves (Charlton et al., 2004). ^{1}H NMR spectra obtained from uniformly grown glasshouse plants revealed differences between transgenic and control groups that exceeded the natural variation of the plants. However, the comparison between the wild-type and null-segregant controls also showed significant differences. These changes due to factors independent of the transgene introduction have larger effects on metabolite profiles than the genetic modification. The authors concluded that the examination of natural variability should provide new methods for rapidly demonstrating that a GM plant destined for human or animal consumption lies within the spectrum of normal variation.

Moreover, this technique has also enabled Sobolev et al. (2007) to examine transgenic lettuce lines expressing the *asparagine synthetase A* (*asnA*) gene from *E. coli*, intended to alter the plant nitrogen status and enhance growth. NMR metabolic profiles showed that the transgenic line and the wild-type counterpart shared the same compounds, but statistical analysis (PCA) revealed side effects on the carbon metabolism following genetic modification. The transgenic samples were characterised by a very large increased accumulation of inulin (up to 30-fold), accompanied by significantly lower concentrations of fructose, sucrose and glucose. It was also shown that the genetic modification affected the Krebs cycle, resulting in decreases of alpha-ketoglutarate, succinate, fumarate and malate and increases of citric and tartaric acids, as well as the alteration of N metabolism (significant changes in Glu and GABA). The increased production of inulins was unexpected but these results indicate that the *pMAC:asnA* lettuce line is of interest for future breeding programs since inulins are believed to be beneficial for human health.

Conclusion

Having considered the targeted and metabolomics approaches to compositional analysis in some detail it is worth considering the role of compositional analysis in the overall safety assessment of GM plants. In fact we have considered two groups of GM plant

- Type 1: plants modified to provide improved agronomic performance, usually by insertion of single genes to confer insect resistance or herbicide tolerance. GM crops of this type are widely grown across many parts of the world.
- Type 2: plants modified by more complex metabolic engineering strategies to provide nutritional or health benefits or to provide greater tolerance to stresses (e.g. drought, high salt). Few GM plants of this type have yet been put forward for regulatory approval.

The scheme shown in Fig. 1 is aimed at demonstrating that the GM plant or food derived from it is *at least as safe* as the parent crop. That is, it is a *relative* safety assessment, since there can be no *absolute* guarantee of safety for any food whatever its origin. Hence knowledge of the parent crop, its history of safe use, and its composition is crucial. Knowledge of its composition guides the choice of compounds for the targeted analysis approach and provides the range of baseline values with which the GM plant composition can be compared. The scheme of Fig. 1 incorporates methods for identifying and characterising hazards associated with both *intended* and *unintended* effects of the modification. With regard to the intended effects tests for possible toxic or allergenic properties of introduced products may be carried out (under the heading 'Characterisation of gene products') using a range of *in silico*, *in vitro* and *in vivo* methods following well established protocols for the testing of single chemicals (EFSA GMO Panel, 2008).

The detection of unintended effects relies on a combined molecular, phenotypic, agronomic and compositional analysis of the GM plant. Based on the outcome of these analyses a decision is taken on whether to proceed to the next stage of the safety assessment which would involve a 90-day animal feeding study (Fig. 1, 'Safety Assessment of New GM Crop/ Food'). Since such studies are testing a whole food rather than a single substance they need to be designed with great care if misleading results are to be avoided (EFSA GMO Panel, 2008; Poulsen et al., 2007). When the results of the combined phenotypic, agronomic and compositional analyses from field trials show equivalence between the GM plant and the conventional comparator it is unnecessary to proceed with a feeding study; however when multiple compositional changes, beyond the intended effects, are observed such a study would be required.

In fact none of the Type 1 GM plants, as described in Table 1, nor any of the currently approved and commercially grown GM crops have shown altered phenotypes or compositional changes (beyond those intended). Nevertheless, as a precautionary measure, feeding trials have frequently been carried out and reported in applications for regulatory approval of Type 1 plants even when compositional substantial equivalence had already been demonstrated: according to the EFSA GMO Panel the 90-day feeding trials contribute little additional useful information towards the safety assessment in these cases (EFSA GMO Panel, 2008). On the other hand animal feeding trials will always have to be seriously considered for testing of food and feed derived from Type 2 plants with claims to provide nutritional or health benefits. Indeed it has been asked whether the safety assessment scheme devised for GMOs should be applied to all novel plant foods even if they are products of modern 'conventional' breeding methods (Kok et al, 2008).

The safety assessment scheme outlined in Fig. 1 has been applied on many occasions to Type 1 GM plants but, as yet, not at all to the newer Type 2 plants which are still at the pre-market stage. The more complex metabolic engineering involved in producing these plants makes the occurrence of unintended effects more probable and modification may lead to intended effects which consist of much more than the introduction of a single gene product. Targeted analysis has proved to be a robust method for determining substantial equivalence in Type 1 crops but examples given in 'Applications of Metabolomics' section show that this method alone may not be adequate for safety assessment of Type 2 GMOs where non-targeted metabolomic and other 'omics analyses could be used advantageously.

Targeted analyses already benefit from availability of validated analytical methods, consensus selection of a set of target compounds for each crop and a growing body of information on the compositional variation of those compounds across many growing locations and seasons. Metabolomics faces a number of challenges if it is to attain the same status and be adopted routinely as part of the safety assessment procedure. The 'Profiling Methods. Metabolomics' section the outlined the many variables that can affect the outcome of a metabolomics experiment and equipment and methods are constantly progressing. However it is a sign of the increasing maturity of the area that detailed experimental procedures are being published in specialist 'protocols' journals and adoption of these will lead to greater standardisation between laboratories. Compounds have been selected for targeted analysis because there is some background knowledge as to why they are important; in metabolomics many additional compounds will be measured for the first time. The implications of these compounds for safety, if any, will be unknown as will the 'normal' concentration range for these compounds in conventional crops.

As pointed out in the 'Application of Potato' section the study of GM potatoes by Catchpole et al., (2005) is the first large scale metabolomic analysis of a field-grown crop. They used non-targeted FIE-MS and GC/MS supplemented by targeted LC/MS. FIE-MS is a rapid method, applicable to large numbers of samples: the data is most suitable for analysis by multivariate methods. NMR spectroscopy has also been a widely used fingerprinting technique. It too relies on multivariate data analysis with good potential for identifying unknowns provided they are present at fairly high concentrations.

The GC/MS technique is much closer to the conventional targeted approach in that individual compounds are quantified (not in terms of absolute concentrations: this makes the transfer and sharing of data more complex although there is an increasing trend to deposit results of metabolomic analyses in publicly available databases). GC/MS results permit the use of both univariate and multivariate statistics: the univariate equivalence or t-test is likely to be required by regulators. Using a single technique, GC/MS enables the measurement of many more compounds (250 in the example quoted) than the usual targeted study. However some important compounds are not covered by GC/MS; many compounds measured by GC/MS are not yet identified chemically; and checking the deconvolution and extraction of peak intensities from the raw data remains a labour intensive task. Nevertheless the study by Catchpole et al. (2005) illustrates the feasibility of supplementing targeted analysis methods with current non-targeted GC/MS technology, particularly for GM crops in which the primary metabolism has been modified; aided by recent LC/MS developments including accurate mass measurements and accompanying databases for secondary metabolites, metabolomics is capable of becoming a practical safety assessment tool for the next generation of novel plant-based foods.

References

Aharoni A, de Vos CHR, Verhoeven HA, et al. (2002) Nontargeted metabolome analysis by use of Fourier Transform Ion Cyclotron Mass Spectrometry. OMICS 6:217–234

Baker JM, Hawkins ND, Ward JL, et al. (2006) A metabolomic study of substantial equivalence of field-grown genetically modified wheat. Plant Biotech J 4:381–392

Catchpole GS, Beckmann M, Enot DP, et al. (2005) Hierarchical metabolomics demonstrates substantial compositional similarity between genetically modified and conventional potato crops. PNAS 102:14458–14462

Cellini F, Chesson A, Colquhoun I, et al. (2004) Unintended effects and their detection in genetically modified crops. Food Chem Toxicol 42:1089–1125

Charlton A, Allnutt T, Holmes S, et al. (2004) NMR profiling of transgenic peas. Plant Biotech J 2:27–35

Colquhoun I J (2007) Use of NMR for metabolic profiling in plant systems J Pestic Sci 32:200–212

De Vos R C, Moco S, Lommen A, et al. (2007) Untargeted large-scale plant metabolomics using liquid chromatography coupled to mass spectrometry. Nat Protocols 2:778–791

Defernez M, Colquhoun IJ (2004) NMR approaches to detect unintended effects of genetic modification in plants. In: Nap J P H, Atanassov A and Stiekema W J (ed) Genomics for biosafety in plant biotechnology, NATO science series, IOS press.

Defernez M, Gunning YM, Parr AJ, et al. (2004) NMR and HPLC-UV Profiling of Potatoes with Genetic Modifications to Metabolic Pathways. J Agric Food Chem 52:6075–6085

Dixon RA (2005) Engineering of plant natural product pathways Curr Opin Plant Biol 8: 329–336

EFSA (2006) Guidance document of the scientific panel on genetically modified organisms for the risk assessment of genetically modified plants and derived food and feed. EFSA J 99:1–100

EFSA GMO Panel (2008) Safety and nutritional assessment of GM plants and derived food and feed: the role of animal feeding trials Food Chem Toxicol 46: S2–S70

Enot D, Beckmann M, Draper J (2007) Detecting a difference – assessing generalisability when modelling metabolome fingerprint data in longer term studies of genetically modified plants. Metabolomics 3:335–347

FAO/WHO (2000) Safety Aspects of Genetically Modified Foods of Plant Origin. Report of a Joint FAO/WHO Expert Consultation on Foods Derived from Biotechnology, Geneva, Switzerland, 29 May-2 June 2000. FAO, Rome http://www.who.

int/foodsafety/publications/biotech/ec_june2000/en/index.html
Fernie AR, Trethewey RN, Krotzky AJ, et al. (2004) Metabolite profiling: from diagnostics to systems biology. Nat Rev Mol Cell Biol 5:763–769
Fiehn O (2002) Metabolomics – the link between genotypes and phenotypes. Plant Mol Biol 48: 155–171
Fiehn O, Kopka J, Dormann P, et al. (2000) Metabolite profiling for plant functional genomics. Nat Biotech 18:1157
Fiehn O, Wohlgemuth G, Scholz M, et al. (2008) Quality control for plant metabolomics: reporting MSI-compliant studies Plant J 53: 691–704
Frenzel T, Miller A, Engel K H (2002) Metabolite profiling – A fractionation method for analysis of major and minor compounds in rice grains. Cereal Chem 79:215–221
George C, Ridley WP, Obert JC, et al. (2004) Composition of grain and forage from corn rootworm-protected corn event MON 863 is equivalent to that of conventional corn (*Zea mays* L.). J Agric Food Chem 52:4149–4158
Grata E, Boccard J, Guillarme D, et al. (2008) UPLC-TOF-MS for plant metabolomics: A sequential approach for wound marker analysis in *Arabidopsis thaliana*. J Chrom B (2008) 871: 261–270
Gullberg J, Jonsson P, Nordström A, et al. (2004) Design of experiments: an efficient strategy to identify factors influencing extraction and derivatization of *Arabidopsis thaliana* samples in metabolomic studies with gas chromatography/mass spectrometry Anal Biochem 331:283–295
Harrigan GG, Ridley WP, Riordan SG (2007) Chemical composition of glyphosate-tolerant soybean 40–3–2 grown in Europe remains equivalent with that of conventional soybean (*Glycine max* L.). J Agric Food Chem 55:6160–6168
Herman RA, Phillips AM, Collins RA, et al. (2004) Compositional equivalency of Cry1F corn event TC6275 and conventional corn (*Zea mays* L.). J Agric Food Chem 52:2726–2734
Herman RA, Storer NP, Phillips AM, et al. (2007) Compositional assessment of event DAS-59122–7 maize using substantial equivalence. Regul Toxicol Pharmacol 47:37–47
Holmes E, Tang H, Wang Y, et al. (2006) The assessment of plant metabolite profiles by NMR-Based methodologies. Planta Medica:771–785
Hothorn LA, Oberdoerfer R (2006) Statistical analysis used in the nutritional assessment of novel food using the proof of safety. Regul Toxicol Pharmacol 44:125–135
Iijima Y, Nakamura Y, Ogata Y, et al. (2008) Metabolite annotations based on the integration of mass spectral information. 54:949–962
Jacobs A, Lunde C, Bacic A, et al. (2007) The impact of constitutive heterologous expression of a moss Na + transporter on the metabolomes of rice and barley. Metabolomics 3:307–317
Katajamaa M, Orešič M (2007) Data processing for mass spectrometry-based metabolomics J. Chromatogr. A 1158:318–328
Kemsley E K, Le Gall G, Dainty J R et al. (2007) Multivariate techniques and their application in nutrition: a metabolomics case study Br J Nutr 98:1–14
Kind T, Fiehn O (2006) Metabolomic database annotations *via* query of elemental compositions: Mass accuracy is insufficient even at less than 1 ppm. BMC Bioinformatics 7:234
Kok EJ, Keijer J, Kleter GA, Kuiper HA (2008) Comparative safety assessment of plant-derived foods Regul Toxicol Pharmacol 50: 98–113
König A, Cockburn A, Crevel RWR, et al. (2004) Assessment of the safety of foods derived from genetically modified (GM) crops. Food Chem Toxicol 42:1047–1088
Kopka J, Schauer N, Krueger S, et al. (2005) GMD@CSB DB: the Golm Metabolome Database, Bioinformatics 21:1635–1638
Krishnan P, Kruger N J, Ratcliffe R G (2005) Metabolite fingerprinting and profiling in plants using NMR J Exp Bot 56:255–265
Kristensen C, Morant M, Olsen CE, et al. (2005) Metabolic engineering of dhurrin in transgenic *Arabidopsis* plants with marginal inadvertent effects on the metabolome and transcriptome. PNAS 102:1779–1784
Le Gall G, Colquhoun I J, Davis A L, et al. (2003) Metabolite Profiling of Tomato (Lycopersicon esculentum) Using ^{1}H NMR Spectroscopy as a tool to detect potential unintended effects following a genetic modification. J Agric. Food Chem 51:2447–2456
Li X, Huang K, He X, et al. (2007) Comparison of nutritional quality between Chinese *Indica* rice with *sck* and *cry1Ac* genes and its nontransgenic counterpart. J Food Sci 72:S420–S424
Limentani GB, Ringo MC, Ye F, et al. (2005) Beyond the t-test: Statistical equivalence testing. Anal Chem 77: 221A–226A
Lisec, J, Schauer N, Kopka J, et al. (2006) Gas chromatography mass spectrometry-based

22

metabolite profiling in plants. Nat Protocols 1: 387–396
Ljung K, Hull AK, Kowalczyk M, et al. (2002) Biosynthesis, conjugation, catabolism and homeostasis of indole-3-acetic acid in Arabidopsis thaliana. Plant Mol Biol 50:309–332
McCann MC, Liu K, Trujillo WA, et al. (2005) Glyphosate-tolerant soybeans remain compositionally equivalent to conventional soybeans (*Glycine max* L.) during three years of field testing. J Agric Food Chem 53:5331–5335
Manetti C, Bianchetti C, Bizzarri M, et al. (2004) NMR-based metabonomic study of transgenic maize. Phytochemistry 65:3187–3198
Manetti C, Bianchetti C, Casciani L, et al. (2006) A metabonomic study of transgenic maize (Zea mays) seeds revealed variations in osmolytes and branched amino acids. J Exp Bot 57:2613–2625
Matsuda F, Yamada T, Miyazawa H, et al. (2005) Characterization of tryptophan-overproducing potato transgenic for a mutant rice anthranilate synthase α-subunit gene (OASA1D). Planta 222:535–545
Mattoo AK, Sobolev AP, Neelam A, et al. (2006) Nuclear magnetic resonance spectroscopy-based metabolite profiling of transgenic tomato fruit engineered to accumulate spermidine and spermine reveals enhanced anabolic and nitrogen-carbon interactions. Plant Physiol. 142:1759–1770
Millstone E, Brunner E, Mayer S (1999) Beyond 'substantial equivalence'. Nature 401:525–526
Oberdoerfer RB, Shillito RD, De Beuckeleer M, et al. (2005) Rice (*Oryza sativa* L.) Containing the *bar* gene is compositionally equivalent to the nontransgenic counterpart. J Agric Food Chem 53:1457–1465
Obert JC, Ridley WP, Schneider RW, et al. (2004) The composition of grain and forage from glyphosate tolerant wheat MON 71800 is equivalent to that of conventional wheat (*Triticum aestivum* L.). J Agric Food Chem 52: 1375–1384
OECD (1993) Safety evaluation of foods derived by modern biotechnology: concepts and principles. OECD, Paris
Oikawa A, Nakamura Y, Ogura T, et al. (2006) Clarification of pathway-specific inhibition by Fourier Transform Ion Cyclotron Resonance/Mass Spectrometry-based metabolic phenotyping studies. Plant Physiol 142:398–413
Parr AJ, Mellon FA, Colquhoun IJ, Davies HV (2005) Dihydrocaffeoyl polyamines (kukoamine and allies) in potato (Solanum tuberosum) tubers detected during metabolite profiling. J Agric Food Chem 53:5461–5466
Poulsen M, Schrøder M, Wilcks A, et al. (2007) Safety testing of GM-rice expressing PHA-E lectin using a new animal test design Food Chem Toxicol 45: 364–377
Renwick AG (2004) Risk characterisation of chemicals in food Toxicol Lett 149: 163–176
Ridley WP, Sidhu RS, Pyla PD, et al. (2002) Comparison of the nutritional profile of glyphosate-tolerant corn event NK603 with that of conventional corn (*Zea mays* L.). J Agric Food Chem 50:7235–7243
Ridley WP, Shillito RD, Coats I, et al. (2004) Development of the international life sciences institute crop composition database. J Food Comp Anal 17:423–438
Roessner U, Willmitzer L, Fernie A (2002) Metabolic profiling and biochemical phenotyping of plant systems. Plant Cell Rep 21: 189–196
Rogan GJ, Bookout JT, Duncan DR, et al. (2000) Compositional analysis of tubers from insect and virus resistant potato plants. J Agric Food Chem 48:5936–5945
Santos FJ, Galceran MT (2003) Modern developments in gas chromatography-mass spectrometry-based environmental analysis. J Chromatogr A 1000:125–151
Schauer N, Steinhauser D, Strelkov S, et al. (2005) GC-MS libraries for the rapid identification of metabolites in complex biological samples, FEBS Lett. 579:1332–1337
Shepherd LVT, McNicol JW, Razzo R, Taylor MA, Davies HV, McNicol JW. (2006) Assessing the potential for unintended effects in genetically modified potatoes perturbed in metabolic and developmental processes. Targeted analysis of key nutrients and anti-nutrients. Transgenic Res 15:409–425
Shepherd T, Dobson G, Verrall SR, et al. (2007) Potato Metabolomics by GC-MS: what are the limiting factors? Metabolomics 3: 475–488
Sidhu RS, Hammond BG, Fuchs RL, et al. (2000) Glyphosate-tolerant corn: the composition and feeding value of grain from glyphosate-tolerant corn is equivalent to that of conventional corn (*Zea mays* L.). J Agric Food Chem 48: 2305–2312
Sobolev A P, Segre A L, Giannino D, et al. (2007) Strong increase of foliar inulin occurs in transgenic lettuce plants (Lactuca sativa L.) Ooerexpressing the asparagine synthetase a gene from Escherichia coli. J Agric Food Chem. 55:10827–10831

Sumner L, Amberg A, Barrett D, et al. (2007) Proposed minimum reporting standards for chemical analysis. Metabolomics 3:211–221

Van Loo J, Cummings J, Delzenne N, et al. (1999) Functional food properties of non-digestible oligosaccharides: a consensus report from the ENDO project (DGXII AIRII-CT94–1095). Br J Nutr 81:121–132

Wakasa K, Hasegawa H, Nemoto H, et al. (2006) High-level tryptophan accumulation in seeds of transgenic rice and its limited effects on agronomic traits and seed metabolite profile. J Exp Bot 57:3069–3078

Wold H (1984) PLS regression. In: Johnson NL, Kotz S (eds) Encyclopedia of statistical sciences. Wiley, New York.

Recent Advances in Traditional Medicines and Dietary Supplements

23

Jeevan Prasain and Stephen Barnes

Abstract Traditional medicines and dietary supplements continue to be used in the search for bioactive agents that maintain health and prevent disease. Individual compounds are being isolated and tested, as is the synergism offered by the variety of compounds in botanicals and supplements. Establishing the composition and quality control of botanicals and supplements is essential in order to carry out clinical trails. Interactions between botanicals and conventinal therapeutics must also be carefully considered. Increasing numbers of investigators are applying–Omic technologies to pursue the mechanisms of action of botanicals.

J. Prasain and S. Barnes
Department of Pharmacology and Toxicology, Birmingham, AL 35294, USA
University of Alabama at Birmingham, and Purdue University-University of Alabama, Botanicals Center for Age-Related Disease, Birmingham, AL 35294, USA
jeevan.prasain@ccc.uab.edu

S. Barnes (✉)
Center for Nutrient-Gene Interaction, Birmingham, AL, 35294, USA
Department of Pharmacology and Toxicology, Room 452, McCallum Research Building, University of Alabama at Birmingham, 1918 University Boulevard, Birmingham, AL, 35294, USA
e-mail: sbarnes@uab.edu

Introduction

Historically, natural products, which are the extremely diverse collection of secondary metabolites produced by plants, marine life, and microorganisms as a result of millions of years of evolution, have been a prolific source of medicinal compounds. These have included the anti-cancer agents, paclitaxel (from the Pacific yew tree, *Taxus brevifolia*), colchicine (from *Colchicum autumnale*) and vinblastine (from Catharanthus sp.) (Fig. 1). It has been reported that 25% of all prescriptions dispensed from community pharmacies in the United States contain active principles that are derived from higher plants (Farnsworth & Morris, 1976). The relevance of traditional medicines can be gauged from the 80% of the people in developing countries of the world who still rely on traditional medicine for their primary health care needs. About 85% of traditional medicine involves the use of plant extracts (Farnsworth, 1988). The continued widespread use of traditional medicines is because they are relatively cheaper than Western medicines, are readily available locally, and are considered to have low or no toxicity. Immigrants to the USA have brought some of these medical traditions with them; mainstream Americans have become aware of the alleged benefits of many of these natural product-based medicines, a process encouraged by the 1994 Dietary Supplement Health and Education Act

A.E. Osbourn and V. Lanzotti (eds.), *Plant-derived Natural Products*,
DOI 10.1007/978-0-387-85498-4_23,

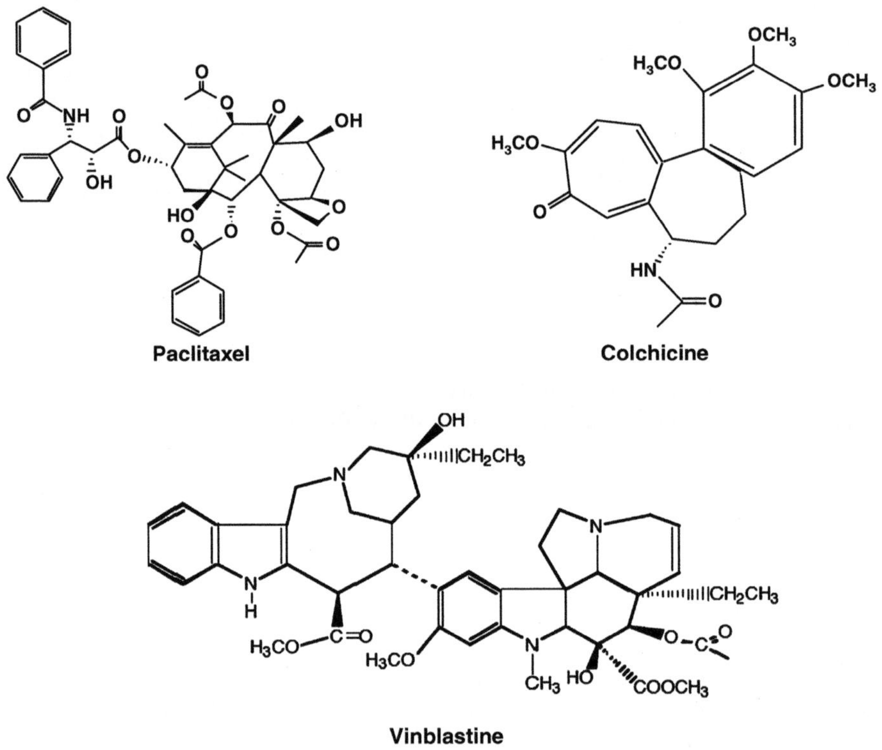

Fig. 1 Structures of some pharmacologically active natural products (paclitaxel, colchicine and vinblastine) isolated from medicinal plants

(DHSEA; http://www.fda.gov/opacom/laws/dshea.html). This legislation, besides allowing existing dietary supplements to be sold in the USA, also contained provisions for NIH Botanicals Centers to carry out research on the composition of botanicals and dietary supplements and the benefits and safety aspects of their use. Six of these centers (Table 1) are currently funded by NIH and carry out research on dietary supplements used to preserve women's health, prevent age-related disease, immunomodulate inflammatory and viral diseases and cancer, prevent the metabolic syndrome, and prevention of atherosclerosis and asthma (see Dietary Supplement Research Centers). A workshop describing the technologies (Barnes et al., 2008) and research carried out in these Botanicals Centers was held at the 2007 Experimental Biology Meetings in Washington, DC; these reports were published in the February 2008 issue of the *American Journal of Clinical Nutrition*.

Natural Products as Sources of Potential Drug Candidates

Following our previous review on the use of traditional medicines and nutraceuticals (Barnes & Prasain, 2005), a comprehensive review of natural products as potential drug targets was published in 2007 (Newman & Cragg, 2007). It emphasized that natural products are the major sources of bioactive molecules and have played a dominant role in the discovery of lead compounds for the development of drugs for the

Table 1 NIH-funded Botanicals Centers

Institution/URL	Focus	Principal botanicals
University of Illinois at Chicago	Women's health	*Black cohosh, red clover, hops*
http://www.uic.edu/pharmacy/centers/uic_nih_botanical_dietary_supplement_research/		
Purdue University-University of Alabama at Birmingham	Age-related disease	*Soy, kudzu root, grapeseed*
http://www.cfs.purdue.edu/fn/bot; http://www.uab.edu/botanicals		
Iowa State University-University of Iowa	Immune modulation	*Ecinachea, Hypericum, Prunella*
http://www.cdfin.iastate.edu/botanical/index.htm		
Pennington Research Center-Rutgers University	Metabolic syndrome	*Russian tarragon, Shilianhua, grape*
http://www.botanical.pbrc.edu		
Wake Forest University-Harvard University	Botanical lipids	*Echium oil; borage seed oil*
http://www.mydietaryfats.org		
Memorial-Sloane-Kettering Medical Center	Immunomodulation	*Echinacea, astragalus, turmeric,*
	Traditional Chinese medicine	*Maitake*

treatment of human diseases (Newman et al., 2003). Among the bioactive secondary metabolites present in medicinal plants, alkaloids, bioflavonoids, isoprenoids and steroids are of high interest. Discoveries of anti-cancer compounds such as taxol, vinblastine and colchicine (Fig. 1) have encouraged the total synthesis of these compounds (Nicolaou et al., 1994; Kuehne et al., 1991; Graening & Schmalz, 2004), or synthesis via natural-product based scaffolds or pharmacophores (Newman & Cragg, 2007). Discovery of lead compounds based on natural products, particularly in the prevention and treatment of cancer, was remarkable during the 1980s to the mid-1990s (Cragg & Newman, 2005).

With the advent of high-throughput screening technology in the 1990s, the demand for the generation of chemical libraries increased tremendously. As a result, a different paradigm based on combinatorial chemistry technology became a major research focus of many pharmaceutical companies, resulting in the synthesis of extremely large numbers of compounds. Despite the generation of impressive, extensive chemical libraries, this changeover from traditional to combinatorial synthesis has been slow to generate any significant increase in the number of new drugs that enter the market (Leach & Hamm, 2000). Indeed, Newman and Cragg (2007) point out that only one anti-cancer drug (sorafenib, or Nexavar) was developed *de novo* during the last 25 years. Using more conventional chemical synthesis and structure optimization methods, an inhibitor of the Bcr-Abl oncogene tyrosine kinase (Gleevec or STI571) was discovered by Druker et al. (1996). However, while it was successfully used in clinical treatment of renal cancer, marked resistance occurred within a year of treatment (Druker et al., 2006).

Because of the slow rate of new drug discovery, a return to biological sources for drug discovery is under way. Why did this happen? Like matter in the universe which is concentrated in just a few places, the parts of chemical space that yield useful compounds may be highly limited. What we see today in the biological resources of this planet is products of selection over the 3.5

billion years of life on Earth. Plant phytochemicals are biologically active both in the plant and often in the life that consumes them. Many enable plants to conduct wars with competitors (other plants, herbivores and microorganisms).

Why Botanicals Are So Attractive

Natural products from botanicals can be regarded as biologically validated structural entities, since they were synthesized by proteins and, therefore, are highly likely to bind to similar motifs again (Schwarz et al., 2007). Phytochemicals are a defensive arsenal produced against environmental stimuli (biotic and abiotic) such as wounding and pathogen attack. Therefore, plant responses result into the biosynthesis of a vast array of phytochemicals. These phytochemicals or secondary metabolites are co-evolved and work in synergy with environmental responses. The synergism is favored by evolution and would enhance the potential of survival of organisms. Plants have developed chemical defenses over millions of years against environmental threats such as UV radiation, reactive oxygen species and microbial attacks. Therefore, phytochemicals are less toxic and biologically active. The practice of traditional herbal medicines is similarly based on multi-component medicines (holistic approach) in contrast to much of modern medicine (single target, single compound).

Mono Drug Therapy Versus Potential Synergism of Multiple Components

A frequent criticism of botanicals is that they are mixtures of bioactive compounds and congeners. In contrast, evaluation of pharmaceutical drugs is usually carried out one at a time. Following the same paradigm, many investigators believe that the bioactive compounds from botanicals should be tested individually to establish their efficacy. However, this approach may miss the whole point about a botanical, namely that it is a mixture of compounds that target more than one receptor, transporter, enzyme site, or signaling pathway. Because of their multiple actions, botanicals can exhibit marked synergism (Schmidt et al., 2007). Also, other components of the botanical or other botanicals may serve to facilitate absorption or to alter the metabolism of the bioactive component. The latter may occur directly by inhibition of an enzyme that otherwise converts the bioactive component to an inactive form, or by preventing the induction of the metabolizing enzyme.

A botanical containing a bioactive amine may be compromised by encountering a monoamine oxidase before it can have its biological effect. However, if a monoamine oxidase inhibitor is present in the botanical or another botanical taken concurrently, then the biological activity of the bioactive amine will be substantially enhanced. Botanicals practitioners are aware of such interactions.

The hallucinogenic beverage *ayahuasca* used in the culture of Amazonian tribes contains both the bioactive amine N,N-dimethyltryptamine (DMT) and monoamine oxidase inhibitors, the beta-carboline alkaloids (McKenna, 2004) (Fig. 2). The presence of the MAO inhibitor causes a profound increase in the oral bioavailability of the DMT and generates a prolonged hallucinogenic event, one well known to the Shamans in these communities (Riba et al., 2003). The soy isoflavone daidzin, previously considered to be an aldehyde dehydrogenase inhibitor (Keung et al., 1997), has been shown to inhibit monoamine oxidases (Rooke et al., 2000) and may alter dopamine-mediated events in susceptible tissues.

Another herbal combination is one in which the second botanical "normalizes" the effect of the first one on the cytochrome P450 system, preventing substantial change in activation of specific isoenzymes. At this time, most herbal–drug interactions are considered to be negative (Gurley et al., 2005, 2008). However, people

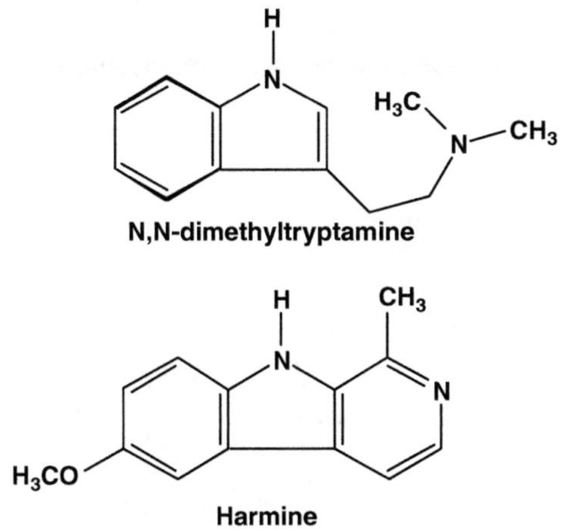

Fig. 2 Structures of N,N-dimethyltryptamine and the monoamine oxidase inhibitor harmine in the hallucinogenic infusion Ayahuasca prepared from the vine *Banisteriopsis caapi*

who are outliers in their response to a given drug compared to the general population may benefit from herbals that "normalize" their own adverse cytochrome P450. This is a possible scenario that may be encountered more often as personalized medicine is introduced into clinical practice.

Dietary Supplements in the Management of Chronic Diseases

Dietary supplements can be distinguished from botanicals because they augment the amount of components that already exist in the *diet*. Examples are minerals and vitamins, They also include foods that are part of the diet, although not necessarily of all peoples. Examples of the latter are soy foods and garlic. A dietary supplement can be a new food formulation that has a high concentration of a presumed bioactive component. Deodorized garlic is a supplement designed to have the bioactive component allicin (Fig. 3). It has been questioned whether these supplements can deliver allicin (Lawson & Wang, 2001). However, allicin is stable when stored at −80°C and enteric coated garlic supplement tablets had similar bioavailability as freshly crushed garlic as measured as allyl methyl sulfide in the breath (Lawson & Gardner, 2005).

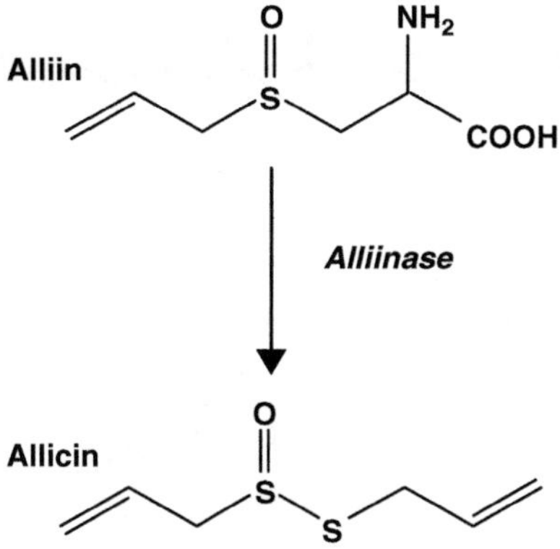

Fig. 3 The allicin chemistry of garlic. Alliin is converted by alliinase to allicin, the bioactive form

Epidemiologic studies have indicated protective effects of plant-based diets on chronic diseases such as cardiovascular disease (CVD) (Hu, 2003), cancer (Béliveau & Gingras, 2007) and diabetes (O'Keefe et al., 2008). Various polyphenols (Fig. 4) present in soy, kudzu root, red clover, flaxseed oil, whole grains, fruits, and vegetables have estrogen-like and antioxidant properties which may be responsible for beneficial effects against CVD (Carlson et al., 2008) and cancer (Khan et al., 2008). Red clover isoflavone supplementation has been extensively studied for decreasing menopausal symptoms (Hidalgo et al., 2005).

Resveratrol (Fig. 4), a bioactive stilbenoid found in nuts and wine from red grapes, has antioxidant, anti-thrombotic, and anti-inflammatory properties (Labinskyy et al., 2006). It also inhibits carcinogenesis in animal models of breast cancer (Whitsett et al., 2006). Muscadine grape skin extract causes phosphatidylinositol 3-kinase Akt-dependent apoptosis in prostate cancer cell lines, but not prostate epithelial cells (Hudson et al., 2007). This is different from resveratrol's effects in the prostate cancer cells which are concentrated on inhibition of the

23

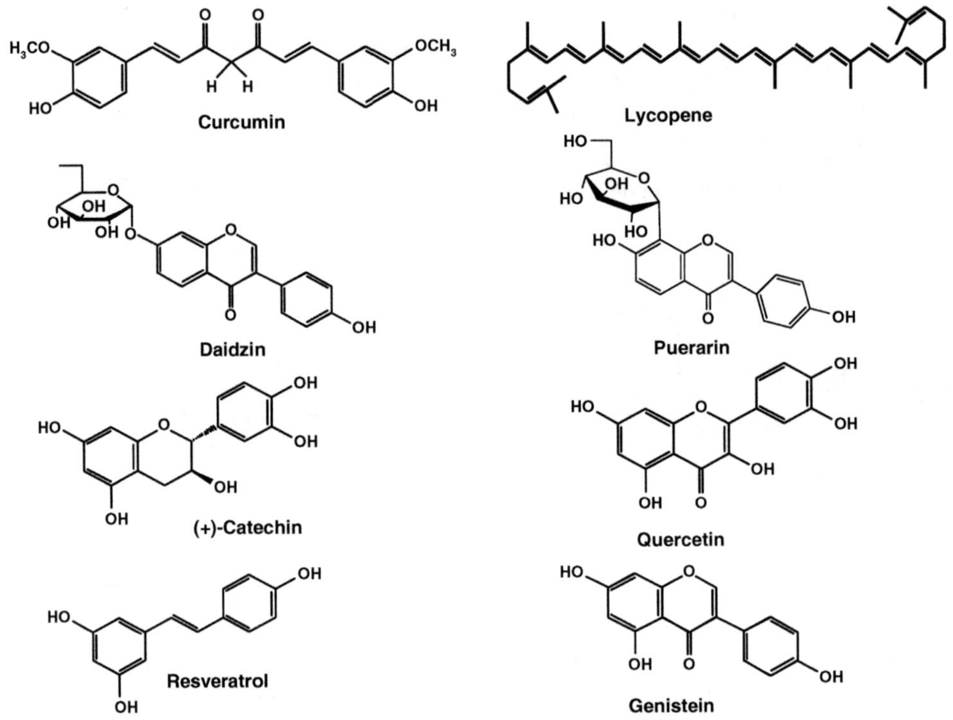

Fig. 4 Structures of phytochemical components in dietary supplements undergoing evaluation of benefits in chronic diseases: (+)-catechin (a flavanol), quercetin (a flavonol), resveratrol (a stilbenoid), lycopene (a carotenoid), genistein (an isoflavone) and puerarin and daidzin (8-C- and 7-O-glucosides of the isoflavone daidzein). They are each drawn so that the A ring is in the same orientation

G_1–S transition in the cell cycle and suggests that muscadine grape skin extracts contain chemopreventive phytochemicals other than resveratrol. Lycopene (Fig. 4), a potent antioxidant carotenoid in tomatoes and other fruits, protects against prostate (Dahan et al., 2008) and other cancers (Seren et al., 2008). Plant sterol-enriched foods are an effective dietary adjuvant in reducing cardiovascular risk by lowering total cholesterol and low density lipoprotein-cholesterol (LDL-C) in serum by up to ~15% (Patch et al., 2006; Jones et al., 2000).

Although there is limited scientific evidence on the effectiveness of dietary supplements on clinical diabetes, a wide range of products claiming to lower blood glucose levels or prevent and treat diabetes complications is marketed to the public (Geil & Shane-McWhorter, 2008). These include aloe vera (Pérez et al., 2007), bitter melon (Shih et al., 2008), chromium (Sahin et al., 2007), cinnamon (Ziegenfuss et al., 2006), fenugreek (Vijayakumar & Bhat, 2008), ginseng (Dascalu et al., 2007), gymnema (Cefalu et al., 2008), and nopal (Yeh et al., 2003). Recent studies using resveratrol and the kudzu isoflavone C-glucoside puerarin (Fig. 4) suggest that these phytochemicals may have therapeutic value in type II diabetes (Milne et al., 2007; Meezan et al., 2005).

The use of dietary supplements in prevention of neurodegenerative diseases is of great interest given the aging of many populations.

Botanicals such as *Ginkgo biloba* xtract (Napryeyenko & Borzenko, 2007), grape seed extract (Deshane et al., 2004; Peng et al., 2005) and coffee (Arendash et al., 2006) are widely considered beneficial in preventing cognitive decline. An extract of *G. biloba* leaves is one of the most popular dietary supplements in the United States and is considered a memory enhancer. In Europe, it is a commonly prescribed drug for treatment of age-related deterioration, including degenerative dementias of the Alzheimer type (AD) (Luo, 2001). A recent clinical study with *G. biloba* highlighted an important point when studying the effects of supplements and botanicals on neurodegenerative disease – when the study results were analyzed on an intent-to-treat basis, there was no significant difference compared to placebo control (Dodge et al., 2008). However, when subject compliance was taken into account, reductions in progression of the clinical dementia rating and in loss of memory were observed. Disturbingly, the occurrence of strokes in *G. biloba* treated subjects was also increased.

Curcumin (Fig. 4), a polyphenol in turmeric, has been studied extensively. It has apparent preventive effects in models of cancer, inflammatory diseases, cystic fibrosis, neurodegenerative disease and atherosclerosis (Strimpakos & Sharma, 2008). Studies in colon cancer patients suggest that doses as high as 8 g/day are safe (Johnson & Mukhtar, 2007). Curcumin has several potential mechanisms, including as an antioxidant, p53- and protein kinase C-mediated signaling (Garg et al., 2008) and ubiquitination (Zhang et al., 2007). In the latter event, breast cancer cells secrete exosomes that fuse with natural killer (NK) cells. These exosomes release tumor proteins into the NK cells and inhibit the activation of NK cells by IL-2. Curcumin at nM concentrations causes increased protein ubiquitination in the exosomes. This is thought to target the exosomal proteins after fusion with the NK cells to the proteasome, thereby restoring NK cell function and recognition of the tumor cells.

Challenge and Opportunities with Dietary Supplements

The estimated prevalence of dietary-supplement use among US adults was 73% in 2002 and this trend is increasing each year. Appropriate use of dietary supplements within the paradigm of evidence-based medicine may be a challenge for medical doctors and non-physician clinicians (Sadovsky et al., 2008). The growing number of dietary supplements available on supermarket shelves and in health food stores are based on previous pre-clinical scientific literature or case reports to back their claims. Only now (in some cases) are they being systematically subjected to prospective randomized, placebo-controlled clinical trials. As noted previously, *G. biloba* is widely marketed as memory enhancer or to treat Alzheimer's disease and other types of dementia. Clinical trials are being reported on its benefits with conflicting results (Dodge et al., 2008; McCarney et al., 2008). Similarly, Echinacea is commonly used as a supplement to stimulate the immune system of the body. Studies in mice support this concept (Zhai et al., 2007). However, a double-blind, placebo-controlled crossover designed clinical trial (Schwarz et al., 2002) found no effect of Echinacea.

Table 2 shows the major herbal dietary supplements available in the USA. It appears that consumers prefer to take dietary supplements with perception that they provide benefits comparable to those found in prescription drugs, with fewer side effects. However, consumers need to make sure that health beneficial claims of dietary supplements have been supported by science-based facts with adequate quality evaluation.

There is also a need for competent monitoring and scientific research to maintain quality assurance (QA) or control (QC) of botanical,

23

Table 2 Top ten botanical dietary supplements in the USA

Herbal	Claimed benefit	Bioactive principle(s)
G. biloba	Enhancement of memory	Terpene lactones, flavonoids
Garlic	Cardiovascular diseases	Allicin, S-allylcysteine
Echinacea	Improve immune system	Alkylamides, flavonoids
Ginseng	Anti-cancer, liver injuries	Ginsenosides
Saw Palmetto	Chronic prostatitis	Acylglycerides
St. John's Wort	Depression	Hypericin, hyperforin
Feverfew	Anti-inflammatory, anti-cancer	Sesquiterpene lactone
Ginger	Anti-inflammatory, anti-cancer	Diarylheptanoids, gingerols
Valerian	Mild sleep disorders, sedative	Iridoids, sesquiterpenoid
Ephedra[a]	Weight loss	Ephedrine

[a]Ephedra is now banned as a dietary supplement by the FDA

chemical, manufacturing, biological, and clinical aspects in order to guarantee value, safety, and efficacy of herbal products. The general public believe that dietary supplements are subject to government regulations similar to those for over-the-counter medications and have undergone extensive review by the Federal Drug Administration (FDA), which is generally not the case. Under the DSHEA Act, companies producing dietary supplements don't have to seek drug approval from the FDA to market their products, but instead are required to submit a safety profile 75 days ahead of the first interstate introduction of the dietary supplement for sale.

Quality Control of Dietary Supplements and Botanicals

In order that reproducible results can be obtained from botanicals in research studies, all investigators should practice good quality control methods both in the selection of the source of botanicals and in the experimental design of their experiments. In a development that began in 2005, investigators seeking funding from the National Center for Complementary and Alternative Medicine (NCCAM), a part of the NIH, are required to provide answers to detailed questions about the nature of the botanical to be used in the proposed study, its origin, chemical composition and stability, and about the analytical methods that will be used to verify its composition and stability throughout the course of the investigation (see http://grants.nih.gov/grants/guide/notice-files/NOT-AT-05–004.html). This thorough product quality evaluation ensures that there is the highest probability that the proposed study will lead to clear outcomes. In the past questions have arisen about which Echinacea sp. were used in clinical trials (Dennehy, 2001). Readers of work on "soy isoflavones" can be confused by failing to realize the source of the isoflavones. When they come from cotyledon-based products, they are predominantly genistein and daidzein *O*-glucosides. Hypocotyl-derived isoflavones are enriched in daidzein and glycitein *O*-glucosides, but not genistein. Many of the "soy isoflavones" sold for women's health are predominantly *C*-glucosides and not *O*-glucosides (Fig. 4) (Prasain et al., 2003). It is intended that the research funded by NCCAM will either place the botanical in a not-ready-for-further research category, or will result in confirmation of its proposed bioactivity. Interestingly, this may lead to other NIH Institutes embracing the botanical as a viable agent that should be included in their research portfolios. It's worth noting that such attention to detail would benefit all of NIH-funded research. Indeed, NIH is now requiring verification of the identity of cell lines to be used in NIH research (http://grants.nih.gov/grants/guide/notice-files/NOT-OD-08–017.html).

Botanical/Drug Interactions

In recent years, multiple case reports of herb – drug interactions have been published (Gardiner et al., 2008), including those that occur in children (Goldman et al., 2008) (Table 3). This should not surprise anyone. Drug-drug interactions occur for many synthetic pharmaceuticals. The statins used in long-term therapy to regulate cholesterol synthesis decrease cytochrome P450 isoenzyme Cyp3A4 activity and thereby prevent the metabolism of other drugs, leading to toxic levels (Igel et al., 2001). Similarly, deaths have occurred in soldiers were being treated for pain associated with their injuries – in one case of a death, 11 prescription drugs were being used – these included "painkillers to treat his physical wounds from an explosion in Iraq and drugs to ease the nightmares, insomnia and memory loss caused by his post-traumatic stress disorder and traumatic brain injury" (National Public Radio, 2008).

Regarding herbal – drug interactions, 32 drugs interacting with herbal medicines in humans have been described (Yang et al., 2006). These drugs mainly include anticoagulants (warfarin, aspirin and phenprocoumon), sedatives and antidepressants (midazolam, alprazolam and amitriptyline), oral contraceptives, anti-HIV agents (indinavir, ritonavir and saquinavir), cardiovascular drugs (digoxin), immunosuppressants (cyclosporine and tacrolimus) and anticancer drugs (imatinib and irinotecan). Most of them are substrates for cytochrome P450s (CYPs) and/or P-glycoprotein (Pgp) and many of which have narrow therapeutic indices (Yang et al., 2006). For example, Kava is a widely consumed psychoactive beverage made from the root of the pepper plant, *Piper methysticum*. It contains herbal ingredients for reliving anxiety and tension. The direct toxicity of kava extracts is quite small; however, they interact with other drugs, in some cases potentiating their hepatotoxicities (Clouatre, 2004).

There are several resources to use to identify potential herb – drug interactions. Cancer patients

Table 3 Drugs subject to negative interactions with herbals/botanicals (Yang et al., 2006)

Herb/botanical	Drug
Betel nut	Procyclidine
Danshen	Warfarin
Donggui	Warfarin
Garlic	Ritinovir; Saquinavir; Warfarin
G. biloba	Aspirin; Digoxin; Trazadone; Warfarin
Ginseng	Digoxin; Warfarin
Kava	Levodopa
Milk Thistle	Indinavir
Piperine	Propanolol
St. John's Wort	Alprazolam; Amitryptyline; Cyclosporine; Dextromorphan; Digoxin; Fexofenadine; Imatinib; Irinotecan; Loperamide; Methadone; Midazolam; Nefazodone; Nevirapine; Omeprazole; Paroxetine; oral contraceptives; Phenprocoumon; Setraline; Simvastine; Tacrolimus; Theophylline; Tolbutamide; Voriconazole; Warfarin

and other interested parties should refer to the web-resource provided by the Integrated Medicine group at Memorial-Sloan-Kettering Cancer Center in New York (http://www.mskcc.org/mskcc/html/11570.cfm). A broader resource has been compiled by Dr. Subhuti Dharmananda at the Institute for Traditional Medicine, Portland, Oregon (http://www.itmonline.org/arts/herbdrug2.htm). It is clear that the medical profession continues to be poorly educated about the use of botanicals and supplements (Xu & Levine, 2008). It is important therefore that medical students in training should be given the opportunity to explore these issues.

Botanicals and Toxicities

Botanical preparations have been used medicinally for thousands of years as traditional medicines. Many commercially available botanical products are being marketed in the United States with little or no publicly available scientific validation of efficacy or consistency. However, there are toxicological concerns that are known to exist for some dietary supplements and they require appropriate attention. Toxicity associated with botanical dietary supplements may result from production problems such as the misidentification or mislabeling of the plant species used in a product; use of incorrect parts of plants; contamination or adulteration with pharmaceutical agents; or contamination with pesticides, herbicides, heavy metals, or microbes (van Breemen et al., 2008). There has been a serious concern about bacterial and fungal contamination of botanicals in dietary supplements or other traditional medicine products. The microbial contamination is most often caused by improper cleaning, drying or storage of the plant material. Therefore, manufacturers should provide quality assurance to consumers regarding the safety of microbial contamination since even very low-level contamination may result in serious infections.

Metals like mercury and arsenic are widely used in Ayurvedic medicines and in high levels, these heavy metals can lead to heart damage, brain damage, kidney failure, reproductive failure, cancer, and death.

Plant sterols and their saturated derivatives, known as stanols, are considered to be cholesterol reducers (Turnbull et al., 1999). Recently, there is an increasing trend in fortification of foods with plant sterols and stanols. However, the consumption of optimal dose of stanols for their beneficial effects in humans is of high importance.

Omic Technologies

Application of omic-technologies (proteomics and metabolomics) is an attractive approach which is in line with the holistic concept and practices of traditional medicines. Metabolomics, particularly, is related with the global or system-wide characterization of small molecule metabolites in biofluids, and tissues using technologies such as nuclear magnetic resonance (NMR), and liquid chromatography tandem mass spectrometry (LC-MS/MS). The importance of "omic" science in human health and nutrition has been realized to identify the function of genes and describe the effects of toxicological, pharmaceutical, nutritional and environmental interventions (Watkins & German, 2002). With the advent of human genome and the concurrent technological developments, genotyping, transcriptomics, proteomics and metabolomics are now available for research in botanicals, dietary supplements and phytomedicine. A comprehensive review on application of the "Omic" technology in phytomedicine has been published by Ulrich-Merzenich et al. (2007) and the term "herbogenomics" has been introduced (Kang, 2008). Furthermore, it has become an important approach to understand mechanisms for drug or xenobiotic actions. A major methodological challenge and first pre-requisite of nutrigenomics is integrating genomics (gene analysis), transcriptomics (gene expression analysis), proteomics (protein expression analysis) and metabonomics (metabolite profiling) to define a "healthy" phenotype (Kussmann et al., 2006). Besides "foods", these technologies can be used to screen the target molecules of the action of traditional medicines (Deshane et al., 2004), and to explore their mechanisms of action. Future research should focus more on the development of multi-target approaches capable to evaluating pharmacological effects of multi-components within a mixture using various 'Omic' technologies. This requires cost effective, sensitive, and robust high-throughput screening techniques. For this, LC-MS/MS has become the most applicable technique for comprehensive analysis of molecules in different stages of systems biology (Feng et al., 2008). LC-MS/MS enables not only profiling of metabolites, but also their quantification which is critical in metabolomic or proteomic studies.

Integrating Traditional and Modern Medicines

With the recognition of the importance of alternative and complementary medicines, the potential of the integrating of traditional and modern medicine in future medical practice is increasing. Traditional medicine relies on the use of natural herbs and herb products for therapeutic measures to boost physical, mental, social and spiritual harmony and improve quality of life. However, the scientific aspects and valuable experiences of traditional medicine should continue to be developed based on modern scientific methodology. A successful, integrated health care system would facilitate more efficient use of traditional medical resources, and enhance overall quality of life not only for poor countries but also industrialized nations. Ethnomedical values of traditional medicine should be evaluated in molecular level with cutting-edge biomedical research. The merger of traditional medicine and modern medicine in a health care system is essential in the twenty-first century.

References

Arendash GW, Schleif W, Rezai-Zadeh K, Jackson EK, Zacharia LC, Cracchiolo JR, Shippy D, Tan J. Caffeine protects Alzheimer's mice against cognitive impairment and reduces brain beta-amyloid production. Neuroscience. 2006;142:941–52.

Barnes S, Prasain JK. Current progress in the use of traditional medicines and nutraceuticals. Curr Opin Plant Biol. 2005;8:324–8.

Barnes S, Birt DF, Cassileth BR, Cefalu WT, Chilton FH, Farnsworth NR, Raskin I, van Breemen RB, Weaver CM. Technologies and experimental approaches at the National Institutes of Health Botanical Research Centers. Am J Clin Nutr. 2008; 87:476S–80S.

Béliveau R, Gingras D. Role of nutrition in preventing cancer. Can Fam Physician. 2007;53:1905–11.

Carlson S, Peng N, Prasain JK, Wyss JM. Effects of botanical dietary supplements on cardiovascular, cognitive, and metabolic function in males and females. Gend Med. 2008;5 Suppl A:S76–90.

Cefalu WT, Ye J, Wang ZQ. Efficacy of dietary supplementation with botanicals on carbohydrate metabolism in humans. Endocr Metab Immune Disord Drug Targets. 2008;8:78–81.

Cragg GM, Newman DJ. Plants as a source of anti-cancer agents. J Ethnopharmacol. 2005;100:72–9.

Clouatre DL. Kava kava: examining new reports of toxicity. Toxicol Lett. 2004;150:85–96.

Dahan K, Fennal M, Kumar NB. Lycopene in the prevention of prostate cancer. J Soc Integr Oncol. 2008;6:29–36.

Dascalu A, Sievenpiper JL, Jenkins AL, Stavro MP, Leiter LA, Arnason JT, Vuksan V. Five batches representative of Ontario-grown American ginseng root produce comparable reductions of postprandial glycemia in healthy individuals. Can J Physiol Pharmacol. 2007;85:856–64.

Dennehy C. Need for additional, specific information in studies with echinacea. Antimicrob Agents Chemother. 2001;45:369–70.

Deshane J, Chaves L, Sarikonda KV, Isbell S, Wilson L, Kirk M, Grubbs C, Barnes S, Meleth S, Kim H. Proteomics analysis of rat brain protein modulations by grape seed extract. J Agric Food Chem. 2004;52:7872–83.

Dietary Supplement Centers: http://ods.od.nih.gov/Research/Dietary_Supplement_Research_Centers.aspx

Dodge HH, Zitzelberger T, Oken BS, Howieson D, Kaye J. A randomized placebo-controlled trial of Ginkgo biloba for the prevention of cognitive decline. Neurology. 2008;70:1809–17.

Druker BJ, Tamura S, Buchdunger E, Ohno S, Segal GM, Fanning S, Zimmermann J, Lydon NB. Effects of a selective inhibitor of the Abl tyrosine kinase on the growth of Bcr-Abl positive cells. Nat Med. 1996;2:561–6.

Druker BJ, Guilhot F, O'Brien SG, Gathmann I, Kantarjian H, Gattermann N, Deininger MW, Silver RT, Goldman JM, Stone RM, Cervantes F, Hochhaus A, Powell BL, Gabrilove JL, Rousselot P, Reiffers J, Cornelissen JJ, Hughes T, Agis H, Fischer T, Verhoef G, Shepherd J, Saglio G, Gratwohl A, Nielsen JL, Radich JP, Simonsson B, Taylor K, Baccarani M, So C, Letvak L, Larson RA, IRIS Investigators. Five-year follow-up of patients receiving imatinib for chronic myeloid leukemia. N Engl J Med. 2006;355: 2408–17.

Farnsworth NR. Screening plants for new medicines. Chapter 9. In: EO Wilson (Ed), *Biodiversity*, Washington, DC: National Academy Press, 1988.

Farnsworth NR, Morris RW. Higher plants – the sleeping giant of drug development. Am J Pharm Sci Support Public Health. 1976;148:46–52.

Feng X, Liu X, Luo Q, Liu BF. Mass spectrometry in systems biology: an overview. Mass Spectrom Rev. 2008 July 17.

Gardiner P, Phillips R, Shaughnessy AF. Herbal and dietary supplement – drug interactions in patients with chronic illnesses. Am Fam Physician. 2008;77:73–8.

Garg R, Ramchandani A, Maru G. Curcumin decreases 12–0-tetradecanoylphorbol-13-acetate-induced protein kinase C translocation to modulate downstream targets in mouse skin. Carcinogenesis. 2008 May 13 [Epub ahead of print]

Geil P, Shane-McWhorter L. Dietary supplements in the management of diabetes: potential risks and benefits. J Am Diet Assoc. 2008;108: S59–65.

Goldman RD, Rogovik AL, Lai D, Vohra S. Potential interactions of drug-natural health products and natural health products-natural health products among children. J Pediatr. 2008;152:521–6.

Graening T, Schmalz H-G. Total syntheses of colchicine in comparison: a journey through 50 years of synthetic organic chemistry. Angew. Chem. Int. Ed. 2004;43;3230.

Gurley BJ, Gardner SF, Hubbard MA, Williams DK, Gentry WB, Khan IA, Shah A. In vivo effects of goldenseal, kava kava, black cohosh, and valerian on human cytochrome P450 1A2, 2D6, 2E1, and 3A4/5 phenotypes. Clin Pharmacol Ther. 2005;77:415–26.

Gurley BJ, Swain A, Hubbard MA, Williams DK, Barone G, Hartsfield F, Tong Y, Carrier DJ, Cheboyina S, Battu SK. Clinical assessment of CYP2D6-mediated herb-drug interactions in humans: effects of milk thistle, black cohosh, goldenseal, kava kava, St. John's wort, and Echinacea. Mol Nutr Food Res. 2008 Jan 23 [Epub ahead of print].

Hidalgo LA, Chedraui PA, Morocho N, Ross S, San Miguel G. The effect of red clover isoflavones on menopausal symptoms, lipids and vaginal cytology in menopausal women: a randomized, double-blind, placebo-controlled study. Gynecol Endocrinol. 2005;21:257–64.

Hu FB. Plant-based foods and prevention of cardiovascular disease: an overview. Am J Clin Nutr. 2003;78:544S–551S.

Hudson TS, Hartle DK, Hursting SD, Nunez NP, Wang TT, Young HA, Arany P, Green JE. Inhibition of prostate cancer growth by muscadine grape skin extract and resveratrol through distinct mechanisms. Cancer Res. 2007;67:8396–405.

Igel M, Sudhop T, von Bergmann K. Metabolism and drug interactions of 3-hydroxy-3-methylglutaryl coenzyme A-reductase inhibitors (statins). Eur J Clin Pharmacol. 2001;57:357–64.

Johnson JJ, Mukhtar H. Curcumin for chemoprevention of colon cancer. Cancer Lett. 2007;255: 170–81.

Jones PJ, Raeini-Sarjaz M, Ntanios FY, Vanstone CA, Feng JY, Parsons WE. Modulation of plasma lipid levels and cholesterol kinetics by phytosterol versus phytostanol esters. J Lipid Res. 2000;41:697–705.

Kang YJ. Herbogenomics: from traditional Chinese medicine to novel therapeutics. Exp Biol Med (Maywood). 2008 June 5 [Epub ahead of print].

Khan N, Afaq F, Mukhtar H. Cancer chemoprevention through dietary antioxidants: progress and promise. Antioxid Redox Signal. 2008;10:475–510.

Keung WM, Klyosov AA, Vallee BL. Daidzin inhibits mitochondrial aldehyde dehydrogenase and suppresses ethanol intake of Syrian golden hamsters. Proc Natl Acad Sci USA. 1997;94:1675–9.

Kuehne Martin E.; Matson Patricia A. Bornmann, William G. Enantioselective syntheses of vinblastine, leurosidine, vincovaline and 20'-epi-vincovaline. J Org Chem 1991;56:513–28.

Kussmann M, Raymond F, Affolter M. OMICS-driven biomarker discovery in nutrition and health. J Biotechnol. 2006;124:758–87.

Labinskyy N, Csiszar A, Veress G, Stef G, Pacher P, Oroszi G, Wu J, Ungvari Z. Vascular dysfunction in aging: potential effects of resveratrol, an anti-inflammatory phytoestrogen. Curr Med Chem. 2006;13:989–96.

Lawson LD, Gardner CD. Composition, stability, and bioavailability of garlic products used in a clinical trial. J Agric Food Chem. 2005;53: 6254–61.

Lawson LD, Wang ZJ. Low allicin release from garlic supplements: a major problem due to the sensitivities of alliinase activity. J Agric Food Chem. 2001;49:2592–9.

Leach AR, Hann MM. The in silico world of virtual libraries. Drug Discov Today. 2000;5:326–336.

Luo Y. Ginkgo biloba neuroprotection: Therapeutic implications in Alzheimer's disease. J Alzheimers Dis. 2001;3:401–407.

McCarney R, Fisher P, Iliffe S, van Haselen R, Griffin M, van der Meulen J, Warner J. Ginkgo biloba for mild to moderate dementia in a community setting: a pragmatic, randomised, parallel-group, double-blind, placebo-controlled trial. Int J Geriatr Psychiatry. 2008 June 9 [Epub ahead of print].

McKenna DJ. Clinical investigations of the therapeutic potential of ayahuasca: rationale and regulatory challenges. Pharmacol Ther. 2004;102:111–29.

Meezan E, Meezan EM, Jones K, Moore R, Barnes S, Prasain JK. Contrasting effects of puerarin and daidzin on glucose homeostasis in mice. J Agric Food Chem. 2005;53:8760–7.

Milne JC, Lambert PD, Schenk S, Carney DP, Smith JJ, Gagne DJ, Jin L, Boss O, Perni RB, Vu CB, Bemis JE, Xie R, Disch JS, Ng PY, Nunes JJ, Lynch AV, Yang H, Galonek H, Israelian K, Choy W, Iffland A, Lavu S, Medvedik O, Sinclair DA, Olefsky JM, Jirousek MR, Elliott PJ, Westphal CH. Small molecule activators of SIRT1 as therapeutics for the treatment of type 2 diabetes. Nature. 2007;450:712–6.

Napryeyenko O, Borzenko I. GINDEM-NP Study Group. Ginkgo biloba special extract in dementia with neuropsychiatric features. A randomised, placebo-controlled, double-blind clinical trial. Arzneimittelforschung. 2007;57:4–11.

Newman DJ, Cragg GM. Natural products as sources of new drugs over the last 25 years. J Nat Prod. 2007;70:461–77.

Newman DJ, Cragg GM, Snader KM. Natural products as sources of new drugs over the period 1981–2002. J Nat Prod. 2003;66:1022–37.

Nicolaou KC, Yang Z, Liu JJ, Ueno H, Nantermet PG, Guy RK, Claiborne CF, Renaud J, Couladouros EA, Paulvannan K, et al. Total synthesis of taxol. Nature. 1994;367:630–4.

O'Keefe JH, Gheewala NM, O'Keefe JO. Dietary strategies for improving post-prandial glucose, lipids, inflammation, and cardiovascular health. J Am Coll Cardiol. 2008;51:249–55.

Patch CS, Tapsell LC, Williams PG, Gordon M. Plant sterols as dietary adjuvants in the reduction of cardiovascular risk: theory and evidence. Vasc Health Risk Manag. 2006;2:157–62.

Peng N, Clark JT, Prasain J, Kim H, White CR, Wyss JM. Antihypertensive and cognitive effects of grape polyphenols in estrogen-depleted, female, spontaneously hypertensive rats. Am J Physiol Regul Integr Comp Physiol. 2005; 289:R771–5.

Pérez YY, Jiménez-Ferrer E, Zamilpa A, Hernández-Valencia M, Alarcón-Aguilar FJ, Tortoriello J, Román-Ramos R. Effect of a polyphenol-rich extract from Aloe vera gel on experimentally induced insulin resistance in mice. Am J Chin Med. 2007;35:1037–46.

Prasain JK, Jones K, Kirk M, Wilson L, Smith-Johnson M, Weaver C, Barnes S. Profiling and quantification of isoflavonoids in kudzu dietary supplements by high-performance liquid chromatography and electrospray ionization tandem mass spectrometry. J Agric Food Chem. 2003;51: 4213–8.

Riba J, Valle M, Urbano G, Yritia M, Morte A, Barbanoj MJ. Human pharmacology of ayahuasca: subjective and cardiovascular effects, monoamine metabolite excretion, and pharmacokinetics. J Pharmacol Exp Ther. 2003;306: 73–83.

Rooke N, Li DJ, Li J, Keung WM. The mitochondrial monoamine oxidase-aldehyde dehydrogenase pathway: a potential site of action of daidzin. J Med Chem. 2000;43:4169–79.

Sadovsky R, Collins N, Tighe AP, Brunton SA, Safeer R. Patient use of dietary supplements: a clinician's perspective. Curr Med Res Opin. 2008;24:1209–16.

Sahin K, Onderci M, Tuzcu M, Ustundag B, Cikim G, Ozercan IH, Sriramoju V, Juturu V, Komorowski JR. Effect of chromium on carbohydrate and lipid metabolism in a rat model of type 2 diabetes mellitus: the fat-fed, streptozotocin-treated rat. Metabolism. 2007;56:1233–40.

Schmidt BM, Ribnicky DM, Lipsky PE, Raskin I. Revisiting the ancient concept of botanical therapeutics. Nat Chem Biol. 2007;3:360–6.

Schwarz E, Metzler J, Diedrich JP, Freudenstein J, Bode C, Bode JC. Oral administration of freshly expressed juice of Echinacea purpurea herbs fail to stimulate the nonspecific immune response in healthy young men: results of a double-blind, placebo-controlled crossover study. J Immunother. 2002;25:413–20.

Schwarz O, Jakupovic S, Ambrosi HD, Haustedt LO, Mang C, Müller-Kuhrt L. Natural products in parallel chemistry-novel 5-lipoxygenase inhibitors from BIOS-based libraries starting from alpha-santonin. J Comb Chem. 2007;9: 1104–13.

Seren S, Lieberman R, Bayraktar UD, Heath E, Sahin K, Andic F, Kucuk O. Lycopene in cancer prevention and treatment. Am J Ther. 2008;15:66–81.

Shih CC, Lin CH, Lin WL. Effects of Momordica charantia on insulin resistance and visceral obesity in mice on high-fat diet. Diabetes Res Clin Pract. 2008 June 10 [Epub ahead of print]

Strimpakos AS, Sharma RA. Curcumin: preventive and therapeutic properties in laboratory studies and clinical trials. Antioxid Redox Signal. 2008;10:511–45.

Turnbull D, Whittaker MH, Frankos VH, Jonker D. 13-week oral toxicity study with stanol esters in rats. Regul Toxicol Pharmacol. 1999;29: 216–26.

Ulrich-Merzenich G, Zeitler H, Jobst D, Panek D, Vetter H, Wagner H. Application of the "-Omic-" technologies in phytomedicine. Phytomedicine. 2007;14:70–82.

van Breemen RB, Fong HH, Farnsworth NR. Ensuring the safety of botanical dietary supplements. Am J Clin Nutr. 2008;87:509S–13S.

Vijayakumar MV, Bhat MK. Hypoglycemic effect of a novel dialysed fenugreek seeds extract is sustainable and is mediated, in part, by the activation of hepatic enzymes. Phytother Res. 2008;22:500–5.

Watkins SM, German JB. Toward the implementation of metabolomic assessments of human health and nutrition. Curr Opin Biotechnol. 2002: 13:512–6.

Whitsett T, Carpenter M, Lamartiniere CA. Resveratrol, but not EGCG, in the diet suppresses DMBA-induced mammary cancer in rats. J Carcinog. 2006;5:15.

Xu S, Levine M. Medical residents' and students' attitudes towards herbal medicines: a pilot study. Can J Clin Pharmacol. 2008;15:e1–4.

Yang XX, Hu ZP, Duan W, Zhu YZ, Zhou SF. Drug-herb interactions: eliminating toxicity with hard drug design. Curr Pharm Des. 2006;12:4649–64.

Yeh GY, Eisenberg DM, Kaptchuk TJ, Phillips RS. Systematic review of herbs and dietary supplements for glycemic control in diabetes. Diabet Care. 2003;26:1277–94.

Zhang HG, Kim H, Liu C, Yu S, Wang J, Grizzle WE, Kimberly RP, Barnes S. Curcumin reverses breast tumor exosomes mediated immune suppression of NK cell tumor cytotoxicity. Biochim Biophys Acta. 2007;1773:1116–23.

Zhai Z, Liu Y, Wu L, Senchina DS, Wurtele ES, Murphy PA, Kohut ML, Cunnick JE. Enhancement of innate and adaptive immune functions by multiple Echinacea species. J Med Food. 2007;10:423–34.

Ziegenfuss TN, Hofheins JE, Mendel RW, Landis J, Anderson RA. Effects of a water-soluble cinnamon extract on body composition and features of the metabolic syndrome in pre-diabetic men and women. J Int Soc Sports Nutr. 2006;3:45–53.

Plant-Derived Natural Products as Leads for Drug Discovery

24

Li Pan, Esperanza J. Carcache de Blanco, and A. Douglas Kinghorn

Introduction

Natural product molecules having their origin from plants, microorganisms, and animals have had an irreplaceable role throughout the last 200 years in treating and preventing diseases, and continue to serve as important leads in modern drug discovery. Several volumes and reviews published recently have focused on this topic [1–6]. A considerable proportion of the natural products used as drugs is derived from terrestrial plants, which offer an invaluable and still incompletely exhausted resource for this purpose. In addition, profound ethnomedical knowledge based on the use of medicinal plants by humans has been accumulated for thousands of years. In the last few decades, pharmaceutical research on plants has been facilitated by the development of relevant technologies including new isolation methods, more sensitive spectroscopic techniques for structural determination, as well as specific high-throughput bioassay systems. Plant-derived bioactive compounds, in addition of being developed directly as drugs, also serve as prototype drug molecules, known as "lead compounds", and as pharmacological probes, to help better understand biochemical and physiological mechanisms.

In this chapter, some successful drugs derived from plant secondary metabolites in their original or modified forms, and other substances, currently under clinical trial as drug candidates, as well as several additional compounds used as biochemical probes afforded from plants, will be described. A general description of the importance and perspective of plants in drug discovery will also be given.

Role of Plants in Drug Development

Medicinal plants have been used as a major source of drugs for thousands of years in human history, and even today they are basis of the systematic traditional medicine practices in many countries all over the world. The first recorded literature on medicinal plants can be traced back to an earlier age of human history, such as the Atharvaveda (2000 BC) in India, the Divine Farmer's Herb-Root Classic (3000 BC) in China, and the Eber Papyrus (1550 BC) in Egypt [7, 8]. It is evident that the modern drug industry has been developed to a considerable degree as a result of plant-based traditional medicines. A review published in 2001 indicated that 88 active compounds isolated from 72 medicinal plants have been introduced into

L. Pan, E.J. Carcache de Blanco, and A.D. Kinghorn (✉)
Division of Medicinal Chemistry and Pharmacognosy, College of Pharmacy, The Ohio State University, Columbus, OH 43210, USA
e-mails: pan.106@osu.edu; carcache-de-blan.1@osu.edu; kinghorn.4@osu.edu

A.E. Osbourn and V. Lanzotti (eds.), *Plant-derived Natural Products*,
DOI 10.1007/978-0-387-85498-4_24,

modern drug therapy, with many of them being considered as the active principle responsible for their ethnopharmacological use [9]. Some of these plant-derived therapeutic agents, such as atropine (anticholinergic), codeine (cough suppressant), colchicine (antigout), ephedrine (bronchodilator), morphine (analgesic), pilocarpine (parasympathomimetic), and physostigmine (cholinesterase inhibitor) are still being widely used today [10].

A whole new era for drug discovery opened up in the early nineteenth century, triggered by the isolation of morphine, a pharmacologically active compound from the plant medicine opium. Later, compounds from plants such as atropine (*Atropa belladonna*), cocaine (*Erythroxylum coca*), ephedrine (*Ephedra* spp.), digitoxin (*Digitalis purpurea*), and quinine (*Cinchona* spp.) were purified and then served as drugs [2, 4, 5]. These discoveries are considered important not only for introducing new single chemical entities as potent medicinal treatments, but also for helping understand human diseases by disclosing the key role that these molecules play, and by promoting the development of pharmacology, medicinal chemistry as well as organic chemistry [8]. As a result of the further purification of drugs such as artemisinin, digoxin, paclitaxel, vinblastine, and vincristine from plants in the twentieth century, the use of plant extractives in prescriptions such as tincture of belladonna was gradually replaced by pure single chemical entities like atropine.

Plants and other organisms may be regarded as libraries of small-molecule secondary metabolite organic compounds with considerable structural diversity, which would otherwise probably be unavailable in a synthetic chemical laboratory [11–14]. As secondary metabolites, these compounds have been elaborated in living organisms by complex enzyme systems developed during a long evolutionary process. It is apparent that these natural products present more "drug-like" or "biologically friendly" molecular qualities than many purely synthetic compounds. These intrinsic properties provide plant-derived small organic molecules and other natural products with an important potential role in modern drug discovery [15, 16].

Although combinatorial chemistry has been employed in the pharmaceutical industry over the last 2 decades to satisfy the need for very large numbers of compounds demanded in high-throughput screens (HTS), the results have not been as promising as expected, and the number of new chemical entities introduced annually as produced by this method has actually declined [3, 17]. A combination of natural products and combinatorial chemistry has been initiated in recent years. In this case the latter serves as a technique to optimize the structure of existing active natural compounds to new agents [17–20].

Active compounds isolated from plants can serve directly as therapies in clinical use, like morphine, atropine, quinine and paclitaxel, or as prototype biologically active "lead" compounds, affording numerous structural analogs as new pharmaceutical agents, such as artemisinin and the opiate derivatives. In addition to their medicinal use, some secondary metabolites from plants have also served as powerful "pharmacological tools" to help explain the mechanisms underlying human diseases [21–24]. Today, drug discovery from plants is based mainly on bioactivity-guided isolation, and groups of scientists with different research backgrounds including botany, biochemistry, pharmacology, pharmaceutics, pharmacognosy, medicinal chemistry, organic chemistry and toxicology are required in this enterprise [12–14, 25, 26].

Plant Natural Products as Drugs

Despite the "synthetic revolution" in the pharmaceutical industry, medicinal plants are still involved in the primary health care of a large proportion of the population in the world, especially in developing countries [9]. Bioactivity-guided fractionation can often lead to the isolation of active principles of these medicinal plants, and some of those chemical entities with acceptable pharmaceutical qualities can be developed as drugs in their original forms directly. These include compounds on the therapeutic market that have been used for many years as important clinical agents mainly in the treatment of cancer, central nervous system disorders, cardiovascular diseases, and infectious diseases [11–14, 20]. In this section, some important plant-derived drugs in their unmodified forms will be mentioned, with a brief description about the plant origin and pharmaceutical use in each case.

Atropine (**1**) [a racemic mixture of (+)- and (-)-hyoscyamine (**2**)] and scopolamine [(-)-hyoscine] (**3**) are tropane-type alkaloids found in certain plants in the Solanaceae (nightshade) family used medicinally for centuries in Europe, such as *Atropa belladonna*, *Hyoscyamus niger*, and *Datura stramonium* [7]. The antispasmodic activities of atropine are due to competitive antagonism of acetylcholine at the muscarinic receptor site. Scopolamine is also an anticholinergic agent, and most commonly used for the prevention of nausea and motion sickness in the form of a transdermal patch. Both of these tropane alkaloids have psychoactive effects as a result of their ability to penetrate the blood-brain barrier [7, 27, 28].

Nicotine (**4**), an agonist on the nicotinic acetylcholine receptor (nAChR) found in *Nicotiana tabacum* (tobacco), is used pharmaceutically for smoking cessation [29]. Analogs of nicotine are considered promising for the treatment of neurodegenerative conditions like Alzheimer's disease [30, 31].

Morphine (**5**) and codeine (methylmorphine) (**6**), two major morphinan-type alkaloids with an isoquinoline skeleton, are extracted from opium, the dried milky sap released from the immature fruits of poppies (*Papaver somniferum*). Morphine and codeine can interact with opioid receptors distributed in brain tissues and the periphery, and are most widely used as narcotic analgesics, with codeine also having an antitussive effect [4, 25, 32].

Galantamine (**7**, Razadyne®, Reminyl®, Nivalin®) is a recently approved drug for the treatment of early-onset Alzheimer's disease [12]. Galantamine (or galanthamine) is an Amaryllidaceae-type alkaloid first purified from the snowdrop (*Galanthus woronowii*) in

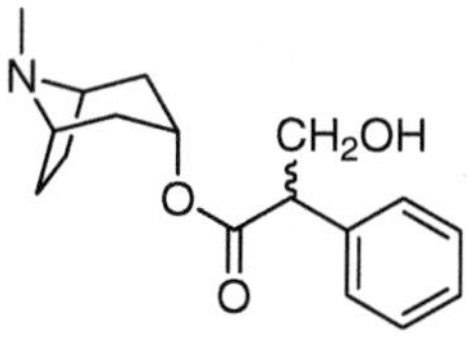

1. atropine

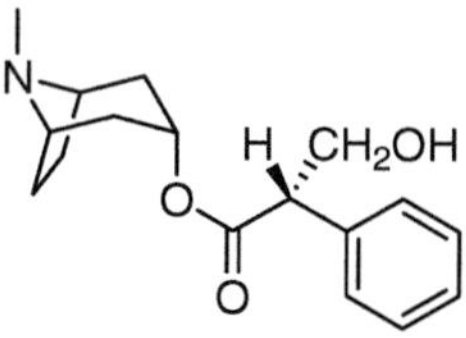

2. (-)-hyoscyamine

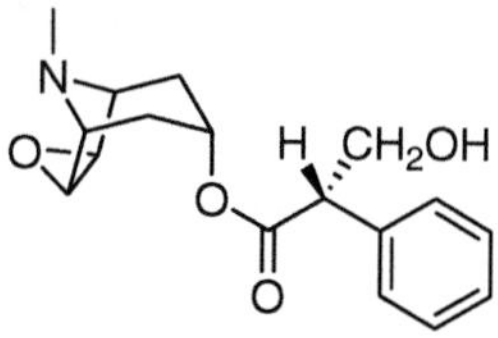

3. scopolamine [(-)-hyoscine]

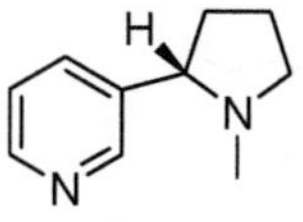

4. nicotine

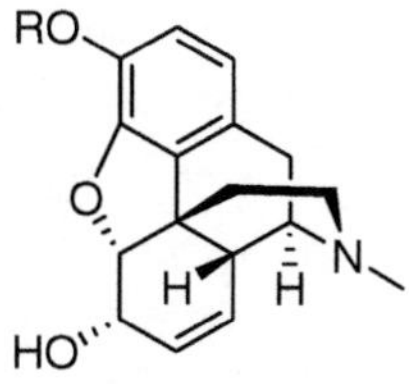

5. R = Me codeine
6. R = H morphine

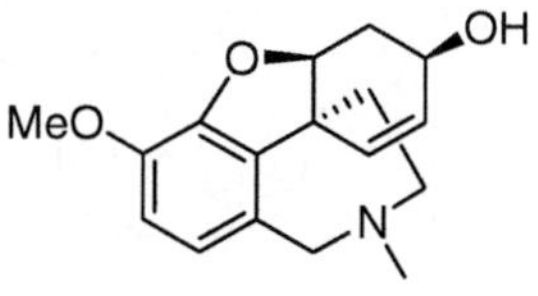

7. galantamine

24

the early 1950s, and later found in other plants of the family Amaryllidaceae [34]. Galantamine can improve cerebral function by acting as a cholinergic agent, and inhibits acetylcholinesterase and modulates nicotinic acetylcholine receptors (nAChRs) [33–36]. The market need for galantamine is now met by the total synthesis of this compound [36].

The demand for the legalization of *Cannabis sativa* (marijuana) for medicinal use has represented an interesting controversy in recent years because of the possibility of abuse [37]. Cannabidiol (CBD, **8**) and Δ^9-*trans*-tetrahydrocannabinol (THC, **9**), two active cannabinoids of marijuana, have been approved recently in Canada as ingredients of an oromucosal spray marketed as Sativex® to alleviate the pain caused by multiple sclerosis (MS). Efforts are being made to introduce Sativex® to other countries in the near future [38, 39].

Quinine (**10**) and quinidine (**11**), two diastereomeric alkaloids containing quinoline rings in their molecules, are obtained from *Cinchona* spp. and other species. Quinine (**10**) was the first effective treatment for falciparum malaria, an often fatal parasitic disease caused by several species of plasmodium. The discovery of quinine relieved European settlers from the harmful effects of this fatal illness, and greatly facilitated colonization in many tropical and subtropical areas of the world [40]. Quinine exerts its activity by inhibiting the heme polymerase of the parasitic host, and still shows some efficacy today as an antimalarial agent in cases where synthetic drugs fail due to parasite resistance [2, 6, 41]. Quinidine (**11**), has some use as a cardiac antiarrhythmic by affecting ion channels [2, 41].

Artemisinin (“qinghaosu”) (**12**), a sesquiterpene lactone possessing an unusual endoperoxide bridge, is a compound discovered in the People’s Republic of China from *Artemisia annua*, which has long been used as a traditional medicinal plant for the treatment of fever. As a naturally occurring antimalarial, artemisinin may be employed as an option for the treatment of chloroquine-resistant malaria in China and some

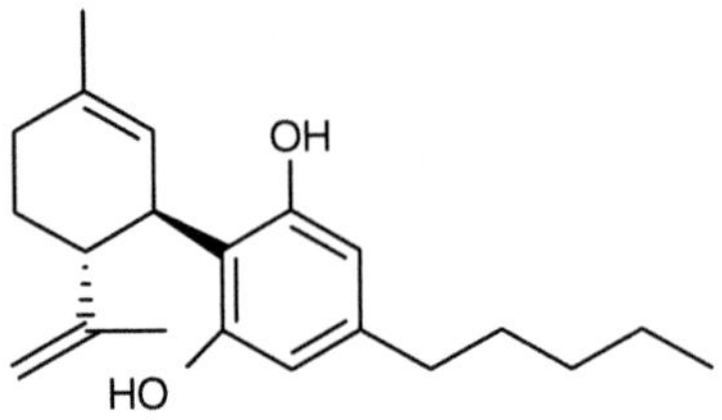

8. cannabidiol (CBD)

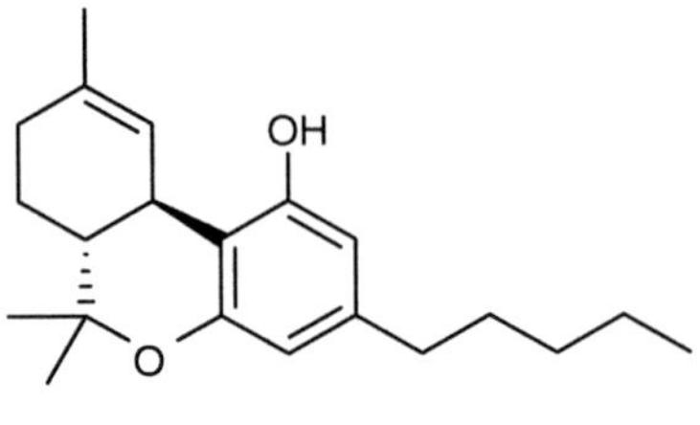

9. Δ^9-*trans*-tetrahydrocannabinol (THC)

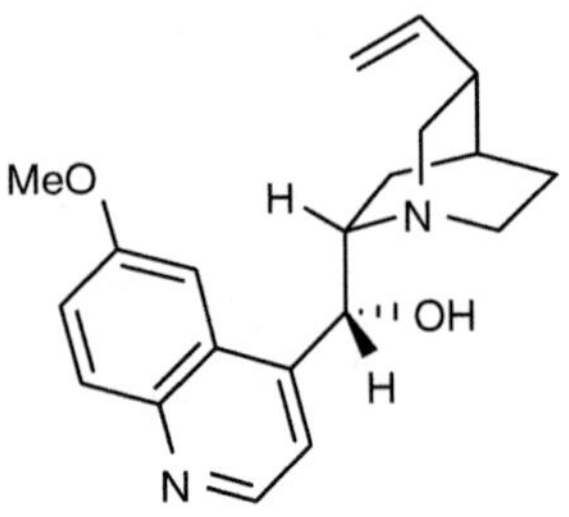

10. quinine

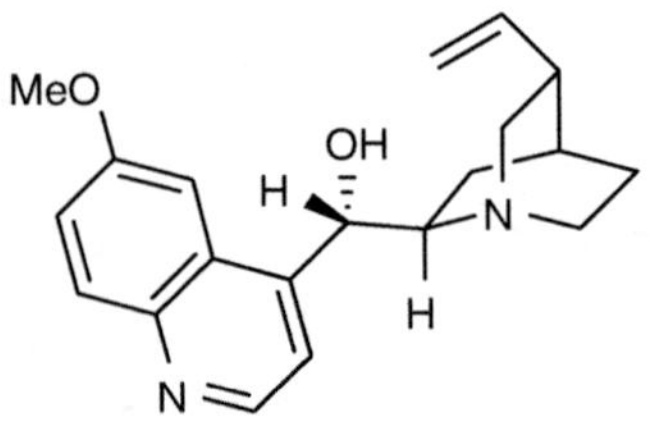

11. quinidine

other countries in Asia [42]. Artemisinin exerts its activity through a unique mechanism by acting on the heme complex [43]. However, the use of artemisinin as a monotherapy antimalarial agent is no longer recommended, since this might lead to parasite resistance to this entire compound class [44].

In the nineteenth century, digitoxin (**13**), a major cardioactive steroid glycoside, was isolated from *Digitalis purpurea*, commonly known as purple foxglove. This plant proved to be an effective treatment for dropsy caused by congestive heart failure in the late eighteenth century in England [45]. Digoxin (**14**) is another active cardiotonic glycoside that was purified from *Digitalis lanata* subsequent to the discovery of digitoxin. These two drugs exhibit a positive inotropic effect by inhibiting the activity of ATPase and cation transport, thus resulting in the increase of Ca^{2+} levels in the myocytes [46].

Paclitaxel (**15**, Taxol®) is a diterpenoid based on the taxane nucleus, possessing an essential oxetane ring and with one of the substituent groups containing a nitrogen atom. This compound was first isolated from the bark of *Taxus brevifolia* (Pacific yew) [47]. As a chemotherapeutic anticancer agent, paclitaxel inhibits mitosis by acting as a microtubule stabilizer and has been used in the clinic primarily for the treatment ovarian cancer, breast cancer, and non-small cell lung cancer [48, 49]. The limited availability of the plant source of paclitaxel was once a considerable obstacle in the development of this drug, until new semi-synthetic and biological methods were developed to solve this problem, as will be discussed later in this chapter.

Vinblastine (**16**, Velban®, Alkaban-AQ®) and vincristine (**17**, Oncovin®) are structurally closely related indole-dihydroindole dimers (bisindolealkaloids), isolated from *Catharanthus*

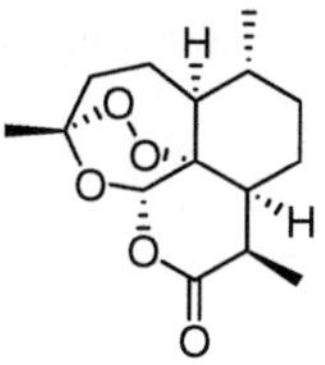

12. artemisinin

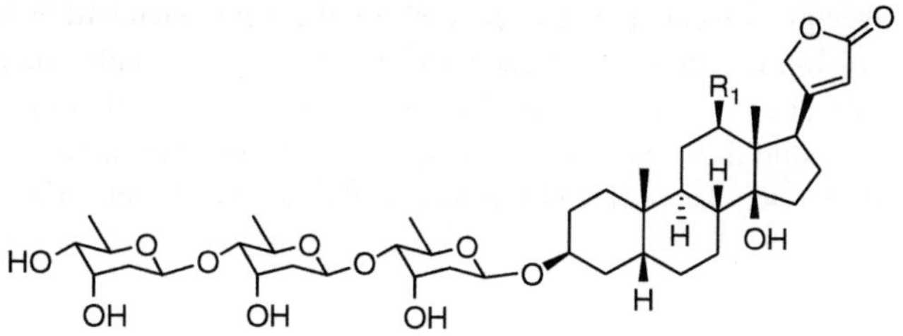

13. R_1 = H digitoxin
14. R_1 = OH digoxin

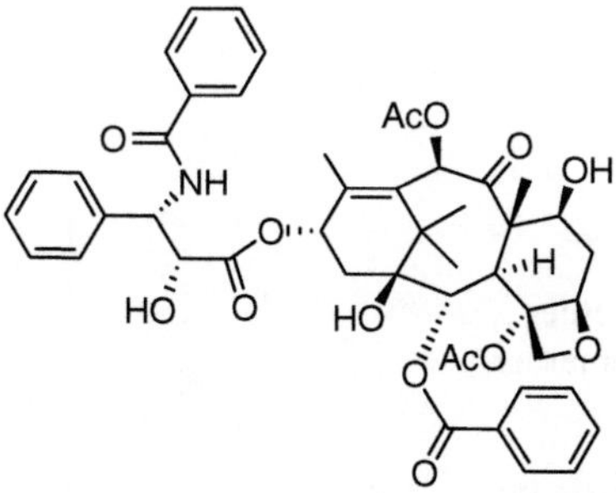

15. paclitaxel (taxol)

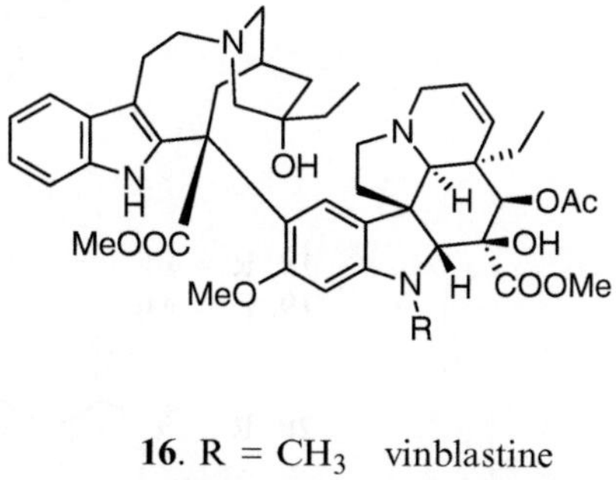

16. R = CH_3 vinblastine
17. R = CHO vincristine

roseus, also known as *Vinca rosea* (Madagascar periwinkle). Vinblastine and vincristine are used mainly to treat different forms of leukemia, although they have efficiency against several other forms of cancer. Those two anticancer agents act as mitotic inhibitors by binding to dimeric tubulin, and then leading to the failure of microtubule assembly in the metaphase stage [50–52].

Semi-synthetic Drugs Based on Plant Secondary Metabolites

The use of semi-synthetic derivatives as drugs developed from plant natural product lead compounds began at the end of the nineteenth century and may be exemplified by drugs like aspirin (acetyl salicylic acid). From then on, the potential of plant-derived compounds as leads in drug discovery has been realized on numerous occasions [8]. A survey released recently revealed that over 40% of the new small-molecular single chemical entity drugs introduced to the market from 1981 to 2006 are natural product derivatives, with 28% of them being chemical semi-synthetic modifications based on natural products, 12% "natural product mimics", and 12% synthetic compounds in which the pharmacophore modeled was that of a natural product [20]. Synthetic optimization of the naturally occurring compounds is often required to enhance therapeutic potency, to increase bioavailability, to remove unpleasant side effects, and to help compensate for a shortage in a natural product drug supply, or to derive different bioactivities. This section will focus on examples of semi-synthetic drugs of plant origin.

A series of artemisinin-based semisynthetic antimalarial derivatives, with all of them maintaining the key endoperoxide bridge, such as arteether (**18**), artemether (**19**), artesunate (**20**), and dihydroartemisinin (**21**), have been designed to improve the water solubility and the metabolic stability of artemisinin [53, 54]. Among them, dihydroartemisinin (artenimol), is considered as a common active metabolite of artemisinin derivatives [53, 54]. Currently, artemisinin-based therapies combined with standard antimalarials such as amodiaquine, sulfadoxine-pyrimethamine, mefloquine, and lumefantrine are recommended by the World Health Organization (WHO) as first-line therapies for malaria [55, 56].

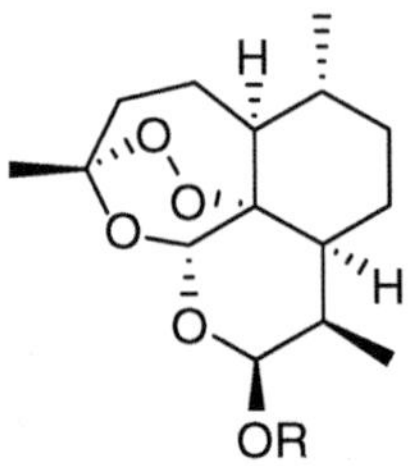

18. R = Et arteether
19. R = Me artemether

20. R = 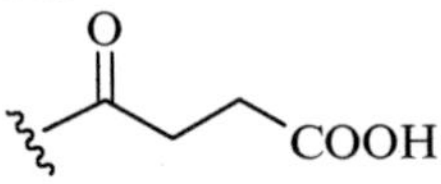

artesunate

21. R = H dihydroartemisinin

As mentioned before, the market demand of the anticancer drug, paclitaxel (**15**, Taxol®) was threatened initially by an unsustainable natural supply, which relied mainly on extracting this compound from the bark of the slow-growing Pacific yew tree, from which it is produced only in a very low yield [57]. Although the total synthesis of paclitaxel has been described, this is still inefficient when considered the extensive market need [58–60]. 10-Deacetylbaccatin III (**22**), which is available with a relatively high yield from the leaves of other renewable yew species, such as *Taxus baccata* L., was introduced as a semi-synthetic precursor compound of paclitaxel by the major pharmaceutical manufacturer, Bristol-Myers Squibb (B-MS) [61]. Docetaxel (**23**), a related taxane anticancer drug can also prepared from 10-deacetylbaccatin III (**22**) [62].

Oseltamivir phosphate (**24**, Tamiflu®), a neuraminidase inhibitor, is administered orally for the treatment and prevention of influenza A and B virus infections, and stockpiling of this drug has been proposed to counter the threat of a pandemic such as avian flu [63, 64]. One synthesis of this compound begins from (−)-shikimic acid (**25**), which serves as the key intermediate in the biosynthesis of a variety of aromatic compounds in plants and microorganisms [2]. Shikimic acid was first isolated from the plant *Illicium anisatum* (Japanese "shikimi") in 1885, and is abundant in the star aniseed (*Illicium verum*) and the plant genus *Liquidambar* ("sweetgum") [65, 66]. In order to overcome the shortage of the natural source of shikimic acid, fermentation of this substance by metabolically engineered *Escherichia coli* strains has been developed successfully as an efficient alternative method of production [67]. Recently, a total synthetic method for oseltamivir phosphate independent of shikimic acid has been developed [68].

Nitisinone (**26**, Orfadin®) is the first drug approved in Europe for the treatment of hereditary tyrosinemia type 1 (HT-1), a rare genetic metabolic disorder caused by a deficiency of fumarylacetoacetate hydrolase (FAH), an enzyme involved in the metabolism of tyrosine [69]. Nitisinone is a derivative of leptospermone (**27**), an effective herbicide present in the bottlebrush

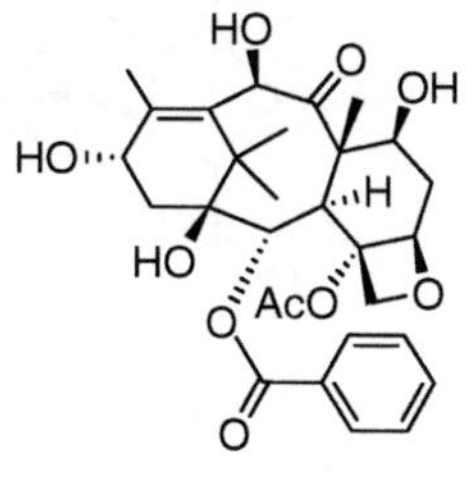

22. 10-deacetylbaccatin III

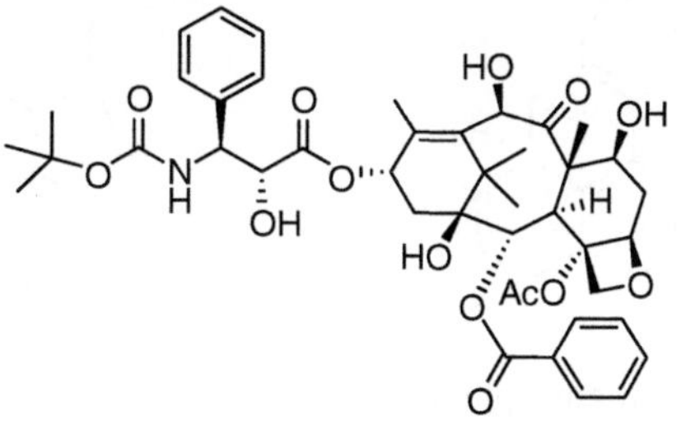

23. docetaxel

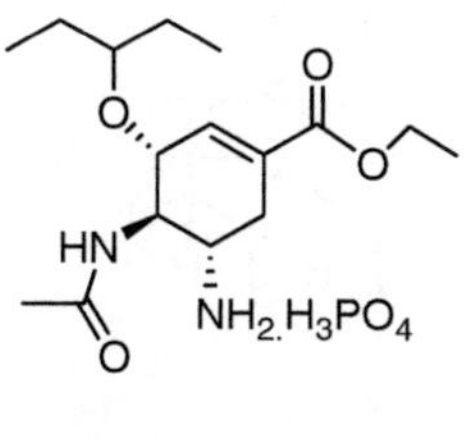

24. oseltamivir phosphate

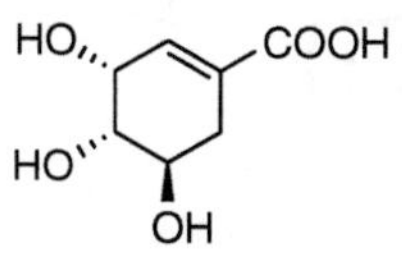

25. (-)-shikimic acid

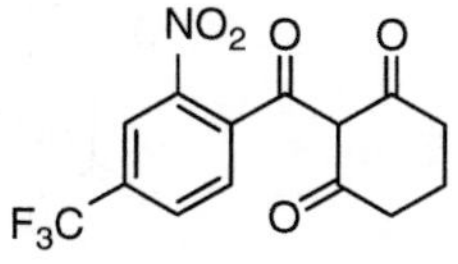

26. nitisinone

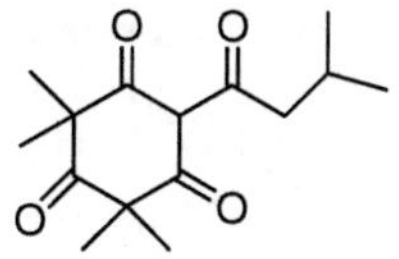

27. leptospermone

plant *Callistemon citrinus* [70]. Nitisinone can interfere with tyrosine catabolism as a competitive inhibitor of 4-hydroxyphenyl pyruvate dioxygenase (HPPD), an upstream enzyme of FAH, to prevent accumulation of toxic tyrosine metabolites that lead to liver and kidney failure [69–71].

Opiates, narcotic compounds extracted or derived from opium, are a remarkable source of lead compounds for their potent pharmaceutical effects such as analgesics, antitussives and ataractics, and of which many synthetic derivatives have been prepared [8, 72]. Apomorphine (**28**), a dopamine agonist derivative from morphine (**5**) but without analgesic properties like morphine, was recently approved as a therapy for Parkinson's disease [73]. Hydrocodone (**30**) is a narcotic agent derived from thebaine (**29**) and is commonly combined with other analgesics such as acetaminophen and ibuprofen as drugs to relieve pain. [74]. Naloxone (**31**) and naltrexone (**32**) are both opioid receptor antagonists. Naloxone is used as a treatment for opioid overdose [75], while naltrexone can be used to treat alcoholism as well as opiate addiction [76]. Dextromethorphan (**33**) is a non-narcotic opiate derivative that is widely used as an over-the-counter cough-suppressant [77].

Podophyllotoxin (**34**), is a lignan constituent of *Podophyllum* resin, produced from the North American mayapple (*Podophyllum peltatum*) and the Himalayan species *Podophyllum hexandrum* [78, 79]. As a result of its considerable toxicity, extensive chemical modification work has been done to improve the pharmaceutical profile of podophyllotoxin. Etoposide (**35**), teniposide (**36**), and etopophos (etoposide phosphate, the prodrug of etoposide) are successful anticancer drugs derived from podophyllotoxin [79]. Instead of acting as microtubule inhibitors in the same manner as the lead compound **34**, derivatives **35** and **36** exert their anticancer activities by acting as inhibitors of the enzyme topoisomerase II, and together are widely used to treat lymphomas, acute leukemia, small cell lung cancer, and testicular cancers [78–80].

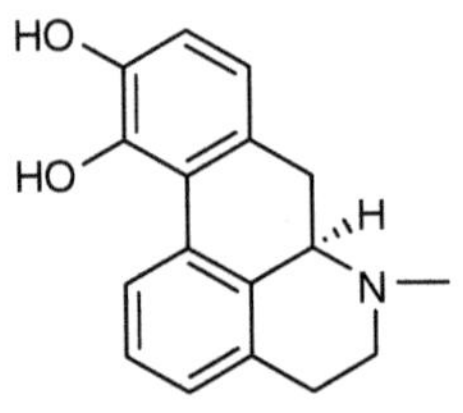

28. apomorphine

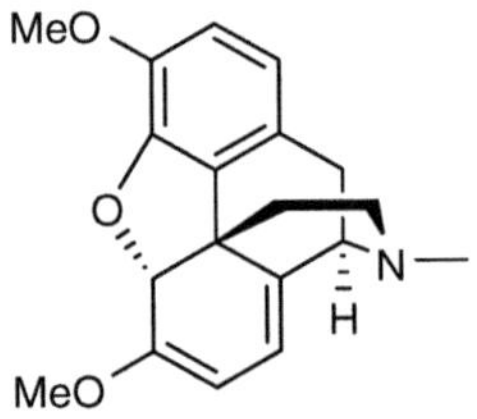

29. thebaine

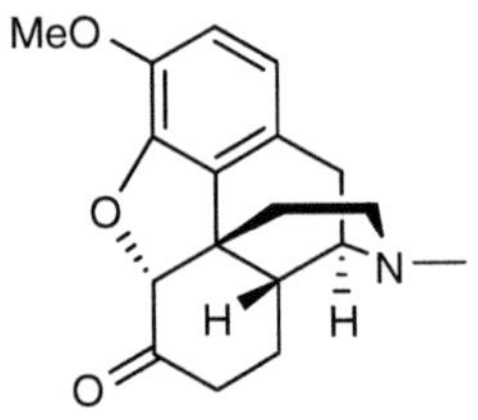

30. hydrocodone

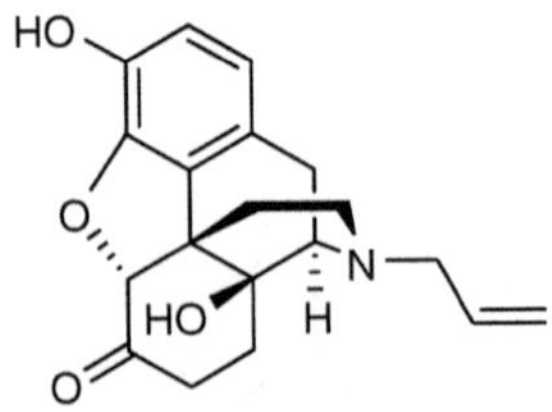

31. naloxone

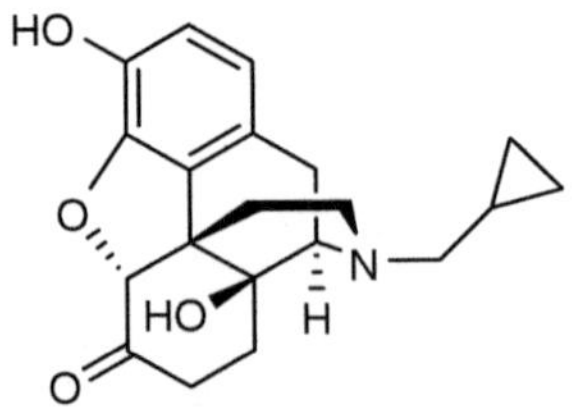

32. naltrexone

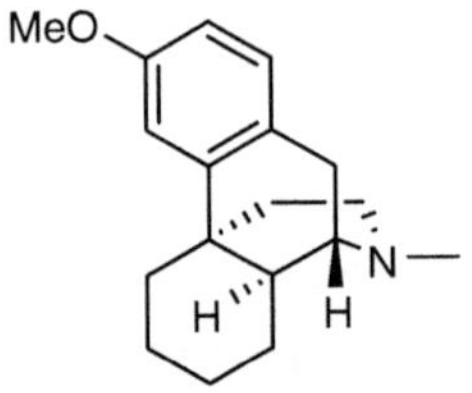

33. dextromethorphan

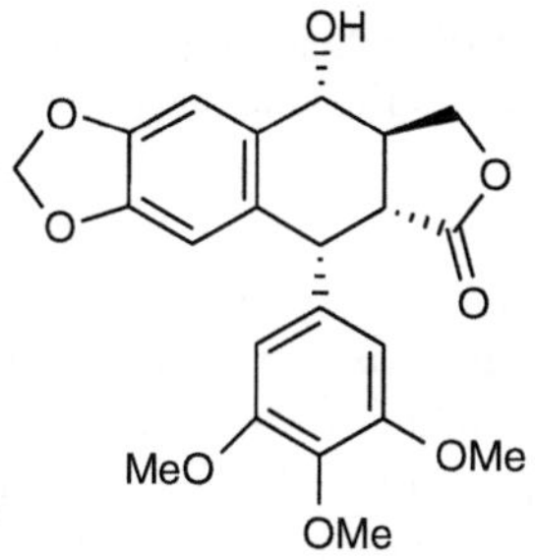

34. podophyllotoxin

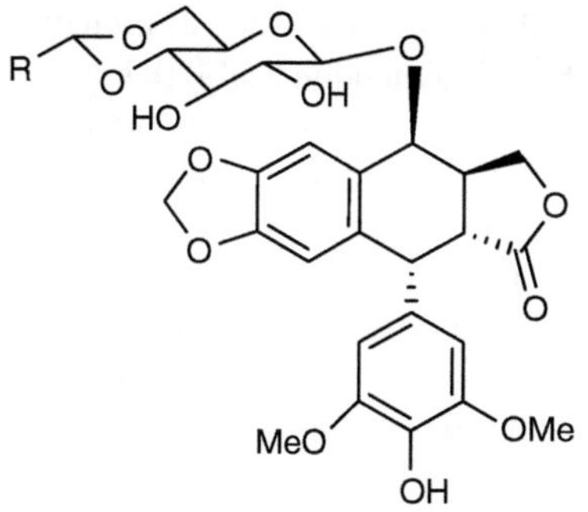

35. R = Me etoposide

36. R = 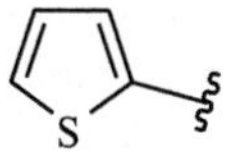 teniposide

Use of Plant-Derived Natural Products as "Pharmacological Probes"

A number of plant secondary metabolites have been found to be valuable "pharmacological probes" or "biochemical tools" to help target various receptors as well as to explain cellular processes and assist with the elucidation of different kinds of molecular targets.

Genistein (**37**), an estrogenic isoflavone present in soybean (*Glycine max*), is a protein tyrosine kinase (PTK) inhibitor, and has been used as a probe to study the interaction between PTK and cyclic nucleotide-gated (CNG) channels [81, 82]. Genistein is currently under clinical trial for its angiogenesis-inhibiting activity [83].

Forskolin (**38**), also known as coleonol, is a labdane diterpene isolated from the roots of *Coleus forskohlii* (Lamiaceae), a traditional medicinal plant used in India for the treatment of heart disease and asthma [84]. As a hypertensive agent with spasmolytic, cardiotonic and anti-platelet-aggregation activity, forskolin can directly activate adenylate cyclase (AC), a target enzyme of multiple G-protein-coupled receptors (GPCRs), leading to the intracellular accumulation of cAMP [85, 86]. This property makes it a potent pharmacological tool widely used in investigating the catalytic mechanism of AC and the regulation of cAMP [86–88].

Quercetin (**39**) is one of the most common natural occurring flavonoids and is found in many fruits and vegetables, such as apples and onions. It is well known for its antiinflammatory, antioxidant, and antiplatelet aggregation activities [89, 91]. Quercetin can induce a wide spectrum of cellular events by interacting with a

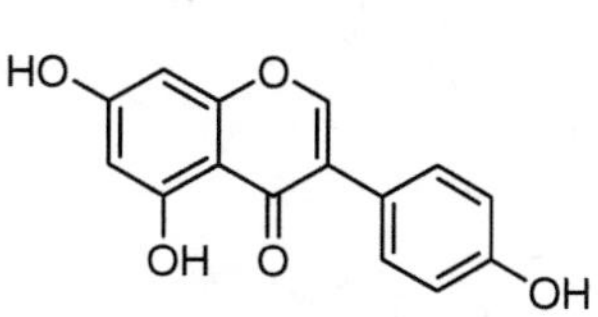

37. genistein

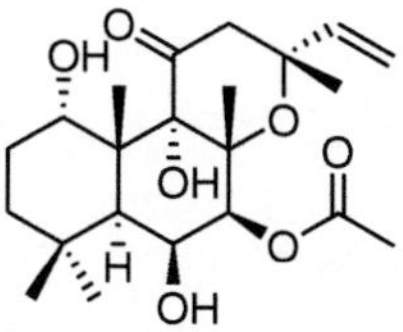

38. forskolin

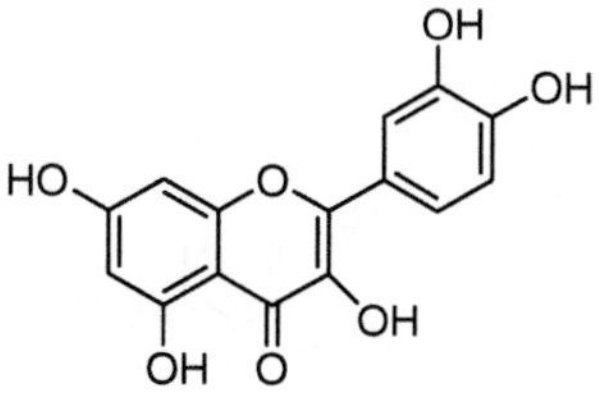

39. quercetin

number of signaling molecules such as protein kinases, NF-κB, and calmodulin (CaM) [89–91]. This flavone can bind to intracellular targets, such as nucleic acids and proteins and exhibits specific fluorescence emission [92, 93]. This property allows quercetin to be a valuable, sensitive, and selective spectroscopic probe to trace phenol-binding complexes in cellular systems and help clarify the relevant molecular functions [92–94].

Phorbol esters, tetracyclic diterpenoids naturally occurring in some plants of the family Euphorbiacaeae, such as *Croton tiglium*, show protein kinase C (PKC) inhibition activity, and are also known for their skin-irritant and tumor-promotion properties [95, 96]. The most abundant *C. tiglium* phorbol ester derivative, 12-*O*-tetradecanoylphorbol-13-acetate (TPA) (**40**), is used as a standard tumor promoter in animal models of full-term carcinogenesis [97].

Some microtubule targeting agents like colchicine (**41**), as well as certain anticancer agents mentioned before, such as podophyllotoxin (**34**), paclitaxel (**15**), vinblastine (**16**) and vincristine (**17**), may be used as biological probes in cancer research. Thus, paclitaxel acts as a promoter of stabilization of microtubules [98–102].

Capsaicin (**42**), a pungent principle in chili peppers (*Capsicum* spp.), as well as the skin irritant, resiniferatoxin (**43**), from the latex of *Euphorbia resinifera*, have been used to probe the transient receptor potential (TRP) channels and the vanilloid receptors responsible for the pain sensation caused by heat. This has led to the successful isolation of the first nociceptive receptor, TRPV1 (transient receptor potential channel, vanilloid subfamily member 1) [103, 104]. Capsaicin-containing creams are now available as nonprescription pain-relievers for the treatment of post-therapeutic neuralgia [105]. The internal use of capsaicin for the treatment of severe post-operative pain, post-traumatic neuropathic pain, and musculoskeletal diseases is currently under clinical trial [106].

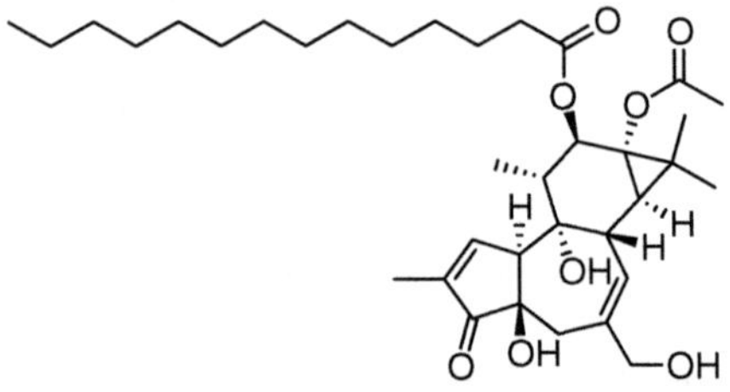

40. 12-*O*-tetradecanoylphorbol-13-acetate

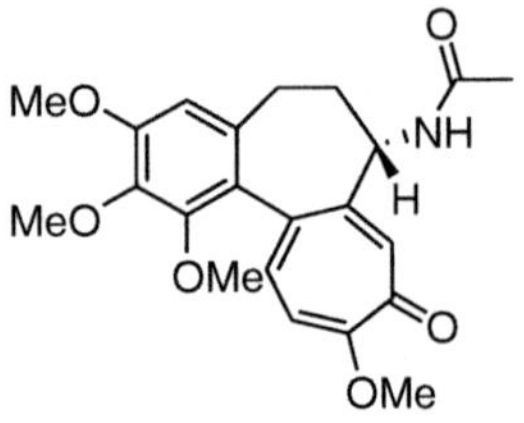

41. colchicine

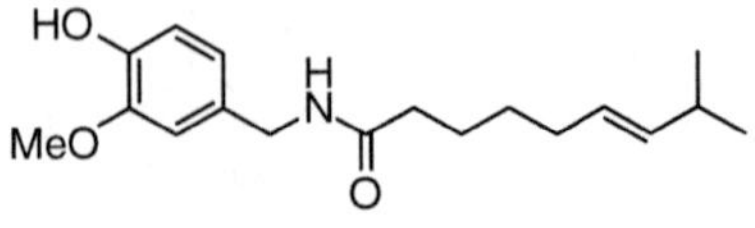

42. capsaicin

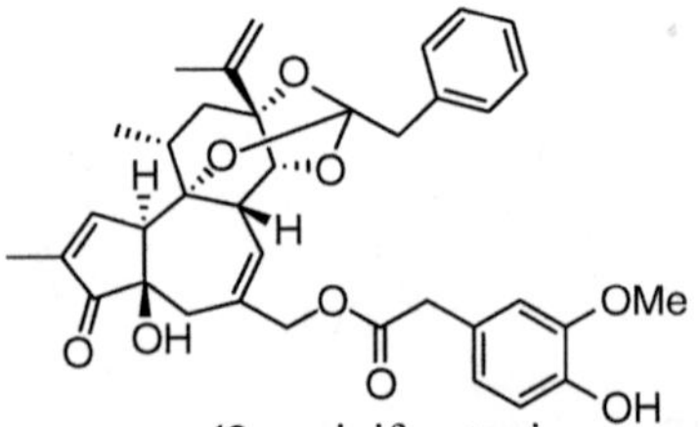

43. resiniferatoxin

Examples of Candidate Drug Molecules from Plants

Although the development of new drugs from plants and other organisms has received less emphasis in the last few decades than previously, plants are still an invaluable resource of therapeutic leads, as judged by the number of chemical entities currently under preclinical or clinical trials [107–109]. A portion of these compounds are analogs developed on the basis of known drugs such as paclitaxel and vinblastine [107] and these will not be covered in the present section. Herein, several other examples of promising candidate drug molecules of plant origin will be described briefly.

Protopanaxadiol (**44**), a triterpene aglycone derived from several saponins of ginseng (*Panax ginseng*) [110], has been demonstrated as a potential anticancer agent with effects on apoptosis induction and cell cycle arrest on cancer cells both *in vitro* and *in vivo*, and has also been reported to show cytotoxicity against multidrug-resistant tumors by blocking P-glycoprotein (P-gp, MDR1) [111–114]. Protopanaxadiol, under the trade name Pandimex®, is now in a Phase I clinical study in the United States for the treatment of lung cancer and other solid tumors [111]. Also, it has been approved conditionally in mainland China for the treatment of advanced cancers of the lung, breast, pancreas, and colon-rectum [114].

Homoharringtonine (Ceflatonin®, HHT) (**45**) is a cytotoxic alkaloid isolated from several plants in the genus *Cephalotaxus*, such as *C. harringtonia*. This compound exerts its potential antineoplastic activity by inhibiting protein synthesis at the ribosome level and inducing the differentiation and apoptosis of cancer cells [115–117]. Homoharringtonine is currently under II/III phase clinical trials for the treatment of patients with chronic myeloid leukemia (CML) in the United States and Europe [117, 118].

Oxymatrine (**46**) is one of the major alkaloids isolated from the aqueous extract of the roots of *Sophora flavescens*, a traditional Chinese medicinal plant mainly used in the treatment of jaundice and other liver disorders [119]. Oxymatrine has been developed in mainland China as an injection for the treatment of chronic hepatitis B virus (HBV) infection [120]. Clinical trials have shown the effect of oxymatrine against chronic hepatitis C (HCV) with the function of inhibiting proliferation, lessening liver fibrosis, regulating the immune reaction though reducing the level of TNFα (tumor necrosis factor alpha) in the serum, and down-regulating the expression of Fas receptor/Fas ligand in the liver tissue [121–123].

Phenoxodiol (**47**), derived from genistein (**48**), a natural occurring soy isoflavone, has been promoted as a new promising drug candidate in oncology for its broad anticancer spectrum and minimal toxicity [124]. Phenoxodiol exhibits its anticancer function by inducing the mitotic arrest and apoptosis of tumor cells through multiple mechanisms such as by decreasing the level of antiapoptotic proteins and acting as a DNA topoisomerase (topo) II

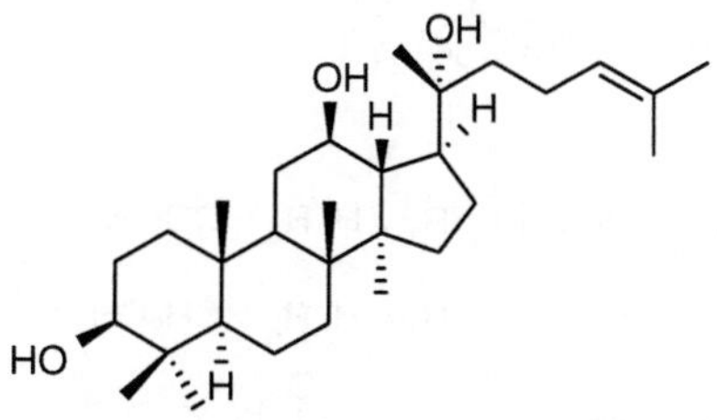

44. protopanaxadiol

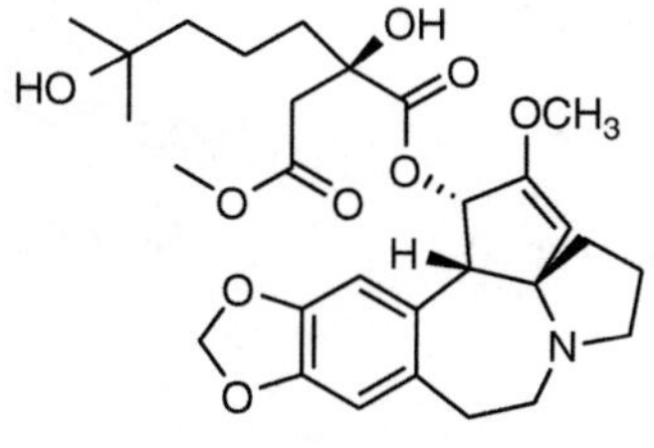

45. homoharringtonine

inhibitor [124, 125]. Phenoxodiol is now undergoing clinical trials as a therapy for ovarian and prostate cancer in the United States, Europe, and Australia [126, 127].

Huperzine A (**49**) is a sesquiterpene alkaloid isolated from a traditional Chinese herbal remedy, *Huperzia serrata* ("Qian Ceng Ta") in 1986 [128]. This compound has been proved to be a potent, selective and reversible acetylcholinesterase (AChE) inhibitor, and demonstrated memory enhancement and neuroprotective functions in clinical trials as a therapeutic against Alzheimer's disease (AD) in the People's Republic of China [129, 130]. In 2004, a phase II clinical trial focused on its cognitive function was initiated by the National Institute on Aging (NIA) in the United States [131]. ZT-1 (**50**), considered more selective than huperzine A, was developed as a semi-synthetic derivative of **49** by cooperation of the Shanghai Institute of Materia Medica and Debiopharm of Switzerland and is currently under phase I/II clinical trials in mainland China and in Europe [132].

Betulinic acid (**51**) and its close structural anolog, betulin (**52**), are lupane triterpenoids widely distributed in the plant kingdom, with the latter especially abundant in the bark of the white birch tree [133]. Betulin can serve as the semi-synthetic precursor of betulinic acid, which is considered biologically more active than betulin. Important biological activities attributed to betulinic acid and its derivatives, are anti-HIV activity and selective cytotoxicity against melanoma cancer cells [133–135]. Betulinic acid can induce apoptosis by acting on the mitochondria of various tumor cells. This compound is currently under phase I/II clinical trials launched by National Cancer Institute (NCI) as a chemotherapeutic agent for the treatment of dysplastic melanocytic nevi, which are considered associated with cutaneous melanomas [136].

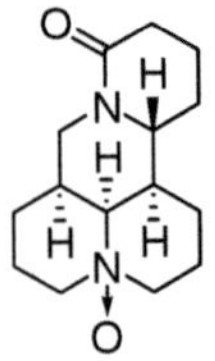

46. oxymatrine

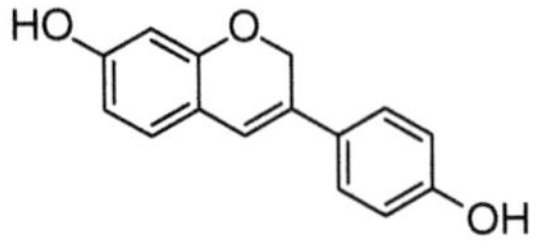

47. phenoxodiol

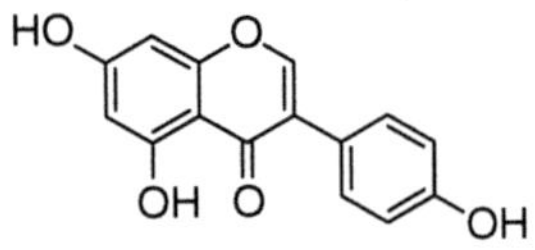

48. genistein

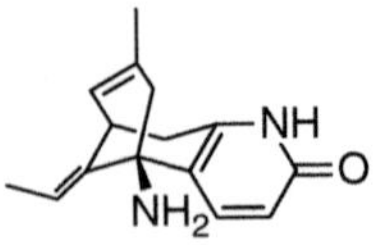

49. huperzine A

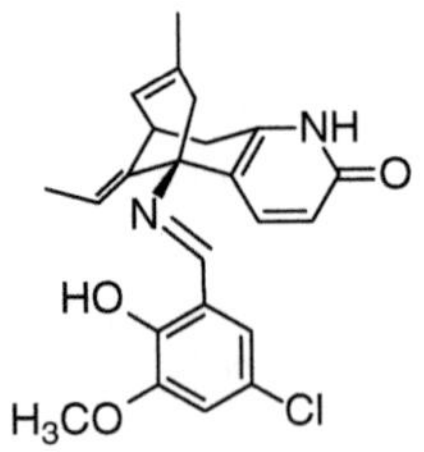

50. ZT-1

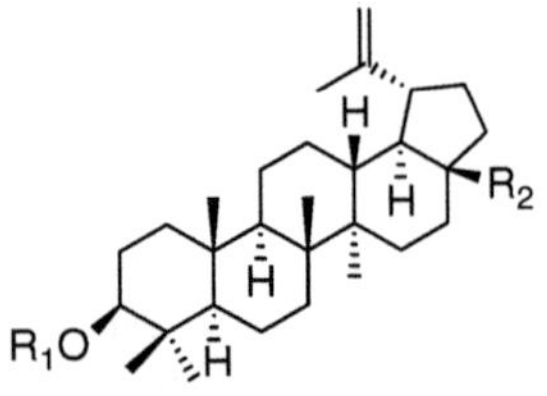

51. betulinic acid $R_1 = H, R_2 = COOH$

52. betulin $R_1 = H, R_2 = CH_2OH$

53. bevirimat $R_1 =$ 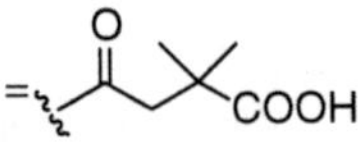

$R_2 = COOH$

Bevirimat (PA-457, **53**) is a semi-synthetic analog of betulinic acid with potent and selective antiretroviral activity. As a first-in-class maturation inhibitor, bevirimat can inhibit the replication of HIV by interrupting the cleavage of the CA-SP1 (capsid-spacer peptide 1) in the Gag processing step, thus leading to defects in viral core condensation [137, 138]. Bevirimat is being developed as a new agent for the treatment of HIV infection, and phase II trials completed recently have suggested a favorable profile of safety and pharmacokinetics that warrant further clinical research [139, 140].

Combretastatin A4 phosphate (**54**, CA4P) is a water-soluble prodrug of the natural stilbenoid, combretastatin A4 (**55**), produced by the South African tree *Combretum caffrum* [141]. CA4P is the first tubulin-binding vascular targeting agent (VTAs) under clinical trial, and can induce morphological changes within endothelial cells and then led to vascular dysfunction in tumors [141, 142]. CA4P as a therapeutic agent for the treatment of anaplastic thyroid carcinoma and myopic macular degeneration is under separate phase II clinical trials [143].

Curcumin (**56**), a yellow polyphenol pigment isolated from turmeric (*Curcuma longa*), commonly used as a spice in cooking, has utility in India not only as a dietary supplement but also for healthcare [144]. It exhibits antioxidant, antiinflammatory, antimicrobial, and immunomodulatory activities as well as other effects germane to cancer. Curcumin regulates numerous intracellular signaling pathways by acting on a broad spectrum of targets including transcription factors, growth factors, inflammatory cytokines, and protein kinases [145]. The potential therapeutic effects of curcumin have raised interest worldwide and multiple clinical trials have been launched in different countries including India, Japan, the People's Republic of China, and the United States, for the treatment of cancer, Alzheimer's disease as well as chronic psoriasis [144].

Future Prospects

The successful track record of natural compounds from plants, microorganisms, and other organisms, has demonstrated amply that these small organic molecules represent a highly useful source of molecular diversity in drug discovery [1, 5, 6, 9, 14]. Although the trend of developing small organic molecules as new drugs in the last 2 decades has been challenged by new synthetic methods such as combinatorial chemistry, natural products have once again come to prominence as sources of libraries of drug-like chemical entities [3, 18, 20]. Many small biotechnology companies as distinct from large pharmaceutical corporations have realized the important role that natural products play in modern drug discovery. Accordingly, new natural product-derived drugs are continually being introduced onto the market at a steady rate and there are numerous drug candidates from plants or other natural sources currently undergoing preclinical or clinical trials [107, 146]. Resourceful interdisciplinary efforts are required

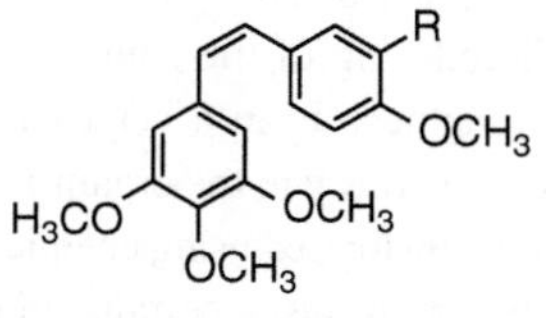

54. combretastatin A4 phosphate R = OPO_3H_2

55. combretastatin A4 R = OH

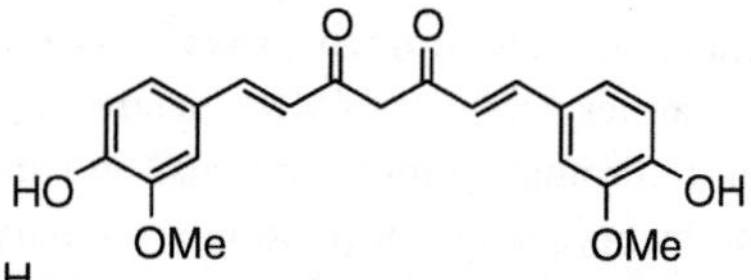

56. curcumin

to ensure that research on natural product drug discovery may keep pace with the many ongoing changes in the pharmaceutical industry.

The investigation of active compounds from plants and other organisms has benefited from many technological breakthroughs over the last 10 years. The utilization of bioassay-guided fractionation has been increased significantly by improving the technologies applied in the processes of compound analysis, purification, and structural identification as well as bioactivity screening. Thus, methods such as HPLC-coupled spectroscopy, higher magnetic field-strength NMR instruments, and robotics to automate high-throughput bioassays have all served to make the lead selection phase of plant-derived drug discovery faster and more reliable.

The combination of HPLC or LC with other techniques such as the diode-array detector (DAD), circular dichroism (CD), mass spectrometry (MS), and nuclear magnetic resonance (NMR) has tremendously increased the ability of analysis and purification of HPLC by providing structural information of compounds on-line with minimum quantities of samples. These upgrades in the use of HPLC have made the structural characterization of compounds in crude natural product mixtures more accurate and their isolation more straightforward [147–149].

Contemporary natural product structure elucidation depends largely on the sensitivity of NMR spectroscopy, which has been increased greatly by recent developments in NMR probe technology. The introduction of microprobes and cryogenically cooled probes in NMR spectroscopy has afforded a considerable increase in sensitivity compared to conventional NMR probes and now enables the structure elucidation of compounds at the microgram level. The development of flow-through probes has provided a seamless link of NMR spectroscopy with liquid chromatography systems, and this promising technique has been adopted in natural products research [150, 151]. Accordingly, improvements to solvent suppression techniques have made it feasible to use non-deuterated solvents instead of expensive deuterated solvents during chromatographic separation and the LC-SPE-NMR technique can also make deuterated solvents unnecessary during the chromatographic separation, using the solid-phase extraction (SPE) technique before NMR analysis [152, 153].

Bioactivity screening is a key step in natural product-derived drug discovery. Over the last decade, numerous efforts have been carried out to develop more efficient screening methods. Thus, improved automated high-throughput techniques have allowed for rapid screening of plant extracts in the same manner as libraries of pure compounds, so the biological assays are no longer a rate-limiting step in the drug discovery process. With advanced data handling systems and robotics, a hundred thousand samples can be tested in just over a week [154]. New technologies based on bioactivity combined with chemo-analytical processes have emerged and some of them have been practiced successfully in plant-based drug research. These new methods, such as bioautography, HPLC-based activity profiling, on-flow bioassays, assays based on capillary electrophoresis, molecular imprinted polymers, biosensors, biological chip-based technologies for affinity separation and expression profiling, and various MS and NMR-based methods, have led to the establishment of effective, flexible and selective approaches for successful secondary metabolite screening from organisms [155, 156].

Biotechnology may be employed to overcome the sourcing problems that are unavoidable obstacles in the process of plant-derived drug manufacturing. A plant natural product can be produced via cell culturing of the source organism or via genetic engineering in a heterologous host. The metabolic pathway in the organism can be modulated and transformed using chemical or biological methods to furnish certain plant secondary metabolites of interest [157–160]. For example, since 2002, Bristol-Myers Squibb has produced paclitaxel using callus cell cultures of the Chinese yew, *Taxus chinensis* [161]. Other plant-derived compounds such as some *Catharanthus* alkaloids, diosgenin from *Dioscorea* species, and the *Panax ginseng* ginsenosides also can be produced by cell culture [162–164]. These biotechnological methods have allowed selected plant natural products to be produced in a relatively controlled manner, and hence provide a supply of plant matrix not limited by sourcing problems, such as environmental, seasonal, geographical, and political factors [160].

The past history of use plant and other natural product-derived drugs in the treatment of many major afflictions like cancer, cardiovascular diseases, and neurological conditions augurs well for their future utilization in this regard. When considering plant sources specifically, no more than 20% of the existing higher plants on earth have been investigated for pharmaceutical purposes [146]. Moreover, even a large proportion of the known compounds with plant origin have never been evaluated in a bioassay [146]. Cooperative efforts from all the technical disciplines related to drug discovery should be continued to make plant-derived natural products research an essential contributor in the future. Towards this end, a greater effort than previously should be made to examine the traditional practices of plants in developing countries, and plant-derived compounds should be examine for efficiency in test systems germane to a wide spectrum of human and animal diseases than previously.

References

1. Kinghorn AD, Balandrin MF (Eds.) (1993) Human medicinal agents from plants, Symposium Series No. 534, American Chemical Society Books: Washington, DC
2. Dewick PM (2002) Medicinal natural products: a biosynthetic approach, 2nd edn. Wiley, New York
3. Newman DJ, Cragg GM, Snader KM (2003) Natural products as sources of new drugs over the period 1981–2002. J Nat Prod 66: 1022–1037
4. Samuelsson G (2004) Drugs of natural origin: a textbook of pharmacognosy, 5th Rev. edn. Swedish Pharmaceutical Press, Stockholm
5. Balunas MJ, Kinghorn AD (2005) Drug discovery from medicinal plants. Life Sci 78: 431–441
6. Paterson I, Anderson EA (2005) The renaissance of natural products as drug candidates. Science 310: 451–453
7. Sneader W (2005) Drug discovery: a history. Wiley, Chichester
8. The history of traditional Chinese medicine of more than 2000 years. Further information available at: http://www.chinese-medicine-guide.org/History-of-traditional-Chinese-medicine.html (accessed November 2007)
9. Fabricant DS, Farnsworth NR (2001) The value of plants used in traditional medicine for drug discovery. Environ Health Perspect 109: 69–75
10. Sneader W (1996) Drug prototypes and their exploitation. Wiley, Chichester
11. Bindseil KU, Jakupovic J, Wolf D, Lavayre J, Leboul J, van der Pyl D (2001) Pure compound libraries: a new perspective for natural product based drug discovery. Drug Discov Today 6: 840–847
12. Butler MS (2004) The role of natural product chemistry in drug discovery. J Nat Prod 67: 2141–2153
13. Vuorelaa P, Leinonenb M, Saikkuc P, Tammelaa P, Rauhad JP, Wennberge T, Vuorela H (2004) Natural products in the process of finding new drug candidates. Curr Med Chem 11: 1375–1389
14. Koehn FE, Carter GT (2005) The evolving role of natural products in drug discovery. Nat Rev Drug Discov 4: 206–220
15. Bajorath J (2002) Chemoinformatics methods for systematic comparison of molecules from natural and synthetic sources and design of hybrid libraries. J Comput Aided Mol Des 16: 431–439
16. Firn RD, Jones CG (2003) Natural products – a simple model to explain chemical diversity. Nat Prod Rep 20: 382–391
17. Lee ML, Schneider G (2001) Scaffold architecture and pharmacophoric properties of natural products and trade drugs: application in the design of natural product-based combinatorial libraries. J Comb Chem 3: 284–289
18. Ganesan A (2004) Natural products as a hunting ground for combinatorial chemistry. Curr Opin Biotechnol 15: 584–590

19. Ortholand JY, Ganesan A (2004) Natural products and combinatorial chemistry: back to the future. Curr Opin Chem Biol 8: 271–280
20. Newman DJ, Cragg GM (2007) Natural products as sources of new drugs over the last 25 years. J Nat Prod 70: 461–477
21. Evans FJ (1991) Natural products as probes for new drug target identification. J Ethnopharmacol 32: 91–101
22. Williamson EM, Okpako DT, Evans FJ (1996) Pharmacological methods in phytotherapy research: Vol. 1. Selection, preparation and pharmacological evaluation of plant material. Wiley, New York
23. Newman DJ, Cragg GM, Holbeck S, Sausville EA (2002) Natural products and derivatives as leads to cell cycle pathway targets in cancer chemotherapy. Curr Cancer Drug Targets 2: 279–308
24. Koehn FE (2006) Therapeutic potential of natural product signal transduction agents. Curr Opin Biotechnol 17: 631–637
25. Kinghorn AD (2001) Pharmacognosy in the 21st century. J Pharm Pharmacol 53: 135–148
26. Jones WP, Chin YW, Kinghorn AD (2006) The role of pharmacognosy in modern medicine and pharmacy. Curr Drug Targets 7: 247–264
27. Hemscheidt T (2000) Tropane and related alkaloids. Topics Curr Chem 209: 175–206
28. Gadzikowska M, Grynkiewicz G (2002) Tropane alkaloids in pharmaceutical and phytochemical analysis. Acta Pol Pharm 59: 149–160
29. Rose JE, Behm, FM, Westman EC, Kukovich P (2006) Precessation treatment with nicotine skin patch facilitates smoking cessation. Nicot Tobacco Res 8: 89–101
30. Clement JA, Yoder BJ, Kingston DGI (2004) Natural products as a source of CNS-active agents. Mini-Rev Med Chem 1: 183–208
31. Anekonda TS, Reddy PH (2005) Can herbs provide a new generation of drugs for treating Alzheimer's disease? Brain Res Brain Res Rev 50: 361–376
32. Snyder SH, Pasternak GW (2003) Historical review: opioid receptors. Trends Pharmacol Sci 24: 198–205
33. Capsoni S, Giannotta S, Cattaneo A (2002) Nerve growth factor and galantamine ameliorate early signs of neurodegeneration in anti-nerve growth factor mice. Proc Natl Acad Sci USA 99: 12432–12437
34. Howes M-JR, Perry NSL, Houghton PJ (2003) Plants with traditional uses and activities, relevant to the management of Alzheimer's disease and other cognitive disorders. Phytother Res 17: 1–18
35. Heinrich M, Teoh HL (2004) Galanthamine from snowdrop – the development of a modern drug against Alzheimer's disease from local Caucasian knowledge. J Ethnopharmacol 92: 147–162
36. Marco-Contelles J, Carreiras MdC, Rodriguez C, Villarroya M, Garcia AG (2006) Synthesis and pharmacology of galantamine. Chem Rev 106: 116–133
37. Frood A (2007) Scientists stir the pot for right to grow marijuana. Nat Med 13: 764
38. Lambert DM, Fowler CJ (2005) The endocannabinoid system: drug targets, lead compounds, and potential therapeutic applications. J Med Chem 48: 5059–5087
39. Barnes MP (2006) Sativex®: clinical efficacy and tolerability in the treatment of symptoms of multiple sclerosis and neuropathic pain. Exp Opin Pharmacother 7: 607–615
40. Cook GC (2006) History of tropical medicine: William Ernest Cooke, FRCSI (1879–1967) and his Asian tour of 1929–1930. Acta Trop 100: 1–10
41. Bruneton J (1995) Pharmacognosy, phytochemistry, medicinal plants, 2nd edn. Lavoisier Publishing, New York
42. Li Y, Wu YL (1998) How Chinese scientists discovered qinghaosu (artemisinin) and developed its derivatives. What are the future perspectives? Med Trop (Mars.) 58: 9–12
43. Golenser J, Waknine JH, Krugliak M, Hunt NH, Grau GE (2006) Current perspectives on the mechanism of action of artemisinins. Int J Parasitol 36: 1427–1441
44. Kerr C (2006) Companies agree to withdraw artemisinin monotherapy. Lancet Infect Dis 6: 397
45. Buchtel L, Ventura HO (2006) Lunar Society and the discovery of digitalis. J La State Med Soc 158: 26–30
46. Warren B (2005) *Digitalis purpurea*. Am J Cardiol 95: 544
47. Wani MC, Taylor HL, Wall ME, Coggon P, McPhail AT (1971) Plant antitumor agents. VI. The isolation and structure of taxol, a novel antileukemic and antitumor agent from *Taxus brevifolia*. J Am Chem Soc 93: 2325–2327
48. Cragg GM, Newman DJ (2004) A tale of two tumor targets: topoisomerase I and tubulin. The Wall and Wani contribution to cancer chemotherapy. J Nat Prod 67: 232–244
49. Kingston DGI, Newman DJ (2007) Taxoids: cancer-fighting compounds from nature. Curr Opin Drug Discov Dev 10: 130–144
50. Lee JC, Harrison D, Timasheff SN (1975) Interaction of vinblastine with calf brain microtubule protein. J Biol Chem 250: 9276–9282

51. Van Der Heijden, R Jacobs DI, Snoeijer W, Hallard D, Verpoorte R (2004) The *Catharanthus* alkaloids: pharmacognosy and biotechnology. Curr Med Chem 11: 607–628
52. Cragg GM, Newman DJ (2005) Plants as a source of anti-cancer agents. J Ethnopharmacol 100: 72–79
53. van Agtmael MA, Eggelte TA, van Boxtel CJ (1999) Artemisinin drugs in the treatment of malaria: from medicinal herb to registered medication. Trends Pharmacol Sci 20: 199–205
54. O'Neill PM (2005) The therapeutic potential of semi-synthetic artemisinin and synthetic endoperoxide antimalarial agents. Expert Opin Investig Drugs 14: 1117–1128
55. Hallett RL, Sutherland CJ, Alexander N, Ord R, Jawara M, Drakeley CJ, Pinder M, Walraven G, Targett GA, Alloueche A (2004) Combination therapy counteracts the enhanced transmission of drug-resistant malaria parasites to mosquitoes. Antimicrob Agents Chemother 48: 3940–3943
56. Facts on acts (Artemisinin-based combination therapies). Further information available at: http://www.searo.who.int/LinkFiles/Drug_Policy_RBMInfosheet_9.pdf (accessed November 2007)
57. Kingston DGI (2000) Recent advances in the chemistry of taxol. J Nat Prod 63: 726–734
58. Nicolaou KC, Yang Z, Liu JJ, Ueno H, Nantermet PG, Guy RK, Claiborne CF, Renaud J, Couladouros EA, Paulvannan K, Sorensen EJ (1994) Total synthesis of taxol. Nature 367: 630–634
59. Holton RA, Somoza C, Kim HB, Liang F, Biediger RJ, Boatman PD, Shindo M, Smith CC, Kim SC, Nadizadeh H, Suzuki Y, Tao CL, Vu P, Tang S, Zhang P, Murthi KK, Gentile LN, Liu JH (1994) First total synthesis of taxol. 1. Functionalization of the B ring. J Am Chem Soc 116: 1597–1598
60. Holton RA, Somoza C, Kim HB, Liang F, Biediger RJ, Boatman PD, Shindo M, Smith CC, Kim SC, Nadizadeh H, Suzuki Y, Tao CL, Vu P, Tang S, Zhang P, Murthi KK, Gentile LN, Liu JH (1994) First total synthesis of taxol. 2. Completion of the C and D rings. J Am Chem Soc 116: 1599–1600
61. Kingston DGI (2006) Taxol and its analogs. In: Cragg GM, Kingston DGI, Newman DJ (eds) Anticancer Agents from Natural Products. Taylor & Francis, Boca Raton, FL, pp 89–122
62. Guénard D, Guéritte-Voegelein F, Potier P (1993) Taxol and Taxotere: discovery, chemistry and structure-activity relationships. Acc Chem Res 26: 160–167
63. De Clercq E (2006) Antiviral agents active against influenza A viruses. Nat Rev Drug Discov 5: 1015–1025
64. Laver G (2006) Antiviral drugs for influenza: Tamifluâ past, present and future. Future Virol 1: 577–586
65. Payne R, Edmonds M (2005) Isolation of shikimic acid from star aniseed. J Chem Ed 82: 599–600
66. Poon T, Enrich L, Mohadjer A (2007) Quantitative analysis and isolation of shikimic acid from *Liquidambar styraciflua*. Abstracts of Papers, 233rd American Chemical Society National Meeting, Chicago, IL, March 25–29
67. Krämer M, Bongaerts J, Bovenberg R, Kremer S, Müller U, Orf S, Wubbolts M, Raeven L (2003) Metabolic engineering for microbial production of shikimic acid. Metab Eng 5: 277–283
68. Yeung YY, Hong S, Corey EJ (2006) A short enantioselective pathway for the synthesis of the anti-influenza neuramidase inhibitor oseltamivir from 1,3-butadiene and acrylic acid. J Am Chem Soc 128: 6310–6311
69. Duke SO, Dayan FE, Romagni JG, Rimando AM (2000) Natural products as sources of herbicides: current status and future trends. Weed Res 40: 99–111
70. McKiernan PJ (2006) Nitisinone in the treatment of hereditary tyrosinaemia type 1. Drugs 66: 743–750
71. Anonymous (2007) Nitisinone: new drug. Type 1 tyrosinemia: an effective drug. Prescrire Int 16: 56–58
72. Booth M (1984) Opium: a history. St. Martin's Press, New York
73. US Food and Drug Administration. CDER new molecular entity (NME) drugs and new biologic approvals in calendar year 2004. Further information available at: http://www.fda.gov/cder/rdmt/nmecy2004.htm (accessed November 2007)
74. Ridgway D (2004) Analgesics for acute pain: meeting the United States Food and Drug Administration's requirements for proof of efficacy. Clin J Pain 20: 123–132
75. van Dorp ELA, Yassen A, Dahan A (2007) Naloxone treatment in opioid addiction: the risks and benefits. Expert Opin Drug Saf 6: 125–132
76. Volpicelli JR, Fenton M (2006) Sustained-release naltrexone formulations for the treatment of alcohol and opioid dependence. Future Neurol 1: 389–398
77. Paul IM, Shaffer ML, Yoder KE, Sturgis SA, Baker MS, Berlin CM Jr (2004) Dose-response

relationship with increasing doses of dextromethorphan for children with cough. Clin Ther 26: 1508–1514
78. Canel C, Moraes RM, Dayan FE, Ferreira D (2000) Molecules of interest: podophyllotoxin. Phytochemistry 54: 115–120
79. Srivastava V, Negi AS, Kumar JK, Gupta MM, Khanuja SPS (2005) Plant-based anticancer molecules: a chemical and biological profile of some important leads. Bioorg Med Chem 21: 5892–5908
80. Martincic D, Hande KR (2005) Topoisomerase II inhibitors. Cancer Chemother Biol Response Modif 22: 101–121
81. Grynkiewicz G, Achmatowicz O, Pucko W (2000) Bioactive isoflavone – genistein. Synthesis and prospective applications. Herba Pol 46: 151–160
82. Molokanova E, Savchenko A, Kramer RH (2000) Interactions of cyclic nucleotide-gated channel subunits and protein tyrosine kinase probed with genistein. J Gen Physiol 115: 685–696
83. Ravindranath MH, Muthugounder S, Presser N, Viswanathan S (2004) Anticancer therapeutic potential of soy isoflavone, genistein. Adv Exp Med Biol 546: 121–165
84. Bhat SV (1993) Forskolin and congeners. Prog Chem Org Nat Prod 62: 1–74
85. de Souza NJ, Dohadwalla AN, Reden J (1983) Forskolin: a labdane diterpenoid with antihypertensive, positive inotropic, platelet aggregation inhibitory, and adenylate cyclase activating properties. Med Res Rev 3: 201–219
86. Tesmer JJ, Sprang SR (1998) The structure, catalytic mechanism and regulation of adenylyl cyclase. Curr Opin Struct Biol 8: 713–719
87. Gilani AUH (1998) Novel developments from natural products in cardiovascular research. Phytother Res 12: S66-S69
88. Iwatsubo K, Okumura S, Ishikawa Y (2006) Drug therapy aimed at adenylyl cyclase to regulate cyclic nucleotide signaling. Endocr Metab Immune Disord Drug Targets 6: 239–247
89. Formica JV, Regelson W (1995) Review of the biology of quercetin and related bioflavonoids. Food Chem Toxicol 12: 1061–1080
90. Moskaug JO, Carlsen H, Myhrstad M, Blomhoff R (2004) Molecular imaging of the biological effects of quercetin and quercetin-rich foods. Mech Ageing Dev 125: 315–324
91. Lakhanpal P, Rai DK (2007) Quercetin: a versatile flavonoid. Internet J Med Update 2(2): Further information available at: http://www.geocities.com/agnihotrimed/paper05_jul-dec2007.htm (accessed December 2007)
92. Rawel HM, Meidtner K, Kroll J (2005) Binding of selected phenolic compounds to proteins. J Agric Food Chem 53: 4228–4235
93. Marinic M, Piantanida I, Rusak G, Zinic M (2006) Interactions of quercetin and its lanthane complex with double stranded DNA/RNA and single stranded RNA: Spectrophotometric sensing of poly G. J Inorg Biochem 100: 288–298
94. Nifli AP, Theodoropoulos PA, Munier S, Castagnino C, Roussakis E, Katerinopoulos HE, Vercauteren J, Castanas E (2007) Quercetin exhibits a specific fluorescence in cellular milieu: a valuable tool for the study of its intracellular distribution. J Agric Food Chem 55: 2873–2878
95. Kazanietz MG, Caloca MJ, Eroles P, Fujii T, Garcia-Bermejo ML, Reilly M, Wang H (2000) Pharmacology of the receptors for the phorbol ester tumor promoters: multiple receptors with different biochemical properties. Biochem Pharmacol 60: 1417–1424
96. Gunjan G, Harinder PSM, George F, Klaus B (2007) Phorbol esters: structure, biological activity, and toxicity in animals. Int J Toxicol 26: 279–288
97. Montesano R, Orci L (1985) Tumor-promoting phorbol esters induce angiogenesis in vitro. Cell 42: 469–477
98. Lee JC, Harrison D, Timasheff SN (1975) Interaction of vinblastine with calf brain microtubule protein. J Biol Chem 250: 9276–9282
99. Cortese F, Bhattacharyya B, Wolff J (1977) Podophyllotoxin as a probe for the colchicine binding site of tubulin. J Biol Chem 252: 1134–1140
100. Horwitz SB, Parness J, Schiff PB, Manfredi JJ (1982) Taxol: a new probe for studying the structure and function of microtubules. Cold Spring Harb Symp Quant Biol 46: 219–226
101. Hearn BR, Shaw SJ, Myles DC (2006) Microtubule targeting agents. Comp Med Chem II 7: 81–110
102. Lampson MA, Kapoor TM (2007) Using small molecules to unravel biological mechanisms. Chem Biol 1: 71–94
103. Szolcsanyi J, Szallasi A, Szallasi Z, Joo F, Blumberg PM (1990) Resiniferatoxin: an ultrapotent selective modulator of capsaicin-sensitive primary afferent neurons. J Pharmacol Exp Ther 255: 923–928
104. Caterina MJ, Schumacher MA, Tominaga M, Rosen TA, Levine JD, Julius D (1997) The capsaicin receptor: a heat-activated ion channel in the pain pathway. Nature 389: 816–824

105. Watson CP, Evans RJ, Watt VR (1988) Postherpetic neuralgia and topical capsaicin. Pain 33: 333–340
106. Anesiva, Inc. Further information available at: http://www.anesiva.com/wt/page/pipeline (accessed November 2007)
107. Butler MS (2005) Natural products to drugs: natural product compounds in clinical trials. Nat Prod Rep 22: 162–195
108. Chin YW, Balunas MJ, Chai HB, Kinghorn AD (2006) Drug discovery from natural sources. AAPS J 8: E239–253
109. Corson TW, Crews CM (2007) Molecular understanding and modern application of traditional medicines: Triumphs and trials. Cell 130: 769–774
110. Tanaka O, Nagai M, Shibata S (1966) Chemical studies on the oriental plant drugs. XVI. The stereochemistry of protopanaxadiol, a genuine sapogenin of ginseng. Chem Pharm Bull 14: 1150–1156
111. Popovich DG, Kitts DD (2002) Structure-function relationship exists for ginsenosides in reducing cell proliferation and inducing apoptosis in the human leukemia (THP-1) cell line. Arch Biochem Biophys 406: 1–8
112. Li G, Wang ZH, Sun YX, Liu K, Wang Z (2006) Ginsenoside 20(*S*)-protopanaxadiol inhibits the proliferation and invasion of human fibrosarcoma HT1080 cells. Basic Clin Pharmacol Toxicol 98: 588–592
113. Liu GY, Bu X, Yan H, Jia WWG (2007) 20*S*-Protopanaxadiol-induced programmed cell death in glioma cells through caspase-dependent and -independent pathways. J Nat Prod 70: 259–265
114. PanaGin Pharmaceuticals, Inc. Further information available at: http://www.panagin.com (accessed November 2007)
115. Smith CR Jr, Powell RG, Mikolajczak KL (1976) The genus *Cephalotaxus*: source of homoharringtonine and related anticancer alkaloids. Cancer Treat Rep 60: 1157–1170
116. Luo, CY, Tang JY, Wang YP (2004) Homoharringtonine: a new treatment option for myeloid leukemia. Hematology 9: 259–270
117. Quintas-Cardama A, Kantarjian H, Garcia-Manero G, O'Brien S, Faderl S, Estrov Z, Giles F, Murgo A, Ladie N, Verstovsek S, Cortes J (2007) Phase I/II study of subcutaneous homoharringtonine in patients with chronic myeloid leukemia who have failed prior therapy. Cancer 109: 248–255
118. The Rocky Mountain blood and marrow transplant program (RMBMTP), Initial patients treated in multinational phase 2/3 study of ceflatonin 09–20–2006. Further information available at: http://www.rockymountainbmt.com/news/Initial-Patients-Treated-in-Multinational-Phase-23-Study-of-Ceflatonin-4682.html (accessed January 2008)
119. Yeung H-C (1985) Handbook of Chinese herbs and formulas. Institute of Chinese Medicine, Los Angeles, CA
120. Further information for oxymatrine available at: http://www.sda.gov.cn/cmsweb/webportal (accessed November 2007)
121. Li J, Li C, Zeng M (1998) Preliminary study on therapeutic effect of oxymatrine in treating patients with chronic hepatitis C. Chin J Integr Tradit West Med 18: 227–229
122. Xiang XX, Wang GJ, Cai X, LI YL (2002) Effect of oxymatrine on murine fulminant hepatitis and hepatocyte apoptosis. Chin Med J 115: 593–596
123. Mao YM, Zeng MD, Lu LG, Wan MB, Li CZ, Chen CW, Fu QC, Wang JY, She WM, Cai X, Ye J, Zhou XQ, Wang H, Wu SM, Tang MF, Zhu JS, Chen WX, Zhang HQ (2004) Capsule oxymatrine in treatment of hepatic fibrosis due to chronic viral hepatitis: a randomized, double blind, placebo-controlled, multicenter clinical study. World J Gastroenterol 10: 3269–3273
124. Constantinou AI, Husband A (2002) Phenoxodiol [2H-1-benzopyran-7-ol, 3-(4-hydroxyphenyl)], a novel isoflavone derivative, inhibits DNA topoisomerase II by stabilizing the cleavable complex. Anticancer Res 22: 2581–2585
125. Choueiri TK, Wesolowski R, Mekhail TM (2006) Phenoxodiol: isoflavone analog with antineoplastic activity. Curr Oncol Rep 8: 104–107
126. Schweigerer L, Christeleit K, Fleischmann G, Adlercreutz H, Wahala K, Hase T, Schwab M, Ludwig R, Fotsis T (1992) Identification in human urine of a natural growth inhibitor for cells derived from solid paediatric tumours. Eur J Clin Invest 22: 260–264
127. Mor G, Fu HH, Alvero AB (2006) Phenoxodiol, a novel approach for the treatment of ovarian cancer. Curr Opin Investig Drugs 7: 542–548
128. Liu JS, Zhu YL, Yu CM, Zhou YZ, Han YY, Wu FW, Qi BF (1986) The structures of huperzine A and B, two new alkaloids exhibiting marked anticholinesterase activity. Can J Chem 64: 837–839
129. Jiang HL, Luo XM, Bai DL (2003) Progress in clinical, pharmacological, chemical and structural biological studies of huperzine A: a drug of traditional Chinese medicine origin for the

treatment of Alzheimer's disease. Curr Med Chem 10: 2231–2252
130. Ma XQ, Tan CH, Zhu DY, Gang DR, Xiao PG (2007) Huperzine A from *Huperzia* species – an ethnopharmacological review. J Ethnopharmacol 113: 15–34
131. US National Institutes of Health. Huperzine A in Alzheimer's disease. Further information available at: http://www.clinicaltrials.gov/show/NCT00083590 (accessed November 2007)
132. Debiopharm. Debio-9902 (formerly ZT-1) A novel treatment for Alzheimer's disease. Further information available at: http://www.debio.com/e/pdf/zt_1_e.pdf (accessed November 2007)
133. Alakurtti S, Mäkelä T, Koskimies S, Yli-Kauhaluoma J (2006) Pharmacological properties of the ubiquitous natural product betulin. Eur J Pharm Sci 29: 1–13
134. Cichewicz RH, Kouzi SA (2004) Chemistry, biological activity, and chemotherapeutic potential of betulinic acid for the prevention and treatment of cancer and HIV infection. Med Res Rev 24: 90–114
135. Liby K, Honda T, Williams CR, Risingsong R, Royce DB, Suh N, Dinkova-Kostova AT, Stephenson KK, Talalay P, Sundararajan C, Gribble GW, Sporn MB (2007) Novel semi-synthetic analogues of betulinic acid with diverse cytoprotective, antiproliferative, and proapoptotic activities. Mol Cancer Ther 6: 2113–2119
136. US National Institutes of Health. Evaluation of 20% betulinic acid ointment for treatment of dysplastic nevi (moderate to severe dysplasia). Further information available at: http://www.cancer.gov/search/ViewClinicalTrials.aspx?cdrid = 494341&version = HealthProfessional&protocolsearchid = 3664931 (accessed November 2007)
137. Temesgen Z, Feinberg JE (2006) Drug evaluation: bevirimat-HIV gag protein and viral maturation inhibitor. Curr Opin Invest Drugs 7: 759–765
138. Adamson CS, Ablan SD, Boeras I, Goila-Gaur R, Soheilian Ferri, Nagashima K, Li F, Salzwedel K, Sakalian M, Wild CT, Freed EO (2006) In vitro resistance to the human immunodeficiency virus type 1 maturation inhibitor PA-457 (bevirimat). J Virol 80: 10957–10971
139. Martin DE, Blum R, Wilton J, Doto J, Galbraith H, Burgess GL, Smith PC, Ballow C (2007) Safety and pharmacokinetics of bevirimat (PA-457), a novel inhibitor of human immunodeficiency virus maturation, in healthy volunteers. Antimicrob Agents Chemother 51: 3063–3066
140. Martin DE, Blum R, Doto J, Galbraith H, Ballow C (2007) Multiple-dose pharmacokinetics and safety of bevirimat, a novel inhibitor of HIV maturation, in healthy volunteers. Clin Pharmacokinet 46: 589–598
141. Young SL, Chaplin DJ (2004) Combretastatin A4 phosphate: background and current clinical status. Expert Opin Investig Drugs 13: 1171–1182
142. Siemann DW, Chaplin DJ, Horsman MR (2004) Vascular-targeting therapies for treatment of malignant disease. Cancer 100: 2491–2499
143. OXiGENE, Inc: Press releases June 5, 2006 and September 12, 2006. Further information available at: http://www.oxigene.com (accessed November 2007)
144. Singh S (2007) From exotic spice to modern drug? Cell 130: 765–768
145. Aggarwal BB, Sundaram C, Malani N, Ichikawa H (2007) Curcumin: the Indian solid gold. Adv Exp Med Biol 595: 1–75
146. Cordell GA, Colvard MD (2005) Some thoughts on the future of ethnopharmacology. J Ethnopharmacol 100: 5–14
147. Bringmann G, Messer K, Wohlfarth M, Kraus J, Dumbuya K, Rückert M (1999) HPLC-CD on-line coupling in combination with HPLC-NMR and HPLC-MS/MS for the determination of the full absolute stereostructure of new metabolites in plant extracts. Anal Chem 71: 2678–2686
148. Niessen WMA, Lin J, Bondoux GC (2002) Developing strategies for isolation of minor impurities with mass spectrometry-directed fractionation. J Chromatogr A 970: 131–140
149. Wolfender JL, Ndjoko K, Hostettmann K (2003) Liquid chromatography with ultraviolet absorbance-mass spectrometric detection and with nuclear magnetic resonance spectrometry: a powerful combination for the on-line structural investigation of plant metabolites. J Chromatogr A 1000: 437–455
150. Reynolds WF, Enriquez RG (2002) Choosing the best pulse sequences, acquisition. parameters, postacquisition processing strategies and probes for natural products structure elucidation by NMR spectroscopy. J Nat Prod 65: 221–244
151. Lewis RJ, Bernstein MA, Duncan SJ, Sleigh CJ (2005) A comparison of capillary-scale LC-NMR with alternative techniques: spectroscopic and practical considerations. Magn Reson Chem 43: 783–789
152. Corcoran O, Spraul M (2003) LC-NMR-MS in drug discovery. Drug Discov Today 8: 624–631

153. Martin GE (2005) Small-volume and high-sensitivity NMR probes. Annu Rep NMR Spectrosc 56: 1–96
154. Anonymous (2007) Assay development. Nat Rev Drug Disc 6: 681
155. Potterat O, Hamburger M (2006). Natural products in drug discovery - concepts and approaches for tracking bioactivity. Curr Org Chem 10: 899–920
156. Roddy TP, Horvath CR, Stout SJ, Kenney KL, Ho P-I, Zhang J-H, Vickers C, Kaushik V, Hubbard B, Wang YK (2007) Mass spectrometric techniques for label-free high-throughput screening in drug discovery. Anal Chem 79: 8207–8213
157. Manchester KL (1995) Louis Pasteur (1822–1895) – chance and the prepared mind. Trends Biotechnol 13: 511–515
158. Turner MK (1995) Biocatalysis in organic chemistry (Part II): present and future. Trends Biotechnol 13: 253–258
159. Rao SR, Ravishankar GA (2002) Plant cell cultures: chemical factories of secondary metabolites. Biotechnol Adv 20: 101–153
160. Fischbach MA, Walsh CT (2006) Biochemistry: directing biosynthesis. Science 314: 603–605
161. Ritter SK (2004) Green innovations. Chem Eng News 12: 25–30
162. Rokem JS, Tal B, Goldberg I (1985) Methods for increasing diosgenin production by *Dioscorea* cells in suspension cultures. J Nat Prod 48: 210–220
163. van Der Heijden R, Jacobs DI, Snoeijer W, Hallard D, Verpoorte R (2004) *Catharanthus* alkaloids: pharmacognosy and biotechnology. Curr Med Chem 11: 607–628
164. Moyano E, Osuna L, Bonfill M, Cusido RM, Palazon J, Tortoriello J, Pinol MT (2005) Bioproduction of triterpenes on plant cultures of *Panax ginseng* and *Galphimia glauca*. In: Pandalai SG (ed) Recent Research Developments in Plant Science, Vol. 3. Research Signpost, Kerala, India, p 195

Speciality Non-food Crops

25

Adrian P. Higson and Alison Hamer

Abstract Potential markets for non-food crops are both large and diverse, ranging from biofuels to performance chemicals and pharmaceuticals. If current consumer demand, high oil prices, and political support are maintained, we can expect to see significant increase in their cultivation. There are hundreds of crops in use or under development for non-food applications around the world. Speciality non-food crops are generally cultivated for specific natural products. They provide an invaluable source of therapeutic agents and high value ingredients including anti-oxidants and essential fatty acids for use in performance chemicals, cosmetics, functional foods, and pharmaceuticals. This chapter looks at the markets for renewable materials from speciality non-food crops and gives examples of crops currently under cultivation.

Current Status for Non-Food Crops

State of the Marketplace

The potential markets for non-food crops are both large and diverse, ranging from biofuels to performance chemicals and pharmaceuticals. Non-food crops are viewed as a means to improve industrial sustainability, improve rural economies and increase crop biodiversity. Globally, governments have developed a raft of policies and strategies to stimulate non-food crop markets and technologies (Alcimed, 2007; Defra and DTI, 2007; Hodsman et al., 2005; Schmitz, 2007; U.S. Department of Energy, 2006).

Comprised of five national bodies representing France, Germany, Netherlands, Belgium and the UK, the European Renewable Raw Materials Association (ERRMA) aims to promote the valorisation of compounds and materials derived from crops for use in the chemical and energy sectors. In the UK, significant progress has been made in the stimulation of markets; the number of farms growing non-food crops under schemes in England rose by 20% between 2003 and 2005. Many of the markets served by high value non-food crops are supply sensitive and can be prone to fluctuations in over and under-supply of material. There is interest and engagement from a widening range of players across the different sectors involved with non-food crops (Hodsman et al., 2005). However, engendering stable and profitable markets for novel crops is not without its challenges and barriers to uptake.

What is a Non-food Crop?

As the name suggests, a non-food crop is a crop used in applications other than for human food or animal feed i.e. the production of renewable

Adrian P. Higson (✉) and Alison Hamer
The National Non-Food Crops Centre, Biocentre, York Science Park, Innovation Way, Heslington, York, Y0105DG, UK
e-mail: a.higson@nnfcc.co.uk, a.hamer@nnfcc.co.uk

A.E. Osbourn and V. Lanzotti (eds.), *Plant-derived Natural Products*,
DOI 10.1007/978-0-387-85498-4_25, © Springer Science+Business Media, LLC 2009

energy, fuels and materials. It is the intended application, or use, that denotes their status, and therefore they are not necessarily distinct from traditional food or feed crops.

From amaranth and artemisia to yarrow and yew, there are hundreds of crops in use or under development for non-food applications around the world. The UK's National Non-Food Crops Centre (NNFCC) holds agronomy and market details on about a hundred of the key European crops.

The majority of non-food crops also have food and/or feed uses, for example wheat and oilseed rape (OSR). The use of existing food crops with well-understood agronomy and developed supply chains can offer an easier route to market over the development of completely novel non-food crops. Current crops dominating the non-food area are those that produce sugar, starch and oils, for use in the production of fuel and bulk chemicals.

Three groups of crops grown solely for industrial, or non-food, purposes are; fibre crops, energy crops and crops grown for their secondary metabolites. In addition to their well-known historical uses, fibre crops such as hemp and flax are used in biocomposites for the automobile industry and in construction materials. Increasingly, crops such as miscanthus and short rotation coppice (SRC) willow or poplar are being cultivated for energy production. Energy crops are grown to maximise yield of biomass per hectare, and developments in the thermal processing of biomass (Bridgewater, 2006) or the manipulation of carbohydrates like starch and cellulose could result in these crops forming the basis of a new economy based on renewable resources ('the bioeconomy') (Morris, 2006). The third type of non-food crop is the low volume, speciality crops grown for components such as secondary metabolites or specific oils. Often with long histories of wild harvesting for traditional use, they are grown to provide plant-derived pharmaceuticals and speciality chemicals for use as personal care and functional food ingredients. Although not always high value, in general, the profit margins available in these high value sectors make the business risk of developing novel non-food crops acceptable. This review looks at activity in the area of speciality non-food crops with respect to secondary metabolites and speciality oils, crops for fibres such as jute and cotton are not considered. Although not strictly 'non-food', crops used to produce ingredients for functional foods have been included as the areas of healthcare and nutrition are closely related, and the movement of functional food ingredients (e.g. omega oils) into personal care products is also evident.

Although a large amount of data is available for large volume crop cultivation, information on the cultivation of speciality crops is less available. The situation is further complicated by the dual use of crops in food and non-food applications. Table 1 serves to give an idea of the cultivation levels of a range of speciality non-food crops in Europe. For comparison, approximately 5 and 25 million hectares (ha) of European land (EU27) are used for OSR and wheat cultivation annually. Land used for cultivation of crops for herbal remedies in Europe amounts to 70,000–100,000 ha (Williamson and MacTavish, 2007).

Table 1 European non-food crop activity (2004) (The Promotion of Non-Food Crops; Hodsman et al., 2005)

Crop	Country	Area (ha)
Caraway	Lithuania	6,500
Chamomile	UK	175
Crambe	UK	1,171
Lavender	France	24,000
Linseed*	Finland, France, Germany, Lithuania, Poland	>7,465
Poppy	France, UK	8,466
St John's Wort	Austria	24
Sunflower	Austria, France, Germany	71,503

* Area not available for all countries

History of Non-food Crops

It is said that "there is nothing new under the sun" and certainly the use of plant products for their functional properties in the home and garden is as old as man himself. Plants have been used for home building, to produce heat, provide decoration and to treat diseases. The first agricultural (Neolithic) revolution some 10,000–12,000 years ago changed man from a nomadic hunter-gatherer to a settled cultivator of crops. The growing of herbs for culinary and medicinal purposes can be traced to ancient Egypt and Mesopotamia. These gardens later appeared as the physic gardens (e.g. Chelsea Physic Garden), of the sixteenth and seventeenth centuries.

Throughout the seventeenth and eighteenth centuries, the understanding of herbal medicines increased to a point in the mid-nineteenth century, where chemists were able to characterise the principle active ingredients in a number of medicines (Fowler, 2006; Sneden, 2005). These active ingredients, included morphine (from *Papaver somniferum*), digitalis (from *Digitalis purpurea*) and quinine (from *Cinchona* tree bark), and provided the impetus for experiments which changed the direction of medicine and gave rise to the chemicals industry. Throughout history, chemicals have been isolated from plants; written records for the use of indigo exists from Mesopotamia (seventh century BC) and by 715 BC wool dying was an established craft in Rome. In the 1850s the paths of herbal medicines and dyes crossed, changing the future development of both products. In 1856, the English chemist William Henry Perkin set out to chemically synthesise the natural product quinine from aniline. Quinine, an expensive plant extract was in demand at the time for the treatment of malaria. Rather than obtain quinine, Perkin synthesised the deep magenta dye '*mauvein'*. This serendipitous discovery paved the way for the modern dyes industry. It also laid the foundations for industrial organic chemistry which would go on to play the fundamental role in shifting western medicine from plant-based herbal extracts to synthetic single entity drugs.

By 1900, purified plant drugs were in use, from these purified compounds came new semi-synthetic drugs and by the 1950s plant compounds were being regarded as starting points for drug development rather than final compounds. From the 1950s fully synthetic drugs (although based on natural product leads) became the preferred route to therapeutic agents. Despite the twenty-first century preference for synthetic drug production, the plant kingdom still provides us with important drug starting materials, and active ingredients competitive with or beyond the commercial reach of synthetic chemistry (Table 2) (Butler, 2005; Ganesan, 2004).

Table 2 Examples of medicinal drugs derived from plants

Crop	Active ingredient	Therapeutic area	Chemical family
Artemisia annua	Artemisinin	Anti-malarial	Sesquiterpene lactone
Catharanthus roseus	Vincristine, vinblastine	Cancer chemotherapy	Bis-indole alkaloids
Cinchona ledgeriana	Quinine	Anti-malarial	Quinoline alkaloid
Digitalis purpurea	Digoxin	Heart conditions	Steroidal glycoside
Narcissus pseudonarcissus	Galanthamine	Alzheimers disease	Isoquinoline alkaloid
Papaver somniferum	Codeine, morphine	Analgesic	Opiate alkaloids
Pilocarpus jaborondi	Pilocarpine	Glaucoma	Imidazole alkaloid
Taxus genus	Paclitaxel	Cancer chemotherapy	Diterpene

25

With the advent of synthetic chemistry, interest in non-food crops waned somewhat. The sector was affected by the availability of cheap oil, resulting in petrochemicals forming the basis of our manufacturing industry. However, as the twenty-first century develops, we see a reawakening of interest in their cultivation. The reasons for the uptake of plant-based materials differ from sector to sector, but in general focus on: mitigation of climate change; security of energy supply; high oil prices; amelioration of environmental contamination; consumer demand for natural ingredients and improved material performance.

While some sectors (e.g. natural dyes) show little sign of recovery, others (e.g. plant-derived fuels named biofuels) are being driven by government incentives and progressed through advances in technology. Other sectors such as plant-derived pharmaceuticals, personal care products, lubricants, and other chemicals are developing based on consumer demand, technical innovation, regulation, and favourable economics.

Crop and Product Development

Overcoming Complexities, Challenges and Opportunities

As with all products, the introduction of new materials from crops requires the development of new supply chains. These supply lines originate at the agricultural/horticultural industry and are often unfamiliar to companies who traditionally work with petrochemical-based feedstocks. Companies will often need to develop new skills and technologies or seek out partner companies to ensure supply chain competency and security. The following areas need to be understood and considered when developing new plant-derived products:

- Knowledge of plants and bioactive contents
- Ability to source sustainable plant material (germplasm)
- Plant breeding, cultivar optimisation
- Agronomy, agricultural scale up, traceability, Good Agricultural Practice (GAP)
- Effects of genotype and phenotype on yield of natural product
- Post-harvest storage, initial extraction
- Separation, purification, standardisation, characterisation
- Synthetic derivatisation
- Regulatory framework; safety, efficacy, clinical trials
- Market demand and acceptance
- Co-ordination of multidisciplinary projects
- Understanding and financial support from private and public sectors

The cultivation of non-food crops can be integrated into the supply chain through a number of routes. Low volume crops can be cultivated by the final product supplier. This production route provides the maximum control over the quality and consistency of cultivation. This control is particularly important when the end product is a botanical extract containing multiple constituents and for use in pharmaceutical applications. The end product manufacturer can contract cultivation to a third party; this approach is often preferable for companies not focused on plant breeding or agronomy. The third supply route is to purchase material from agricultural contractors who have detailed knowledge of agricultural markets and existing supply chains. This route is more common for larger scale production of food crops and crops for biofuels, but is also well demonstrated by plant oil supply chains. For example, in the UK, contracts are available to farmers for borage, echium, and camelina, details of such contracts are held by The National Non-Food Crops Centre.

The complexity of supply chains and the need to develop collaborations is seen in the cultivation of sweet gale (*Myrica gale*) for use as a cosmetic ingredient (Galley and Simpson,

2007). The deciduous shrub sweet gale (also known as bog myrtle or Dutch myrtle) has been used for flavouring beer, as an insect repellent and an herbal remedy. More recent analysis showed extracts displayed interesting anti-bacterial activity against acne-causing skin bacteria. Although harvested from the wild for thousands of years, sweet gale had never been commercially propagated or cultivated. To develop cosmetic products containing sweet gale, a consortium capable of the following was required; identifying the best plant material, developing the propagation and agronomy, designing bespoke harvesting machinery, developing bioactive extraction processes, and performing product formulation and safety tests.

Analysis of wild sweet gale showed considerable variation in composition from plant to plant and also by location. The Scottish Agricultural College (SAC) developed tests allowing standardisation of extracts and developed propagation methods. SAC also developed an understanding of the agronomic requirements for commercial production. Highland Natural Products Ltd. performed environmental impacts assessments on commercial production and looked at economic effects of land use. Two critical technologies needed to be developed. Cranfield University (UK) designed and built a harvester suitable for the wet upland conditions. Critical Processes Ltd. assessed extraction methods and optimised a steam distillation process. Product formulation, efficacy, and safety tests were performed by Boots. The development project resulted in new products forming part of the Boots 'Botanics' range. Sweet gale oil is the first essential oil to be developed in 40 years and plans to cultivate 5,000 ha are underway.

Development of New Crop Varieties

Developments in high-throughput screening technologies now allow the rapid and detailed analysis of genes, transcripts, proteins and metabolites. Studied in combination, 'systems biology' creates new opportunities for both conventional development of new crop varieties and transgenic modification (or genetic modification, GM) (Miflin, 2000; Smallwood, 2006). This understanding can be used to improve both input (inputs required for successful cultivation, e.g. water, fertiliser, fungicides and pesticides) and output (quality and quantity of the desired products) traits in food and non-food crops. Although resistance to the cultivation of GM crops in Europe is strong, it has been estimated that worldwide, GM crops cover almost 4% of total arable land (WHO, 2005). GM strains of maize, soybeans, OSR and cotton are grown commercially. Also, GM papaya, potato, rice, squash, sugar beet and tomato are available. In the EU, GM 'Bt maize' – a variety containing a gene that allows the maize to defend itself against the European corn borer – was grown on more than 100,000 ha (250,000 acres) in 2007 (EuropaBio, 2007). Currently focused on agronomic trait developments which minimise economic damage in crops, such as increasing pest or disease resistance, pesticide tolerance and crop yield, the potential exists through metabolic engineering to optimise the production of specific fatty acids and secondary metabolites in crops (Kinney, 2006; Smallwood, 2006). An active area of research and development is in the production of biological therapeutic compounds (biologics) in plants and is discussed further in Section 3.2.

Biorefineries

Biorefineries are large integrated facilities, which, by 'biorefining', provide a range of products including fuel, chemicals, and power from bio-based feedstocks. Key to the successful development of biorefineries is the availability of technologies to derive maximum value from all constituents of a very high volume of biomass, thus it is not currently clear to what degree they will drive market uptake of speciality crops. Driven by a high volume chemical or biofuel, the co-processing of extraction products such as waxes, terpenes, proteins, sterols and antioxidants plus the efficient use of other polymers such as lignin for heat and power generation or chemical production will be required to derive maximum economic returns (Tamutech Consultancy, 2007). Through the operation of biorefineries, even materials that are present in low concentrations may become commercially viable.

There is a very wide range of potentially valuable extracts from plant biomass. Many of these are specific to particular genera or species. Others are found quite widely across the plant kingdom. Ingredients from a specific type of plant, e.g. a natural product pharmaceutical ingredient, will normally be extracted in a dedicated facility, which is unlikely to form part of a biorefinery. However, more generic materials found in many types of plant; particularly crops associated with biofuel feedstocks e.g. maize, wheat and OSR could be extracted.

The extraction of squalene is a notable example of a terpene found in low concentration in a range of plants. It is normally produced from shark liver oil, but during the process of refining low grade olive oil, squalene is concentrated in a fraction called the deodorizer distillate. Ten percent or more of this fraction can consist of squalene, and extraction becomes cost-effective. The squalene is hydrogenated to give squalane, a high value ingredient in demand for cosmetic and personal care products (Bondioli et al., 1993; Fernández-Bolaños et al., 2006).

Non-food Sectors

Pharmaceuticals

Natural products and their derivatives are an invaluable source not only of lead compounds for pharmaceutical development but also therapeutic agents themselves (Table 2) (Fowler, 2006). In terms of production cost, the chemical complexity of these agents places them beyond the reach of synthetic chemistry and natural product extraction, and derivatisation is the preferred production route (Butler, 2005; Fowler and Law, 2006; Koehn and Carter, 2005; Srivastava et al., 2005). This approach is exemplified by the manufacture of the taxanes, paclitaxel and docetaxel. These compounds form the basis of the blockbuster drugs Taxol® and Taxotere® which are used to treat many types of cancer. Taken from yew species (*Taxus* spp.) both natural paclitaxel and synthetic intermediates suitable for conversion to paclitaxel and docetaxel are extracted.

First performed in 1804, the extraction of morphine from opium poppies (*Papaver somniferum*) still provides an important source of opiate analgesics and a starting material for codeine synthesis. These alkaloids are widely used in the treatment of severe pain. As for the taxanes mentioned above, extraction from the poppy remains the preferred route for the production of morphine and codeine. Globally, opium poppies are legally cultivated on around 40,000 ha annually. The major supplier of opium poppies is Tasmania, supplying around 40% of the market. GlaxoSmithKline, Tasmanian Alkaloids and Johnson-Matthey operate fully integrated supply chains for opiate production. Levels of production are controlled by the UN Single Convention on Narcotic Drugs, which limits production to reasonable saleable quantities with allowance made for contingency stocks (Fowler and Law, 2006).

Plants are an important source of new drug leads (Fowler, 2006; Kong et al., 2003). Developments continue across a number of scientific areas (e.g. plant molecular biology, extraction and separation technology and molecular characterisation tools). Plus, there is increasing concern about the lack of new pharmaceuticals entering the marketplace. Therefore, interest in the cultivation of high value non-food crops for pharmaceutical applications is re-emerging. The reasons for this renewed interest have been reviewed in detail (Fowler and Law, 2006) and are summarised below:

- Lack of delivery at the expected levels of synthetic and combinatorial chemistry.
- Recognition that natural product-derived drugs occupy a very different ‘chemical space’ to synthetically derived drugs, providing an important extra dimension of chemical diversity - this is especially true of plant systems.
- Significant advances have been made in technologies for preparation of plant extracts and the identification and rapid isolation of specific active entities (de-replication).

A number of companies hold valuable libraries of natural products (Fowler and Law, 2006). Phytopharm Plc and GW Pharmaceuticals Plc are developing new drug pipelines based on medicinal plants. In collaboration with Unilever, Phytopharm are developing a functional food targeted at obesity sufferers. The extract from the South African *Hoodia* plant, is claimed to contain a novel satiety stimulator that reduces calorie intake in overweight subjects. In addition to detailed agronomy studies, Phytopharm are developing supply chains and undertaking clinical and consumer studies.

Cannabis (*Cannabis sativa*) has a long history of medicinal and non-medicinal use and records date back 4,000 years. Naturally occurring cannabis contains a unique class of compounds known as cannabinoids. Over 60 members of this family have been identified but only a few have been well studied. The two best known and most studied cannabinoids are delta-9-tetrahydrocannabinol (THC) and cannabidiol (CBD) (Fig. 1). These two compounds show markedly different pharmacological activity. THC is best known for its psycho-activity but also displays analgesic, anti-spasmodic, anti-tremor, anti-inflammatory, and anti-emetic activity. CBD

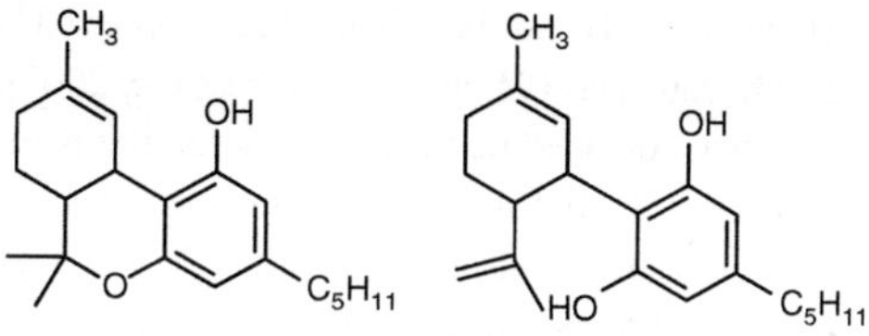

1 2

Fig. 1 Cannabinoids delta-9-tetrahydrocannabinol (1) and cannabidiol (2)

has anti-inflammatory, anti-convulsant, and anti-psychotic activity as well as neuroprotective and immunomodulatory effects. Critically, CBD is not psychoactive or intoxicating and may play a role in reducing the unwanted side effects of THC.

GW Pharmaceuticals are leading research into the use of cannabinoids for treatment of a range of illnesses such as Multiple Sclerosis symptoms including muscle spasticity, spasms, bladder dysfunction and pain control. The company states that precise pharmacological activity can be achieved by blending different cannabinoids in defined ratios. In addition to the ratio of cannabinoids; GW are interested in the action of other plant components modulating pharmacological activity.

By its nature, the cultivation of *Cannabis sativa* plants requires additional control beyond Good Agricultural Practice (GAP) and Good Manufacturing Practice (GMP) requirements for plant-derived pharmaceutical production. GW cultivates *Cannabis sativa* under licence from the UK Home Office and the crop is grown under a highly secure and controlled glasshouse environment. Computer control allows temperature, humidity, air change, and photoperiod to be optimised to give a controlled year round supply of material. Initial crops were grown from seed but subsequent chemovars (varieties) have been grown as cuttings to give genetically identical clones. The use of cuttings for propagation allows the generation of crops with very narrow ratios of cannabinoid content. GW’s first product SativexTM, an oromucosal spray composed

25

primarily of THC and CBD, has been approved by Health Canada (GW Pharmaceuticals Plc, 2007).

When considering bioactive molecules occupying novel chemical space the anti-malarial compound artemisinin stands out. Artemisinin (also known as Qinghaosu) is an extract of *Artemisia annua*, and is a complex 1,2,4-trioxane tetracyclic sesquiterpene (Fig. 2).

Shown to be effective for treatment of malaria by Chinese scientists in the 1970s, artemisinin, as part of an artemisinin based combination therapy (ACT), is now considered to be the leading treatment for malaria in chloroquine resistant areas (WHO, 2006). Large quantities of *Artemisia* are required to meet clinical need. The number of companies extracting artemisinin in China rose from 10 to 80 between 2003 and 2005 and from 3 to 20 in Vietnam. Commercial cultivation also began in East Africa and Madagascar (Kindermans et al., 2007). However, the commercial market for artemisinin is extremely volatile with large fluctuations in price and recently prices have dropped to $200/kg (down from a high of over $1,000/kg in 2004); such low prices caused some growers to switch to more profitable crops (CNAP, 2007; Kindermans et al., 2007). The Centre for Novel Agricultural Products (CNAP) at the University of York (UK) is using fast track breeding techniques to increase yields of artemisinin. With funding from the Bill and Melinda Gates Foundation (CNAP, 2007), they have developed new high-throughput screening methods, which can extract and measure the artemisinin content of plants in minutes. The team eventually aims to screen 25,000 plants generating a collection of individuals with valuable traits.

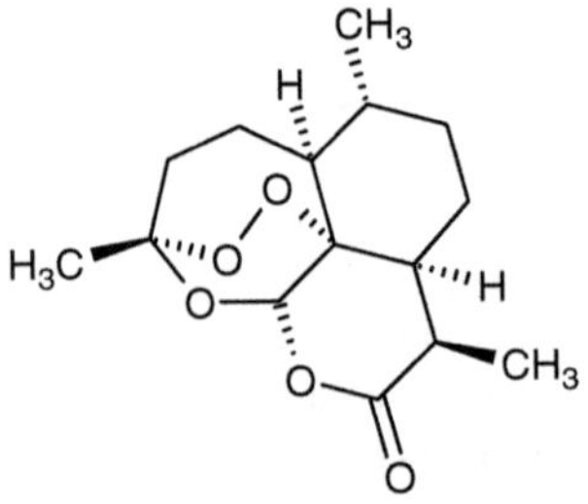

Fig. 2 Artemisinin extracted from *Artemisia annua*

Artemisia annua grows wild in northern Europe. Commercial scale cultivation is being studied by European companies looking at germplasm, agronomy, and requirements for harvesting, processing, and delivery (Botanical Developments, 2006).

Recombinant Proteins and Molecular Farming

Over the last decade, the possibility of using plants as vectors for the large scale production of recombinant proteins and vaccines has moved from development to commercial activity. So called 'molecular farming' is seen as an opportunity to overcome the problems of scale-up associated with fermentation or transgenic animal systems. However, over 80% of the production cost of these proteins is associated with downstream isolation and purification of the protein and these downstream costs vary little with production method. Therefore, to compete with simpler rival production methods, the increased cost of plant production must be offset by higher yields of expressed protein (Fowler and Law, 2006; Fisher et al., 2004; Twyman et al., 2003).

Due to its ease of manipulation and significant potential to produce biomass, the tobacco plant (*Nicotiana tabacum* L.) has received a great deal of attention as a candidate for production of recombinant proteins (Epobio, 2007a; Twyman et al., 2003). The European biotechnology company Meristem has developed systems for recombinant protein expression in whole leaf biomass. Proteins under development include gastric lipase and lactoferrin. The Large Scale Biology Corporation also chose tobacco plants for production of a range of therapeutic vaccines and proteins. They have adopted a new approach using viral vectors with the capacity to express genes of interest in non-transgenic plants. These tobacco mosaic virus and related vectors move rapidly throughout the entire plant, causing protein synthesis and accumulation in most tissues. Over the last 50 years, safflower (*Carthamus tinctorius*) has been cultivated mainly for its vegetable oil. Recently, in addition to the production of nutritional oils, SemBioSys Genetics has developed systems for production of insulin and proteins in safflower for use in cardiovascular therapy. These products are currently undergoing clinical trials. The potential for in-plant production of recombinant proteins is clear; however, it remains to be seen whether this production system can compete with microbial fermentation and whether the required GM technology will be accepted by society.

Herbal Medicines

Although single active ingredient pharmaceuticals dominate western medicine, in many parts of the world herbal or traditional medicine forms the mainstay of healthcare. The use of herbal medicine is also growing in both Europe and the US. The global herbal medicine market was worth over €45 billion per year in 2003 and is growing steadily (WHO, 2003). Although often not supported by extensive clinical data, herbal medicines are commonly used for treatment of a wide range of complaints from depression to hypertension. Some preparations have been well studied and have proven clinical effects; these include treatment of moderate depression with St John's Wort extract, garlic for decreasing high blood pressure, the use of *Ginkgo biloba* extract for the treatment of dementia (Bardia et al., 2007; Mar and Bent, 1999). The herbal medicines market is largely served through the wild harvesting of medicinal plants (Williamson and MacTavish, 2007).

There are, however, concerns over the quality and authenticity of some herbal medicines. These concerns have resulted in the European Traditional Herbal Medicinal Products Directive (2004/24/EC) (THMPD) (European Union, 2004). Coming into force in 2011, the Directive will bring consistency and regulation to the herbal medicines market across Europe. In additional to concerns over safety, the increased amount of wild crafting is raising sustainability questions. Of the 40,000–50,000 plant species used in herbal medicines, 70% are wild crafted and 4,000 of those face extinction. The opportunity for commercial cultivation of medicinal herbs has been reviewed (Williamson and MacTavish, 2007). The study identified up to 36 species of plants suitable for cultivation in Northern European agroclimatic conditions or under controlled conditions. The authors reviewed potential volumes and prices and the species with greatest potential for economic development are shown in Table 3.

Whether the commercial cultivation of medicinal plants increases will be largely dependent on how the industry and consumers respond to new regulations such as the THMPD. The costs associated with the Directive may appear to be off-putting to producers, but on the other hand the reassurance provided to consumers and medical practitioners may further stimulate the sector. In addition, the need to grow crops under GAP may help prevent the exploitation of endangered wild species.

Nutrition and Personal Care

In addition to the long standing uses of natural products in pharmaceuticals and herbal medicines, new developments in nutrition and cosmetics continue to provide new opportunities for cultivation of non-food crops. Providing the momentum for developments such as sweet gale is a strong customer drive for products derived from natural sources. This can be seen in the personal care sector in products from shampoo to skin care. The natural cosmetic market is believed to be near the €1billion mark in Europe (Organic Monitor, 2006). Another

25

Table 3 Medicinal plant species offering opportunities for commercial cultivation in Northern Europe

Species	Common name	Note	Type of bioactive
Arctostaphylos uva-ursi	Bearberry	Endangered and in demand	Hydroquinones, arbutin, tannis, phenolic glycosides and flavonoids
Atropa belladonna	Deadly Nightshade, Belladonna	Controlled cultivation necessary	Tropane alkaloids
Baptista tinctorum	Wild Indigo	Immune system herb increasingly in demand	Arabinogalactan proteins
Cetraria islandica	Iceland Moss	Endangered species, sustainable cultivation required	Lichen acids (usnic acid) and polysaccharides
Crataegus oxyacantha	Hawthorn	Long term crop with increasing demand	Bioflavonoids, triterpenes, proanthocyanins, polyphenols, coumarins
Gentiana lutea	Yellow Gentian	Endangered, requires cultivation research	Bitter principles (gentiopicriside, amarogentin), gentainose, inulin, phenolic acids
Hydrastis canadensis	Goldenseal	Endangered with increasing demand	Isoquinoline alkaloids.
Orchis mascula and *Orchis morio*	Purple and Green Winged Orchid	Endangered. Sustainable supplies required	
Primula veris	Cowslip, Oxlip	Endangered	Triterpene saponins, flavonoids, phenols
Ruscus aculeatus	Butchers Broom	Endangered	Saponin glycosides
Saponaria officinalis	White Soapwort	Endangered with increasing demand	Saponins
Scutellaria baicalensis	Baical Skullcap	Potential wider applications	Baicalin, baicalein, tenaxin, skullcap flavones
Viola odorata	Sweet Violet	Many application with history of UK cultivation	Phenolic glycosides, saponins, flavonoids, alkaloids, mucilage

new and growing market is functional foods; in 2006, the US functional food market was estimated to value $36 billion (Sloane, 2006). Evidence from epidemiological studies increasingly demonstrates the benefits of plant-rich diets. Identifying the bioactive compounds responsible for these benefits is an active area of research. Compound groups under examination include polyphenols, flavonoids, lignans, stilbenes, salicylates, and sterols (Hooper and Cassidy, 2006).

Many of these compounds are being analysed for antioxidant activity. Antioxidants are an important and growing area in functional foods and the potential for novel antioxidants from plant sources has not gone unnoticed. Tocopherol and ascorbic acid, either synthetic or natural are the most widely used biological antioxidants; most natural tocopherols are isolated from soybean oil whereas ascorbic acid is prepared through a double fermentation process. The antioxidant activity of rosmarinic acid is well known and understood (Chang et al., 1977). Rosemary (*Rosmarinus officinalis* L.) is a member of the Labiateae or mint family (Petersen and Simmonds, 2003). It is a slow growing, cold sensitive, woody perennial cultivated for its aromatic foliage. The crop is widely grown on a commercial scale in Spain, France, Italy, Croatia, Tunisia, and Morocco for its high-value essential oils. The antioxidant potential of rosemary has been recognised since the 1950s. A range of phenols with antioxidant activity has been isolated from rosemary leaves including rosmarinic acid and carnosic acid (Fig. 3). Carnosic acid is known to undergo degradation and rearrangement to a number of derivatives (carnosol, rosmanol, rosmariquinone and methyl carnosate) all with antioxidant activity.

Polyunsaturated fatty acids (PUFAs) such as α-linolenic (ALA) and γ-linolenic (GLA) (Fig. 4) are in high demand by both the nutraceutical and cosmeceutical markets.

Although not produced by the human body, these Essential Fatty Acids (EFAs) are required for normal brain function, growth, development, bone health, stimulation of skin and hair growth, regulation of metabolism, and maintenance of reproductive processes (University of Maryland Medical Centre, 2008). The health benefit provided by PUFAs is dependent on the precise PUFA being taken. The health benefits of the fish derived PUFAs eicosapentanoic acid and docosahexanoic acid are well described (Sanderson et al., 2002). There is interest in whether plant produced PUFAs such as ALA and GLA offer the same benefits. The metabolic profile of GLA and its efficacy in the treatment of several diseases has been reviewed (Fan and Chapkin, 1998). The efficacy of GLA has been noted for the treatment of pain, stiffness, and tender joints in rheumatoid arthritis sufferers (Soeken et al., 2003). A number of non-food crops are cultivated to meet the demand for both healthcare and personal care applications. Oil from hemp (*Cannabis sativa)* seed contains approximately 56% linoleic acid, 22% ALA and 4% GLA (Callaway, 2004). Only cultivars with less than 0.2% of the psychotropic agent THC may be grown for fibre and seed oil production in the EU. Camelina or Gold of Pleasure (*Camelina sativa*) produces an oil high in polyunsaturates (>50%). In addition to high levels of ALA (~40%) and GLA (~15%), the oil also contains antioxidants such as tocopherols. Adaptable to both climate and soil, Camelina has been grown in several European countries and is attracting interest from US growers (IENICA, 2002). When in flower, borage (*Borago officinalis*) is a distinctive blue crop. It is grown for its high concentrations of GLA, which accounts for 20–23% of its oil content. The oil, branded as Starflower Oil can be found in health foods, skin care creams, and cosmetics. The global requirement for borage oil has been estimated at around 1,500 t.

Croda Chemicals Europe Ltd. has developed Echium oil from *Echium plantagineum* for use as

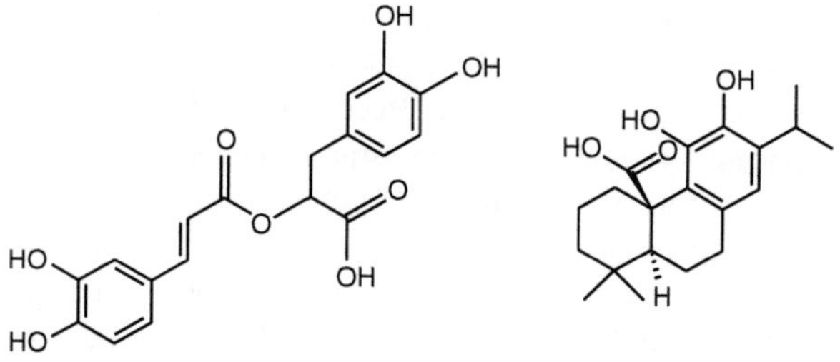

1 2

Fig. 3 Antioxidants, rosmarinic acid (1) and carnosic acid (2)

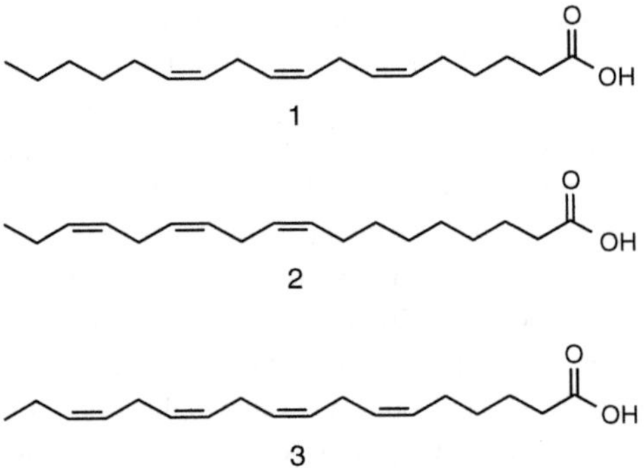

Fig. 4 Polyunsaturated fatty acids, γ-linolenic acid (1), α-linolenic acid (2) and stearidonic acid

a functional food. Echium contains high levels of several EFAs including GLA (9–12% of oil content). Despite the interest in GLA, it is the high levels of stearidonic acid (Fig. 4), with demonstrated anti-inflammatory properties, which make the extract of particular interest. The use of stearidonic acid in anti-wrinkle creams and sun aftercare make the oil attractive to the personal care market. Studies on stearidonic acid as treatment for eczema, acne and other skin disorders may lead to its medicinal use.

Many of the ingredients used in cosmetics and household cleaning products are products of the oleochemical industry. Of particular importance for the cosmetic sector is coconut and palm kernel oil because of their high percentage of fatty acids with short to medium chain length (mainly lauric acid C12 and myristic acid C14). These fatty acids are particularly suitable for conversion to surfactants for cosmetic and washing agents.

Also used in surfactant production, betaine is isolated from sugar beet as a value added co-product from sugar refining. Betaine is used in skincare formulations as a moisturiser and stabiliser for its water binding and cell membrane protecting properties.

Behenic acid (Fig. 5) is a major component of Ben oil (or behen oil), from the seeds of the Ben oil tree (*Moringa oleifera*). The Ben oil tree is widely cultivated in tropical Africa and America, Sri Lanka, India, Mexico, Malabar, Malaysia and the Philippine Islands as a food crop. Although behenic acid is also present in other oil-bearing plants, including OSR and peanut, it is produced commercially using erucic acid (Fig. 5) from high erucic acid varieties of OSR. Behenic acid is converted to a salt and used as a conditioner in hair products.

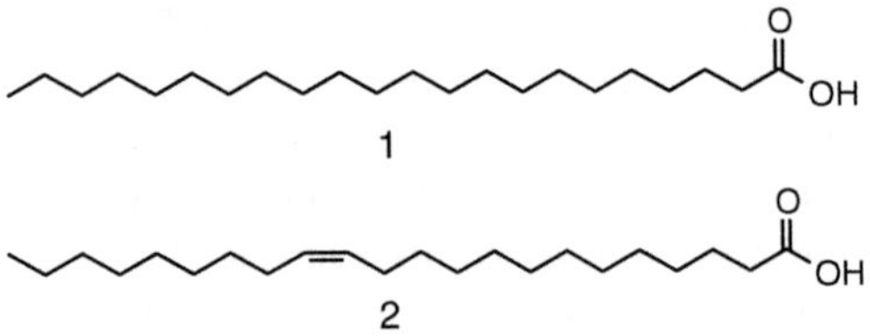

Fig. 5 Fatty acids, behenic acid (1) and erucic acid (2)

The use of natural oils as cosmetic ingredients is increasing globally. Apricot oil, yuzu (a Japanese citrus fruit) and raspberry extracts form part of a lipstick formulation developed by Kanebo which is claimed to stimulate microcirculation giving a deeper, redder colour. Extracts from orange, almond, carrot and many other plants are finding their way into a range of personal care products. Cognis Care Chemicals have launched a new hair product based on vegetable derived proteins. The complex includes wheat protein Gluadin W 40, Gluadin Soy and wheat micro protein Gluadin® WLM, which due to its low molecular weight (below 1 kDa) is claimed to penetrate, strengthen, and protect hair.

Plant Protection Products

Many secondary metabolites are produced in response to insect challenge. Identification and extraction of these natural products, many of which act by altering insect feeding patterns, presents opportunities for the development of crop protection products. Extracts of the neem tree (*Azadirachta indica*) containing azadirachtin have been studied and used extensively for crop protection. A wide range of plant extracts with insect anti-feedant activity are known including thyme (*Thymus vulgaris*), ashwagandha (*Withania somnifera*) and checkerberry (*Gaultheria procumbens*) (Isman, 2002). Rotenone is a natural insecticide obtained from the roots of several tropical and subtropical plant species belonging to the genera *Lonchocarpus* or *Derris*. A regulated plant protection product, rotenone can be used in organic farming practice. Carvone, a volatile terpenoid component of caraway oil (*Carum carvi*), has been shown to suppress sprouting of potatoes during storage, being marketed in the Netherlands for this purpose as the product 'Talent'. Large scale development beyond niche markets for plant-derived plant protection products is hindered by development costs and regulatory requirements.

Oil Crops for Industrial Applications

Oil crops are not restricted to nutritional or personal care applications. Driven by environmental regulations plant-based oils have found applications in areas such as cutting fluids, chainsaw lubricants, metal working fluids, hydraulic oils and 2-stroke engine oils. Around 90% of industrial and commercial lubricant applications can be satisfied by the newer generations of biolubricants derived from crops like OSR and sunflowers (Schneider, 2006). To overcome technical issues with OSR biolubricants, new cultivars with specific fatty acid profiles have been developed. High oleic acid rape varieties containing low linolenic acid levels have been developed, known as HO, LL varieties. This oil has improved oxidative stability allowing its use as a lubricant in higher temperature applications.

OSR varieties (HEAR) containing high levels of erucic acid have also been developed. Erucic acid is used in numerous applications (Gunstone and Hamilton, 2001) but its primary use is through conversion to erucamide for use as a plastic slip agent. HEAR varieties contain 40–50% erucic acid as a percentage of the oil. Crambe (*Crambe abyssinica*) also known as Abyssinian mustard, is a new introduction to UK agriculture (Epobio, 2007b). Prevalent across Asia and Western Europe, it requires moderate rainfall and warm temperate conditions. Crambe is grown for its high content of erucic acid (~56% of oil content) and provides advantages over HEAR for production of erucamide.

Concluding Remarks

From fatty acids to steroids and proteins, crops can provide us with a wide range of ingredients for healthcare, personal care and industrial materials. Increasing consumer demands for natural ingredients, concerns over environmental issues and the increasing price of crude oil is creating a commercial climate ripe for the development of renewable and sustainable non-food crops. Hurdles do need to be overcome in order to develop stable markets and secure supply chains. Ranging from improved understanding of non-food crop agronomy through to regulatory requirements for production and use, the sector requires coordinated development. Food crop yield and quality has increased dramatically throughout the second half of the twentieth century, whether this trend will be sustained and whether this success can be transferred to non-food crops remains to be seen. Other issues relevant for the sector will be developments on yield and natural product output plus the routes to development of optimised crop varieties. Sustainable non-food crops can be used to deliver industrial sustainability, cut our greenhouse gas emissions, improve rural economies, and increase crop biodiversity. If current consumer demand, high oil prices, and political support are maintained we can expect to see a further and significant increase in the cultivation and use of materials from non-food crops in everyday life.

References

Agro Business Consultants (2006) The Agricultural Budgeting & Costing Book. Agro Business Consultants, Melton Mowbray

Alcimed (2007) Marché actuel des Bioproduits et Biocarburants & Evolutions Prévisibles à Echéance 2015/2030. L'Agence de l'Environnement et de la Maîtrise de l'Energie, Angers

Bardia A, Nisly N L, Zimmerman M B, et al. (2007) Use of Herbs Among Adults Based on Evidence-Based Indications: Findings From the National Health Interview Survey. Mayo Clin Proc 82: 561–566

Bondioli P, Mariani C, Lanzani A, Fedeli E, Muller A (1993) Squalene recovery from olive oil deodorizer distillates. J Am Oil Chem Soc 70: 763–766

Botanical Developments, East Malling Research, NIAB (2006). Field cultivation of Artemisia annua and enhanced extraction of artemisisin used in novel anti malarial treatments. Department for Food Environment and Rural Affairs: Available via http://www2.defra.gov.uk/research/project_data/More.asp?I = NF0613 Cited Aug 17, 2007

Bridgewater T (2006) Biomass for energy. J Sci Food Agric 86:1755–1768

Butler M S (2005) Natural products to drugs: natural product derived compounds in clinical trials. Nat Prod Rep 22:162–195

Callaway J C (2004) Hempseed as a nutritional resource: An overview. Euphytica 140:65–72

Chang S S, Ostric-Matijasevic B, Hsieh O A, et al. (1977) Natural antioxidants from rosemary and sage. J Food Sci 42:1102–1106

CNAP (2007) The CNAP Artemisia Research Project. The Centre for Novel Agricultural Products. Available via http://www.york.ac.uk/org/cnap/artemisiaproject. Cited 7 Jan 2008

Defra DTI (2007). Creating value from renewable resources. Response to 2 year progress report on strategy for non-food crops and uses. Available via Defra: http://www.defra.gov.uk/ENVIRONMENT//climatechange/uk/energy/renewablefuel/pdf/non-foodstrat-2yearreview.pdf. Cited Oct 19, 2007

Epobio (2007a) Industrial Crop Platforms for the Production of Chemicals and Biopolymers. CPL Press, Newbury

Epobio. (2007b). Oil Crop Platforms for Industrial Uses. CPL Press, Newbury

EuropaBio. (2007, October 29). Press Release. EuropaBio. Available via http://www.europabio.org/GBE_media/GMOfigures/PR_Industry_Biotech%20Figures_FINAL29102007.pdf. Cited Jan 3, 2008

European Union. (2004). Directive 2004/24/EC of the European Parliament and of the Council. Official J EU, 47(L136):86–90

Fan Y-Y, Chapkin R (1998) Importance of Dietary Gamma-Linolenic Acid in Human Health and Nutrition. J Nutr 128:1411–1414

Fernández-Bolaños J, Rodríguez G, Rodríguez R, et al. (2006) Potential use off olive by-products – Extraction of interesting organic compounds from olive oil waste. Grasas Aceites 57:95–106

Fisher R, Stoger E, Schillberg S, et al. (2004) Plant-based production of biopharmaceuticals. Curr Opin Plant Biol 7:152–158

Gunstone F, Hamilton R J (2001) Oleochemical manufacture and application. CRC Press, Boca Raton, FL

Fowler M W (2006) Plants, medicines and man. J Sci Food Agric 86:1797–1804

Fowler M W, Law I (2006) Plant-based pharmaceuticals. A strategic study relating to UK activity and interests. The National Non-Food Crops Centre, New York

Galley E, Simpson M (2007). From Scottish Bog to International Beauty Counter – the Story of Bog Myrtle. Cosmetic Science Technology

Ganesan A (2004) Natural products as a hunting ground for combinatorial chemistry. Curr Opin Biotech 15:584–590

GW Pharmaceuticals Plc. (2007). News and Media., GW Pharmaceuticals. Avilable via http://www.gwpharm.com/news_press_releases.asp?id = /gwp/pressreleases/currentpress/2007–08–07/. Cited 17 Aug 2007

Hodsman L, Smallwood M, Williams D (2005) The Promotion of Non-Food Crops. A report for the European Parliament Committee on Agriculture and Rural Development Available via http://www.biomatnet.org/secure/other/s2025.htm

Hooper L, Cassidy A (2006) A review of the health care potential of bioactive compounds. J Sci Food Agric 86:1805–1813

IENICA. (2002). Camelina. The National Non-Food Crop Centre. Available via http://www.nnfcc.co.uk. Cited 28 Aug 2007

Isman M (2002) Insect antifeedants. Pestic Outlook 13:152–156

Kindermans J-M, Pilloy J, Olliaro P, et al. (2007) Ensuring sustained ACT production and reliable artemisinin supply. Malaria J 6:125

Kinney A J (2006) Metabolic engineering in plants for human health and nutrition. Curr Opin Biotech 17:130–138

Koehn F E, Carter G T (2005) The evolving role of nautral products in drug discovery. Nat Rev Drug Discov 4:206–220

Kong J-M, Goh N-K, Chia L-S, et al. (2003) Recent advances in traditional plant drugs and orchids. Acta Pharm Sinic 24:7–21

Mar C, Bent S (1999) An evidence-based review of the 10 most commonly used herbs. Western J Med 171:1–5.

Miflin B (2000) Crop improvement in the 21st century. J Exp Bot 51:1–8

Morris D (2006) The next economy: from dead carbon to living carbon. J Sci Food Agric 86:1743–1746.

National Institute of Agricultural Botany (NIAB), Botanical Developments Ltd. (2006) Field Cultivation of Artemisia annua in the UK and Enhanced Extraction of Artemisisin Used in Novel Anti-malarial Treatment. National Non-Food Crops Centre. Available via http://www.nnfcc.co.uk. Cited 7 Jan 2008

NNFCC. (2007) NNFCC Crop Factsheet: Borage. The National Non-Food Crops Centre. Available via http://www.nnfcc.co.uk. Cited 16 August 2007

Organic Monitor (2006) The European market for natural cosmetics. Organic Monitor, London.

Petersen M, Simmonds M S (2003) Rosmarinic acid. Phytochemistry 62:121–125

Sanderson P, Finnegan Y E, Williams C M et al. (2002) UK Food Standards Agency α-linoleic acid workshop report. Brit J Nutr 88:573–579

Schneider M (2006) Plant-oil-based lubricants and hydraulic fluids. J Sci Food Agric 86:1769–1780

Sloane A E (2006) Top 10 functional food trends. In Food Technology, Apr 2006 Institute of Food Technology

Smallwood M (2006) The impact of genomics on crops for industry. J Sci Food Agric 86:1747–1754

Schmitz N (2006) Marktanalyse Nachwachsende Rohstoffe. Fachagentur Nachsende Rohstoffe e. V., Gülzow

Sneden A T (2005) Natural Products as Medicinally Useful Agents. Available via http://www.people.vcu.edu/~asneden/MEDC%20310%20Intro.htm. Cited 17 Aug 2007

Soeken K L, Miller S A, Ernst E (2003) Herbal medicines for the treatment of rheumatoid arthritis: A systematic review. Rheumatology 42:652–659

Srivastava V, Negi A S, Kumar J K, et al. (2005) Plant-based anticancer molecules: A chemcial and biological profile of some important leads. Bioorg Med Chem 13:5892–5908

Tamutech Consultancy (2007) Mapping the Development of UK Biorefinery Complexes. NNFCC Available via http://www.nnfcc.co.uk. Cited 14 Sep 2007

The Advisory Committee on Novel Foods and Processes. (2006) Refined Echium Oil. The Advisory Committee on Novel Foods and Processes. Available via http://www.acnfp.gov.uk/assess/fullapplics/refechiumoil. Cited 17 Aug 2007

Twyman R M, Stoger E, Schillberg S, et al. (2003) Molecular farming in plants: host systems and expression technology. Trends Biotechnol 21 570–578

U.S. Department of Energy (2006). Breaking the Biological Barriers to Cellulosic Ethanol: A Joint Research Agenda. U.S. Department of Energy Office of Science and Office of Energy Efficiency and Renewable Energy. Available via www.doegenomestolife.org/biofuels/. Cited 14 Jan 2008

University of Maryland Medical Centre (2008) Complementary and Alternative Medicine Index. University of Maryland Medical Centre. Available via http://www.umm.edu/altmed/. Cited 5 Jan 2008

WHO (2006) Facts on ACTS (Artemisinin based combination therapies). World Health Organisation. Available via http://www.searo.who.int/LinkFiles/Drug_Policy_RBMInfosheet_9.pdf. Cited 17 Aug 2007

WHO (2003) Traditional Medicine. World Health Organisation. Available via http://www.who.int/mediacentre/factsheets/fs134/en/. Cited 28 Aug 2007

WHO (2005) Modern food biotechnology, human health and development:an evidence-based study. World Health Organisation, Geneva

Williamson E, MacTavish H (2007) An investigation into the potential market and feasibility of introducing new medicinal plant crops into the UK and Europe. The National Non-Food Crops Centre, York. Available via http://www.nnfcc.co.uk. Cited 28 Aug 2007

Index

H

I

J

K

L

M

O

P

Q

R

S

T

V

X

Z

GPSR Compliance

The European Union's (EU) General Product Safety Regulation (GPSR) is a set of rules that requires consumer products to be safe and our obligations to ensure this.

If you have any concerns about our products, you can contact us on ProductSafety@springernature.com

In case Publisher is established outside the EU, the EU authorized representative is:

Springer Nature Customer Service Center GmbH
Europaplatz 3
69115 Heidelberg, Germany

Batch number: 10755310

Printed by Printforce, the Netherlands